SPACEFLIGHT MECHANICS 2003
Part II

AAS PRESIDENT
    Dr. Robert E. Lindberg   National Institute of Aerospace

VICE PRESIDENT - PUBLICATIONS
    Dr. Robert G. Melton   Pennsylvania State University

EDITORS
    Dr. Daniel J. Scheeres   University of Michigan
    Dr. Mark E. Pittelkau   Applied Physics Laboratory
    Dr. Ronald J. Proulx   Charles Stark Draper Laboratory
    Dr. L. Alberto Cangahuala   Jet Propulsion Laboratory

SERIES EDITOR
    Robert H. Jacobs   Univelt, Incorporated

Special thanks are due Dr. Mark E. Pittelkau for preparing the CD ROM supplement to the hard copy version of the proceedings. Thanks are also due Bonita Roach, Univelt, Incorporated, for final preparation work done on the manuscript.

Front Cover Illustration:

Photo of the Arecibo Observatory, located in Puerto Rico. The Arecibo Observatory is part of the National Astronomy and Ionosphere Center (NAIC), a national research center operated by Cornell University under a cooperative agreement with the National Science Foundation (NSF). As the site of the world's largest single-dish radio telescope, the Observatory is recognized as one of the most important national centers for research in radio astronomy, planetary radar and terrestrial aeronomy. Photo courtesy of the NAIC - Arecibo Observatory, a facility of the NSF.

American Astronautical Society

SPACEFLIGHT MECHANICS 2003

Volume 114
Part II
ADVANCES IN THE ASTRONAUTICAL SCIENCES

Edited by
Daniel J. Scheeres
Mark E. Pittelkau
Ronald J. Proulx
L. Alberto Cangahuala

Proceedings of the AAS/AIAA Space Flight Mechanics Meeting held February 9-13, 2003, Ponce, Puerto Rico.

*Published for the American Astronautical Society by
Univelt, Incorporated, P.O. Box 28130, San Diego, California 92198
Web Site: http://www.univelt.com*

FOREWORD

This volume is the thirteenth of a sequence of Spaceflight Mechanics volumes which are published as a part of *Advances in the Astronautical Sciences*. Several other sequences or subseries have been established in this series. Among them are: Astrodynamics (published for the AAS every second year), Guidance and Control (annual), International Space Conferences of Pacific-Basin Societies (ISCOPS, formerly PISSTA), and AAS Annual Conference proceedings. Proceedings volumes for earlier conferences are still available either in hard copy or in microfiche form. The appendix at the end of Part III of this volume lists proceedings available through the American Astronautical Society.

Spaceflight Mechanics 2003, Volume 114, *Advances in the Astronautical Sciences*, consists of three parts totaling about 2,300 pages, plus a CD ROM supplement which contains all the available papers in digital format. Papers which were not available for publication are listed on the divider pages of each section in the hard copy volume. A chronological index and an author index are appended to the third part of the volume.

In our proceedings volumes the technical accuracy and editorial quality are essentially the responsibility of the authors. The session chairs and our editors do not review all papers in detail; however, format and layout are improved when necessary by our editors.

We commend the general chairs, technical chairs, session chairs and the other participants for their role in making the conference such a success. An extra special word of thanks is extended to Dr. Mark E. Pittelkau for preparing the CD ROM for publication. We would also like to thank those who assisted in organizational planning, registration and numerous other functions required for a successful conference.

The current proceedings are valuable to keep specialists abreast of the state of the art; however, even older volumes contain some articles that have become classics and all volumes have archival value. This current material should be a boon to aerospace specialists.

AAS/AIAA SPACEFLIGHT MECHANICS VOLUMES

Spaceflight Mechanics 2003 appears as Volume 114, *Advances in the Astronautical Sciences* (Including CD ROM). This publication presents the complete proceedings of the AAS/AIAA Spaceflight Mechanics Meeting 2003.

> **Spaceflight Mechanics 2002,** Volume 112, *Advances in the Astronautical Sciences*, Eds. K.T. Alfriend et al., 1570p, two parts.
>
> **Spaceflight Mechanics 2001,** Volume 108, *Advances in the Astronautical Sciences*, Eds. L.A. D'Amario et al., 2174p, two parts.

Spaceflight Mechanics 2000, Volume 105, *Advances in the Astronautical Sciences*, Eds. C.A. Kluever et al., 1704p, two parts.

Spaceflight Mechanics 1999, Volume 102, *Advances in the Astronautical Sciences*, Eds. R.H. Bishop et al., 1600p, two parts.

Spaceflight Mechanics 1998, Volume 99, *Advances in the Astronautical Sciences*, Eds. J.W. Middour et al., 1638p, two parts; Microfiche Suppl., 2 papers (Vol. 78 *AAS Microfiche Series*).

Spaceflight Mechanics 1997, Volume 95, *Advances in the Astronautical Sciences*, Eds. K.C. Howell et al., 1178p, two parts.

Spaceflight Mechanics 1996, Volume 93, *Advances in the Astronautical Sciences*, Eds. G.E. Powell et al., 1776p, two parts; Microfiche Suppl., 3 papers (Vol. 73 *AAS Microfiche Series*).

Spaceflight Mechanics 1995, Volume 89, *Advances in the Astronautical Sciences*, Eds. R.J. Proulx et al., 1774p, two parts; Microfiche Suppl., 5 papers (Vol. 71 *AAS Microfiche Series*).

Spaceflight Mechanics 1994, Volume 87, *Advances in the Astronautical Sciences*, Eds. J.E. Cochran, Jr. et al., 1272p, two parts.

Spaceflight Mechanics 1993, Volume 82, *Advances in the Astronautical Sciences*, Eds. R.G. Melton et al., 1454p, two parts; Microfiche Suppl., 2 papers (Vol. 68 *AAS Microfiche Series*).

Spaceflight Mechanics 1992, Volume 79, *Advances in the Astronautical Sciences,* Eds. R.E. Diehl et al., 1312p, two parts; Microfiche Suppl., 11 papers (Vol. 65 *AAS Microfiche Series*).

Spaceflight Mechanics 1991, Volume 75, *Advances in the Astronautical Sciences,* Eds. J.K. Soldner et al., 1353p, two parts; Microfiche Suppl., 15 papers (Vol. 62 *AAS Microfiche Series*).

All of these proceedings are available from **Univelt, Inc., P.O. Box 28130, San Diego, California 92198** (Web Site: *http://www.univelt.com*), publishers for the American Astronautical Society.

Robert H. Jacobs
Series Editor

PREFACE

The 13th Annual Space Flight Mechanics Meeting was held February 9-13, 2003 in Ponce, Puerto Rico at the Ponce Hilton Hotel. Hosted by the American Astronautical Society and co-sponsored by the American Institute of Aeronautics and Astronautics, and organized by the AAS Space Flight Mechanics Technical Committee and the AIAA Astrodynamics Technical Committee, this meeting brought together 160 participants to hear 130 presentations in the 20 scheduled technical sessions. There were three concurrent sessions held each morning and afternoon for three days, and two concurrent sessions held on the last morning of the conference. Technical sessions were held on the following topics: Attitude control, attitude determination, attitude dynamics, formation flight, mission design, mission operations, navigation, orbit determination, orbital debris, orbital dynamics, orbital transfers, space-based interferometry, spacecraft constellations, trajectory control, trajectory optimization, and tethers. Of the 130 papers presented, 26 were lead-authored from researchers representing 9 foreign countries, including Canada, France, Italy, Japan, Korea, the Netherlands, Poland, Spain, and the United Kingdom.

The conference included several special events. First, a memorial to the Columbia astronauts was set up in the registration and paper sales room. We appreciate Ron Proulx and Draper Lab for conceiving of and creating this fitting tribute. On Monday evening the Brouwer Award Lecture was given by Prof. Roger Broucke, 2002 AAS Dirk Brouwer Prize Awardee, entitled "A Half Century of Astrodynamics." On Tuesday evening Dr. Daniel R. Altschuler, the Director of the Arecibo Observatory, gave a lecture entitled "The Arecibo Observatory – History and Science." On Thursday afternoon, after the conference ended, a group of almost 100 conference attendees and guests made a special trip to the Arecibo Observatory, where they were given a guided tour of the large radio dish and after-hours access to the education center. This was an impressive trip, and enjoyed by all. The talk by Dr. Altschuler and the visit to the observatory was the inspiration for the cover art for these proceedings.

We would like to thank the session chairs for all of their efforts. The original session chairs were Dennis Byrnes, Paul Cefola, Shannon Coffey, Prasun Desai, Chris Hall, Felix Hoots, David Hyland, Ron Lisowski, Thomas Alan Lovell, Kim Luu, Don Mackison, Jim McAdams, Craig McLaughlin, Jay Middour, Beny Neta, Jon Sims, David Spencer, Al Treder, Bobby Williams, and Michael Zedd. Special appreciation to Terry Alfriend and Mike Ross who substituted for Craig McLaughlin and Prasun Desai who were not able to attend the conference. We greatly appreciate the donation of the printing of the program booklet by The Charles Stark Draper Laboratory, Inc. We also appreciate the support provided by AAS and AIAA staff and by secretarial staff at our business locations. We thank Shannon Coffey for placing program material on the AAS web site (www.space-flight.org).

We also would like to thank the numerous volunteers who staffed the registration and paper sales table. Most especially, we would like to express our appreciation to the authors for their efforts in performing the research and preparing the papers and presentations for the meeting. Finally, we wish to express our appreciation to the AAS Publications Office for producing this volume.

L. Alberto Cangahuala
AIAA General Chair

Mark E. Pittelkau
AIAA Technical Chair

Ronald J. Proulx
AAS General Chair

Daniel J. Scheeres
AAS Technical Chair

CONTENTS

Page

FOREWORD   vii

PREFACE   ix

Part I

SPECIAL LECTURES   1

Brouwer Award Lecture: A Half a Century of Astrodynamics (Summary)
(AAS 03-148)
    Roger A. Broucke 3

The Arecibo Observatory—History and Science (Abstract) (AAS 03-256)
    Daniel R. Altschuler 5

ATTITUDE CONTROL   7

Integrated Orbit and Attitude Control for a Nanosatellite With Power
Constraints (AAS 03-100)
    Bo J. Naasz, Matthew M. Berry, Hye-Young Kim and Christopher D. Hall . . 9

Integrated Structural and Control Optimization Approach for a Large Space
Station Design (AAS 03-101)
    Ijar M. Fonseca and Peter M. Bainum 27

Cost Functions for Bang-Off-Bang Control of Axisymmetric Spacecraft
(AAS 03-102)
    Robert A. Hall and Nathan C. Lowry 45

Structured Model Reference Adaptive Control for Vision Based Spacecraft
Rendezvous and Docking (AAS 03-103)
    Puneet Singla, Kamesh Subbarao, Declan Hughes and John L. Junkins . . . 55

The Effect of Averaging in Matched Basis Function Real Time Repetitive
Control (AAS 03-104)
    Masaki Nagashima and Richard W. Longman 75

Creating a Short Time Equivalent of Frequency Cutoff for Robustness in
Learning Control (AAS 03-105)
    Kenneth Chen and Richard W. Longman 95

Adaptive Inverse Iterative Learning Control (AAS 03-106)
    Richard W. Longman, Yen-Tun Peng, Taekjoon Kwon, Hilmi Lus,
    Raimondo Betti and Jer-Nan Juang 115

Repetitive Control to Eliminate Periodic Measurement Disturbances:
Application to Disk Drives (AAS 03-107)
    Yi-Ping Hsin and Richard W. Longman 135

FORMATION FLIGHT I   **151**

Nonlinear Modeling and Control of Spacecraft Relative Motion in the
Configuration Space (AAS 03-108)
    Pini Gurfil and N. Jeremy Kasdin 153

Linear and Nonlinear Control Laws for Formation Flying (AAS 03-109)
    S. S. Vaddi and S. R. Vadali 171

Deployment of Spacecraft Large Formations (AAS 03-111)
    Giovanni B. Palmerini 189

Control Strategies for Formation Flight in the Vicinity of the Libration
Points (AAS 03-113)
    K. C. Howell and B. G. Marchand 197

Suppose $A(t)$ Isn't Constant (AAS 03-114)
    William E. Wiesel 231

Incorporating Secular Drifts into the Orbit Element Difference Description
of Relative Orbits (AAS-03-115)
    Hanspeter Schaub 239

OPTIMIZATION   **259**

Probabilistic Optimization Applied to Spacecraft Rendezvous and Docking
(AAS 03-116)
    Jeff M. Phillips, Lydia E. Kavraki and Nazareth Bedrossian 261

Minimum-Time Orbital Phasing Maneuvers (AAS 03-117)
    Christopher D. Hall and Victor Collazo Perez 275

Optimal Rephasing Problem (AAS 03-118)
    Dario Izzo and Chiara Valente 293

Trajectory Optimization of a Constrained Mission About a Libration Point
(AAS 03-119)
    S. Infeld and W. Murray 307

Trajectory Optimization for a Mission to NEOS, Using Low-Thrust
Propulsion and Gravity Assist (AAS 03-120)
    Mauro Massari, Franco Bernelli-Zazzera and Massimiliano Vasile 317

Computation of Optimal Mars Trajectories Via Combined Chemical/
Electrical Propulsion, Part 2: Minimum Time Solutions With Bounded
Thrust Direction (AAS 03-121)
    A. Miele, T. Wang and P. N. Williams 331

Fuel-Optimal Orbital Transfers for Variable Specific Impulse Powered
Spacecraft (AAS 03-122)
Hans Seywald, Carlos M. Roithmayr and Patrick A. Troutman 347

Factors Influencing Solar Electric Propulsion Vehicle Payload Delivery for
Outer Planet Missions (AAS 03-123)
Michael Cupples, Shaun Green and Victoria Coverstone 365

ATTITUDE DETERMINATION AND CONTROL HARDWARE 383

Integrated Power and Attitude Control for a Spacecraft With Flywheels
and Control Moment Gyroscopes (AAS 03-124)
Carlos M. Roithmayr, Christopher D. Karlgaard, Renjith R. Kumar
and David M. Bose 385

Historical Review of Spacecraft Simulators (AAS 03-125)
Jana L. Schwartz, Mason A. Peck and Christopher D. Hall 407

An Airbearing-Based Testbed for Momentum Control Systems and
Spacecraft Line of Sight (AAS 03-127)
Mason A. Peck, Les Miller, Andrew R. Cavender, Mario Gonzalez
and Tim Hintz 427

User Interface Design for Moon-and-Star Night-Sky Observation
Experiments (AAS 03-128)
Christian Bruccoleri and Daniele Mortari 447

Innovative Spacecraft Sun Acquisition Algorithm Using Reaction Wheels
(AAS 03-130)
Che-Hang Charles Ih and Richard A. Noyola 461

Non-Dimensional Star Identification for Un-Calibrated Star Cameras
(AAS 03-131)
Malak A. Samaan, Daniele Mortari and John L. Junkins 477

FORMATION FLIGHT II 491

Flying a Four-Spacecraft Formation by the Moon . . . Twice (AAS 03-132)
José J. Guzmán and Ariel Edery 493

Preliminary Planer Formation-Flight Dynamics Near Sun-Earth L2 Point
(AAS 03-133)
A. M. Segerman and M. F. Zedd 511

Relative Trajectory Analysis of Dissimilar Formation Flying Spacecraft
(AAS 03-134)
Shankar K. Balaji and Adrian R. Tatnall 531

Geopotential and Luni-Solar Perturbations in the Satellite Constellation and
Formation Flying Dynamics (AAS 03-136)
Edwin Wnuk and Justyna Golebiewska 545

Preliminary Assessment of Interferometric SAR Baseline Determination
Using DGPS (AAS 03-137)
    Frederic J. Pelletier, Stefano Casotto, Alberto Zin and Boris Padovan .   .   .   561

Meet the Cluster Orbits With Perturbations of Keplerian Elements
(COWPOKE) Equations (AAS 03-138)
    Chris Sabol, Craig A. McLaughlin and K. Kim Luu   573

Analysis of the Reconfiguration and Maintenance of Close Spacecraft
Formations (AAS 03-139)
    T. A. Lovell and S. G. Tragesser   595

OPTIMIZATION AND ORBITAL TRANSFERS   **611**

Guess Value for Interplanetary Transfer Design Through Genetic Algorithms
(AAS 03-140)
    Paola Rogata, Emanuele Di Sotto, Mariella Graziano and Filippo Graziani .   .   613

A Global Approach to Optimal Space Trajectory Design (AAS 03-141)
    Massimiliano Vasile   629

Adaptive Grids for Trajectory Optimization by Pseudospectral Methods
(AAS 03-142)
    I. Michael Ross, Fariba Fahroo and Jon Strizzi   649

Design of a Multi-Moon Orbiter (AAS 03-143)
    S. D. Ross, W. S. Koon, M. W. Lo and J. E. Marsden   669

Lunar Transfer Trajectory Design and the Four-Body Problem (AAS 03-144)
    James K. Miller   685

A Systematic Method for Constructing Earth-Mars Cyclers Using Direct
Return Trajectories (AAS 03-145)
    Ryan P. Russell and Cesar A. Ocampo   697

Planar High-Thrust and Low-Thrust Orbital Transfers From Earth to Europa
(AAS 03-146)
    Hans Seywald, Carlos M. Roithmayr, Daniel D. Mazanek,
    Frederic H. Stillwagen, Patrick A. Troutman and Sang-Young Park .   .   717

Desensitized Optimal Trajectories With Control Constraints (AAS 03-147)
    Hans Seywald   737

Part II

                                                            Page

ATTITUDE DETERMINATION AND CONTROL – MISSIONS   **745**

Messenger Spacecraft Pointing Options (AAS 03-149)
Daniel J. O'Shaughnessy and Robin M. Vaughan   747

CONTOUR Phasing Orbits: Attitude Determination and Control Concepts
and Flight Results (AAS 03-150)
Jozef van der Ha, Gabe Rogers, Wayne Dellinger and James Stratton . . . 767

Acquisition, Tracking and Pointing of Bifocal Relay Mirror Spacecraft
(AAS 03-151)
Brij N. Agrawal   783

In-Flight Performance and Calibration of the Autonomous Minisatellite
PROBA (AAS 03-152)
Jean de Lafontaine, Jean-Roch Lafleur, Isabelle Jean,
Pieter Van den Braembussche and Pierrik Vuilleumier 801

Attitude Sensor Alignment and Calibration for the Timed Spacecraft
(AAS 03-153)
Mark E. Pittelkau and Wayne F. Dellinger.   821

Modeling and Analysis for Nutation Time Constant Determination of
On-Axis Diaphragm Tanks on Spinners: Application to the Deep Space One
Spacecraft (AAS 03-155)
Marco B. Quadrelli 835

Modeling of MARSIS Segmented Booms and Prediction of In-Flight
Dynamics of the Mars Express Spacecraft (AAS 03-156)
Edward Mettler and Marco B. Quadrelli   855

NAVIGATION AND CONTROL   **873**

Numerical Solution to the Small-Body Hovering Problem (AAS 03-157)
S. Broschart and D. J. Scheeres 875

Development of Spacecraft Orbit Determination and Navigation Using Solar
Doppler Shift (AAS 03-159)
Andrew J. Sinclair, Troy A. Henderson, John E. Hurtado
and John L. Junkins   895

Mutibody Parachute Flight Simulations for Planetary Entry Trajectories
Using "Equilibrium Points" (AAS 03-163)
Ben Raiszadeh 913

ORBIT DETERMINATION 925

Real-Time Estimation of Local Atmospheric Density (AAS 03-164)
James R. Wright 927

Comparison of MSIS and Jacchia Atmospheric Density Models for Orbit
Determination and Propagation (AAS 03-165)
Keith A. Akins, Liam M. Healy, Shannon L. Coffey
and J. Michael Picone 951

Implementing the MSIS Atmospheric Density Model in OCEAN
(AAS 03-166)
Lisa A. Policastri and Joseph M. Simons 971

A Validation of Atmospheric Density Determination From LORAAS
Ultraviolet Spectra By Comparison With HASDM Satellite Drag
Measurements (AAS 03-167)
S. H. Knowles, A. C. Nicholas, S. E. Thonnard, J. M. Picone,
K. F. Dymond and S. McCoy 983

Brightness Loss of GPS Block II and IIA Satellites on Orbit (AAS 03-169)
Frederick J. Vrba, Henry F. Fliegel and Lori F. Warner 995

Comparison of Accuracy Assessment Techniques for Numerical Integration
(AAS 03-171)
Matthew M. Berry and Liam M. Healy 1003

SPACE-BASED INTERFEROMETRY 1017

Formation Path Planning for Optimal Fuel and Image Quality for a Class of
Interferometric Imaging Missions (AAS 03-172)
I. I. Hussein, D. J. Scheeres, D. C. Hyland 1019

Worst Case and Mean Squared Performance of Imaging Systems:
A Feature-Based Approach (AAS 03-173)
Suman Chakravorty, Pierre T. Kabamba and David C. Hyland 1039

Interferometric Observatories in Earth Orbit (AAS 03-174)
I. I. Hussein, D. J. Scheeres, D. C. Hyland 1057

Design of Spacecraft Formation Orbits Relative to a Stabilized Trajectory
(AAS 03-175)
F. Y. Hsiao and D. J. Scheeres 1075

Optimal Coordination of Mobile Agents: Application to Space-Based
Interferometers (AAS 03-176)
Venkatesh G. Rao and Pierre T. Kabamba 1095

Probabilistic Controller Analysis and Synthesis: The Method of HPD
Inscription (AAS 03-177)
Hiroaki Fukuzawa and Pierre T. Kabamba 1115

CONSTELLATIONS 1125

Walker Constellations to Minimize Revisit Time in Low Earth Orbit
(AAS 03-178)
    Thomas J. Lang 1127

Daily Repeat-Groundtrack Mars Orbits (AAS 03-179)
    Gary Noreen, Stuart Kerridge, Roger Diehl, Joseph Neelon, Todd Ely
    and Andrew E. Turner 1143

Single Satellite Orbital Figure-of-Merit (AAS 03-180)
    John L. Young III, David W. Carter, John E. Draim and Paul J. Cefola . . 1157

Drag and Stability of a Low Perigee Satellite (AAS 03-182)
    Joseph R. Schultz and Mark J. Lewis 1179

ORBITAL DEBRIS 1195

Improved Analytical Expressions for Computing Spacecraft Collision
Probabilities (AAS 03-184)
    Ken Chan 1197

Risk of Collision for the Navigation Constellations: The Case of the
Forthcoming Galileo (AAS 03-185)
    A. Rossi, G. B. Valsecchi and E. Perozzi 1217

The 2002 Italian Optical Observations of the Geosynchronous Region
(AAS 03-186)
    Manfredi Porfilio, Fabrizio Piergentili and Filippo Graziani 1237

A Geometrical Approach to Determine Blackout Windows at Launch
(AAS 03-187)
    V. Rabaud and B. Deguine 1253

Orbital Debris Analysis of Timed Spacecraft Mission (AAS 03-188)
    S. S. Badesha, S. K. Dion and R. E. O'Hara 1267

ATTITUDE DETERMINATION AND DYNAMICS 1283

Conformal Mapping Among Orthogonal, Symmetric, and Skew-Symmetric
Matrices (AAS 03-190)
    Daniele Mortari 1285

Spacecraft Angular Rate Estimation Algorithms for Star Tracker-Based
Attitude Determination (AAS 03-191)
    Puneet Singla, John L. Crassidis and John L. Junkins 1303

Autonomous Artificial Neural Network Star Tracker for Spacecraft Attitude
Determination (AAS 03-192)
    Aaron J. Trask and Victoria L. Coverstone 1317

An Analysis of the Quaternion Attitude Determination Filter (AAS 03-194)
    Mark E. Pittelkau 1337

Attitude Interpolation (AAS 03-197)
    Sergei Tanygin   1353

NAVIGATION AND ORBIT DETERMINATION – OPERATIONS I   **1371**

Reconstruction of the Voyager Saturn Encounter Orbits in the ICRF System
(AAS 03-198)
    Robert A. Jacobson   1373

Satellite Ephemerides Update Schedule for the Cassini Mission
(AAS 03-199)
    Ian Roundhill and Duane Roth   1391

Cassini Navigation During Solar Conjunctions Via Removal of Solar Plasma
Noise (AAS 03-200)
    P. Tortora, L. Iess, J. J. Bordi, J. E. Ekelund and D. Roth .   .   .   1407

Interplanetary Navigation During ESA's *BepiColombo* Mission to Mercury
(AAS 03-201)
    Rüdiger Jehn, Juan L. Cano, Carlos Corral and Miguel Belló-Mora .   . 1425

Genesis Trajectory and Maneuver Design Strategies During Early Flight
(AAS 03-202)
    Roby S. Wilson and Kenneth E. Williams   1439

Orbit Determination Support for the Microwave Anisotropy Probe (MAP)
(AAS 03-203)
    Son H. Truong, Osvaldo O. Cuevas and Steven Slojkowski.   .   .   1457

Navigating CONTOUR Using the Noncoherent Transceiver Technique
(AAS 03-204)
    Eric Carranza, Anthony H. Taylor, Dongsuk Han, Cliff E. Helfrich,
    Ramachand Bhat and Jamin S. Greenbaum   1473

Estimating General Relativity Parameters From Radiometric Tracking of
Heliocentric Trajectories (AAS 03-205)
    Ryan S. Park, Daniel J. Scheeres, Giacomo Giampieri,
    James M. Longuski and Ephraim Fischback   1493

Part III

	Page
MISSION DESIGN I	1513

SIRTF Mission Design (AAS 03-207)
Eugene P. Bonfiglio and Mark D. Garcia 1515

Design and Implementation of CONTOUR's Phasing Orbits (AAS 03-208)
David W. Dunham, Daniel P. Muhonen, Robert W. Farquhar,
Mark Holdridge and Edward Reynolds 1535

MESSENGER Mercury Orbit Trajectory Design (AAS 03-209)
James V. McAdams 1549

Options for a Mission to Pluto and Beyond (AAS 03-210)
Massimiliano Vasile, Robin Biesbroek, Leopold Summerer,
Andres Galvez and Gerhard Kminek 1569

Trajectory Design for the Mars Reconnaissance Orbiter Mission
(AAS 03-211)
C. Allen Halsell, Angela L. Bowes, M. Daniel Johnston,
Daniel T. Lyons, Robert E. Lock, Peter Xaypraseuth,
Shyam K. Bhaskaran, Dolan E. Highsmith and Moriba K. Jah 1591

Primary Science Orbit Design for the Mars Reconnaissance Orbiter Mission
(AAS 03-212)
Angela L. Bowes, C. Allen Halsell, M. Daniel Johnston,
Daniel T. Lyons, Robert E. Lock, Peter Xaypraseuth,
Shyam K. Bhaskaran, Dolan E. Highsmith and Moriba K. Jah 1607

| **TETHERS** | 1625 |

Libration Control of Electrodynamic Tethers in Inclined Orbit (AAS 03-214)
J. Peláez and E. C. Lorenzini 1627

Damping in the Dynamic Stability of Deorbiting Bare Tethers (AAS 03-215)
J. Peláez and M. Lara 1647

Modeling and Estimation of ProSEDS Decay (AAS 03-216)
Mario L. Cosmo, Joshua Ashenberg and Enrico C. Lorenzini 1667

Dynamics of a Multi-Tethered Satellite System Near the Sun-Earth
Lagrangian Point (AAS 03-218)
Brian Wong and Arun K. Misra 1675

Flexibility Effects on Non-Planar Spin-up Dynamics of
Artificial-Gravity-Generating Tethered Satellite System (AAS 03-219)
Andre P. Mazzoleni and John H. Hoffman 1695

Control of a Rotating Variable-Length Tethered System (AAS 03-220)
Mischa Kim and Christopher D. Hall 1713

Retargeting Dynamics of a Linear Tethered Interferometer (AAS 03-221)
Claudio Bombardelli, Enrico C. Lorenzini and Marco B. Quadrelli . . . 1733

Analysis and Damping of Lateral Vibrations in a Linear Tethered
Interferometer (AAS 03-222)
Enrico C. Lorenzini, Claudio Bombardelli and Marco B. Quadrelli . . . 1749

ORBIT DETERMINATION II   **1767**
Autonomous Landmark Based Spacecraft Navigation System (AAS 03-223)
Yang Cheng and James K. Miller 1769

Optical Landmark Detection for Spacecraft Navigation (AAS 03-224)
Yang Cheng, Andrew E. Johnson, Larry H. Matthies and Clark F. Olson . . 1785

A Model for Satellite Position and Velocity With Respect to Ground
Topography (AAS 03-226)
Stefano Casotto 1805

Contributions of Individual Forces to Orbit Determination Accuracy
(AAS 03-227)
Justin E. Register 1819

Error Assessment of a Low-Budget Precision Orbit Determination Program
for a Low-Earth-Orbit Satellite Using GPS Observation Data (AAS 03-228)
Jung Hyun Jo, Nammi Jo Choe and John E. Cochran, Jr. 1839

Precise Real-Time Orbit Estimation Using the Unscented Kalman Filter
(AAS 03-230)
Deok-Jin Lee and Kyle T. Alfriend 1853

Determination of Satellite Formation Geometry and Phasing From Range
Data (AAS 03-231)
Trevor Williams, David Thompson and Anees Syed 1873

ORBITAL DYNAMICS   **1885**
Long and Short-Period Librations and Some Mechanical Analogies
(AAS 03-232)
Roger A. Broucke 1887

Phase Space Structure for Three-Dimensional Motion Around Europa
(AAS 03-233)
Martín Lara and Juan F. San Juan 1907

Numerical Investigation of Perturbation Effects on Orbital Classifications in
the Restricted Three-Body Problem (AAS 03-235)
H. Yamato and D. B. Spencer 1923

Comparison of the DSST and the USM Semi-Analytical Orbit Propagators
(AAS 03-236)
Paul J. Cefola, Vasiliy S. Yurasov, Zachary J. Folcik, Eric B. Phelps,
Ronald J. Proulx and Andrey I. Nazarenko 1943

Long-Term Evolution of the KOMPSAT-1 Orbit (AAS 03-237)
Byoung-Sun Lee, Jeong-Sook Lee and Jae-Hoon Kim 1985

Motions in a Central Force Field With a Quadratic Drag Model
(AAS 03-238)
Mayer Humi and Thomas Carter 2005

Formulas for the Drag Constant and Time of Flight in the Two-Body
Problem With Quadratic Drag (AAS 03-239)
Thomas Carter and Mayer Humi 2019

Clohessy-Wilshire Equations Modified to Include Quadratic Drag
(AAS 03-240)
Thomas Carter and Mayer Humi 2025

MISSION DESIGN II **2041**
Mission Planning for the Space Maneuver Vehicle (AAS 03-241)
William E. Wiesel 2043

Outer-Planet Mission Analysis Using Solar-Electric Ion Propulsion
(AAS 03-242)
Byoungsam Woo, Victoria L. Coverstone, John W. Hartmann
and Michael Cupples 2051

Heliocentric Earth Trailing Orbit Design for a Small Probe Concept
(AAS 03-243)
Emanuele Di Sotto, Lorenzo Tarabini and Mariella Graziano 2063

Preliminary Design of Earth-Mars Cyclers Using Solar Sails (AAS 03-244)
Robert Stevens and I. Michael Ross 2075

Designing Phase 2 for the Double-Lunar Swingby of the Magnetospheric
Multiscale Mission (MMS) (AAS 03-245)
Ariel Edery 2089

Study on Recovery of Escape Missions (AAS 03-246)
Stefania Cornara, Miguel Belló-Mora and Martin Hechler 2101

The CloudSat Mission: A Virtual Platform (AAS 03-247)
Ronald J. Boain 2121

NAVIGATION AND ORBIT DETERMINATION – OPERATIONS II   **2139**

Characterization of Space Surveillance Sensors Using Normal Places
(AAS 03-248)
    John H. Seago, Mark A. Davis and Anne E. Reed 2141

Navigation for the Mars Premier Netlander Delivery (AAS 03-249)
    S. Delavault, L. Francillout, D. Carbonne, H. Fraysee,
    P. D. Burkhart, D. Craig, J. Guinn 2163

Approach Navigation for the 2009 Mars Large Lander (AAS 03-250)
    P. Daniel Burkhart 2183

Orbit of Mars Explorer NOZOMI and its Determination by Delta-VLBI
Technique (AAS 03-251)
    Makoto Yoshikawa, Jun'ichiro Kawaguchi, Hiroshi Yamakawa,
    Takaji Kato, Tsutomu Ichikawa, Takafumi Ohnishi and Shiro Ishibashi . . 2199

Radiometric Orbit Determination Activities in Support of Navigating Deep
Space 1 to Comet Borrelly (AAS 03-252)
    Brian M. Kennedy, Shyam K. Bhaskaran, Joseph E. Riedel
    and Mike Wang. 2217

GRACE Precise Orbit Determination (AAS 03-253)
    Z. Kang, B. Tapley, S. Bettadpur, P. Nagel and R. Pastor 2237

The Trajectory of a Photon: General Relativity Light Time Delay
(AAS 03-255)
    James K. Miller and Slava G. Turyshev 2245

APPENDICES   **2257**

Publications of the American Astronautical Society. 2258
    Advances in the Astronautical Sciences 2259
    Science and Technology Series 2267
    AAS History Series. 2274

INDEX   **2277**

Numerical Index 2279

Author Index 2287

ATTITUDE DETERMINATION AND CONTROL - MISSIONS

SESSION 7

Chair: Al Treder
                                                  Dynacs Inc.

The following paper was not available for publication:

AAS 03-154
   "Center of Mass Calibration of GRACE Mission," by Furun Wang, Larry Romans,
   Byron D. Tapley, Srinivas V. Bettadpur, University of Texas (Paper Withdrawn)

MESSENGER SPACECRAFT POINTING OPTIONS

Daniel J. O'Shaughnessy[*] and Robin M. Vaughan[*]

Planning is now underway for the MESSENGER mission to Mercury. Scheduled for launch in March 2004, MESSENGER will orbit the planet for one Earth-year beginning in April 2009. A variety of different spacecraft attitudes are needed to support science observations, permit communication with Earth, and perform engineering activities. With the diverse suite of instruments and constant thermal constraints, the commanded attitude algorithms are an important part of MESSENGER's guidance and control (G&C) system design. This paper presents the basic architecture of the on-board guidance system being implemented to generate desired (or commanded) spacecraft attitude and rate. Algorithms are given for the suite of base pointing commands and scan pattern combinations. The strategy for enforcing thermal safety constraints in the commanded attitude computation is described. Additionally, guidance commands during ΔV maneuvers are briefly discussed.

INTRODUCTION

Planning is now underway for the MESSENGER (MErcury Surface, Space ENvironment, GEochemsitry, and Ranging) mission. As part of NASA's Discovery program, MESSENGER will be the first spacecraft to closely observe the planet Mercury since the Mariner 10 flybys of the mid-1970s. This scientific investigation of Mercury will provide insight into the formation and evolution of all the terrestrial planets.[1] Scheduled for launch in March 2004, MESSENGER will make two flybys of Venus and two of Mercury prior to orbiting the planet for one Earth-year beginning in April 2009. These flybys will assist in developing the focused science gathering of the year-long orbit phase of the mission. MESSENGER will carry a diverse suite of miniaturized science instruments to globally characterize the planet.[2] A variety of different spacecraft attitudes are needed to support the different observation strategies of these instruments, communicate with Earth, and perform engineering activities such as trajectory correction maneuvers.

Due to the extreme environment at Mercury, thermal and radiation concerns are the main factor driving the spacecraft design.[3] Protection from this environment is accomplished with a large sunshade mounted on the −Y body axis, which shields the spacecraft components from direct exposure to the Sun, as shown in Figure 1. Nominally, MESSENGER is flown with this sunshade centered on the sunline, and the Sun vector must remain near the −Y axis at all times while in orbit about Mercury. Power generation is handled with solar panels that point out the +/-X body axes and are capable of rotating about the X axis to track the Sun. MESSENGER carries 17 thrusters onboard for trajectory corrections, attitude control (nominally during burns only), and momentum offloads. The main engine for large maneuvers is mounted on the top deck of the spacecraft, along the −Z direction. Additional thrusters are mounted around the spacecraft structure to provide redundant force/torque capability about all body axes. Most of the science

---

[*] Space Department, Mission Concept and Analysis Group, The Johns Hopkins University, Applied Physics Laboratory, 11100 Johns Hopkins Road, Laurel, Maryland 20723-6099.

instruments are co-boresighted along the +Z axis and are mounted inside the launch vehicle adapter ring. Two sets of high- and medium-gain antennas are mounted on opposite sides of the spacecraft for communication with Earth.

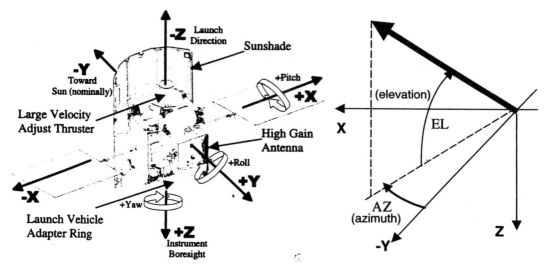

Figure 1. Current Configuration of the MESSENGER Spacecraft and Body Frame Definition

The MESSENGER guidance and control system maintains a 3-axis stabilized spacecraft using reaction wheels as the primary actuators for attitude control. Inertial reference for attitude and body rotation rates is provided through two co-boresighted star trackers (with one acting as a cold spare) and an inertial measurement unit with four gyroscopes. MESSENGER also carries a set of six Sun sensors to provide Sun-relative attitude knowledge if there is a failure in the primary attitude sensors. Guidance and control also uses the propulsion system and thrusters for attitude control during trajectory maneuvers and momentum dumps and may also use the thrusters as a backup system for attitude control in the event of multiple wheel failures.

MESSENGER operates in three distinct spacecraft modes. The first mode is termed "Operational" and represents the normal mode for science collection and engineering activities. In this mode, all the pointing options described in this paper are available for use. Restrictions on the pointing are imposed in either of the safing modes. The intermediate level safing mode is "Safe Hold" mode, and in this mode MESSENGER is restricted to an attitude that permits communication with the Earth via one of its phased-array antennas. In the event that there is not sufficient information onboard to establish communication with the Earth, the spacecraft goes into its lowest level safing mode, "Earth Acquisition" mode. In this mode, the spacecraft attempts to use the available information to search for the Earth (and possibly the Sun, as needed). Promotion to a higher mode can only occur via ground command, although demotion may occur autonomously.

With the diverse suite of instruments and constant thermal constraints, the commanded attitude algorithms are an important part of MESSENGER's guidance and control system design. This paper presents the basic architecture of the on-board guidance system used to generate desired (or commanded) spacecraft attitude. First, the overall system architecture and general formulation of the commanded attitude is described. This includes the underlying ephemeris and celestial body models used in computing commanded attitude, as well as a general algorithm for computing attitude and rate commands. Then, the suite of pointing options available to mission operators and science planners is discussed. Algorithm details are given for special science scenarios such as atmospheric limb scans and engineering attitudes such as data downlink via the spacecraft's phased-array antennas. Key system features such as the interfaces with the antennas and science cameras for setting boresight positions, overlay of scan patterns on the base pointing, and the strategy for enforcing Sun-related attitude safety constraints are discussed. Lastly, the guidance strategy for maneuvers is briefly described.

GUIDANCE SYSTEM ARCHITECTURE

The guidance system architecture for the MESSENGER mission has its roots in the structure of the Near Earth Asteroid Rendezvous (NEAR) mission flight code.[4] MESSENGER's onboard guidance system computes commanded spacecraft attitude and rotation rate. Attitude is represented by the transformation from the EME2000 inertial reference frame to the spacecraft body axes defined in Figure 1. Rotation rate is given as an angular velocity vector in the body frame. These commands are computed at a rate of 1 Hz by the main background guidance task and propagated at the attitude control frequency of 50 Hz. The system first determines the pointing option to be used based on operator request and the current spacecraft operational mode. Certain pointing options are automatically chosen regardless of operator requests when MESSENGER is in one of its two safe modes. The quaternion describing the body frame orientation relative to the inertial frame and the associated rotation rate are then computed assuming no restrictions from the various pointing constraints. Next the result is checked for violations of safety constraints and modified if a violation is detected. Positions for the science camera or phased-array antenna boresights are computed if these are included in the target specification.

Ephemeris models and models for the shape, size, and rotation of a target planet are available to the guidance system when needed to formulate the commanded attitude. Ephemerides stored in the flight processor memory give the position and velocity of the Sun, the Earth, a target planet, and the spacecraft referenced to the solar system barycenter. The target planet is either Venus or Mercury depending on the mission phase. Venus is only used around the two flybys while Mercury is used for the flybys and during the orbital phase. The on-board ephemerides are derived from NAIF SPK kernels supplied by the Jet Propulsion Laboratory navigation team.[5] Additional parameters are stored in the flight processor memory for the standard International Astronomical Union (IAU) model giving target planet body-fixed frame orientation relative to the inertial frame and for the triaxial ellipsoid approximation of the planet's shape and size. This IAU rotation model is included to enable precise pointing on/around Mercury when in the orbital phase of the mission.

For each G&C science or engineering activity, there is a set of commands used to define the desired attitude and rotation rate. Each activity involves setting a basic pointing command, which can then be supplemented with a scan pattern command. These two commands specify the unconstrained pointing, while an additional command determines whether or not to apply the Sun safety constraints.

All of the inertial pointing commands share a common framework, which uses four vectors to define the desired attitude. Two of these vectors are specified in the spacecraft body frame, and two must lie in the external (inertial) frame, as the targets for the body axes. Though the external targets must be expressed in terms of the inertial coordinate axes, they need not be specified in the inertial frame, as long as the spacecraft has the requisite transformation matrices to establish the target's state in the inertial frame. For instance, the target may be provided as a latitude and longitude on a specified planet, as long as rotation models for that planet are maintained to transform that planetary position into an inertial position.

As mentioned above, the spacecraft basebody attitude is specified via 4 vectors. The two external vectors provide targets for the two specified axes in the spacecraft body frame. The primary external target is termed the aimpoint and designated by the vector **A**, and the secondary external target is called the external roll reference vector (ERRV) and is designated by the vector **E**. In the body frame, the axis to be aligned with the primary target is called the boresight designated by the vector **B**, and the axis to be aligned with the external roll reference vector is called the body roll vector (BRV) and is designated by the vector **R**. The external target vectors are specified in terms of the EME2000 frame components and in general, can contain nonzero velocity components. The body vectors are specified in terms of the spacecraft body axes, and these vectors may also be time varying. Based on this nomenclature, the attitude is formulated as:

$$\hat{x} = unit(\mathbf{A}) \tag{1}$$

$$\hat{z} = unit(\mathbf{A} \times \mathbf{E}) \tag{2}$$

$$\hat{y} = \hat{z} \times \hat{x} \tag{3}$$

where $[\hat{x}\ \hat{y}\ \hat{z}]$ are the unit vector components of the axes of an intermediate frame (designated the "prime" frame) expressed in terms of the inertial frame axes. This intermediate frame has been adopted purely as a mathematical convenience and is physically meaningless. These direction cosines can easily be transformed into a quaternion, $q_{I/P}$, representing the transformation from the inertial frame to the prime frame. Likewise a second transformation from another intermediate frame to the body frame can be created as:

$$\hat{x} = unit(\mathbf{B}) \tag{4}$$

$$\hat{z} = unit(\mathbf{B} \times \mathbf{R}) \tag{5}$$

$$\hat{y} = \hat{z} \times \hat{x} \tag{6}$$

where $[\hat{x}\ \hat{y}\ \hat{z}]$ are the unit vector components of the axes of the prime frame expressed in terms of the body axes. These direction cosines can easily be transformed into a quaternion, $q_{B/P}$, representing the transformation from the inertial frame to the prime frame. Equating the prime frames from Eqs. (1)-(3) and Eqs. (4)-(6), and by adopting the notation in Wertz[6], the quaternion from the inertial frame to the basebody frame is:

$$q_{B/I} = (q_{I/P} *) q_{B/P} \tag{7}$$

where the * operation indicates the quaternion inverse, and the result is the quaternion product of $q_{P/I}$ and $q_{B/P}$. The flight code passes this quaternion directly to the attitude control algorithms for use in the quaternion feedback controller. From Eqs. (1) and (4) it is easy to see that this aligns the boresight with the aimpoint, and then the spacecraft is rolled around the boresight axis until the body roll vector lies in the same half-plane as the external roll reference vector. Note that in general, it is not possible to achieve simultaneous precision pointing of two targets, especially when both body axes are fixed in the body frame.

The angular velocity components of the body frame can readily be expressed as functions of the attitude vectors and their first derivatives. Again, an intermediate frame is adopted for convenience and the equations are formulated with respect to this prime frame. The rates of the prime frame unit vectors in the inertial frame can be expressed as:

$$\omega_{P/I_x} = -\frac{[(\mathbf{A} \times \mathbf{E}) \times \mathbf{A}] \cdot (\mathbf{A} \times \dot{\mathbf{E}} + \dot{\mathbf{A}} \times \mathbf{E})}{|\mathbf{A} \times \mathbf{E}|^2 |\mathbf{A}|} \tag{8}$$

$$\omega_{P/I_y} = -\frac{(\mathbf{A} \times \mathbf{E}) \cdot [(\mathbf{A} \times \dot{\mathbf{A}}) \times \mathbf{A}]}{|\mathbf{A} \times \mathbf{E}||\mathbf{A}|^3} \tag{9}$$

$$\omega_{P/I_z} = \frac{[(\mathbf{A} \times \mathbf{E}) \times \mathbf{A}] \cdot [(\mathbf{A} \times \dot{\mathbf{A}}) \times \mathbf{A}]}{|\mathbf{A} \times \mathbf{E}||\mathbf{A}|^4} \tag{10}$$

where $\dot{\mathbf{A}}$ and $\dot{\mathbf{E}}$ are the first time derivatives of the aimpoint position vector and the external roll reference position vector, respectively. Similarly, the expression for the prime frame unit vector angular rates expressed in the body frame is given by:

$$\omega_{P/B_x} = -\frac{[(\mathbf{B} \times \mathbf{R}) \times \mathbf{B}] \cdot (\mathbf{B} \times \dot{\mathbf{R}} + \dot{\mathbf{B}} \times \mathbf{R})}{|\mathbf{B} \times \mathbf{R}|^2 |\mathbf{B}|} \tag{11}$$

$$\omega_{P/B_y} = -\frac{(\mathbf{B} \times \mathbf{R}) \cdot [(\mathbf{B} \times \dot{\mathbf{B}}) \times \mathbf{B}]}{|\mathbf{B} \times \mathbf{R}| |\mathbf{B}|^3} \tag{12}$$

$$\omega_{P/B_z} = \frac{[(\mathbf{B} \times \mathbf{R}) \times \mathbf{B}] \cdot [(\mathbf{B} \times \dot{\mathbf{B}}) \times \mathbf{B}]}{|\mathbf{B} \times \mathbf{R}| |\mathbf{B}|^4} \tag{13}$$

where $\dot{\mathbf{B}}$ and $\dot{\mathbf{R}}$ are the first time derivatives of the boresight position vector and the body roll vector expressed in terms of the body axes, respectively. From Eqs. (8)-(10) and (11)-(13) the desired body rates with respect to the inertial frame can be found from the component equations:

$$\omega_{B/I_x} = \omega_{P/I_x} - \omega_{P/B_x} \tag{14}$$

$$\omega_{B/I_y} = \omega_{P/I_y} - \omega_{P/B_y} \tag{15}$$

$$\omega_{B/I_z} = \omega_{P/I_z} - \omega_{P/B_z} \tag{16}$$

This vector is expressed in terms of the inertial axes, and must be rotated into the body frame using the current spacecraft commanded attitude to find the rate command. The above procedure provides the unconstrained attitude and rate commands to the control system. These commands may be supplemented with a scan pattern for imaging mosaics or could also be modified to accommodate the Sun-safety constraints. Both of these additions require modifications to the commanded attitude and/or rate and will be discussed in subsequent sections. Having completed the definition for the attitude and rate commands, the collection of options for each of the attitude vectors will be presented.

INERTIAL POINTING OPTIONS

The guidance system has one "generic" pointing option that requires all four attitude vectors to be specified independently. Each vector may be specified "by hand," which allows for the ground to command any constant inertial pointing. For many of the fixed boresight instruments, it is not necessary to specify a time-varying vector. Frequently, it is of interest to specify external targets that move in the inertial frame. To prevent the complicated upload of a series of inertially-fixed pointing schemes to approximate the path of a time varying external target, the capability to compute the inertial position of several common targets is built into the software. The options for setting each attitude vector individually are described. This includes many static options, as well as vectors with some velocity in their respective frames. For external targets with some motion in the inertial frame, the target motion must be known or modeled in the software. These mobile targets are limited to solar system bodies and surface points on those bodies.

Ephemeris models for the spacecraft as well as other solar system bodies are made available to the guidance system when needed to formulate the commanded attitude. Each body for which onboard ephemeris is necessary has both a precise and coarse ephemeris model. The precision models are used under normal spacecraft science gathering operations and are necessary to meet the pointing requirements of the science team. The coarse models are lower precision models used as a contingency for spacecraft safing operations. When MESSENGER is in its intermediate safing mode, the spacecraft will continue to use the precise models until they become unavailable, at which point it will autonomously switch into the coarse models. In the lowest level safing mode, the spacecraft assumes that the precise models have failed, and use is restricted to the coarse models.

The precision models use Chebyshev polynomial fits generated by ground software that accommodates the desired accuracy and storage/uplink requirements. These polynomial coefficients are then uplinked to the spacecraft and the guidance software evaluates the polynomials and their time derivatives to extract the necessary positions and velocities for MESSENGER or a celestial body. The precise models are checked with the Sun sensor data as a check on the ephemeris data. If the ephemeris data are not consistent with the Sun sensor readings, the precise models are declared invalid, and the spacecraft begins to use the coarse models as a backup.

The coarse models use 4th-order polynomial fits for the classical Keplerian elements, and the stored coefficients are valid for the entire mission duration. When necessary, these polynomials are evaluated and converted to a Cartesian state vector. The coarse models are only available in the intermediate and lowest-level safing modes and are not suited for the pointing precision required in operational mode.

It is also of interest to model the inertial motion of points on the surface of the target planet. This allows the spacecraft and instruments to track surface features and target planet body-fixed positions in a more intuitive manner. For this purpose, the software maintains a simplified IAU rotation model for the target planet. The standard IAU model consists of three polynomials, two for the planet pole direction and one for the rotation angle about its rotation axis as a function of time. Because the rates of change of the right ascension and declination of the pole are very slow, this computation is eliminated, and the pole direction is assumed fixed in inertial space. This approximation to the actual pole direction simplifies the algorithm, and does not compromise the required pointing accuracy. This allows a transformation matrix to be computed at any instant to rotate body-fixed positions into the inertial frame for computation in the attitude algorithms.

Boresight Options

The boresight vector defines the primary spacecraft body axis for pointing at an external target. The complete set of options for setting this vector is summarized in Table 1. The boresight vector is typically defined as a unit vector, and as such can easily be specified via its Cartesian components, or azimuth and elevation angles in the spacecraft body frame. The convention for these angles is defined in Figure 1. These options are adequate to specify any vector in the body frame, but to simplify the interface for the movable-boresight devices, additional means of setting this vector in the attitude algorithms are provided.

Besides providing attitude and rate commands to the attitude controller, the guidance sets the boresight directions for the Mercury Dual Imaging System (MDIS) and the phased-array antennas. MDIS is a gimbaled camera with a rotational degree-of-freedom about the spacecraft X axis. This imager is capable of scanning 90° and is configured to scan 40° sunward and 50° anti-sunward, as depicted in Figure 2. When pointing the MDIS camera for imaging, the ground operators can simply specify the boresight as an angle relative to the +Z spacecraft axis in the instrument scan plane.

The phased-array antennas are the primary antennas for science downlinks and have a 12° electronically steerable beam over a range of 90°. There are two antennas mounted in the spacecraft X-Y plane, covering the +X, +Y quadrant and the –X, –Y quadrants. When pointing these antennas, ground operators can simply specify the boresight as an azimuth angle in the spacecraft body plane. If the azimuth angle specified is not in the valid range for the antennas, or the MDIS angle is specified outside its valid range, the flight code defaults to the nearest bound of the valid antenna range. Figure 3 illustrates the phased-array antenna fields-of-view in the spacecraft body frame.

Table 1
Summary of Boresight Vector Options

Vector Option	Reference Frame	Number of Parameters
(Unit) vector	Spacecraft body frame	3
Azimuth and elevation angles	Spacecraft body frame	2
Fixed angle in MDIS scan plane	MDIS scan range	1
Fixed angle in phased-array scan plane	Phased-array scan plane	1

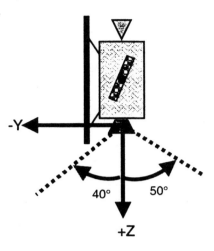

Figure 2. MDIS Scan Range in Spacecraft Body Frame

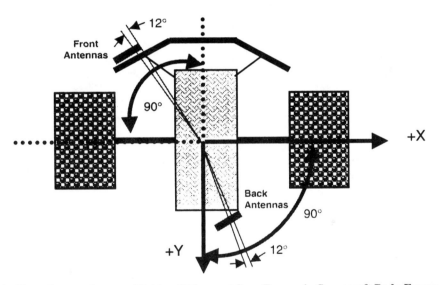

Figure 3. Phased-array Antenna Fields-of-View and Scan Ranges in Spacecraft Body Frame

Aimpoint Vector Options

The aimpoint vector specifies the primary target in the inertial frame. The complete set of options for setting this vector is summarized in Table 2. This target usually defines the science observation point but also can be used for engineering and communications activities. The aimpoint may be specified manually by setting an inertial direction via a unit vector in the inertial frame or right ascension (RA) and declination (DEC) angles referenced to the EME2000 inertial axes. This vector may also be specified by a position in the inertial frame relative to the Earth (ECI frame), Sun (SCI), or the target planet (TPCI). Other alternatives are to specify this vector in the target planet-rotating frame or as latitude and longitude angles and height above the surface. This requires the IAU model for the target planet rotation in the inertial frame, to transform body fixed positions into the inertial frame.

753

Table 2
Summary of Aimpoint Vector Options

Vector Option	Reference Frame	Number of Parameters
(Unit) vector	Inertial (EME2000) frame	3
Right ascension and declination angles	Inertial (EME2000) frame	2
Earth-centered inertial position	Inertial (EME2000) frame	3
Sun-centered inertial position	Inertial (EME2000) frame	3
Target planet-centered inertial position	Inertial (EME2000) frame	3
Target planet body-fixed vector	Target planet rotating frame	3
Target planet LAT/LONG and height	Target planet rotating frame	2
AZ and EL in LVLH frame (AZ reference Is downtrack direction)	LVLH frame	2
AZ and EL in LVLH frame (AZ reference is target planet to Sun vector in LH frame)	LVLH frame	2
Spacecraft-to-Sun vector	N/A	0
Spacecraft-to-Earth vector	N/A	0
Spacecraft-to-target planet vector	N/A	0
Target planet subsolar point	N/A	0
Target planet optimal lighting condition	N/A	0

The aimpoint may also be specified in rotating frames centered at the spacecraft, termed local-vertical, local-horizontal (LVLH) frames. There are two such frames defined in the MESSENGER flight software. For both LVLH frames, the local vertical direction is aligned with the target planet-spacecraft vector. In the first LVLH frame, the azimuth reference in the local horizontal frame is along the flight path (the velocity vector projection in the local horizontal plane). The second frame uses the target planet-Sun vector projection in the local horizontal frame as the azimuth reference. Figure 4 illustrates the two LVLH frames.

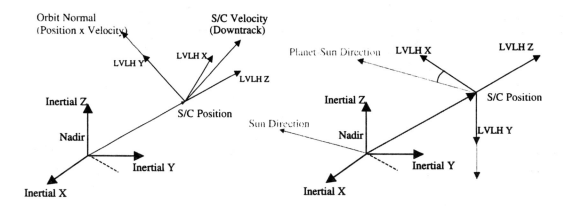

Figure 4. LVLH frames, in which Azimuth Reference is the Spacecraft Downtrack Direction (left) and the Target Planet-Sun Direction (right)

The above methods for describing the aimpoint allow pointing at locations that are fixed in the inertial frame or any of the additional frames modeled onboard the spacecraft. The aimpoint may also be specified as one of a collection of targets whose motion in the inertial frame is modeled. By using the ephemeris models, any of the modeled bodies may be specified as an aimpoint. This includes the spacecraft-Sun vector, the spacecraft-Earth vector, or the spacecraft-target planet vector. Additionally, the science team has specified two specialized viewing geometries of interest that optimize the lighting conditions of the surface point on the target planet. For both options, it is convenient to define the instrument incidence angle (I) and the instrument emission angle (E). The incidence angle is defined as the angle between the target planet surface normal and the Sun direction, in the spacecraft-target planet-Sun plane. The emission angle is the angle between the surface normal and the spacecraft-to-surface point vector. The first pointing scenario aims the spacecraft boresight to the surface location that minimizes the incidence angle or (expressed another way) as close to the planet subsolar point as possible. The intent of the second scenario is to target the instrument boresight close to the subsolar point without creating excessively oblique viewing conditions for the instrument. The science team created an approximate mathematical formulation to quantify their required optimal viewing geometry. Because the mathematical expression is not exact, a tight tolerance on the solution is (in general) not necessary, and simplifying assumptions in the solution are allowed. The desired viewing conditions are achieved by finding the surface point that maximizes the following function:

$$F(E, I) = \cos(E) * \cos(I) \qquad (17)$$

$$E = \pi - \cos^{-1}\left(\frac{r^2 - r_t^2 - R_M^2}{-2r_t R_M}\right) \qquad (18)$$

$$r_t = \sqrt{r^2 + R_M^2 - 2rR_M \cos(\alpha - I)} \qquad (19)$$

where $\mathbf{F}$ is the function to be maximized, r is spacecraft to nadir distance, R_M is central body radius, r_t is the distance from spacecraft to surface point with incidence angle I, E is instrument emission angle, α is spacecraft-nadir-Sun angle. Although Eq. (17) has been expressed above as a function of both E and I, it may be reduced to a function of only the incidence angle by performing the requisite substitution Eq. (19) into Eq. (18), and the result into Eq. (17). This reduces the problem to solving a transcendental function of one variable. The direct approach to solving this equation requires computation of the function's derivative, which is computationally expensive. In order to avoid computation of this derivative, the maximum of the function can easily be found using the family of 1st-order methods, which allow the extremum to be found via function evaluations only. By inspection of Figure 5, it is easy to see that the solution can quickly be bounded to a region between 0 emission and 0 incidence, which assists in the numerical

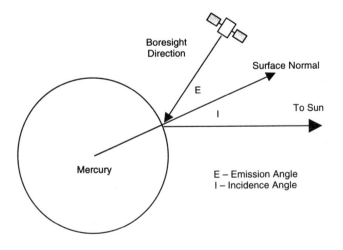

Figure 5. Geometry of Instrument Emission Angle and Incidence Angle for a Given Surface Point

solution. The Golden Section algorithm[7] is used as it is guaranteed to converge for a bounded solution, and the algorithm has the added benefit of giving the number of iterations required for a specified solution tolerance. The method assumes a spherical planet (R_M is constant), which is a reasonable approximation as the mathematical formulation of the lighting condition is not exact and there are no reliable measurements of the oblateness of Mercury. The current standard IAU size model for Mercury assumes a spherical shape. Note that the incidence angle for a surface point on the night side of the planet is meaningless. The algorithm will continue to converge to a solution in this case, although this scenario is meaningful only when the instrument can see the day side of the planet. There is no attempt by the flight code to recognize or avoid the use of this scenario on the night side of the central body, and it is left to the science planning tools and mission operations to avoid this situation.

Body Roll Vector Options

The body roll vector specifies the secondary vector in the body frame to be aligned with the secondary target in the inertial frame. The complete set of options for setting this vector is summarized in Table 3. This vector is typically specified as a unit vector (dimensional vectors in the body frame can be reduced to unit vectors without loss of generality). All possible vectors can be specified with Cartesian components for this unit vector in the body frame or equivalently, with azimuth and elevation angles.

Table 3
Summary of Body Roll Vector Options

Vector Option	Reference Frame	Number of Parameters
(Unit) vector	Spacecraft body frame	3
Azimuth and elevation angles	Spacecraft body frame	2

External Roll Reference Vector Options

This vector specifies the secondary external target in the inertial frame. The complete set of options for setting this vector is summarized in Table 4. The suite of secondary targets is less extensive then the options for the aimpoint vector. These options include inertial directions specified as a unit vector in the inertial frame or right ascension (RA) and declination (DEC) angles referenced to the EME2000 axes. Additionally, the external roll reference vector can be specified as any of the ephemeris bodies: the Sun, Earth, or target planet.

Table 4
Summary of External Roll Reference Vector Options

Vector Option	Reference Frame	Number of Parameters
(Unit) vector	Inertial (EME2000) frame	3
Right ascension and declination angles	Inertial (EME2000) frame	2
Spacecraft to Sun vector	N/A	0
Spacecraft to Earth vector	N/A	0
Spacecraft to target planet vector	N/A	0

INERTIAL POINTING SCENARIOS

In the prior section, the options for setting each of the attitude vectors were described. In some cases this involves setting the vector explicitly, and in other cases time-varying positions are computed onboard to track the motion of celestial bodies. As a further convenience to mission planners and the ground operations team, common sets of these vectors have been packaged together as scenario "shortcuts." These sets of attitude vectors can wholly specify the spacecraft attitude and body rates, without the ground manually setting each vector individually. Additionally, a scenario has been added at the request of the science team to conduct atmospheric limb scans. Rather than manually setting the attitude vectors that accomplish this motion, this special case has been included as its own scenario. The scenario descriptions follow and are summarized in Table 5.

Table 5
Summary of Scenario Shortcut Options

Pointing Shortcut Name	Aimpoint	Boresight	BRV	ERRV
+Z pointing	User defined	+Z body	-Y	MESSENGER to Sun
Nadir pointing	Target Planet Nadir	+Z body	-Y	MESSENGER to Sun
MDIS pointing	User defined	MDIS boresight	-Y	MESSENGER to Sun
Double target pointing	User defined/User defined	+Z body/MDIS boresight	-Y	MESSENGER to Sun
Downlink pointing	Spacecraft to Earth	Phased-array boresight	-Y	MESSENGER to Sun
Limb pointing (terminator case)	Surface point on target planet limb and terminator plane	+Z body	-Y	MESSENGER to Sun
Limb pointing (generic case)	Surface point on target planet limb with user defined azimuth	+Z body	-Y	MESSENGER to Sun

Plus Z Pointing

Most of the instruments onboard MESSENGER are mounted on the instrument deck of the spacecraft and are co-boresighted along the +Z spacecraft body axis. Consequently, it is convenient to define a pointing scenario where the aimpoint can be supplied as any target, but the boresight is selected from the suite of instruments looking along the +Z axis. By specifying the secondary target as the Sun and the secondary body axis as –Y, the +Z axis may be pointed at the primary target, without compromising Sun safety. This primary target (aimpoint) may be any (possibly time-varying) inertial target. Allowances have been made for variability in the direction of the boresight of each instrument on the instrument deck. This scenario requires only that the user select the instrument and its intended aimpoint (target).

Nadir Pointing

This scenario is similar to that described in Plus Z Pointing, but the primary target is specified as nadir of the target planet. Under this scenario shortcut, the user need only specify the instrument used for observation (from the suite that is co-boresighted along the +Z axis).

MDIS Pointing

This scenario is used to point the MDIS camera. In addition to achieving viewing geometries not possible with the fixed boresight instruments, the MDIS imager makes it possible to off point the spacecraft under certain geometries in order to use solar torque to passively dump accumulated momentum. In this scenario, the BRV and ERRV are the –Y axis and spacecraft-Sun direction, respectively, which help ensure Sun safety. The aimpoint is the only attitude vector which must be specified by the ground.

Double Target Pointing

The double targeting strategy allows the science team to accommodate different science goals by collecting data with one of the fixed boresight instruments and the MDIS imager. This scenario points a fixed boresight instrument at its desired target, and then uses the scan platform in the MDIS imager to point its camera at a second target, all while maintaining Sun safety. This allows the MDIS instrument team to obtain images while viewing geometries are optimized for any of the other fixed-boresight instruments. This scenario enforces Sun safety as the top priority and always emphasizes pointing the fixed boresight instrument over the MDIS imager. Said another way, this scenario is much like the +Z pointing while using the MDIS flexibility to obtain additional images opportunistically. The reverse heirarchy, emphasizing MDIS targeting over the fixed instruments is not supported.

Downlink Dump Pointing

MESSENGER designates 8-hours every other orbit (centered at apoapsis) for downlinking collected science data. This scenario requires that one of the two fanbeam antennas be pointed at the Earth while simultaneously maintaining Sun safety. The fanbeam antennas each have a 15° by 4° FOV but are electronically steerable along their primary axis up to 90°, so their coverage nominally spans an entire quadrant of the X-Y spacecraft body plane and extends above and below this quadrant by 2°. A sketch of the FOV of the fanbeam antennas is provided in Figure 3. This antenna configuration allows communication with the Earth at all times despite the Sun-safety constraint, although there are short durations during cruise where Sun occultations prevent ground communication. The guidance uses the variability in the boresight for these antennas to dump momentum passively by off pointing the spacecraft-Sun line whenever possible. This passive momentum dump is only constrained when the Earth-spacecraft-Sun geometry is nearly collinear. This passive momentum dumping strategy is an important part of the guidance design, as it reduces the number of momentum offloads from the thrusters and permits better science observations as the orbit remains unperturbed for long stretches. The mathematical formulation for attitudes that use the Sun as a means of passively dumping momentum while in the downlink configuration are described in Vaughan et. al.[8], and are not repeated here.

Limb Pointing/Scans

Another scenario of interest to the science team is an atmospheric limb scan shown in Figure 6. In these scenarios, MESSENGER points an instrument at the central body limb and scans up and down along the radial direction. This scenario is divided into two separate categories, one where the point on the central body's surface must coincide with the terminator plane, and another more general case, where this point is defined by an azimuth reference in the spacecraft's local horizontal plane (LH).

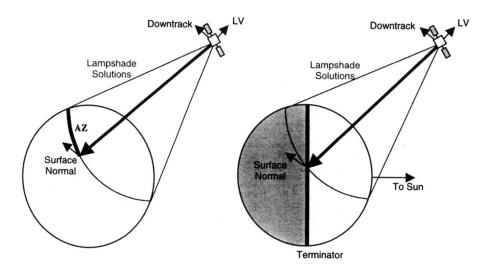

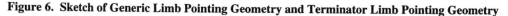

Figure 6. Sketch of Generic Limb Pointing Geometry and Terminator Limb Pointing Geometry

In order for such a limb point to exist, it must meet three conditions: (1) it must lie on the surface of the central body ellipsoid, (2) it must intersect the spacecraft lampshade cone, and (3) it must lie in the plane defined by the azimuth angle (or defined by the terminator plane). Notice that for the generic case, there is always a unique solution, however, for the terminator case there can be 0, 1, or 2 solutions. Mathematically, the three conditions are:

$$\frac{x^2}{a^2} + \frac{y^2}{b^2} + \frac{z^2}{c^2} = 1 \tag{20}$$

$$\frac{r_x}{a^2} x + \frac{r_y}{b^2} y + \frac{r_z}{c^2} z = 1 \tag{21}$$

$$S_x x + S_y y + S_z z = 0 \tag{22}$$

where [x y z] is the location of the limb solution, $[r_x \; r_y \; r_z]$ is the spacecraft position in the central body rotating frame, $[S_x \; S_y \; S_z]$ is the Sun position vector in the central-body rotating frame, and a, b, and c are the central-body ellipsoid radii. Because Eq. (20) is nonlinear, the simultaneous solution yields a quadratic, and thus 0, 1 or 2 solutions, as expected. For the 2-solution case, the solution that is closer to the spacecraft downtrack direction is selected, and for the case without a real solution (imaginary roots in the resulting quadratic) the algorithm defaults to nadir pointing, [x y z] = [0 0 0]. Atmospheric scans are accomplished by linearly varying the limb height over time and adding this additional term along the radial direction of the current limb point. Note that the limb point does creep along the surface during the scan and that the scan is conducted along the radial direction, which for the triaxial ellipsoid is generally not the surface normal direction.

SCAN PATTERNS

One other aspect of the guidance system is the ability to superimpose scan patterns on any of the inertial pointing scenarios. These scans are used to design mosaics or continuous scans that enable target motion in an instrument FOV. Each ground command that sets the spacecraft pointing option has a second component, which indicates whether or not a scan pattern is to be used with the commanded pointing. Scans are parameterized by tuning values in the flight code that regulate the durations, rates, and reversals of the scan motion, as well as specifying the reference frame for the scan commands.

Scan patterns can be executed in several reference frames to achieve a variety of boresight motions relative to a target. Scans may be specified in the spacecraft body frame, or in four different ways when referenced to an external frame. Scans may be commanded as inertial rotations, inertial translations, planet body-fixed translations, or rotations in a local-vertical, local-horizontal frame (LVLH). The guidance system enforces certain compatibility restrictions between the scan frame and the base pointing option, and scan options are not allowed in the downlink pointing scenario. Scans referenced to the body axes may be supplemented to any pointing option (except downlink), but scans constructed in any of the external frames must be consistent with the base pointing option. For instance, if the command was given to point an instrument at a location on the celestial sphere, imposing an inertial translation scan is meaningless, as the target is considered to be infinitely far from the spacecraft. Thus to move this target along the inertial axes by some translational rate is physically meaningless. If incompatible scan motions are requested, the software defaults to the basic pointing and ignores the requested scan.

Each scan pattern is specified as a series of motions at a constant rate, pauses, and reversals along each of the three orthogonal axes of the specified frame. Each axis of the scan frame may have a different commanded scan rate, duration, and (possibly) intervening pauses. This allows the ground team easily to construct intricate boresight tracks with a single algorithm. For scans commanded in the body frame, the algorithm modifies the nominal boresight direction in the body frame with the commanded scan motion. The spacecraft must compensate for the mobile boresight direction in the body frame by rotating the spacecraft in the opposite direction to keep the target in the sensor FOV. For scans referenced to an external frame, it is easy to see that the aimpoint direction in the inertial frame can be modified to include the desired motion, and the nominal spacecraft commands will track this mobile target. So, instead of modifying the attitude and rate command directly with a complex algorithm that models target/body motion, the guidance modifies the attitude vectors prior to computing attitude and body rates. These vectors include the scan motion by changing either the boresight and its rate in the body frame (if the scan pattern command is in the body frame) or the aimpoint and its rate in the inertial frame (if the scan pattern command is in any external frame). This allows the algorithm that computes the commanded attitude and body rates Eqs. (1)-(16) to remain unchanged.

Note that the scan pattern algorithm requires some level of intelligent use by the ground team. At each discrete change in scan rate (at a reversal or a pause), the control system will experience some overshoot in both the attitude and rate profiles. The magnitude and duration of this overshoot is dependent on the desired scan rates, and the torque authority of the actuators (in this case, the reaction wheels). This requires judicious use by the ground operators not to generate scan patterns that exceed the capability of the spacecraft hardware. Likewise, it requires careful planning on the part of the science team to account for the physical limitations of the hardware and to allow for the control system response during imaging sequences.

Pointing Transitions

Figures 7 and 8 show a typical transition between pointing scenarios for the commanded attitude and the estimated attitude. The simulation begins in a downlink attitude, where the Earth appears in the X-Y body plane and inside the +X, +Y quadrant, as this corresponds to one of the phased-array antenna FOVs (EL=0°, AZ=-33°). The Sun direction in the body frame is centered on the −Y axis (0° AZ and EL). The commanded slew begins at 500 seconds, and the spacecraft response takes about 70-100 seconds to settle to the new commanded attitude and rate. The second scenario command is MDIS instrument to Mercury nadir, and this is consistent with the azimuth and elevation for Mercury in the body frame, as it lies in the MDIS scan range. The commanded attitude is discontinuous across transitions between scenarios, and it is left to the control law to handle this discontinuity. There is no attempt in the guidance algorithm to smooth this transition between pointing scenarios, even for large slews. Despite this, the estimated attitude never violates the Sun keep-in zone (SKIZ). In fact, for this case, the Sun never leaves the spacecraft −Y body axis (0° AZ and EL).

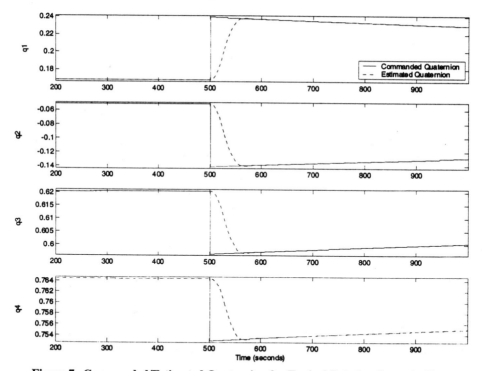

Figure 7. Commanded/Estimated Quaternion for Typical Pointing Scenario Change

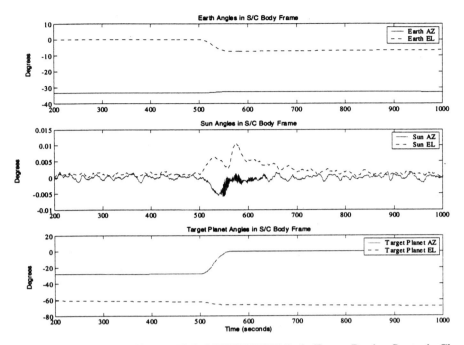

Figure 8. Celestial Body AZ and EL in MESSENGER Body Frame During Scenario Change

CONSTRAINED POINTING

The prior commands specify the attitude vectors that define the pointing during unconstrained operation. With careful planning, it is possible that the attitude safety constraints are never violated, but the guidance continually monitors the commanded attitude for violation of these constraints. If a violation is detected, the command is forced to a valid attitude that both satisfies the constraint and continues to get as close as possible to the intended target.

Sun Safety Constraint

As a result of the extreme thermal environment MESSENGER will experience when in orbit around Mercury, an integral part of the spacecraft design is the thermal sunshade. It is mission critical that this sunshade isolate the instruments and other spacecraft components from direct exposure to the Sun. Consequently, Sun safety is the primary constraint monitored by the guidance algorithm. A constraint boundary may be placed around any spacecraft axis to define the SKIZ. In the event that the ground has determined that the Sun safety constraints are active, the computed attitude must place the spacecraft-Sun line inside this SKIZ. This constraint checks the azimuth and elevation of the spacecraft-Sun line in the attitude command and verifies that it lies within some uploadable bounds on both angles. Nominally, for Mercury orbit conditions, the axis is chosen as the spacecraft -Y body axis (center of the sunshade), and thus the Sun azimuth and elevation must satisfy:

$$-15° < AZ_{Sun} < 15° \tag{23}$$
$$-13° < EL_{Sun} < 13° \tag{24}$$

If the nominal desired attitude places the Sun direction outside this SKIZ, the guidance recomputes an attitude that places the Sun at the constraint boundary. This is accomplished inside the guidance algorithm by first computing the point along the SKIZ boundary that is as close as possible to the Sun vector from the original attitude command. This becomes the new boresight vector, and its corresponding aimpoint is the spacecraft-Sun direction. This will place the Sun vector at the boundary to the SKIZ, and the spacecraft is rotated around this line until the original aimpoint and the original boresight lie as close as possible to one another. By swapping the vectors in the attitude algorithms, the same algorithm can be used to compute the safe attitude and resulting body rates. The external/body vector switching is described in Table 6.

Table 6
Summary of Attitude Vectors in Before/After SKIZ Constraint Applied

	Before SKIZ Constraint	After SKIZ Constraint
Aimpoint	A	Spacecraft to Sun
External Roll Reference	E	A
Boresight	B	Body vector to SKIZ boundary
Body Roll Vector	R	B

Placing the Sun at the SKIZ boundary will no longer allow for precise pointing at the designated aimpoint, but this new attitude will place the original boresight as close as possible to the original aimpoint. The guidance never issues a command outside of the SKIZ boundary, although any attitude may be achieved by changing the limits of the SKIZ or by switching off the constraint entirely. Because of the discontinuous nature of the guidance commands and the resulting tracking errors, it is possible for the control system response to violate the SKIZ. These errors are generally due to overshoot in the control response and as such are of limited magnitude and duration. In addition, the anticipated overshoot can be accounted for in the parameterization of the attitude constraint boundaries, so that Sun safety violations are not triggered. In other words, the software constraint boundaries can be set slightly inside the actual constraint boundary imposed by the physical dimensions of the sunshade.

Hot-pole Keep-Out-Zone
The second constraint monitored by the guidance is also the result of the thermal conditions when in orbit at Mercury. When MESSENGER passes within a small distance of the subsolar point at low altitudes, the top deck (-Z body direction) must be pointed away from the planet. This region is termed the "hot-pole keep-out" (HPKO) zone, and is designed to prevent the top deck from damage due to radiation from Mercury. The geometry for this event will be known well ahead of its occurrence, and under normal circumstances mission planners will avoid generating any commands that violate the HPKO. Guidance must monitor this constraint to prevent inadvertently commanding a violation in the event that the spacecraft is in an intermediate safing mode during a hot-pole crossing. In this event, guidance computes an attitude that is at the HPKO constraint boundary, while as close as possible to the downlink configuration. This modification to satisfy the constraint only persists as long as the constraint persists. Once the spacecraft passes out of the defined hot-pole region, normal downlink pointing is automatically reestablished.

In the event that either of these constraint violations persists for some short period of time, a constraint monitor (external to the guidance algorithm) will force MESSENGER to execute a safing turn. This safing turn interrupts the current observation and rotates the spacecraft to a Sun- (or planet-) safe attitude and demotes the spacecraft to the next lowest safing mode. Due to the drastic response of a constraint violation, the tolerances on the boundaries of the constraint monitor are designed so that they are not triggered by small violations such as overshoot in the control system response. This design prevents unnecessary safing turns without sacrificing spacecraft safety.

Safe Modes
The wide range of options described above is available in normal operation; MESSENGER's operation is limited in both of its lower-level safing modes. In Safe Hold mode, the spacecraft is in an attitude that maintains communication with Earth and enforces strict Sun safety. From a commanded attitude standpoint, if the spacecraft is demoted to Safe Hold mode the guidance logic will accept pointing commands only for downlink attitude. All other commands during this time are ignored. The flight code maintains a separate set of parameters for this scenario (apart from the operational set of downlink parameters) that allows for different parameterization of Sun keep-in-zones. This mode never allows scan patterns and always enforces the Sun safety constraints.

In Earth Acquisition mode, inertial reference has been lost, and the spacecraft is assumed to have only Sun sensor and gyro data. In this mode, the guidance commands the spacecraft into a slow rotation about the Sun direction in the body frame, while the antennas attempt to reestablish communication with the Earth. This safing mode uses a special algorithm in the flight code, as there are not enough vectors to specify completely a unique attitude. In this case, in lieu of giving the controller a commanded attitude and rate, the guidance passes the quaternion error defined by:

$$q_{err} = \begin{bmatrix} unit(\mathbf{S} \times \mathbf{B}) \cdot \sin\left(\frac{\theta}{2}\right) \\ \cos\left(\frac{\theta}{2}\right) \end{bmatrix} \tag{25}$$

$$\theta = \cos^{-1}(\mathbf{S} \cdot \mathbf{B}) \tag{26}$$

where the vector $\mathbf{S}$ is the sensed Sun direction in the spacecraft body frame, θ is the angle between the sensed Sun vector and the spacecraft boresight, and the first three elements of q_{err} denote the vector portion of the quaternion error and the last element the scalar component. The control algorithm uses this attitude error instead of computing it from the commanded/estimated attitudes. The rate command for this mode is the desired rotation rate about the boresight direction.

In the event that the Sun direction in the body frame is lost, due to a loss of Sun sensors and a failure of the coarse ephemerides (an unlikely event), the guidance commands a slow rotation about a sequence of body axes that will allow the Sun sensors to reacquire the Sun vector. This condition is another special case, as there is no attitude command computed in the guidance algorithm for use in the controller. In this case, the control algorithm is in "rate-only" mode, and the guidance software is only creating a body rate command. This rate command begins by commanding a rotation about one of the spacecraft principal axes for a specified duration. If, after some duration, the Sun hasn't returned, the algorithm begins commanding a new rate about a second principal axis. This process of cycling through spacecraft axes is repeated, in an attempt to search randomly the sky for the Sun, until a valid Sun direction reading is obtained from the Sun sensors? or the star trackers reacquire inertial reference, at which point the ephemeris models can be used to obtain the Sun direction.

ΔV GUIDANCE

MESSENGER carries a set of thrusters which are available for attitude control, momentum dumps, and ΔVs. There are three different size thrusters, with each set primarily used for differnent functions. The large ΔV manuevers are handled with the bi-propellant Large Velocity Adjust (LVA) 660-N thruster. This is mounted on the top deck of the spacecraft (-Z face) and imparts a force along the +Z direction, as depicted in Figure 1. Surrounding the LVA are the C-cluster 22-N thrusters, all pointed along the +Z direction and mounted at the corners of the spacecraft –Z face, that are used for smaller manuevers and attitude control when using the LVA. There is also a set of eight 4.4-N thrusters mounted in the corners of the spacecraft for momentum dumps and attitude control during burns. Ther are four additional 4.4-N thrusters that allow for small manuevers in the +Y and –Y when the spacecraft is unable to burn in these body directions due to Sun-safety constraints. All of the 4.4-N and 22-N thrusters are hydrazine monopropellant units, and they can operate in a pressure regulated mode off one of the main fuel tanks, or can operate in blowdown mode when using the auxiliary fuel tank.

The spacecraft relies on the ground commands for manuever execution. The guidance software requires a desired trajectory for the burn to follow and a set of thrusters to use for the burn. In order to follow this desired velocity-space trajectory, the software computes the aimpoint vector that points the thrust vector along the desired ΔV path. The boresight vector for this scenario is the estimated thrust vector direction in the spacecraft body frame. The ERRV and BRV are assumed to be the spacecraft-Sun direction and –Y axis, respectively. The mission design has incorporated the Sun safety constraints into the manuever design, so the actual burns are not expected to violate the SKIZ. Thus for the ΔV guidance, the secondary pair of vectors serves only to specify completely the attitude and is not relied on for Sun safety, as is the case with all other pointing scenarios.

The desired trajectory is given in terms of the velocity profile the spacecraft is to follow (minus any gravitational effects). This trajectory is linear for turn and burn manuevers and can be up to a cubic polynomial for the powered turn used for the Mercury orbit-insertion manuever. A nominal desired trajectory in two-dimensional velocity space is shown in Figure 9. The discussion here is limited to this 2-D case, but can easily be extended to the full three-dimensional case.

Beginning on the dashed desired reference trajectory at Point *1*, the aimpoint is targeted towards a future point on the reference trajectory at Point *2*. Because the thrust vector has some alignment uncertainty and the control system has some imprecision, steering with this aimpoint lands the spacecraft at the off-nominal Point *3*. This has produced some angular error, θ. In order to drive the spacecraft back onto the desired trajectory at some future point, corrections need to be applied to the steering direction computed from the reference profile.

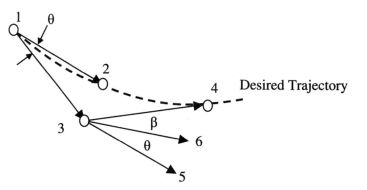

Figure 9. ΔV Guidance Correction Angles

The nominal aimpoint would lie along the direction from Point *2* to Point *4* for the second stage in the burn. Noting that the segments *2-4* and *3-6* are parallel, steering with this aimpoint (derived from the reference trajectory) will cause further perturbations from the desired trajectory. Again this steering strategy will force the spacecraft to drift by angle θ from the original direction, to Point *5*. Correcting for this angular drift in Stage 2 will correct for some of the errors in the actual trajectory, and the spacecraft will end up at Point *6*. However, simply compensating for the misalignment does not allow the full trajectory error to be removed, as there is an accumulated offset due to prior pointing/sensor errors. By applying a second correction, given by angle β, the accumulated offset can be eliminated and the spacecraft will land back on the reference curve at Point *4*. Thus the algorithm has two corrections: θ for thrust vector misalignments and β for accumulated deviations from the reference curve.

The description above is idealized, for even by applying both corrections in Stage 2, it is unlikely that the spacecraft will land back on the trajectory at Point *4*. Misalignments will vary slightly as the center of mass wanders while fuel is consumed. It is also unlikely that controller errors or accelerometer noise are consistent over constant time intervals in the burn. The β correction automatically compensates for these new offsets and will always help to drive the spacecraft back onto the reference curve. However, β is a reactive correction as it can only eliminate accumulated errors. By making the predictive correction, θ, an accumulated correction, that is: by letting $\theta_1 = \theta_0 + \theta_1$, this correction is driven to the thrust vector offset from nominal. This procedure significantly reduces errors due to thrust vector misalignments, and this correction accurately models the thrust vector in the body frame, which is useful for designing future burns. This correction is computed once per second during the burn, and so it is possible that each burn can have hundreds or even thousands of such corrections. This procedure of repeatedly making this predictive correction allows the actual burn to mirror closely the desired trajectory, even in light of measurement noise and center-of-mass motion.

Figure 10 shows the quaternion command and control response during a sample 10 m/s burn. The associated spacecraft velocity errors between the actual (simulated) trajectory and the desired reference trajectory are provided in Figure 11. Although the burn is a simulation of a turn-and-burn maneuver, the

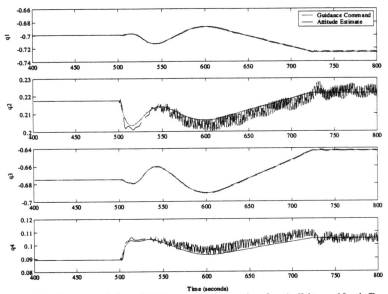

Figure 10. Commanded and Estimated Quaternion for Aribitray 10m/s Burn

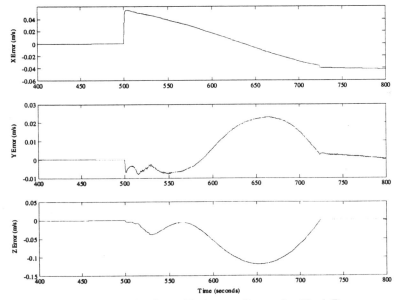

Figure 11. Velocity Space Trajectory Errors for 10m/s Burn

Time (seconds)

quaternion command does wander in order to keep the burn along the desired inertial direction. This variability in the attitude command removes errors due to the thruster performance variations and misalignments as well as accelerometer bias, drift, and noise. The simulation was terminated on accumulated magnitude of the spacecraft velocity change, and results in very small errors in the burn magnitude (~2mm/s) and direction (~0.2°).

CONCLUSION

The guidance algorithms and architecture for the flight code for the MESSENGER mission emphasize computational efficiency and robustness and offer a great deal of flexibility to mission planners to achieve a variety of anticipated pointing conditions. While system accuracy is the chief concern, developing a suite of pointing options that have a high degree of autonomy was an additional motivation. The collection of base pointing options presented, can, in general, be commanded with a single ground command to determine the commanded attitude and rate. This reduces costly communication time with the spacecraft, and the extensibility of the pointing options allows for many viewing geometries to be accommodated without the need for complex ground commands. The base pointing options can be modified with scan patterns that enable complex imaging mosaics to be easily executed. An additional strength of the architecture presented is that the same algorithm is used to compute commanded attitude with and without scan patterns or attitude constraints. Likewise, the burn scenarios are handled with a single algorithm, both in the turn-and-burn case, and the more specialized powered-turn. Development time and costs were also reduced by using flight heritage[4] whenever possible.

ACKNOWLEDGEMENT

The work described in this paper was performed at The Johns Hopkins University Applied Physics Laboratory, under contract (NAS5-97271) with the National Aeronautics and Space Administration, Discovery Program Office.

REFERENCES

1. S. C. Solomon et. al., "The MESSENGER Mission to Mercury: Scientific Objectives and Implementation," *Planetary and Space Science*, Vol 46, Issues 14-15, pp. 1445-1465, December 2001.

2. R. E. Gold et. al., "The MESSENGER Mission to Mercury: Scientific Payload," *Planetary and Space Science*, Vol 46, Issues 14-15, pp 1467-1479, December 2001.

3. A. G. Santo et. al., "The MESSENGER Mission to Mercury: Spacecraft and Mission Design," *Planetary and Space Science*, Vol 46, Issues 14-15, pp. 1481-1500, December 2001.

4. G. A. Heyler and A. P. Harch, "Guidance and Control Aspects of the Mathilde Fast Flyby," AAS Guidance and Control Conference, AAS-98-071, Breckenridge, CO, Feb. 4-8, 1998.

5. C. Acton, *SPICE Overview*, http://pds.jpl.nasa.gov/New_SPICE_Overview.pdf , 2001.

6. J. R. Wertz, *Spacecraft Attitude Determination and Control*, D. Reidel Publishing Co., Dordrecht, Holland, 1978.

7. G. N. Vanderplaats, *Numerical Optimization Techniques for Engineering Design*, Vanderplaats Research and Development, Inc., Colorado Springs, CO, 1998, pp. 49-54.

8. R. M. Vaughan et. al., "Momentum Management for the MESSENGER Mission," Paper AAS 01-380, AAS/AIAA Astrodynamics Specialists Conference, Quebec City, Quebec, Canada, July 30-August 2, 2001.

CONTOUR PHASING ORBITS: ATTITUDE DETERMINATION AND CONTROL CONCEPTS AND FLIGHT RESULTS

Jozef van der Ha,[*] Gabe Rogers,[*] Wayne Dellinger[*] and James Stratton[*]

The CONTOUR spacecraft was launched on July 3, 2002 and placed in an Earth phasing orbit that lasted about 6 weeks. The spacecraft was kept in a spin-stabilized configuration throughout this period. The main objective during this phase was to achieve the proper orbit and attitude parameters for the injection (by means of a Solid Rocket Motor) into a heliocentric trajectory to Encke's comet. The paper describes the main characteristics of the attitude determination and control concepts behind the design of the CONTOUR spin mode. The Earth Sun Sensor operating concept, attitude determination algorithms, as well as attitude accuracy estimates are addressed. Next, the execution of attitude maneuvers, the maneuver calibration concept, and the design of the propulsion system are discussed. Finally, flight results such as the performances of the attitude maneuvers are presented.

INTRODUCTION

The CONTOUR (Comet Nucleus Tour) mission is part of NASA's Discovery Program and aimed at performing imaging and other science explorations of at least two comet nuclei: Encke and Schwassmann-Wachmann-3. The CONTOUR spacecraft was launched on July 3, 2002 and placed in an Earth phasing orbit with a period of 1.75 days by a Delta-7425. The injection into a heliocentric trajectory on course to Encke's comet took place on August 15. Unfortunately, a mishap occurred near the end of the STAR-30BP Solid Rocket Motor (SRM) firing and contact with the spacecraft could not be re-established[1].

During its 6 weeks of Earth phasing orbits, CONTOUR was kept in a spin-stabilized mode with nominal spin rates of 20 and 60 rpm. A large number of orbit and attitude maneuvers were executed with the objective to achieve the most favorable orbit and attitude parameters as well as hydrazine mass at the time of the SRM burn. The spacecraft performance was practically flawless throughout the phasing orbits. This paper summarizes the pertinent design concept as well as the observed in-flight performances of the guidance and control capabilities that supported the spin mode used throughout the phasing orbits.

ATTITUDE DETERMINATION CONCEPT

Earth-Sun Sensor

The orientation of the spacecraft spin axis in inertial space was determined by means of measurements generated by an integrated Earth-Sun Sensor (ESS) unit manufactured by Galileo Avionica of Florence, Italy. This sensor has been used extensively in geo-stationary transfer orbit operations for over 25 years and has shown excellent reliability. The Sun sensor produces pulses when the Sun crosses over the meridian and skew slits during each spin revolution (Figure 1). The spin rate follows from the time difference between two successive meridian pulses, and the Sun angle θ is calculated from the delay between the meridian and skew slit pulses.

* Applied Physics Laboratory, The Johns Hopkins University, 11100 Johns Hopkins Road, Laurel, Maryland 20723-6099.
E-mail: JvdHa@aol.com; Gabe.Rogers@jhuapl.edu; Wayne.Dellinger@jhuapl.edu; James.Stratton@jhuapl.edu.

The Earth sensor has two static pencil-beams oriented at angles μ_i (with $\mu_1 = 60$ and $\mu_2 = 65$ degrees) relative to the positive spin axis. The fundamental measurements consist of the Space/Earth (S/E) and Earth/Space (E/S) crossing times of the Earth's Infra-Red horizon. Figure 2 shows that these crossings (in combination with the Sun sensor's meridian slit crossing time) result in measurements of the Sun-Earth Azimuth Angles (SEAA) α_i and the Half-Chord Angles (HCA) κ_i for each of the pencil-beams ($i = 1, 2$). The SEAA angle α represents the dihedral rotation angle that is formed by the spacecraft spinning (about its Z axis) from the Sun's meridian up to the meridian containing the center of the Earth.

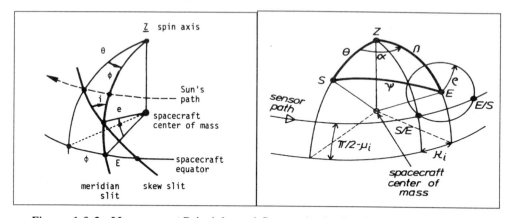

Figures 1 & 2 - Measurement Principles and Geometries for Sun Sensor & Earth Sensor

The design of the Earth sensor has been customized for the CONTOUR-specific phasing orbits with perigee altitudes near 200 km and apogees of about 18 Earth radii (see Ref. 1). It delivers its most accurate performance in the altitude range from about 50000 to 60000 km altitude, which corresponds to the location of the Earth sensor coverage intervals for a spin axis attitude close to the SRM firing direction.

Figure 3 illustrates that the sensor coverage intervals occur in the region after the apogees of the phasing orbits under the selected pencil beam alignment conditions and for the nominal SRM firing attitude direction. The geometrical conditions of the Sun and Earth vectors with a Sun-Earth angle ψ of about 55 degrees in mid August are fairly favorable in terms of expected attitude determination accuracy.

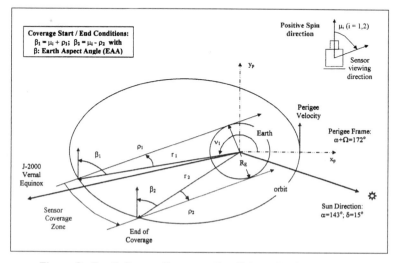

Figure 3 - Earth Sensor Coverage Conditions (Projected View)

Attitude Determination Principles

The most critical requirement for the attitude pointing is imposed by the precision of the thrust direction of the Solid Rocket Motor (SRM): i.e., a half-cone angle of less than 0.75 degrees (including knowledge and control). A pointing error of this magnitude would require a correction of at least 25 m/sec in terms of delta-v that would need to be delivered by the hydrazine thrusters during the heliocentric trajectory.

Figure 4 illustrates the nominal evolution of the measured half-chord angles κ_i (for the two pencil-beams i = 1, 2) as a function of the decreasing Earth Aspect Angle (EAA) β over the coverage intervals. Also shown is the evolution of the apparent Earth radius $\rho(\beta)$ as seen from the spacecraft's position in its orbit.

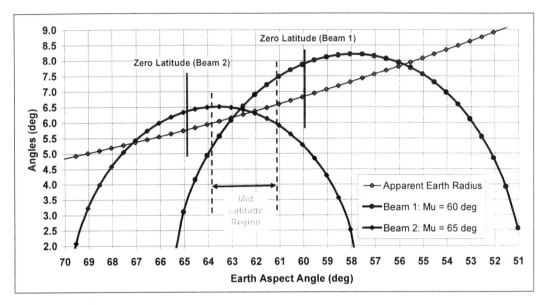

Figure 4 - Half-Chord Angles and Apparent Earth Radius vs. Earth Aspect Angle

The half-chord angles κ_i shown in Figure 4 are derived from the familiar measurement equations for an Earth sensor with nominal pencil-beam mounting alignments of $\mu_1 = 60$ and $\mu_2 = 65$ degrees (Figure 2):

$$\cos\mu_i \cos\beta + \sin\mu_i \sin\beta \cos\kappa_i = \cos\rho \quad (i = 1, 2) \tag{1}$$

It should be noted that (during in-flight operations) the chord angles κ_i form the measurements from which the EAA β needs to be determined. The quality and usefulness of the Earth sensor data vary substantially during the few hours of Earth sensor coverage in accordance with the lengths of the scanned chords and the geometrical sensitivity conditions (Ref. 2). The smaller chords are not useful because of the relatively high systematic errors due to the 'near-tangential' Earth crossings. The larger chords, on the other hand, have a favorable crossing geometry but the attitude determination error may be relatively large because of the low measurement sensitivity when scanning in the equatorial region (near the zero latitudes shown in Figure 4): a small error in the measured chord angle results in a large change in the calculated Earth aspect angle.

The best conditions in terms of the resulting attitude determination accuracy occur when the pencil-beams scan the so-called 'mid-latitude' regions of the Earth corresponding to latitude bands between about 10 and 40 as well as -10 and -40 degrees. It may be noted that the 'latitudes' as used here refer to characteristics of the pencil-beam scans and do not necessarily correspond to the familiar geographical Earth latitudes. The combined mid-latitude regions for both pencil-beams under the nominal SRM attitude conditions would correspond to Earth aspect angles in the range from about 61 to 64 degrees as shown in Figure 4.

Two different software programs were employed for attitude determination support during the phasing orbits while CONTOUR was kept in spin-stabilized mode:

1. **Equal-Chord Method (ECM):** this method makes use of a single Earth aspect angle, namely at the time when the chords produced by each of the two pencil-beams are exactly equal; an attitude estimate is obtained by means of a geometrical method involving also the Sun and SEAA angles

2. **Fine Attitude Determination (FAD):** this method uses a batch of Sun and Earth sensor data collected over about an half-hour interval (within the combined mid-latitude region); it produces a precise attitude estimate by means of a 'Weighted Least-Squares' estimation algorithm.

Equal Chord Method

Even though this method is not widely used, it has a number of favorable characteristics (first mentioned in Ref. 3) and a relatively straightforward mathematical formulation. It uses a short interval of Earth sensor measurements around the crossing of the chords produced by the two pencil-beams. The actual crossing time t_e is determined from the quadratic fits of the two arcs of Half-Chord Angle measurements κ_i that are produced by the two pencil beams (i = 1, 2): the time t_e provides immediately the reference Earth vector $\underline{E}_e$.

The Earth Aspect Angle (EAA) β_e at the crossing time t_e can readily be established by subtracting the functional relationships for the two half-chord angles κ_i given in equation (1) and illustrated in Figure 4:

$$\beta_e = \arctan \{\tan\mu / \cos\kappa_e\} \cong \mu + \tfrac{1}{4} \kappa_e^2 \sin(2\mu) + O(\kappa_e^4) \tag{2}$$

Here, μ is the mean value of the two pencil-beam mounting angles relative to the spin axis (i.e., $\mu = 62.5°$ for CONTOUR's design values) and κ_e is the measured half-chord angle at the time t_e. It can be seen from the simulated results shown in Figure 4 that $\kappa_e \cong 6.4$ degrees. It follows immediately from equation (2) that the Earth aspect angle β_e at the time of equal chords will be $\beta_e \cong 62.6$ degrees with very good accuracy.

Measurement errors in the half-chord-length κ_e do not have an appreciable effect on the resulting β_e since the sensitivity of the EAA β_e to errors in κ_e is small: $|\partial\beta_e/\partial\kappa_e| \cong 0.05$. Furthermore, bias errors in the Earth's Infra-Red horizon have largely been eliminated since the apparent Earth radius does not appear in equation (2) due to the subtraction of the two chords. There may of course be errors that are induced by the non-uniformity of the Earth's Infra-Red radiance profile (since the two pencil-beams are scanning over different parts of the Earth) but these error effects will be fairly small. Therefore, sensor mounting misalignments and/or spacecraft balancing imperfections will most likely be the main contributors to β_e errors.

In addition to the EAA β_e calculated at the time of equal chords, the Equal Chord Method makes use of the Sun Aspect Angle (SAA) ϑ_e and the Sun-Earth Azimuth Angle (SEAA) α_e which is the mean value of the SEAA measurements delivered by the two beams. The complete set of measurement equations is given by:

$$\underline{Z} \bullet \underline{S}_e = \cos\vartheta_e; \quad \underline{Z}_e \bullet \underline{E}_e = \cos\beta_e; \quad \underline{Z} \bullet (\underline{S}_e \times \underline{E}_e) = \sin\vartheta_e \sin\beta_e \sin\alpha_e \tag{3}$$

As long as the Sun and Earth vectors are not co-linear, a unique attitude vector $\underline{Z}$ can be derived from these three equations (note: $\underline{Z}$ must be normalized because of measurement errors, Ref. 4). In geometrical terms, the first two equations provide two solutions representing the intersections of the two cones with half-cone angles ϑ_e and β_e and centered around the instantaneous Sun and Earth vectors $\underline{S}_e$ and $\underline{E}_e$. The measurement α_e provides the resolution of the two-fold ambiguity in the attitude solution through the third equation.

Thus, the ECM attitude determination method does not require any a priori attitude knowledge and is relatively straightforward in terms of implementation effort. Furthermore, the method is extremely 'robust' in terms of sensitivity to chord-length errors and to variations in the apparent Earth radius induced for instance by variations in the Earth's Infra-Red radiance profile.

Fine Attitude Determination

The Fine Attitude Determination (FAD) method uses a batch of Sun and Earth sensor measurements collected over an interval of about 0.5 hr duration while the pencil-beams are scanning over the combined mid-latitude region. The estimation algorithm uses a system of measurement equations similar to equations (3) but expressed in terms of the batch of observation angles ϑ_k, β_k, α_k collected at the instants t_k ($k = 1,...,$ K) with K of the order of 1000 to 2000 at the 60 rpm spin rate. Whereas the motion of the Sun's position in inertial space is practically negligible over a 0.5 hr interval, the changing Earth vector positions $\underline{E}_k$ must be taken into account. A representative Sun-Earth Azimuth Angle measurement α_k at time t_k is provided by the mean value of the two individual SEAA measurements $\alpha_i(t_k)$ delivered by the two pencil-beams.

The model for the EAA measurements $\beta_k = \beta(t_k)$ is more complicated since there are two solutions $\beta_k^{(+)}$ and $\beta_k^{(-)}$ that satisfy a given half-chord angle measurement, i.e. the scans above and below the 'equatorial' Earth crossing, as can be seen from equation (1) and Figure 2. In order to eliminate this ambiguity an a priori reference attitude $\underline{Z}_{ref}$ will be introduced. The attitude established by the ECM method is obviously a suitable candidate for $\underline{Z}_{ref}$. The differences $\delta\kappa_i(t_k)$ between the actual half-chord measurements $\kappa_i(t_k)$ and the predicted (on the basis of the reference attitude) measurements $\kappa_{i,ref}(t_k)$ will be used in the measurement model. After linearization of the two measurement equations in (1), the differences in the Earth Aspect Angles $\delta\beta_i$ can be expressed in terms of $\delta\kappa_i$ at the times t_k for each pencil-beam i = 1, 2:

$$\delta\beta_i(t_k) = f_i(t_k)\, \delta\kappa_i(t_k) \qquad (i = 1,2;\ k = 1, ... , K) \qquad (4)$$

The functions $f_i(t_k)$ represent complicated expressions of the known quantities μ_i, $\kappa_{i,ref}(t_k)$, and $\beta_{ref}(t_k)$. It is important to recognize that the functions $f_i(t_k)$ vary in accordance with the varying sensitivities of the EAA $\beta_i(t)$ as a function of the chord-length angles $\kappa_i(t)$ over the coverage intervals. A single solution $\delta\beta(t)$ may be established by using a linear weighted combination of the individual results for the two pencil-beams:

$$\delta\beta(t_k) = w(t_k)\, f_1(t_k)\, \delta\kappa_1(t_k) + [1-w(t_k)]\, f_2(t_k)\, \delta\kappa_2(t_k) \qquad (5)$$

It makes sense to select the weights $w_k = w(t_k)$ in such a way that the resulting variance of $\beta_k = \beta(t_k)$ will be minimal: the 'optimal' (in a minimum-variance sense) solution $\delta\beta_k^*$ can be established by requiring that $\partial\{\sigma_\beta^2\}/\partial w$ vanishes at every point t_k. It may be noted that the variance $\sigma_\beta^2 = E\{\beta^2\} = E\{(\delta\beta)^2\}$ since the values of the reference Earth aspect angle $\beta_{ref}(t_k)$ are known a priori on the basis of the selected $\underline{Z}_{ref}$. For convenience, it will be assumed that the chord measurements are uncorrelated and will have equal variances σ_κ^2 for the two pencil-beams throughout the mid-latitude region of the coverage interval.

The solutions for the weights w_k can be shown to be equal to $w = f_2^2/(f_1^2 + f_2^2)$ at every point t_k. It follows that the weights of the 60-degree beam will dominate those of the 65-degree beam during the first part of the coverage interval (i.e., for EAA > 62.35 degrees) and vice versa for the second part. The differences are of the order of 2 to 3 at the start and end of the half-hour data interval centered around the time of equal chords. This result can be shown to be consistent with the nature of the evolution of the sensitivity functions $|\partial\beta/\partial\kappa_i|$ for the two pencil-beams.

Finally, the resulting minimum-variance Earth aspect angle and the associated variance are as follows:

$$\beta^*(t_k) = \beta_{ref}(t_k) + \{\delta\kappa_1(t_k)/f_1(t_k) + \delta\kappa_2(t_k)/f_2(t_k)\}/F(t_k); \qquad \sigma_\beta^2(t_k) = \sigma_\kappa^2/F(t_k) \qquad (6)$$

Here, the auxiliary function $F(t_k)$ denotes $1/f_1^2 + 1/f_2^2$ at time t_k. A representative worst-case value for the CONTOUR coverage conditions is given by $F \cong 0.5$ so that $\sigma_\beta^2 \cong 2\sigma_\kappa^2$.

A weighted least-squares solution $\underline{Z}^*$ for the attitude vector can be established on the basis of the batch of 'observation angles' ϑ_k, β_k^*, and α_k at the times t_k ($k = 1, ..., K$) in the following form (see, for instance, Ref. 4, equations 45-48):

$$\underline{\mathbf{Z}}^* = [P]\, \Sigma_k \{\cos\vartheta_k\, \underline{\mathbf{S}}_k / \sigma_S^2 + \cos\beta_k\, \underline{\mathbf{E}}_k / \sigma_E^2 + \sin\vartheta_k\, \sin\beta_k\, \sin\alpha_k\, (\underline{\mathbf{S}}_k \times \underline{\mathbf{E}}_k)/\sigma_N^2\} \tag{7}$$

Here, [P] denotes the covariance matrix of the attitude vector $\underline{\mathbf{Z}}^*$ and σ_S^2, σ_E^2, σ_N^2 represent the variances of the observations (note that these variances have been assumed to remain constant over the data interval):

$$\sigma_S^2 = \sigma_\theta^2 \sin^2\theta; \quad \sigma_E^2 = \sigma_\beta^2 \sin^2\beta; \quad \sigma_N^2 = \sigma_N^2(\theta, \beta, \alpha, \sigma_\theta^2, \sigma_\beta^2, \sigma_\alpha^2) \tag{8}$$

After selecting a reference frame for the attitude vector, it will be possible to establish explicit expressions for the state covariance matrix [P] in terms of the results of the three variances given in (8).

Attitude Determination Accuracy

It is important to distinguish between random and systematic errors in the attitude determination process: typical random errors in the amplitudes of the ESS measurements are less than 0.001 degree for the Sun aspect angle and about 0.02 degrees for the Half-Chord and Sun-Earth Azimuth Angles measured by the Earth sensor. It is obvious that random errors have no appreciable effect on the accuracy of the attitude estimate, in particular when using a batch of hundreds or more individual measurements. Systematic errors (also known as 'biases'), on the other hand, usually have significant impacts on the achievable attitude accuracy. Furthermore, their adverse effects can not be attenuated by using more measurements.

Systematic errors may be induced for instance by any of the following sources:

- Local variations in the Earth's Infra-Red horizon: 0.3 degrees (North-South, worst-case)
- Spacecraft dynamic imbalance (including fuel imbalance): 0.1 degrees (worst-case)
- Residual sensor calibration errors: 0.05 degrees (worst-case)
- ESS mounting and spacecraft alignment errors: 0.03 degrees (worst-case)
- Earth and Sun ephemeris errors: 0.02 degrees (worst-case)
- Thermal effects (structural distortions and electronic effects): 0.02 degrees (worst-case)

Of particular interest are the effects induced by spacecraft balancing errors. The in-flight 'dynamical' spin axis (which corresponds to the actual maximum principal inertia axis) will be 'tilted' with respect to the designated 'geometrical' spin axis (which has been the reference for the mounting of the ESS unit). There are two main contributors to the dynamic imbalance:

1. Measurement threshold of the equipment used for the pre-Launch balancing operations: this may result in a worst-case spin axis tilt of the order of 0.04 degrees;
2. Imbalance induced by the asymmetric use of fuel in the two tanks: this imbalance can be kept well below 1 kg by an appropriate strategy for switching between the two tanks in-between maneuvers: the 1 kg fuel imbalance would correspond to a tilt angle of about 0.08 degrees.

Realistic predictions for the effects of the systematic errors on the actual numerical values of the angles used in the measurement model can not easily be established but conservative limits may be the following:

$$\Delta\theta < 0.1°; \quad \Delta\kappa < 0.2°; \quad \Delta\alpha < 0.3°; \quad \Delta\mu < 0.1°; \quad \Delta\rho \text{ and } \Delta\psi < 0.02° \tag{9}$$

The effect of a particular systematic error on the resulting attitude accuracy may be evaluated by means of a covariance analysis for the single-frame attitude solution based on these input biases (e.g., Ref. 2). A representative explicit formula for the resulting 'attitude error' (i.e., the angular deviation of the unit-vector along the spin axis) may be represented by the trace of the covariance matrix [P] introduced in equation (7):

$$\sigma_{attitude} \cong \{\sigma_S^2 + \sigma_E^2 + \sigma_N^2\}^{1/2} / \sin\psi \tag{10}$$

After substitution of the relevant input parameters, an expected worst-case attitude error of the order of 0.5 degree follows for the CONTOUR-specific conditions prior to the firing of its SRM in mid August 2002.

ATTITUDE CONTROL CONCEPT

Attitude slew maneuvers were performed by a set of 4 hydrazine thrusters (see Figure 10 below) firing in pulsed mode to deliver a precession torque that changes the direction of the angular momentum vector (and thus also the orientation of the spacecraft spin axis) in inertial space. The selected thrusters form a 'balanced' set, which implies that the resulting perturbing effects on the orbit will be negligible. Also the capability for spin control maneuvers was provided and was actually used in-flight to perform the spin-down from 60 to 20 rpm shortly after Launch and the spin-up back to 60 rpm prior to the SRM firing.

When executing attitude maneuvers it is possible to control the direction of the spin axis motion by selecting the appropriate timing of the thrust pulse initiation within each spin revolution. This can be achieved by introducing a constant delay time for these initiations with respect to the occurrence of the Sun sensor meridian pulse. The required delay time is calculated on-ground beforehand and up-linked to the spacecraft along with the number of thrust pulses and the firing duration of the pulses. The resulting motion of the spin axis on the celestial sphere (under a constant delay angle) follows a so-called rhumb path as shown in Figure 5. More details on rhumb-line maneuvers are provided in Wertz[5] (pp. 651-654).

In order that the deviation of the attitude pointing direction during a long attitude slew maneuver can be kept within acceptable bounds, it is necessary to perform careful calibrations of the effective thrust level, the rhumb angle (or heading direction) as well as spin rate. An elaborate maneuver calibration scheme was implemented as part of the first 180-degree slew (i.e., the so-called 'Flip' maneuver) that was required for establishing the proper attitude for executing an orbit maneuver at the time of the second perigee (Ref. 1).

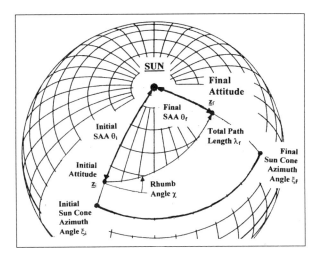

Figure 5 – Geometry of a Rhumb-Line Attitude Maneuver

Figure 6 illustrates the Flip maneuver strategy. The first leg of the Flip maneuver is about 19 degrees long and has a rhumb angle of (nominally) 90 degrees so that the attitude vector will move along a meridian circle in the direction towards the Sun. This results in a 'pure' Sun angle variation so that a very precise calibration of the maneuver path-length λ_f can be established on the basis of the measured change in Sun aspect angle over this maneuver leg. This is equivalent to the calibration of the effective thrust level performance factor P_{thrust} (to a precision of the order of 0.5 %):

$$P_{thrust} = \lambda_{f, measured} / \lambda_{f, prepared} \cong \Delta\theta_{measured} / \Delta\theta_{prepared} \tag{11}$$

The second leg of the Flip maneuver is about 57 degrees long and follows a direction that is perpendicular to the first leg. The (nominal) rhumb angle that will be maintained throughout this leg is 180 degrees so the attitude will move along the Sun cone, which means that the Sun aspect angle should remain constant during this maneuver leg. Therefore, any observed change in Sun aspect angle over the maneuver leg must be caused by an error in the effective rhumb angle χ. This is equivalent to the calibration of the effective mean centroid delay angle of the thrust pulses of the selected set of thrusters (to a precision of about 0.1°):

$$\Delta\chi = \chi_{measured} - \chi_{prepared} \cong \Delta\theta_{measured} / \lambda_{f,\ prepared} \qquad (12)$$

It has been assumed here that the thrust level calibration result in (11) has already been incorporated in the prepared path length $\lambda_{f,\ prepared}$ (otherwise, a more intricate formula should be used).

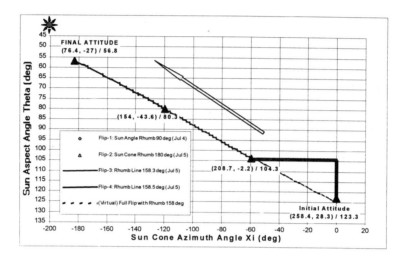

Figure 6 – Design of 180-degree Flip Maneuver, Including Calibration Strategy

The nominal length of the second leg was selected so that its end-point lies precisely on the great circle of the full 180-degree Flip maneuver. Furthermore, the Flip maneuver strategy included a useful independent confirmation of the achieved attitude pointing at this time with the aid of an Earth sensor coverage interval that occurred shortly after conclusion of the second leg. By employing the calibration results as well as the observed spin variations established over the first two maneuver legs, it was possible to reduce the expected error in the final attitude pointing after the remaining two legs (each about 64 degrees long) to less than 3 degrees. The established calibration knowledge was employed for the benefit of all attitude maneuvers throughout the phasing orbits (this implies of course that the same set of thrusters needed to be used).

SOFTWARE DESIGN & IMPLEMENTATION

The software used for ground based attitude determination and control was generated using The Mathworks® Matlab program. The Equal Chord Method and Fine Determination Method algorithms, as well as the algorithms developed to calculate the thruster commands for the maneuvers, were implemented in m-files that were placed under configuration control on a PC in the CONTOUR mission operations center. These m-files would read an input ASCII input file that contained data on the current state of the spacecraft, calibration coefficients, alignment parameters, and desired states following a maneuver.

To determine the attitude of the spin axis, telemetry from the spacecraft was processed and loaded into a Matlab workspace, where the m-files processed it to determine an attitude. The results of each run were recorded in an attitude history file that was used by mission operations as well as by the navigation team. Maneuvers were generated using the current and desired states of the spacecraft. The end result of this process was a maneuver file containing all information needed to change the rate or orientation of the spin axis, or the orbit of the spacecraft. The resulting maneuver files were then transferred to the mission operations team for testing and upload to the spacecraft.

SPIN MODE HARDWARE DESIGN

The hardware implementation for the CONTOUR spin mode was relatively straightforward. The Earth Sun Sensor (ESS) was the source of information for all attitude measurements during this mode as explained above. Figure 7 shows a photograph of the ESS sensor unit taken during spacecraft integration. The unit consists of two Earth sensor pencil-beams and a V-slit Sun sensor. The 'meridian' Sun sensor slit is aligned with the spin axis of the spacecraft and is used to determine the spin rate. This slit serves also as reference for the Earth sensor measurements, which allows the calculation of the Sun-Earth Azimuth Angle (as shown in Figure 2). The 'skew' Sun sensor slit, corresponding to the canted slit in Figure 7, is used along with the meridian slit to determine the Sun angle.

The V-slit Sun sensor is equipped with eight silicon photo-detectors (which are redundant within the principal range of Sun angles from about 60 to 120 degrees) operating in the visible band of the spectrum (from 0.3 to 1.1 μm). Each pencil-beam of the Earth sensor has an immersed thermistor bolometer Infra-Red detector operating in the CO_2 spectral band between 14 and 16.25 μm. The crossing times of the Earth's Infra-Red horizon are obtained through (on-board) processing of the detector measurements.

Output of the ESS sensor consists of six pulses on separate wires corresponding to: the Sun crossing the meridian slit, the Sun crossing the skew slit, space-to-Earth crossing detected by the upper Earth sensor pencil-beam, Earth-to-space crossing detected by the upper Earth sensor, and similar detections by the lower Earth sensor beam.

Figure 7 – Earth Sun Sensor Unit during Spacecraft Integration

Figure 8 shows a block diagram of the hardware configuration used for spin mode. The ESS pulses are detected by the Thruster / Attitude Control Card (TAC). Upon detection of the meridian pulse, the TAC counts the number of one-microsecond tics that occur from the time the meridian pulse is detected to the time each of the other five pulses are detected. This timing information is packed for distribution onto the '1553 bus' for receipt by the Command and Data Handling (C&DH) processor and for subsequent transmission to the ground. The timing data are de-commutated by the ground system and then processed by the various attitude estimation algorithms.

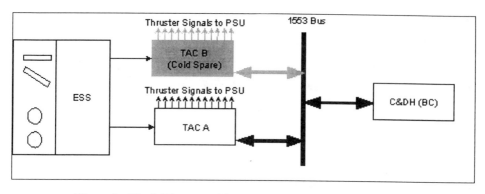

Figure 8 – Block Diagram of Spin Mode Hardware Configuration

Thruster control is equally straightforward. The required thruster firing parameters (consisting of delay time, on-time, number of pulses, etc.) are calculated on-ground and up-linked to the TAC via the C&DH. The TAC then commands the individual thrusters based on the parameter values. The time resolution for commanded thruster firings is one millisecond and the minimum firing duration is 0.125 sec.

PROPULSION SYSTEM DESIGN

A schematic of the CONTOUR Liquid Propulsion System (LPS) is shown in Figure 9. The CONTOUR LPS includes a total of sixteen thrusters: fourteen small 0.2 lb_f thrusters, and two large 5 lb_f thrusters. Orbit delta-v maneuvers (see Ref. 1) are normally performed by means of the two 5-lbs thrusters firing in continuous mode. Orbit correction maneuvers in a radial direction may be delivered by a set of 4 or more radial 0.2 lbs thrusters firing in pulsed mode with prescribed delay angles relative to the Sun meridian pulse (in a similar manner as described above for the attitude reorientation maneuvers).

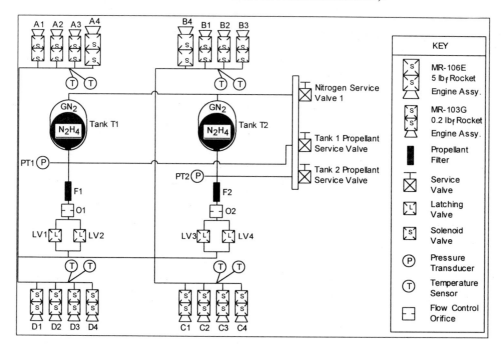

Figure 9 - CONTOUR Liquid Propulsion System Schematic

The thrusters use high purity hydrazine (N_2H_4) and are fed propellant from two propellant/pressurant tanks by simple blow-down. Each thruster contains a series-redundant solenoid valve, redundant valve heaters and redundant catalyst bed heaters. These thrusters are mounted in four different Rocket Engine Modules (REM's), which are positioned as shown in Figure 10. The design provides 'near-pure' couple torques about all three axes as well as both positive and negative ΔV corrections in all three directions. The adjective 'near-pure' refers to the resulting undesirable effects on the orbit: these are typically not more than a few mm/sec after a 180-degree attitude maneuver.

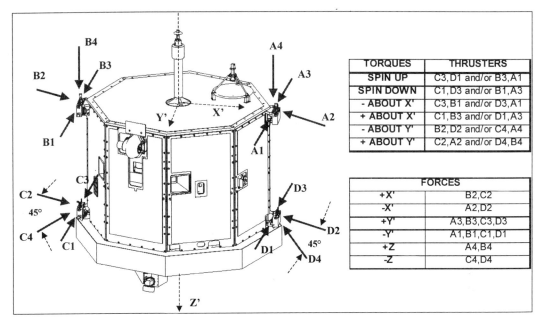

TORQUES	THRUSTERS
SPIN UP	C3,D1 and/or B3,A1
SPIN DOWN	C1,D3 and/or B1,A3
- ABOUT X'	C3,B1 and/or D3,A1
+ ABOUT X'	C1,B3 and/or D1,A3
- ABOUT Y'	B2,D2 and/or C4,A4
+ ABOUT Y'	C2,A2 and/or D4,B4

FORCES	
+X'	B2,C2
-X'	A2,D2
+Y'	A3,B3,C3,D3
-Y'	A1,B1,C1,D1
+Z	A4,B4
-Z	C4,D4

Figure 10 – Overview of CONTOUR Thruster Utilization

The LPS tanks are internally pressurized using Nitrogen, and have their ullages manifolded together to minimize the risk of propellant migration due to thermal and pressure variations in the tanks. The ullages of both tanks are pressurized through a single pressurant service valve. Each tank has its own individual pressure transducer, service valve, flow-control orifice, propellant filter, and a parallel redundant set of two latch valves downstream of the filter. Propellant flow paths have been designed such that both tanks are capable of providing propellant for all sixteen thrusters, and switching between propellant tanks (for maintaining the fuel imbalance below the required 1 kg) is possible by using the latch valves. The propellant flow path layout and the selective matching of components result in a nearly identical pressure drop from each tank to the thrusters. Latch valves were also required to meet the Range Safety requirement that there are at least three valve seats between the propellant in the tanks and the outlets of the thrusters.

SUMMARY OF IN-FLIGHT PERFORMANCES

Initial Nutation Damping Performance

The passive nutation damping induced by the energy dissipation effects of the hydrazine fuel was much better than expected. It had been predicted before launch that the Nutation Time Constant (NTC) of the spacecraft would be of the order of 15000 seconds. However, following separation and power up of the ESS the NTC was measured at 500 seconds for a spin rate of 49.5 RPM. At 20 RPM this number rose to 1500 seconds, and at 60 RPM the number dropped to around 300 seconds. Figure 11 shows the nutation damping performance of the spacecraft immediately after separation from the Delta Launcher's third stage.

777

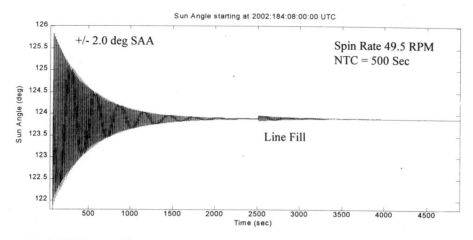

Figure 11 - ESS Measured Sun Angle Following Separation from Delta third Stage at 49.5 rpm.
(Note: the small increase at 2500 seconds was caused by the initial priming of the fuel lines)

Summary of Attitude Maneuvers

During the six weeks of phasing orbits, a total of 12 attitude maneuvers, 7 orbit and 4 spin correction maneuvers were performed. The total length of all slews combined was over 500 degrees and the longest attitude maneuver performed was about 95 degrees. Precise calibrations of thrust-level as well as rhumb angle (including spin effects) were instrumental in achieving adequate pointing precision prior to the execution of the orbit delta-v maneuvers.

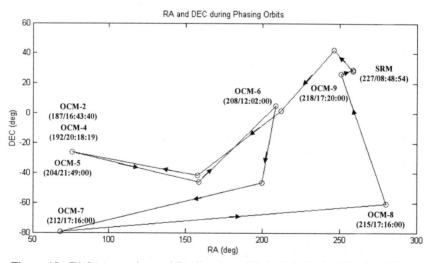

Figure 12 - Right Ascension and Declination of Spin Axis during Phasing Maneuvers

Figure 12 shows the progress of the right ascension (RA) and declination (DEC) of the spin axis during the phasing orbits (it may be noted that the RA and DEC angles are defined with respect to the Earth's equatorial plane within the inertial J2000 reference frame). The objective of each of the attitude maneuvers was to orient the spacecraft in preparation for an Orbit Correction Maneuver (OCM). The separation attitude was very close to the SRM attitude, which is why the spin axis motion appears to create a loop.

Table 1 - Simulated and Observed Data from Attitude Control Maneuvers

ACM	Starting RA/DEC (deg)	End RA/Dec (deg) Observed	End RA/Dec (deg) Simulated	Delta (%)	End Sun Angle (deg) Observed	End Sun Angle (deg) Simulated	Delta (deg)	Tank Used	Com-ments
1	258.40 / 28.30	No ESS Data	246.24 / 42.18	Unknown	106.18	104.3	-(1.83)	1	(1), (2)
2	No ESS Data	211.76 / 1.87	211.80 / 1.65	0.29	105.36	106.13	-(0.77)	1	(1)
3	211.76 / 1.87	No ESS Data	157.70 / -41.86	Unknown	80.29	80.46	0.17	1	(2)
4	No ESS Data	76.2 / -26.22	77.50 / -27.90	1.54	56.6	56.74	0.14	1	(3)
5	76.06 / -26.08	No ESS Data	158.50 / -46.32	Unknown	73.56	73.3	0.26	1	(2)
6	No ESS Data	208.48 / 4.80	208.81 / 4.89	0.25	83.01	82.88	0.13	1	None
7	208.25 / 5.05	No ESS Data	199.52 / -46.70	Unknown	90.78	90.78	0	1	(2)
8	No ESS Data	No ESS Data	68.28 / -79.25	Unknown	103.21	103.12	-(0.09)	1	(2)
9	No ESS Data	No ESS Data	280.32 / -60.38	Unknown	131.8	131.89	0.09	2	(2)
10	No ESS Data	250.65 / 26.39	251.3 / 26.10	0.27	103.87	103.74	0.13	1	(4)
11	250.71 / 25.07	258.56 / 29.10	258.69 / 29.39	3.81	105.69	105.93	-(0.24)	2	(5)
12	258.56 / 29.10	258.66 / 29.21	258.72 / 29.28	36.42	103.73	103.66	-(0.07)	2	(5)

Notes: (1) Simulation deltas are based on distance traveled between ESS measurements.
(2) Simulation accounted for linear spin rate changes during burn.
(3) End RA/DEC based on ESS data post maneuver.
(4) Observed sun angle calibrated for ESS misalignment.
Comments: (1) Calibration Maneuver 10.9% underperformance from model. 37 msec thrust pulse centroid delay.
(2) No ESS data following this maneuver before start of next ACM.
(3) OCM between end of ACM and ESS measurement, which could shift spin axis by +/- 0.2 deg.
(4) SCM between end of ACM and ESS measurement.
(5) Small maneuver. Attitude knowledge uncertainty is large portion of delta uncertainty.

Table 1 provides an overview of the performances of the attitude control maneuvers. It can be seen that after the completion of the two calibration maneuvers (ACM-1 and 2), the difference between the simulated and observed Sun angle was never greater than 0.26 degrees. Most of the difference was due to unanticipated changes in the spin rate. Following a maneuver these changes were substituted into the simulation and the result was a solution that very closely matched truth. Note that the larger percentage deltas for ACM-11 and -12 were due to the fact that the maneuver lengths were relatively small (i.e., only 8 and 0.3 degrees respectively) so that attitude uncertainties have a large effect on the apparent performances.

Thruster Performances & Calibrations

Following each maneuver, and in preparation for each future maneuver, the thrusters were calibrated using observed maneuver performances. Thruster performance was measured against blow-down predictions based on thruster acceptance test calibration data and measurements of the current propellant tank pressure. The thruster performance curves generated during thruster acceptance testing were used throughout the mission to predict thruster performance. Performance variations were modeled as changes in the thruster feed (inlet) pressure. This feed pressure was modeled relative to the measured tank pressure, so that estimates of future feed pressure could be made. This technique provided very accurate and consistent estimates of thruster performances throughout the phasing orbits as shown in Table 1.

Attitude Determination Results

In order to be able to achieve the required attitude determination accuracy use was made of the on-ground 'calibration data' that were measured by the manufacturer prior to delivery of the ESS unit. These data model the delays between the actual ESS pulses and the expected geometric crossing times. In particular, the calibrations of the Earth sensor S/E and E/S crossings are critical since they lead to corrections of the measured chord centers (i.e., SEAA angle) of as much as $1.51°$ and $1.31°$ for the two beams. Furthermore, the best available knowledge of the ESS mounting angles as measured during spacecraft alignment tests needed to be incorporated. This resulted in a change of almost 14 arc-min in the Sun sensor elevation angle (i.e., $\Delta e = -0.226°$ with nominal design value $e = 0$) as well as in the pencil-beam mounting angles $\Delta\mu_1 = \Delta\mu_2 = +0.226°$ relative to their nominal design values of 60 and 65 degrees (see Figures 1 to 3).

Table 2 – Final Attitude Determination Residuals (in Degrees) Prior to SRM Firing on August 15

Solution	Attitude Pointing RA & DEC		Residuals over 0.5 hour near Equal Chords SAA	EAA	HCA-1 & 2		SEAA-1 & 2		Distance to SRM Attitude	Attitude Consistency
ThBe	258.62	29.27	0.005	0.008	0.23	0.22	0.03	0.22	0.09	0.125
ThBeAl	258.60	29.18	0.023	0.008	0.22	0.21	0.03	0.16	0.16	0.036
RES	258.53	29.10	0.003	0.007	0.20	0.19	0.09	0.10	**0.26**	0.086
ECM	258.64	29.13	0.078	0.008	0.22	0.21	0.07	0.12	0.18	0.093
FAD	258.61	29.15	0.046	0.008	0.22	0.21	0.06	0.13	0.17	Reference
Averages:			0.031	0.008	0.22	0.20	0.06	0.14	0.17	0.085

The attitude determination results obtained by the software algorithms were validated by means of a scrutiny of the residuals in the Sun and Earth sensor measurement angles ϑ, κ_1, κ_2, α_1, α_2. These residuals represent the differences between the 'predicted' measurement angles (resulting from simulations based on the applicable attitude estimate) and the actual down-linked sensor measurements over the relevant interval.

Table 2 summarizes the results for the residuals (i.e., their average values over the half hour of sensor data centered on the time of equal chords) during the final coverage interval before the SRM firing. A total of five different attitude determination results are shown in Table 2: the 'spread' or consistency between these attitude solutions provides an indication of the likely remaining error. The best possible estimate is most likely delivered by the Fine Attitude Determination (FAD) solution since it is based on the largest set of measurements. The following list describes the five solutions in increasing order of 'sophistication':

1. **ThBe**: represents the geometric attitude solution based only on the SAA θ_e and the EAA β_e measured at the time of equal chords
2. **ThBeAl**: represents the geometric solution obtained from all three angular observations, namely the SAA θ_e, the EAA β_e and the SEAA α_e all taken at the time of equal chords; the deviation of this attitude result from **ThBe** under point 1 provides an indication of the consistency of the SEAA measurement which is likely the 'weakest' entry due to its sensitivity to the North-South IR effects.
3. **RES**: is the solution that aims at reconciling the observed residuals as well as possible; it is obtained from ThBeAl in an iterative manner by (visually) minimizing the evolution of the differences between the simulated and measured SEAA's over the interval of interest
4. **ECM**: is the result of the ECM software (described in detail above): the main difference with the **ThBeAl** approach lies in the quadratic fitting of the chords over the interval around the equal chords
5. **FAD**: is the result of the FAD software (described in detail above): this is expected to be the most accurate result available as it is based on all available measurements with their applicable weightings over the half hour data interval.

Figure 13 illustrates the locations of these attitude solutions relative to each other as well as the designated SRM attitude in a (linearized) Right Ascension (RA) versus Declination (DEC) diagram. It is seen from Figure 13 and Table 2 that all 5 solutions are contained within a circle of diameter of less than 0.2 degrees, which is indicative of the excellent consistency between all possible solutions. Furthermore, all of the solutions (except for RES) are expected to be within 0.2 degrees from the target SRM attitude as shown in Table 2 above.

Because of the mishap that occurred near the end of the SRM firing, it has unfortunately not been possible to reconstruct the actual attitude error on the basis of the observed delta-v direction delivered during the SRM firing to a very good precision. The available knowledge of the orbits of the three pieces that have been observed would indicate that the error in the attitude orientation during the SRM burn has in any case been below 0.44 degrees. This result represents the maximum of the three reconstructed attitude errors resulting from the analyses of the post-SRM orbits as outlined in Refs. 1 and 7.

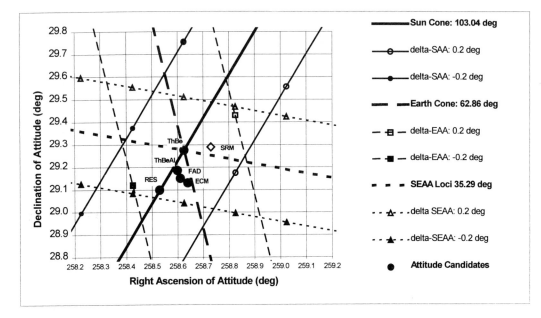

Figure 13 – Attitude Solutions Relative to SAA, EAA & SEAA Loci (August 15)

CONCLUSION

The paper has outlined the concepts underlying the design of CONTOUR's spinning mode that was used throughout the 6 weeks phasing orbits preceding its ill-fated SRM firing. The in-flight results confirm that the minimal and relatively straightforward hardware and software capabilities that have been implemented were able to support all operational requirements with excellent reliability and accuracy performances.

REFERENCES

1. D. W. Dunham et al., "Design and Implementation of CONTOUR's Phasing Orbits", *AAS/AIAA Space Flight Mechanics Meeting*, Ponce, Puerto Rico, February 9-13, 2003, Paper AAS 03-208.

2. J. C. van der Ha, "Attitude Determination Covariance Analysis for Geo-stationary Transfer Orbits", Journal of Guidance, Control, and Dynamics, Volume 9, Number 2, March-April 1986, pp. 156-163.

3. A. S. Fagg, & L. van Holtz , "Application of Simplified Spin Axis Attitude Determination Techniques to ESA Telecom Satellites, *Proceedings of the CNES Symposium on 'Space Dynamics for Geostationary Satellites'*, Cepadues Editions, Toulouse, France, October 1985, pp. 219-226.

4. M. D. Shuster, "Efficient Algorithms for Spin-Axis Attitude Estimation", *The Journal of the Astronautical Sciences*, Vol. XXXI, April-June 1983, pp. 237-249.

5. J. R. Wertz (Editor), *Spacecraft Attitude Determination and Control*, Kluwer, 1978, Section 19.3.

6. Stratton, J., Engelbracht, C., Deboer, J., and J. Morris, "Description of the Liquid Propulsion System of the Comet Nucleus Tour Spacecraft", Paper AIAA-2001-3391.

7. E. Carranza et al., "Navigating CONTOUR using the Noncoherent Transceiver Technique", AAS 03-204, *AAS/AIAA Space Flight Mechanics Conference*, Ponce, Puerto Rico, February 9-12, 2003.

ACQUISITION, TRACKING AND POINTING OF BIFOCAL RELAY MIRROR SPACECRAFT

Brij N. Agrawal[*]

This paper presents the attitude dynamics and control of bifocal relay mirror spacecraft. The spacecraft consists of two optically coupled telescopes used to redirect the laser light from ground-based, aircraft-based, or spacecraft based lasers to distant points on the earth or in space. The spacecraft has very tight pointing and jitter requirements to meet the performance of the optical payload. Because the two large telescopes are gimbaled, the spacecraft inertia has a large variation during the large angle maneuvers. The attitude control system consists of reaction wheels, star trackers, gyros, sun sensors, magnetometers and magnetic rods. Feed forward control and quaternion formulation are used. Kalman filter is used to update the rate gyros biases and the attitude angles during star tracker measurements. Fast steering mirrors are used for fine beam control. The simulation results show that the fine tracking requirements can be achieved. However, the feed forward control and integrated spacecraft and fast steering mirror control are critical to meet these requirements.

1. INTRODUCTION

For the current spacecraft attitude control designers, a very challenging problem is to meet performance requirements of fine tracking, pointing, and low jitter for several current and future spacecraft with optical payloads (Refs. 1 and 2). These spacecraft include optical communications relay link satellites, imaging satellites, space telescopes to view distant galaxies and stars, and space based laser satellites. In order to meet these requirements, some of these satellites need to use fast steering mirrors for fine beam control, jitter sensors, active vibration suppression and isolation control, and precision sensors. This requires integration of several different controls. The problem is further complicated because several of these missions require large apertures, in the range of 10-30m diameter, resulting in interaction of structures and control interaction. Therefore, significant research effort is needed to develop acquisition, tracking, and pointing (ATP) technologies for these spacecraft.

At the Spacecraft Research and Design Center, Naval Postgraduate School, a research project has been undertaken to develop ATP technologies for Bifocal Relay Mirror Spacecraft. The spacecraft is composed of two optically coupled telescopes used to redirect the laser light from ground-based, aircraft-based or spacecraft based lasers to

---

[*] Distinguished Professor and Director, Spacecraft Research and Design Center, Naval Postgraduate School, Monterey, California 93943. E-mail: agrawal@nps.navy.mil.

distant points on the earth or in Space. The receiver telescope captures the incoming laser beam and the transmit telescope directs the beam to the desired point.

A Laser and Space Optical Systems (LASSOS) study (Ref. 3) commissioned by Air Force Research Laboratory (AFRL) in 1996 explored the missions that could be supported by space-based laser or space relay mirrors and identified a large number of operational concepts and non-weapon missions that can be supported by such systems. At the Naval Postgraduate School (NPS), students under a spacecraft design course completed a preliminary design of a demonstration Bifocal Relay Mirror spacecraft in the summer of 2000. The spacecraft, as shown in Figure 1, consists of single axis gimbaled receive and transmit telescopes with 1.64 meter diameter primary mirrors and fast steering mirrors for fine beam control. The transmit telescope has a majority of the spacecraft bus subsystems including the attitude control sensors and actuators. The spacecraft mass is 3300 kg at launch and the spacecraft orbit altitude is 715 km with an inclination of 40 degrees. The mission requirements are for a 3 meters spot beam on the ground, jitter less than 144 nanoradians rms and mean dwell duration per pass of 250 seconds. The acquisition sequence consists of pitch and roll motion of transmit telescope to acquire the transmit point. The yaw motion of the transmit telescope and the one axis motion of the receive telescope is used to acquire the receive point.

There are several unique multi-body pointing and control problems in the Bifocal Relay Mirror (BRM) spacecraft. The spacecraft has two large inertia telescopes that are gimbaled. This results in continual change in the spacecraft dynamics and spacecraft inertia during the operation. These telescopes must slew the line of sight at a fast rate, as much as 180 degrees in a few seconds so as potentially to look at another relay mirror spacecraft. Simultaneously, the vibrations and jitter must be reduced to optical tolerances. The mission of BRM is quite different from Hubble Space Telescope (HST). The dynamics required to track an HST object are very benign, being constantly pointing in an inertial frame. The HST stabilization system is based upon relatively low bandwidth systems without a need for high bandwidth stabilization or tracking system. HST does not require fast steering mirror architectures as required for BRM. Therefore significant effort is required in the development of ATP technology for BRM spacecraft over the ATP technology for HST.

In December 2000, NPS and AFL team received a research award for initial development of the acquisition, tracking, and pointing technologies for bifocal relay mirror spacecraft and experimental demonstration. This paper presents the analytical work performed on the dynamics and control of the bifocal relay mirror spacecraft.

2. DYNAMICS OF BIFOCAL RELAY MIRROR SPACECRAFT

This analytical model of BRM (Ref. 4) is shown in Figure 2. It consists of two rigid bodies, transmit telescope and receive telescope. The receive telescope rotates with respect to transmit telescope about a single axis. The c.m. of the receive telescope is on the rotation axis. Therefore, the c.m. of the system is fixed during the relative motion of the receive telescope.

The coordinate system x, y, z is fixed in the transmit telescope with x-axis parallel to rotation axis of the receive telescope, z as telescope axis and y is normal to x and z such that the x, y, z coordinate system is right handed mutually orthogonal frame. The origin of the coordinate system is at c.m. of the transmit telescope. Unit vectors along x, y, and z are $\bar{i}, \bar{j},$ and $\bar{k}$ respectively. The coordinate system x', y', z', is fixed in the receive system with x' axis as the rotation axis, parallel to x axis, z' as telescope and y' is normal of x' and z' such that x', y', z' coordinate system is right handed mutually orthogonal frame with origin at the c.m. of the receive telescope. The coordinate system x', y', z' are obtained from the coordinate system x', y', z' by rotation α about x-axis. The relative motion of the transmit fast steering mirror with respect to transmit telescope is represented by β_x^T rotation about x axis and β_y^T rotation y-axis. The relative motion of the receive fast steering mirror with respect to receive telescope is represented by $\beta_{x'}^R$ rotation about x' axis and $\beta_{y'}^R$ rotation about y' axis. The fast steering mirror rotations are assumed to be small. The equations of motion of the system are written in the coordinate frame x, y, z.

2.1 Equations of Motion

In a general case, rotational equations of a body about an arbitrary point P is given by

$$M_p = \dot{H}_p - \dot{\rho}_c \times m \ \dot{r}_c \tag{1}$$

Where
M_p = total sum of external forces about P
H_p = angular momentum of the body about P
ρ_c = total vector from P to c.m. of the system
m = total mass of the body
$\dot{r}_c$ = velocity of center of mass of the body

If the point P is cm, then $\rho_c = 0$, then

$$M = \dot{H}\,|_I \tag{2}$$

The angular momentum of the spacecraft, H_S can be written as follows:

$$H_S = H + H_{rel} + H + H_{FS}^T + H_{FS}^R \tag{3}$$

where
H_S = total angular momentum of the spacecraft.

785

H = total angular momentum by neglecting the contribution by the relative motion of the receive telescope and reaction wheels with respect to transmit telescope

H_{rel} = angular momentum due to relative motion of the receive telescope

H_w = angular momentum due to relative motion of the reaction wheels.

H^T_{FS} = angular momentum of transmit fast steering mirror relative to transmit telescope

H^R_{FS} = angular momentum of receive fast steering mirror relative to receive telescope

The next step in derivation is to determine these angular momentums.

Inertia

Let I^T be the inertia matrix of the transmit telescope about its c.m. in coordinate frame x, y, and z

$$I_T = \begin{bmatrix} I^T_{xx} & -I^T_{xy} & -I^T_{xz} \\ -I^T_{yx} & I^T_{yy} & -I^T_{yz} \\ -I^T_{zx} & -I^T_{zy} & I^T_{zz} \end{bmatrix} \tag{4}$$

Let the vector from c.m. of the transmit telescope to the c.m. of the spacecraft is given by

$$r_T = x_T i + y_T j + z_T k \tag{5}$$

Let I'_R be the inertia matrix of the receive telescope about its c.m. in coordinate frame x', y', and z'.

It is given by

$$I'_R = \begin{bmatrix} I^R_{x'x'} & -I^R_{x'y'} & -I^R_{x'z'} \\ -I^R_{y'x'} & I^R_{y'y'} & -I^R_{y'z'} \\ -I^R_{z'x'} & -I^R_{z'y'} & I^R_{z'z'} \end{bmatrix} \tag{6}$$

Next we transform I'_R to x', y', z' frame. The transformation matrix from x, y, z to x', y', z', $^RC^T$ is given by

$$^RC^T = \begin{bmatrix} 1 & 0 & 0 \\ 0 & Cos\alpha & Sin\alpha \\ 0 & -Sin\alpha & Cos\alpha \end{bmatrix} \tag{7}$$

The transformation matrix $^TC^R$ from x', y', z' to x, y, z is transpose of $^RC^T$. The I'$_R$ in x', y', z' system, I$_R$, is given by

$$I_R = {}^TC^R I'_R \ {}^RC^T \tag{8}$$

Let the vector from c.m. of the receive telescope to the c.m. of the system is given by

$$r_R = x_R i + y_R j + z_R k \tag{9}$$

The inertia matrix of the spacecraft about its c.m. in coordinate frame x, y, z is given by

$$I_S = I_T + I_R + m_T \begin{bmatrix} y_T^2 + z_T^2 & -x_T y_T & -x_T z_T \\ -x_T y_T & z_T^2 + x_T^2 & -y_T z_T \\ -x_T z_T & -y_T z_T & x_T^2 + y_T^2 \end{bmatrix}$$

$$+ m_R \begin{bmatrix} y_R^2 + z_R^2 & -x_R y_R & -x_R z_R \\ -x_R y_R & z_R^2 + x_R^2 & -y_R z_R \\ -x_R z_R & -y_R z_R & x_R^2 + y_R^2 \end{bmatrix} \tag{10}$$

Angular Velocities

The angular velocity of the transmit telescope

$$\omega_T = \begin{bmatrix} \omega_x \\ \omega_y \\ \omega_z \end{bmatrix} \tag{11}$$

The relative angular velocity of the receive telescope with respect to transmit telescope ω_{rel} is given by

$$\omega_{rel} = \begin{bmatrix} \dot{\alpha} \\ 0 \\ 0 \end{bmatrix} \tag{12}$$

The angular velocity of the receive telescope, ω_R, is

$$\omega_R = \omega_T + \omega_{rd} = \begin{bmatrix} \omega_x + \dot{\alpha} \\ \omega_y \\ \omega_z \end{bmatrix} \tag{13}$$

787

The angular velocity of the transmit fast steering mirror (TFSM) relative to transmit telescope is

$$\omega_{FS}^T = \dot{\beta}_x^T \vec{i} + \dot{\beta}_y^T \vec{j} = \begin{bmatrix} \dot{\beta}_x^T \\ \dot{\beta}_y^T \\ O \end{bmatrix} \tag{14}$$

The angular velocity of the receive fast steering mirror (RFSM) relative to receive telescope is

$$W_{FS}^R = \dot{\beta}_{x'}^R \vec{i}^1 + \dot{\beta}_{y'}^R \vec{j}^1 = \begin{bmatrix} \dot{\beta}_{x'}^R \\ \dot{\beta}_{y'}^R \\ 0 \end{bmatrix} \tag{15}$$

Angular Momentums

The angular moment H is given by

$$H = I_s \omega_T \tag{16}$$

The relative angular momentum H_{rel} is given by

$$H_{rel} = I_R \omega_{rel} \tag{17}$$

The angular momentum H_{FS}^T is given by

$$H_{FS}^T = I_{FS}^T \ \omega_{FS}^T \tag{18}$$

where I_{FS}^T is the inertia matrix of TFSM in x, y, z coordinate frame. The angular momentum H_{FS}^R is given by

$$H_{FS}^R = {}^T C^R \ I_{FS}^R \ \omega_{FS}^R \tag{19}$$

where I_{FS}^R is the inertia matrix of RFSM in x', y', z' coordinate frame.
Substituting eqs. (16), (17), (18), and (19) into Eq. (3), we get

$$H_S = I_s \omega_T + I_R \omega_{rel} + H_w + I_{FS}^T \omega_{FS}^T + {}^T C^R I_{FS}^R \omega_{FS}^R \tag{20}$$

Equations of Motion

Using Eq. (2), the equation of motion of the spacecraft is given by

$$M = \dot{H}_s\big|_I$$
$$= \dot{H}_s\big|_T + \omega_T \times H_s \tag{21}$$

where

$\dot{H}_s\big|_I$ = rate of change in inertial frame

$\dot{H}_s\big|_T$ = rate of change in transmit telescope frame

Substituting Eq. (15) into Eq. (16) we get

$$M = \dot{I}_S\omega_T + I_S\dot{\omega}_T + I_R\dot{\omega}_{rel} + \dot{I}_R\omega_{rel} + \dot{H}_W + I_{FS}^T\dot{\omega}_{FS} + {}^T\dot{C}{}^R I_{FS}^R \omega_{FS}^R$$
$$+ {}^T C{}^R I_{FS}^R \dot{\omega}_{FS}^R + \omega_T \times \left[I_S\omega_T + I_R\omega_{rel} + H_W + I_{FS}^T\omega_{FS} + {}^T C{}^R I_{FS}^R \omega_{FS}^R \right] \tag{22}$$

It should be noted that I_S is function of α, and therefore time dependent. Eq. (17) can be rewritten as

$$\frac{d}{dt}\left[I_S\omega_T + I_R\omega_{rel} + H_W + I_{FS}^T\omega_{FS} + {}^T C{}^R I_{FS}^R \omega_{FS}^R \right] = M - \omega_T \times$$
$$\left[I_S\omega_T + I_R\omega_{rel} + H_W + I_{FS}^T\omega_{FS} + {}^T C{}^R I_{FS}^R \omega_{FS}^R \right] \tag{23}$$

Fast Steering Mirrors

The fast steering mirrors are supported on soft springs and are represented by spring mass system. The actuators re voice coils. The equations of motion of a spring mass system is given by

$$\ddot{\beta} + 2\,\xi\,\omega_n\,\dot{\beta} + \omega_n^2\,\beta = \frac{T}{I} - a_{base} \tag{24}$$

where β =   relative angle of the spring mass system
   ξ =   damping ratio
   ω_n =   natural frequency
   T   =   applied torque
   a_{base} =   base acceleration.

For the TFSM, the base acceleration a^T is given by

$$a^T = \dot{\omega}_T = \begin{bmatrix} \dot{\omega}_x \\ \dot{\omega}_y \\ \dot{\omega}_z \end{bmatrix} \tag{25}$$

The equations of motion for TFSM is

$$\ddot{\beta}_1^T + 2\,\xi_1^T\,\omega_{n1}^T\,\dot{\beta}_1^T + \left(\omega_{n1}^T\right)^2 \beta_1^T = \frac{T^T}{I_{11}^T} - a_1^T \tag{26}$$

where subscript 1 is substituted for x and y for axis x and y, respectively. For RFSM, the base angular velocity in x′, y′, and z′ coordinate is given by

$$\omega_R' = {}^R C^T \omega^R \tag{27}$$

The base acceleration a^R is given by

$$
\begin{aligned}
a^R &= {}^R\dot{C}^T \omega_R + {}^R C^T \dot{\omega}_R \\
&= \dot{\alpha} \begin{bmatrix} 0 \\ -\sin\alpha\,\omega_y + \cos\alpha\,\omega_z \\ -\cos\alpha\,\omega_y - \mathrm{Sin}\,\alpha\,\omega_z \end{bmatrix} + \begin{bmatrix} \dot{\omega}_x + \ddot{\alpha} \\ \cos\alpha\,\dot{\omega}_y + \sin\alpha\,\dot{\omega}_y \\ -\sin\alpha\,\dot{\omega}_y + \cos\alpha\,\dot{\omega}_z \end{bmatrix}
\end{aligned} \tag{28}
$$

The equations of motion for RFSM is

$$\ddot{\beta}_1^R + 2\,\xi_1^R\,\omega_{n1}^R\,\dot{\beta}_1^R + \left(\omega_{n1}^R\right)^2 \beta_1^R = \frac{T^R}{I_{11}^R} - a_1^R \tag{29}$$

where the subscript 1 is substituted for x′ and y′ for axes x′ and y′, respectively.

2.2 Gravity Gradient Torque

The gravity gradient torque M_G is given by

$$M_G = \frac{3\mu}{R_0^3} \begin{bmatrix} C_{13} \\ C_{23} \\ C_{33} \end{bmatrix} \times \begin{bmatrix} I_{xx} & -I_{xy} & -I_{sz} \\ -I_{xy} & I_{yy} & -I_{yz} \\ -I_{xz} & -I_{yz} & I_{zz} \end{bmatrix} \begin{bmatrix} C_{13} \\ C_{23} \\ C_{33} \end{bmatrix} \tag{30}$$

2.3 Magnetic Moments

Magnetic moment is given by

$$M_m = m \times B \tag{31}$$

where

M_m = magnetic moment (n.m)
m = spacecraft magnetic dipole (a.m^2)
B = earth's magnetic field vector (n/(a.m))

The earth's magnetic field is given by

$$B = \frac{k}{R^3}[3(M.R)R - M] \tag{32}$$

where

$K = 7.943 \times 10^{15}$ n.m^2/a^2
R = earth c.m. to spacecraft c.m (m)
M = unit dipole vector

The earth's magnetic field component in orbit coordinates is given by

$$B_{xo} = \frac{k}{R^3}[\cos \alpha(\cos \varepsilon \sin i - \sin \varepsilon \cos i \cos u) - \sin \alpha \sin \varepsilon \sin u]$$

$$B_{yo} = \frac{k}{R^3}[-\cos \varepsilon \cos i - \sin \varepsilon \sin i \cos u] \tag{33}$$

$$B_{zo} = \frac{k}{R^3}[2 \sin \alpha(\cos \varepsilon \sin i - \sin \varepsilon \cos i \cos u) + 2 \cos \alpha \sin \varepsilon \sin u]$$

where

α = true anomaly of the spacecraft
E = magnetic dipole tilt from north pole

i = orbit inclination
u = right ascension angle of magnetic dipole with respect to right ascension of the orbit normal.

The earth's magnetic field vector in body frame is given by

$$B_B = {}^BC^O B_O \tag{34}$$

where ${}^BC^O$ is transformation matrix from orbit frame to body frame.

Magnetic Disturbance Moment

The magnetic disturbance moment is given by

$$M_{md} = M \times B \tag{35}$$

where M is the spacecraft magnetic dipole

3. SPACECRAFT CONTROL

For the spacecraft attitude control, reaction wheels are used as actuators with magnetic torque rods to desaturate the wheels. Fast steering mirrors are used for fine beam pointing. The primary spacecraft sensors are rate gyros to determine spacecraft angular rates and star trackers to determine spacecraft attitude. The rate gyros have bias errors and star trackers have measurements gaps. During the measurement gap for the star trackers, rate gyros are used to determine angular rates and angular position. When the star trackers measurements are taken, using Kalman Filter, the angular position is corrected and rate gyro biases are updated. During the simulations, disturbance torques are assumed from gravity gradient and earth's magnetic field. The spacecraft also has target trackers for receive and transmit trackers for both receive and transmit telescopes. The trackers determine line of sight errors and feed back to fast steering mirrors.

The block diagrams of the control approaches used are shown in Figures 3 and 4. In Figure 3, independent spacecraft and fast steering mirror control is shown. The angular motion of the fast steering mirrors (FSM) is limited to only few degrees. In order to avoid reaching the limits for the FSM, the spacecraft and FSM controls are integrated as shown in Figure 4. The symbols are defined as follows:

$\theta_{S/C}$ = true attitude of the spacecraft. In the reality this quantity is not known. In the simulations it is given by the integration of the equations of motion;

$\theta_{S/C-M}$ = measured attitude of the spacecraft;

$\theta_{C-S/C}$ = reference attitude of the spacecraft;

θ_{Target} = attitude of the spacecraft at which the transmitter telescope axis hits the target. In the presented simulations this quantity is considered equal to $\theta_{S/C}$

$\theta_{e-transm} = \theta_{Target} - \theta_{S/C}$ = pointing error;

$\theta_{uncertainties}$ = line of sight uncertainties. It can derive from different causes: motion of the target, imperfect knowledge of the reference attitude, LOS jitter. In our we considered uncertainties given by a low frequency pseudorandom signal,

θ_{e-LOS} = total pointing error, including the uncertainties and the correction from the fast steering mirror.

θ_{fsm} = fast steering mirror angular motion

3.1 Magnetic Control Moment

The desaturation of the reaction wheel is performed by magnetic control. The command for magnetic dipole of the torque rods or coils is given by

$$m_c = -k (B \times h) \tag{36}$$

where $\quad$ h = angular momentum of the wheel
$\qquad$ k = gain

3.2 Wheel Control Laws

The wheel control laws are as follows

$$
\begin{aligned}
\dot{H}_{W1} &= 2 \ k_1 q_{1E} q_{4E} + k_{1d} \omega_{1E} \\
\dot{H}_{W2} &= 2 \ k_2 q_{2E} q_{4E} + k_{2d} \omega_{2E} \\
\dot{H}_{W3} &= 2 \ k_3 q_{3E} q_{4E} + k_{3d} \omega_{3E}
\end{aligned}
\tag{37}
$$

where ω_E is the error between commanded angular rate and measured angular rate. As shown in Fig. 4, the FSM angular motion is also fed into the wheel control for integrated spacecraft and fast steering mirror control.

3.3 Feed Forward control

The feed forward control torque, M_{fd}, by neglecting the effect of fast steering mirrors because of small inertia and motion is given by

$$M_{fd} = I_S \omega_T + I_S \dot{\omega}_T + I_R \dot{\omega}_{rel} + \omega_T \times \left[I_S \omega_T + I_R \omega_{rel} + H_W \right] \tag{38}$$

4. SIMULATION RESULTS

The spacecraft dynamics and control laws are implemented on MATLAB/SIMULINK. The following parameters are used for the simulation The simulation time period is 500 seconds. The simulation solver method is ode5 (Dormand-Prince), and the solver fixed step size is 0.005 seconds. Altitude of the circular orbit: 715 Km. Mass of the transmitter telescope = 2267 Kg, mass of the receiver telescope = 972 Kg. Transmit telescope inertia: $I_{xx} = 2997$ kgm^2, $I_{yy} = 3164$ kgm^2, and $I_{zz} = 882$ kgm^2; receiver telescope inertia: $I_{x'x'} = 1721$ kgm^2, $I_{y'y'} = 1560$ kgm^2, and $I_{z'z'} = 183$ kgm^2. For both the transmitter and receiver fast steering mirrors $I_{xx} = I_{yy} = 0.01$ kgm^2. Natural frequency around both x and y-axes, $\omega_n = 10$ Hz. Damping ratio around both x and y-axes $\xi = 0.01$.

Magnification factor of both telescopes is 8.2. Field of views for both telescopes is 5e-4 rad. This field of view corresponds to a circular footprint area with radius of 375 m on the earth surface (when the telescope axis is perpendicular to the earth surface). The secular torques magnitude is 1e-4 Nm. The control law delay for initial determination errors is 30 seconds. A star tracker measurement gap is assumed between 100 and 300 sec. The rate gyros static rate biases are 1e-4*[-1,1.5,1] rad/sec. The initial errors are: for quaternion [0.008, 0.012, -0.008] and for angular rate [-0.001,0.001,0.002] rad/sec.

The control parameters are as follows: Gains for the PD control of the spacecraft are k = [1500, 3500, 2250] and k_d = [1000, 2000, 1000]. For the reaction wheels, the maximum allowable angular momentum is 1 Nm and the maximum torque is 10Nm/sec. Gains for the PID control of both the transmitter and the receiver fast steering are k= [160,160], k_d = [10,10], k_i = [15,15].

The command attitude profile is shown in Figure 5. The left figure shows the transmit telescope rotation in quaternion and the right figure shows the relative angle rotation of the receive telescope. The profile represents the maneuver required to maintain transmit and receive telescopes pointing during an overhead pass to conduct laser relay operations. The majority of the maneuver is performed in the spacecraft pitch axis, q_2, as both telescopes orient to point at fixed ground sites. Based on the ground site separation distance and orbital altitude, the largest relative angle is about 30 degrees during a near overhead pass between the uplink and downlink ground sites.

Figure 6 shows the quaternion errors with star tracker and assuming fixed fast steering mirrors. During the star tracker measurement gap, the error reaches up to 8×10^{-4}. However, during star tracker measurement, the errors are less than 2×10^{-4}. Figure 7 shows the pseudo-random signal added to the target sensor output during the simulations with uncertain knowledge of the target position. Figure 8 shows the simulation results for the line of sight errors for independent spacecraft and fast steering mirror control. The left figure shows the error when the FSM is not operational. The right figure shows the error when the FSM control is operational. Figure 9 shows the line of sight error for integrated spacecraft and fast steering mirror control. It is clear that feed forward and integrated spacecraft fast steering mirror control are necessary to meet the performance requirements of the mission.

5. CONCLUSIONS

The attitude dynamics and control simulation of the bifocal relay mirror spacecraft have been successfully implemented on MATLAB/SIMULINK. Based on the simulation results, the wheel control, the feed forward control and Kalman filter for estimating attitude and attitude rates have been effective to meet the performance of the spacecraft. The feed forward control is critical to meet the performance requirements, as there are two large telescopes gimbaled about one axis having large angle maneuvers. The integration of spacecraft and fast steering mirror control is also necessary to meet the mission performance requirements. Further work is required to include flexibility of the telescopes and experimental validation of control laws. A test bed for experimental validation is under development.

6. ACKNOWLEDGMENTS

The authors would like to acknowledge the significant contribution by Lt. William J. Palermo and Dr. Marcello Romano to this work.

7. REFERENCES

1. G. Baister and P. V. Gatenby, "Pointing, Acquisition and Tracking for Optical Space Communications," Electronics and Communications Engineering Journal, Vol. 6, pp. 271-280, December 1994.

2. D. Miller, O. de Weck, S. Uebelhart, R. Grogan, and I. Basdogan, "Integrated Dynamics and Controls Modeling for the Space Interferometery Mission," IEEE Aerospace Conference, Vol. 8, Big Sky, Montana, March 2001.

3. Laser and Space Optical Systems (LASSOS) Study, AFRL-DE-TR-199-1072, Vol. I.

4. B. Agrawal and C. Senenko, " Attitude Dynamics and Control of Bifocal Relay Mirror Spacecraft", AAS Paper No. 01-418, AAS/AIAA Astrodynamics Specialist Conference, Quebec City, Quebec, Canada, July 30-August 2, 3001

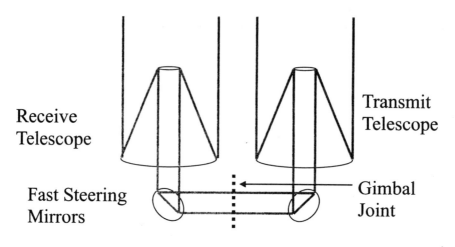

Figure 1 Spacecraft optical payload

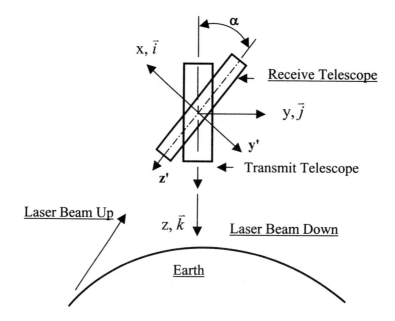

Figure 2 Bifocal Relay Mirror Spacecraft dynamic model

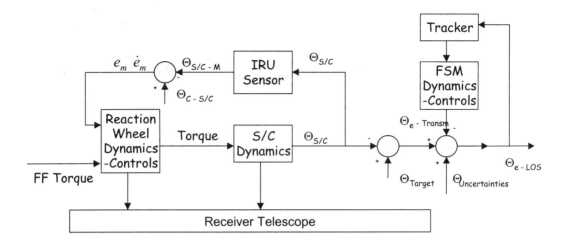

Figure 3 Independent spacecraft and fast steering mirror control

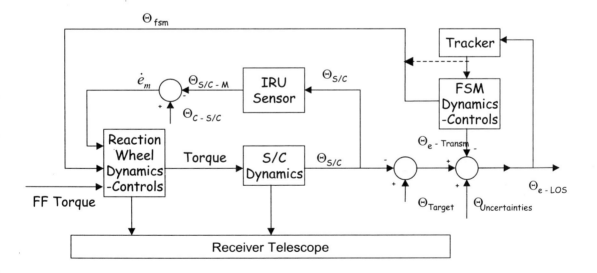

Figure 4 Integrated spacecraft and fast steering mirror control.

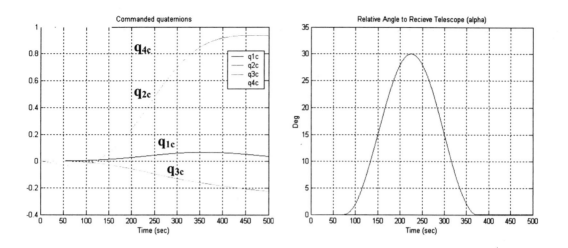

Figure 5 Command attitude profile during tracking maneuver.

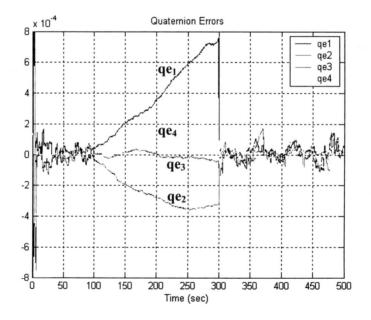

Figure 6 Quaternion errors with fixed fast steering mirror.

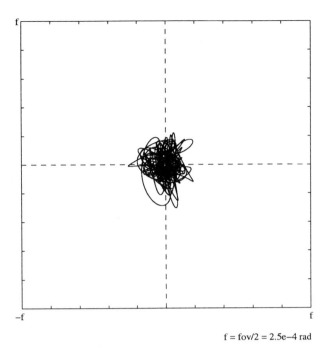

f = fov/2 = 2.5e−4 rad

Figure 7 Pseudorandom signals added to the target sensor output to simulate the uncertain knowledge of the target position.

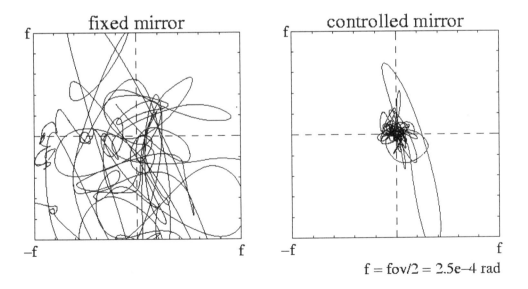

f = fov/2 = 2.5e−4 rad

Figure 8 Performance with and without fast steering control for independent spacecraft and fast steering mirror control.

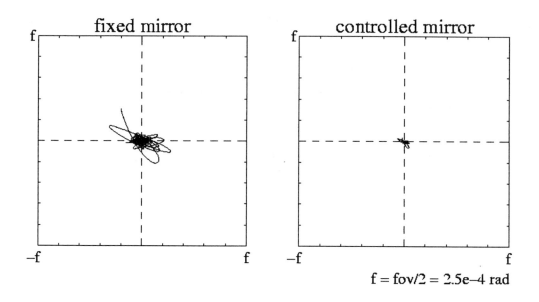

Figure 9 Performance with and without using fast steering mirror control for integrated spacecraft and fast steering mirror control.

IN-FLIGHT PERFORMANCE AND CALIBRATION OF THE AUTONOMOUS MINISATELLITE PROBA

Jean de Lafontaine,[*] Jean-Roch Lafleur,[*] Isabelle Jean,[*]
Pieter Van den Braembussche[†] and Pierrik Vuilleumier[‡]

In the context of its PRoject for On-Board Autonomy (PROBA), the European Space Agency is currently supporting the development of autonomy-enabling technologies for on-board mission management and payload operation. With the successful launch of the PROBA spacecraft on 22 October 2001, these technologies are now being evaluated in orbit. This paper presents the in-flight performance of the autonomous navigation, guidance and control functions of PROBA. It also presents the algorithms for the in-flight inertia tensor determination and payload alignment calibration. Due to the particular nature of the PROBA design, these algorithms required novel developments, presented here.

INTRODUCTION

It is often said that autonomy on-board Earth satellites could bring benefits in terms of reduced operational costs, an improved efficiency in the use of limited on-board resources and a higher mission availability due a faster reaction to on-board anomalies [1-2]. It has also been argued that more autonomy could imply a higher cost for the on-board software development and testing and a higher complexity of the on-board software, leading to poor operator visibility of the decisions taken on board by the software. In order to assess in a realistic scenario the potential benefits and complexity of on-board autonomy, the European Space Agency has initiated the **PR**oject for **On**-Board Autonomy (PROBA) [2]. The main objective of the project is to demonstrate in orbit new technologies and techniques that will be required in future space missions.

On 22 October 2001, the first demonstration satellite of the series, PROBA-1, was launched successfully from Shriharikota, India. The mini-satellite, built by Verhaert Design & Development (Belgium), was designed to conduct an Earth-observation mission in a completely autonomous way : the Users would command the spacecraft, via Internet, by simply providing the coordinates of the Earth target to image (longitude, latitude, altitude). The on-board systems would plan the operations, predict the target overfly, prepare the observation, and manoeuvre the instrument to acquire the images, all without any ground assistance. To accomplish these autonomous operations, PROBA-1 featured a highly autonomous guidance, navigation and control system, the Attitude Control and Navigation System (ACNS). The ACNS software was designed and developed at the Université de Sherbrooke, Canada.

An overview of the ACNS system and its on-board software design were presented in an earlier conference (Ref. 3). The autonomous guidance function was described in Ref. 4. The purpose of

---

[*] Université de Sherbrooke, Canada.

[†] Verhaert Design and Development nv, Belgium.

[‡] ESA/ESTEC, The Netherlands.

the present paper is to present some of the flight results – and lessons learned – gathered during commissioning and more than one year of successful operations of the PROBA-1 spacecraft. During that period, additional developments were conducted in order to calibrate in orbit the alignment of the main payload, relative to the star sensors, and to estimate the inertia matrix of the spacecraft. The theoretical basis for these calibrations, as well as the numerical results, will also be discussed. A brief overview of the PROBA-1 mission and of the ACNS design will first be presented to a level of detail compatible with the subsequent discussion their flight performance. The algorithms for calibrations will then be described followed by conclusions on the lessons learned.

THE PROBA-1 MISSION
The PROBA-1 mission is summarised in terms of mission objectives, orbit, spacecraft, payloads and their requirements on the ACNS operation and autonomy. Details can be found in Ref. 3.

Mission Objectives
The objectives of the PROBA-1 mission are:

- to successfully demonstrate on-board autonomy in orbit
- to successfully demonstrate new space technologies and tools
- to successfully carry out a hyper-spectral Earth-observation mission.

From the ACNS point of view, the main autonomous functions include:

- the on-board determination of the orbit and the attitude
- the on-board prediction of orbital events (entry/exit into/from eclipse, ground station overfly, Earth imaging target overfly over the next 24 hours, etc)
- the on-board generation of the reference attitude profile (quaternion, angular rate and angular acceleration, in 3 axes) required to meet the imaging requirements
- the on-board control of the manoeuvres and of the instrument pointing
- the detection and identification of on-board failures (FDI).

As for the 2nd mission objective, new space technologies and tools are demonstrated in the use of:

- a high-performance ERC-32 RISC processor, a space version of a standard commercial processor, whose development has been initiated by ESA [1],
- an autonomous star sensor, capable of acquiring inertial attitude without a priori information,
- a GPS receiver, capable of determining on board the orbit knowledge, and
- the use of Computed-Aided Software Engineering (CASE) tools for the development, test and automatic coding of the flight software, including that of the ACNS.

PROBA's Earth-observation mission was specially selected to impose realistic and challenging requirements on the autonomy of the ACNS. It will be described below with the discussion on the payloads.

The Orbit and the Spacecraft
The PROBA spacecraft was launched directly into its final slightly eccentric, Sun-synchronous orbit with a perigee of 560 km and an apogee of 680

Fig. 1 : PROBA-1 on the PLSV Launcher

km. At the beginning of life, its passage at the descending node was at 10:30 AM local time. An eclipse of 35 minutes occurs during its 100-minute orbit period. A single ground station provides coverage of the spacecraft. The PROBA spacecraft is a 100-kg mini-satellite built in a honeycomb structure with dimensions 600 x 600 x 800 mm (Fig. 1). Gallium-arsenide solar cells mounted on the external panels provide electrical power to a protected 28V regulated power bus. A 9Ah Lithium-Ion battery takes over during eclipse. Thermal control is passive. A S-band communication system provides a 4kbit/s uplink and a 1Mbit/s downlink. A redundant ERC-32 central processor performs all the primary computations and data processing required by the PROBA commanding, house-keeping and telecommunication functions. On-board propulsion is not baselined.

The Payloads and their Requirements

In order to impose realistic constraints on the on-board autonomous operation of the spacecraft, PROBA carries various payloads that have been specially selected for their demanding operational requirements. As far as pointing strategy is concerned, these requirements can be categorised in terms of the fraction $1/N$ ($N \geq 1$) between the required ground speed of the instrument footprint and the nominal push-broom speed of the sub-satellite point (Figs. 2):

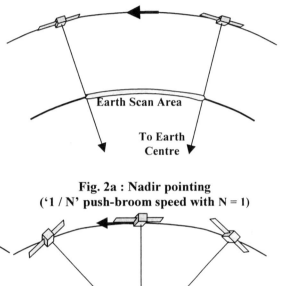

Fig. 2a : Nadir pointing
('1 / N' push-broom speed with N = 1)

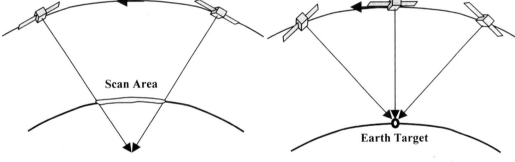

Fig. 2b : Reduced-speed imaging scan
('1 / N' push-broom speed with 1<N<∞)

Fig. 2c : Fixed Target Pointing
('1 / N' push-broom speed with N = ∞)

- *for a nadir pointing or offset-nadir pointing*:
  ground speed = push-broom speed/N with N = 1 (nominal push-broom speed)
- *for an Earth-fixed target pointing*:
  ground speed = push-broom speed/N with N = ∞ (zero ground speed of the footprint)
- *for a "1/N ground speed"* :
  ground speed = push-broom speed/N with 1 < N < ∞ (reduced ground speed by factor N).

The *Space Radiation Environment Monitoring* (SREM) instrument and the *Debris In-orbit Evaluator* (DEBIE) instrument do not impose special constraints and requirements on the ACNS design except for a nadir pointing attitude (N=1).

The *High-Resolution Camera* (HRC) has a spatial resolution of a few meters and requires a highly stable and accurate pointing of its two-dimensional CCD to an Earth-fixed target (N = ∞).

The primary PROBA payload is the *Compact High-Resolution Imaging Spectrometer* (CHRIS). This imager operates in the spectral range 450 nm to 1050 nm with a spatial resolution of 25m at nadir and a spectral resolution of 5 nm to 12 nm. Its task is to acquire 5 images of the same User-selected land area by scanning a single line of detectors over a 20-km Earth target. The ground speed of the line of detectors is reduced by a factor of 5 relative to the push-broom speed (N = 5) in order to increase the radiometric resolution and/or reduce the sampling distance. Nadir-pointing operation of the CHRIS (N = 1) has also been implemented. In order to acquire BRDF information on the target, the 5 images are taken from different viewing angles in the same pass in a 55-deg co-elevation cone (Fig. 3). The main imaging payload – the Compact High-Resolution Imaging Spectrometer or CHRIS – has a field of view to Earth along the yaw axis, through the interface ring with the launcher. Since CHRIS is rigidly fixed to the spacecraft body, this instrument imposes specific attitude and rate profiles on the spacecraft during imaging.

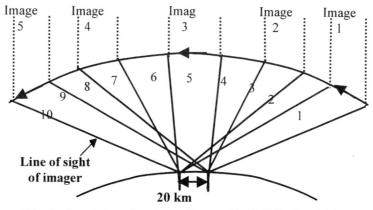

Fig. 3 : The 5 Imaging Scans for the CHRIS Payload (N = 5)

The ACNS Operational Modes and Pointing Requirements
Given the above operational requirements, the PROBA ACNS was designed to operate into three broad classes of modes (Fig. 4) : the *terrestrial modes*, the *celestial modes* and the *magnetic mode*. The terrestrial modes require on-board knowledge of the Earth attitude and position which, for in the latter case, is equivalent to a knowledge of the orbit. The terrestrial modes include:

- the **IMAGING** mode, where five reduced-speed scans of the target by the CHRIS payload are performed at the fraction 1/N (N = 5) of the nominal push-broom speed;
- the **ETARGET** mode (Earth target mode), where the ground speed of the HRC two-dimensional detector relative to the target is zero (N = ∞);
- the **RELORB** mode, where the attitude of the spacecraft is fixed relative to the orbital frame, thereby including the nadir pointing and offset-nadir pointing attitude (N = 1);
- the **HPOWER** mode (high power generation mode) which is basically a nadir-pointing attitude with an additional cyclic roll-yaw rotation to maximise the Sun energy input on the body-fixed solar cells.

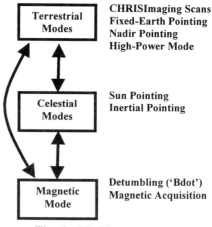

Fig. 4 : PROBA ACNS Modes

804

The celestial modes include the **SUN** mode where the spacecraft attitude is defined relative to the Sun frame and the **INERTIAL** mode where it is defined relative to the International Celestial Reference System. The magnetic class of modes included only a single mode, the **BDOT** mode which uses magnetometers and magneto-torquers to damp the excess spacecraft angular momentum acquired at separation from the launcher. The BDOT mode has also been used as a safe mode and a low-energy "week-end" mode. As a background mode to all the above primary modes, the control of the *total* spacecraft angular momentum was achieved with the magneto-torquers in order to prevent reaction wheel saturation.

The pointing requirements associated with the celestial and terrestrial modes are given in Table 1. The APE and RPE requirements for terrestrial pointing are applicable to the following index:

$$Index = \sqrt{\varepsilon_{roll}^2 + \varepsilon_{pitch}^2 + H\varepsilon_{yaw}^2}$$

where H is the payload field of view (1.5 deg) and ε is the pointing error.

Table 1 : PROBA-1 Pointing requirements

Errors at 95% confidence level	Celestial Pointing	Terrestrial Pointing
Absolute Pointing Error (APE)	100 arcsec per axis	150 arcsec (index)
Relative Pointing Error (RPE)	5 arcsec over 10 s per axis	5 arcsec over 10s (index)
Absolute Measurement Error (AME)	10 arcsec per axis	20 arcsec per axis
Position knowledge accuracy		200 m (3σ)

The Autonomy Scenario

As indicated earlier, one of the unique features of the PROBA mission is the fact that all mission planning and payload operations are managed by the on-board software.

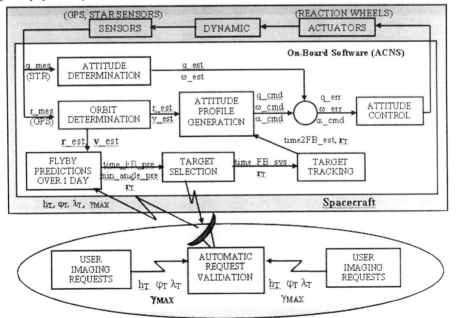

Fig. 5 : The PROBA Autonomy Scenario

805

As illustrated on the PROBA autonomy scenario of Fig. 5, the Users that require images of particular Earth targets of interest simply have to transmit their request in terms of the target coordinates (longitude λ_T, latitude φ_T and altitude h_T) and a 'figure of merit' for the quality of the image, expressed as the maximum allowable minimum co-elevation angle to the target (γ_{max}) beyond which it is preferable to wait for the next orbit to obtain a better line of sight to the target. All the operations required to predict the 'best' conditions for target overfly, to plan for payload operations and attitude manoeuvres are executed on-board, within the ACNS software. The functions and operations illustrated in Fig. 5 have been described in an earlier paper [4] and are not further discussed here.

THE PROBA AUTONOMOUS ACNS

In order to better appreciate the in-flight results of the PROBA spacecraft, an overview of the PROBA ACNS is required. The ACNS software is decomposed into four top-level functions : *navigation*, *guidance*, *control* and *failure detection & identification* (FDI). Each is briefly described next along with an overview of the PROBA-1 ACNS hardware.

The ACNS Hardware

During the celestial and terrestrial modes, a set of four all-skewed reaction wheels provide attitude control of the PROBA-1 spacecraft. Control during the magnetic mode and angular momentum management is ensured by magneto-torquers. In order to reduce hardware costs, attitude and position measurements with respect to the stars, the Sun, the Earth and Earth target are provided by only two sensors: two autonomous star trackers and a Global Positioning System (GPS) receiver. In addition, wheel-mounted tachometers provide a knowledge of the angular rate of each reaction wheel. The functional equivalence to the gyro, the Sun and the Earth sensors is ensured by software. For the magnetic mode, magnetometers complete the set of sensors.

Navigation

The spacecraft attitude and angular rate are estimated continuously at 6Hz with a 10th-order Extended Kalman Filter (EKF) using as inputs the quaternions from the two star sensors and the four wheel tachometer measurements. Since one or both of the star sensors may be blinded by the Earth during manoeuvres, only the available measurements within each of the 1/6 s cycles are used to update the estimated state in a sequential process. The outputs of the EKF module are the estimated (filtered) inertial attitude and the inertial angular velocity expressed in body axes. Since the delay in the delivery of the star sensor measurements (more than 1s latency) was incompatible with the absolute pointing requirements applicable in the high-rate IMAGING and ETARGET modes, a *delay-recovery algorithm* was implemented. It effectively propagates the measurements in the future, so the estimated attitude state would become contemporary with the point in time when they are needed to compute the control actions. This delay-recovery algorithm is described in Ref. 5. 'Blind' operation (i.e. without star sensor measurements) of the state estimator is allowed for a given (operator defined) period beyond which the outputs of the estimator are declared invalid. A second, more robust and simpler state estimation algorithm was also implemented in the ACNS in order to allow background (i.e. not-in-the-loop) execution of the attitude EKF for troubleshooting, if required.

The spacecraft inertial position and velocity are estimated continuously at 6Hz with a 6th-order Extended Kalman Filter (EKF) using as inputs the GPS position fixes provided at 1 Hz. 'Blind' operation (i.e. without GPS measurements) of the state estimator is allowed for a given (operator defined) period beyond which the outputs of the estimator are declared invalid. Such blind operation was required during eclipse (35 minutes) so the GPS could be switched off to save power. The outputs of the orbit navigation module are the estimated (filtered) inertial position and velocity of the spacecraft. Here as well, the delay in the delivery of the GPS measurements was incompatible with the absolute position requirements (Table 1) and the same *delay-recovery algorithm* was implemented. A second, more robust and simpler state estimation algorithm was also implemented in the ACNS in order to allow background (i.e. not-in-the-loop) execution of the orbit EKF for troubleshooting, if required. A third orbit estimation algorithm, the SGP4 algorithm that uses the NORAD Two-Line Elements as inputs, was also implemented in the ACNS navigator as a back-up to the previous two algorithms.

The navigation module also includes an accurate model of the Earth motion (precession, nutation, sidereal rotation, polar motion) and Sun ephemeris in order to support the autonomous on-board prediction of ground station overfly's, imaging target overfly's and entry/exit into/from eclipse.

Guidance

The guidance module [4] generates on-board the reference attitude quaternions, angular rates and angular acceleration, in 3 axes, that are required to ensure that the ground trace of a given instrument field of view has the desired geographical coordinates and relative speed (with the commanded 1/N ratio) as a function of time. Using as input the User's target coordinates, the model of the Earth's rotation and the estimated satellite orbit, the next flyby of the target is predicted. When the spacecraft is nearing the entry into the 55-deg observation cone centred at the target zenith, the ACNS initiates autonomously the generation of the attitude profiles required by the IMAGING, ETARGET, RELORB and HPOWER modes. The detailed procedure on how the profiles are generated on board is described in Ref. 4.

Control

Since PROBA is an experimental and technology-demonstration mission, three different control algorithms were implemented on-board the ACNS for nominal pointing modes. One controller is always in operation when large-angle manoeuvres are required. It is based on an optimal sliding-mode controller that operates either at the maximum torque that the wheels can deliver, or at the maximum angular momentum the spacecraft can receive from the wheels (near wheel saturation). Given the all-skewed configuration of the wheels, algorithms were developed to compute the effective maximum torque and angular rate that are available about any arbitrary spacecraft rotation axis.

For fine-pointing attitude control, two different algorithms are available and can be selected by ground command. One is based on a classical phase-advance-plus-integral control algorithm, the other is based on a full-state feedback structure where the feedback gain matrix was computed by pole placement.

All three of the above controllers are executed at 6 Hz. In addition, two other non-pointing control algorithms, using magnetic sensors and actuators, are executed at 1 Hz. The BDOT mode, as the name indicates, is a simple algorithm that estimates the current angular rate of the spacecraft relative to the magnetic field using successive magnetometer measurements. Once this pseudo angular velocity is determined, magnetic torques are applied in a direction opposite to this angular velocity until it is reduced to zero. At that point, the spacecraft is "attached" to the magnetic field of the Earth, executing at each orbit two full rotations around the orbit normal.

Reaction wheel desaturation is also controlled with the magneto-torquers. Because of the high-rate manoeuvres required by the PROBA mission, the *total* angular momentum of the spacecraft is monitored. A parallelepiped has been computed in body axis such that, whenever the total angular momentum vector lies within this volume, none of the wheels will saturate. Due to the all-skewed configuration of the wheels, this allowable region is quite complex to compute and has been simplified to the largest sphere that fits within it. This sphere is offset from the spacecraft centre of mass so the large manoeuvres required during IMAGING and ETARGET will not trigger the angular-momentum control. A hysteresis has also been designed around the sphere to avoid multiple momentum management control near the boundary of the sphere.

Failure Detection and Identification (FDI)
Because of the large size and complexity of the on-board ACNS software, a structure has been designed into it to detect problems, identify them and avoid their propagation to other modules downstream. Basically, each ACNS software module generates at its output two logical flags: a FDI flag and an OUT flag. These flags are computed as follows. Within each module, a series of consistency tests, validity tests and numerical tests are executed every cycle on the data the module

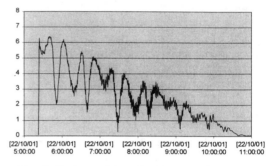

Fig. 6 : Estimated Angular Rate in BDOT Mode

generates. If all the computations within the module are successful and all the data it generates are valid, the OUT flag is unity and the FDI flag is zero. Whenever there is a problem, the OUT flag is zero, the FDI flag is positive and the actual value of the FDI flag is coded to indicate which test(s) failed, thereby identifying the source of the problem. Conversely, each module that requires as inputs the outputs from other modules upstream must also ensure that all these input data are valid before the module uses them. To verify that, each module also has as input an INP flag. This INP flag is unity when all the OUT flags of the upstream modules are true (unity) and it is false (zero) whenever one of them is false. In the latter case, the module does not execute and generates nominal (or dummy) data as output, with its OUT flag set to zero. Details on this FDI structure are provided in Ref. 6.

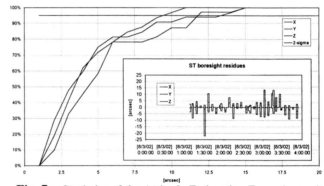

Fig. 7a: Statistics of the Attitude Estimation Error (arcsec)

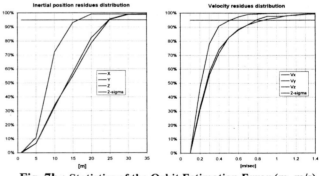

Fig. 7b : Statistics of the Orbit Estimation Error (m, m/s)

808

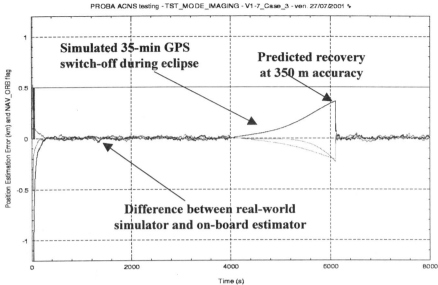

Fig. 8a : Orbit Knowledge Recovery after Simulated GPS Switch-Off

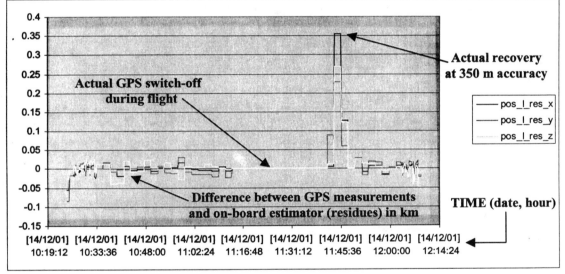

Fig. 8b : Orbit Knowledge Recovery after In-Flight GPS Switch-Off

ACNS FLIGHT PERFORMANCE

Initial Acquisition

The BDOT mode was initiated after separation from the launcher. Figure 6 shows the telemetry of the pseudo angular rate, as estimated from the measured magnetic field. From the initial angular rate of about 6.5deg/s, the BDOT control reduced the rate to a near-zero value in less than 6 hours. This near-zero rate relative to the magnetic field corresponds, in an inertial frame, to an average of 0.12 deg/s (720 deg/orbit). The BDOT was shown to be robust and reliable and was used later in the mission as a safe mode and during weekends.

Navigation Performance

During the commissioning activities, the residues of the attitude and orbit estimators (difference between estimated and measured states) were collected in order to compile statistics on the navigation performance. They are shown in Figs. 7. The 2-sigma pointing knowledge error is around 10 to 12 arcseconds (Fig. 7a) and the orbit knowledge is around 15 to 25 m in position and 0.5 to 0.8 m/s in velocity (Fig. 7b).

Figures 8 compare the predicted and actual performance of the navigator when it propagates the orbit state with the GPS switched off during the 35-min eclipse to save power. Figure 8a was obtained by simulation before launch and Fig. 8b gives the telemetry obtained during commissioning. Both show a recovery of the orbit knowledge with a residual error of 350 m.

Control Performance

Figure 9 shows a typical tracking error in degrees in the RELORB mode (nadir pointing attitude in this case) over a period of about 9 hours. The tracking error is the difference between the commanded attitude and the estimated attitude. As shown in the figure, the pointing performance is well within the specification of 150 arcseconds.

Autonomy Demonstration

One of the main features to demonstrate during commissioning was the autonomous operation of the ACNS. To achieve this, a large number of pictures, both in ETARGET mode (for HRC pictures) and in IMAGING mode (for CHRIS imaging scans) were commanded by specifying only the coordinates of the Earth target, often the day before the picture was to be taken. Many weekend pictures were taken and the commands were sent to the spacecraft on the Friday before.

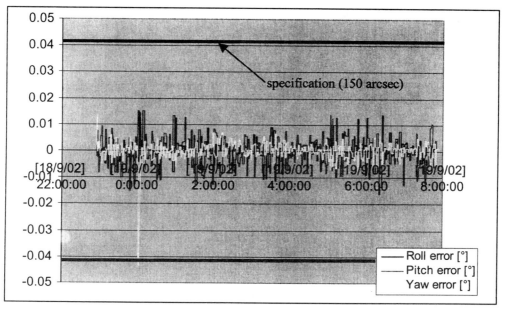

Fig. 9 : Tracking Error During Nadir Pointing

Figure 10 shows the telemetry of a series of autonomous manoeuvres executed by the spacecraft in order to acquire a picture with the HRC in the ETARGET mode. The 3-axis inertial angular velocity of the spacecraft in body coordinates is shown. The spacecraft was initially in a nadir-pointing attitude. At the transition to the ETARGET mode, a large-angle sliding-mode manoeuvre is commanded at the largest available torque and angular velocity that the wheels can produce (here a forward pitch rate around 0.6 deg/s). The spacecraft acquires the target and tracks it by pitching backward until it comes almost directly above the target (minimum co-elevation angle). At that time, the negative pitch rate is maximum (shortest range to the target) and the picture is taken. Compensation for the Earth rotation can be seen in roll and yaw. As the target disappears behind the spacecraft, a return to nadir pointing is commanded, again with the maximum positive pitch rate that the wheels can deliver in that new inertial attitude (0.9 deg/s).

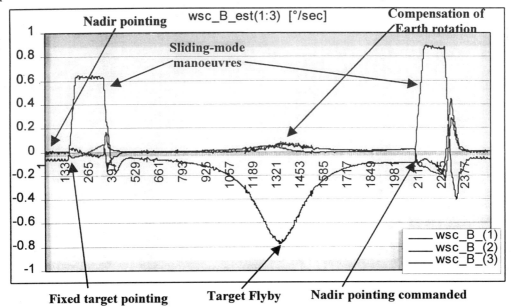

Fig. 10 : Angular Velocity (deg/s) during Fixed Target Pointing

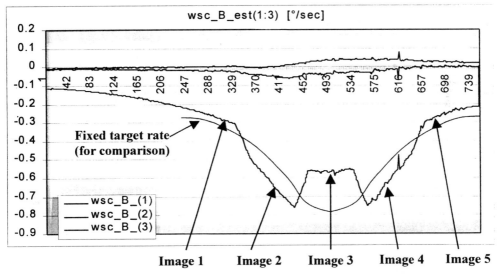

Fig. 11 : Angular Velocity (deg/s) during CHRIS Imaging Scans

811

Figure 11 gives the telemetry of a similar manoeuvre, this time in IMAGING mode. Compared with the ETARGET mode (with relative ground speed of 1/N = 0, which has been added on the graph for illustration), the modulation of the pitch rate to achieve the relative ground speed of 1/N = 1/5 can be seen. The odd-numbered images scan the target forward (pitch rate "less negative" than ETARGET) and the even-numbered images scan backward (pitch rate "more negative").

Figure 12 illustrates the relative mapping accuracy in IMAGING mode. After having acquired images in the IMAGING mode, their ground trace was projected on an existing map of the area. As can be verified in Fig. 12, the odd-numbered images shown overlap relatively accurately. Images no 1 and 5 are larger than the central image no 3, as can be expected from the fact that the field of view of the imager intersects a larger area on the Earth when the line of sight makes a shallower angle with the surface. A similar performance (not shown) was also demonstrated for the even-numbered images.

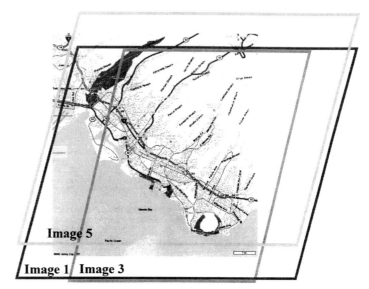

Fig. 12 : Relative Mapping Accuracy in Imaging Mode

Fig. 13a : CHRIS Image of San Francisco (left)

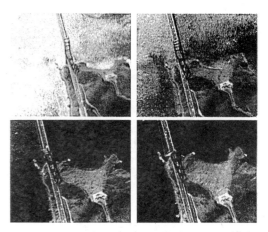

Fig. 13b : Zoom into HRC picture showing Northern tower of the Golden Gate bridge

On Figs. 13 (taken from Ref. 6), a CHRIS picture of San Francisco is shown along with a zoom on part of the image, extracted from an HRC picture of the same area. Figure 14 is a CHRIS picture of the Mount Etna eruption that took place some months ago.

Fig. 14 : A CHRIS Picture of Mount Etna

CALIBRATION OF THE PAYLOAD ALIGNMENT

Basic Principle

Calibrating the alignment of a payload consists in measuring the 3-axis orientation of the payload – the CHRIS payload in the case of PROBA – relative to the primary sensors of the spacecraft – the star sensors for PROBA. An accurate knowledge of this alignment is important since pointing the payload to an Earth target is achieved on PROBA by pointing the sensor to the appropriate inertial direction. Knowledge of this alignment was first acquired on ground during integration. However, the launch stress, the zero-g relaxation, outgassing and thermal gradients may alter the physical alignment of the payload and a new measurement is required during flight. Looking at the first pictures acquired by PROBA, it was concluded that the alignment knowledge was indeed no longer accurate and had to be corrected.

The in-flight measurement principle is based on the fact that the inertial coordinates of targets identified in a CHRIS image can be computed in two ways: (1) using a knowledge of these coordinates in the payload axes, then spacecraft axes and finally with respect to the stars (via the star sensors), and (2) using a knowledge of the target coordinates on Earth (longitude, latitude, altitude), then translated into inertial coordinates using a knowledge of the Earth inertial attitude. In this matrix equality, there is one unknown: the alignment matrix of the payload relative to the spacecraft axes. The equation can then be solved for this unknown.

Given the geocentric position vector $\vec{r}_i$ of a known Earth target i, the spacecraft geocentric position vector $\vec{r}$ at the time of imaging and the position of the Earth target $\vec{r}_{rel,i}$ relative to the spacecraft at that time, the following equality holds:

$$\vec{r}_i = \vec{r} + r_{rel,i}\,\vec{d}_i$$

where $r_{rel,i}$ and $\vec{d}_i$ are the magnitude of $\vec{r}_{rel,i}$ and the unit vector along it, respectively. Expressing this vector equation in inertial coordinates, via (1) the Earth frame and (2) via the spacecraft body frame, we get:

$$r_i^I = C_{TI}^{-1}\,r_i^T = r^I + r_{rel,i}\,C_{SI}^{-1}\,C_{SP}\,d_i^P$$

where superscripts I, T, S and P denote vector components expressed in inertial, terrestrial, sensor and payload frames respectively. The 3x3 matrices C_{MN} denote the rotation matrices from frame N to frame M. The inverse notation has been used on these matrices, instead of the transpose notation, to avoid confusion with the notation for the terrestrial frame. In the above equation, C_{TI} is known from a model of the Earth nutation, precession and rotation, r_i^T is known from the geographic coordinates of target i recognised in the image, the inertial spacecraft position r^I is obtained from the GPS, the range from spacecraft to target $r_{rel,i}$ is computed from the magnitude of $r_i^I - r^I$, the inertial attitude of the star sensor C_{SI} is known from its measured quaternion, and the unit vector d_i^P to the Earth target i in payload axes can be read from its position on the image. The only unknown in this equation is the rotation matrix C_{SP}, the desired alignment matrix between payload and sensor axes. For N Earth targets in the image, the equation can be re-arranged into the compact form:

$$A = C_{SP}\, D$$

where matrix A, composed of N column-matrices A_i, is a matrix of *known* inertial measurements :

$$A = [A_1, A_2, A_3, \cdots, A_N] \qquad \text{with} \qquad A_i = C_{SI}\left(C_{TI}^{-1}\,r_i^T - r^I\right)/\,r_{rel,i}$$

and matrix D is a collection of N unit vectors pointing to the N identified Earth targets in the image and measured in image axes i.e. payload axes:

$$D = \left[d_1^P, d_2^P, d_3^P, \cdots, d_N^P\right].$$

Solving this equation for the alignment matrix C_{SP} is the well-known Wahba problem [7] and many solutions have been proposed to solve it [8].

Application to the Calibration of the CHRIS Payload

So far, the description of the alignment problem is conventional. Its application to PROBA is slightly more complex. Indeed, during the push-broom motion, the CHRIS payload takes the image one line at a time. Supposing that the payload boresight is along the P_z axis, the line of detectors is along the P_y axis and the push-broom motion is along the P_x axis, then all the unit vectors to the Earth targets will have the general format: $d_i^P = \begin{bmatrix} 0 & y_i & z_i \end{bmatrix}$ i.e. all the measurements are taken in the Y-Z plane, along the line of detectors, since the spacecraft provides the motion along P_x. Therefore, even though there may be many Earth targets on a single line of the image, payload rotations around the P_y axis cannot evidently be detected from them and thus cannot be calibrated.

The solution to this particular problem was to perform a *dynamic* calibration in which the matrix A_i above varies with time, as the payload acquires the thousand of lines constituting the image. Therefore, for each of the N targets in the image, the on-board time (to compute C_{TI}), the star sensor quaternion (C_{SI}) and the spacecraft position r^I were extracted from the telemetry at the time the line containing each target was imaged. The matrix A_i is then constructed and the equation solved for C_{SP}. Since at least three targets were available in each calibration image, a solution based on the singular value decomposition of the matrix of measurements was used [8].

Numerical Results

Using a CHRIS picture of New York, it was possible to identify a number of features with well-known geographical coordinates and the algorithm was applied. Two alignment matrices were computed, C_{1SP} and C_{2SP}, one for each of the two star sensors on board PROBA. By comparing these alignment matrices with those measured on ground before launch, it could be observed that important discrepancies (of the order of 1 deg) were observed.

In order to have a measure of the relative accuracy of the calibration, the equation $A_c = C_{SPc} D$ was used to re-compute the matrix of inertial measurements A_c using the calibrated alignment matrix C_{SPc} just computed and the original payload measurements matrix D. Next, by computing the *calibration error matrix* $E_c = A - A_c$, between the initial set of inertial measurements A used in the calibration and that computed from the re-mapped payload measurements after calibration A_c, a measure of the consistency in the data is obtained. The error matrices for both alignment matrices is given below :

$$E_{1c} = \begin{bmatrix} 0.00000 & 0.00022 & -0.00028 \\ -0.00015 & 0.00021 & -0.00001 \\ 0.00039 & -0.00044 & 0.00000 \end{bmatrix} \quad E_{2c} = \begin{bmatrix} -0.00039 & 0.00001 & 0.00032 \\ 0.00021 & -0.00014 & -0.00001 \\ 0.00031 & -0.00029 & -0.00000 \end{bmatrix}$$

The last column in these matrices effectively give the error in the direction of the payload boresight axis P_z expressed in the sensor reference frame. From that information, it can be computed that the calibration error in the boresight direction is 0.017 deg for star sensor 1 and 0.019 deg for star sensor 2. As for the rotational error around the boresight axis, the computations give around 0.031 deg for sensor 1 and 0.031 deg for sensor 2. The fact that the relative accuracy around the boresight is not as good as that for the boresight direction is a consequence of the geometric dilution of precision caused by the limited field of view of the instrument (all targets are within a degree or so of the boresight). These relative accuracies are much smaller than the corrections in the C_{SP} matrices that were required (about 1 deg). It can thus be concluded with some confidence that the in-flight calibration improved the alignment knowledge significantly.

IN-FLIGHT MEASUREMENT OF THE SPACECRAFT INERTIA MATRIX

As noted earlier in this paper, the knowledge of the spacecraft angular rates, required for the stable control of the attitude during imaging and fixed-target pointing, is not obtained from gyros but from a Kalman state estimator. In addition, the algorithm that generates the attitude control actions during these fast manoeuvres uses a feedforward term based on the commanded acceleration computed in the guidance function. Both these algorithms require a knowledge of the spacecraft inertia matrix. The in-flight measurement of the effective inertia matrix can improve the accuracy and performance of the above two algorithms.

Basic Principle

The measurement of the inertia matrix for a spacecraft equipped with reaction wheels relies on the principle of conservation of angular momentum [9]. For short-duration manoeuvres, the effect of external perturbing torques is assumed negligible and the total angular momentum of the spacecraft $\mathbf{h}_{tot}$ remains constant in inertial coordinates. During the manoeuvre, the total angular momentum is thus distributed between the reaction wheels and the spacecraft body and the momentum exchanged between them is proportional to their respective inertias. Since the inertia of the reaction wheels J_w is typically known to high accuracy and does not change much during launch and flight, J_w is assumed *known* and the only unknown left in this momentum exchange is the spacecraft inertia $\mathbf{J}$. The translation of the above principle in mathematical terms follows.

In body axes, the total angular momentum of the spacecraft is given by:

$$h_{tot} = J\omega_B + \sum_i J_{wi}\,\omega_{rel,i}\,a_i$$

where $\mathbf{J}$ is the 3x3 inertia matrix of the spacecraft plus stationary wheels, J_{wi} is the (scalar) inertia of wheel i around its spin axis, ω_B is the 3x1 inertial angular velocity of the spacecraft in body axes, $\omega_{rel,i}$ is the angular rate of wheel i relative to the spacecraft body and the a_i are the 3x1 coordinates of the wheel rotation axis in body axes. Re-arranging the wheel inertias into a diagonal matrix $\mathbf{J}_w$, the wheel angular rates in a column matrix ω_{rel} and the spin-axis unit vectors in a transformation matrix $\mathbf{A}$, the equation can be rewritten as:

$$h_{tot} = J\omega_B + AJ_w\omega_{rel}\,.$$

As explained earlier, during a manoeuvre from time t_0 to time t_1, the total angular momentum is constant in inertial coordinates, thus:

$$C_{BI}^{-1}(t_0)\big[J\omega_B(t_0) + AJ_w\omega_{rel}(t_0)\big] = C_{BI}^{-1}(t_1)\big[J\omega_B(t_1) + AJ_w\omega_{rel}(t_1)\big]$$

where C_{BI} is the rotation matrix from inertial to spacecraft frames (the spacecraft attitude). When the spacecraft angular velocity ω_B is known (e.g. from gyros), the wheel angular rates ω_{rel} are known (from the wheel tachometers), the wheel inertias $\mathbf{J}_w$ and alignment $\mathbf{A}$ are well known (from ground measurements before launch) and the spacecraft attitude C_{BI} is known (e.g. from star sensors), the only unknown left in this equation is the spacecraft inertia $\mathbf{J}$.

In order to solve the above equation for $\mathbf{J}$, Ref. 9 assumes that the spacecraft angular velocity at the beginning of the manoeuvre $\omega_B(t_0)$ is zero. The above equation can thus be simplified to:

$$J\omega_B(t_1) = C_{BI}(t_1)C_{BI}^{-1}(t_0)AJ_w\omega_{rel}(t_0) - AJ_w\omega_{rel}(t_1) = q(t_1)$$

where the right-hand side has been compressed into the 3x1 column-matrix $q(t_1)$. The left-hand side of this equation can be re-arranged so that all 6 unknowns in the inertia matrix are organised into a 6x1 column-matrix $\mathbf{J}^* = [J_{xx}, J_{yy}, J_{xx}, J_{xy}, J_{xz}, J_{yz}]^T$:

$$\omega_B^*(t_1)\mathbf{J}^* = q(t_1)$$

where the components of the angular velocity ω_B have also been re-arranged accordingly into the 3x6 matrix ω_B^*. Finally, by taking N measurements at times t_i, $i = 1, 2, 3, ...N$ during the manoeuvre, a $3N$ x 6 matrix of angular velocity measurements $W = [\omega_B^*(t_1); \omega_B^*(t_2); ... \omega_B^*(t_N)]$ (where the semi-colon denotes arrangement row by row) and a $3N$ x 1 column-matrix of right-hand sides $Q = [q(t_1); q(t_2);... q(t_N)]$ can be compiled, leading to the $3N$ equations:

$$WJ^* = Q.$$

A least-square solution for the inertia matrix $\mathbf{J}$ can be obtained with the pseudo-inverse of W:

$$\mathbf{J}^* = \left[W^T W\right]^{-1} W^T Q .$$

The above equation is a generalisation of that derived in Ref. 9 to the case where all the elements of the inertia matrix are estimated at the same time using measurements from manoeuvres in all 3 axes. Reference 9 estimated one element of $\mathbf{J}$ at a time with a single-axis manoeuvre at a time.

Application to PROBA

The above results cannot be directly applied to the PROBA spacecraft and, for that matter, to any spacecraft not initially at rest at the start of the manoeuvre. Indeed, since PROBA is nominally in a nadir-pointing attitude, the spacecraft body is constantly rotating inertially at the orbital rate. Therefore, the assumption made in Ref. 9 that $\omega_B(t_0) = 0$ cannot be made. Looking at the general equation in this case:

$$C_{BI}^{-1}(t_0)\left[J\omega_B(t_0) + AJ_w\omega_{rel}(t_0)\right] = C_{BI}^{-1}(t_1)\left[J\omega_B(t_1) + AJ_w\omega_{rel}(t_1)\right],$$

it can be seen that the inertia matrix cannot be isolated on the left-hand side as was done in Ref. 9. The best we can achieved is:

$$J\omega_B(t_1) - C_{BI}(t_1)C_{BI}^{-1}(t_0)J\omega_B(t_0) = C_{BI}(t_1)C_{BI}^{-1}(t_0)AJ_w\omega_{rel}(t_0) - AJ_w\omega_{rel}(t_1) = q(t_1)$$

A further generalisation of Ref. 9 is required. Re-arranging as before the unknowns into the 6x1 column-matrix $\mathbf{J}^*$, it can be shown after a lengthy by simple development that the equations can be re-organised into a similar format:

$$V J^* = Q$$

where this time, the $3N$ x 6 matrix V contains not only the measured angular velocities at time t_i, $i = 1, 2, ..., N$, like W, but also that at t_0 plus the elements of the rotation matrices $C_{BI}(t_i)$, $i = 0$, 1,..., N. The least-squares solution has the same form as above, with W replaced by V.

817

Numerical Results

The above algorithm was validated by simulation on the PROBA simulator. Multiple simultaneous manoeuvres around each axis were generated and noisy measurements of the attitude and angular rates were simulated. The algorithm was applied and the simulated inertia matrix was recovered to an accuracy compatible with the measurement accuracies.

In order to apply the algorithm on the real spacecraft, a problem arises due to the fact that the angular velocity of the spacecraft is not directly measured by gyros but estimated via the extended Kalman Filter which, itself, uses a knowledge of the inertia matrix. In order to avoid a circular mathematical development, the angular velocity of the spacecraft must be obtained by differentiation of the quaternions. Although differentiating noisy measurements is not the best approach, it is the only one available for the present case. Such an algorithms was fortunately implemented on PROBA.

CONCLUSIONS – LESSONS LEARNED

This paper has described some of the flight results that demonstrated the performance and the autonomy of the PROBA guidance, navigation and control system. In addition, the procedures and algorithms for the in-flight calibration of the payload alignment and inertia matrix have been described with supporting numerical results. The paper now concludes with some "lessons learned", some aspects being positive (+), some others to be improved (o).

o   The performance and usefulness of the sensor delay-recovery algorithm was demonstrated but it was also shown to be very sensitive to timing and synchronisation errors (e.g. when a bad time correlation was performed on board or when the GPS time had jumps).

o   Despite the fact that the star sensor was designed to flag to the ACNS any measured quaternions that were considered invalid, invalid quaternions did make it to the attitude state estimator, causing jumps and noise in the attitude estimates. A bad-data rejection algorithm should have been implemented at the input of the filter. Such a rejection algorithm was proposed and discussed during the development but was eventually not implemented on PROBA.

o   As explained earlier, the quaternion measurements coming from the two star sensors are processed in sequence. When both star sensors deliver valid quaternions, that approach increases considerably the computing load on the on-board processor. A fusion of the two quaternions before they enter the state estimator, using techniques such as those described in Ref. 8, would have been preferable. This alternative was considered too late in the design (and too close to the launch) to be implemented.

+   The process by which the imaging of targets could be accomplished autonomously, by specifying only its geographic coordinates, was demonstrated to be effective and simple for the ground operators and the Users.

+   The performance and usefulness of the state estimation using a sensor delay-recovery algorithm was demonstrated although the delay-recovery algorithm must be made robust to timing and synchronisation errors.

+   The open-loop estimation of the orbit during eclipse with negligible degradation was demonstrated.

+   The execution of fast and accurate attitude manoeuvres without gyros and the accurate pointing of the spacecraft to Earth targets without Earth sensors was demonstrated using their functional equivalent obtained via software (i.e. *software* or *virtual* sensors) and star sensors.

+   The optimal large-angle, sliding-mode manoeuvres with skewed reaction wheels was demonstrated.

+ The autonomous management of *total* spacecraft angular momentum was shown to be effective.

+ The use of a network of FDI checks and data-validity propagation within the software was shown to be very useful not only during flight but also during development, since the FDI flags were used during simulations and ground validations to find out the origin of problems.

+ The use of the MATRIX-X / SystemBuild / Autocode family of software design tools was shown to provide not only an immense reduction in the time required for software development and validation, it would have been nearly impossible to develop such a complex on-board software with the time and budget in effect for the PROBA project.

REFERENCES

1. F. Teston, R. Creasey, J. Bermyn, K. Mellab, **PROBA: ESA's Autonomy and Technology Demonstration Mission**, *Proceedings of the 48th International Astronautical Congress* (1997), Paper IAA-97-11.3.05.

2. R. Creasey, **Project for On-Board Autonomy : PROBA. An ESA Technological Microsat Mission to Demonstrate Autonomous Spacecraft Operations**, *Workshop on On-Board Autonomy*, Noordwijk. The Netherlands, 17-19 Oct 2001.

3. J. de Lafontaine, P. Vuilleumier, P Van den Braembussche, **Autonomous Control of the PROBA Spacecraft,** Paper AAS 99-375, in *Astrodynamics 1999*, K.C. Howell et al., eds., *Advances in the Astronautical Sciences*, Vol. 103-II, pp. 1131-1144, 2000 (paper presented at the AAS/AIAA Astrodynamics Specialists Conference, Girdwood, Alaska, 16-19 August 1999).

4. J. de Lafontaine, P. Vuilleumier, P Van den Braembussche, **Autonomous Navigation and Guidance On-Board Earth Observation Mini-Satellites**, Paper AAS 01-421, in *Astrodynamics 2001*, D.B. Spencer et al., eds., *Advances in the Astronautical Sciences*, Vol. 109-III, pp. 1733-1748, 2002 (paper presented at the AAS/AIAA Astrodynamics Specialists Conference, Québec, Canada, 30 July – 2 August 2001).

5. J. de Lafontaine, Annie Robillard, **Autonomous Failure Detection and Identification for Future Space Missions**, *Workshop on On-Board Autonomy*, Noordwijk. The Netherlands, 17-19 Oct 2001.

6. P. Van den Braembussche, J. de Lafontaine, P. Vuilleumier, **PROBA Attitude Control and Guidance In-Orbit Performance**, *5th ESA Guidance, Navigation & Control Conference*, Frascati, Italy, 22-25 November 2002.

7. G. Wahba, **A Least Squares Estimate of Spacecraft Attitude**, *SIAM Review*, Vol.7, No.3, July 1965.

8. F.L. Markley, D. Mortari, **How to Estimate Attitude from Vector Observations**, Paper AAS 99-427, in *Astrodynamics 1999*, K.C. Howell et al., eds., *Advances in the Astronautical Sciences*, Vol. 103-II, pp. 1979-1996, 2000 (paper presented at the AAS/AIAA Astrodynamics Specialists Conference, Girdwood, Alaska, 16-19 August 1999.

9. J.A. Wertz, A.Y. Lee, **In-Flight Estimation of the Cassini Spacecraft's Inertia Tensor**, Paper AAS-01-179, in *Spaceflight Mechanics 2001*, L.A. D'Amario et al., eds., *Advances in the Astronautical Sciences*, Vol. 108-I, pp. 1087-1102, 2001 (paper presented at the AAS/AIAA Space Flight Mechanics Meeting, Santa Barbara, California, 11-15 February 2001).

ATTITUDE SENSOR ALIGNMENT AND CALIBRATION FOR THE TIMED SPACECRAFT[*]

Mark E. Pittelkau[†] and Wayne F. Dellinger[‡]

This paper presents the attitude sensor alignment and gyro calibration for Thermosphere-Ionosphere-Mesosphere Energetics and Dynamics (TIMED) and demonstrates the effectiveness of an Alignment Kalman Filter and a composite calibration algorithm. The on-orbit attitude determination performance prior to and after calibration is presented. The optical alignment performed prior to launch is compared to the estimated alignment parameters. Various means of validating the alignment calibration are discussed.

INTRODUCTION

The Thermosphere-Ionosphere-Mesosphere Energetics and Dynamics (TIMED) spacecraft was built by the Applied Physics Laboratory and was launched in December 2001. Its science objective is to perform an exploratory study of the physical and chemical processes acting within and upon the coupled lower-thermosphere/ionosphere system between 60 and 180 km. The science instrument suite comprises the SEE (Solar EUV Experiment), SABER (Sounding of the Atmosphere using Broadband Emission Radiometry), GUVI (Global Ultraviolet Imager), and TIDI (TIMED Doppler Imager). The spacecraft flies in an orbit of 625 km altitude and 74.1 degrees inclination and performs a 180 degree yaw maneuver approximately every 60 days. It maintains attitude control to the orbital reference frame within the requirement of 0.5 degrees 3σ per axis and the pointing knowledge within the requirement of 0.03 degrees 3σ per axis (36 arcsec, 1σ). The attitude sensors are two Lockheed-Martin AST-201 star trackers (designated AST-1 and AST-2) and a Honeywell YG9666C ring-laser gyro (IRU).

This paper presents the attitude sensor alignment and gyro calibration for TIMED and demonstrates the effectiveness of the alignment calibration algorithm (the Alignment Kalman Filter or AKF) described in [1] and the composite calibration algorithm (LSE) described in [2]. The calibration parameter vectors are the non-orthogonal gyro axis misalignment, ξ; symmetric scale factors of the gyro, λ; asymmetric scale factors of the gyro, μ; and misalignment of two star trackers, δ_{s1} and δ_{s2}.

It was shown in [3] that if all attitude sensors are modeled with misalignment parameters, there will be three independent and unobservable degrees of freedom of attitude unless one or more payload sensors can provide measurements of attitude. The payload then effectively defines the body reference frame. This is known as absolute alignment calibration [4]. Without payload measurements of attitude, the error in the three unobservable degrees of freedom is limited only by the

---

[*]   Copyright © 2003 by The Applied Physics Laboratory. Published by the American Astronautical Society with permission.

[†]   The Johns Hopkins University, Applied Physics Laboratory, 11100 Johns Hopkins Road, Laurel, Maryland 20723. E-mail: mark.pittelkau@jhuapl.edu.

[‡]   The Johns Hopkins University, Applied Physics Laboratory, 11100 Johns Hopkins Road, Laurel, Maryland 20723. E-mail: wayne.dellinger@jhuapl.edu.

a priori knowledge of alignment, which is based on ground measurements of alignment corrupted by orbit insertion errors.

Sufficient payload measurements of attitude are not available for alignment calibration, so absolute alignment calibration is not possible. As shown in [1, 5], a relative alignment calibration can be performed by choosing one of the attitude sensors to define the body reference coordinates. The choice should be the sensor for which the payload-to-sensor alignment is the best known and most stable and for which the sensor data is always available. On TIMED, this sensor is the IRU. Neither of the star trackers is the best choice because they are subject to dropouts due to moon interference, and the alignment of their focal plane to their external optical cubes is not accurately known.

Ground-measured alignments of the star trackers were not used as a priori data in the alignment estimation because they were not consistent with initial alignment estimates. The nominal mounting alignments were used instead. The a priori alignment errors were assumed to be zero with a 1 degree (3600 arcsec) uncertainty, which is large compared to the error in the nominal mounting data and the optical alignment data, and is large compared to the desired final uncertainty. The SRS frame, which is defined by a pair of optical flats, is taken to be the body reference frame. The optically measured IRU-to-SRS alignment was used to define the IRU mounting, and so the SRS is effectively the body frame. Because the IRU-to-SRS transformation is not observable, it must be assumed to be (sufficiently) accurately known. Fortunately the optically measured IRU-to-SRS alignments were consistent. The uncertainty in the final alignment and gyro calibration estimates depends on the angular rate of the spacecraft during the time in which star tracker and gyro data is collected for calibration. A conjecture in [3], which is empirically true, is that the angular rates in each body axis have to be such that a certain matrix integral over the calibration interval is positive definite. Non-harmonically related angular rates are sufficient; harmonically related and identical rates in each axis are not sufficient for full observability of the calibration parameters. TIMED was not designed to maneuver for the purpose of calibration. It is fortunate, however, that the yaw turn maneuver required every 60 days causes motion in all axes that provides almost enough observability. The algorithm in [2] is used to combine the calibration estimates from several maneuvers to obtain a composite calibration that should be more accurate than any individual calibration.

Ring laser gyros such as the Honeywell YG9666C inherently have symmetric scale factors, so the initial estimate of the asymmetric scale factor was set to zero with an uncertainty of only 1 part per million or 10^{-6}. The scale factor of the gyro is inherently very stable and well-characterized by the manufacturer over a wide range of angular rates. Nevertheless we attempted to estimate the symmetric scale factor despite the limited calibration maneuvers. The symmetric scale factor estimate was initialized to zero with a 1500 part per million (ppm) uncertainty in each axis.

The initial attitude estimate in the AKF is the quaternion measurement from one of the star trackers. Because the error in this initial estimate contains star tracker misalignment error, it is vital that the covariance between the attitude estimate and the alignment estimate be initalized accordingly. The covariance of the initial attitude estimate must also account for the misalignment error as well as the measurement noise. This was demonstrated in [1].

Hardware Layout

The two star trackers are mounted on an aluminum honeycomb composite laminate optical bench with the four TIDI telescopes and the SEE instrument as shown in Figure 1. The optical bench is kinematically mounted to the top deck of the bus. The gyros are mounted on the inside of the top deck. The SABER and GUVI instruments are mounted on the opposite end of the bus.

A Master Reference Cube (MRC) is mounted to the top deck to serve as an optical alignment reference. Two MRC frames (MRC-A and MRC-B) are defined by the MRC since opposite faces are not perfectly parallel. Optical measurements were taken with respect to both MRC-A and MRC-B. The Spacecraft Reference Surfaces (SRS) are a pair of nearly orthogonal flat surfaces located near the opposite end of the spacecraft. The SRS are used to define the body reference frame. Optical alignment measurements with respect to the MRC are referenced to the SRS. Since the alignment

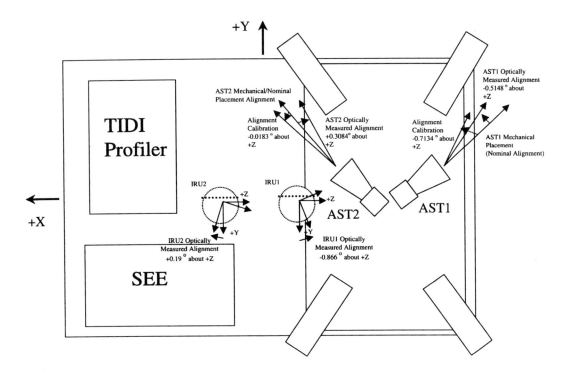

Figure 1: Attitude sensing hardware locations and orientations.

calibration is with respect to a fixed and assumed known body-to-gyro transformation matrix, the body (SRS) reference frame is known to within the accuracy of the optically measured alignment of the gyro with respect to the SRS. This may be corrupted by launch shock or other effects. The axes shown in Figure 1 are defined to be parallel to the body reference frame axes defined by the SRS.

Although optical alignments were not used as a priori data in the alignment calibration, it was necessary to accept the alignment between the IRU and the SRS as truth because the IRU is chosen as the reference sensor in the relative calibration algorithm. In addition, there is no indication of ambiguity in the optical measurements of the IRU orientation with respect to the SRS. The IRU manufacturer's measurement of external-to-internal orthogonal misalignment of the IRU axes is used to complete the SRS-to-IRU mounting transformation.

After shipment of TIMED to the launch facility, optical alignment measurements were made with respect to an Instrument Bench Cube (IBC); the IRU was not accessible for measurement at the launch facility. The IBC to SRS alignment changed by $(1.08, 4.68, 1.44)$ arcsec $((x, y, z)$ axes) and the MRC to SRS alignment changed by $(-8.64, 20.52, \ll 1)$ arcsec after shipment. This is larger than the alignment change from before to after thermal vacuum and vibration testing and it appears that the optical bench buckled during shipment. If the top deck warped during shipment, the IRUs would be misaligned, but this could not be measured after shipment. The alignment change expected due to orbit insertion is expected to be smaller than the change due to shipment.

RESULTS

Alignment Calibration Results From Day 142 Telemetry

The Alignment Kalman Filter results for the data from Day 2002-142 are shown in Figure 2 through Figure 5. The gyro scale factors are scaled to parts per million (ppm) and the misalignments ξ, δ_{s1}, δ_{s2} are scaled to arcseconds (arcsec). Angular rate is shown in radians/second. The standard deviation of error shown in the figures is from the square root of the diagonal of the AKF covariance matrix.

Figure 2 shows the AKF residuals (essentially the measurement minus prediction) in arcsec and the chi (χ) values for each star tracker. The χ^2 parameter is given by $\chi^2 = \nu^T V^{-1} \nu$, where ν is the $m \times 1$ residual vector and V is the residual covariance computed by the filter. Ordinarily, χ^2 is chi-distributed with m degrees of freedom. The AKF is designed to update with one measurement component at a time, and so three χ values each with one degree of freedom are computed. Thus χ should be less than one 68% of the time, less than two 95% of the time, and less than three 99.7% of the time.

The angular rate over the calibration interval on Day 142 is shown in Figure 3. The angular rate during the yaw maneuver peaks at -0.015 and $+0.01$ rad/sec or -1 and $+0.6$ deg/sec in both roll and yaw (x and y) and in pitch the peak angular rate reaches only 0.1 deg/sec. Comparing Figure 2c and Figure 2d, it is evident that the residuals are large when the angular rate changes sign due to increased error from the star tracker. The measurement errors appear to agree with the model variance in the AKF when the slope of the angular rate is constant.

Figure 3 shows the convergence of the calibration parameter estimates. These parameters change almost immediately from their initial value of zero and then are essentially constant until maneuvering comences. Figure 4 shows the convergence of the corresponding standard deviations, which are the square root of the diagonal of the covariance matrix. Notice that the standard deviations of the attitude estimation error and the star tracker alignment estimation errors in Figures 4a,c,d converge from the initial 3600 arcsec to about 2000 arcsec almost immediately and then do not converge further until the maneuver takes place. This is because the relative alignment between the two trackers is easily observable but the alignment between the trackers and the gyro is not observable until sufficient maneuvering occurs. The standard deviation of the gyro bias estimate shown in Figure 4b also does not converge until maneuvering takes place. Figure 4e and Figure 4f shows that the non-orthogonal alignment and the symmetric scale factors of the gyro also do not converge until sufficient maneuvering occurs.

The standard deviation of the steady-state attitude estimation error, which is shown on a smaller scale in Figure 5, is approximately $(7, 7, 25)$ arcsec in the (x, y, z) body axes. This steady-state accuracy is determined by the accuracy of the calibration parameters, whose accuracy is limited by the sufficiency of the calibration maneuver.

Composite Calibration Results

At the time this work was performed, high rate telemetry during the yaw maneuvers was available only for Days 142 and 197 in the year 2002. The LSE composite estimator [2] was used to combine the calibration estimates from those two days. The individual and composite calibration results are shown in Table 1. Recall that the asymmetric scale factors of the gyro $\boldsymbol{\mu}$ were essentially not estimated by setting their a priori values to zero and the a priori covariance to 1 part per million (ppm) or 10^{-6}. The symmetric scale factors $\boldsymbol{\lambda}$ of the gyro are scaled to ppm and the misalignments δ_{s1} and δ_{s2} are scaled to arcsec. The LSE composite estimator detected that the Day 142 and Day 197 scale factor estimates in the x and z axes are not consistent to within 4 standard deviations (and in fact have opposite signs) and so the LSE algorithm eliminated the Day 197 scale factors λ_x and λ_z from the composite estimate. The scale factor estimate then comprises the Day 142 estimate. It was ultimately decided to set the scale factor correction to zero in the on-orbit filter. The non-orthogonality of the gyro axes, $\boldsymbol{\xi}$, are also inconsistent between the two days' estimates in the y and z axes. Although the gyro calibration parameters were estimated with relatively large

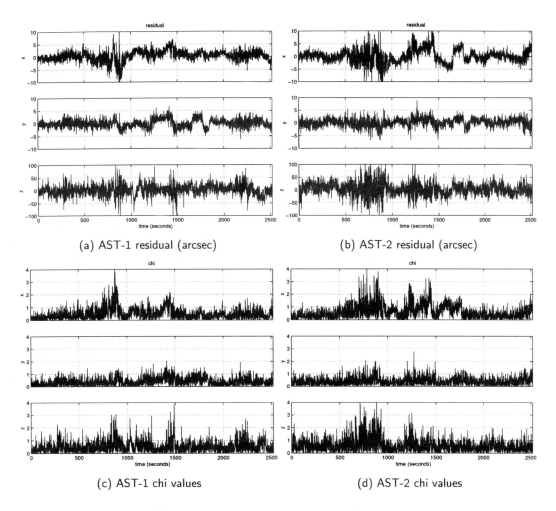

(a) AST-1 residual (arcsec)

(b) AST-2 residual (arcsec)

(c) AST-1 chi values

(d) AST-2 chi values

Figure 2: AKF performance for Day 142 data: Residuals and chi values.

uncertainty, it is important to realize that the normal mission operation of TIMED is to point to the local-level/local-horizonal coordinate system (z-axis to zenith and the y-axis to orbit normal or anti orbit normal). Because the angular rate is small during normal mission operation, the gyro calibration uncertainty does not degrade the pointing accuracy during that time. The attitude error during a yaw maneuver is not important because accurate pointing is not required during that time. The calibration does, however, prevent the measurement residuals from being too large during the maneuver and removes coupling of the orbital angular rate from the gyro bias estimates in the x and z axes.

A comparison of the estimated attitude before and after calibration shows that calibration changes the attitude by $(-0.2441, 0.0817, 0.3268)$ degrees in the (x, y, z) body axes. A large part of the correction about the z-axis is due to the use of the optically-measured IRU-to-SRS mounting matrix instead of the nominal mounting.

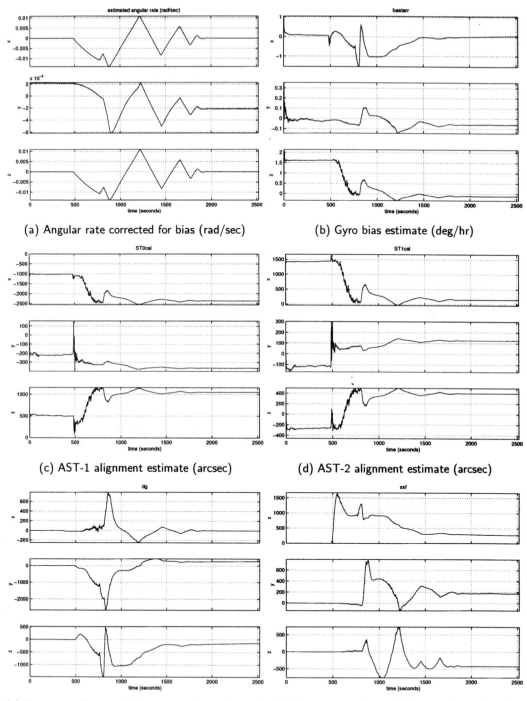

(a) Angular rate corrected for bias (rad/sec)

(b) Gyro bias estimate (deg/hr)

(c) AST-1 alignment estimate (arcsec)

(d) AST-2 alignment estimate (arcsec)

(e) Gyro nonorthogonal alignment estimate (arcsec)

(f) Gyro symmetric scale factor estimate (ppm)

Figure 3: AKF convergence for Day 142 data: Estimated parameters.

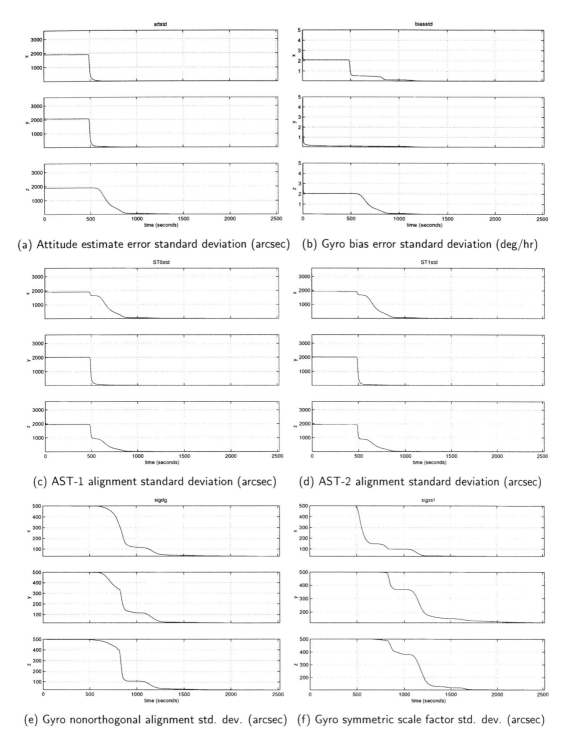

(a) Attitude estimate error standard deviation (arcsec)

(b) Gyro bias error standard deviation (deg/hr)

(c) AST-1 alignment standard deviation (arcsec)

(d) AST-2 alignment standard deviation (arcsec)

(e) Gyro nonorthogonal alignment std. dev. (arcsec)

(f) Gyro symmetric scale factor std. dev. (arcsec)

Figure 4: AKF convergence for Day 142 data: Standard deviation of error from the covariance matrix.

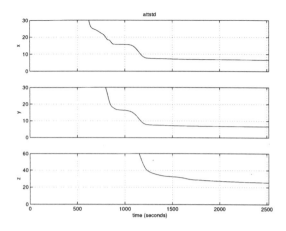

Figure 5: AKF convergence for Day 142 data: Standard deviation of attitude error (arcsec) from the filter covariance matrix (magnification of Figure 4a).

VALIDATION

The validity of the alignment calibration was checked in various ways, including comparison of measurements of the two star tracker, evaluation of the estimated gyro bias after calibration, examination of the measurement residuals in the on-orbit filter, and comparison of the alignment estimates with optically-measured alignments.

Measurement Residual and Gyro Bias Check

Telemetry from Day 154, for which there was no maneuver, was run through an attitude determination filter using the calibration parameters. The residuals revealed a 2 arcsec relative misalignment error. The star tracker misalignment estimates were adjusted accordingly and the filter was run again. The resulting residuals are shown in Figure 6a,b. The residuals show no bias, which means that the relative alignment is good. The χ values in Figure 6c,d are characteristic of a properly operating filter. Residuals greater than 4σ appear on occasion and are detected by a residual edit test. The standard deviation of the attitude error is shown in Figure 6e. This indicates that the KF attitude error is about 1.5 arcsec per axis. The KF doesn't "know" about the uncertainty due to miscalibration (i.e., calibration error). The attitude determination error is more accurately given by the steady-state standard deviation in Figure 5, which is obtained from the AKF covariance. The AKF accuracy in Figure 5 is (7,7,25) arcsec in the (x, y, z) body frame axes. This reflects both the random error and the calibration error in the attitude estimate. The estimated gyro bias is shown in Figure 7 and its standard deviation of error is shown in Figure 6f to be about 0.02 deg/hr. Before calibration, the gyro bias estimate from telemetry was $(3.245, 0.0622, -0.5795)$ deg/hr in the (x, y, z) IRU axes, which are nearly coaligned with the body axes. The bias estimate essentially changes sign after a yaw maneuver. To a large extent, this bias was due to the omission of the optically-measured alignment of the SRS-to-IRU from the on-orbit filter and the resultant orbital-rate coupling into the x and z axes of the gyro. After calibration and uploading the SRS-to-IRU mounting transformation to the on-orbit filter, the estimated gyro bias vector was $(0.15, 0.05, -0.18)$ deg/hr, which is well within the bias performance of the IRU. It is expected that a small part of this bias is due to residual misalignment, including axis northogonality.

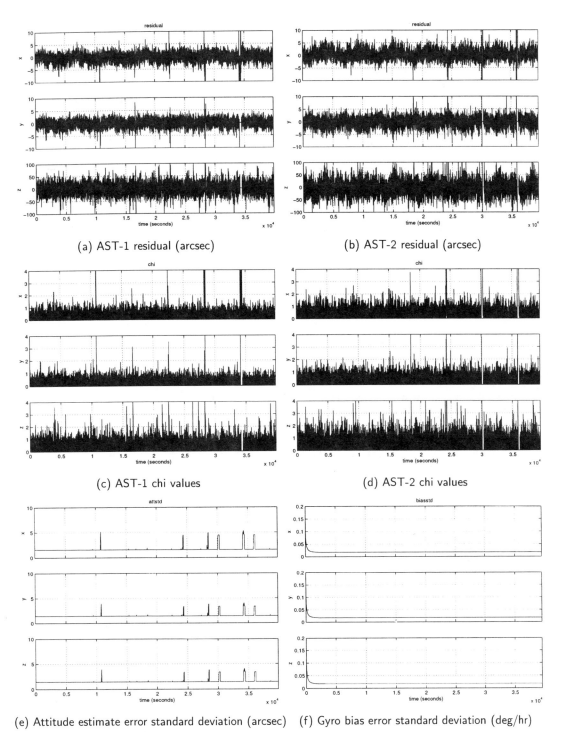

(a) AST-1 residual (arcsec)

(b) AST-2 residual (arcsec)

(c) AST-1 chi values

(d) AST-2 chi values

(e) Attitude estimate error standard deviation (arcsec)

(f) Gyro bias error standard deviation (deg/hr)

Figure 6: KF performance for Day 154 data: Residuals and chi values.

Table 1: Individual and Composite Calibration Estimates

parameter	axis	Day 142 estimate	Day 142 σ	Day 197 estimate	Day 197 σ	Composite estimate	Composite σ	units
λ	x	270.2	31.6	−213.6	31.2	270.3	21.0	
	y	173.8	120.7	13.4	88.4	57.2	69.4	ppm
	z	−424.5	102.9	163.2	98.2	−425.7	65.5	
μ	x	0.0	0.0	0.0	0.0	0.0	0.0	
	y	0.0	0.0	0.0	0.0	0.0	0.0	ppm
	z	0.0	0.0	0.0	0.0	0.0	0.0	
ξ	x	−0.7	33.3	29.3	27.7	17.8	20.0	
	y	254.7	22.6	406.6	21.6	256.1	14.4	arcsec
	z	−148.0	26.4	−328.1	19.6	−146.7	15.2	
δ_{s1}	x	−2364.5	22.5	−2362.5	16.2	−2362.4	12.9	
	y	−364.9	6.5	−360.7	6.4	−363.1	4.3	arcsec
	z	1049.7	13.7	1047.0	11.0	1048.1	8.1	
δ_{s2}	x	153.6	22.5	158.4	16.2	156.7	12.9	
	y	125.1	6.5	125.0	6.4	124.8	4.3	arcsec
	z	399.6	13.7	401.0	11.1	399.4	8.1	

Comparison Of Star Tracker Measurements

One check of the validity of the star tracker alignment parameters is performed by converting the AST-1 quaternion measurements for Day 154 to the AST-2 frame by using the calibrated mounting quaternions, and then by computing the rotation error between the two quaternion measurements. The rotation error is plotted in Figure 8. This shows only measurement error; relative alignment error, if any remained after calibration, would appear as a bias in this data. There is no discernable bias in the graphs, but a small cyclical component of error at orbital rate is evident. The mean rotation error vector was found to be $(0.42, 0.16, -2.96)$ arcsec and the rms value of the rotation vector was found to be $(10.9, 15.0, 23.7)$ arcsec in the x and y (cross-boresight) and z (boresight) axes of AST-2. This may be compared with the expected covariance C, which is given by

$$C = \mathbf{T}_{s1}^{s2} R \mathbf{T}_{s2}^{s1} + R \tag{1}$$

where $R = \mathrm{diag}(2.7^2, 2.7^2, 32^2) = \mathrm{diag}(7.3, 7.3, 1024)$ arcsec^2 is the manufacturer-specified measurement error covariance of the star trackers (cross-correlations are unknown), $\mathbf{T}_{s1}^{s2} = \mathbf{T}_b^{s2}(\mathbf{T}_b^{s1})^T$ and $\mathbf{T}_{s2}^{s1} = (\mathbf{T}_{s1}^{s2})^T$, where $\mathbf{T}_b^{s1}$ and $\mathbf{T}_b^{s2}$ are the body-to-tracker mountings for AST-1 and AST-2. We obtain $\sqrt{\mathrm{diag}(C)} = (18.7, 26.4, 32.1)$ arcsec. The standard deviation computed from the data was $(10.9, 15.0, 23.7)$ arcsec with a sample mean of $(0.4, 0.2, -3.0)$ arcsec. The standard deviation of error computed from equation (1) is greater than the standard deviation computed from the data by a factor of $(1.72, 1.76, 1.35)$ in the (x, y, z) axes of AST-2. This may be because the expected error levels are based on end-of-life performance and on errors due to angular rates up to 0.2 deg/sec. The Day 154 data was taken near beginning-of-life at the orbital angular rate of 0.064 deg/sec.

Comparison With Optically-Measured Alignments

The estimated alignment rotation vectors transformed to body coordinates are shown in Table 2. The difference between the rotation vectors can also be computed since they are in the same coordinate system, and these are also shown in Table 2 in the rows labeled "AST-1 to AST-2". The optically-measured alignment correction to the nominal star tracker alignment quaternions are also shown in the table. The z-axis alignment data in the table is shown in Figure 1. The last column

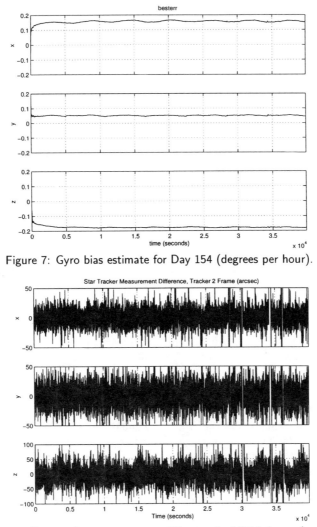

Figure 7: Gyro bias estimate for Day 154 (degrees per hour).

Figure 8: Star tracker measurement difference in AST-2 frame (arcsec).

of data shows the rotation between the optically-measured alignment and the AKF-estimated align-ment. The first column of data is approximately equal to the sum of the second and third columns of data.

The optical alignment includes the internal-to-external alignment of the star trackers as measured by the manufacturer. The optically-measured alignments differ from the estimated alignments by error in the optical measurements, internal-to-external alignment measurement error of the trackers, and launch shock and other orbit insertion effects. The last column of Table 2 shows that the optically measured and estimated alignments are comparable except that the z-axis component of the difference (-1176 arcsec or -0.3267 degrees) is a bit large. The change in relative alignments (last column of the "AST-1 to AST-2" row) indicates that this difference is due mainly to a common shift of both star trackers, the optical bench, or of the gyro, all of which would manifest themselves the same way.

Table 2: Alignment Comparisons. All rotation vectors are in body coordinates.

Star Tracker	Body Axis	Nominal to Estimated	Nominal to Optical	Optical to Estimated	Units
AST-1	x	14.0	−212.3	226.2	arcseconds
	y	−452.2	−310.3	−143.3	
	z	−2568.2	−1853.3	−714.7	
AST-2	x	418.8	372.3	46.1	
	y	140.9	133.9	8.1	
	z	−65.7	1110.3	−1176.0	
AST-1 to AST-2	x	404.8	584.6	−180.2	
	y	593.0	444.2	151.4	
	z	2502.5	2963.6	−461.3	
AST-1	x	0.0039	−0.0590	0.0628	degrees
	y	−0.1256	−0.0862	−0.0398	
	z	−0.7134	−0.5148	−0.1985	
AST-2	x	0.1163	0.1034	0.0128	
	y	0.0391	0.0372	0.0023	
	z	−0.0183	0.3084	−0.3267	
AST-1 to AST-2	x	0.1124	0.1624	−0.0500	
	y	0.1647	0.1234	0.0421	
	z	0.6951	0.8232	−0.1281	

CONCLUSION

The results presented demonstrate the relative alignment calibration algorithm and the composite calibration algorithm developed by the author. The yaw maneuver executed by the TIMED spacecraft every two months is sufficient to produce an accurate alignment calibration, but insufficient to accurately estimate gyro calibration parameters. The combined estimate of more than one calibration is an improvement. Although the gyro calibration parameters cannot be estimated with high accuracy, this does not degrade the pointing accuracy during normal nadir-pointing mission operation because the angular rate is small during that time. Various means of validating the estimated calibration parameters were presented, and all indicate that the calibration is satisfactory.

Like many spacecraft, TIMED was not designed and implemented with calibration requirements explicitly stated. An alignment calibration plan should be developed early in the design cycle (Phase A/B) of a spacecraft and be factored into the system error budget and operational plan. An alignment calibration plan should require that we:

1. Choose the calibration parameters;

2. Specify data needed for calibration;

3. Specify pointing maneuvers and pointing directions;

4. Determine how accurately the calibration parameters can be estimated within operational constraints—and sometimes imposing operational requirements to achieve a required accuracy;

5. Determine attitude error and attitude knowledge error due to miscalibration (residual calibration error);

6. Define mechanical placement, optical alignment measurement before and after environmental testing, thermal distortion analysis, and analyze orbit insertion effects (launch shock and gravity offset predictions, prediction of moisture desorption effects, etc.).

This is a tall order for a Phase A/B effort, but is feasible if an Alignment Calibration Filter or similar estimator is available to aid in evaluating calibration maneuver options and attitude sensor performance.

REFERENCES

1. PITTELKAU, M. E., "Everything is Relative in System Alignment Calibration", *AIAA Journal of Spacecraft and Rockets*, Vol. 39, No. 3, 2002, pp. 460–466.

2. PITTELKAU, M. E., "Composite Estimate of Spacecraft Sensor Alignment Calibrations", *AIAA Journal of Guidance, Control, and Dynamics*, Vol. 26, No. 2, 2003.

3. PITTELKAU, M. E., "Kalman Filtering for Spacecraft System Alignment Calibration", *AIAA Journal of Guidance, Control, and Dynamics*, Vol. 24, No. 6, 2001, pp. 1187–1195.

4. SHUSTER, M. D., PITONE, D. S., and BIERMAN, G. J., "Batch Estimation of Spacecraft Sensor Misalignments, I. Relative Alignment Estimation", *Journal of the Astronautical Sciences*, Vol. 39, No. 4, 1991, pp. 519–546.

5. SHUSTER, M. D. and PITONE, D. S., "Batch Estimation of Spacecraft Sensor Misalignments, II. Absolute Alignment Estimation", *Journal of the Astronautical Sciences*, Vol. 39, No. 4, 1991, pp. 547–571.

MODELING AND ANALYSIS
FOR NUTATION TIME CONSTANT DETERMINATION OF
ON-AXIS DIAPHRAGM TANKS ON SPINNERS:
APPLICATION TO THE DEEP SPACE ONE SPACECRAFT[*]

Marco B. Quadrelli[†]

This paper describes the modeling, analysis, and testing done to determine the DS1 nutation time constant. The significance of this analysis and testing program is that Deep Space One was the first spacecraft flown by JPL with a diaphragm tank located on the spin axis. First, modeling considerations are made, and in order to simulate the nonlinear behavior of the spacecraft containing the liquid (Hydrazine), the dynamics of a rigid body coupled through a universal joint to a pendulum mass representing the liquid slosh has been analyzed. This nonlinear simulation is done in order to estimate the nutation time constant with a pendulum slosh model. Second, some considerations follow on the modeling of tanks with diaphragms in spinner tanks. Third, the actual spin drop tests are described. These tests also confirm a nutation time constant in excess of 1000 seconds, both in the case of tests done for the Xenon tank only and in the case of tests done for the Hydrazine tank only. The large value of the nutation time constant computed and verified by testing for this system ensured that it remained well above the required values of 150 seconds at ignition, and 50 seconds at burnout.

INTRODUCTION

The Nutation Time Constant (NTC) estimate is required for launch vehicle stability analysis. It involves any launch with a spinning upper stage for injection, namely STAR motors, and is supplied by the spacecraft customer. The divergence time constant is a significant input to the Nutation Control System analysis. It establishes initial cone angle at third stage ignition, it contributes to the velocity pointing and loss during motor burn, and establishes cone angle at the start of spacecraft separation. The Deep Space One (DS1) spacecraft utilized a monopropellant hydrazine blow-down propulsion system for attitude control and trajectory correction maneuvers. This system incorporates the use of one spherical, 16.5 inches diameter titanium propellant tank. This tank is mounted on the +Z axis of the spacecraft, and incorporates a semi-rigid elastomeric diaphragm for propellant control/management. Because of the configuration of the tank on the spacecraft, because the spacecraft is spinning at 62.8 rpm at thruster ignition, and because of the nature of the diaphragm, it is important to evaluate the influence of this particular topology on the nutation divergence time constant of the spacecraft when energy dissipation due to liquid motions is possible. This paper describes the challenges of the problem under investigation, discusses several modeling approaches,

---

* The U.S. Government has a royalty-free license to exercise all rights under the copyright claimed for Government purposes. All other rights are reserved by the copyright owner.

† Jet Propulsion Laboratory, California Institute of Technology, Mail Stop 198-326, 4800 Oak Grove Drive, Pasadena, California 91109-8099. E-mail: marco@grover.jpl.nasa.gov.

and presents a slosh pendulum analysis done to support the study of the effect of an elastomeric diaphragm located inside an on-axis spinning tank. The paper also describes the testing done to support the modeling and analysis. The drop tests for the hydrazine tank and for the Xenon tank give separately two different time constants, τ_{Xe} and τ_{Hy}, which are combined to give the final system's nutation time constant as $\tau = 1/\tau_{Xe} + 1/\tau_{Hy}$. The forces and torques acting on the structural system depend on the pressure distribution on the liquid container walls. On the other hand, the current distribution of vorticity in the body of fluid depends on the current distribution of linear and angular velocities to which the base vehicle is subjected. The wobble amplification characteristics depend on the geometric distribution of the liquid with respect to the vehicle's principal axes. The nutation characteristics of the whole system also depend on the dissipation induced by the liquid viscosity, as well as on the presence and damping characteristics of propellant management devices (baffles, membrane diaphragms, or others). Energy dissipation in spinning bodies containing liquids generally depends on: tank shape and geometry (Spherical, Cono-spherical, Cylindrical, Isotensoid), tank Fill Fraction (FF), Propellant Management Device (PMD) shape & viscoelastic properties (shape, orientation in tank, material constants), PMD orientation in on-centerline (CL) or off-CL tank, tank location with respect to vehicle centerline and center of mass, interaction with moving center of mass (during burn), and vehicle inertia ratio β (spin to transverse inertia).

It is well known that a prolate spinner is unstable, i.e., if the ratio of the spin to transverse moment of inertia is less than one, and the body is spinning about the axis of minimum moment of inertia, an initially small nutation angle θ_0 will increase in a finite time Δt to θ_f according to the approximate law $\theta_f = \theta_0 \exp(\Delta t/\tau)$ where τ represents the nutation divergence time constant. The case of the Exosat spacecraft (early eighties) [16] which was a spinner with an on-axis diaphragm tank, is significant: analysis showed that adding the diaphragm decreased the divergent nutation time constant by a factor of 6-7 relative to a bare tank. However, the diaphragm configuration was different than in DS1, and more prone to static and dynamic instability. In [23], the conclusions of [16] concerning the responsibility of inertial wave resonances on the nutation instability of Exosat are questioned based on the fact that there is no correlation between the Exosat test results and the pre-test predictions (also because the predictions did not take into account the increased damping effects due to the presence of a flexible bladder). Recent studies of the nutation time constant of the Deep Impact spacecraft [11], with two on-axis Hydrazine tanks of different dimensions, and to be launched in 2003, also conclude that the liquid induced resonances (and perhaps the nutation synchronous mode described below) may also appear during the pre-burn and post-burn phases.

SPACECRAFT MASS PROPERTIES AND DERIVATION OF SLOSH PARAMETERS

The configuration of the DS1 spacecraft is shown in Figure 1. The STAR 37 FM motor and spacecraft are spin stabilized (spin rate=62.8 rpm); the spherical hydrazine tank (radius = 8.25 in = 0.209 m) is located on the spin axis, directly above the Xenon tank; the tank contains 27.8 kg of hydrazine (fill factor = 87.5 percent); the hydrazine tank contains a flexible diaphragm as propellant management device (thickness = 0.06 in, more properties below); the required minimum nutation time constants: 150 seconds at ignition, and 50 seconds at burnout. At launch, the fill level of the equivalent clean tank, i.e., a tank of the same radius as the tank in DS1, but without diaphragm, is 87.5 percent. The liquid mass in the tank is 27.8 kg. The component mass properties were (mass in kg, position in m, inertias in kg-m²): The mass properties of the whole third stage vehicle (DS1 attached to the STAR37 solid rocket motor) are given in Table 1 [18]:

Table 1

	Pre-burn	Pre-burn	Post-burn	Post-burn
DS1 s/c cm location (m)	0.265	0.261	0.265	0.261
mass (slugs)	117.7	117.7	44.1	44.1
cm location, s/c coords (ft)-	2.87	-2.91	-0.425	-0.522
transverse inertia (slug-ft2)	925	893	399	388
spin inertia (slug-ft2)	206	206	128	128
inertia ratio	0.223	0.231	0.321	0.330
tank center/cm offset (in)	54.73	55.20	25.4	26.6

The tank center location is at 20.29 inches from the reference plane, on the spin axis. The tank diameter is 16.5 inches, the fuel is Hydrazine at a 87.5 percent fill level. The Hydrazine tank contains a diaphragm of thickness 0.065 inches, made of an elastomer with Young's modulus equal to 1200 psi. Finally, the nominal spin rate is 60 rpm.

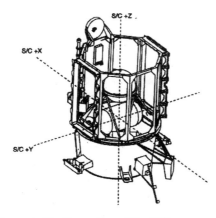

Figure 1. Configuration of the DS1 spacecraft.

DIAPHRAGM PROPERTIES AND DERIVATION OF SLOSH MODEL PARAMETERS

The diaphragm is made of AF-E-332, which is a relatively stiff elastomer. While this material can be stretched to 300 percent of its original length in a uniaxial direction, in its three-dimensional usage the diaphragm can only stretch 10 percent before failing. The diaphragm in this state provides a semi-stiff control surface and is resistant to deformation. Filling of the tank occurs with the diaphragm originally pulled completely toward the propellant side of the tank (i.e., completely evacuated). As the propellant enters the tank, the diaphragm buckles into its reversed position. This buckling can occur in a random direction, and leads to some uncertainties in the diaphragm ultimate shape. The diaphragm is stiff enough, however, to prevent the fluid from settling into its lowest energy state in the static 1-g environment. The properties of the diaphragm are as follows: tensile strength = 1650 psi, elongation = 260 %, tensile modulus (psi, at 100 elongation) = 12000, hardness = 90 points, tear strength = 300 (lb/in), Poisson's ratio = unknown, damping properties = unkown, tensile strength change = -20 %, and elongation change = -20 %. The last two properties represent the fuel resistance to hydrazine, after 96 hours of exposure at 160 degrees F. In the preliminary evaluation phase of the spin drop test, the scaling of the tensile and bending stiffness of the model diaphragm had to be considered. The spin drop test will be conducted on a 1:3.5 reduced model of the Star37FM motor attached to the casing containing the DS1 vehicle with tank full. The thickness of the diaphragm inside the hydrazine tank needs to be scaled from 1/16th of an inch down to 19/1000th of an inch. The scaled diaphragm needs to be much stiffer than the full scale version, also because a very thin membrane is difficult to manufacture using an injection molding technique. The Young's modulus for the membrane material is 1170 psi for the full scale diaphragm, and 25000 psi for the scaled diaphragm. The bladder is flexible, but nearly inextensible in its reversed position, i.e., at maximum liquid fill level (87.5 percent). The equilibrium shape of the membrane at maximum liquid fill level is axi-symmetrical about the +Z axis. This axi-symmetry is close to perfect the more the tank is full. The equally important issue of scaling down the diaphragm damping properties remains open, although its effect is unknown and intuitively of less magnitude than that of the stiffness.

From [21] and [22], the damping ratio can be expressed as a non-dimensional function of diaphragm stiffness and tank diameter for small oscillations. This fact was used in [20] to predict the nutation time constant for Mars Pathfinder. Since DS1 uses exactly the same type of tank as Mars Pathfinder, with the same diaphragm, except that the tank is located on-axis, the extrapolation of the model in [20] to the DS1 model is well justified. As in [20], we can make the assumption to use data developed for clean tanks (without diaphragm) of diameter d to obtain the fixed pendulum mass m_f, the moving (slosh) mass m_s, and the pendulum length l. We also use the data from [21] obtained for a 20.5 inches diameter tank with 0.010-0.030 inch thick diaphragms to extrapolate and to determine the diaphragm stiffness and damping for the DS1 diaphragm. Since damping increases with increasing diaphragm thickness for spherical tanks [21], these estimates of damping are conservative. The pendulum equation of motion in an acceleration field $g=\omega_{spin}^2 d/2$, where d is the tank diameter, and denoting by $(.)'$ a time derivative, is:

$$m_s\, l_s^2\theta''+c\theta'+k\theta+m_sgl_s\sin(\theta)=0$$

The pendulum frequency and damping, after linearization, are $\omega_s^2 = (g/(l_s)+k/(m_sl_s^2)$ and $\xi = c/(2\omega_sm_sl_s^2)$, respectively. From [5,6], the stiffness and damping coefficients for the slosh pendulum can be computed as follows $K_\theta=0.30m_sl^2\omega_{1g}^2$ and $C_\theta=0.22m_sl^2\omega_s$, where ω_{1g} is the slosh frequency in rad/sec for one-g conditions. We derive the slosh parameters, i.e., slosh mass, slosh pendulum length, and slosh pendulum hinge point, based on data obtained [1] for a spherical tank of radius a, with liquid level h (hence fill factor $h/(2a)$), and fluid density ρ_f with no propellant management devices. An important assumption to keep in mind is that the charts from [1] are derived assuming a horizontal acceleration field, instead of a centrifugal acceleration field. There are no slosh analysis charts reported in the literature assuming as external excitation a uniform rotational acceleration field. Given the 87.5 percent fill factor, the chart in [1], page 206, provides the following parameters: mass of full tank $m_{full}=(4/3)\pi\rho_f a^3=28.70$kg; effective mass $m_t=\rho_f\pi h^2(d-(h/3))$; undamped frequency $\omega_0=\sqrt{(g/d)}=6.8511$rad/sec; $\omega_s=2.335\omega_0=15.997$rad/sec; $m_s=0.075m_t=2.0686$kg; $l_s=0.335d=0.07002$m; hinge point location from tank bottom $l_h=0.99d$; rest mass $m_0=m_{full}-m_s=26.631$kg; and rest mass location $l_0=0.99d$. Note that the rest mass is so high because the fill factor is very high, approximating the problem to that of a fully filled spherical tank. Consequently, we assumed the rest mass and the pendulum hinge point to be located at the tank center. The frequency parameter used in Stofan's chart [21,22] turns out to be $\omega_s\sqrt{(d/g)}=2.335$. As a function of diaphragm thickness, linear extrapolation shows the damping ratio to be as in Table 2. Consequently, the stiffness and damping coefficients for this configuration become k=3.403 N/m and c=0.03165 Ns/m, respectively.

Table 2

Thickness [in]	Damping ratio
.01	.27
.02	.38
.03	.42
.06	.73

PHYSICS OF LIQUID-WALL INTERACTION WITH AND WITHOUT DIAPHRAGM

In a fluid-structure interaction problem there are three different modes of fluid motion. See [1], [10] and [8] for a more detailed description of this phenomenon. We refer to Figure 2, and proceed to summarize the essential features of this dynamic interaction.

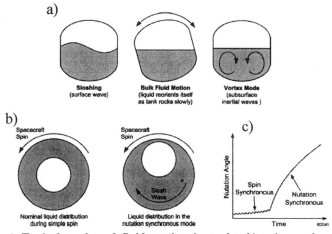

Figure 2. a) Typical modes of fluid motion in tanks. b) spin-synchronous vs. nutation-synchronous mode, c) nutation angle of spin-synchronous vs. nutation-synchronous mode vs. time. (courtesy of Dr. Carl Hubert of Hubert Astronautics Inc.).]

Bulk Motion Mode

In the bulk motion mode (Figure 2a), the liquid rotates as a rigid body, following the rotational motion of the walls. Bulk fluid motion is characterized by liquid reorienting itself within a tank in response to changes in the tank's alignment relative to the acceleration field and/or inertial space. For example, when a half-filled coffee cup is slowly tilted, the free surface remains horizontal as the liquid repositions itself to minimize its gravitational potential energy. On a spinning vehicle, bulk fluid motion is similar to that of the coffee cup example except that the shape of the free surface or diaphragm is governed by centrifugal force instead of gravity. Nutation produces a cyclic bulk motion that dissipates energy due to viscous drag as the liquid flows past the tank walls and diaphragm. Unlike free-surface sloshing, bulk motion affects the entire body of the liquid. This means that it can produce much more flow past the tank walls and diaphragm than simple surface wave sloshing. This, in turn, means that under non-resonant conditions bulk motion can dissipate far more energy than sloshing. For a simple spherical tank, bulk motion can be modeled by treating the liquid as a solid body that pivots about the center of the tank. Accurate mechanical analogs are more difficult to develop for non-spherical and diaphragm tanks. Figure 2c) shows that the nutation synchronous mode is excited in a different manner than the conventional spin synchronous mode. The bulk motion in tanks with internal devices may dissipate a lot more energy than in bare tanks. In fact, the diaphragm causes the fluid-filled spherical cavity to become non-spherical, resulting in a vigorous fluid motion initiated by the viscous boundary layer at the tank walls. In the case of a partially filled cavity, the diaphragm assumes a grossly wrinkled shape, which differs from the bare tank free surface. Because the diaphragm causes the fluid-filled cavity to be non-spherical, the liquid motion will dissipate more energy on account of a more turbulent velocity distribution than in a bare tank.

Sloshing Mode

The sloshing mode (Figure 2a) occurs when the liquid has a free surface, and involves liquid motion in a direction perpendicular to the free surface with very little or no circulation.

The determination of the sloshing oscillations of a fluid in an axi-symmetric tank consists of a solution for the velocity potential described by Laplace's equation in three dimensions, which satisfies the prescribed tank wall and boundary conditions and free-surface boundary conditions. From this solution, the equations for the hydrodynamic pressures, forces, moments, surface waveforms, and natural frequencies can be obtained. A simple mechanical analogy in the form of a pendulum or equivalent spring mass will exactly duplicate the forces and moments produced in the fluid oscillations. Physically, the motion of the fluid can be interpreted as follows. A certain portion of the liquid can be considered as rigid and not participating in the sloshing motion, while the fundamental mode reacts as a pendulum and can oscillate as such. The higher modes are not important, as the sloshing masses participating in the motion become smaller as the natural frequency increases. Since the fluid is generally assumed to be incompressible and inviscid, the mechanical analogy does not apply well in situations where the predominant source of internal recirculation is viscous friction at the tank walls or at the diaphragm surface.

In a spherical diaphragm tank with ullage over the liquid, the diaphragm reduces the wavelength of the slosh mode, increases the frequency, and results in a similar mode shape to the case with the bare tank. The slosh mode frequency and damping ratio increase with diaphragm thickness, and because of the diaphragm viscoelasticity, the free-surface slosh mode is highly damped. The determination of diaphragm viscoelasticity and its effects on this mode are very important.

Nutation Synchronous Mode

The nutation synchronous mode [17, 2, 3] depicted in Figures 2b) and 2c), involves a unidirectional (non-cyclic) surface wave that revolves around the spin axis at the nutation frequency. On-axis tanks can also support this type of non-oscillatory liquid motion. With an inertia ratio of about 0.2, with a spin rate of 60 rpm, this nutation synchronous wave would circulate around the spin axis at over 40 rpm in a direction opposite to the spin. Such a high velocity flow can dissipate significant amounts of energy and cause rapid nutation growth. A vehicle that experienced this type of behavior is the IMAGE spacecraft, which had no liquid propellant but did have a passive viscous-ring nutation damper that contained 1.2 kg of mercury. The passive damper was for use after separation from the spinning upper stage. The device acted as a de-damper during third stage operations. Pre-flight analysis indicated that this small amount of mercury could enter the nutation synchronous mode and that if this happened it would require vigorous activity by the NCS. The behavior did occur in flight and Boeing reported that there was more NCS thruster firing during the IMAGE launch than during any other Delta flight.

A good equivalent mechanical model for nutation synchronous motion is a lumped mass traveling on a circular path that is centered on the spin axis. This is essentially the model that was used to predict the performance of the IMAGE damper. Although this model is easy to study by simulation and/or closed form analysis, it is unclear how to select or even bound key model parameters for a spacecraft with an on-axis tank containing a diaphragm. For IMAGE this was simple because the motion involved flow thought a pipe, which is well understood. To excite nutation synchronous behavior requires a minimum nutation amplitude. This critical angle is a function of several parameters, including spin rate, inertia ratio, tank radius, distance from the vehicle cm to the tank center, amount of mobile liquid, and effective damping coefficient. If the effective damping coefficient is low enough and/or the mobile mass is high enough then the critical angle for exciting nutation synchronous motion will be within the range of angles that could be experienced in flight. As indicated by the IMAGE experience, nutation synchronous motion can cause very rapid nutation growth. The difference in nutation growth characteristics under the influence of simple oscillatory spin-synchronous motion and the more vigorous nutation-synchronous motion is shown qualitatively in Figures 2b) and 2c). Because nutation synchronous behavior is nonlinear, it cannot be analyzed with eigenvalue methods resulting from pendulum models.

Inertial Waves

A rotating fluid supports inertial waves if it is Rayleigh stable [8], i.e. if the angular momentum per unit mass of the circular flow lines increases with the radius from the spin axis. The Rayleigh criterion asks whether the force due to inward radial pressure gradient is adequate to maintain an inward centripetal acceleration for a generic element of the fluid in that rotating flow. If the criterion is satisfied, i.e. if the flow is stable to perturbations, then fluid elements moving radially out or radially inward will tend to return to their initial radius when the perturbation has ceased. Unless the flow is heavily damped (i.e. if baffles are present or the fluid is highly viscous), the element of perturbed fluid will overshoot from the undisturbed radial position, and initiate an oscillatory wave motion called the inertial wave.

Inertial waves (Figure 2a) correspond to the natural resonant modes for the current tank geometry. They are traveling and also dispersive solutions of a hyperbolic equation [8], which implies that discontinuities can occur in the fluid across characteristic surfaces (cones). These waves die out in a time of the order of the spin-up time (roughly proportional to the square root of the Reynolds number). In a sphere, for example, the first inertial mode is a spin of the interior fluid about a single axis as if it were a rigid body. Higher modes are obtained by seeking smaller subdivisions of the sphere with corresponding motions that satisfy both continuity and momentum balance. Inertial wave modes are a dissipative mechanism, which can occur only in spinning tanks, are independent of the presence of a free surface, and can couple with nutation motion. Inertial waves can be excited in non-spherical tanks, and are the result of redistribution of the fluid initiated at the tank walls, progressively diffusing inside the spinning liquid, and these modes have a sectorial character. Inertial waves are driven by the cyclic pressure waves from nutation-induced, angular motions of the tank wall coupled with Coriolis forces inside the liquid. Inertial wave resonances can occur within a bandwidth of two times the spin rate for a prolate spinner, and they can be identified by producing one or more sharp peaks in a plot of the energy dissipated by the liquid in a spinning spacecraft vs. inertia ratio. Because of the sharp energy dissipation that they can produce, the nutation time constant can be drastically reduced if the configuration of the spacecraft is such that the resonance can occur. In a clean fully filled spherical tank, the only fluid-structure interaction mechanism is at the boundary layer at the tank wall, sustained by viscous shear. Inertial waves can occur in fully loaded tanks, with 100% fill factor. As stated in [24], slosh resonances are generally excited in (forced motion spin table) tests driven at low spin rate (<30 rpm) and high nutation rate (>80 rpm). Conversely, inertial wave resonances, if they exist, should generally be excited in tests driven at large spin rates (>60 rpm) and small nutation rates (<40 rpm). [24] uses a rotor model with a viscous damper to model the boundary layer dissipation and the inertial wave resonance.

Unlike the modal decomposition of an elastic member, a spinning fluid does not have a lowest frequency mode and a modal structure in which the frequencies progressively march up from the lowest value to infinity. Rather, as described in [15], the eigenvalues tend to cluster in the interval $\{-2\Omega, 2\Omega\}$, where Ω is the spin rate with numerous excursions outside this interval when a free surface is present. Hence, the fluid modes cannot be ordered according to the magnitude of the eigenvalue. Instead, this ordering is based on the number of nodal surfaces that a particular mode has.

In order to excite these inertial wave modes, energy must be introduced into the flow. For a clean sphere, the excitation may originate at the viscous boundary layer with the tank walls. For a nonspherical surface, the forcing function also includes normal velocities of the boundaries relative to the fluid mass. When there is structural symmetry, solutions may be expressed in terms of Legendre polynomials or Bessel functions, which typically represent axisymmetric expansions. Once excited, the inertial waves move both axially and radially, producing conical shear surfaces, which can be visualized in the laboratory using transparent tank walls and tracers in the illuminated liquid. In general, one could say that inertial resonances are extremely sensitive to changes in system geometry.

In a clean spherical tank, the only fluid-structure interaction mechanism is at the boundary layer at the tank wall, sustained by viscous shear. Previous experiments done with water and mercury on a non-spinning tank with 17.5 percent ullage (83.5 percent filled) [19] show that the slosh response was found to be very dependent on the stiffness and deformed shape of the tank positive expulsion diaphragm, for tests done under horizontal excitation. The diaphragm altered the free surface of the fluid by supporting the fluid away from the tank walls. As a result, the classical first free surface slosh mode, which is represented by a fluid pendulum motion does not occur. Therefore, the system stiffness appeared to be almost entirely controlled by the stiffness and shape of the diaphragm. In addition, in [19] it was also established that a single multiple degree of freedom pendulum model could not be assumed across the entire frequency range of operation for the tests, rather different models had to be used in different frequency ranges.

A report by F. Dodge on the NEAR spacecraft [4], which had a spinning tank with a similar on-axis tank configuration with diaphragm, presents some of the issues of interest in the model. Since the tanks are located on the spin axis, spinning simply causes the propellant and diaphragm to be oriented in an axi-symmetric shape about the spin axis. No significant lateral shifting of the spacecraft center of mass from the axis was expected. Given the absence of internal hardware, the viscous stresses at the walls are the only available mechanisms to cause the liquids to rotate. There is a very complicated transient liquid flow pattern during spin up, caused by growing boundary layers, internal recirculation flows, and tank end effects.

In general, the presence of a diaphragm inside a fuel tank adds a significant uncertainty to the vehicle dynamics of a spinner spacecraft. One remarkable case was EXOSAT (another spinner spacecraft with on-axis tank with diaphragm), in which tests showed that the major sources of dissipation were the internal modes and not the free surface slosh modes, that the effect of the diaphragm orientation inside the tank is crucial for determining the fluid regime, and that adding diaphragms to the propellant tanks decreased the divergent nutation time constant by a factor of 6 or 7 relative to the bare tank. Generally, to ensure success, the configuration to be flown must lie within the parameter space of previously tested or flown configurations, the reason being that there is a significant sensitivity to parameter variations (fill fraction, tank shape, inertia ratio, etc). Figure 3 depicts the grossly deformed diaphragm of a Hydrazine tank used for the Mars Polar Lander spacecraft at different fill factors. Figure 3 shows some modes of deformation of an elastomeric diaphragm inside a draining tank, obtained from a recent report [14]. The analysis in [14], although quite detailed and promising, is limited to small deformations of an elastomeric bladder, under gravity, and the fluid contribution is obtained from pressure equilibrium in a one g field only.

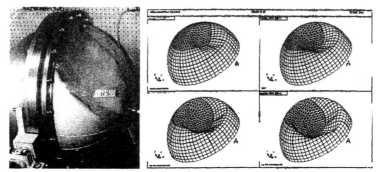

Figure 3. Mars Polar Lander test of tank with diaphragm (left). Modes of deformation of an elastomeric diaphragm inside a draining tank, from [14] (right).

PENDULUM MODELS

The pendulum model is routinely used by the aerospace industry to provide a solution to the slosh problem to first order of approximation. Unfortunately, it can be useful only to roughly estimate the nutation divergence time constant, and it is not useful for accurate predictions. Traditionally, analytical (pendulum) models often grossly underestimate the severity of the resonances occurring in the fluid-structure interaction process. The presence of a propellant management device (PMD) could be incorporated into the pendulum model via a rotary stiffness and damping coefficient, thereby providing a viscoelastic model of a pendulum as in Figure 4. One the one hand (like the spring-mass-damper oscillator) it provides some back of the envelope estimates. On the other hand, the true physics ends up not being correctly included in the model mainly because it is more of an art to determine the parameters to be used in the model.

As stated above, the determination of the forced oscillations of a fluid in a tank of axi-symmetric shape consists of a solution for the velocity potential described by the Laplace equation in three dimensions, which satisfies the prescribed tank wall and boundary conditions and free-surface boundary conditions. Once this solution, has been obtained, the equations for the hydrodynamic pressures, forces, moments, surface waveforms, and natural frequencies are combined to compute the effects of the variation in velocity potential. The mechanical analogy in the form of a pendulum or equivalent spring mass will exactly duplicate the forces and moments produced in the fluid oscillations, up to any order in a series expansion of the velocity potential. The physical equivalence consists in having a certain portion of the liquid that can be considered rigid and not participating in the sloshing motion, while the fundamental sloshing mode reacts as a pendulum and can oscillate as such. As in any modal analysis, the higher sloshing modes participate in the motion with increasingly smaller amplitudes as the natural frequency increases. One drawback of the pendulum models for sloshing only is its ineffectiveness in capturing internal recirculation dynamics since the fluid is generally assumed to be incompressible and inviscid. Consequently, the mechanical analogy does not apply well in situations where the predominant source of internal recirculation is viscous friction at the tank walls or at the diaphragm surface.

One mechanical model which could be used for cases in which internal recirculation may be important is a conical pendulum held by a spherical joint at the tank center. The angular motion parallel to the spin axis is restrained by a torsional spring and damper representing the viscoelasticity of the diaphragm, whereas the motion about the spin axis could be modeled by the third rotational degree of freedom enabled by the spherical joint, which is restrained only by some form of damper representing the friction of the walls. In this way, the Coriolis coupling which becomes the source of the internal inertial waves is more closely captured.

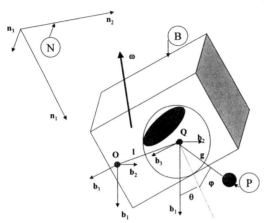

Figure 4. One mechanical model for nonlinear time simulation.

DYNAMIC ANALYSIS AND NUTATION TIME CONSTANT PREDICTION

A nonlinear simulation was carried out of the slosh pendulum model mounted on a rigid spacecraft (as shown in Figure 4) for the purpose of determining the nutation time constant explicitly. Since the nutation time constant is an output of a linearized analysis, it is estimated by means of exponential fitting to the output of the nonlinear program. The nonlinear model represents the problem as a free-floating rigid body connected to a point mass by a universal joint. The system is initially spinning about one axis at 62.3 rpm, and an initial nutation angle is given. The simulation is run for the two cases of pre-ignition and burn-out mass properties, and the trend of nutation angle is observed in time. Details of the model are given in [18].

The system is simulated starting from an initial nutation angle of 0.1 degrees. The results show that, for both the pre-ignition and the burn-out cases, a good exponential fit of the divergent nutation angle θ in degrees is provided by $\theta_{ignition}=0.0996\exp(t/1590)$ and by $\theta_{burnout}=0.0996\exp(t/1990)$. The two nutation time constants are 1590 seconds for pre-ignition, and 1990 seconds for burnout, for the given parameters of the spacecraft and Hydrazine slosh model. Figure 5, 7, 6, 8 depict, respectively, the nutation angle at ignition and burnout vs. time showing the sensitivity to different parameters (center of mass location, diaphragm stiffness and damping). It is interesting to note that these numbers are significantly different than those that the energy sink analysis predicted without including the effect of the diaphragm. An energy sink analysis gives an initial estimate for the nutation time constant of $\tau=25354$ seconds at ignition, and of $\tau=14660$ seconds at burn-out. Therefore, we conclude that the results of the slosh pendulum model show no danger of instability. Because the nonlinear slosh dynamic model is more detailed than the energy sink analysis, it is also more realistic. The results shown in Figure 5 assumed a perfectly balanced spacecraft, and the center of mass location (in spacecraft coordinates) was at 10.13 inches along the Z axis (no X or Y components present). However, subsequent mass properties changes resulted in a slight center of mass offset. The new center of mass location is at 10.59 inches along the Z axis, -0.16 inches along the X axis, and -0.37 inches along the Y axis. Some simulations have been run incorporating this new change in the pre-burn and burnout cases, and the results are shown in Figure 5 and Figure 6. The new (computed) nutation time constant is about 900 seconds (pre-burn) and 1450 seconds (burnout). What we can infer from these figures is that the presence of an offset of the center of mass along the X and Y axes starts a wobbling motion of amplitude less that 1 degree, superimposed on the nutational instability typical of a prolate spacecraft. The wobbling remains superimposed, since it is not damped, being equivalent to spinning about a different geometrical axis. Eventually, but only over a long time, is the nutation growth visible in the plot of nutation angle vs. time. Note that in the previous case, i.e. when the center of mass is on the Z axis, the wobble angle is exactly zero, and the attitude motion is governed by the nutational instability, induced by the slosh motion, alone. In the new case, the estimated wobble angle is 0.30 degrees in the pre-burn case, and 0.24 in the burnout case. We see that, besides the wobble angle induced by the center of mass offset, there is an additional contribution to the amplitude of the nutation angle which is due to the wobble amplification induced by the motion of the slosh mass: its effect is similar to having one flexible mode. If in addition, we assume that the spacecraft has been balanced, but imperfectly, so that a residual offset is left along the X and Y axes (in this case, an offset of 3.2e-2 inches along the X and of -3.2e-2 inches along the Y axes), the result is still different, implying that a very slight offset has also the potential of initiating a wobbling motion. In this case, this wobbling motion is of much smaller amplitude. The nutation time constants are practically the same as in the perfectly balanced case, but with some slight wobbling superimposed (1500 seconds for pre-burn, 1900 seconds for burnout). The estimated wobble angle is 0.035 degrees for the pre-burn case, and 0.028 for the burnout case. In this case, the wobble amplification due to slosh motion is much less than in the previous case. Figure 8 depicts the nutation angle vs. time at ignition for the cases in which the stiffness and damping coefficients are equal to 10 times larger, and 10 times smaller than the nominal values, respectively. From these plots, we see that although the magnitude remains confined to approximately the same value and that the behavior is qualitatively different, an exponential fit would provide in all three cases a very large value of nutation time constant (at least 1000 seconds). At this point we deem necessary to quote [7]: *Using an equivalent dynamic pendulum model...is is impossible to achieve such a simplified equivalence for rotating tanks where Coriolis accelerations and vorticity prevail. Therefore, a pendulum model can only give an incomplete fluid behavior representation.* Table 3 summarizes the results.

Table 3

Configuration	wobble angle [deg]	time constant [s]
pre-burn (no offset)	0	1590
burnout (no offset)	0	1990
pre-burn (-0.16,-0.37,10.59)	0.30	900
burnout (-0.16,-0.37,10.59)	0.24	1450
pre-burn (+0.03,-0.03,10.59)	0.035	1560
burnout (+0.03,-0.03,10.59)	0.028	1960

In conclusion, the effect of having a center of mass off the Z axis is to reduce the time constant (even a slight offset causes this reduction, showing how sensitive the spacecraft is to imbalance), and to induce a wobbling motion of max amplitude of 0.6 degrees in the pre-burn case, and a maximum amplitude of 0.4 degrees in the burnout case. The wobbling motion is also amplified to some degree by the inertial loading imposed by the moving slosh mass, which, from a dynamical point of view, is equivalent to having an additional flexible mode in the system. Nevertheless, these nutation time constants are still well above the minimum requirements, and the wobbling motion is of very small amplitude.

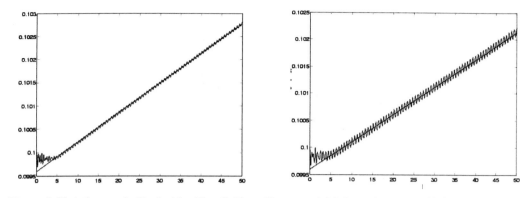

Figure 5. Nutation angle [deg] at ignition (left) and burnout (right) vs. time. com=[0;0;10.13] in.

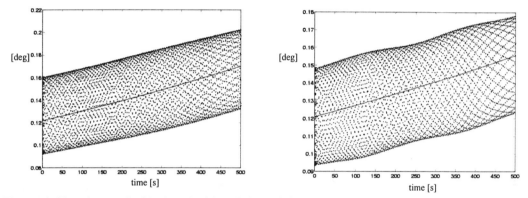

Figure 6. Nutation angle [deg] at ignition (left) and burnout (right) vs. time. com=[3.2e-2;-3.2e-2;10.59] in.

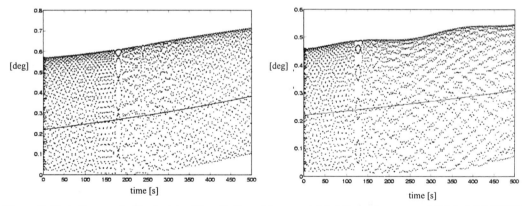

Figure 7. Nutation angle [deg] at ignition (left) and burnout (right) vs. time. com=[-0.16;-0.37;10.59] in

845

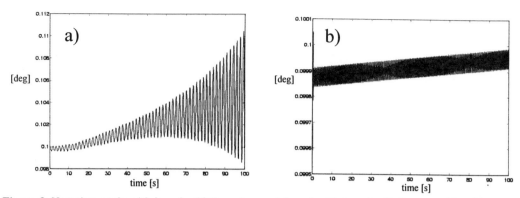

Figure 8. Nutation angle with k and c 10 times larger (a), and with k and c 10 times smaller (b).

SPIN DROP TESTS

The drop test facility used by JPL and run by Dr. Jon Harrison is at Applied Dynamics Laboratories. The facility is located in a three story building especially designed for all aspects of drop test model research, design, construction, testing, and data reduction and analysis. The basement area is used for storage of model construction materials, the first floor is used as a machine shop, and the second floor is the office, data reduction, and computer area. The drop tower is located at one end of the building. The hardware specifically related to drop testing includes the following: spinup/release mechanism; tower and catch box; telemetry and data processing system. Details of the facility are given in [18] and in [9]. Figure 9 shows details of the scaled tank and diaphragm.

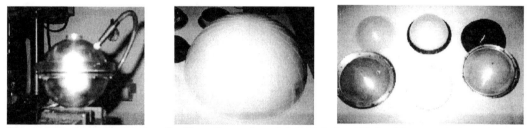

Figure 9.Photograph showing the Deep Space 1 scaled tank being filled with water before the spin drop test (left). Photograph showing the scaled diaphragm to be placed inside the Deep Space 1 scaled tank (center). Photograph showing the components of the Deep Space 1 model tank with diaphragm (right).

Phase I: DS1 Xenon Tank Drop Test Program

The drop tests for the Xenon tank use a drop test model with a 1/5 scale model of the Xenon tank filled with an appropriately scaled liquid. The nominal spin rate of the full scale vehicle is 60 rpm, and the nominal inertia ratios are 0.222 (pre-burn) and 0.296 (post-burn). Along with predicting the nutation time constant for the nominal parameters, the main effort of the test program is to uncover any fluid resonances that may exist. The search was limited to inertia ratios approximately 10% above and below the nominal values with an increment of approximately 0.01. Ballast weights on the model are adjusted to change the inertia ratio. The time constant for the full-scale vehicle is computed by scaling the model results using scaling laws based on the Energy Sink assumptions. The spacecraft/upper stage mass properties used in the determination of the Xenon tank time constant are as in Table 4.

846

Table 4

	Pre-burn	Post-burn
Spin inertia (slug ft^2)	200	121
Transverse inertia (slug-ft2)	897	409
Inertia Ratio	0.222	0.296

The interior of the Xenon tank is a 19.3 inch long cylinder with identically domed ends. The volume is 50 liters. The tank load is between 63 and 83 kg. Since the tank volume is 50 liters, the fluid density is approximately 1.2 and 1.7 kg/liter. The full scale test results are based on the higher density since that gives the shortest time constant. The kinematic viscosity for any load is approximately 0.054 centistokes. The design of the model is driven by the need to accurately simulate the Xenon fluid. The two scaling laws that are relevant are the Mach number ($d\omega/V$) and Reynolds number ($d^2\omega/v$), where d=tank diameter, ω is the spin rate, V is the velocity of sound in fluid, and v is the kinematic viscosity. The full scale Mach number is between 0.0055 and 0.0124. This is so much less than one that compressibility effects are of much less importance than viscosity effects. Therefore, the model, which uses a liquid to represent the Xenon, is designed to accurately match Reynolds number. The Xenon fluid has a very low viscosity which in turn requires a very low viscosity model liquid. The liquid used is a 1,1,2 Trichloroethane. It has a density of 1.44 kg/liter and a kinematic viscosity of 0.83 centistokes (1/12 that of water). With a length scale of 1/5, the Reynolds numbers exactly match at a model spin rate of 2300 rpm. This is used for all tests. No data were found for the velocity of sound in 1,1,2 Trichloroethane but, based on data for other similar liquids, it is estimated that the value is approximately 1500 meters/second. The Mach number for the model is, therefore, approximately 0.01 which is close to the full scale value. Since the Mach numbers for the model and the full scale vehicle are 1) so much less than one, and 2) so closely match, it is felt that the model liquid accurately simulates the Xenon fluid, a compressed gas. The model tank is constructed in two halves which are sealed with an O-ring and bolted together. The tank is an exact 1/5 scale replica of the full scale tank. The cylinder and domed ends are machined on a computer controlled lathe with the R-S coordinate points connected by a smooth curve. The main body of the drop test model is a 6 inches diameter, 12 inch long PVC tube with aluminum end caps. The tank is mounted to the tube with screws through each end of the tank. One of these screws serves as a fill/drain port. Lead weights are added around the circumference of the tube to achieve the desired inertia ratio. The following scaling law, which is based on the Energy Sink assumptions, is used to scale the data:

$$\rho d_0 \omega \tau / J = \text{constant}$$

for the model and full scale vehicle, where d=tank diameter, ω is the spin rate, ρ is the liquid density, τ is the nutation time constant, J is the spin moment of inertia. Using the full scale (FS) values for moment of inertia, the scaling laws become $\tau_{FS}=9628\tau_M/J_M$ in the pre-burn case, and $\tau_{FS}=5825\tau_M/J_M$ in the post-burn case, where the subscript M refers to the model, and FS to the full scale vehicle. The model spin inertia J_M changes somewhat as ballast is added or removed to alter the inertia ratio. The results from the tests are summarized in Figure 10.

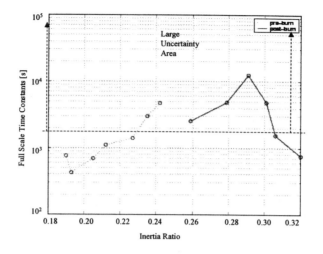

Figure 10. Full scale time constants in the pre- and post-burn cases vs. inertia ratio for the Xenon tank.

The data can be corrected for the effects of air damping and parasitic energy dissipation by using an appropriate tare value. The nutation angles at the beginning and at the end of the drop are shown as well of the inertia ratio (as determined by the spin speed and nutation frequency) and the model time constant. The latter is computed by finding the exponential curve which best fits the data. Some tests were tare tests. These test are run without liquid, with rigid ballast replacing the tank and liquid, to determine the effect of air drag and parasitic structural energy dissipation. The results show extremely long time constants. This means that these effects are negligible. In the rest of the tests, the tank is completely filled with liquid. In one test, a liquid with a viscosity 7.7 times higher than the regular test liquid (but the same density) is used. This was done to see the effect of the Reynolds number error.

The tests results indicate full scale pre-burn time constants in excess of 1000 seconds for all inertia ratios above 0.21. Below 0.21 inertia ratio, the data indicate that a liquid resonance mode exists at approximately 0.19 inertia ratio. The pre-burn and post-burn results are plotted in Figure 10. Note that there is significant uncertainty in the data for model time constants above about five seconds (area enclosed within dotted lines). This is the reason for the scatter in the long time constant pre-burn data, as well in the post-burn data. Even though all of the post-burn time constants are long, the data suggests that another resonance mode may exist above inertia ratio 0.33. In conclusion, the nutation time constant considering the presence of the Xenon tank only is in excess of 1000 seconds, which means that the Xenon gas behaves almost like a solid lump, rigidly connected to the tank walls.

Phase IIa: DS1 Hydrazine Tank Drop Test Program

Phase IIa of the Hydrazine tank drop test program was completed with all of the original objectives met. Analytical scaling laws were developed which define the parameters of the model diaphragm. Based on the assumptions and depth of the analysis, a high confidence exists that the model diaphragm was accurately simulating its full-scale counterpart. Three different elastomeric compounds were developed and several diaphragms were made from each type of material. Depending on the material used, the correctly scaled model spin rate is between 870 and 1300 rpm. The diaphragm was successfully installed in the tank, moved into the reversed position, and the tank filled. Many drop tests were performed without any case of leakage or diaphragm breakage. Even though the model did not represent DS1 in terms of mass center location or inertia ratios, it is worthwhile commenting on the nutation behavior. With the tank fill fraction approximately 85 percent, the diaphragm appeared to be slightly stretched and quite rigid. The measured time constants were extremely long. Some tests were run at lower fill levels, and these resulted in shorter time constants. Presumably, this was due to wrinkles or flexing in the diaphragm when it was not stretched. It should be emphasized that the model tested had the mass center relatively close to the tank center. For DS1, particularly in the pre-burn state, the mass center to tank center distance is quite large. There was a strong possibility that this lever arm would powerfully drive the liquid/diaphragm system and cause short nutation time constants even at high fill levels. The remainder of this section will discuss the scaling laws and model diaphragm in more detail.

A detailed analytical study of the equations governing the flexible diaphragm deflection due to the spinning liquid was performed. This analysis resulted in the following diaphragm scaling laws which are dimensionless ratios which must be preserved for both the model and the full-scale vehicle: h/R, defines the model diaphragm thickness; $E/\rho R^2 \omega^2$, defines the relationship between model spin rate and diaphragm modulus of elasticity, where h is the diaphragm thickness, R is the tank radius, E is the modulus of elasticity of the diaphragm material, ρ is the liquid density, and ω is the spin speed. It was assumed that the Poisson's ratio is reasonably well matched if the model and the full-scale diaphragm materials are similar. The full-scale parameters are as follows: R=8.25 inches, h=0.065 inches, ρ=0.0364 lb/in^3, ω=60 rpm, E=1200 lb/in^2. The values for E (full-scale and model) are obtained by measuring the deflection (d) of a sample of material (thickness h, width b). The sample is suspended on two knife-edge supports (separation l), and a force (W) is applied. The modulus of elasticity is calculated as: $E=Wl^3/4dbh^3$. A model tank of radius 2.375 inches is used with water (.0361 lb/in^3) simulating the fuel. The model parameters based on the scaling laws are therefore: h=0.187 inches, ω/E=36.5 (ω in rp, E in lb/in^2).

The design of the model diaphragm was driven by two requirements: 1) accurate geometric scale, including thickness and details of the ribs; 2) and the modulus of elasticity sufficiently high to allow testing in the spin speed range of 1000-2000 rpm. Spin speeds of this magnitude usually give enough nutation cycles for accurate drop test results. The decision was made early on in the program to develop an injection mold. The mold consists of male and female hemispherical surfaces, which are moved into close proximity, separated by the desired diaphragm wall thickness of 0.0187 inches. Plastic pellets are placed in a hopper, heated to the melting point, and the fluid injected under high pressure. The material enters the hemispherical cavity at the pole, flows around the surface and exits at the rim of the equator. The excess material (flash) is later trimmed off. The desirable aspect of injection molding is that there is a wide range of plastics available. Therefore, at least potentially, diaphragms of virtually any desirable stiffness could be manufactured. Another important reason that an injection mold was chosen is that it can also be used for compression molding using rubber. In compression molding the raw rubber material is inserted between the two mold halves and they are squeezed together. This compresses the rubber and forces it to fill the mold cavity. The part is then removed and cured at high temperature. This vulcanizing process changes the internal structure of the material and increases the stiffness. The first diaphragms made used low density polyethylene in an injection process. It proved to be extremely difficult to produce a whole diaphragm using this process, although eventually about six accurate parts were made. Hundreds of parts were discarded because only part of the mold was filled. It was difficult to keep the two halves of the mold in the desired relative orientation as the molten plastic was injected due to hydraulic action shifting the mold and tight tolerances dictated by the thin (0.0187 inch) desired wall section. About three parts were made using a higher density polyethylene but these were even more difficult to obtain and were so stiff that the diaphragm could not be reversed (turned inside out) without breaking. There were no breakage problems with those made from the low density material, and these were used for all of the drop tests. They have a modulus of elasticity of 40,000 psi, resulting in a model spin speed of 1200 rpm. After the injection molded parts were made, the compression molding process was used using rubber compounds specially blending for this project. Different proportions of styrene and natural rubber are blended to create materials of varying stiffness. Two such blends were used for this project, resulting in values for E of 20,800 and 44,800 psi. The respective spin speeds are 870 rpm and 1300 rpm. Higher stiffness rubber diaphragms for subsequent tests for the following reasons: 1) the compression molded rubber units were easier to manufacture in case more were needed; 2) for a given modulus of elasticity, the rubber diaphragm appeared to be deformed more readily than the polyethylene diaphragm without breaking; 3) since the full-scale diaphragm was made out of a rubber compound, it was assumed that the rubber model diaphragm would be more likely to match Poisson's ratio.

Scaling of fuel-membrane-gas system in a rotating tank

This analysis describes in some more detail the steps that have led to the numbers indicated above. The analysis was done to identify the way to correctly scale the elastic properties of the diaphragm, i.e. axial and bending stiffness, and see if the Poisson ratio needs to be scaled as well. The analysis considered a spinning rigid tank, an incompressible inviscid fluid in the tank, and an elastic membrane modeled as a shell. Since the output of this analysis would have identified the correct stiffness of the scaled membrane, before manufacturing, and there was uncertainty on the manufacturing process, the possibility existed of having to deal with a composite shell, made of two layers. The quantities involved in this scaling procedure are: ω_0 the initial angular speed of the tank, E the shell elastic modulus, h the shell wall thickness, R the tank radius, I the central inertia dyadic for the rigid portion of the vehicle, ρ the fuel density, p the fuel pressure, E_1 (E_2) the modulus of the inner (outer) layer of the composite shell, h_1 (h_2) thickness of inner (outer) layers of composite shell. Then, the tensile stiffness becomes $E_F h_F = E_1 h_1 + E_2 h_2$, and the bending stiffness becomes $E_M h_M^3 = E_2 h_2 [h_2^2 + 3h_1(h_1+h_2)] + E_1 h_1^3$. Introducing the ratios $\alpha = R_{test}/R_{fullscale}$, $\beta = \omega_{0test}/\omega_{0fullscale}$, $\gamma = \rho_{testliquid}/\rho_{fuel}$, the ratios (or nondimensional numbers) that must remain unchanged for the governing equations to remain unchanged are: $I/(\rho R^5)$ to satisfy the momentum equilibrium of the rigid body plus the tank and fluid, $p/(\rho R^2 \omega_0^2)$ to satisfy the Euler equation for the incompressible, inviscid fluid in the tank, $pR/(E_F h_F)$ to satisfy the radial equilibrium of a shell element, $(E_M h_M^3)/(R^2 E_F h_F)$ to satisfy the moment equilibrium of a shell element. The last two also reduce to the following ones (after recombination with the first two): $(E_F h_F)/(\rho R^3 \omega_0^3)$ and $(E_M h_M^3)/(\rho R^5 \omega_0^2)$.

With these definitions, the tank diameter D in the test model becomes $D_T = \alpha D$, the spin rate ω in the test model becomes $\omega_{0T} = \beta \omega_0$, the fuel density ρ in the test model becomes $\rho_T = \gamma \rho$, the rigid inertia dyadic I in the test model becomes $I_T = \alpha^5 \gamma I$, the membrane tensile stiffness Eh in the test model becomes $E_F h_F = \alpha^3 \beta^2 \gamma E$ h, and the bending stiffness E h^3 in the test model becomes $E_M h^3_M = \alpha^5 \beta^2 \gamma E$ h^3. Given the full scale values of E=1200 psi, v=0.49, h=0.065 inches, D=16.5 inches, $\omega 0$=60 rpm, ρ= 0.0361 lbm/in³, the scaling can be done following three approaches:

- Increase the Young's modulus, so that both the tensile and bending stiffnesses scale properly. In this case, if D_{test}=4.75 inches, ω_{0T}=750 rpm, α=4.75/16.5, β=750/60, h_T=0.0187 inches, E_T=15540 psi, and if D_{test}=3.6 inches, ω_{0T}=750 rpm, α=3.6/16.5, β=750/60, h_T=0.0142 inches, E_T=8926 psi.
- Consider a composite shell, for which one obtains: D_{test}=4.75 inches, ω_{0T}=58 rpm, h_1=0.013 inches, E_1=12000 psi, h_2=0.019 inches, E_2=1200 psi.
- Consider the shell material to be the same as in the full-scale tank, scale the thickness so that the bending stiffness is properly scaled, and accept the improperly scaled tensile stiffness. Then, with D_{test}=4.75 inches, ω_{0T}=750 rpm, E=1200 psi, h_T=0.044 inches, E h_T=52.8 lbf/in, and $\alpha^3 \beta^2 E$ h = 291 lbf/in, while with with D_{test}=3.6 inches, ω_{0T}=750 rpm, E=1200 psi, h_T=0.0277 inches, E h_T=33.2 lbf/in, and $\alpha^3 \beta^2 E$ h = 126.6 lbf/in.

The strain components in the plane tangent to the middle surface of the shell have the form $\varepsilon_{ij} = A_{ij} + B_{ij}(z/R)$ where A_{ij} and B_{ij} are dimensionless parameters, and z is the thickness variable. Assuming negligible stresses through the thickness (i.e., σ_{zz}), the stress components have the form $\sigma_{ij} = (E/(1-v^2))[C_{ij} + D_{ij}(z/R)]$ where C_{ij} and D_{ij} are non-dimensional coefficients depending on the Poisson's ratio. With these expressions for strain and stresses, the internal force resultants take the form $F_{ij} = ((Eh)/(1-v^2))C_{ij}$ and the internal moment resultants take the form $M_{ij} = ((Eh)/(1-v^2))D_{ij}$. As before, C_{ij} and D_{ij} are non-dimensional coefficients depending on the Poisson's ratio, but not on the size of the deformed middle surface. Therefore, if the size is scaled and Poisson's ratio is held fixed, E, h, and the other quantities of the system such as spin rate may be scaled in such a way that the shape of the deformed middle surface is the same in the test as in the actual vehicle.

Phase IIb-DS1 Hydrazine Tank Drop Test Program

Phase IIb of the DS1 Hydrazine tank test program was also successfully completed. A newly developed rubber diaphragm was used for all the tests. This was easier to stretch than the plastic version used in Phase IIa, allowing the desired fill level of 87.5 percent to be reached (with the plastic diaphragm, the maximum fill level was 85 percent). Models representing the pre- and post-burn state were tested at the nominal inertia ratio and for values approximately 10 percent higher and lower than the nominal. All results show very long time constants.

The parameters of the drop test model are as follows: the tank diameter is 4.75 inches, the liquid is Water, at 87.5 percent fill level, the diaphragm thickness is 0.019 inches, the diaphragm Young's modulus is 44,750 psi, the diaphragm material is a blend of natural rubber and styrene, and the nominal spin rate is 1300 rpm. The 4.75 inch model tank defines the length scale to be 1/3.47, which then defines the diaphragm thickness to be 0.019 inches. The diaphragm stiffness scaling law $E / \rho R^2 \omega^2$ discussed in the section describing Phase 1 defines the model spin rate of 1300 rpm. The drop test model is composed of six parts: a 24 inch long, 6-inch diameter PVC tube; a fuel tank; two aluminum end caps (the left cap is considered the forward end, and the right end is considered the aft end, closest to solid motor, and carries the model telemetry electronics); a 22 pound solid steel weight fastened to the tube (this represents the solid motor); a lead ring fastened to the outer surface of the tube to alter the inertia ratio. The pre-burn and post-burn models differ only in tank location. The tank center is 2.8 inches from the left end of the tube for pre-burn, and 11.8 inches for post-burn. In each case, the distance between the tank center and the cm of the model (cm offset) is correctly scaled.

The first test was a tare test (no liquid in tank) to verify the lack of parasitic energy dissipation in the dry model. The model constant was an extremely long 45 seconds. The full scale time constant was computed by inertia scaling using the standard scaling law $\rho R_0 \omega \tau / J$, where ρ is the liquid density, R is the tank radius, ω is the spin speed, τ is the time constant, and J is the spin inertia. The model spin inertia varies slightly when inertia ratio weights are added but the scaling is approximately $\tau_{fullscale} = 98 \tau_{model}$ for the pre-burn case, and $\tau_{fullscale} = 59 \tau_{model}$ for the post-burn case. The actual model spin inertias are used to compute the full-scale time constant.

Figure 11 (left) shows a plot of the time constants for the test results in the pre-burn and post-burn cases. For this test program, a high confidence level exists for model time constants shorter that about 5 seconds. For values longer than 5 seconds (all test results), there is uncertainty in the results, which explains the scatter in the data. When dealing with very long time constants, as in this test program, it would also be more meaningful to speak in terms of inverse time constant, since that reflects the energy dissipation rate more accurately. Finally, Figure 11 (right) depicts the sensitivity of the nutation time constant to the liquid fill fraction.

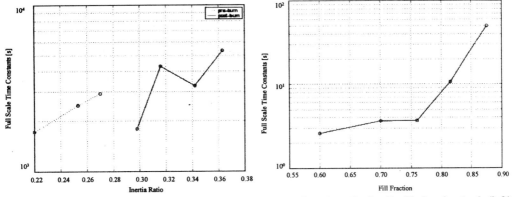

Figure 11. Full scale pre- and post-burn time constants vs. inertia ratio for the Hydrazine tank (left). Full scale time constants vs. fill fraction for the Hydrazine tank (right).

FLIGHT TEST RESULTS

Figure 12 shows the pitch acceleration, thrust acceleration, NTC tank pressure, and filtered rate gyro vs. time during Star-38 ignition and burnout, respectively (courtesy of James Corbo, Boeing Astronautics).

These plots confirm the results of the analysis and testing in that the nutation barely developed in these phases of the DS1 flight.

CONCLUSIONS

The conclusions of this paper are as follows:
- Several modeling issues are described, particularly concerning the effect of the membrane in the Hydrazine tank. An energy sink analysis gives an initial estimate for the nutation time constant of $\tau=25354$ seconds at ignition, and of $\tau=14660$ seconds at burn-out. We observe that, assuming a perfect mass balance, liquid slosh is probably not an issue in this type of diaphragm tank configuration. The only uncertain dynamic response was in terms of liquid resonances (inertial wave modes) which might have been excited inside the Xenon and the Hydrazine tanks.
- In order to simulate the nonlinear behavior of the spacecraft containing the liquid (Hydrazine), the dynamics of a rigid body coupled through a universal joint to a pendulum mass representing the liquid slosh has been analyzed. Despite the uncertainty in diaphragm viscoelastic properties, the results of the analysis show that the nutation time constant is $\tau=1590$ seconds at ignition, and $\tau=1990$ seconds at burn-out. The fact that these two numbers are not very different can be intuitively explained because of the very large tank fill factor, which approximates the behavior of this tank to that of a fully filled spherical tank.
- Spin drop tests also confirm a nutation time constant in excess of 1000 seconds, both in the case of tests done for the Xenon tank only and in the case of tests done for the Hydrazine tank only.
- Flight data reports a small number of Nutation Control System firings, and Nutation Time Constants of 1055 seconds (pre-burn) and 605 seconds (post-burn), confirming the analysis and test results.

ACKNOWLEDGMENTS:

The research described in this paper was carried out at the Jet Propulsion Laboratory, California Institute of Technology, under a contract with the National Aeronautics and Space Administration. The author is grateful to Dr. Fred Hadaegh, Dr. Sam Sirlin, Dr. Dankai Liu, and Mr. Mike Davis of JPL for financial and intellectual support, to Dr. Jon Harrison of Applied Dynamics Laboratories, Inc. for conducting the drop tests, to Prof. Charles Smith, retired from Oregon State University, for providing some of the membrane scaling analysis, to Dr. Carl Hubert of Hubert Astronautics, Inc. for providing useful information on the nutation synchronous mode, and to Mr. James Corbo of Boeing Astronautics for providing the flight test data.

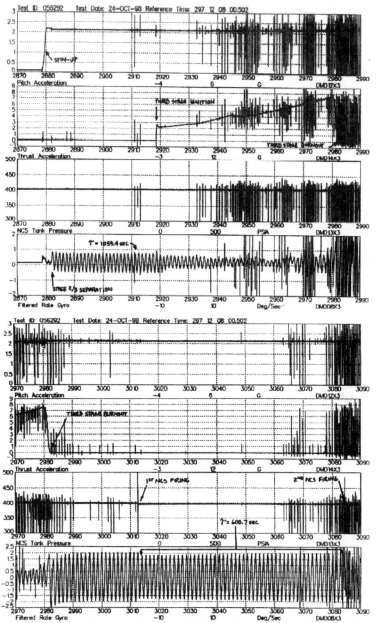

Figure 12. Pitch acceleration, thrust acceleration, NTC tank pressure, and filtered rate gyro vs. time during Star-38 ignition (top) and burn-out (bottom). (courtesy of James Corbo, Boeing Astronautics).

REFERENCES

1. Abramson, H.N.: *The Dynamic Behavior of Liquids in Moving Containers*, NASA SP-106, 1966.
2. Alfriend, K.T.: *Partially Filled Viscous Ring Nutation Damper*, Journal of Spacecraft and Rockets, vol. 11, no. 7, 1974, pp. 456-462.

3. Cartwright, Massingill, and Trueblood: *Circular Constraint Nutation Damper*, AIAA Journal, vol. 1, no. 6, 1963, pp. 1375-1380.

4. Dodge, F.: *Propellant Dynamics and PMD Design For The Near Earth Asteroid Rendezvous (NEAR) Spacecraft*, SWRI Project 04-6297, Southwest Research Institute, April 1994.

5. Dodge, F.T. and Kana, D.D.: *Dynamics Of Liquid Sloshing In Upright And Inverted Bladdered Tanks.* 1985

6. Dodge, F.T., : *Slosh Investigation*, Final Report to RCA, SWRI Project 06-1378-101, April 20, 1987.

7. El-Raheb, M., and Wagner, P.: *Vibration of a Liquid With A Free Surface In A Spinning Spherical Tank*, Journal of Sound and Vibration, Vol. 76, Part 1, 1981, pp. 83-93.

8. Greenspan, H.P.: *The Theory of Rotating Fluids*, Cambridge University Press, 1969.

9. Harrison, J.: *Analysis of spacecraft nutation dynamics using the drop test method*, Space Communications and Broadcasting, vol. 5, 1987, pp. 265-280.

10. Hubert, C.: *Assessment of MSP Lander Propellant Dynamics During Spinning Operations*, Hubert Astronautics Report, 30 January 1997.

11. Hubert, C.: *Deep Impact Nutation Assessment*, Hubert Astronautics Report, 3 July 2002.

12. Kana, D.D., :*Validated Spherical pendulum Model for Rotary Liquid Slosh*, Journal of Spacecraft and Rockets, vol.26, no.3, pp.188-195.

13. Kana and Dodge: *Preliminary Study of Liquid Slosh in the Tracking and Data Relay Satellite Hydrazine Tanks*, Final Report of Project 02-5887, Southwest Research Institute, San Antonio, TX, December 7, 1979.

14. Kreis, A., Kurz, A., Klein, M., Deloo, Ph.: *Static and Dynamic Modeling of Diaphragm Tanks*, Proc. Conference on Spacecraft Structures, Materials, and Mechanical Testing, Noordwijk, The Netherlands, 27-29 March, 1996.

15. McIntyre, J. E. and Tanner, T. M.: *Fuel Slosh in a Spinning On-Axis Propellant Tank: An Eigenmode Approach*, Space Communication and Broadcasting, no. 5, 1987, pp. 229-251.

16. Marce, J. et al.: *Energy Dissipation In Exosat Tanks - Test Results*, CNES Technical Note, CNES-NT-99, March 1981.

17. Mingori L. and Harrison J.: *Circularly Constrained Particle Motion in Spinning and Coning Bodies*, AIAA Journal, vol. 12, no. 11, 1974, pp.-1553-1558.

18. Quadrelli, M.: *Final Report of Deep Space 1 Nutation Time Constant Analysis and Testing*, Internal JPL Engineering Memorandum EM-3455-98-01, February 18, 1998.

19. Ross, R.G., and Womack, J.R.: *Slosh Testing of a Prototype Electric Propulsion Mercury Propellant Tank with Positive Expulsion Diaphragm*, AIAA paper 73-1120, presented at the 10th Electric Propulsion Conference, Lake Tahoe, NV, October 31-Nov. 2, 1973.

20. Sirlin, S.W.: *Mars Pathfinder Launch Vehicle Nutation Analysis*, IOM-3456-96-032, JPL, 19 July 1996.

21. Stofan, A. and Pavli, A.: *Experimental Damping Of Liquid Oscillations In A Spherical Tank By Positive Expulsion Bags And Diaphragms*, NASA TN-D-1311, July 1962.

22. Stofan, A.J. and Sumner, I.E.: *Experimental Investigation Of The Slosh-Damping Effectiveness Of Positive Expulsion Bags And Diaphragms In Spherical Tanks*, NASA TN-D-1712, June 1963.

23. Wood, J., *Mars '98 Lander Nutation Time Constant Estimate Issues*, VSG-ML-M-0897-003R1, 18 August 1997.

24. Zedd, M.F., and Dodge, F.T, :*Energy dissipation of liquids in nutating spherical tanks measured by a forced motion spin table*, AIAA paper 84-1842, 1984.

MODELING OF MARSIS SEGMENTED BOOMS AND PREDICTION OF IN-FLIGHT DYNAMICS OF THE MARS EXPRESS SPACECRAFT[*]

Edward Mettler[†] and Marco B. Quadrelli[‡]

In this paper we provide an independent modeling and dynamic analysis of the MARSIS Antenna segmented booms deployed on the Mars Express Spacecraft. The Mars Express Mission is a joint NASA/ESA Cooperative project. The Mars Express Spacecraft being built by ESA will be launched in June 2003 and arrive at Mars in December 2003 to begin a four year study of the planet's atmosphere, surface, and subsurface. The Mars Advanced Radar for Subsurface and Ionospheric Mapping (MARSIS), provided by NASA and managed by JPL, is a key instrument in the search for water on Mars. Our objective was to determine the antenna's dynamic interaction with the spacecraft bus (or central rigid body). Static and modal analyses make use of boom material parameters, mass properties, and laboratory test results provided by TRW Astro Aerospace, Goleta, California, in addition to data on spacecraft mass properties and orbital parameters from Astrium, Toulouse, France. Solutions are derived involving coupled equations of motion for vehicle orbital mechanics, rigid body spacecraft bus, and flexible-appendages dynamics. Numerical simulations were performed of a "flying model" of the spacecraft in the perigee phase of its elliptical Mars orbit, with all MARSIS booms fully deployed and Reaction Wheels used to both disturb and control the spacecraft attitude. The spacecraft's Reaction Wheel model and Attitude Controller were designed by the authors and is not optimized nor based on any specific information provided by Astrium. Realistic excitations of the bus and boom appendages were imposed by short reaction wheel torque-time profiles that were constructed to maximize excitation of the system fundamental vibration mode. To clearly identify only the MARSIS interaction with the bus, the solar panels were modeled as rigid elements, and the attitude sensors and reaction wheels were assumed to be free from noise and other errors. The simulation results verify the Antenna system modeling fidelity and provide data on proximate "worst case" dynamic interactions between the flexible booms and the spacecraft bus that demonstrate the MARSIS dynamic compatibility with the spacecraft.

[†]   Jet Propulsion Laboratory, California Institute of Technology, Mail Stop 198-138, 4800 Oak Grove Drive, Pasadena, California 91109-8099. Email: edward.metter@jpl.nasa.gov.

[‡]   Jet Propulsion Laboratory, California Institute of Technology, Mail Stop 198-326, 4800 Oak Grove Drive, Pasadena, California 91109-8099. E-mail: marco@grover.jpl.nasa.gov.

INTRODUCTION

This paper provides an independent modeling and dynamic analysis of the MARSIS Antenna segmented booms deployed on the Mars Express Spacecraft. This work was done on behalf of the JPL MARSIS Experiment Project Office to establish the highest confidence for the Antenna System in-flight dynamics compatibility with the Mars Express mission ([1], [2], [4]).

The Mars Express Mission is a joint NASA/ESA Cooperative project. The Mars Express Spacecraft being built by ESA will study Mars for about four years from a highly elliptical orbit (15,000 km apoapsis and 250 km periapsis altitude) with seven major science instruments on the spacecraft plus a small lander vehicle, named Beagle 2. The mission will be launched about June 2003 and arrive at Mars in December 2003. The Mars Advanced Radar for Subsurface and Ionospheric Mapping (MARSIS), provided by NASA and managed by JPL, is a key instrument in the search for water on Mars. It is a four frequency band Synthetic Aperture Altimeter with penetration capability for subsurface and atmospheric sounding from 40 meter long dipole antennas oriented parallel to the planet surface and a 7 meter long monopole along the nadir for clutter cancellation.

Our objective was to determine the antenna's dynamic interaction with the spacecraft bus (or central rigid body). Static and modal analyses make use of boom material parameters, mass properties, and laboratory test results provided by TRW, in addition to Astrium data on spacecraft mass properties and orbital parameters. Solutions are derived involving coupled equations of motion for vehicle orbital mechanics, rigid body spacecraft bus, and flexible-appendages dynamics. Numerical simulations were performed of a "flying model" of the spacecraft in the perigee phase of its elliptical Mars orbit, with all MARSIS booms fully deployed and Reaction Wheels used to both disturb and control the spacecraft attitude. The spacecraft's Reaction Wheel model and Attitude Controller were designed by the authors and is not optimized nor based on any specific information provided by Astrium. Realistic excitations of the bus and boom appendages were imposed by short reaction wheel torque-time profiles that were constructed to maximize excitation of the system fundamental vibration mode. To clearly identify only the MARSIS interaction with the bus, the solar panels were modeled as rigid elements, and the attitude sensors and reaction wheels were assumed to be free from noise and other errors. The simulation results verify the Antenna system modeling fidelity and provide data on proximate "worst case" dynamic interactions between the flexible booms and the spacecraft bus that demonstrate the MARSIS dynamic compatibility with the spacecraft.

Approach to the Dynamic Analysis and Modeling

Clearly, the MARSIS uniquely segmented booms present a dynamic modeling challenge and a special problem in predicting the on-orbit deployed dynamic behavior. For the Mars Express mission, where a precise attitude pointing stability prediction is required, the solution needs to include the segmented dipole and monopole antennas flexible boom appendages, the attachment boundary conditions, and the rigid body inertial dynamics of the spacecraft in its Martian elliptical orbit with orbital mechanics, gravity gradient torques, and closed-loop reaction wheel control. The following modeling, analysis, and simulations were performed with this approach. From Figure 1 we observe that, in the deployed state, the appendage model is far from being a uniform, homogeneous beam. Rather, it consists of 13 segments hinged together by some kinematic constraints. This motivates the multi-body analysis to be discussed next.

Assumptions of the Dynamic Model

A dynamic model of the deployed dipole and monopole booms was derived with the following assumptions:

- Only a small portion of the orbit is simulated, i.e., 100 seconds representing the spacecraft's periapsis passage in its elliptical Mars orbit.

- The spacecraft bus is modeled as a rigid body with three Reaction Wheels aligned along the principal axes.

- The spacecraft Attitude Control System is deliberately modeled with perfect position and rate reference sensors. Reaction Wheels are modeled with realistic rotor inertia, momentum storage, and torque capabilities.

- The solar panels are deliberately modeled as rigid fixed elements, and their inertia contribution is included in the moment of inertia matrix of the bus about its center of mass. This is done to avoid additional complications in the model, but primarily because the solar panels are much more stiff than the antennas.

- The vehicle dynamics is coupled to the orbital dynamics, but the spacecraft undergoes only small rotational motions of less than one degree.

- The monopole and dipole booms were first modeled as cantilever beams of uniform and homogeneous material properties along their length. However, we were not able to come close to the experimental results provided by TRW.

- An improved equivalent model was derived assuming that each boom is modeled as a serial chain of hinged flexible beams, with rotational springs at the "root" (attachment point to the bus) and between the segments.

- The root spring constants are obtained from the TRW static deflection water tank (zero-g) tests

- The rotational spring constants between each hinged segment are kept as free parameters in order to match the first mode of the tested article. The hinges are modeled as perfect spherical joints.

Boom-Bus Attachment Boundary Condition Influence

Comparative information is presented below that shows the dynamic behavior that supports the conservatism of using fixed - free, or various degrees of fixity such as "root" hinge stiffness, versus the actual in-space condition of quasi free - free boundaries seen by the MARSIS booms. The term "quasi" is used to express that the spacecraft central body mass and rotational inertia will impose force and moment constraints on the boom root interface. If the central body were massless then a true free - free boundary condition would be present for the booms.

The natural frequencies (in [rad/s]) of a homogeneous cantilever beam can be written from Ref. ([3]) as $f_j = \frac{a_j^2}{L^2}(\frac{EI}{m})^{\frac{1}{2}}$ where: j = 1, 2, 3, ...n; a_j = Non-rigid body mode coefficients (iteratively derived from transcendental hyperbolic equations for homogeneous cantilever beams); L = length of beam; E = modulus of elasticity; I = area moment of inertia of beam about neutral axis; m = mass per unit length of beam. Then, comparing the two extreme cantilever beam boundary conditions of Fixed-Free vs Free-Free, the first four mode coefficients, a_j, are:

Non-Rigid Body Mode Number	Fixed (or Clamped) - Free	Free - Free
1	1.8751	4.7300
2	4.6941	7.8532
3	7.8547	10.9956
4	10.9955	14.1371

One can readily see the ratio of the a_j^2 terms for two boundary conditions provides modal frequency scaling, e.g., the first modes will have $(4.7300)^2 / (1.8751)^2 = 6.364/1$. Thus the pure free - free beam fundamental frequency will be 6.364 times higher than the fixed - free beam fundamental frequency. While these facts provide a conservative direction for TRW's results, they do not allow accurate prediction of the actual in-flight dynamics of the segmented MARSIS booms. It is clear, however, that the finite stiffness of the "root" hinge is not a fixed or clamped case, and *tends towards a free-free* case or higher frequency. Mitigating that are the segment "spliced-joints" which tend to lower the overall boom stiffness. Prediction of the resulting global dynamics of the booms requires much further modeling and analysis, and makes careful use of the TRW test results.

Boom Stiffness Symmetry

TRW conducted static load-deflection water tank (~zero-g) tests of the fully deployed 20-meter boom to determine the global bending transverse stiffness, the stiffness of the boom "root" in the two transverse directions, K_{ry} and K_{rz}, and also to obtain data on the stiffness of the "spliced" joints. The x direction is defined as the roll axis (see Figure 1), whereas the y and z axes are the yaw and pitch axes respectively. The "root" stiffness K_{ry} was found to be 926.47 N-m/radian (or 8200 in-lb/rad), and K_{rz} was 206.76 N-m/radian (or 1830 in-lb/rad). Thus, the "root" section was 4.48 times stiffer in the K_{ry} direction. This is not surprising since the slotting of the boom tube at the segment fold points creates an obvious asymmetry. The test data also indicated the stiffness of the typical "spliced" section was 2.6 times greater in the z direction then the y direction. This result is opposite to the "root" stiffness asymmetry, and leads to the expectation that the combined segmented boom deflections due to "root and splices" will tend to cancel out local differences in directionality for the global boom behavior. These static deflection tests also indicated reasonable transverse symmetry for the global stiffness, that was later confirmed by the symmetric frequencies obtained from the TRW ten-meter Free-Fixed boom scaled impulse response dynamic tests. We have therefore taken this into account in this analysis and detailed modeling of the segments and joints, and assumed approximately the same global stiffness and natural frequencies in both transverse directions of the fully deployed booms.

EQUATIONS OF MOTION

The equations of motion of the entire system will be derived in this section. Following Figure 2, an inertial reference frame $\mathcal{F}_I$ is defined by the X-axis along the vernal equinox, the Z-axis along the direction of the system's angular momentum, and the Y-axis completes the right-handed triad. The origin of $\mathcal{F}_I$ is placed at the center of a Mars geocentric frame. The position and velocity of the center of mass of bodies i and j (also representing nodes i and j of an extended finite element body) is given by vectors $\mathbf{r}_i$ and $\mathbf{r}_j$, and $\dot{\mathbf{r}}_i$ and $\dot{\mathbf{r}}_j$ respectively measured from the origin of $\mathcal{F}_I$. Similarly, the attitude of the reference frames $\mathcal{F}_i$ and $\mathcal{F}_j$ of bodies i and j with respect to the inertial frame $\mathcal{F}_I$ is described by tensors $\mathbf{A}_i$ and $\mathbf{A}_j$, and their angular velocity by vectors ω_i and ω_j, respectively. We parameterize the translation of body i by the components of vectors $\mathbf{r}_i$ and $\dot{\mathbf{r}}_i$ in $\mathcal{F}_I$, and its rotation wrt. $\mathcal{F}_I$ by the quaternion parameters $\mathbf{q}_i$ and the angular velocity ω_i. It is also useful to introduce the orbiting reference frame $\mathcal{F}_{ORF}$, which we use to describe the near field dynamics of the spacecraft relative to its orbit. This reference frame is attached to a point that follows a Keplerian orbit around the primary body. $\mathcal{F}_{ORF}$ is defined by the direction of the orbital velocity vector (x-axis), the local vertical (z-axis), and the orbit normal (y-axis). The orbit of the origin of $\mathcal{F}_{ORF}$ is defined by the six orbital elements a (semimajor axis), e (eccentricity), i (inclination),

Ω_l (longitude of ascending node), ϖ (argument of perigee), ν (true anomaly), and time of passage

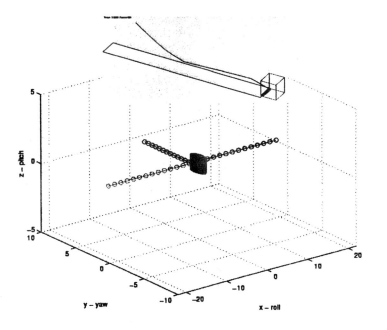

Figure 1: Finite element model of vehicle with deployed MARSIS antennas, showing a snapshot of the deploying boom in vacuo.

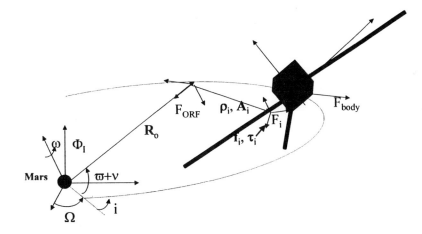

Figure 2: MARSIS orbital geometry model.

through perigee. The orbital radius is R_0, and the orbital angular velocity vector is denoted by Ω. The transformation between $\mathcal{F}_{ORF}$ and $\mathcal{F}_I$ is given by $\mathcal{F}_{ORF} = \mathbf{R}\mathcal{F}_I$ with

$$\mathbf{R} = \begin{bmatrix} 1 & 0 & 0 \\ 0 & -1 & 0 \\ 0 & 0 & -1 \end{bmatrix} \mathbf{R}_3\left(\varpi + \nu\right) \mathbf{R}_2\left(i\right) \mathbf{R}_3\left(\Omega_l\right) \tag{1}$$

where $\mathbf{R}_i\left(\cdot\right)$ denotes a rotation matrix of $\left(\cdot\right)$ around the direction specified by the subscript. It is useful to refer the translational dynamics of body i to the origin of $\mathcal{F}_{ORF}$. Therefore, we have $\mathbf{r}_i = \mathbf{R}_0 + \boldsymbol{\rho}_i$. We define the state vector as $\mathbf{X} = (\mathbf{x}, \dot{\mathbf{x}})$, where $\mathbf{x} = (\mathbf{R}_0, \boldsymbol{\rho}, \mathbf{q}, \sigma_1, \sigma_2, \sigma_3, \boldsymbol{\rho}_1, \boldsymbol{\theta}_1, ..., \boldsymbol{\rho_N}, \boldsymbol{\theta_N})$. Since we decide to work with the near field dynamics, the translational and rotational kinematics equations become $\mathbf{v}_i = \dot{\boldsymbol{\rho}}_i$, and $\boldsymbol{\omega}_i = 2\mathbf{G}\left(\mathbf{q}_i\right)\dot{\mathbf{q}}_i$, respectively, where $\mathbf{G}\left(\mathbf{q}_i\right)$ denotes the transformation between angular velocities and rotation parameters. Consequently, we have that $\dot{\mathbf{r}}_i = \dot{\mathbf{R}}_0 + \dot{\boldsymbol{\rho}}_i + \Omega \times \boldsymbol{\rho}_i$. We also use the notation $\tilde{\boldsymbol{\omega}}$ for the cross-product operator. Measuring translational quantities with respect to $\mathcal{F}_{ORF}$, the translational and rotational dynamics equations of the spacecraft bus become:

$$m_i\ddot{\boldsymbol{\rho}}_i = -m_i\ddot{\mathbf{R}}_0 - m_i\Omega \times \Omega \times \boldsymbol{\rho}_i - 2m_i\Omega \times \dot{\boldsymbol{\rho}}_i + m_i\mathbf{R}\ddot{\mathbf{r}}_i + \mathbf{f}_e \tag{2}$$

with

$$\ddot{\mathbf{r}}_i = \left(\frac{f_S}{m_i} - \mu_S\right)\frac{\mathbf{r}_i}{|\mathbf{r}_i|^3} - \mu_E\frac{(\mathbf{r}_i - \mathbf{r}_E)}{|\mathbf{r}_i - \mathbf{r}_E|^3} + \mathbf{A}_i\frac{\mathbf{f}_a}{m_i} \tag{3}$$

and

$$\mathbf{J_i}\dot{\boldsymbol{\omega}}_i + \boldsymbol{\omega}_i \times \left(\mathbf{J}_i\boldsymbol{\omega}_i + \mathbf{h}_i\right) = \mathbf{r}_{cp2cm} \times f_S\frac{\mathbf{r}_i}{|\mathbf{r}_i|^3} + \boldsymbol{\tau}_a + \boldsymbol{\tau}_e \tag{4}$$

$$\dot{\mathbf{h}}_i = -\boldsymbol{\tau}_a \tag{5}$$

where m_i and $\mathbf{J_i}$ are the mass and moment of inertia matrix of the i-th body, $\mathbf{f}_S$ is the solar force acting on the i-th body, μ_S and μ_E represent the solar and Earth gravitational parameters, $\mathbf{h}_i$ represents the internal angular momentum distribution for the i-th body (originated by reaction wheels), $\mathbf{r}_{cp2cm}$ represents the vector from center of mass to center of pressure of the i-th body, $\mathbf{f}_a$ and $\boldsymbol{\tau}_a$ represent actuator control forces and torques, and $\mathbf{f}_e$ and $\boldsymbol{\tau}_e$ represent the generalized structural reaction forces and moments at the root of each appendage. The combined model of the bus plus appendages is based on eliminating these constraint reaction forces and torques via finite element assembly. The disturbance models that we consider acting on the spacecraft bus are: gravity, solar pressure, and actuator forces and torques. These perturbations are already included

in equation (3) and in equation (4). Control inputs are reaction wheel forces and torques. For the reaction wheel dynamic model, we adopt an idealized reaction wheel model. Three reaction wheels are located and centered along the spacecraft principal axes. Each reaction wheel dynamics is as described in equation (5).

CONSTRAINED FLEXIBLE BODY DYNAMICS

Hamilton's Principle states that, for any kinematically admissible variation of the displacement and rotation fields, i.e. allowed by the geometry of the motion, the following stationarity condition holds for a system $\mathcal{S}$ of nb bodies:

$$\sum_{i=1}^{nb} \int_i^{i+1} \left[\delta\mathcal{L}\left(\boldsymbol{\eta}, \dot{\boldsymbol{\eta}}, t\right) + F \cdot \delta\boldsymbol{\eta}\right] dt = \left[\boldsymbol{\sigma} \cdot \delta\boldsymbol{\eta}\right]_i^{i+1} \tag{6}$$

where $\mathcal{L}(\boldsymbol{\eta}, \dot{\boldsymbol{\eta}}, t) = \mathcal{T}(\boldsymbol{\eta}, \dot{\boldsymbol{\eta}}) + \mathcal{U}(\boldsymbol{\eta}, t)$ is the Lagrangean of $\mathcal{S}$, $\mathcal{T}(\boldsymbol{\eta}, \dot{\boldsymbol{\eta}})$ is the kinetic energy of $\mathcal{S}$ and $\mathcal{U}(\boldsymbol{\eta}, t)$ the potential energy of $\mathcal{S}$, $\boldsymbol{\eta}$ and $\dot{\boldsymbol{\eta}}$ are the vectors of generalized coordinates and speeds, $\boldsymbol{\sigma}$ is the vector of generalized momenta, and F is the vector of generalizes forces. For simplicity, let us impose that $\delta\boldsymbol{\eta} = \mathbf{0}$ at the boundaries of the time interval. The vectors $\boldsymbol{\eta}$, $\dot{\boldsymbol{\eta}}$, and F are defined as follows: $\boldsymbol{\eta} = (\mathbf{r}_i, \mathbf{q}_i)$, $\delta\boldsymbol{\eta} = (\delta\mathbf{r}_i, \boldsymbol{\theta}_{\delta i})$, $\dot{\boldsymbol{\eta}} = (\dot{\mathbf{r}}_i, \boldsymbol{\omega}_i)$, and $F = (\mathbf{f}_g + \mathbf{f}_i, \boldsymbol{\tau}_i)$, where $\boldsymbol{\theta}_{\delta i}$ stands for a virtual variation of the quasicoordinate describing the rotation. The vectors $\mathbf{f}_i$ and $\boldsymbol{\tau}_i$ include external perturbation and control forces, and $\mathbf{f}_g$ represents the gravitational force $-\mu m_i \frac{\mathbf{r}_i}{|\mathbf{r}_i|^3}$. Therefore $\mathbf{A}_i = \mathbf{A}_i(\mathbf{q}_i)$. From eq. (6), we obtain

$$\sum_{i=1}^{nb} \delta\mathbf{r}_i \cdot \left(m_i \ddot{\mathbf{r}}_i + \mu m_i \frac{\mathbf{r}_i}{|\mathbf{r}_i|^3} - \mathbf{f}_i \right) + \tag{7}$$
$$\boldsymbol{\theta}_{\delta i} \cdot (\mathbf{J}\dot{\boldsymbol{\omega}}_i + \tilde{\boldsymbol{\omega}}_i \mathbf{J}\boldsymbol{\omega}_i - \boldsymbol{\tau}_i)$$
$$= 0$$

The virtual displacements $\delta\mathbf{r}_i$ and virtual rotations $\boldsymbol{\theta}_{\delta i}$ in eq. (7) are kinematically admissible as they satisfy any constraint equation imposed on body i, namely if $\boldsymbol{\Phi}_{\mathbf{r}_i} \cdot \delta\mathbf{r}_i + \boldsymbol{\Phi}_{\theta_i} \cdot \boldsymbol{\theta}_{\delta i} = 0$, where $\boldsymbol{\Phi} = [\boldsymbol{\Phi}_{\mathbf{r}_i}, \boldsymbol{\Phi}_{\theta_i}]$ represents the Jacobian of a certain algebraic equation $\Psi = \Psi(\boldsymbol{\eta}, \dot{\boldsymbol{\eta}}, t) = 0$. Therefore, there exists a vector of Lagrange multipliers $\boldsymbol{\lambda}$ such that the new equations of motion become:

$$\sum_{i=1}^{nb} \delta\mathbf{r}_i \cdot \left(m_i \ddot{\mathbf{r}}_i + \mu m_i \frac{\mathbf{r}_i}{|\mathbf{r}_i|^3} - \mathbf{f}_i - \boldsymbol{\Phi}_{\mathbf{r}_i}^T \boldsymbol{\lambda} \right) +$$
$$\boldsymbol{\theta}_{\delta i} \cdot \left(\mathbf{J}_i \dot{\boldsymbol{\omega}}_i + \tilde{\boldsymbol{\omega}}_i \mathbf{J}_i \boldsymbol{\omega}_i - \boldsymbol{\tau}_i - \boldsymbol{\Phi}_{\theta_i}^T \boldsymbol{\lambda} \right) = 0 \tag{8}$$

Finally, for arbitrary admissible $\delta\mathbf{r}_i$ and $\boldsymbol{\theta}_{\delta i}$, we obtain the following equations of motion for body i:

$$m_i \ddot{\mathbf{r}}_i = -\mu m_i \frac{\mathbf{r}_i}{|\mathbf{r}_i|^3} + \mathbf{f}_i + \boldsymbol{\Phi}_{\mathbf{r}_i}^T \boldsymbol{\lambda} \tag{9}$$

$$\mathbf{J}_i \dot{\boldsymbol{\omega}}_i + \tilde{\boldsymbol{\omega}}_i \mathbf{J}_i \boldsymbol{\omega}_i = \boldsymbol{\tau}_i + \boldsymbol{\Phi}_{\theta_i}^T \boldsymbol{\lambda} \tag{10}$$

The set of equations for body i are eq.(9), eq. (10), and eq. (5).

The equations of motion for the system formed by the i-th and j-th bodies can now be written in matrix form as:

$$\begin{bmatrix} \mathbf{M} & \boldsymbol{\Phi}^T \\ \boldsymbol{\Phi} & \mathbf{0} \end{bmatrix} \begin{pmatrix} \ddot{\boldsymbol{\eta}} \\ \boldsymbol{\lambda} \end{pmatrix} = \begin{pmatrix} \mathbf{g} \\ \mathbf{0} \end{pmatrix} \tag{11}$$

where $\ddot{\boldsymbol{\eta}} = \begin{pmatrix} \ddot{\mathbf{r}}_i & \dot{\boldsymbol{\omega}}_i & \ddot{\mathbf{r}}_j & \dot{\boldsymbol{\omega}}_j \end{pmatrix}^T$, $\boldsymbol{\lambda} = \begin{pmatrix} \boldsymbol{\lambda}_r & \boldsymbol{\lambda}_\theta \end{pmatrix}^T$, $\mathbf{M}$ is a block diagonal matrix for each body, $\mathbf{g}$ is the vector of external and nonlinear terms, and $\boldsymbol{\Phi} = [\boldsymbol{\Phi}_i \quad \boldsymbol{\Phi}_j]$.

APPENDAGE EQUATIONS

We assume linearized structural dynamics of each appendage with respect to the bus body frame. If the appendage were a continuous member (a cantilevered beam, for instance) the equations of motion for each appendage in global coordinates (subscript $_g$) can be written as follows:

$$M_g \ddot{\mathbf{q}} + K_g \mathbf{q} = \mathbf{f}_R \tag{12}$$

where the vector $\mathbf{q}$ (of dimension $n_g \times 1$) contains the nodal displacements and rotations of each node in global coordinates, and $\mathbf{f}_R$ represents the vector of generalized external forces on the appendage nodes (damping terms can be added later).

In our analysis, we noticed that a better correlation with the experimental modal frequencies of each appendage (monopole and dipole booms) can be achieved if the appendage is modeled as a series of elastic Bernoulli-Euler beams connected serially to each other by spherical joints, each joint supporting a rotational spring (in each transverse direction). Instead of the fixed boundary condition at the root, a spherical joint with the root hinge spring constants K_{ry} and K_{rz} provided by the TRW report ([1]) was used. The model of the appendage then becomes:

$$M_t \ddot{\mathbf{q}}_t + K_t \mathbf{q}_t = \mathbf{f}_t \tag{13}$$

where the vector $\mathbf{q}_t$ (of dimension $n_t \times 1$) contains the nodal displacements and rotations of each node in global coordinates, and $\mathbf{f}_t$ represents the vector of generalized external forces on the appendage nodes. Here, n_t is the total number of degrees of freedom of each elastic segment times the number of segments. The model is in block diagonal form as

$$M_t = \begin{bmatrix} M_1 & 0 & \cdots & 0 \\ 0 & M_2 & \cdots & 0 \\ \vdots & \vdots & \ddots & \vdots \\ 0 & 0 & \cdots & M_n \end{bmatrix} \tag{14}$$

$$K_t = \begin{bmatrix} K_{11} & -K_{12} & \cdots & 0 \\ -K^T{}_{12} & K_{22} & \cdots & 0 \\ \vdots & \vdots & \ddots & \vdots \\ 0 & 0 & \cdots & K_n \end{bmatrix} \tag{15}$$

where K_{ij} includes the effect of the rotational spring at each hinge and root joints.

For an ideal spherical joint, the constraint Jacobian is $\mathbf{\Phi} = [\mathbf{\Phi}_{\mathbf{r}_i}, \mathbf{0}, \mathbf{\Phi}_{\mathbf{r}j}, \mathbf{0}]$, with $\mathbf{\Phi}_{\mathbf{r}_i} = -\mathbf{1}_3$ and $\mathbf{\Phi}_{\mathbf{r}j} = \mathbf{1}_3$.

COMPONENT MODEL REDUCTION

In this section, we obtain an expression of the reduced set of multiple flexible body dynamics equations.

The algorithm makes use of the Singular Value Decomposition (SVD) to project the equations of motion of the constrained system into the tangent subspace of the motion (eliminates reaction forces and torques between pairs of interacting bodies) [4].

The equations of motion with the constraints may be written as

$$\mathbf{M}\ddot{\boldsymbol{\eta}} + \mathbf{K}\boldsymbol{\eta} + \mathbf{\Phi}_q^T \boldsymbol{\lambda} = \mathbf{G}\mathbf{u} + \mathbf{Q} \tag{16}$$

together with the constraints $\mathbf{\Phi}_q \dot{\boldsymbol{\eta}} = \boldsymbol{\nu}(\mathbf{t})$ and $\mathbf{\Phi}_q \ddot{\boldsymbol{\eta}} = \boldsymbol{\gamma}(\mathbf{t})$. By introducing a coordinate transformation $\mathbf{P}$ such that

$$\boldsymbol{\eta} = \mathbf{P}\mathbf{q}_{\mathbf{r}} = \begin{pmatrix} P_1 & P_2 \end{pmatrix} \begin{pmatrix} \mathbf{q}_{r1} & \mathbf{q}_{r2} \end{pmatrix}^T \tag{17}$$

where $\mathbf{P}_{1[n \times m]} = \mathbf{orth}\left(\mathbf{\Phi}_q^T\right)$ and $\mathbf{P}_{2[n \times (n-m)]} = \mathbf{null}\left(\mathbf{\Phi}_q\right)$, one obtains a projection of the dynamics of the constrained system in a direction tangent to the constraint manifold. The matrix $\mathbf{P}$ maps the minimal system state $\mathbf{q}_r$ into the global system state $\boldsymbol{\eta}$. This means that the projected system moves in the direction of the kinematically admissible displacements, and the effect of the constraints on the balance of forces vanishes. This transformation is equivalent to the one obtained via a singular value decomposition of the constraint jacobian, i.e. $[P, \Sigma, V] = \mathbf{svd}\left(\mathbf{\Phi}_q^T\right)$ such that $\mathbf{\Phi}_q\mathbf{P}_1$ is invertible and $\mathbf{P}_2^T\mathbf{\Phi}_q^T = 0$. Therefore, by premultiplying the equations of motion of each appendage by $\mathbf{P}_2$, we have a way to eliminate the reaction forces from the equations of motion. This elimination process is exact, however it requires some extra computation at each integration time since the algebraic operations required by the SVD may be time consuming. This is a marginal problem, since this computation is carried out off-line, before the dynamic simulation is carried out. Inserting eq.(17) into the equations of motion eq.(16), we obtain:

$$P_2^T M P_2 \ddot{q}_r + P_2^T K P_2 q_r + \boxed{P_2^T \Phi_q^T \lambda} = P_2^T G u + P_2^T Q \tag{18}$$

where the enclosed term vanishes because of the projection.

The reduced model of the appendage then becomes:

$$M_r \ddot{q}_r + K_r q_r = P_2^T G u + P_2^T Q = f_r \tag{19}$$

where the vector q_r (of dimension $n_r \times 1$) contains the reduced nodal displacements and rotations of each node in global coordinates, and f_r represents the vector of generalized external forces on the appendage nodes.

COUPLING THE BUS EQUATIONS WITH THE APPENDAGE MULTIBODY EQUATIONS

It is important to accurately capture the interaction between the rigid, elastic, and gyroscopic energy flowing throughout the system. With this in mind, the wheel can be seen as a freely rotating body which is coupled to the base structure via a revolute joint. Assume that there exists a finite element model of the base structure, where the wheel is mounted on. The wheel can be modeled with a localized inertia at a particular node, where the degree of freedom corresponding to the wheel rotation is left free. The equations of motion of the rigid spacecraft bus (superscript 1) and of node w (location of reaction wheel) are as follows:

$$M_{bus}^i \ddot{d} + S_{bus}^i \dot{\omega} = f^i \tag{20}$$

$$S_{bus}^i \ddot{d} + J_{bus+w}^i \dot{\omega} + J_w^w \dot{\Omega} + G^i(\mathcal{H}^w)\omega + \omega \times (J_{bus+w}^i \dot{\omega}) = \tau^i - \tau_w^w \tag{21}$$

$$J_w^w \dot{\omega} + J_w^w \dot{\Omega} = \tau_w^w \tag{22}$$

where d is the nodal displacement, ω is the nodal angular velocity, and Ω is the wheel rate vector, τ_w^w is the vector of reaction wheel actuation torque, f^i is the vector of external forces at node i, and τ^i is the vector of external torques at node i. M_{bus}^i is the mass matrix of node i, S_{bus}^i the first moment of inertia matrix, J_{bus+w}^i the second moment of inertia matrix, J_w^w the diagonal matrix of wheel axial inertia, and $G^i(\mathcal{H}^w)$ is the skew-symmetric gyroscopic matrix, which depends on the relative angular momentum $\mathcal{H}^w$ present at node w.

The equations of motion for the spacecraft in global coordinates (subscript $_g$), including the gyroscopic wheel effect, can now be written as follows:

$$M_g \ddot{q} + (G_g + D_g)\dot{q} + K_g q = u \tag{23}$$

where the vector q (of dimension $(n_g + 1) \times 1$) contains the nodal displacements and rotations of each node in global coordinates plus the reaction wheel rotation angles plus the degrees of freedom of the bus, and u is the vector of nodal external forces and moments (including external perturbations and control forces and moments) on the bus node, and the reaction wheel torques. Some nodes are artificially defined to be non-structural (for example, the nodes of an optical prescription), and they do not have any mass or stiffness properties associated with them. Hence, the global equations need to be reduced from the global set n_g to a set of independent degrees of freedom n_e. This is done by the transformation $\mathbf{q} = T\mathbf{q_e}$, where T is of dimension $n_g \times n_e$.

Splitting the equations in elastic (e) and rigid (r) coordinates, we have:

$$M_{ee}\ddot{\mathbf{q}}_e + M_{er}\dot{\Omega} + (G_{ee} + D_{ee})\dot{\mathbf{q}}_e + K_{ee}\mathbf{q}_e = f_e \tag{24}$$

$$M_{re}\ddot{\mathbf{q}}_e + M_{rr}\dot{\Omega} = f_r \tag{25}$$

where now $M_{ee} = T^T M_g T$, and so on. With the assumptions stated above, i.e., by assuming small angular rates (so that the nonlinear terms are negligible, and the modes are still the mass-normalized undamped modes), we can impose the modal transformation $\mathbf{q}_e = \phi\eta$, and rewrite the equations as:

$$\ddot{\eta} + \{\phi^T[G(\mathcal{H}^w)]\phi + 2\Lambda\xi\}\dot{\eta} + \Lambda^2\eta + \phi^T M_{er}\dot{\Omega} = \phi^T T^T b\mathbf{f_r} \tag{26}$$

$$M_{re}\phi\ddot{\eta} + M_{rr}\dot{\Omega} = \mathbf{f_r} \tag{27}$$

where ϕ is the modal matrix, Λ is the diagonal matrix of natural frequencies, and ξ is the modal damping coefficient. We assume some percentage of modal damping in the computations, between 2 and 5 percent of critical. Clearly, $M_{rr} = J_w^w$, $M_{re} = M_{er}{}^T$ is the coupling inertia term, and $\mathbf{f_r}$ is the vector of the wheel disturbance forces and torques on the bus, plus the reaction wheel actuation torques.

Introducing the state vector as $\mathbf{x} = (\eta, \beta, \dot{\eta}, \Omega)^T$, where β represents the reaction wheel angle, the state space model becomes as follows:

$$\dot{\mathbf{x}} = A\mathbf{x} + B\mathbf{u} \tag{28}$$

$$\mathbf{y} = C\mathbf{x} + D\mathbf{u} \tag{29}$$

where $\mathbf{u}$ is the vector of inputs (reaction wheel torque at location w), and is the observation matrix which reads all the finite element state vector. The individual state matrices are as follows:

$$A = \begin{bmatrix} \mathbf{0} & \mathbf{I} \\ -M^{-1}K & -M^{-1}(G + D) \end{bmatrix} \tag{30}$$

$$M = \begin{bmatrix} \mathbf{I} & \phi^T M_{er} \\ M_{er}\phi & M_{rr} \end{bmatrix} \tag{31}$$

$$K = \begin{bmatrix} \mathbf{K}_{ee} & \mathbf{0} \\ \mathbf{0} & \mathbf{0} \end{bmatrix} \tag{32}$$

$$G + D = \begin{bmatrix} \phi^T[G(\mathcal{H}^w)]\phi + 2\Lambda\xi & \mathbf{0} \\ \mathbf{0} & \mathbf{0} \end{bmatrix} \tag{33}$$

$$B = \begin{bmatrix} \mathbf{0} \\ -\phi^T T^T b \\ \mathbf{0} \quad \mathbf{I}_{3x3} \end{bmatrix} \tag{34}$$

$$b = \begin{bmatrix} \mathbf{0} & \mathbf{I}_{3\times3} & -\mathbf{I}_{3\times3} \end{bmatrix} \tag{35}$$

$$C = \begin{bmatrix} T\phi & 0 & 0 & 0 \\ 0 & \mathbf{I} & 0 & 0 \\ 0 & 0 & T\phi & 0 \\ 0 & 0 & 0 & \mathbf{I} \end{bmatrix} \tag{36}$$

CONTROL LAWS

The control laws applied to the spacecraft during the maneuvers are of the feedback (proportional-derivative) plus feedforward type. The translation control actually implemented on the spacecraft is of the form

$$\mathbf{f} = K_p(\mathbf{s}_{Cmd} - \mathbf{s}_{Est}) + K_v(\dot{\mathbf{s}}_{Cmd} - \dot{\mathbf{s}}_{Est}) + M^i\ddot{\mathbf{s}}_{Cmd} \qquad (37)$$

where $\mathbf{s}$ represents the position vector of the center of mass, K_p and K_v are translation control gain matrices, M is the spacecraft mass matrix, q_{Est} and q_{Cmd} represent the estimated and commanded translation state, respectively. The rotational control instead is of the following form

$$\boldsymbol{\tau} = \Gamma_p(\lambda\theta_{err}) + \Gamma_v(\omega_{Cmd} - \omega_{Est}) + J\ddot{\alpha}_{Cmd} \qquad (38)$$

where Γ_p and Γ_v are rotational control gain matrices, J is the spacecraft moment of inertia matrix, λ is the eigenaxis of rotation, and θ_{err} is the magnitude of rotation corresponding to the difference between the commanded and the estimated quaternions.

As per equation (5), the torques in equation (38) are applied with negative sign to the reaction wheels. The control gains were chosen as follows:

direction	proportional gain [Nm/rad]	derivative gain [Nms/rad]
roll	1.7028e2	3.9582e+2
yaw	1.7028e2	3.9582e+2
pitch	4.7300e1	1.0995e2

On account of the closed reaction wheel loops, the open and closed loop eigenstructure for a model with 4 modes becomes as follows:

mode	open loop pole [Hz]	closed loop pole [Hz]	experimental [Hz]
7	0	5.9197e-2	0
8	0	8.3560e-2	0
9	0	8.3605e-2	0
10	8.3560e-2	9.2141e-2	7.80e-2
11	8.3605e-2	9.2277e-2	8.10e-2
12	9.2141e-2	1.4400e+0	N/A
13	9.2277e-2	4.2351e+0	N/A

The comparison of the open loop frequencies obtained with the multibody model described in this paper (the eigenspectrum of the whole vehicle, derived with the dynamical model described above, includes the effect of the rigid rotors) with the experimental data reported in the JPL Memorandum (Ref. [2]) and in the third column of the last table is excellent. Figure 3 shows open loop and closed loop transfer functions from pitch reaction wheel torque to bus angular rates. One can notice the significant energy content around 0.1 Hz, and that the higher boom modes do not contribute significantly to the bus dynamics.

NUMERICAL SIMULATION

A numerical simulation has been carried out to show the performance of the model. The spacecraft bus is given an excitation by a reaction wheel torque pulse of 0.33Nm for a duration of 6 second about

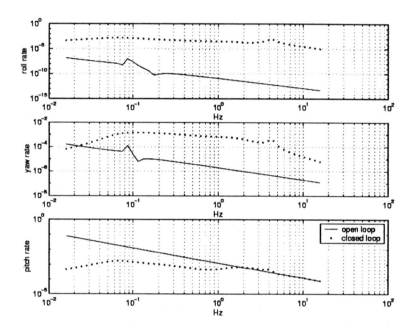

Figure 3: Bode plots of transfer functions from pitch reaction wheel input torque to bus roll, yaw, and pitch rate.

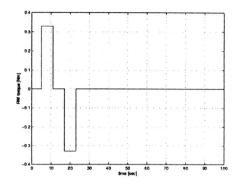

Figure 4: Applied pitch reaction wheel torque vs. time.

the bus Z-axis (pitch). Axes X (roll) and Y (yaw) are not actuated. A closed loop proportional-derivative controller from spacecraft angles and angular rates to spacecraft body torques is applied to maintain the monopole antenna and Y axis always pointed along the Nadir direction (local Mars vertical). Figure 4 depicts the torque profile applied to the pitch axis, designed to excite the 0.083 Hz mode. Figures 5, 6, and 7 depict the open loop response of the system (roll, yaw, and pitch angle) to the torque input with 2% damping. Figure 8 shows the effect of 2% and 5% damping on the yaw rate. Figures 9, 10, 11, show the roll, yaw, and pitch angles of the bus when 2% structural damping is applied to the monopole and dipole booms. Finally, Figure 12 shows the pitch axis reaction wheel speed during the transient maneuver, in which approximately 475 rpm are reached (the spin inertia of each wheel is 0.2 kgm^2), for a change in angular momentum of 1.66 Nms. The maximum momentum storage capability of each reaction wheel is 10 Nms.

With the Reaction Wheel control loops closed, the disturbed response of the spacecraft bus to a Pitch axis symmetric torque pulse waveform of +/-2 Nms (+/-0.33 Nm for 6 seconds) was settled within 10 seconds following the end of the symmetric input disturbance. This disturbance torque profile was deliberately designed to excite the 20 m boom fundamental mode of 0.0835 Hz (measured with the attitude loops open) and also have a net zero momentum transfer to the bus and zero steady state rate. The residual settling oscillations after 10 seconds had amplitudes of sub-arcseconds with a frequency of 0.125 Hz. The residual oscillations had damping decrements of 40 seconds for 50% peak amplitude decay with the 5% structural damping factor, and 55 seconds with the 2% damping factor. The peak bus motion for both dampings was +/- 0.2 degrees in pitch motion. Apparently, the Reaction Wheel closed loop damping of the disturbed bus was far more effective in controlling its settling motion than the structural damping. This is as it should be for such low frequency appendage dynamics.

CONCLUSIONS

As predicted from boundary condition theory for cantilevered beams, the resulting open-loop oscillatory dynamics of the spacecraft central body are significantly (~12%) higher in frequency (0.0835 Hz) then the individual 20-meter boom modal dynamics of 0.0746 Hz. Two levels (2% and 5%) of boom structural damping were used in the simulations. Following the imposed disturbance, the open-loop settling time with 2% damping to a peak-peak oscillation of ~0.002 degrees was ~77 seconds. Importantly, with the attitude control system (ACS) Reaction Wheels in a closed-loop mode,

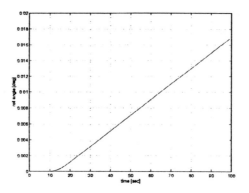

Figure 5: Roll angle vs. time with no feedback (2% damping).

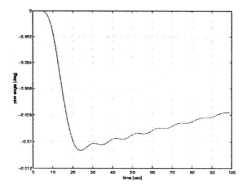

Figure 6: Yaw angle vs. time with no feedback (2% damping).

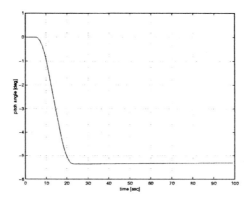

Figure 7: Pitch angle vs. time with no feedback (2% damping).

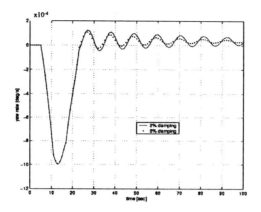

Figure 8: Yaw rate vs. time with no feedback.

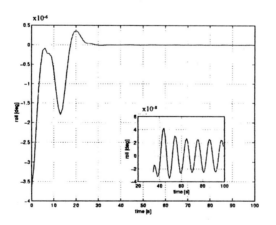

Figure 9: Roll angle with 2% damping and feedback.

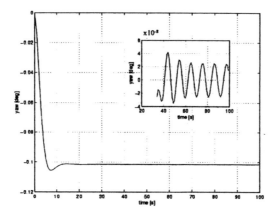

Figure 10: Yaw angle with 2% damping and feedback.

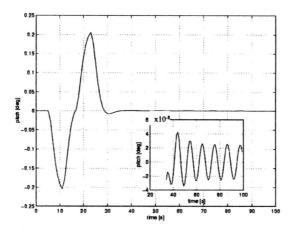

Figure 11: Pitch angle with 2% damping and feedback.

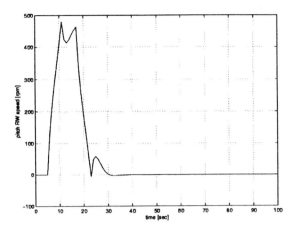

Figure 12: Pitch Reaction Wheel speed in rpm during maneuver with feedback.

the corresponding settling time to sub-arcsecond stability in the local-vertical local-horizontal Mars centered frame was very short (~10 seconds). With the system in a closed-loop mode, the effective inertial stiffness was increased and the bus settling oscillation frequency was about 50% higher at 0.125 Hz compared to the open-loop condition. From this analysis we can conclude that the MARSIS booms have open-loop dynamic properties compatible with the Mars Express Spacecraft. Furthermore, we can predict that a realistically sized Reaction Wheel Controller will be able to suppress the antenna residual transient vibration dynamics so that it contributes negligible error to the mission science pointing accuracy and jitter (line-of-sight stability) capabilities.

Acknowledgements: The research described in this paper was carried out at the Jet Propulsion Laboratory, California Institute of Technology, under a contract with the National Aeronautics and Space Administration. The authors wish to gratefully acknowledge the detailed technical information expeditiously provided by personnel at TRW Astro Aerospace, Goleta and Carpenteria, CA, and with the Mars Express Project at Astrium, Toulouse, France. We also wish to extend our appreciation to the JPL MARSIS Experiment project managers who encouraged our independent examination of the design, modeling, and testing of the antenna booms to establish the highest confidence in its deployment and in-flight dynamics compatibility with the Mars Express mission. Finally, we wish to acknowledge our JPL colleague, Jeff Umland, for his helpful technical discussions.

References

[1] TRW Astro Aerospace Memorandum, April 2, 2002.

[2] Umland J. and Mettler E., *MARSIS Antenna Element Frequency Estimate Review Findings*, JPL Interoffice Memorandum 352B-JWU-02-IOM, February 25, 2002.

[3] Harris, C.M. editor: *Shock and Vibration Handbook*, 4th Edition, Chapter 7, McGraw Hill, 1996.

[4] Mettler, E. and Quadrelli, M., *Modeling and Analysis of MARSIS Boom Dynamics*, JPL Engineering Memorandum 3457-02-005, and JPL Document D-23738, May 6, 2002

NAVIGATION AND CONTROL

Chair: Bobby Williams
                                                              KinetX, Inc.

The following papers were not available for publication:

AAS 03-158

"Almost Geostationary Almost Periodic Orbits With Low Control," by F. Blesa, A. Elipe, M. Lara (Paper Withdrawn)

AAS 03-160

"MER Entry, Descent, and Landing Reconstruction Using an Unscented Kalman Filter," by G. G. Wawrzyniak (Paper Withdrawn)

AAS 03-161

"An Extended Kalman Filter (EKF) to Reconstruct Mars Exploration Rover (MER) Entry and Descent," by M. Lisano (Paper Withdrawn)

AAS 03-162

"A Low Cost Pseudo-Galileo System for Civil Aircraft," by Jianping Yuan, Xingand Li, Northwestern Polytechnical University, P.R. China (Paper Withdrawn)

NUMERICAL SOLUTION TO
THE SMALL-BODY HOVERING PROBLEM

S. Broschart[*] and D. J. Scheeres[†]

The small gravitational forces associated with a small celestial body, such as an asteroid or comet nucleus, allow visiting spacecraft to implement active control strategies to improve maneuverability about the body. One form of active control is hovering, where the nominal forces on the spacecraft are cancelled by a nearly-continuous control thrust. We investigate the stability of hovering station-keeping maneuvers in the body-fixed and inertial reference frames. We also investigate using these hovering control laws for descent maneuvers to the body's surface. Due to the irregular shape and mass distributions typical of small-bodies, stability of spacecraft trajectories in their vicinity is not determined trivially. We have developed a detailed numerical simulation to aid us in our analysis. With the help of this simulation, we find that body-fixed hovering is stable inside a region roughly approximated by the body's resonance radius and inertial frame hovering is stable in all regions outside the resonance radius.

MOTIVATION

Mission proposals currently exist to visit small celestial bodies such as comets, asteroids, and planetary satellites. Due to the small size of such objects, the gravitational forces that these bodies exert on a nearby spacecraft are relatively weak. This allows us to consider active spacecraft control strategies, where thrusters are used in a near-continuous manner to maneuver about the body. This is as opposed to a more traditional orbital approach, where natural dynamics govern the motion of the spacecraft, with control thrust generally being limited to orbital maintenance. Hovering is a particular active control strategy where the nominal forces on the spacecraft are cancelled by the spacecraft's thrusters, creating an equilibrium point at the nominal position. A hovering trajectory could be advantageous to some missions, allowing the spacecraft to spend more time investigating areas of interest on the surface of the body. In cases where the target body is particularly small, orbital control strategies become unstable due to the perturbations of the solar radiation pressure [1]. Hovering control strategies are more likely to remain stable under these types of perturbations, as nominal solar radiation pressures can be negated by the hovering thrust. Hovering in the body-fixed frame also allows for simplified descent maneuvers, which may be very advantageous to a small-body sample return mission.

A mission planner would have to implement a control law to govern the spacecraft's hovering. One type of controller that has been proposed is a combination of an open-loop controller to cancel

[*]   Graduate Student, Department of Aerospace Engineering, University of Michigan, Ann Arbor, Michigan 48109-2140.
       E-mail: sbroscha@engin.umich.edu.

[†]   Associate Professor, Department of Aerospace Engineering, University of Michigan, François-Xavier Bagnoud
       Building, 1320 Beal Avenue, Ann Arbor, Michigan 48109-2140. E-mail: scheeres@umich.edu.

nominal acceleration and a dead-band control on altitude to control perturbataions from the nominal point [2]. [Sawai, Scheeres, and Broschart] went on to develop analytical criteria for stabilty of a spacecraft subject to this control for hovering in the body-fixed frame. For this solution to be useful in mission planning, it must be shown to be valid under realistic circumstances; this necessitates the use of numerical simulation to determine stability characteristics. We have developed such a simulation and can quantify the stability of a given trajectory under realistic conditions. Using this simulation, we can demonstrate the validity of the analytical criteria, showing under which circumstances the criteria hold and where the assumptions may lead to a misleading result. We will also compute numerical results for a slight variation of the controller in the hopes of improving performance. Using the Japan Space Agency's MUSES-C mission as an example[3], we will investigate the stability of hovering in the inertial frame as well. Lastly, we will examine the performance of a hovering descent controller.

PROBLEM FORMULATION

The small-body hovering problem is a form of the two-body gravitational problem, where the two masses are the spacecraft and the small-body. The spacecraft is regarded as having negligible mass relative to the small-body. Because we are concerned with maneuvering very close to the body, we can neglect the third body effect of the Sun. For our analysis, we will use the two-body equations of motion in the body-fixed, rotating frame. The only forces present in the system are gravitational forces, inertial forces, and forces from the spacecraft's propulsion system. In this analysis, we will ignore forces of smaller magnitude, such as the solar radiation pressure. It is assumed that the asteroid has a constant density and is rotating uniformly about a fixed axis (z) corresponding to its maximum moment of inertia. The equations of motion in the body fixed Cartesian coordinate frame with origin at the body's center of mass are defined as follows:

(1)
$$\ddot{x} - 2\omega\dot{y} = -\frac{dV}{dx} + T_x + \omega^2 x$$

(2)
$$\ddot{y} + 2\omega\dot{x} = -\frac{dV}{dy} + T_y + \omega^2 y$$

(3)
$$\ddot{z} = -\frac{dV}{dz} + T_z$$

Here, ω represents the rotation rate of the small body, V is its gravitational potential, and T_x, T_y, and T_z represent the component of the spacecraft thrust in each respective direction. For this analysis, we assume that the thrusters are one-dimensional and are fixed to the body of the spacecraft. The thrusters produce a constant magnitude force when they are enabled and come in 'pairs', allowing thrusting in either the positive or negative sense.

PREVIOUS WORK

In [Sawai, Scheeres, and Broschart], they were able to define an analytical set of sufficiency criteria to characterize the stability of a hovering, station-keeping trajectory in the body-fixed frame. Through their analysis of the problem, they determined that the Jacobian matrix of the gravitational potential at the nominal point determines the stability of hovering, subject to their particular controller, over an arbitrarily shaped body. The sufficiency conditions they developed for stability are listed below:

(4)
$$3\omega^2 v_{3z}^2 + \omega^2 - (\alpha_1 + \omega^2) - (\alpha_2 + \omega^2) \geq 0$$

(5)
$$(\alpha_1 + \omega^2)(\alpha_2 + \omega^2) - \omega^2 v_{1z}^2(\alpha_2 + \omega^2) - \omega^2 v_{2z}^2(\alpha_1 + \omega^2) \geq 0$$

(6)
$$(\alpha_1 - \alpha_2 - \omega^2)^2 + 3\omega^4 v_{3z}^2(v_{3z}^2 + 2) - 8\omega^2 v_{3z}^2\{(\alpha_1 + \omega^2) + (\alpha_2 + \omega^2)\} \geq 0$$

where ω is the body's rotation rate, α_1, α_2, α_3 refer to the three eigenvalues of the Jacobian matrix at the nominal point, and v_1, v_2, v_3 are the eigenvectors of the same matrix (subscripts x, y, and z refer to the respective components of these eigenvectors). The eigenvectors and eigenvalues are arranged such that the third set refers to the eigenvector 'nearly' aligned with the gravitational direction. Of the remaining two, the eigenvector/eigenvalue pair with the largest eigenvalue is considered the first set.

These criteria were developed by linearizing the equations of motion about the nominal hovering point and assuming an infinitely tight dead-band in the control direction. They allowed [Sawai, Scheeres, and Broschart] to compute the theoretical stability of a given trajectory based on a given nominal spacecraft position. A region of stability around the body, roughly corresponding to the area inside the body's resonance radius, is defined by these three inequalities. The resonance radius is defined as the radial distance from the body's rotation axis in the equatorial plane at which centripetal acceleration is equal and opposite to gravitational acceleration.

The gravitational potential field around an arbitrarily-shaped small body is highly nonlinear. The assumption of infinitely tight control is also not realistic, as motion through a finite dead-band and dead-band overshoot may have significant effects on system dynamics. This necessitates an analysis using the full nonlinear equations of motion to validate these results. We further this work by numerically simulating the nonlinear equations of motion with a realistic controller to obtain the actual trajectory of the hovering spacecraft. Having the entire trajectory defined, we are able to quantify and measure its stability. These numerical results will allow us to more accurately predict the level of stability a spacecraft will have at a given postion near a small body. We can also compare and contrast these results with the previous findings to determine in which regions the sufficiency conditions are accurate.

DESCRIPTION OF THE HOVERSIM PROGRAM

The Hoversim simulation package was developed to simulate hovering trajectories over an arbitrarily shaped, rotating body. The code for the simulation is written using Matlab and Simulink software.

The simulation was developed to work with two different asteroid shape and gravity models: the tri-axial ellipsoid and the polyhedron. The tri-axial ellipsoid is a three-dimensional ellipsoid parameterized by its three semi-major axes. The body-fixed coordinate frame is such that the largest dimension of the ellipsoid is along the x-axis, the intermediate dimension aligned with the y-axis, and the z-axis aligned with the smallest semi-major axis. It is assumed that the body is rotating uniformly about the z-axis. Tri-axial ellipsoids are often good approximations of real small-body shapes. Simulations using this shape can be used for initial mission planning, since calculations are much simpler for this shape than for the polyhedral shape model.

More complex geometries can be specified to the simulation as n-faced polyhedrons. The polyhedron can model a much wider range of shapes than the ellipsoidal model, allowing depressions, ridges, cliffs, caverns, and holes on the body. Such models currently exist for a number of asteroids. This shape model can be very accurate, with resolution increasing with the number of faces used. As with the ellipsoid model, it is assumed that the asteroid rotates uniformly about the z-axis of the body-fixed frame, aligned along the maximum moment of inertia.

Two input files describe the shape of the polyhedron [4]. The vertex file contains the Cartesian coordinates in the body-fixed frame of each vertex on the polyhedron. The face file contains information about which vertices connect to form faces on the surface of the body. It is assumed that all faces are triangular, i.e. described by exactly three vertices. The verticies are ordered such that the vector normal to the surface can be determined.

Calculation of the Gravitational Potential

To simulate the equations of motion (Equations 1-3), we must know the rotation rate and the gravitational potential field around the asteroid. The rotation rate is constant and aligned with the

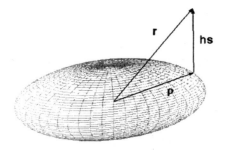

Figure 1: Vector Diagram

body-fixed z-axis. Its magnitude is specified as a simulation input. The gravitational potential is somewhat more complicated to compute, but previous works exist that exactly define the potential around elliptical[5] and polyhedral[6] shaped, constant density bodies. The methods described in these papers are implemented into code to allow calculation of gravitational potential, attraction, the Jacobian, and the Laplacian. For the case of the polyhedral shape model, the Laplacian is used to determine whether the spacecraft has impacted the body.

Control Law

The spacecraft thrust (T_x, T_y, T_z) is determined by the control law chosen by the user. The simulation includes three different types of control that can be applied to the spacecraft. These include a position dead-band controller, a velocity dead-band controller, and an open-loop, constant thrust controller. The simulation can be run using any combination of these three controllers. It is also possible to specify the sensing and control directions. The sensing direction is the direction in which altitude is measured and the control direction is the direction in which thrust is applied.

When using the position dead-band controller, the user specifies a target altitude for the spacecraft and a tolerance factor. If the spacecraft altitude is within the tolerance factor of the target altitude, no thrust is added. If the spacecraft leaves this region, a thrust is applied along the specified control direction, either positively or negatively, to return the spacecraft to the target altitude band. The user specifies the magnitude of this thrust. It is possible to implement only one 'side' of the dead band controller, i.e. thrusters are only fired when the spacecraft altitude is less than (greater than) the desired altitude.

In order to implement this dead-band controller, it was necessary to develop a means for determining altitude above both the elliptical and polyhedral shape models. For this simulation, it is assumed that some attitude controller exists which fixes the spacecraft's attitude with respect to the body, so that the altitude sensing direction is constant. On an actual spacecraft, altitude would be given by the spacecraft altimeter mounted in the appropriate direction.

For the sake of simulation, we can use our knowledge of the spacecraft position in the body-fixed frame to determine the altitude. The problem geometry is specified in Figure 1.

Here **r** is the spacecraft position (given by the simulation's integrator), ρ specifies the point on the surface that the altimeter intersects, h is the spacecraft altitude, and **s** is the normalized sensing direction. From this diagram, we can derive the following relation:

(7) $$\rho = \mathbf{r} - h\mathbf{s}$$

For the ellipsoidal case, we know that ρ must lie on the surface of the ellipsoid and therefore is a solution to the equation:

(8)
$$\rho E \rho^T = 1$$

where E is the symmetric matrix describing the ellipsoid shape. Substituting equation 7 into equation 8 yields:

$$(\mathbf{r} - h\mathbf{s})E(\mathbf{r} - h\mathbf{s})^T = 1$$
$$\mathbf{r}E\mathbf{r}^T - h\mathbf{r}E\mathbf{s}^T - h\mathbf{s}E\mathbf{r}^T + h^2\mathbf{s}E\mathbf{s}^T = 1$$
$$(\mathbf{s}E\mathbf{s}^T)h^2 - (2\mathbf{s}E\mathbf{r}^T)h + (\mathbf{r}E\mathbf{r}^T - 1) = 0$$

This quadratic yields two solutions, as we would expect (one intersection on either side of the ellipsoid).

$$h = \frac{\mathbf{s}E\mathbf{r}^T \pm \sqrt{(\mathbf{s}E\mathbf{r}^T)^2 - (\mathbf{s}E\mathbf{s}^T)(\mathbf{r}E\mathbf{r}^T - 1)}}{\mathbf{s}E\mathbf{s}^T}$$

Assuming that we chose our sensing direction properly such that both solutions for h are positive, we use the smallest solution of the two, corresponding to the side of the ellipsoid closest to the spacecraft.

(9)
$$h = \frac{\mathbf{s}E\mathbf{r}^T - \sqrt{(\mathbf{s}E\mathbf{r}^T)^2 - (\mathbf{s}E\mathbf{s}^T)(\mathbf{r}E\mathbf{r}^T - 1)}}{\mathbf{s}E\mathbf{s}^T}$$

The calculation of the altitude above the polyhedral shape is fundamentally different, as the surface is described by not one, but many equations. To determine the correct altitude, we calculate the altitude over a number of facets, then sort the results.

We begin with the same vector relation as above (Equation 7). We also know that the groundtrack vector, ρ, must lie on the surface of the plane defined by a particular facet. This plane is defined by its normal vector $\mathbf{n}_i$. Thus, the following equation must hold:

(10)
$$\mathbf{n}_i \cdot \rho = C_i$$

where C_i is the scalar constant which defines the facet's position in space. The value of C_i is obtained simply by substituting the vector pointing to one of the face's vertices for ρ in the above equation. Again, we combine equations 7 and 10 to find:

$$\mathbf{n}_i \cdot (\mathbf{r} - h_i\mathbf{s}) = C_i$$

Thus,

(11)
$$h_i = \frac{\mathbf{n}_i \cdot \mathbf{r} - C_i}{\mathbf{n}_i \cdot \mathbf{s}}$$

The number of faces in a polyhedral model can be very large, necessitating a heavy computational load to perform this altitude computation for all the body's faces. This load is lessened by creating a temporary, revised face file which includes only faces that are in the vicinity of the current spacecraft groundtrack and have surface normals in the proper direction. Because the spacecraft moves in a continuous manner, we can assume that subsequent altitude measurements will fall on this same set of faces. If the simulation is unable to find a satisfactory altitude from this set of faces, which could occur if the spacecraft has traveled outside this area, a new set of faces based on the spacecraft's current position are chosen.

We have defined the altitude above any of the shape's infinitely extended planar facets (Eqn. 11). Next we must determine which of these solutions intersect the planar sections on the actual surface of the body. Using our knowledge of h_i and $\mathbf{r}$, equation 7 allows us to compute the 'groundtrack' vector on the surface of the plane, ρ. For the altitude to be valid, ρ must lie inside the area defined by that face's three vertex vectors. Now, we have only valid intersections with facets facing in the proper direction. Because the polyhedral shape can be quite arbitrary, this set can include many altitudes. As in the previous case, the lowest altitude from this set is the correct one.

For both the ellipsoidal and polyhedral altitude measurement calculations, it is possible that there may be no valid altitude, i.e. the sensing direction does not intersect the body. In this case, the thrust produced by the position dead-band controller is set to zero. This means that the spacecraft has drifted 'off' the body and indicates a major failure of the hovering control law.

The simulation also allows the user to implement a velocity dead-band if desired. This type of controller is active only when the spacecraft is inside the position dead-band and fires the spacecraft thrusters when the spacecraft changes altitude faster than the specified rate. This controller adds a damping effect, reducing large oscillations about the altitude dead-band. Using knowledge of the spacecraft velocity (from the simulation integrator), position, and surface geometry, we can determine the spacecraft velocity in the altitude sensing direction. This altitude-based velocity includes components arising from both the spacecraft velocity and changes in the surface of the asteroid. An actual spacecraft would obtain this rate by comparing successive altimeter readings.

The velocity in the altitude sensing direction for the polyhedron shape model is determined in the following way. We use the same geometry presented in Figure 1, but vary these nominal quantities to their values after a small time step, Δt. Values previous to the time step are noted using a subscript 0. From the figure, we know that:

$$\mathbf{r}_0 - h_0\mathbf{s} = \rho_0$$

We also know that the change in groundtrack ρ will be in the plane of the surface and therefore:

$$(\rho - \rho_0) \cdot \mathbf{n} = 0$$

After a small timestep Δt:

$$(\mathbf{r}_0 + \mathbf{v}\Delta t) - (h_0 + \Delta h)\mathbf{s} = \rho$$

where $\mathbf{v}$ is spacecraft velocity. Subtract ρ_0 from both sides and dot both sides with $\mathbf{n}$:

$$[(\mathbf{r}_0 + \mathbf{v}\Delta t) - (h_0 + \Delta h)\mathbf{s} - \rho_0] \cdot \mathbf{n} = (\rho - \rho_0) \cdot \mathbf{n} = 0$$
$$[(\rho + h\mathbf{s} - \rho_0) + \mathbf{v}\Delta t + \Delta h\mathbf{s}] \cdot \mathbf{n} = 0$$
$$\Delta t(\mathbf{v} \cdot \mathbf{n}) + \Delta h(\mathbf{s} \cdot \mathbf{n}) = 0$$

Thus,

$$(12) \qquad \frac{\Delta h}{\Delta t} = -\frac{\mathbf{v} \cdot \mathbf{n}}{\mathbf{s} \cdot \mathbf{n}}$$

For the ellipsoidal shape model, the velocity of the spacecraft in the altitude direction is determined similarly. We can find the instantaneous altitude-based velocity of the spacecraft using equation 12 by simply substituting the normal vector of the ellipsoid at the groundtrack point, ρ. The vector normal to an ellipsoid at point ρ on the surface is:

$$(13) \qquad \mathbf{n}_{ellipse} = \nabla(\rho E \rho) = 2E\rho$$

As with the position dead-band, if no valid altitude is found, this controller is turned off and creates no thrust.

The final controller used by Hoversim is an open-loop controller. This controller creates a constant thrust that nulls gravitational and centripetal accelerations at the prescribed hovering point.

User Interface

In Simulink, the HoverSim controller includes a front-end user interface. This allows the user to easily specify the various parameters of the system. These include: small-body parameters such as the shape model, density, and rotation rate; spacecraft parameters including initial position, thrust level, and errors in position and velocity; and control law parameters such as which control type to use and tolerances associated with them.

Simulation Output

Upon completion of the simulation, a number of outputs are given to the user. The simulation returns to the Matlab workspace raw data giving the time history of the system. These include histories of spacecraft position, altitude, altitude-based velocity, thrust, and simulation time steps. A Matlab script is run by the simulation upon completion which processes this data and creates a number of plots. These plots include: position vs time, thrust magnitude vs time, altitude vs time, altitude-based velocity vs time, spacecraft groundtrack (θ vs ϕ), radial distance from the origin vs time, and deviation from the prescribed hovering point vs time. This script also creates variables in the Matlab desktop describing these plots.

Iterative Capabilities

This simulation also has the capability to be run iteratively to obtain results for a range of initial conditions. This iterative capacity allows us to more easily judge the overall quality of various control laws. This is done using the same Simulink code, but instead of the initial conditions being specified within Simulink, the user creates a Matlab script to describe the parameters of the system, initial conditions, and number of iterations to the simulation. The creation of such scripts is described in the HoverSim User's Guide [4]. An iterative run returns a Matlab workspace file containing the system parameters and information gathered from each simulation, which includes: initial position, position of maximum deviation, maximum distance from hovering point, and maximum angular deviation from the hovering point.

The data obtained from iterative runs can be compiled and presented in graphical form using a script included with the HoverSim package. This script creates a plot of the area surrounding the asteroid with a mark at each initial position simulated. The specific mark corresponds to the stability of the spacecraft at that initial position subject to the system parameters. Plots are created that use both the maximum distance from the hovering point and the angular deviation from the hovering point as measures of stability.

Verification of Program

In order to assure that the simulation works as intended, we ran a series of tests with known results and compared the simulation output to the expected answer.

First, we confirmed that the spacecraft's angular momentum and energy were conserved throughout the simulation above a spherical body if no thrust was applied. The equations for energy and angular momentum are given below:

(14)
$$E = \frac{|\mathbf{v}_i|^2}{2} - \frac{\mu}{|\mathbf{r}|}$$

(15)
$$\mathbf{h}_i = \mathbf{r} \times \mathbf{v}_i$$

where,
$$\mathbf{v}_i = \mathbf{v}_r + \omega \times \mathbf{r}$$

where μ is the small-body's gravitational parameter. The subscript i denotes a property of the inertial frame and the subscript r represents a property of the rotating frame. For our test case, our spacecraft has initial position $[25, \pi/6, 2\pi/3]$ (radius(km), θ(rad), ϕ(rad)) and is initially at rest in the rotating frame. θ and ϕ represent longitude and latitude respectively. Our small-body is a sphere with a radius of 15 km, has a 10 hour period, and a density of 3000 kg/m^3. We can calculate the initial energy and angular momentum as follows:

$$\omega = \begin{bmatrix} 0 & 0 & \frac{2\pi}{T} \end{bmatrix}^T = \begin{bmatrix} 0 & 0 & 1.7453 \end{bmatrix}^T E-4\,rad/s$$

$$\mathbf{r} = \begin{bmatrix} r\cos\theta\cos\phi & r\sin\theta\cos\phi & r\sin\phi \end{bmatrix}^T = \begin{bmatrix} -10825.3 & -6250.0 & 21650.6 \end{bmatrix}^T m$$

$$\mathbf{v}_i = \mathbf{v}_r + \omega \times \mathbf{r} = \begin{bmatrix} 1.0908 & -1.8894 & 0 \end{bmatrix}^T m/s$$

$$\mu = \frac{4\pi}{3}G\rho r_b^3 = 2.8297\,E6\,m^3/s^2$$

where G is the Universal Gravitational Constant, T is the body's rotation period, ρ is the body's density, and r_b is the radius of the spherical body. Equations 14 and 15 tell us:

(16)
$$E = \frac{|\mathbf{v}_i|^2}{2} - \frac{\mu}{r} = -110.81\frac{m^2}{s^2}$$

and

(17)
$$\mathbf{h}_i = \mathbf{r} \times \mathbf{v}_i = \begin{bmatrix} 4.0906 & 2.3617 & 2.7271 \end{bmatrix}^T E4\,m^2/s$$

We then ran the simulation for a time of 600 seconds with these initial conditions. At this time, the spacecraft's position and velocity in the rotating body-fixed frame were as follows:

$$\mathbf{r}_f = \begin{bmatrix} -1.0517 & -0.60993 & 2.0938 \end{bmatrix} E4\,m$$

$$\mathbf{v}_{f,r} = \begin{bmatrix} 1.0592 & 0.4742 & -2.4014 \end{bmatrix} m/s$$

where the subscript f denotes values at the final time. This tells us that:

$$\mathbf{v}_{f,i} = \mathbf{v}_{f,r} + \omega \times \mathbf{r}_f = \begin{bmatrix} 2.1238 & -1.3614 & -2.4041 \end{bmatrix}^T m/s$$

Therefore,

(18)
$$E_f = -110.81\frac{m^2}{s^2}$$

and

(19)
$$\mathbf{h}_{f,i} = \mathbf{r}_f \times \mathbf{v}_{f,i} = \begin{bmatrix} 4.3151 & 1.9212 & 2.7271 \end{bmatrix}^T E4\,m^2/s$$

We can see that energy is in fact conserved. We can also see that the magnitude of the angular momentum vector is conserved. We know that the angular velocity vector should be rotated from its initial direction by $\omega\Delta t$ around the rotation axis. It can be shown that the final angular momentum vector is indeed obtained by rotating the initial vector by this amount.

Our next test was to verify that the simulation was computing the correct body forces on the spacecraft. This was done by solving analytically for the forces on the spacecraft at a given point, then adding a open-loop thrust of equal magnitude and opposite direction to the spacecraft. If the simulation is computing the forces on the spacecraft correctly, this open-loop thrust should cancel those forces and the spacecraft should not move. We will use the same central body as in the previous case (spherical, 15km radius, 3000 $\frac{kg}{m^3}$). The spacecraft will begin at rest at position [19 $\frac{-\pi}{2}$ $\frac{-\pi}{6}$] (radius(km), θ(rad), ϕ(rad)). Since the gravitational field is spherical, we know that the spacecraft's acceleration in the body fixed frame is:

$$\mathbf{a} = -\frac{\mu}{r^3}\mathbf{r} + \omega \times (\omega \times \mathbf{r})$$

The spacecraft's position is given by:

$$\mathbf{r} = \begin{bmatrix} r\cos\theta\cos\phi & r\sin\theta\cos\phi & r\sin\phi \end{bmatrix}^T$$
$$= \begin{bmatrix} 0 & -16454 & 9500 \end{bmatrix}^T m$$

From here, we can calculate the acceleration on the spacecraft:

$$\mathbf{a} = \begin{bmatrix} 0 & 0.0073 & 0.0039 \end{bmatrix}^T m/s^2$$

We then put the negative of this acceleration into the simulation as an open-loop control thrust and ran the simulation with the same initial conditions. As we would expect, the spacecraft did not move. This demonstrates that the body forces on the spacecraft are being calculated properly.

Our final verification test was done to ensure that the dead-band controller was working as intended. We can determine this by looking at the thrust and altitude vs time plots. Thrust should be enabled in the proper direction whenever the altitude dead-band is violated. We performed this test for a number of initial conditions and found the dead-band controller to be working as expected.

STABILITY OF BODY-FIXED HOVERING

We implemented the iterative capabilities of the HoverSim program to examine stability of hovering in the body-fixed frame. We have obtained results for two different control laws and compared them with the region of stability suggested by Sawai's sufficiency criteria (Eqns. 4-6).

Since we are trying to characterize stability from numerical results, we must define the measure of stability that we will use. We will be using a modified form of Lyapunov stability suitable to evaluate our finite-time trajectories. Beginning with a perturbation from equilibrium of a given magnitude, we will gauge stability by the size of the three-dimensional spheroid centered at the initial hovering point that contains the entire trajectory for a given simulation time. The size of this sphere is measured in two ways, distance from the initial point and angular deviation from the initial point (measured from the body's center of mass). Angular deviation from a given point may be more important to a mission than absolute distance if the spacecraft is trying to point at something on the surface of the body.

Numerical Simulation of Previous Work

The first hovering, station-keeping controller we will evaluate is the controller used in the previous work by [Sawai, Scheeres, and Broschart]. Recall that this controller implements an open-loop control to cancel nominal forces on the spacecraft and a dead-band control on altitude. Both the sensing and control directions are aligned with the third eigenvector of the Jacobian matrix of the potential.

Figure 2 shows stability results for body-fixed hovering subject to Sawai's controller. Each mark on the plot corresponds to an initial hovering point simulated. The type of mark depends on the size of the smallest solid angle (measured from the origin) such that the entire trajectory (for the duration of the simulation) is contained within it. The different marks are defined in the figure's legend. For this analysis, we will consider angular deviations of less than 0.5° to be stable. The area satisfying Sawai's sufficiency criteria (Eqns 4-6) is outlined. The spacecraft was initially placed at the specified hovering point with a random initial velocity error of up to 1 cm/s in each direction. The small-body used for this result was a 15x7x6 km ellipsoidal body with a density of 3000 kg/m^3 and a ten hour rotation period. The size of this body is a rough estimate of the asteroid Eros, which was visited during NASA's NEAR mission.

Each simulation was run for 20000 sec (about 5.5 hrs). This duration is appreciably less than the body's rotation period. We justify this relatively short simulation time by noting that a spacecraft operating in the body-fixed frame very near the surface would have little reason to remain in one position for long periods of time. Thus, we have concluded that stability over relatively short periods of time is of greater importance to mission planners.

As we can see in the figures, the numerical results are in fair agreement with the sufficiency criteria developed previously. We can note, however, that the agreement is certainly not exact. There are regions above the body's leading edge in the XY plane that satisfy the sufficiency criteria, but are not stable. We can also see that these criteria are overly conservative in some regions, with stability extending well beyond the area covered by the sufficiency criteria. From this we can conclude that the assumptions presented in the previous analytical analysis can lead to incorrect stability results. However, overall, this controller does produce a large region for stable hovering and presents a feasible option for mission planners.

Alternate Control Law

We also performed numerical simulations using a slightly different controller than used in the previous paper. We still use a dead-band controller based on altimeter readings, but we eliminate the open-loop part. By eliminating the open-loop controller, we leave the component of spacecraft acceleration normal to the gravitational vector uncontrolled. To account for this, we change the control direction from being aligned with a particular eigenvector of the Jacobian of the gravitational potential to being aligned with the total nominal acceleration vector at the desired hovering point. We also change the direction of altitude measurement from along this particular eigenvector to measure altitude in the direction normal to the surface at the hovering point. This controller may offer fuel savings over the previous controller since the open-loop controller is eliminated. This controller may also offer advantages in a surface descent scenario. There will exist some angle between the thruster plume and the surface normal. This angle could help to avoid contamination of the surface regolith by the thrusters in a sample collecting mission.

With some thought, we can expect that this controller will catastrophically fail near the resonance radius of the body. This is because as you approach the resonance radius, the initial acceleration vector will turn 180° as the nominal acceleration transitions from pointing toward the body to away.

Given this constraint, we found the controller to work very well at altitudes inside the resonance radius. We also found this controller to perform more predictably than Sawai's controller, creating more clearly discernable regions of stability and instability. This controller also seems to be less affected by 'gravitational wake' effects associated with the rotation of the ellipsoidal shape than the previous controller. Figure 3 shows results of simulation at a number of initial conditions for the same body and system parameters used in figure 2.

STABILITY OF INERTIAL HOVERING

A spacecraft visiting a small-body may also be interested in hovering in the inertial frame. Inertial frame hovering may be ideal for a spacecraft to map the body's surface and gravitational field from.

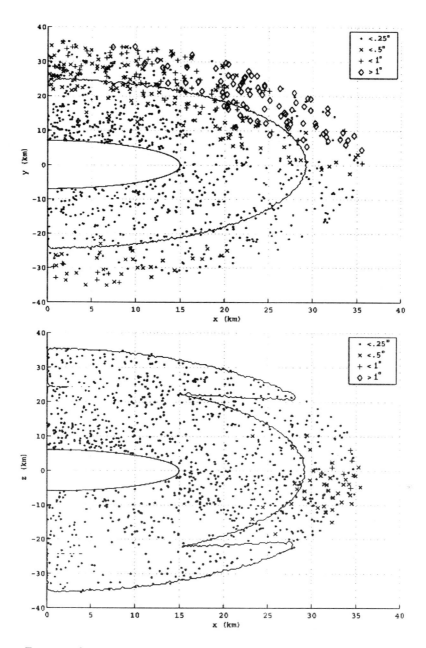

Figure 2: Stability Results for Sawai's Controller (XY and XZ planes)

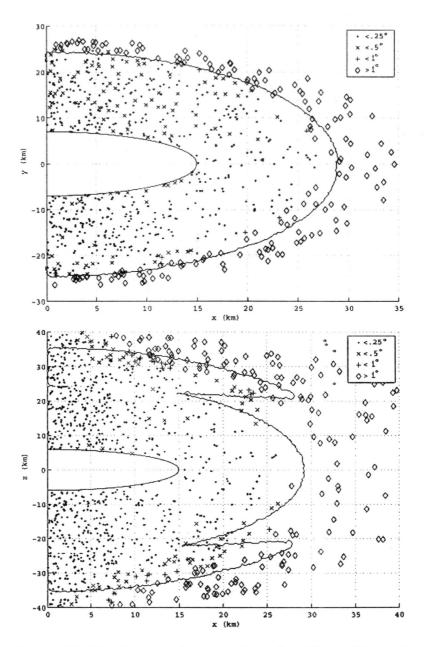

Figure 3: Stability Results for Alternative Controller (XY and XZ planes)

We have extended our work to examine the stability of hovering in a rotating, arbitrary potential field. For the purpose of analysis, we will look at the equations of motion for this problem in the body-fixed frame, where the spacecraft is effectively in a retrograde, circular, constant latitude orbit with angular rate equal to the negative of the body's rotation rate. Here the equations of motion are:

$$(20) \qquad \ddot{x} + 2\omega_0 \dot{y} - \omega_0^2 x = -\frac{dV}{dx} + T_x$$

$$(21) \qquad \ddot{y} + 2\omega_0 \dot{x} - \omega_0^2 y = -\frac{dV}{dy} + T_y$$

$$(22) \qquad \ddot{z} = -\frac{dV}{dz} + T_z$$

where ω_0 is the magnitude of the orbital velocity vector (aligned with the z-axis) of the spacecraft with respect to the asteroid.

We set the nominal thrust value as that necessary to make the retrograde, constant latitude orbit a solution to these equations. This solution is:

$$x_0(t) = r\cos\omega_0 t, \qquad y_0(t) = -r\sin\omega_0 t, \qquad z_0(t) = z(0)$$

where r is the orbital radius.

Linearizing the equations of motion (20 - 22) about this solution we obtain the perturbation equations:

$$(23) \qquad \ddot{\Delta x} + 2\omega_0 \dot{\Delta y} - \omega_0^2 \Delta x = -\left.\frac{d^2V}{dx^2}\right|_0 \Delta x - \left.\frac{d^2V}{dxdy}\right|_0 \Delta y - \left.\frac{d^2V}{dxdz}\right|_0 \Delta z$$

$$(24) \qquad \ddot{\Delta y} + 2\omega_0 \dot{\Delta x} - \omega_0^2 \Delta y = -\left.\frac{d^2V}{dxdy}\right|_0 \Delta x - \left.\frac{d^2V}{dy^2}\right|_0 \Delta y - \left.\frac{d^2V}{dydz}\right|_0 \Delta z$$

$$(25) \qquad \ddot{\Delta z} = -\left.\frac{d^2V}{dxdz}\right|_0 \Delta x - \left.\frac{d^2V}{dydz}\right|_0 \Delta y - \left.\frac{d^2V}{dz^2}\right|_0 \Delta z$$

This is a periodic, time-varying linear system. We can analyze the stability of this system using Floquet theory, which suggests that the state transition matrix is of the form $I = P(t)exp(Mt)$, where P(t) is a periodic matrix and M is a constant matrix. We can find the stability of the system by looking at the eigenvalues of M. This is done by evaluating the state transition matrix after one period of motion, i.e when P is the identity matrix. We can numerically calculate the state transition matrix using the potential code that was developed for the HoverSim program. Using this data, we can compute the eigenvalues after one period.

We also can see that this is a Hamiltonian system, and thus, the eigenvalues must come in complex conjugate and inverse pairs. This means that for stability, the eigenvalues of the system must all lie on the unit circle in the complex plane.

Without doing any calculations, we would imagine that the system will have at least two unstable eigenvalues corresponding to motion along the gravitational attraction vector. If there are some perturbations in the radial direction, the gravitational attraction the spacecraft feels will either increase or decrease in the same manner that the nominal thrust will be inaccurate, resulting in exponential growth of the perturbations. In this analysis, we will consider an infinitely tight, dead-band controller aligned in the radial direction to eliminate this instability. Thus, this analysis is a precursor to a more extensive analysis to come.

We have found that inertial-frame hovering is stable at most radial distances and inclinations. The instabilities that exist lie in the regions very close to the body and near harmonics of the body's

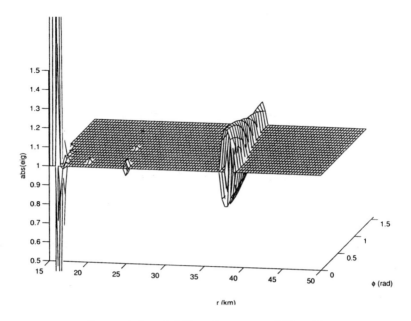

Figure 4: Inertial Hovering over an Ellipsoid

resonance radius. Hovering is stable beyond the body's resonance radius up to the point where our assumptions break down and third-body effects of the sun become an issue. Latitude has little qualitative effect. As latitude increases from the equator to the pole, its effect to is move the radii of unstable hovering to slightly shorter distances and to marginally decrease the magnitude of the instability. Another effect is that hovering near the body at low latitudes is highly unstable but as latitude increases, this instability quickly disappears. Below are the results of this analysis for hovering above a rotating ellipsoid (Fig.4). The figure shows the magnitude of the eigenvalues after one rotation period for nominal trajectories of varying radius and latitude, ϕ. The ellipsoid used was 15x7x6 km, had density of 2300 kg/m^3, and a rotation period of 20 hours. This corresponds to a resonance radius of 37.6 km.

In this figure, we can see that the instabilities of the greatest magnitude exist very close to the surface of the body. This region would have to be avoided by an inertially hovering spacecraft. There also exists an instability at the resonance radius of the body. This is similar to the instability of orbiting a rotating body at resonant frequencies. If you look closely at this case, a weak instability can be seen at the radius corresponding to orbiting at three times the resonant frequency. This plot also shows that the latitude, ϕ, has little effect on the stability of hovering, except very close to the body, where hovering becomes stable at higher latitudes. The properties of this ellipsoidal example are reflective of the other ellipsoidal shapes we have looked at.

It should be noted that the magnitude of the instabilities that exist near the resonance radius and its harmonics is fairly small (<2). This would suggest that transition across these regions while inertially hovering is likely possible. This analysis is left to future research.

Inertial Hovering in the MUSES-C Mission

For a portion of the Japan Space Agency's upcoming MUSES-C mission, their spacecraft will maintain a hovering trajectory in the inertial frame above asteroid SF36 at an altitude of approximately 20 kilometers[7]. During this period, instrumentation aboard the spacecraft will be used to map the

Inclination	i	1.7228315°
Long. of ascending node	Ω	71.1396491°
Eccentricity	e	0.2804327
Orbital period	T	1.52404 Julian yrs
Periapsis passage	t_p	May 4.5893954,2001
Argument of periapsis	ω	160.7980749°
Semi-major axis	a	1.324318555 AU
Time of arrival (s/c)	t_a	6-15-05
Time of departure (s/c)	t_d	11-2-05

Table 1: SF36 Parameters, *Courtesy of "Horizons" Website[9]. Japan Institute of Space and Astronautical Science[10]*

surface and accurately determine asteroid parameters. The mission designers plan to implement a three-dimensional 'box' type controller, which consists of a three position dead-band type controllers each aligned with a local orthogonal coordinate axis, to maintain the spacecraft's approximate position. They have performed some analysis of this control and set the parameters of control as to minimize the usage of fuel necessary to maintain position.

Calculation of Hovering Latitude

As an example of inertial hovering over a polyhedral shaped body, we have simulated the conditions of the MUSES-C mission to asteroid SF36 (asteroid shape model courtesy of NASA-JPL [8]). To do this accurately, we must first calculate the latitude at which the spacecraft will be hovering. The spacecraft will be hovering along the Earth-Asteroid line during its mapping phase, which is well approximated by inertial hovering. The parameters for the orbit of asteroid SF36 are included in Table 1.

The planned time of arrival (t_a) at the asteroid for the spacecraft is June 15, 2005 and the planned time of departure (t_d) is November 2, 2005. From these times, we can calculate the range of time past asteroid periapsis during which the spacecraft will be visiting the asteroid.

$$\left[\begin{array}{c} t_1 \\ t_2 \end{array} \right] = \left[\begin{array}{c} t_a \\ t_d \end{array} \right] - \left[\begin{array}{c} t_p \\ t_p \end{array} \right] = \left[\begin{array}{c} 1.0653 \\ 1.4445 \end{array} \right] yrs$$

We can also calculate the mean motion of the asteroid around the sun.

$$n = \sqrt{\frac{\mu}{a^3}} = 1.3064\,E - 7\,rad/s$$

We orient our coordinate frame such that the x-y plane is the plane of the Earth's ecliptic with the positive x axis pointing to the Earth's vernal equinox and the origin at the center of the sun. In this coordinate frame, the position vector of the asteroid can be found for any time through the following analysis.

$$M = n\delta t = E - e\sin E$$

where M is the mean anomaly, δt is the time past periapsis, and E is the eccentric anomaly. The above transcendental equation can be used to calculate the eccentric anomaly, E, at a given time past periapsis. From this information, the following equations can be used to calculate the true anomaly, f.

$$\cos f = \frac{\cos E - e}{1 - e\cos E}$$

$$\sin f = \frac{\sqrt{1 - e^2} \sin E}{1 - e \cos E}$$

From here, we can calculate the asteroid's position in this reference frame using the following equation.

(26)
$$\mathbf{r} = r \begin{bmatrix} \cos f + \omega \cos \Omega - \sin f + \omega \sin \Omega \cos i \\ \cos f + \omega \sin \Omega - \sin f + \omega \cos \Omega \cos i \\ \sin f + \omega \sin i \end{bmatrix}$$

where $r = \frac{a(1-e^2)}{1+e \cos f}$ is the orbital radius. This analysis yields the position of SF36 at the time of arrival and departure. This can also be used to determine the position of the Earth during the rendezvous period, since we are concerned with the direction of Earth from the asteroid. The positions of the Earth and SF36 at the time of arrival and departure are given in the table below.

$$r_{sf36,a} = \begin{bmatrix} -1.5126 \, E11 & -1.8064 \, E10 & 4.1297 \, E9 \end{bmatrix} \, km$$
$$r_{sf36,d} = \begin{bmatrix} -1.4079 \, E11 & -4.5766 \, E10 & 3.5624 \, E9 \end{bmatrix} \, km$$
$$r_{Earth,a} = \begin{bmatrix} 1.2695 \, E10 & 1.5144 \, E11 & 0 \end{bmatrix} \, km$$
$$r_{Earth,d} = \begin{bmatrix} -1.1213 \, E11 & -9.7208 \, E10 & 0 \end{bmatrix} \, km$$

We want to find the range of latitudes on asteroid SF36 that the Earth moves across during the given period with respect to the rotation pole of the asteroid. To do this, we must find the direction of the asteroid's rotation pole, which is fixed in inertial space. The asteroid's pole is at located at 320 deg longitude (L) and -75 deg latitude (λ). We can calculate this unit vector in our coordinate frame as follows:

$$\hat{\mathbf{z}} = \begin{bmatrix} \cos L \cos \lambda \\ \sin L \cos \lambda \\ \sin \lambda \end{bmatrix}$$

The angle between the pole and the Earth-asteroid line is given by:

$$\cos \Theta = \hat{\mathbf{z}} \cdot \hat{\mathbf{r}}_{\mathbf{E/A}}$$

where $\hat{\mathbf{r}}_{\mathbf{E/A}} = \mathbf{r}_{Earth} - \mathbf{r}_{sf36}$ and $\hat{\mathbf{z}}$ is the unit vector normal to the ecliptic. The latitude of the Earth-asteroid line with respect to the asteroid's rotation pole is 90 degrees - Θ. This latitude is 2.0156^o at the time of arrival and 17.4417^o at the time of the departure of the MUSES-C spacecraft. Because the motion of the Earth and the asteroid are both continuous in time, we know the Earth-asteroid line will move continuously across this range of latitudes during the rendezvous time.

Simulation Results for MUSES-C Mission

We choose a latitude of 10 degrees at which to simulate hovering over SF36, corresponding to a time somewhere near the middle of the mission. Figure 5 shows the magnitudes of the eigenvalues obtained for hovering at 10 degrees latitude over asteroid SF36 versus radial distance of hovering. This plot is interpreted the same as the previous result for inertial hovering. If eigenvalues of magnitude different than unity exist, then hovering is unstable.

The results here are exactly what we would expect. The resonance radius for asteroid SF36 with a density of 2300 kg/m^3 (estimate) and a rotation period of 12.12 hours is calculated to be 551 m. We can see in this plot that the primary instability exists in the region surrounding this radius. Again, we can note that although this is the largest instability present, its magnitude is reasonably small. Beyond this region of instability, all inertial hovering is stable. The MUSES-C mission plans to inertially hover at an altitude of 20 km. This is well within the region of stable hovering.

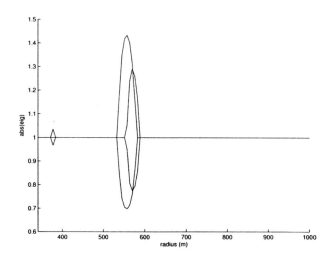

Figure 5: Stability of Inertial Hovering over SF36, 10° Latitude

CONTROL OF A HOVERING DESCENT

Once we can stably hover in the body-fixed frame, it is a natural progression to descend under the same control law. The desire to return samples drives many of today's proposed small-body missions. In order to collect a surface sample, the spacecraft will at least have to descend to very near the surface. We have begun a preliminary analysis of descent trajectories subject to the body-fixed, station-keeping controller which used control thrust in the direction of initial acceleration. This controller offers some advantage in that the sensing and control directions are different. This means that as the spacecraft descends, the thruster plume will extend at some angle from the bottom of the spacecraft. This may prevent contamination of the surface regolith by the thruster outgassing in sample return missions.

Our strategy for descent to the surface will be as follows. Using the same dead-band controller used for station-keeping, we will vary the target altitude linearly down to the surface. During ascent, the altitude will initially be put at some distance off the surface to give the spacecraft an initial thrust so that oscillations around the dead-band do not cause the spacecraft to reimpact the surface. Because this initial period of thrusting will cause large oscillations in distance from the desired hovering point (the linearized system has no damping effect), we implement the velocity dead-band controller discussed previously during ascent. Only the part of the velocity controller regulating negative altitude rates will be implemented, as not to impede the initial thrusting off the surface.

It is important that the controller keep the velocity of the spacecraft at impact as small as possible to protect the spacecraft. Touchdown should occur very close to the target landing site. The spacecraft should also ascend to a point near where it began.

Derivation of Necessary Control Thrust

In the descent maneuver, deviations from the prescribed hovering point have more serious consequences than in station-keeping maneuvers. The larger the deviations from the desired altitude, the larger the maximum velocity obtained by the spacecraft is. As mentioned previously, it is important that the spacecraft impact the surface at a reasonable velocity. This impact can occur at any point in the dead-band, so it is important to keep the spacecraft's velocity low at all times. It is also important to have tight control on the spacecraft to avoid unwanted collisions with the surface when operating at very low altitudes. To maintain this tight control on the spacecraft, we must make the width of the dead-band controller very small and we must limit overshoot beyond the deadband.

The way we can limit overshoot outside the deadband is through the magnitude of our control thrust. At very low altitudes, as during a descent/landing maneuver, it is a reasonable approximation to model the spacecraft's motion as one-dimensional motion in a constant gravity field.

Let us suppose that the spacecraft begins at rest some distance h above the bottom of the dead-band. The acceleration while within this dead-band is equal to the constant gravitational acceleration, g. When below the band, the spacecraft has constant upward acceleration, a. Since the spacecraft is subject to constant acceleration, we can define the spacecraft's position and velocity by:

$$(27) \qquad\qquad x = x_0 + v_0 t - \frac{1}{2} g t^2$$

and

$$(28) \qquad\qquad v = v_0 - gt$$

when above the dead-band and

$$(29) \qquad\qquad x = x_0 + v_0 t + \frac{1}{2} a t^2$$

$$(30) \qquad\qquad v = v_0 + at$$

when below the dead-band. Here x refers to spacecraft position, v refers to velocity, and the subscript 0 indicates initial quantities. Beginning from $x_0 = h$, equations 27 and 28 tell us that when the spacecraft reaches the dead-band boundary,

$$\Delta t_1 = \sqrt{\frac{2h}{g}} \qquad and \qquad v_1 = -g\Delta t_1 = -\sqrt{2gh}$$

Using these initial conditions, equations 29 and 30 tell us that

$$\Delta t_2 = -2\frac{v_1}{a} = 2\frac{\sqrt{2gh}}{a} \qquad and \qquad v_2 = -v_1$$

are the conditions when the spacecraft returns again to the dead-band boundary. Since the velocity of the spacecraft when it returns to the dead-band is equal and opposite to when it leaves, it can easily be seen that the spacecraft will again reach a height of h with zero velocity and will experience periodic motion.

We want to keep the overshoot, or the distance that the spacecraft travels outside of the dead-band, to a minimum. Using equation 29 with $t = \sqrt{2gh}/a$ and $v_0 = -\sqrt{2gh}$, we find that the overshoot distance is given by:

$$(31) \qquad\qquad d = \frac{gh}{a}$$

Equation 31 tells us that the overshoot is proportional to the magnitude of gravitational acceleration divided by the control acceleration, a. Since for descent, we seek to minimize overshoot, this tells us that we should select g/a to be very small, thus selecting a relatively large control thrust. Overshoot can be kept small (1-2% of the dead-band width) by selecting a magnitude of control thrust between 50 and 100 times the nominal gravitational acceleration. This thrust should not be selected too large, as the larger it becomes, the more trouble small errors in the thruster on/off times become.

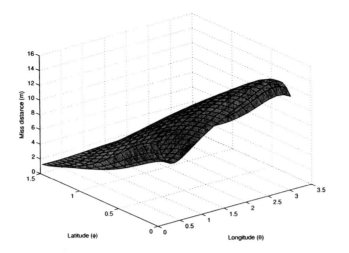

Figure 6: Drift for Descent/Ascent Maneuver

Results of Descent Simulations

We have conducted some preliminary numerical simulations of descent trajectories over an ellipsoidal shape that roughly approximates asteroid SF36. A 50 meter descent to the surface and ascent was simulated for a range of latitudes and longitudes. The descent was designed to take 1200 seconds. An ellipsoidal shape model was used (0.3x0.18x0.17 km) with a density of 2300 kg/m^3. A two meter altitude dead-band was implemented with a 0.1 m/s velocity deadband. Control thrust was set to roughly 60 times the local gravitational acceleration.

For this particular case, we found the average impact velocity to be 3.24±1.75 cm/s, which should be acceptable for most missions. Figure 6 shows the lateral drift from the initial position for the complete descent/ascent cycle. These results seem to suggest that this drift is a regular and predictable phenomenon, dependent on the rotation rate and shape of the body. It can be seen that the greatest drift occurs on the trailing edge of the body. The drift from the initial groundtrack point to the touchdown point shows similar regularity, but 180° out of phase with the drift distance after ascent; that is, the maximum drift distance at touchdown occurs on the body's leading edge and the minimum value occurs on the trailing edge. For both the drift at touchdown and the drift after ascent, values are at a minimum at 90° latitude, i.e. the poles. This is expected since no centripetal acceleration exists at these points.

The regularity we see in the drift values suggests that a nominal solution for this distance can be found. If we can find this nominal solution, then we can quantify the accuracy with which a spacecraft can land at a target site under this control. This analysis is a topic of our ongoing research.

CONCLUSIONS

We have examined the stability of hovering trajectories over small-bodies in both the body-fixed and inertial frames. Our analysis was aided by the development of a numerical simulation package, HoverSim, whose functionality is outlined here.

Previous analytical results were numerically simulated to check their validity under realistic assumptions. We found the correlation between numerical and analytical results to be fair, but not exact. We numerically investigated a variation of the previously used control law and found it created more predictable, though smaller, regions of stability. This controller would likely be more useful to a mission planner than the previous one due to its more consistent performance.

Analytical analysis of hovering in the inertial frame, supported by numerical tools from the HoverSim program, was also performed. We found that inertial hovering is stable in the region outside the body's resonance radius up to the point that third-body effects become important. Inertial hovering near the equatorial plane close to the body is very unstable. We also found that hovering at distances corresponding to orbital frequencies that are harmonics of the body's rotation rate is also unstable. This results holds for both elliptical and the more general polyhedral shape models.

Finally, we presented some preliminary ideas and simulation results using our hovering controller for descent/ascent trajectories, as may be necessary in a sample return mission. It is suggested that our hovering control law will extend well to descent maneuvers.

ACKNOWLEDGEMENTS

The research described in this paper was sponsored by the IPN Technology Program by a grant to The University of Michigan from the Jet Propulsion Laboratory, California Institute of Technology which is under contract with the National Aeronautics and Space Administration.

References

[1] D.J. Scheeres and F. Marzari, "Temporary Orbital Capture of Ejecta from Comets and Asteroids", *Astronomy and Astrophysics* 356, 747-756.

[2] S. Sawai, D.J. Scheeres, and S. Broschart, "Control of Hovering Spacecraft using Altimetry", *Journal of Guidance, Control, and Dynamics*, 25(4): 786-795.

[3] T. Kubota, T. Hashimoto, M. Uo, M. Maruya, and K. Baba, "Maneuver Strategy for Station Keeping and Global Mapping around an Asteroid", *Spaceflight Mechanics 2001: Advances in the Astronautical Sciences*, Volume 108, 2001, pp. 769-779

[4] S. Broschart, "HoverSim User's Guide", Unpublished

[5] D.J. Scheeres, "Satellite Dynamics About Tri-Axial Ellipsoids", presented at "Advances in Non-Linear Astrodynamics", University of Minnesota, Nov 8-10, 1993

[6] R.A. Werner, D.J. Scheeres, "Exterior Gravitation of a Polyhedron Derived and Compared with Harmonic and Mascon Gravitation Representations of Asteroid 4769 Castalia", *Celestial Mechanics and Dynamical Astronomy*, Volume 65, 1997, pp. 313-344

[7] T. Hashimoto, T. Kubota, J. Kawaguchi, M. Uo, K. Baba, and T. Yamashita, "Autonomous Descent and Touch-down via Optical Sensors", *Spaceflight Mechanics 2001: Advances in the Astronautical Sciences*, Volume 108, 2001, pp. 469-490

[8] E-mail correspondence with S.J. Ostro, NASA-JPL, January 19, 2003.

[9] Website: JPL's Horizons System, http://ssd.jpl.nasa.gov/horizons.html

[10] E-mail correspondence with S. Sawai, Japan Institute of Space and Astronautical Science, October 2, 2002.

DEVELOPMENT OF SPACECRAFT ORBIT DETERMINATION AND NAVIGATION USING SOLAR DOPPLER SHIFT

Andrew J. Sinclair,[*] Troy A. Henderson,[†]
John E. Hurtado[‡] and John L. Junkins[**]

In using the Sun as a navigation reference body, we can obtain autonomous (self-contained) spacecraft navigation anywhere that the Sun is the dominant visible body. The use of solar radial velocity knowledge is included in Extended Kalman Filter and Gaussian Least Square Differential Correction implementations and in both rectangular coordinate and orbital element representations. These estimation methods were used to evaluate the sensitivity of spacecraft navigation to the measurement accuracy. An improvement in spacecraft state knowledge resulting from solar radial velocity measurement was demonstrated.

INTRODUCTION

Currently, the two most common methods for performing orbit determination, and therefore navigation, (in earth orbit) use radar measurements and/or optical measurements taken from Earth or radio frequency measurements taken from GPS satellites. One disadvantage of these methods is their reliance on artificial signals. Optical measurements are typically very brief and require complex and expensive ground facilities. A disadvantage of using GPS receivers is the limitation to low Earth orbits— the GPS signal is only broadcast inside of a specified altitude making it useless on interplanetary spacecraft missions. While radar measurements can also be performed on satellites in high orbits and interplanetary spacecraft, this requires complex and expensive ground facilities such as the Deep Space Network (DSN). Satellites that are dependent on artificial signals can be rendered useless by obstruction of those signals. This is of particular concern for military missions.

The hardware and associated software algorithms are being developed at Texas A&M University for a novel method to perform autonomous spacecraft navigation and control taking passive, unalterable measurements. This method uses measurements of the

* Graduate Student, Department of Aerospace Engineering, Texas A&M University, College Station, Texas 77843-3141. E-mail: sinclair@tamu.edu. AIAA and AAS Student Member.

† Graduate Research Assistant, Department of Aerospace Engineering, Texas A&M University, College Station, Texas 77843-3141. E-mail: troy_henderson@tamu.edu. AIAA and AAS Student Member.

‡ Assistant Professor, Department of Aerospace Engineering, Texas A&M University, College Station, Texas 77843-3141. E-mail: jehurtado@tamu.edu. AIAA Member.

** George J. Eppright Chair, Distinguished Professor, Department of Aerospace Engineering, Texas A&M University, College Station, Texas 77843-3141. E-mail: junkins@tamu.edu. AIAA and AAS Fellow.

radial velocity of the spacecraft relative to the Sun to perform orbit determination analysis.

The development of a sensor to precisely measure radial velocity from a spacecraft to the Sun is underway using a tunable Fabry-Perot interferometer (FPI). An FPI consists of two partially transparent, highly reflective, parallel plates or mirrors. When light is introduced into the mirror cavity, multiple beams interfere according to the spacing between the mirrors. A tunable FPI is one in which one of the mirrors can be very precisely positioned allowing a single wavelength to be measured. The large number of interfering rays produced in the cavity gives extremely high wavelength resolution. By measuring the output fringe spacing and intensity, and knowing the distance between the mirrors in the FPI, the wavelength of the spectral line can be determined precisely. The complete theory behind the FPI is well documented by many resources.[1,2,3]

With knowledge of the measured wavelength and reference wavelength, the simple Doppler shift formula can be applied to obtain the radial velocity of the instrument relative to the Sun. The utility of solar radial velocity knowledge for spacecraft navigation has been demonstrated. The observability of this type of system was initially investigated by Guo.[4] The feasibility of spacecraft navigation using solar radial velocity and various line of sight measurements was further demonstrated by Yim.[5,6]

The current development investigates several options for incorporating radial velocity measurements into spacecraft state estimations. Two options studied are implementations of the Extended Kalman Filter and the Gaussian Least Square Differential Corrector estimator. Another choice exists in the selection of the spacecraft state representation on which the estimator operates. In the paper, representations in rectangular coordinates and orbital elements are compared.

ORBIT DYNAMICS

In the study, the motion of an interplanetary spacecraft was modeled as a Keplerian orbit about the Sun. No perturbations were modeled. The orbital elements for this reference orbit were based on the Mars Global Surveyor's transfer orbit. The following elements were used: a=185,106,991.1 km, e = 0.207228763, i = 1.667986 deg, Ω = 45.534369 deg, ω = 19.445019 deg, f_0 = 0 deg. Additionally, because the measurements depended upon the position of the Earth at each time step, the Earth's orbit was also modeled. The true anomaly at the initial time was selected to bring the spacecraft near the Earth at the initial time (within 1,700,000 km). The resulting trajectories are shown in Figure 1.

Rectangular Coordinate Dynamics

One expression of the state of the spacecraft is rectangular coordinates. In this system the state vector is given by the following.

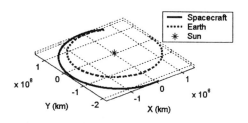

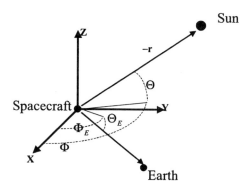

Figure 1 Interplanetary trajectory   **Figure 2 Measurement geometry**

$$[\mathbf{x}_{RC}] = \begin{bmatrix} \mathbf{r} \\ \mathbf{v} \end{bmatrix} = \begin{bmatrix} x & y & z & \dot{x} & \dot{y} & \dot{z} \end{bmatrix}^T \tag{1}$$

The dynamical model for a Keplerian orbit expressed in rectangular coordinates is given by the following state equation.

$$[\dot{\mathbf{x}}_{RC}] = [\mathbf{f}_{RC}(\mathbf{x}_{RC})] = \begin{bmatrix} \dot{\mathbf{r}} \\ \dot{\mathbf{v}} \end{bmatrix} = \begin{bmatrix} \mathbf{v} \\ -\dfrac{\mu}{r^3}\mathbf{r} \end{bmatrix} \tag{2}$$

In order to perform estimation of the system, the Jacobian of the state derivatives relative to the states is needed.

$$\begin{bmatrix} \dfrac{\partial \mathbf{f}_{RC}}{\partial \mathbf{x}_{RC}} \end{bmatrix} = [\mathbf{F}_{RC}] = \begin{bmatrix} 0_{3\times 3} & I_{3\times 3} \\ G_{3\times 3} & 0_{3\times 3} \end{bmatrix} \tag{3}$$

Where G is the gravity gradient matrix.

$$[G] = \frac{\mu}{r^5} \begin{bmatrix} -r^2 + 3x^2 & 3xy & 3xz \\ 3xy & -r^2 + 3y^2 & 3yz \\ 3xz & 3yz & -r^2 + 3z^2 \end{bmatrix} \tag{4}$$

Orbital Element Dynamics

An alternative state representation for the spacecraft is to use the six orbital elements.

897

$$[\mathbf{x}_{OE}] = [a \quad e \quad i \quad \Omega \quad \omega \quad f]^T \tag{5}$$

For a Keplerian orbit a, e, i, Ω, and ω are constants with respect to time. Thus the only nonzero component of the state equation is the derivative of the true anomaly.

$$[\dot{\mathbf{x}}_{OE}] = [\mathbf{f}_{OE}(\mathbf{x}_{OE})] = [0 \quad 0 \quad 0 \quad 0 \quad 0 \quad \dot{f}]^T \tag{6}$$

The state equation for the true anomaly can be found directly from Kepler's second law. Here however, the famous relation between the true anomaly and the eccentric anomaly and Kepler's equation were used.

$$\tan\frac{f}{2} = \sqrt{\frac{1+e}{1-e}} \tan\frac{E}{2} \tag{7}$$

$$E - e\sin E = \sqrt{\frac{\mu}{a^3}}(t - \tau) \tag{8}$$

These give the following expression for the derivative of the true anomaly.

$$\dot{f} = \frac{\sqrt{\frac{\mu}{a^3}(1-e^2)}}{\left\{1 - e\cos\left[2\tan^{-1}\left(\sqrt{\frac{1-e}{1+e}}\tan\frac{f}{2}\right)\right]\right\}^2} \tag{9}$$

Again, to operate the estimation procedures the Jacobian of the state derivatives with respect to the states. Clearly the only nonzero components of this Jacobian are the partial derivatives of $\dot{f}$ relative to a, e, and f. These are straightforward to compute, and are not reproduced here.

$$\left[\frac{\partial \mathbf{f}_{OE}}{\partial \mathbf{x}_{OE}}\right] = [\mathbf{F}_{OE}] = \begin{bmatrix} & & \mathbf{0}_{5\times 6} & & \\ \frac{\partial \dot{f}}{\partial a} & \frac{\partial \dot{f}}{\partial e} & 0 & 0 & 0 & \frac{\partial \dot{f}}{\partial f} \end{bmatrix} \tag{10}$$

MEASUREMENT MODELS

It has been shown that observability of the interplanetary spacecraft system can be achieved by using measurements of the radial velocity relative to the Sun as well as line of sight measurements to the Sun and Earth.[5] Therefore these measurements were used in the current work. The geometry of these measurements is shown in Figure 2. To perform estimation, models of these measurements as functions of the state variables of both expressions and the partial derivatives relative to the states are needed.

$$[\mathbf{y}] = [\mathbf{h}_{RC}(\mathbf{x}_{RC})] = [\mathbf{h}_{OE}(\mathbf{x}_{OE})] = [\dot{r} \quad \Phi \quad \Theta \quad \Phi_E \quad \Theta_E]^T \tag{11}$$

Rectangular Coordinate Measurements

The different measurements can each be modeled as functions of the rectangular coordinates. The radial velocity is found from the dot product of the position and velocity vectors.

$$\dot{r} = \frac{\mathbf{r}^T \mathbf{v}}{r} \tag{12}$$

The azimuth and elevation angles defining the line of sight from the spacecraft to the Sun can be found from the components of the position vector.

$$\Phi = \tan^{-1}\left(\frac{y}{x}\right) \ , \quad \Theta = \sin^{-1}\left(\frac{-z}{r}\right) \tag{13}$$

The line of sight to Earth angles can be similarly found with knowledge of the Earth's position.

$$\Phi_E = \tan^{-1}\left(\frac{y_E - y}{x_E - x}\right) \ , \quad \Theta_E = \sin^{-1}\left(\frac{z_E - z}{r_{SE}}\right) \tag{14}$$

The partial derivatives of these measurements with respect to the rectangular coordinate states can also be found.

$$\frac{\partial \dot{r}}{\partial \mathbf{x}_{RC}} = \frac{1}{r}\left[\frac{\partial \mathbf{r}}{\partial \mathbf{x}_{RC}}\dot{\mathbf{r}} + \mathbf{r}\frac{\partial \dot{\mathbf{r}}}{\partial \mathbf{x}_{RC}} - \frac{\partial r}{\partial \mathbf{x}_{RC}}\dot{r}\right] \tag{15}$$

The partials of the line of sight angles are computed as follows.

$$\frac{\partial \Phi}{\partial \mathbf{x}_{RC}} = \frac{1}{1 + \left(\frac{y}{x}\right)^2}\frac{\partial}{\partial \mathbf{x}_{RC}}\left(\frac{y}{x}\right) \ , \quad \frac{\partial \Theta}{\partial \mathbf{x}_{RC}} = \frac{1}{\sqrt{1 - \left(\frac{-z}{r}\right)^2}}\frac{\partial}{\partial \mathbf{x}_{RC}}\left(\frac{-z}{r}\right) \tag{16}$$

Similarly, the partials of the line of sight to the Earth are shown below.

$$\frac{\partial \Phi_E}{\partial x} = \frac{1}{1+\left(\dfrac{y_E-y}{x_E-x}\right)^2} \frac{\partial}{\partial x}\left(\frac{y_E-y}{x_E-x}\right) \quad , \quad \frac{\partial \Theta_E}{\partial x} = \frac{1}{\sqrt{1-\left(\dfrac{z_E-z}{r_{SE}}\right)^2}} \frac{\partial}{\partial x}\left(\frac{z_E-z}{r_{SE}}\right) \qquad (17)$$

These derivatives form the measurement sensitivity matrix.

$$[\mathbf{H}_{RC}] = \left[\frac{\partial \mathbf{h}_{RC}}{\partial \mathbf{x}_{RC}}\right] = \left[\left(\frac{\partial \dot{r}}{\partial \mathbf{x}_{RC}}\right)^T \left(\frac{\partial \Phi}{\partial \mathbf{x}_{RC}}\right)^T \left(\frac{\partial \Theta}{\partial \mathbf{x}_{RC}}\right)^T \left(\frac{\partial \Phi_E}{\partial \mathbf{x}_{RC}}\right)^T \left(\frac{\partial \Theta_E}{\partial \mathbf{x}_{RC}}\right)^T\right]^T \qquad (18)$$

During simulation it was found that $\mathbf{H}_{RC}$ had a condition number on the order of 10^{22} when the spacecraft was at the beginning of its trajectory, near the Earth. A condition number on the order of 10^{25} was found at the end of the trajectory, when the spacecraft was at a large distance from the Earth. The condition number of the matrix is the ratio of the largest and smallest singular values. The large condition number reflects the low observability of the system.

Orbital Element Measurements

Additionally, the radial velocity and line of sight measurements can be modeled as functions of the orbital elements. The radial velocity in terms of the orbital elements is one component of the hodograph plane and is given below.

$$\dot{r} = e\sqrt{\frac{\mu}{a\left(1-e^2\right)}}\sin f \qquad (19)$$

The line of sight from the spacecraft to the Sun depends on the normalized position vector in inertial coordinates. This vector can be found from the normalized position vector in the orbital plane and the rotation matrix from the orbital plane to the inertial coordinates.

$$\begin{bmatrix} \hat{x} \\ \hat{y} \\ \hat{z} \end{bmatrix} = [R(\Omega, i, \omega)] \begin{bmatrix} \cos f \\ \sin f \\ 0 \end{bmatrix} \qquad (20)$$

The line of sight to the Sun angles can then be found.

$$\Phi = \tan^{-1}\left(\frac{\hat{y}}{\hat{x}}\right) \quad , \quad \Theta = \sin^{-1}(-\hat{z}) \qquad (21)$$

From these models, the measurement sensitivity matrix, $\mathbf{H}_{OE}$, is found. Computing the partial derivatives of these measurements with respect to the orbital element states is a straightforward process, applying the chain rule to Eqs. (19), (20), and (21).

Finding the line of sight to Earth angles as functions of the orbital elements is slightly more complicated. This is done by writing the rectangular coordinates as functions of the orbital elements and using the same measurement models as in the previous section. Then, to find the partial derivatives of these angles with respect to the orbital elements the chain rule must be used.

$$\frac{\partial \Phi_E}{\partial \mathbf{x}_{OE}} = \frac{\partial \Phi_E}{\partial \mathbf{x}_{RC}} \frac{\partial \mathbf{x}_{RC}}{\partial \mathbf{x}_{OE}} \quad , \quad \frac{\partial \Theta_E}{\partial \mathbf{x}_{OE}} = \frac{\partial \Theta_E}{\partial \mathbf{x}_{RC}} \frac{\partial \mathbf{x}_{RC}}{\partial \mathbf{x}_{OE}} \tag{22}$$

To use these expressions the Jacobian of the rectangular coordinates relative to the orbital elements must be evaluated. The measurement sensitivity matrix can then be assembled. This matrix was found to have a condition number on the order of 10^{16} at the beginning of the simulated trajectory and 10^{17} at the end.

Measurement Accuracy

An important aspect of the current study was to evaluate the estimate accuracies achievable for various accuracies of the radial velocity and line of sight measurements. The estimation methods were evaluated for radial velocity accuracies of 1 m/s, 20 cm/s, and 1 cm/s and line of sight accuracies of 10^{-4} rad, 10^{-5} rad, and 10^{-6} rad. These line of sight accuracies roughly correspond with the accuracies of Sun sensors, star trackers, and next-generation star trackers, respectively. A radial velocity accuracy of 1 m/s requires measuring Doppler shifts to an accuracy of approximately 2×10^{-6} nm. An accuracy of 20 cm/s reflects the size of errors that can be caused by not accounting for the presence of sunspots.[7] An accuracy of 1 cm/s represents an upper goal of what can be achieved by Doppler shift measurements. The measurement errors were modeled as Gaussian white noise. The set of measurements was taken every 100 minutes over the entire trajectory.

ESTIMATION METHODS

Rectangular Coordinate Extended Kalman Filter

One method that has been evaluated for determining the orbit of the spacecraft is to use an Extended Kalman Filter to estimate the rectangular coordinate states (RCEKF). To implement this, the dynamics and measurement models described above in terms of the rectangular coordinates are used. The dynamics are propagated between measurement updates by numerically integrating the nonlinear model. The predicted error covariance is propagated between measurements by numerically integrating a Lyapunov differential equation.

$$\dot{\mathbf{P}} = \mathbf{FP} + \mathbf{PF}^{T} + \mathbf{GQG}^{T} \tag{23}$$

When new measurements are received from the sensors, the nonlinear measurement model is evaluated at the propagated states. A correction is then made to the propagated states based on the measurement residual.

$$\hat{\mathbf{x}}_{k}^{+} = \hat{\mathbf{x}}_{k}^{-} + K_{k}\left[\mathbf{y}_{k} - h\left(\hat{\mathbf{x}}_{k}^{-}\right)\right] \tag{24}$$

The Kalman gain is computed from the following.

$$\mathbf{K}_{k} = \mathbf{P}_{k}^{-}\mathbf{H}_{k}^{T}\left[\mathbf{H}_{k}\mathbf{P}_{k}^{-}\mathbf{H}_{k}^{T} + \mathbf{R}\right]^{-1} \tag{25}$$

At measurement update time steps, the predicted covariance is also updated.

$$\mathbf{P}_{k}^{+} = \left[\mathbf{I} - \mathbf{K}_{k}\mathbf{H}_{k}\right]\mathbf{P}_{k}^{-}\left[\mathbf{I} - \mathbf{K}_{k}\mathbf{H}_{k}\right]^{T} + \mathbf{K}_{k}\mathbf{R}\mathbf{K}_{k} \tag{26}$$

For the simulations performed in this study, the initial error covariance was based on the true error in the initial guess of the state estimates. The measurement noise was also matched to the noise added to the true measurement models. Because there were no unmodeled dynamics, the process noise was set to zero.

Orbital Element Extended Kalman Filter

An alternative method that has been developed is to use an Extended Kalman Filter to estimate the spacecraft's orbital elements (OEEKF). The operation of this filter is identical to the RCEKF, except that the dynamics, measurement models, and partial derivatives used are all in terms of the orbital elements. The same methods were also used to select the initial covariance, measurement noise, and process noise.

Gaussian Least Squares Differential Correction

The Gaussian Least Squares Differential Correction estimator (GLSDC) is essentially an $n \times m$ extension of Newton's root solving method. The estimation problem is recast to iteratively find the initial orbital conditions which produce a trajectory that best matches the true measurements taken over the path of the spacecraft. The measurements are weighted according to their accuracies, and no errors in the dynamical model are considered.

An initial guess for the initial conditions is propagated forward using the orbital dynamics model. Simultaneously, the measurement model and the measurement sensitivity matrix are evaluated along the propagated trajectory. Here though, the measurement sensitivity matrix gives the partial derivative of the measurements at each time step with respect to the initial condition vector.

$$\mathbf{H}_k = \frac{\partial \mathbf{y}(t_k)}{\partial \mathbf{x}(t_0)} = \frac{\partial \mathbf{y}(t_k)}{\partial \mathbf{x}(t_k)} \Phi(t_k, t_0) \tag{27}$$

The state transition matrix is propagated as follows.

$$\dot{\Phi}(t_k, t_0) = \frac{\partial \mathbf{f}_k}{\partial \mathbf{x}_k} \Phi(t_k, t_0) \quad , \quad \Phi(t_0, t_0) = \mathbf{I} \tag{28}$$

The measurement residuals are calculated at each time step.

$$\Delta \mathbf{y}_k = \mathbf{y}_k - h(\hat{\mathbf{x}}_k) \tag{29}$$

The residuals and measurement sensitivity matrices from each time step are then assembled into global matrices.

$$[\Delta \mathbf{Y}] = \begin{bmatrix} \Delta \mathbf{y}_1 \\ \Delta \mathbf{y}_2 \\ \vdots \\ \Delta \mathbf{y}_n \end{bmatrix} \quad , \quad [\mathbf{H}] = \begin{bmatrix} \mathbf{H}_1 \\ \mathbf{H}_2 \\ \vdots \\ \mathbf{H}_n \end{bmatrix} \tag{30}$$

The global weight matrix is based on the covariance matrix for the set of measurements taken at each time step.

$$[\mathbf{W}] = \begin{bmatrix} \mathbf{R}_1^{-1} & \mathbf{0} & \cdots & \mathbf{0} \\ \mathbf{0} & \mathbf{R}_2^{-1} & \cdots & \mathbf{0} \\ \vdots & \vdots & \ddots & \vdots \\ \mathbf{0} & \mathbf{0} & \cdots & \mathbf{R}_n^{-1} \end{bmatrix} \tag{31}$$

A numerically accurate method for calculating the corrections to the initial condition estimate is given by the following.

$$\Delta \mathbf{x}(t_0) = pinv\left(\sqrt{\mathbf{W}}\mathbf{H}\right)\sqrt{\mathbf{W}}\Delta \mathbf{Y} \tag{32}$$

The new estimate for the starting guess is then computed.

$$\hat{\mathbf{x}}_{l+1}(t_0) = \hat{\mathbf{x}}_l(t_0) + \Delta \mathbf{x}(t_0) \tag{33}$$

The process is then repeated using the new initial condition estimate until the cost function, $\mathbf{J}_l$, converges to satisfy $\left|\dfrac{\mathbf{J}_l - \mathbf{J}_{l-1}}{\mathbf{J}_l}\right| < 10^{-6}$, with

$$\mathbf{J}_l = \left(\sqrt{\mathbf{W}}\Delta\mathbf{Y}\right)^T \sqrt{\mathbf{W}}\Delta\mathbf{Y} \tag{34}$$

While a predicted error covariance is not explicitly involved in the estimation process of the GLSDC, it is simple to post-compute this information. The covariance of the measurements at each time step is mapped into the states at the initial time.

$$\mathbf{P}_0 = \left(\mathbf{H}^T\mathbf{W}\mathbf{H}\right)^{-1} \tag{35}$$

The initial covariance can then be mapped into the covariance at any time step.

$$\mathbf{P}(t_k) = \Phi(t_k, t_0)\mathbf{P}_0\Phi^T(t_k, t_0) \tag{36}$$

Similar to the Extended Kalman Filter, the GLSDC can be implemented with either rectangular coordinate or orbital element representations. For the rectangular coordinate implementation, $\dfrac{\partial\mathbf{y}(t_k)}{\partial\mathbf{x}_{RC}(t_k)}$ and $\dfrac{\partial\mathbf{f}}{\partial\mathbf{x}_{RC}}$ are used to compute the measurement sensitivity matrix and propagate the state transition matrix. The orbital element implementation uses $\dfrac{\partial\mathbf{y}(t_k)}{\partial\mathbf{x}_{OE}(t_k)}$ and $\dfrac{\partial\mathbf{f}}{\partial\mathbf{x}_{OE}}$.

RESULTS

In order to evaluate the different estimation methods, simulations of the system were performed. There are several possible methods for comparing the accuracy of the estimators. While they provide state estimates in two different representations, the rectangular coordinates can be converted to orbital elements and vice versa. This allows computation of the sample error covariance resulting from the error between the estimated states and the true states in either state representation for both filters.

Additionally, the predicted error covariances provided by each estimation method corresponds with the state representation of the estimator. These predicited covariances, though, can also be converted to provide a predicted error covariance for the other state representation using the Jacobian of the two state representations.

$$\mathbf{J} \equiv \frac{\partial\mathbf{x}_{RC}}{\partial\mathbf{x}_{OE}} \tag{37}$$

The transformation of the covariance uses a linear mapping. In order for the mapping to be computed in practice, the linearization is evaluated at the estimated state, not the true state.

$$\mathbf{P}_{RC} = \mathbf{J}\mathbf{P}_{OE}\mathbf{J}^{T} \quad , \quad \mathbf{P}_{OE} = \mathbf{J}^{-1}\mathbf{P}_{RC}\mathbf{J}^{-T} \tag{38}$$

An alternative to using the inverse formulation to convert from rectangular coordinate to orbital element covariances is to directly compute the Jacobian of the orbital elements relative to the rectangular coordinates. However, this was not done here.

Another option for comparing the accuracies of the estimation processes is to look at the covariances, either predicted or sampled, as averaged over the steady-state or at the final time. The former gives information about the accuracy of the estimation over the entire trajectory, while the latter would be important for a vehicle that is intended to rendezvous with a celestial body or another spacecraft at the final time.

Figures 3 through 10 show examples of the simulations that were conducted. The case with radial velocity measurement accuracy of 20 cm/s and line of sight accuracy of 10^{-5} rad is shown. The figures represent the time histories of the errors in the state estimates produced by the RCEKF and the rectangular coordinate GLSDC. The OEEKF and the orbital element GLSDC produced results nearly identical to the rectangular coordinate versions of the two methods and are not shown. The errors are shown in both rectangular coordinate states (Figures 3 through 6) and the orbital elements (Figures 7 through 10). Also shown is a 3σ-bound computed from the predicted error covariances.

The plots show how the Kalman Filter updates the estimate at each time step to match the measurements, creating errors that jump in time. However, the GLSDC only uses the dynamics model to propagate the measurements. Therefore the errors vary smoothly with time (and are constant for the first five orbital elements). Also, the figures depicting the rectangular coordinate errors show how the accuracy varies as the spacecraft moves through its trajectory. At different points in the orbit, the measurements vary in their sensitivity to the X and Y coordinates.

Additionally, the errors tend to grow as the spacecraft moves along its trajectory. As time passes, the spacecraft moves away from both the Earth and the Sun. Therefore, while the angular accuracy of these line of sight measurements remains constant, the value of the information they provide about the spacecraft position is degraded. In general the standard deviations in the true anomaly and the position at the final time were approximately 1.5 to 2 times greater than the steady-state average. This indicates that using a nearby body as a line of sight reference provides better estimation than a distant body. For a spacecraft such as the Mars Global Surveyor, that is intended to rendezvous with another planet, using that planet as a reference would be preferable.

Figures 11 through 14 show the steady-state standard deviations computed from the predicted covariances for the RCEKF and rectangular coordinate GLSDC. Again, the

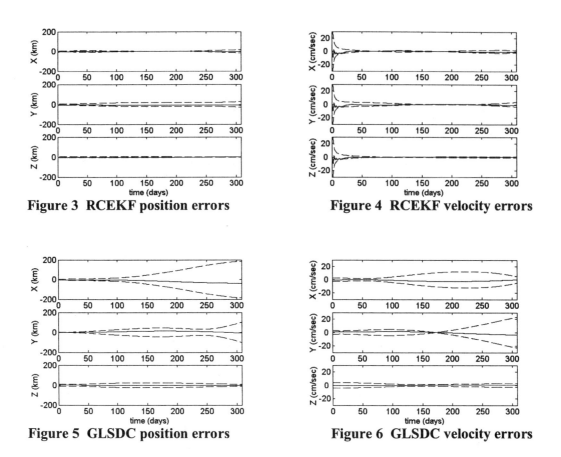

Figure 3 RCEKF position errors

Figure 4 RCEKF velocity errors

Figure 5 GLSDC position errors

Figure 6 GLSDC velocity errors

results for the OEEKF and the orbital element GLSDC were nearly identical to the rectangular coordinate versions and are not shown. The results are shown for standard deviations in both rectangular coordinates and orbital elements. Each plot depicts the standard deviation predicted in a particular state for the various measurement accuracies.

Several trends can be seen in these data. The first trend is the reduction of standard deviation size with the improvement in measurement accuracy as expected. This trend is apparent in both the RCEKF and GLSDC and in both the rectangular coordinate states and the orbital elements.

Another trend dealing with the estimation accuracy achieved for various measurement accuracies has to do with the improvements seen while holding one measurement accuracy constant and improving the accuracy of the second measurement type. If the radial velocity measurement accuracy is held constant at 1 cm/s, then the improvement in estimation seen by improving line of sight accuracy from 10^{-4} rad to 10^{-5} rad to 10^{-6} rad is not as great as the improvement seen if the radial velocity accuracy is

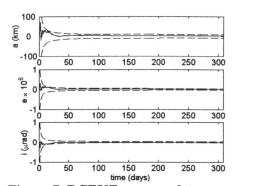

Figure 7 RCEKF a **,** e **, and** i **errors**

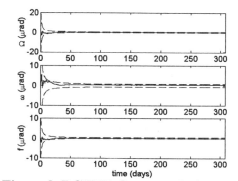

Figure 8 RCEKF Ω **,** ω **, and** f **errors**

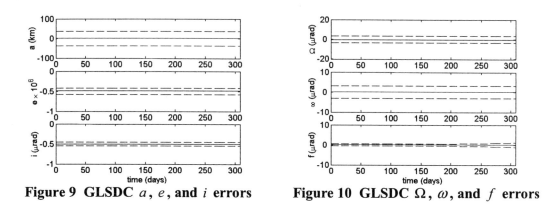

Figure 9 GLSDC a **,** e **, and** i **errors**   **Figure 10 GLSDC** Ω **,** ω **, and** f **errors**

held at 1 m/s. The converse is also true for holding line of sight accuracy constant at 10^{-6} rad versus 10^{-4} rad while varying radial velocity measurement accuracy. This trend is also expected because the accuracy of one measurement type is not as critical if the other measurement type is providing very accurate information.

Finally, the results of the Extended Kalman Filter can be compared to the GLSDC. The GLSDC produced estimates comparable to the RCEKF for the cases with very accurate measurements. But for cases with low accuracy measurements, the GLSDC gave higher standard deviations than the RCEKF. This trend is repeated when looking at either the orbital elements or rectangular coordinates.

While not shown, the final-time standard deviations followed the same trends. And while detailed Monte Carlo simulation has not been performed, the standard deviations from the sampled covariances agree well with the predicted covariances and also follow the same trends.

CONCLUSIONS

The analysis of the Extended Kalman Filter and the Gaussian Least Square Differential Correction show that the measurement of radial velocity to the Sun can be a useful source of information in estimating the spacecraft state. Developing a spectrometer instrument to measure solar Doppler shift to the accuracies necessary to determine radial velocity to within 1 cm/s, 20 cm/s, or 1 m/s would be a beneficial development for interplanetary navigation.

While this paper compared the Kalman Filter and GLSDC for various measurement accuracies, an important factor in evaluating these methods is the accuracy of the dynamical model that can be incorporated in the estimation process. In this study no dynamical model errors were considered. Though there are several significant perturbation to the dynamics considered here, many of these are also well modeled. Therefore, future work could include more sophisticated dynamics models in the estimation process.

Additionally, while not shown here it was found that the Kalman Filter exhibits a preferential coordinate system when process noise is included. This indicates that for estimation with the Kalman Filter it may be preferable to operate as closely as possible on the desired information, rather than estimating some alternate states and then transforming the estimate via post-computation. An area of future work is to further investigate this coordinate system dependence.

Finally, the utility of solar radial velocity measurements for spacecraft navigation has been further demonstrated. The accuracy of the state estimates that can be achieved for various measurement accuracies has been shown. This motivates continuing development in the design of an instrument to produce these measurements. A critical aspect of this development will be to determine the accuracy that can be achieved with the FPI. Also, establishing the cost and complexity associated with achieving the measurement accuracy will be important. This will allow design choices to be made on the use of solar Doppler shift for spacecraft navigation.

ACKNOWLEDGMENTS

The authors are pleased to acknowledge the support of the State of Texas Advanced Technology Program. We are also pleased to acknowledge our productive interaction with Dr. J. Yim whose previous work on related orbit determination problems was very useful.

REFERENCES

1. Nave, R., "Fabry-Perot Interferometer", http://hyperphysics.phy-astr.gsu.edu/hbase/phyopt/fabry.html

2. Born, M. and Wolf, E., *Principles of Optics: Electromagnetic Theory of Propagation, Interference and Diffraction of Light,* Pergamon Press, New York, 1965.

3. Vaughn, J. M., *The Fabry-Perot Interferometer: History, Theory, Practice, and Applications,* Institute of Physics, Philadelphia, 1989.

4. Guo, Y. P., "Self-contained Autonomous Navigation System for Deep Space Mission," *AAS/AIAA Space Flight Mechanics Meeting,* Breckenridge, CO, Feb. 1999, AAS-99-177.

5. Yim, J. R., Crassidis, J. L., and Junkins, J. L., "Autonomous Orbit Navigation of Interplanetary Spacecraft," *AIAA Astrodynamics Specialist Conference*, Denver, CO, Aug. 2000, AIAA Paper 2000-3936.

6. Yim, J. R., *Autonomous Spacecraft Orbit Navigation*, Ph.D. dissertation, Texas A&M University, 2002.

7. Gabriel, A. H., "Global Oscillations at Low Frequency from the SOHO Mission (GOLF)," Solar Physics, Vol. 162, 1995, pp. 61-99.

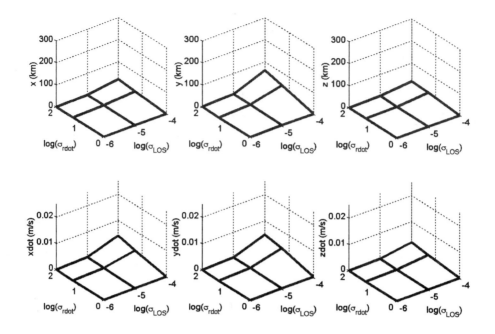

Figure 11 RCEKF predicted standard deviations for rectangular coordinates

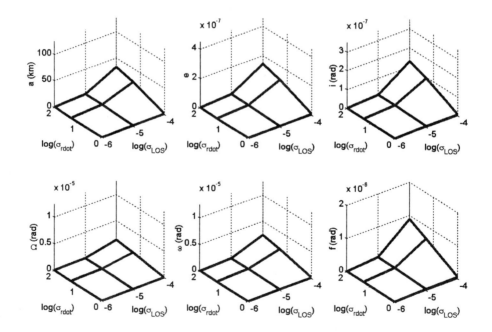

Figure 12 RCEKF predicted standard deviations for orbital elements

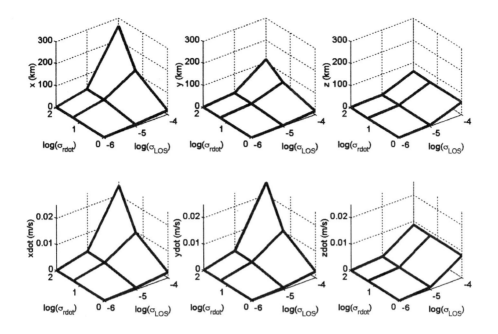

Figure 13 GLSDC predicted standard deviations for rectangular coordinates

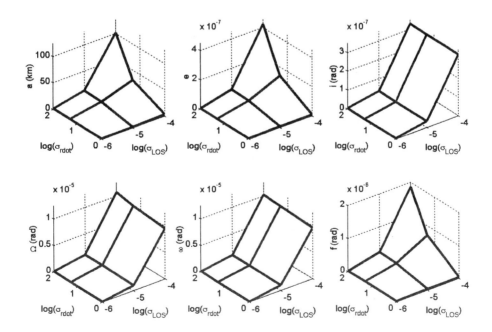

Figure 14 GLSDC predicted standard deviations for orbital elements

MUTIBODY PARACHUTE FLIGHT SIMULATIONS FOR PLANETARY ENTRY TRAJECTORIES USING "EQUILIBRIUM POINTS"

Ben Raiszadeh[*]

A method has been developed to reduce numerical stiffness and computer CPU requirements of high fidelity multibody flight simulations involving parachutes for planetary entry trajectories. Typical parachute entry configurations consist of entry bodies suspended from a parachute, connected by flexible lines. To accurately calculate line forces and moments, the simulations need to keep track of the point where the flexible lines meet (confluence point). In previous multibody parachute flight simulations, the confluence point has been modeled as a point mass. Using a point mass for the confluence point tends to make the simulation numerically stiff, because its mass is typically much less that than the main rigid body masses. One solution for stiff differential equations is to use a very small integration time step. However, this results in large computer CPU requirements. In the method described in the paper, the need for using a mass as the confluence point has been eliminated. Instead, the confluence point is modeled using an "equilibrium point". This point is calculated at every integration step as the point at which sum of all line forces is zero (static equilibrium). The use of this "equilibrium point" has the advantage of both reducing the numerical stiffness of the simulations, and eliminating the dynamical equations associated with vibration of a lumped mass on a high-tension string.

INTRODUCTION

Many planetary entry systems employ a parachute system as a decelerator device from supersonic to subsonic velocities. Simulating the parachute portion of the planetary entry often involves modeling the parachute and the suspended bodies as individually at-

---

* NASA Langley Research Center, Hampton, Virginia 23681-2199. E-mail: b.raiszadeh@larc.nasa.gov.

tached rigid Six-Degree-Of-Freedom (6 DOF) bodies. The parachute and the suspended bodies are connected by flexible lines on these planetary entry configurations. In this model, the suspension lines meet at a confluence point. The Mars Exploration Rover (MER) mission uses a parachute system to slow down its descent and after chute deployment has one such confluence point (Fig. 1). In current MER simulations, confluence points are modeled using Three-Degree-Of-Freedom (3 DOF) point mass bodies. Due to its small mass, the confluence point is subject to rapid accelerations. The rapid accelerations make the simulations numerically stiff, requiring very small time increments on the order of 0.0001 second. This causes the simulation run times to significantly increase.[1] This is undesirable in Monte-Carlo simulations where fast turn-around is required to support trade studies. In previous multibody entry models,[1-3] line masses are lumped into the confluence points. In reality, this mass is distributed evenly along the suspension lines. Therefore, the vibrational modes will be those of a distributed mass system, as in a vibrating string. Thus, the vibrational modes introduced by the confluence point mass may not be realistic.

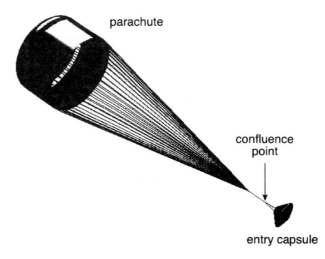

Figure 1   Vehicle entry configuration.

This paper outlines a method to eliminate confluence point masses, and introduces the "equilibrium point". The "equilibrium point" is defined as the point at which the sum of the line forces is zero, the point where the line forces are in equilibrium. This has the same effect as using a confluence point with an infinitely small mass, and allowing all the transients to die out. Instead of integrating the confluence point mass, the "equilibrium point" location is calculated at every time step. The solution is then used to calculate the line forces. This allows the simulations to run at a larger time step on the order of 0.01 second (100 times larger than without), and the high frequency vibrations caused by the small confluence mass are eliminated. This also reduces the number of differential equations to integrate.

SYMBOLS AND ABBREVIATIONS

MER	Mars Exploration Rover
POST	Program to Optimize Simulated trajectories
DOF	Degree of Freedom
$\hat{p}_i$	Position vector of the attach point per line
$\hat{p}_{ep}$	Position vector of the equilibrium point
e_i	Line strain per line
$\dot{e}_i$	Line strain rate per line
L_{0i}	Free length per line
$\hat{u}_i$	Unit vector in direction of the line
f_i	Magnitude of line force per line
$\hat{f}_i$	Line force vector per line
$\hat{f}_{ki}$	Line force vector due to stiffness per line
$\hat{f}_{ci}$	Line force vector due to damping per line
L_t	Line length in current time step
L_{t-1}	Line length in previous time step
$[x_i \quad y_i \quad z_i]$	Attach point coordinates
$[x_{ep} \quad y_{ep} \quad z_{ep}]$	Equilibrium point coordinates
Ixx	Moment of inertia about the roll axis
Iyy	Moment of inertia about the pitch axis
Izz	Moment of inertia about the yaw axis
C_D	Drag coefficient
Cp	Center of pressure
S_{ref}	Reference surface area
Ω	Mars gravitational constant
R_e	Mars equatorial radius
R_p	Mars polar radius
ω	Mars rotation rate
J	Mars gravity zonal harmonics

BACKGROUND

The underlying simulation used to test the "equilibrium points" concept is the Program to Optimize Simulated Trajectories II (POST II). POST II is the latest major upgrade to POST.[4] POST was originally developed for the Space Shuttle program to optimize ascent and entry trajectories. Over the years it has been upgraded and improved to include many new capabilities. POST II relies on most of the technical elements established by POST, but the executive structure has been reworked to take advantage of today's faster computers. The new executive routines allow POST II to simulate multiple bodies simultaneously, and to mix Three-Degree-Of-Freedom (3DOF) bodies with Six-Degree-Of-Freedom (6DOF) bodies in a single simulation. The already established and verified multiple body capability using confluence point masses allows POST II to simulate parachutes. This is done by connecting the spacecraft and the parachute by massless spring-dampers. The springs can be attached at any point on the body. No moments are applied except those due to force application away from the center of mass. Each line connects an attach point on one body to an attach point on another body and provides a tension-only force. When the lines are stretched, tension in the lines is determined as function of strain and strain rate. In the current multiple body POST II, the differential equations of motion for all bodies are explicitly integrated numerically, and the line forces are calculated based on relative position and velocity of the bodies attached.

APPROACH

First, the algorithm to calculate the "equilibrium point" is described. This algorithm needs to be robust since it has to solve the "equilibrium point" at every time step, and also be able to account for special cases, such as when the lines are slack. Although this capability has been added to POST II, it could be used with any multiple body simulation programs such as ADAMS.[5] To verify the current multiple body capability of POST II (without "equilibrium points"), a series of more than 35 tests of increasing complexity were performed. These tests were intended to prove that the POST II model was implemented correctly by evaluating its performance on problems that could be verified by other means. The test cases started with a simple vertical drop from rest of the fully deployed parachute and entry capsule and gradually increased in complexity to include parachute opening, non-zero initial conditions, line deployments, wind gusts, and other effects. The POST II model was compared to both MATLAB-based and ADAMS-based multi-degree-of-freedom simulations. In each case the agreement between the simulations was excellent. These tests have validated the general parachute model within POST II. Some of the validation is reported more fully in Reference 1. To verify the method using "equilibrium point" some of the same test cases are used, except the confluence point mass is replaced by the "equilibrium point", and the results are compared.

ALGORITHM

The "equilibrium point" is defined as the point at which the sum of all line forces is zero. Note that in the general case an arbitrary number of lines can be connected at the equilibrium point. In this algorithm the "equilibrium point" position is the unknown.

$$\hat{p}_{ep} = [x_{ep} \quad y_{ep} \quad z_{ep}]$$

The position of the attach point is fixed in each time step.

$$\hat{p}_i = [x_i \quad y_i \quad z_i]$$

Line strain is found by subtracting the position vectors, calculating the magnitude, and dividing by free length. The force is in the direction of the line connecting the end points.

$$e_i = \frac{|p_i - p_{ep}| - L_{0i}}{L_{0i}}$$

The unit vector in the direction of the line force is found by subtracting the position vectors and dividing by the magnitude.

$$\hat{u}_i = (\hat{p}_i - \hat{p}_{ep}) / |\hat{p}_i - \hat{p}_{ep}|$$

Force vector due to stiffness is found by the following equation:

$$\hat{f}_{ki} = L_{0i} K e_i \hat{u}_i$$

In calculating the damping force we do not have direct access to velocity of the "equilibrium point" in inertial space. Thus strain rate of the line is not readily available. Instead, the strain rate is approximated numerically by subtracting the line length from the previous time step and dividing by time increment. The direction of damping force is along the line, and opposite to the strain rate of the line.

$$\dot{e}_i = \frac{\dfrac{L_{t-1} - L_t}{dt}}{L_{0i}}$$

Force due to damping is then found using the following relationship:

$$\hat{f}_{ci} = L_{0i} C_i \dot{e}_i \hat{u}_i$$

The net line force is the sum of force due to stiffness and damping force

$$\hat{f}_i = L_{0i} K_i e_i \hat{u}_i + L_{0i} C_i \dot{e}_i \hat{u}_i$$

$$\hat{f}_i = L_{0i} (K_i e_i + C_i \dot{e}_i) \hat{u}_i$$

At the "equilibrium point" sum of all the line forces is equal to zero.

$$\sum_{i=1}^{n} \hat{f}_i = 0$$

At the initial guess for the "equilibrium point" there will most likely be a non—zero resultant force vector. The objective is to find a unique location in space where the summation all line forces is zero. The only variable here is the position of the "equilibrium point", and the output is the resultant force vector. This problem can be solved using one of many available methods for solving system of equations. Here, the Newton method is chosen to solve this system. In summary, the Newton method uses the slope of the output variable (in this case force vector), by varying the input variable (in this case the "equilibrium point" position vector) to converge to a point where the output is zero (static equilibrium).

RESULTS

As mentioned earlier, some of the same test cases used previously to verify multiple body capability of POST are used again to evaluate the "equilibrium point" solution. For a detailed description of the test cases see Reference 1. Test cases 2 and 3 are chosen for this study. Comparison is made between the solution using the "equilibrium position", and the solution utilizing the confluence point mass. All test cases are performed in a Mars environment. But this method can be applied to parachute entry simulations on any planet. A constant atmospheric density of 0.0135 kg/m^3 is assumed for all runs. Aerodynamic drag acts on the parachute only. Mars gravity and an oblate planet model have been used. The planet is assumed to be non-rotating. All simulations start at zero latitude and zero longitude at a height of approximated 8.4 kilometers. Tables 1, 2 and 3 summarize the inputs used.

Table 1
LINE PROPERTIES

Line	Parameter	Value	
Single riser	L_0	1.832	m
	K	60,000	N/m
	C	600	N/(m/s)
Triple risers	L_0	0.71524	m
	K	47,000	N/m
	C	470	N/(m/s)

Table 2
PLANET MODEL

Parameter	Value	
Ω	4.2828286853e^{13}	m^3/s^2
R_c	3.393940e^6	m
R_p	3.376780e^6	m
ω	0.0	rad/s
J terms	0.0	

Table 3
PARACHUTE AND ENTRY CAPSULE INPUTS

Body	Parameter	Value	
Parachute	DOF	6	
	Mass	16.0	kg
	Ixx	253.7	kg.m^2
	Iyy	1126.5	kg.m^2
	Izz	1126.5	kg.m^2
	C_D	0.46	
	Cp	1.57	m
	S_{ref}	178.47	m^2
Backshell/lander	DOF	6	
	Mass	761	kg
	Ixx	238.02	kg.m^2
	Iyy	179.13	kg.m^2
	Izz	212.51	kg.m^2

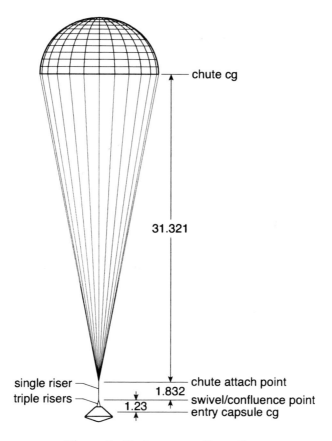

Figure 2 Test cases configuration.
Note: Dimensions are typical (not based on any specific Mars mission).

Test case 2 is repeated here with the "equilibrium point" and results are compared. In this test case a parachute system is dropped from rest. To introduce dynamics into the system we included a slack of one centimeter in the Single Riser (see Figure 2). Note that the aerodynamic drag acts on the parachute only. The entry capsule initially drops faster than the parachute. Eventually, the single riser runs out of slack, thus exciting the system. In this simulation it takes the system about sixty seconds to reach terminal velocity. It takes approximately two seconds for vibrational dynamics to damp out. Figure 3a is the plot of force in the Single Riser. Note that the simulation started with a one-centimeter slack in the Single Riser. It takes about 0.7 second for the slack to run out. The entry bodies then undergo an oscillatory motion. The oscillations damp out approximately 0.5 second after they start (Figure 3b). After this point, the line force gradually builds up until it reaches a steady state value. Note that in Figure 3a the curves are indistinguishable. In Figure 3b the differences become more visible. In the solution with the "equilibrium point", an integration time step of 0.01 second was used as opposed to 0.0001 second for the simulation using confluence point mass.

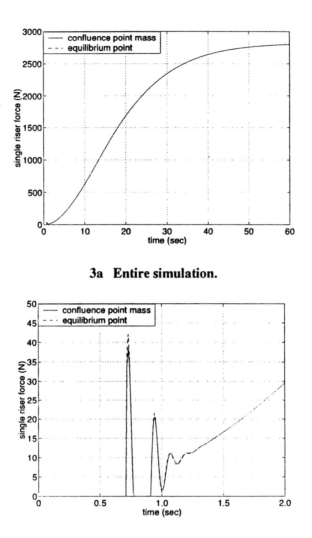

3a Entire simulation.

3b Initial two seconds.

Figure 3 Single Riser force.

An interesting set of results is obtained by zeroing out line damping, and leaving all other inputs unchanged. The comparison of the Single Riser force is presented in Figure 4. In Figure 4a, the overall force-time history of the line appears to produce a good comparison. But a closer inspection shows that the model with confluence point mass has an additional high frequency mode (Figures 4b and 4c). This is caused by the confluence point mass vibrating back-and-forth in between the lines in tension. This example illustrate how the confluence point mass tends to introduce unwanted frequency contents into the simulation, and how it can be eliminated using the "equilibrium point".

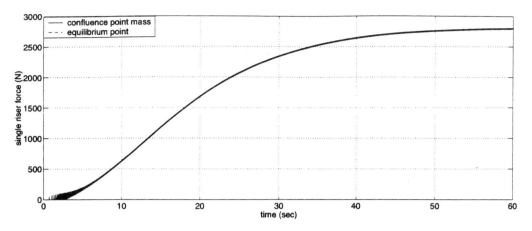

4a   Entire simulation.

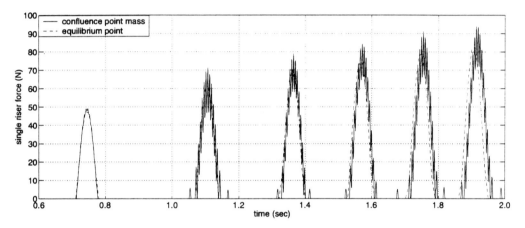

4b   Initial two seconds.

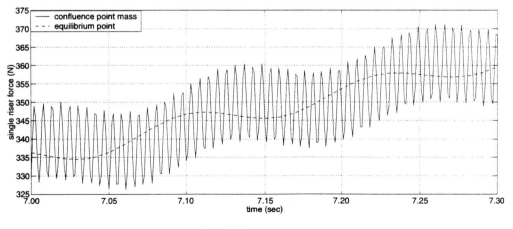

4c   Midway range.

Figure 4   Single Riser force, no line damping.

The next test case is identical to the previous with the exception that the entry capsule is given an initial horizontal velocity of 1 m/s. In this test case, more degrees of freedom are excited, as opposed to the previous test case where the bodies were excited in the vertical direction only. Figure 5 shows the plot of the Single Riser line force as a function of time. The overall force curve shows good agreement (Figure 5a), but when observed in detail the model using confluence point mass has additional frequency contents caused by the point mass (Figure 5b). The simulation with confluence point mass ran using integration time step of 0.0001 second, and the simulation using "equilibrium point" used an integration time step of 0.01 second, resulting in substantial savings in run time (at least by a factor of ten).

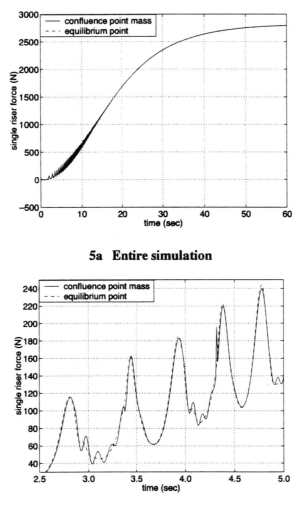

5a Entire simulation

5b Midway range

Figure 5 Single Riser, non-zero initial conditions

CONCLUSION

Modeling the confluence points using a point mass is a numerical challenge for all the multibody parachute entry simulations (past and present). The method described in this paper reduces numerical stiffness associated with modeling the confluence point as a point mass. This method also eliminates unwanted high frequency oscillations caused by the confluence point mass. Reduced stiffness is desirable because it allows larger integration time steps which should lead to shorter run times. Also, by eliminating the confluence point mass, the simulation has fewer equations of motion to integrate. These speed enhancements are somewhat offset by the need to numerically solve for the "equilibrium point" position. Effectiveness of this scheme can be improved by refining numerical techniques used to solve for the "equilibrium point."

REFERENCES

1. Ben Raiszadeh, Eric M. Queen, Partial Validation of Multibody Program to Optimize Simulated Trajectories II (POST II) Parachute Simulation With Interacting Forces, NASA/TM-2002-211634, April 2002

2. Eric M. Queen, Ben Raiszadeh, Mars Smart Lander Parachute Simulation Model, AIAA Paper 2002-4616, August 2002

3. Kenneth S. Smith, Chia-Yen Peng, Ali Behboud, Multibody Dynamic Simulation of Mars Pathfinder Entry, Descent and Landing, JPL D-13298, April 1995

4. Program to Optimize Simulated Trajectories: Volume II, Utilization Manual, prepared by: R.W. Powell, S.A. Striepe, P.N. Desai, P.V. Tartabini, E.M. Queen; NASA Langley Research Center, and by: G.L. Brauer, D.E. Cornick, D.W. Olson, F.M. Petersen, R. Stevenson, M.C. Engel, S.M. Marsh; Lockheed Martin Corporation, Version 1.1.1.G, May 2000

5. ADAMS, Software Package for Simulating Force and Motion Behavior of Mechanical System, Property of Mechanical Dynamics Inc.

ORBIT DETERMINATION

Chair:
Paul Cefola
Massachusetts Institute of Technology

The following papers were not available for publication:

AAS 03-168

"Preliminary Validation of Atmospheric Neutral Density Derived form Ultraviolet Airglow Observations," by S.E. Thonnard, A.C. Nicholas, J.M. Picone, K.F. Dymond, S.A. Budzien, Naval Research Laboratory; S.H. Knowles, Raytheon Technical Corp.; E.E. Henderlight, Praxis, Inc.; R.P. McCoy, Office of Naval Research (Paper Withdrawn)

AAS 03-170

"Adaptive Stepsize Control for a Family of Chebyshev Methods for the Integration of First-Order Ordinary Differential Equations," by J.W. Mitchell and D.L. Richardson (Paper Withdrawn)

REAL-TIME ESTIMATION OF
LOCAL ATMOSPHERIC DENSITY[*]

James R. Wright[†]

A method is presented for the real-time estimation of atmospheric density, locally along the trajectory of a low altitude spacecraft. Atmospheric density is estimated simultaneously with six parameters of the spacecraft orbit, and with other parameters that are both observable and unknown. Atmospheric density is derived by estimating local corrections directly to a global atmospheric density model. These corrections are derived, in part, from real-time range and/or Doppler tracking data. They are also derived from $F_{10.7}$ and a_p measurement values. $F_{10.7}$ and a_p measurements are used conventionally to drive the global atmospheric density model. But they are also employed by two new *stochastic* atmospheric density *error* models: A baseline error model and a dynamic error model. The baseline error model is derived from the historical record of $F_{10.7}$ and a_p measurements across multiple eleven year solar cycles. The dynamic error model is an extension to the baseline model, and is derived from current $F_{10.7}$ and a_p measurements. This provides a new physical connection between the physics of atmospheric density and atmospheric density estimation. *Real-time* here means that the timelag for estimation of the local atmospheric density is less than one second following the arrival of new range and/or Doppler tracking data. Atmospheric density estimation is demonstrated with real LEO tracking data acquired during July 2000 - at solar maximum.

1 Introduction

A method is presented for the real-time estimation of atmospheric density, locally along the trajectory of a low altitude spacecraft. Atmospheric density is estimated simultaneously with six parameters of the spacecraft orbit, and with other parameters that are both observable and unknown. Atmospheric density is derived by estimating local atmospheric density corrections to a global atmospheric density model. These corrections are derived, in part, from real-time range and/or Doppler tracking data. They are also derived from $F_{10.7}$ and a_P measurement values. $F_{10.7}$ and a_P measurements are used conventionally to drive the global atmospheric density model. But they are also employed here by two new *stochastic* atmospheric density *error* models: A baseline error model and a dynamic error model.

Let ρ denote atmospheric density, and $\bar{\rho}$ an estimate of ρ derived from a global a priori atmospheric density model. Define $\Delta\rho = \rho - \bar{\rho}$ the error in $\bar{\rho}$, $\Delta\hat{\rho}$ a real-time estimate of $\Delta\rho$, and $\hat{D} = \Delta\hat{\rho}/\bar{\rho}$.

1.1 Baseline Atmospheric Density Error Model

The baseline error model is derived from the *historical record* of F_{10} and a_P measurements across multiple eleven year solar cycles. An exponential Gauss-Markov sequence is employed to propagate relative atmospheric density error estimates $\hat{D}$ at perigee height, and to add appropriate baseline

---

† Senior Engineer, Analytical Graphics, Inc., 40 General Warren Blvd., Malvern, Pennsylvania 19355.

error process noise variance $q_{\Delta\rho/\bar{\rho}}$ for propagations during quiet solar weather. In the absence of measurements the relative atmospheric density error variance $\sigma^2_{\Delta\rho/\bar{\rho}}$ is a time constant (stationary). But during the processing of each observable measurement, $\sigma^2_{\Delta\rho/\bar{\rho}}$ is reduced (i.e., $\sigma^2_{\Delta\rho/\bar{\rho}}$ is non-stationary during measurement processing).

A transformation is defined to relate the atmospheric density error estimate at perigee height to that at current spacecraft height.

1.2 Dynamic Atmospheric Density Error Model

The dynamic error model is an extension to the baseline model. It simultaneously invokes the atmospheric density due to *current* $F_{10.7}$ and a_P measurements, and the atmospheric density due to global *mean* values of $F_{10.7}$ and a_P measurements, to form a scaling ratio R. The scaling ratio is used to boost the baseline error variance $\sigma^2_{\Delta\rho/\bar{\rho}}$ to derive $Q_{\Delta\rho/\bar{\rho}} = R^2 \sigma^2_{\Delta\rho/\bar{\rho}}$ dynamically. With every significantly increasing sequence of atmospheric density $\bar{\rho}$ at perigee height, the relative atmospheric density error $\Delta\rho/\bar{\rho}$ is modeled as a modified random walk sequence. When the increasing sequence is terminated, the relative atmospheric density error variance propagation returns to baseline Gauss-Markov propagation. The rate of return is defined by the density of range and/or Doppler tracking measurements processed, and by exponential half-life of the Gauss-Markov sequence. This provides a new physical connection between the physics of atmospheric density and atmospheric density estimation.

The dynamic error model is particularly useful during solar maximum with volitile increasing measurements of $F_{10.7}$ and a_P that predict associated increases in atmospheric density and atmospheric density error magnitude. This enables the error process noise variance $Q_{\Delta\rho/\bar{\rho}}$, and the sequential filter gain on atmospheric density error $\Delta\rho/\bar{\rho}$, to be opened appropriately at exactly the right time.

1.3 Real-Time

The new method is distinguished from existing and previous methods in that atmospheric density error correction $\Delta\hat{\rho}/\bar{\rho}$ is estimated directly in *real-time*, in association with optimal sequential orbit determination [11]. *Real-time* here means that the time-lag for estimation of the local atmospheric density is less than one second following the arrival of new tracking data[1]. This performance is, of course, impossible using batch least squares techniques.

1.4 Real Data Results

The Bastile Day geomagnetic storm began with a solar flare and solar coronal mass ejection (CME) on 14 July 2000 that was observed by the SOHO spacecraft coronagraph (J. Burch[15]). The CME slammed into the earth late on 15 July 2000, accompanied by a sharp decrease (-300 nanoteslas) in the strength of the geomagnetic field at the earth's surface. And a_P rose to 400 (max a_P). We acquired real range tracking data covering the interval from 1 July 2000 through 26 July 2000 for two near-circular LEO spacecraft $LEOh497$ and $LEOh780$. Spacecraft heights were roughly $497km$ and $780km$ respectively. According to Tom Gidlund [18], the estimation of ballistic coefficient time-constants using least squares orbit determination, for these LEOs, presented significant orbit determination problems during that time at solar maximum.

The results of both sequential filtering and sequential smoothing are presented for time-varying atmospheric density estimation for $LEOh497$ and $LEOh780$. The times of significant corrections to atmospheric density, due to both direct solar radiation and geomagnetic disturbances, are shown to

---

[1]The term *near real-time* has been used recently to refer to a time-lag of up to eight hours. Were it not for this, I would have used *near real-time* in place of *real-time*.

be in agreement with the effective lags, 1.70^d for $F_{10.7}$ and 0.279^d for K_P, expected by the Jacchia 1971 (J71) global atmospheric density model[1].

1.5 Minimization of Orbit Error Magnitudes

The new method should prove to be useful for minimization of orbit error magnitudes for real-time and predicted LEO spacecraft trajectories. This is due to the use of an optimal orbit determination method[11], where the local time-varying atmospheric density modeling error is appropriately absorbed by the local time-varying atmospheric density error estimate. Otherwise atmospheric density error would be aliased into correlated orbit states (mean motion and mean longitude most significantly), thereby degrading the orbit estimate.

Least squares orbit determination methods minimize the sum of squares of measurement residuals, not orbit error magnitudes. And least squares methods estimate a ballistic coefficient time constant, not the local time varying atmospheric density. It is thus not surprizing that least squares methods frequently do not estimate atmospheric density errors in response to significant increases in a_P (see Owens[10]). Unestimated atmospheric density modeling errors are unfortunately aliased into the least squares orbit estimate.

1.6 Predictions

The new method employs a serially correlated stochastic sequence, a Gauss-Markov model, to represent atmospheric density estimates. These time-varying estimates are derived from range and/or Doppler tracking measurements by the filter measurement update function. They are predicted forward by the filter time update function. And they are returned to zero exponentially with time, according to the half-life specified for the Gauss-Markov model.

There are at least two sources of air-drag acceleration error, an error in atmospheric density and an error in the ballistic coefficient (see Eq. 10). Both errors must be estimated and thereby removed. Given the technique identified below to estimate and remove the observable part of any constant mean ballistic coefficient, an optimal predicted estimate of the local atmospheric density is achieved.

This is in sharp contrast to the use of an estimated time-constant ballistic coefficient for least squares predictions, while ignoring the time-varying error in atmospheric density.

2 Global Atmospheric Density Model

A global atmospheric density model refers to a capability to model atmospheric density at any LEO position and time, whether past, current, or in the future. Global models are driven deterministically by $F_{10.7}$ and a_P (or K_P) measurement values. Global atmospheric density estimates are accompanied by significant atmospheric density modeling errors.

LEO least squares orbit determination programs use global atmospheric models, and estimate local ballistic coefficient *time constants* in an effort to absorb the effect of *time-varying* atmospheric density modeling errors. For least squares this is necessary, but it is far from optimal. Atmospheric density modeling errors are time variable, not time constant, and during solar maximum relative error magnitudes can, and do, exceed 100% (see the filter-smoother results presented herein).

2.1 Jacchia 1971

The Jacchia 1971 (J71) global atmospheric density model[1] was adopted in 1972 as the COSPAR international Reference Atmosphere (CIRA72)[2]. It models atmospheric density ρ from a lower height of 110 km to an upper height of 2000 km, referred to the earth's surface. J71 has been used operationally for thirty years. I have selected it for use in the real-time estimation of atmospheric density.

2.1.1 K_P vs a_P

K_P and a_P are used interchangably in the liturature on atmospheric density modeling. K_P and a_P refer to the same geomagnetic measurement information, and can be related[2] approximately with the empirical nonlinear transformation (compare Eqs. 21 and 22 from J70 [3]):

$$28°K_P + 0.03° \exp(K_P) = 1.0°a_P + 100° [1 - \exp(-0.08a_P)]$$

where the temperature unit is in degrees Kelvin. Rearrange the equation above to define the non-linear function $F(K_P, a_P)$:

$$F(K_P, a_P) = \{1.0°a_P + 100° [1 - \exp(-0.08a_P)]\} - \{28°K_P + 0.03° \exp(K_P)\}$$

where:

$$F(K_P, a_P) = 0$$

Given a value for K_P (or a_P), then solve the function $F(K_P, a_P)$ for a_P (or K_P) iteratively (e.g., with Newton-Raphson). Although this seems a trivial matter, it is necessary to be able to discuss K_P and a_P interchangably.

2.1.2 $F_{10.7}$, $\bar{F}_{10.7}$ and K_P (or a_P) in J71

$F_{10.7}$ is the $10.7cm$ daily solar flux, a daily average of measured values, reported daily. $\bar{F}_{10.7}$ is an average of $F_{10.7}$ over six solar rotations, reported daily. $F_{10.7}$ measures particular active regions of the solar disk, but $\bar{F}_{10.7}$ is associated with the entire solar disk. Measured values of the geomagetic index K_P are averaged across three hour intervals to provide a single K_P (and a_P) constant for each three hour interval. Atmospheric density is extremely sensitive to realizable variations in input values for K_P and $F_{10.7}$, according to J71 [1]. Atmospheric temperatures define the bridge between K_P and $F_{10.7}$ measurement values and atmospheric density. From Jacchia Eq. 14:

$$T_c = 379° + 3.24° \bar{F}_{10.7} + 1.3° \left(F_{10.7} - \bar{F}_{10.7}\right), \qquad \text{for } K_P = 0 \tag{1}$$

where T_c (degrees Kelvin) denotes the night time minimum of the global asymptotic exospheric temperature T_∞ for $K_P = 0$ (and $a_P = 0$). T_∞ is a function of T_c, then T_∞ is augmented with K_P information according to Jacchia Eq. 18:

$$\Delta T_\infty = 28°K_P + 0.03° \exp(K_P) \tag{2}$$

Then T_∞ drives the calculation of atmospheric density ρ.

Thus according to Jacchia, atmospheric density is driven by both $F_{10.7}$ and K_P. Correlations between ρ, T_∞, $\bar{F}_{10.7}$, $F_{10.7}$, and K_P are presented graphically in Jacchia Figures 6 and 7 [1].

---

[2] K_P is a three-hour average of irregular geomagnetic range disturbances in two horizontal field components (H and D), measured at thirteen mid-latitude stations by ground based magnetometers. At a particular station, the natural logarithm of the largest excursion in H or D over a three hour period is recorded on a scale from 0 to 9. K_P measurements are defined only at thirds of each unit (e.g., $0, 1/3, 2/3, 3/3$), where $0 \leq K_P \leq 9$. a_P was defined to present geomagnetic range disturbances on a linear scale, as a table of transformed values $0 \leq a_P \leq 400$, with a one-to-one discreet mapping with K_P, and boundaries defined by $(K_P = 0) \Leftrightarrow (a_P = 0)$, and $(K_P = 9) \Leftrightarrow (a_P = 400)$. The transformation given by $F(K_P, a_P)$ provides a continuous map between K_P and a_P, but where $(K_P = 0) \Leftrightarrow (a_P = 0)$ is not satisfied exactly.

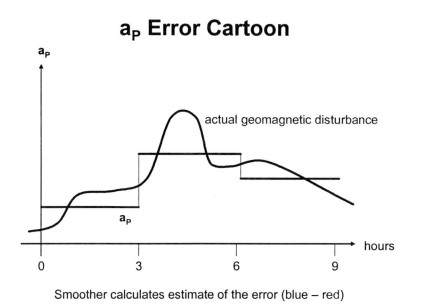

a_P Error Cartoon

actual geomagnetic disturbance

a_P

a_P

hours

0   3   6   9

Smoother calculates estimate of the error (blue − red)

Figure 1: Geomagnetic Disturbance vs a_P

And according to Jacchia, the effect of changes of $F_{10.7}$ on the atmosphere lags the $F_{10.7}$ measurement time by 1.7 days (40.8 hours). And the effect of changes in the K_P measurement on the atmosphere lags the K_P measurement time by 0.279 days (6.7 hours). Given current values for $F_{10.7}$ and K_P, these lags provide *predictive power* for real-time atmospheric density estimation and for real-time orbit determination.

3   Global Atmospheric Density Modeling Errors

3.1   Constants and Discontinuities in a_P

On 15 July 2000, values for a_P did jump from 32 over the interval $(9^h, 12^h)$ to 207 at 12^h for the interval $(12^h, 15^h)$. These a_P values are given as constants across three hour time intervals. But in fact the discontinuity at 12^h is fictitious, and it generates a huge error in the value for atmospheric density ρ calculated by J71. Another huge error in ρ is incurred because the actual profile of geomagnetic disturbance with time is absent across each three-hour a_P interval. The a_P peak variation magnitude is obviously significantly larger than any average. It is useful to view a sequence of a_P three-hourly values begining at nine hours on 15 July 2000: $\{\dots, 32, 207, 300, 400, 300, 179, 80, 32, \dots\}$, a sequence that invokes huge modeling errors correlated in part with discontinuities between the three-hour constants. I treat all increasing sequences in a_P as modified random walk functionals. That is, given the value $a_P = a_P(t, t + 3h)$, then define the predicted estimate $\hat{a}_P$:

$$\hat{a}_P(t+3h, t+6h) = a_P(t, t+3h)$$

Then the error $\delta a_P(t+3h, t+6h)$ in $\hat{a}_P(t+3h, t+6h)$ is given by:

$$\delta a_P(t+3h, t+6h) = a_P(t+3h, t+6h) - \hat{a}_P(t+3h, t+6h)$$

$$= a_P(t+3h, t+6h) - a_P(t, t+3h)$$

This generates a realistic mechanism for modeling errors due to discontinuities in a_P between the three-hourly time constants in a_P. These errors map to evaluations of atmospheric density ρ according to Eq. 2.

The cartoon defined by Fig. 1 presents a fictitious explanatory example: Let d denote the difference between actual geomagnetic disturbance and a_P. For those times between discontinuities in a_P, d is a continuous function because both the geomagnetic disturbance and a_P are continuous functions. But at each discontinuity in a_P, the difference d is also discontinuous. Then given real-time range tracking data that measures the actual geomagnetic disturbance, one should expect that the sequentially smoothed corrections to modeled atmospheric density would present continuous functions between discontinuities in a_P, but would present discontinuities at the times of discontinuities in a_P. Also, this cartoon suggests the existence of both negative and positive values in the difference d. Thus one should also expect both negative and positive values in the smoothed corrections to modeled atmospheric density. Actual variations of the smoothed atmospheric density estimation results presented herein are thus explained, and an important necessary condition for validation of the method is established.

During quiet geomagnetic weather the maximum values of a_P are smaller than at solar maximum, and the three-hourly variation magnitudes in a_P are also smaller. Consequently the associated values of ρ at fixed spacecraft height are smaller than at solar maximum, and variation magnitudes in ρ at fixed spacecraft height are also smaller. But during solar maximum the maximum values of a_P are large, and the three-hourly variation magnitudes in a_P are also large. Consequently the associated values of ρ at fixed height are relatively large, and variation magnitudes in ρ at fixed height are also large.

3.2   Constants and Discontinuities in $F_{10.7}$

A similar phenomenan exists with the daily mean values of $F_{10.7}$. I treat increasing daily mean value sequences of $F_{10.7}$ as modified random walk functionals. During quiet solar weather the maximum values of $F_{10.7}$ are relatively small, and the daily variation magnitudes in $F_{10.7}$ are also small. Consequently the associated values of ρ at fixed spacecraft height are relatively small, and variation magnitudes in ρ at fixed spacecraft height are also small. But during solar maximum the maximum values of $F_{10.7}$ are large, and the daily variation magnitudes in $F_{10.7}$ are also large. Consider the sequence beginning with 6 July 2000: $\{\ldots, 174, 187, 210, 211, 244, 241, 315, 232, 204, 213, 219, 228, \ldots\}$. Consequently the associated values of ρ at fixed spacecraft height are relatively large, and variation magnitudes in ρ at fixed spacecraft height are also large.

3.3   Estimation Errors in Global Models

A review of the development of J71 [1][2] reveals the source of significant atmospheric density modeling errors that are independent of discontinuities in $F_{10.7}$ and a_P.

h (km)	$F_{10.7}$	a_P	$\rho \left(kg/m^3 \right)$	Ratio
497	150	20	1.1×10^{-12}	
497	213	400	5.9×10^{-12}	5
780	150	20	2.6×10^{-14}	
780	213	400	4.3×10^{-13}	17

Table 1: Height Dependent Density Ratios

4 Adopted Principles

4.1 The Fixed-Height Principle

A mechanism is required to increase the filter baseline error variance σ_ρ^2 on atmospheric density due to the effects of significant increases in $F_{10.7}$, $\bar{F}_{10.7}$ and a_P, particularly during solar maximum. This requirement derives especially from the use of time constants and discontinuities in $F_{10.7}$ and a_P discussed above. For fixed spacecraft height, I identify a *fixed height principle* between atmospheric density ρ and its error variance σ_ρ^2: A significant increase in atmospheric density ρ, at fixed spacecraft height, implies a significant increase in estimated atmospheric density error variance σ_ρ^2 at that height.

4.1.1 A Simple Experiment with Height Variations Using J71

The *fixed height principle* suggests the need for a definition for mean atmospheric density derived from mean values $\langle F_{10.7} \rangle$, $\langle \bar{F}_{10.7} \rangle$, and $\langle a_P \rangle$ for comparison to current atmospheric density derived from current values of $F_{10.7}$, $\bar{F}_{10.7}$ and a_P, all at fixed spacecraft height. But the *fixed height principle* leads to a surprizing result. The description of the simple experiment that follows illuminates this result.

I have selected $\langle F_{10.7} \rangle = \langle \bar{F}_{10.7} \rangle = 150$ and $\langle a_P \rangle = 20$ as mean values for $F_{10.7}$, $\bar{F}_{10.7}$ and a_P across multiple eleven year solar cycles. Associate density $\langle \rho(h) \rangle$ with $\langle F_{10.7} \rangle$, $\langle \bar{F}_{10.7} \rangle$, and $\langle a_P \rangle$. Evaluate atmospheric density at spacecraft heights $h = 497km$ and $h = 780km$, for mean values $\langle F_{10.7} \rangle$, $\langle \bar{F}_{10.7} \rangle$, $\langle a_P \rangle$, and for[3] $F_{10.7} = 213$ and $a_P = 400$. Denote the density function $\rho = \rho \left(h, F_{10.7}, \bar{F}_{10.7}, a_P \right)$, and define the ratio:

$$\text{Ratio} = R \left(h, F_{10.7}, \bar{F}_{10.7}, a_P \right) = \frac{\rho \left(h, F_{10.7}, \bar{F}_{10.7}, a_P \right)}{\langle \rho(h) \rangle} \tag{3}$$

Then evaluate the ratio of densities at each height using J71:

Then for $F_{10.7} = 213$ and $a_P = 400$ on 15 July 2000, referred to mean values $\langle F_{10.7} \rangle$ and $\langle a_P \rangle$, the density is increased by a factor of five at $h = 497km$, but is increased by a factor of seventeen at $h = 780km$. This is an example of a general property that I shall refer to as the *ratio-height principle*.

4.2 The Ratio-Height Principle

The *Ratio Height Principle*: The ratio of atmospheric density during solar maximum to mean atmospheric density, both at the same height, increases significantly when spacecraft height is increased.

---

[3] At 18 hours on 15 July 2000, the record shows $F_{10.7} = 213$ and $a_P = 400$. July 2000 was at solar maximum, and the explosion in a_P was due to earth impact of a coronal mass ejection.

5 Baseline Model for Atmospheric Density Error Variance

Define:

$$\sigma_{\Delta\rho/\bar{\rho}}(h) = \sqrt{E\left\{\left(\frac{\Delta\rho(h)}{\bar{\rho}(h)}\right)^2\right\}} = \frac{\sqrt{E\left\{(\Delta\rho(h))^2\right\}}}{\bar{\rho}(h)} = \frac{\sigma_{\Delta\rho}(h)}{\bar{\rho}(h)} \tag{4}$$

where:

$$\Delta\rho(h) = \rho(h) - \bar{\rho}(h) \tag{5}$$

where $\rho(h)$ is true atmospheric density at height h (in km), and $\bar{\rho}(h)$ is the associated value of estimated atmospheric density according to J71. The *baseline error variance model* was derived as a height dependent function by evaluating first and second moments on relative atmospheric errors across multiple solar cycles. An estimated graph of the root-variance $\sigma_{\Delta\rho/\bar{\rho}}(h)$ is presented by Figure 2.

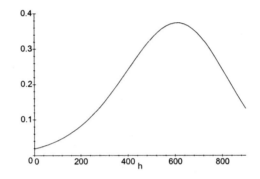

Figure 2: Sigma for Relative Error in Air Density

The ordinate displays $\sigma_{\Delta\rho/\bar{\rho}}(h)$ as a function of spacecraft height h in kilometers. Denote this graph with the function:

$$f(h) = \sigma_{\Delta\rho/\bar{\rho}}(h) \tag{6}$$

Then $f(h)$ can be sampled simultaneously at two heights, say h at current spacecraft height and h_P at orbit perigee height, to define a ratio $f(h)/f(h_P)$ of relative atmospheric density error in terms of averages across multiple solar cycles:

$$\frac{f(h)}{f(h_P)} = \frac{\sigma_{\Delta\rho/\bar{\rho}}(h)}{\sigma_{\Delta\rho/\bar{\rho}}(h_P)} = \frac{\sigma_{\Delta\rho}(h)/\bar{\rho}(h)}{\sigma_{\Delta\rho}(h_P)/\bar{\rho}(h_P)} \tag{7}$$

See Eq. 4.

6 Air Drag Acceleration Error Model

The perturbative air-drag acceleration $\ddot{z}_D$ has the form:

$$\ddot{z}_D = -\frac{1}{2}B\rho\dot{s}^2 K \,, \qquad B = \frac{C_D A}{m} \tag{8}$$

where $\ddot{z}_D$ is a 3×1 perturbative drag acceleration matrix with inertial components, C_D is the unitless drag coefficient, A is the spacecraft area projection onto a plane orthogonal to the spacecraft velocity vector $\dot{s}$ referred to a rotating Earth, $\dot{s}$ is the length of $\dot{s}$, m is spacecraft mass, ρ is atmospheric density, and K is a 3×1 unit matrix that contains inertial components of $\dot{s}/\dot{s}$.

Differentiate Eq. 8 to derive random errors in $\ddot{z}_D$ from random errors in C_D, A, ρ, and m:

$$\Delta\ddot{z}_D = \left(\frac{\Delta C_D}{C_D} + \frac{\Delta A}{A} + \frac{\Delta\rho}{\rho} - \frac{\Delta m}{m}\right)\ddot{z}_D \tag{9}$$

Suppose $\Delta C_D/C_D$, $\Delta A/A$, $\Delta m/m$, and $\Delta\rho/\rho$ are all unknown. If they are observable, optimal orbit determination[11] would generate useful estimates of all four parameters. Typically, they are not all observable. But if there exists a useful mean value of B that is a time constant, then the differentiation of Eq. 8 provides:

$$\Delta\ddot{z}_D = \left(\frac{\Delta B}{B} + \frac{\Delta\rho}{\rho}\right)\ddot{z}_D \tag{10}$$

noting that $\Delta\rho/\rho = \Delta\rho(t)/\rho(t)$ is always time variable. If one has sufficient tracking data to observe both $\Delta B/B$ and $\Delta\rho/\rho$, then one could estimate them simultaneously. But for now I shall take a different approach.

Denote $D(t) = \Delta\rho(t)/\rho(t)$. If one at first sequentially estimates only $D(t)$, and not $\Delta B/B$, using a biased a priori estimate $\bar{B}$ in Eq. 8, and ignores the error $\Delta B = B - \bar{B}$ in $\bar{B}$, then the estimated sequence $\bar{D}(t) = D(t) - \Delta D$ will produce a biased graph. Then modify the a priori estimate $\bar{B}$ experimentally so as to generate an unbiased graph $\hat{D}(t)$, and an associated estimate $\Delta\hat{B}$, where $\hat{B} = \bar{B} + \Delta\hat{B}$. Henceforth use $\hat{B}$ in Eq. 8. Then one derives unbiased estimates for the sequence $D(t) = \Delta\rho(t)/\rho(t)$ because useful mean values of B and ΔB are time constants and the bias ΔB has been removed. How shall we model $D(t)$?

7 Gauss-Markov Sequence

Define:

$$D(t) = \frac{\Delta\rho_{h_P}(t)}{\bar{\rho}_{h_P}(t)} \tag{11}$$

at mean perigee height h_P, where $D(t)$ satisfies the equation:

$$D(t_{k+1}) = \Phi(t_{k+1}, t_k) D(t_k) + \sqrt{1 - \Phi^2(t_{k+1}, t_k)}\, w(t_k) \quad , \quad k \,\epsilon\, \{0, 1, 2, \ldots\} \tag{12}$$

where $w(t)$ is a Gaussian white random variable with mean zero and variance σ_w^2, where:

$$D(t_0) = w(t_0) \tag{13}$$

$$\Phi(t_{k+1}, t_k) = e^{\alpha|t_{k+1} - t_k|} \tag{14}$$

$$\text{constant } \alpha < 0 \tag{15}$$

$$\sigma_{\Delta\rho/\bar{\rho}}^2(t_k) = E\left\{D^2(t_k)\right\}, \text{ for each } k, \tag{16}$$

and where $D(t_k)$ is Gauss-Markov, and:

$$\sigma^2_{\Delta\rho/\bar{\rho}}(t_k) = E\left\{D^2(t_k)\right\} = \sigma^2_w, \text{ for each } k \tag{17}$$

Variance $\sigma^2_{\Delta\rho/\bar{\rho}}(t_k)$ varies significantly with height h. Thus it is necessary to choose a height for $\Delta\rho/\bar{\rho}$ that is fixed. Therefore I have anchored the relative air-density error to mean perigee height h_P.

7.1 Propagation of State Estimate

Let $\hat{D}_{n|m}$ denote an optimal estimate of $D(t_n)$, where t_n is the epoch for $\hat{D}_{n|m}$ and t_m is the time of last measurement. Then according to Sherman's Theorem [6][7][9]:

$$\hat{D}_{n|m} = E\left\{D(t_n)|y_m\right\} \tag{18}$$

Apply Eq. 18 to Eq. 12, where y_k at time t_k was the last measurement processed:

$$\hat{D}_{k+1|k} = \Phi(t_{k+1}, t_k)\,\hat{D}_{k|k} \tag{19}$$

Eq. 19 is the filter state estimate propagation equation for the filter time update. Given measurement y_{k+1} at time t_{k+1} use Kalman's filter measurement update theorem, derived from application of Eq. 18 to $D(t_{k+1})$, for the representation:

$$\hat{D}_{k+1|k+1} = E\left\{D(t_{k+1})|y_{k+1}\right\} \tag{20}$$

Propagation of $\hat{D}_{k+1|k+1}$ to time t_{k+2}:

$$\hat{D}_{k+2|k+1} = \Phi(t_{k+2}, t_{k+1})\,\hat{D}_{k+1|k+1}$$

7.2 State Estimate Error

Define the error in $\hat{D}_{n|m}$ by:

$$\delta\hat{D}_{n|m} = D_n - \hat{D}_{n|m} \tag{21}$$

Insert Eqs. 19 and 12 into Eq. 21:

$$\delta D_{k+1|k} = \Phi(t_{k+1}, t_k)\,\delta D_{k|k} + \sqrt{1 - \Phi^2_{k+1,k}}\,w_k \tag{22}$$

7.3 Base-Line Process Noise Model for Filter Time Update

Square Eq. 22 and apply the expectation operator to get:

$$E\left\{\left(\delta D_{k+1|k}\right)^2\right\} = \Phi^2(t_{k+1}, t_k)\,E\left\{\left(\delta D_{k|k}\right)^2\right\} + \left(1 - \Phi^2_{k+1,k}\right)\sigma^2_w \tag{23}$$

Notice that:

$$E\left\{\left(\delta D_{k|k}\right)^2\right\} < \sigma^2_w \tag{24}$$

due to the processing of measurements by the optimal filter. Thus the stochastic sequence defined by Eq. 22 is not stationary.

7.3.1 Baseline Algorithm

The second term in the right-hand side of Eq. 23 is the base-line process noise covariance for deweighting prior estimates of D:

$$q_{k+1,k}^{\Delta\rho/\rho} = \left(1 - \Phi_{k+1,k}^2\right)\sigma_w^2 \tag{25}$$

Refer to $[t_k, t_{k+1}]$ as the propagation time interval. For long propagation time intervals the factor $\left(1 - \Phi_{k+1,k}^2\right)$ tends to unity, and so $q_{k+1,k}^{\Delta\rho/\rho}$ tends to σ_w^2. For short propagation time intervals the factor $\left(1 - \Phi_{k+1,k}^2\right)$ tends to zero, and so $q_{k+1,k}^{\Delta\rho/\rho}$ tends to zero. Thus the factor $\left(1 - \Phi_{k+1,k}^2\right)$ drives the variance $E\left\{\left(\delta D_{k+1|k}\right)^2\right\}$ toward σ_w^2 in the absense of measurements, but adds little or nothing during dense measurements. When mean values $\langle a_P \rangle$ and $\langle F_{10} \rangle$ are experienced, this is appropriate for the base-line Gauss-Markov model.

But when a_P and F_{10} are much larger than $\langle a_P \rangle$ and $\langle F_{10} \rangle$, then an important model extension is called for. The Gauss-Markov sequence must be immediately interrupted to open the filter gain, particularly during dense measurements.

7.4 Dynamic Process Noise Model for Filter Time Update

The J71 atmospheric density model $\bar{\rho}$ is a function of several arguments. It will suffice here to write:

$$\bar{\rho} = \bar{\rho}\left(h, F_{10}, \bar{F}_{10}, a_P, t_{k+1}\right) \tag{26}$$

Define the unitless ratio:

$$R = \frac{\bar{\rho}\left(h_P, F_{10}, \bar{F}_{10}, a_P, t_{k+1}\right)}{\bar{\rho}\left(h_P, \langle F_{10}\rangle, \langle F_{10}\rangle, \langle a_P\rangle, t_{k+1}\right)} \tag{27}$$

Define and initialize R_{max}:

$$R_{\max} = 1 \tag{28}$$

Define and set ϵ to an appropriate small constant positive value. The value of ϵ quantifies what is meant by the least *significant increase* in atmospheric density at perigee height.

7.4.1 Dynamic Algorithm

If $R > R_{\max} + \epsilon$, set $R_{\max} = R$ and define:

$$Q_{k+1,k}^{\Delta\rho/\rho} = R^2\sigma_w^2 \tag{29}$$

else, set $R_{\max} = R$ and define:

$$Q_{k+1,k}^{\Delta\rho/\rho} = q_{k+1,k}^{\Delta\rho/\rho} \tag{30}$$

Use $Q_{k+1,k}^{\Delta\rho/\rho}$ for air-density error process noise covariance deweighting. Effect: When $\bar{\rho}$ is increasing at h_P due to $F_{10.7}$, $\bar{F}_{10.7}$, and a_P, then the air density variance and filter gain are immediately opened to enable tracking measurements to estimate significant increases in atmospheric density. But when $\bar{\rho}$ is decreasing at h_P due to $F_{10.7}$, $\bar{F}_{10.7}$, and a_P, then the air density variance and filter gain begin their return to the baseline model.

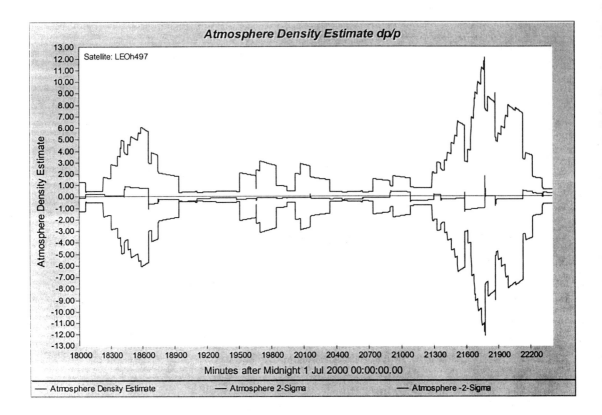

Figure 3: Filtered Atmospheric Density Estimates

8 Baseline Transform From Perigee Height to Spacecraft Height

Recall Eqs. 6 and 4 to write:

$$\frac{\sigma_{\Delta\rho}(h)}{\bar{\rho}(h)} = f(h)\frac{f(h_P)}{f(h_P)} = \frac{f(h)}{f(h_P)}f(h_P) = \frac{f(h)}{f(h_P)}\frac{\sigma_{\Delta\rho}(h_P)}{\bar{\rho}(h_P)} \tag{31}$$

Eq. 31 can be derived from:

$$\frac{\Delta\rho(h)}{\bar{\rho}(h)} = \frac{f(h)}{f(h_P)}\frac{\Delta\rho(h_P)}{\bar{\rho}(h_P)} \tag{32}$$

which, with the aid of Eq. 11, can be written:

$$D(h) = \frac{f(h)}{f(h_P)}D(h_P) \tag{33}$$

From Eq. 5:

$$\rho(h) = \bar{\rho}(h) + \Delta\rho(h)$$

$$= \bar{\rho}(h)\left[1 + \frac{\Delta\rho(h)}{\bar{\rho}(h)}\right] \tag{34}$$

938

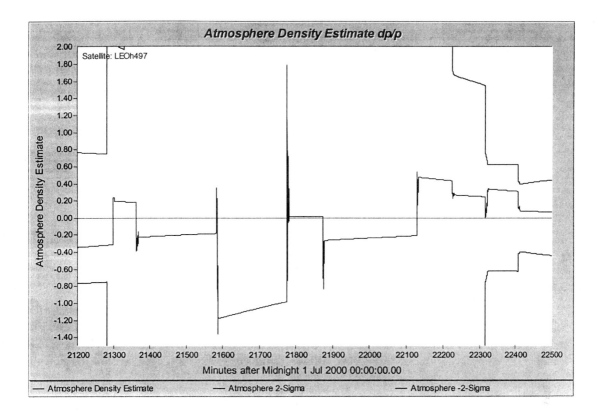

Figure 4: Filtered Geomagnetic Response to CME

Insert Eq. 32 into Eq. 34:

$$\rho(h) = \bar{\rho}(h) \left[1 + \frac{f(h)}{f(h_P)} \frac{\Delta\rho(h_P)}{\bar{\rho}(h_P)} \right]$$

and apply Sherman's Theorem to get:

$$\hat{\rho}(h) = \bar{\rho}(h) \left[1 + \frac{f(h)}{f(h_P)} \frac{\Delta\hat{\rho}(h_P)}{\bar{\rho}(h_P)} \right]$$

or:

$$\hat{\rho}(h) = \bar{\rho}(h) \left[1 + \frac{f(h)}{f(h_P)} \hat{D}(h_P) \right] \tag{35}$$

Eq. 35 defines my method to map the filter estimate $\hat{D}$ at perigee height h_P to the estimate $\hat{\rho}(h)$ of atmospheric density, at current height h, for use in trajectory propagation.

8.1 Discussion

The baseline height transformation factor $f(h)/f(h_P)$ is used in Eq. 32, as well as in Eq. 31, to guarantee consistency between the height dependent *stochastic error* transform model and the

939

height dependent *error estimate* transform model for atmospheric density. This frees the height transform from wild perturbations that would be suffered due to the use of local time-constants and discontinuities in $F_{10.7}$ and a_P, and provides stability to the estimate $\hat{D}(h_P)$ at perigee height. The spacecraft does sample its own perigee height once per orbit, and perigee height is most significant with respect to atmospheric density modeling errors.

9 Real Data Results

The new method for real-time atmospheric density estimation has been implemented and tested in STK/OD[4] with simulated tracking data and real tracking data. The figures for real data results display atmospheric density estimation response to range tracking data, from fourteen ground stations, to two spacecraft in LEO. The sequential filter began processing range data early on 1 July 2000, and terminated late on 26 July 2000 due to the beginning of a six day gap in the archived tracking data.

$F_{10.7}$ began with 163.7 on 1 July, had a global peak at 314.6 on 12 July, receded to 203.9 on 14 July, had a minor peak of 261.9 on 18 July, and receded to 174.6 on 26 July.

a_P began with 4.0 on 1 July, had a global peak of 400.0 at 18^h on 15 July, and experienced ten other minor peaks during the 26 day July interval.

The absissa (X Axis) for each figure is given in units of minutes after 1 July 2000 0^h UTC, denoted hereafter as MAE (Minutes After Epoch). The ordinate (Y Axis) for each figure presents the change in atmospheric density relative to J71. The time varying estimate of relative atmospheric density $\Delta\hat{\rho}/\bar{\rho}$ is displayed in blue, and the associated $\pm 2\sigma$ error envelope is displayed in black. Each unit on the ordinate is associated with a 100% change in atmospheric density.

9.1 Filter Update Functions

The sequential filter performs time update and measurement update functions recursively. The time update function has two vital activities: (1) It accumulates atmospheric density error variance with time, and (2) It propagates the atmospheric density estimate and its error variance across time intervals between measurements. Time intervals between station passes are long, and time intervals between range measurements within each station pass are short. The Gauss-Markov exponential half-life used by the filter sends the estimate (blue line) toward zero, in these time intervals, at a rate defined by the value used for half life. Here the exponential half life constant was set to 700 minutes.

It is important to note that significant step changes in modeled atmospheric density are induced by changes in $F_{10.7}$, $\bar{F}_{10.7}$, and a_P according to J71. These changes are always *wrong* because they are modeled as time constants, and have significant fictitious discontinuities at time constant boundaries, as discussed above. The associated errors in $F_{10.7}$ or a_P may thus be positive or negative.

Each significant step change in the sequential atmospheric density estimate is associated with the commencement of range tracking data[5] processing by the filter measurement update function after a long time interval between station passes. These step changes may be positive or negative because the associated atmospheric density errors may be positive or negative.

---

[4] A new orbit determination product from Analytical Graphics, Inc.

[5] It would be preferable to process dense GPS range and/or Doppler data completely around the orbit so as to reduce or eliminate this step change.

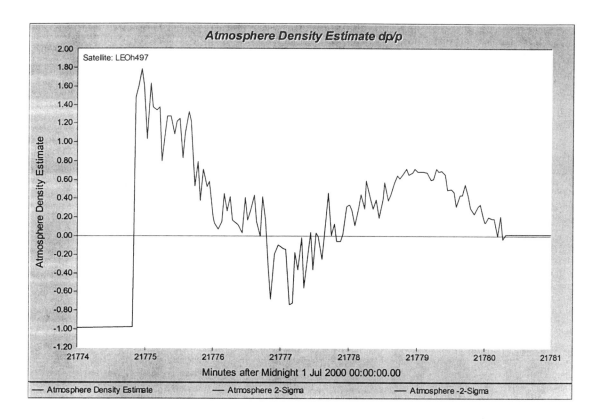

Figure 5: Filter Response to Range Measurements at Geomagnetic Peak

9.2 LEO $h = 497km$

9.2.1 Sequential Filter

Fig. 3 displays the filter response to range data from 13 July 2000 $12^h\ 0^m$ (18000 MAE) through 16 July 2000 $13^h\ 20^m$ (22400 MAE). The filter was initialized on 1 July 2000 0^h. The baseline error variance model is clearly distinguished here from the dynamic error variance model.

The large positive step in the estimate at approximately MAE 18400 (13 July 2000 $18^h\ 42^m$) is correlated with a peak of $F_{10.7} = 314.6$ for 12 July 2000. This is the global peak in $F_{10.7}$ across the entire July interval. The Jacchia defined $F_{10.7}$ lag for the direct solar effect on atmospheric density is 1.70 days, or 40.8 hours. This corresponds to the interval from 16.8 hours on 13 July 2000 to 16.8 hours on 14 July 2000.

It is significant to note that the global peak in a_P follows, and is clearly detached from, the global peak in $F_{10.7}$. The global peak in a_P is associated with the largest estimate magnitudes for atmospheric density across the entire processing interval.

Fig. 4 displays the filter response, between MAE 21200 and MAE 22500. These times are associated with 15 July 2000 $17^h\ 20^m$ and 16 July 2000 $15^h\ 0^m$ respectively. The peak STK/OD filtered estimate in atmospheric density is seen to be about 1.8 (180%), of the density modeled by J71, at approximately 16 July 2000 2.92 hours.

The $a_P = 400$ peak occurs on 15 July 2000 across the interval $(18, 21)$ hours. The Jacchia defined

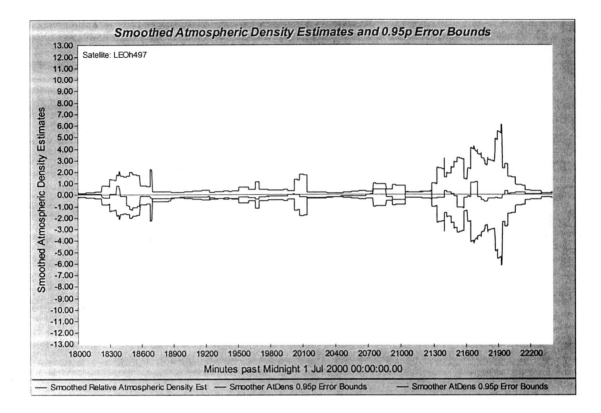

Figure 6: Smoothed Atmospheric Density Estimates

a_P lag for the geomagnetic effect on atmospheric density is 0.279 days, or 6.696 hours. Then the *effect* of the $a_P = 400$ peak on atmospheric density occurs on 16 July 2000 across the interval $(0.696, 3.696)$ hours. Note that the peak STK/OD estimate in atmospheric density falls within this interval, but that most of the four minute positive estimate sequence lies to the right of interval $(0.696, 3.696)$ hours.

Fig. 5 magnifies the effective a_P peak on 16 July 2000 2.92 hours, and shows the detailed response to filtering range measurements.

9.2.2 Sequential Smoother

Fig. 6 displays the filter-smoother response to range data from 13 July 2000 $12^h\ 0^m$ (18000 MAE) through 16 July 2000 $13^h\ 20^m$ (22400 MAE). It is scaled as the filtered response of Fig. 3 so as to enable easy comparison.

The filter responds only to range data prior to the time of response, whereas the smoother responds to range data both prior to and after the time of response. The smoothed estimate uses more local information than the filtered estimate. It is thus not surprizing that smoothed estimates and root-variances are smaller than those from the filter. But the filtered estimate function has been reshaped by the smoother.

Fig. 7 focuses the smoother response on the CME interval, between MAE 21200 and MAE 22500, and is comparable to the filtered response of Fig. 4. The *effective* $a_P = 400$ peak on atmospheric

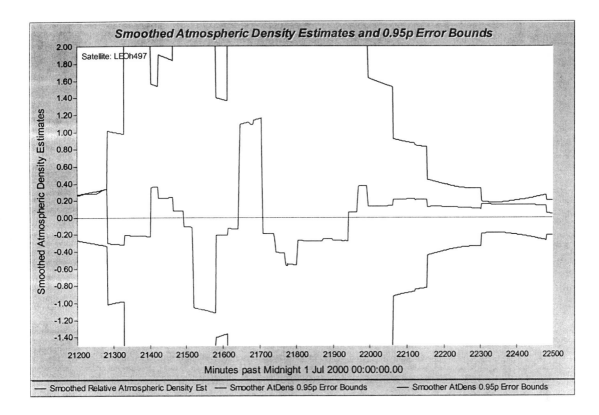

Figure 7: Smoothed Geomagnetic Response to CME

density occurs on 16 July 2000 across the interval $(0.696, 3.696)$ hours. Note that the entire one hour smoothed estimate peak in atmospheric density falls within this interval. The smoother has shifted the filtered estimate function to the left by 110 minutes (bringing it entirely within the interval predicted by Jacchia), has reshaped it, and has reduced the filtered peak from 1.8 to less than 1.2.

9.3   LEO $h = 780km$

9.3.1   Sequential Filter

Fig. 8 displays the filter response to range data from 13 July 2000 $12^h\ 0^m$ (18000 MAE) through 16 July 2000 $16^h\ 40^m$ (22600 MAE). The filter was initialized on 1 July 2000 0^h. The baseline error variance model is not so clearly distinguished from the dynamic error variance model.

The large positive step in the estimate at approximately MAE 18400 (13 July 2000 $18^h\ 42^m$) is correlated with a peak of $F_{10.7} = 314.6$ for 12 July 2000. This is the global peak in $F_{10.7}$ across the entire July interval. The Jacchia defined $F_{10.7}$ lag for the direct solar effect on atmospheric density is 1.70 days, or 40.8 hours. This corresponds to the interval from 16.8 hours on 13 July 2000 to 16.8 hours on 14 July 2000.

The step at 18400 MAE of Fig. 8 is correlated with a peak of $F_{10.7} = 314.6$ for 12 July 2000. This is the global peak in $F_{10.7}$ across the entire July interval. 18400 MAE is associated with 18.7 hours on 13 July 2000. The Jacchia defined $F_{10.7}$ lag for the direct solar effect on atmospheric density is

943

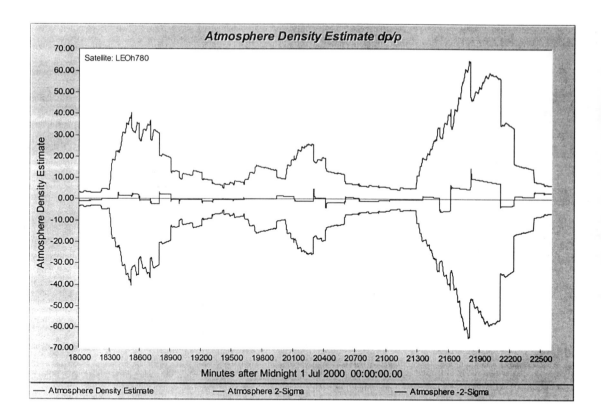

Figure 8: Filtered Atmospheric Density Estimates

1.70 days, or 40.8 hours. This corresponds to the interval from 16.8 hours on 13 July 2000 to 16.8 hours on 14 July 2000.

Fig. 9 displays the filter response, between MAE 21200 and MAE 22600. These times are associated with 15 July 2000 $17^h 20^m$ and 16 July 2000 $16^h 40^m$ respectively. The peak STK/OD filtered estimate in atmospheric density is seen to be about 14.7 (1470%), of the density modeled by J71, at approximately 16 July 2000 $3^h 40^m$.

The $a_P = 400$ peak occurs on 15 July 2000 across the interval $(18, 21)$ hours. The Jacchia defined a_P lag for the geomagnetic effect on atmospheric density is 0.279 days, or 6.696 hours. Then the *effect* of the $a_P = 400$ peak on atmospheric density occurs on 16 July 2000 across the interval $(0.696, 3.696)$ hours. Note that the peak STK/OD estimate in atmospheric density falls within this interval.

Fig. 10 displays the filtered response to range measurements for the peak at 21816 MAE (16 July 2000 $3^h 36^m$).

9.3.2 Sequential Smoother

Fig. 11 displays the filter-smoother response to range data from 13 July 2000 $12^h 0^m$ (18000 MAE) through 16 July 2000 $13^h 40^m$ (22600 MAE). It is scaled as the filtered response of Fig. 8 so as to enable easy comparison.

Fig. 12 magnifies the smoothed response between 15 July 2000 $17^h 20^m$ (MAE 21200) and 16

944

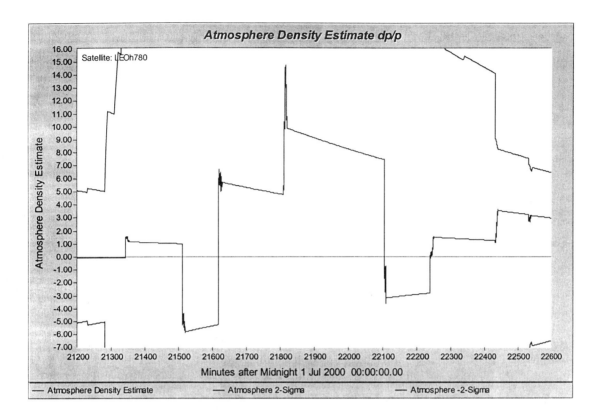

Figure 9: Filtered Geomagnetic Response to CME

July 2000 16^h 40^m (MAE 22600). The filtered peak in the atmospheric density estimate is reduced by the smoother from 14.7 to 10.0, is shifted to the left by about 50 minutes, and is broadened and reformed.

9.4 The Smoother Shift to the Left

The peak filtered estimate in atmospheric density was shifted to the left by the smoother for both *LEOh*497 and *LEOh*780. The following explanation is offered.

Range measurements sit in position space, two time-integrals above the air-drag accelerations. These integrals are time lags. Thus the sensing of atmospheric density from range measurements by the filter lags the time of actual change in the atmospheric density.

The optimal filtered estimate is derived only from information that sits to the left of the filter epoch. On the other hand, the optimal smoothed estimate is derived from information that sits both to the left *and to the right* of the smoother epoch. This provides the smoother with a significant advantage in estimation of the atmospheric density error function, relative to the filter. Having used information from the right of a peak, and from the right of the integral lag, the smoother is enabled to more accurately estimate the atmospheric density error function.

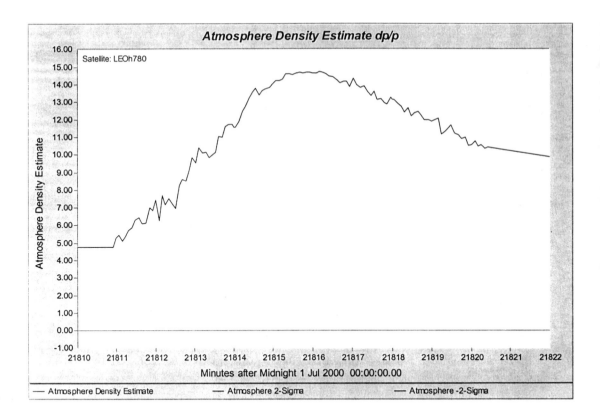

Figure 10: Filter Response to Range Measurements at Geomagnetic Peak

9.4.1   Use the Filtered Estimate for Predictions

The last filtered estimate is the first smoothed estimate, and subsequent smoothed estimates are calculated backwards with time. The last filtered estimate is the optimal estimate for use in trajectory prediction. Observable atmospheric density modeling error is absorbed by the atmospheric density error filter state parameter. Otherwise the atmospheric density modeling error would be aliased into the orbit estimate.

10   Optimality Validation

Optimality validation consists in demonstrating that range residuals are white, and that the McReynolds filter-smoother concistency test is satisfied at the 0.99 probability level. The first test is satisfied, and the second test is satisfied most of the time. An explanation for failures in the second test has been identified.

10.1   White Noise Range Residuals

Inspection of range residual graphics shows consistency with white noise.

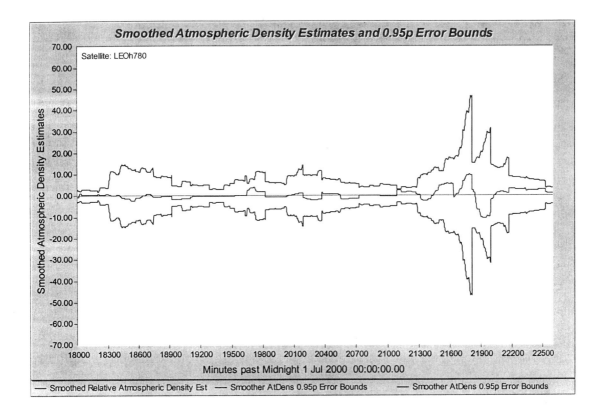

Figure 11: Smoothed Atmospheric Density Estimates

10.2 McReynolds' Filter-Smoother Consistency Test

Inspection of filter-smoother consistency test graphics shows consistency most of the time. However there are time intervals during which this test clearly fails. These failures have been identified to coincide with the commencement of range data, after long prediction time intervals, when filtered atmospheric density estimates make very large corrections. These failures would be eliminated by taking tracking data densely about the orbit; e.g., by using an onboard GPS receiver.

11 Summary

Using real range tracking data at solar maximum for two spacecraft in LEO, the new method for real-time estimation of atmospheric density has been shown to estimate significant time-varying corrections to local atmospheric density that are closely correlated in time to the *effective* times for observed values of both $F_{10.7}$ and a_P. Atmospheric density is demonstrated to be consistently driven by significant disturbances in the geomagnetic field. The atmospheric density estimation algorithm has new features:

- *Sequential* processing of range and/or Doppler measurements using an optimal filter-smoother

- Simultaneous estimation of local atmospheric density for multiple spacecraft

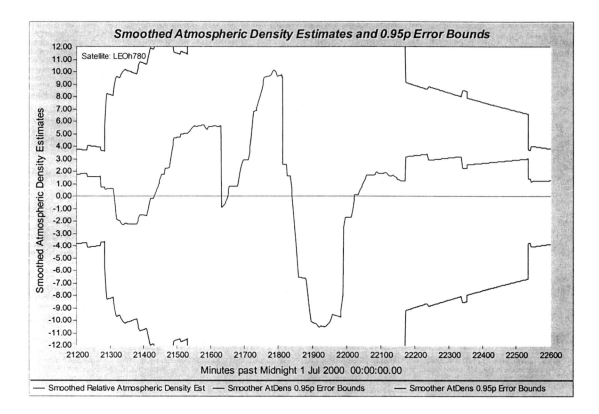

Figure 12: Smoothed Geomagnetic Response to CME

- Use of the historical record on $F_{10.7}$ and a_P to define a baseline stochastic atmospheric density error model

- Use of current values of $F_{10.7}$ and a_P to define a dynamic stochastic atmospheric density error model

12 Real-Time Global Atmospheric Density Estimation

Looking forward, it is appropriate to consider again[14] the sequential real-time *global* estimation of atmospheric density. What is the relation between *local* and *global* estimation of atmospheric density?

Given tracking measurements from an ensemble of LEO spacecraft, consider global estimation without local estimation. Then locally there would always exist significant *observable* atmospheric density modeling errors. But by definition, optimal orbit determination[11] requires that every unknown observable effect must have a place in the state estimate structure. Therefore local estimation is required *simultaneously* with global estimation.

In 1990 I proposed[14] the transformation of appropriate parameter time constants in J71 to serially correlated stochastic parameter sequences. The stochastic parameters were to be estimated in real time, their error variances were to be constant (stationary) during propagation, and reduced appropriately due to measurements. This part of the 1990 proposal is proposed again.

13 Acknowledgements

13.1 Tom Gidlund

Tom Gidlund [18], with U. S. Air Force support, provided a substantial sequence of real range and Doppler tracking data from two LEOs at solar maximum, from the summer of year 2000, for use in the validation of our new sequential orbit determination method STK/OD. The LEO orbits are near circular, and are separated in height by roughly $280km$. The effects of a huge coronal mass ejection impacted the earth's geomagnetic field with a_P rising to 400 at 18^h on 15 July, where $F_{10.7}$ had peaked at 314.6 on 12 July. This provided two very useful LEO test cases.

This tracking data was also used in the development of our new *baseline and dynamic atmospheric density error models*. The separation in height of the two LEOs presented useful cases in relating volatile changes in atmospheric density to specification of filter process noise error variance at significantly different spacecraft heights.

13.2 Hujsak, Cottam, and Pechenick

Dependence of our new *baseline atmospheric density error model* on the historical record of $F_{10.7}$ and a_P was first suggested by Richard Hujsak [17]. A related study was assigned to Russ Cottam and Kay Pechenick under ATA contract to AFSPACECOM in 1986. Unfortunately the AFSPACECOM contract was cancelled before an atmospheric density error model could be completed, applied, or demonstrated.

13.3 Analytical Graphics, Inc.

Conversion of the MACH 10 prototype to STK/OD was performed by members of the AGI development team, especially Kevin O'Brien, Jim Woodburn, Jim Fields, Matt Amato, Kevin Murray, Jeff Gassert, Jim Wilson, Dave Holland, and Ralph Patrick. I wish to thank Paul Graziani for his patient and sustained support during my development of the MACH 10 prototype.

References

[1] L. G. Jacchia, *Revised Static Models of the Thermosphere and Exosphere with Empirical Temperature Profiles*, Smithsonian Astrophysical Observatory, Special Report 332, 1971

[2] The Committee for the CIRA Working Group 4, *COSPAR International Reference Atmosphere 1972 (CIRA 1972)*, Akademie-Verlag, Berlin, 1972

[3] L. G. Jacchia, *New Static Models of the Thermosphere and Exosphere with Empirical Temperature Profiles*, Smithsonian Astrophysical Observatory, Special Report 313, 1970

[4] L. G. Jacchia, *Thermospheric Temperature, Density, and Composition: New Models*, Smithsonian Astrophysical Observatory, SAO Special Report 375, 1977

[5] Sherman, S., *A Theorem on Convex Sets with Applications*, Ann. Math. Stat., 26, 763-767, 1955.

[6] Sherman, S., *Non-Mean-Square Error Criteria*, IRE Transactions on Information Theory, Vol. IT-4, 1958.

[7] Kalman, R. E., *New Methods in Wiener Filtering Theory*, Proceedings of the First Symposium on Engineering Applications of Random Function Theory and Probability, edited by J. L. Bogdanoff and F. Kozin, John Wiley & Sons, New York, 1963.

[8] Rauch, H. E., *Solutions to the Linear Smoothing Problem*, IEEE Trans.Autom. Control, vol. AC-8, p. 371, 1963

[9] Meditch, J. S., *Stochastic Optimal Linear Estimation and Control*, McGraw-Hill, New York, 1969.

[10] J. K. Owens, *NASA Marshall Engineering Thermosphere Model – Version 2.0*, NASA/TM-2002-211786, June 2002, Section 3.2.3 Geomagnetic Activity: "Although high-latitude ionospheric current fluctuations drive the magnetic field fluctuations observed at these stations, the magnetic field fluctuations do not drive the thermosphere."

[11] James R. Wright, *Optimal Orbit Determination*, Paper AAS02-192, AAS/AIAA Space Flight Mechanics Meeting, San Antonio, Texas, 27-30 January, 2002

[12] James R. Wright, *Sequential Orbit Determination with Auto-Correlated Gravity Modeling Errors*, AIAA, Journal of Guidance and Control, Vol 4, No. 2, May-June 1981, page 304.

[13] James R. Wright, *Orbit Determination Solution to the Non-Markov Gravity Error Problem*, AAS/AIAA Paper AAS 94-176, AAS/AIAA Spaceflight Mechanics Meeting, Cocoa Beach, FLA, Feb., 1994.

[14] James R. Wright, *Near Real Time Atmospheric Density Estimation*, Applied Technology Associates, Inc. (ATA), 11 August 1990. This informal ATA company paper proposed a near real time global method for estimation of atmospheric density.

[15] James L. Burch, *The Fury of Space Storms*, Scientific American, April 2001, page 86.

[16] Meditch, J. S., Personal Communications, 1974. A private lecture on the importance of Sherman's Theorem

[17] Richard Hujsak, Personal Communications, 1980-1996.

[18] Tom Gidlund, Personal Communications, 2000-2001.

[19] L. G. Jacchia, Personal Communication via telecom, 1978

COMPARISON OF
MSIS AND JACCHIA ATMOSPHERIC DENSITY MODELS
FOR ORBIT DETERMINATION AND PROPAGATION

Keith A. Akins,[*] Liam M. Healy,[†] Shannon L. Coffey[‡] and J. Michael Picone[**]

Two atmospheric density model families that are commonly chosen for orbit determination and propagation, Jacchia and MSIS, are compared for accuracy. The Jacchia 70 model, the MSISE-90 model, and the NRLMSISE-00 model may each be used to determine orbits over fitspans of several days and then to propagate forward. With observations kept over the propagation period, residuals may be computed and the accuracy of each model evaluated. We have performed this analysis for over 4000 cataloged satellites with perigee below 1000 km for September-October 1999, and the 60 HASDM calibration satellites with a large observation set for February 2001. The purpose of this study is to form a picture of the relative merits of the drag models in a comprehensive view, using all satellites in a manner consistent with the operational practice of US space surveillance centers. A further goal is to refine this knowledge to understand the orbital parameter regions where one of the models may be consistently superior.

INTRODUCTION

Precise prediction of satellite motion in low-earth orbits requires significant knowledge of the effects of the atmospheric drag force. This force has two components which are usually difficult to ascertain: the ballistic coefficient of the satellite, and the atmospheric density at a particular point in its orbit. The problem of atmospheric density has been studied for decades, and two of the most widely-used atmospheric density models are Jacchia 70 and MSIS.

This paper presents a comparison of orbit determination and propagation results between three atmospheric density models for over 4000 satellites in low-earth orbit. These satellites represent all the unclassified satellites with perigees below 1000 km tracked by Naval Network and Space Operations Command (NNSOC, formerly Naval Space Command) in the period of September-October 1999. Further analysis is conducted using the HASDM satellite and observation data sets. This data includes 60 primary, or calibration, objects for which a very large number of observations were accumulated starting in early 2001.

[*]   Pennsylvania State University, Aerospace Engineering, University Park, Pennsylvania 16802; and Naval Research Laboratory, Code 8233, Washington, DC 20375-5355. E-mail: Keith.Akins@nrl.navy.mil.

[†]   Naval Research Laboratory, Code 8233, Washington, DC 20375-5355. E-mail: Liam-Healy@nrl.navy.mil.

[‡]   Naval Research Laboratory, Code 8233, Washington, DC 20375-5355.

[**] Naval Research Laboratory, Code 7643, Washington, DC 20375-5355.

DRAG MODELS

An important component of low-earth motion is atmospheric drag, and an important part of determining this force is knowledge of the atmospheric density. There are two major families of atmospheric density models based on empirical data collection in the 1960s, 70s, and 80s. These models provide estimates of the statistical mean temperature, total mass density, and number density of each species as a function of position, local time, universal time, and solar and geomagnetic indices. Of greatest interest to the astrodynamics community is the total mass density.

The two most widely used families of density models are *Jacchia*, based on the work of Luigi Jacchia, and *MSIS* (Mass Spectrometer — Incoherent Scatter), based on the work of Alan Hedin and collaborators (Ref. 1). The operational orbit determination community primarily uses the Jacchia 70 (J70) model (Ref. 2) while the atmospheric physics community has preferred the MSIS-class models, of which the latest is NRLMSISE-00 (N00) (Ref. 3). To compare the performance of J70 model and N00 for drag estimation, a discussion of the respective underlying observational data and the methods of generating the models is necessary. For a complete comparison, one must include the preceding MSIS-class model, MSISE-90 (M90) (Ref. 4). The two MSIS-class models extend from the ground ($z = 0$ km) to beyond the exosphere ($z \approx 500$ km) while the Jacchia models apply only above 90 km, unless the programmer has augmented the model at lower altitude with either the U. S. Standard Atmosphere (Ref. 5), as is done in the present analysis, or an MSIS representation, as is done in many space surveillance centers.

At thermospheric altitudes ($z > 90$ km), all three models primarily represent parameterized solutions of the equations of diffusive equilibrium for the individual chemical constituents of the atmosphere, such as molecular nitrogen (N_2), molecular oxygen (O_2), and atomic oxygen (O). The key altitude profile parameters for thermospheric variables are the temperature and the temperature gradient near the inflection point of temperature as a function of altitude ($120 - 125$ km), the exospheric temperature, and the total mass density and individual species densities at prescribed lower altitudes. In J70, the important parameters are temperature at 90 km and the mixing ratio 90 km $< z < 105$ km; for the MSIS models they are composition and temperature at 120 km. These profile parameters are, in turn, decomposed into harmonic, spherical harmonic, and other terms such as the dependence of the 10.7 cm solar flux index, $F_{10.7}$, on density, each with a coefficient to be determined from data. These terms capture dependencies on characteristic temporal and spatial scales, e.g., local time (diurnal-scale tides), UT, day of year (annual-scale cycles), latitude, longitude (not explicit in J70), solar declination (not explicit in M90, N00). The generation of a model consists of computing the atmospheric densities from observations over a range of the model's parameters and then doing a nonlinear least squares fit to this density data to derive the coefficients. These coefficients then become the core of the model; they allow the estimation of density given appropriate input parameters. For a realistic model, the number of coefficients can be quite large; N00, for example, has over 2000 nonzero coefficients.

The oldest of the three models, Jacchia 70, is a fit to a set of density values derived from orbit observations of satellites covering the time period 1961–1970, falling short of a complete solar cycle. The characteristic fitspans for retrieving density were 1–4 days. Because the region of the perigee dominates the integrated drag over an orbit, Jacchia was able to assign local geophysical and temporal variables (e.g., solar zenith angle) to the density values. Jacchia then inferred the various coefficients in his analytic density model to evaluate the temperature and density profiles as functions of the subroutine arguments (time and location variables, solar index, geomagnetic index). Based on our reading of Jacchia's reports, his inferences were not the result of a formal multivariate fitting procedure, but were instead based on a filtering or separation of the individual terms (semiannual variation, diurnal variation, etc.) from the data. It has been found (Ref. 3) that J70 does not produce an acceptable representation of his set of data for particular subsets of the model arguments. Given the computing capability during the 1960s, this is neither an indictment of Jacchia's work nor a surprise. Rather, one respectfully notes the utility of J70 in operational orbital drag estimation.

In the thermosphere, 90 km $< z < 500$ km, the MSISE-90 model is a Levenberg-Marquardt (nonlinear least-squares) fit to a data set consisting primarily of individual species density values and temperature data, as measured by ground-based and satellite-borne systems. The vast majority of the data covers the two decades prior to 1983, but the fit excluded the total mass density derived from satellite-borne accelerometer data, and orbit determination that Jacchia used. The explicit fit to temperature and individual species density observations has rendered the MSIS-class temperature and composition estimates demonstrably superior in agreement with direct atmospheric observations to those of the Jacchia models. An important contributing factor is the use of optimal multivariate assimilation techniques in generating the model; Jacchia apparently did not do this. The atmospheric physics community has validated the model by direct measurement of the density, for example by rocket flights with mass spectrometers, primarily against short time-scale, local measurements, but has not performed similar studies of orbital drag data. These data usually resolve timescales no shorter than a day and have similarly broader spatial scales, as compared to the N00 input parameters temperature and number density. We hope that this paper will provide the first step in that process.

The primary M90 deficiency is the exclusion of data on thermospheric total mass density. The NRLMSISE-00 model now removes this deficiency by adding extensive data from orbit determination (including Jacchia's data set) and from accelerometers (Ref. 3). From the standpoint of the model generation data, the N00 model therefore represents an improvement over both M90 (total mass density) and J70 (composition and temperature). In addition, the superior representation of the temperature function should translate to superior extrapolation or interpolation of the N00 total mass density measurements to the conditions of orbital observations. On the other hand, strong temporal and spatial filtering of density estimates by the orbit determination process could mitigate any potential advantage of N00 to the operational user, depending on the orbit sampling rate and accuracy.

Our motivation in undertaking this study is primarily to see if the MSIS atmospheric model, with its better representation of the atmosphere, can improve orbit propagation over Jacchia model still widely used in the astrodynamics community. Our intent is to examine a broad spectrum of satellites, in fact all relevant satellites, and not just a few specially selected ones.

Furthermore, we anticipate, in the net few years, the capability of real-time revision of atmospheric density. Research is presently underway examining the effectiveness of modifying the N00 model with near real-time, location-specific constituent data to provide a more accurate model. Both the former LORAAS instrument, onboard the ARGOS spacecraft, and the future SSULI instrument, planned for launch on the the next DMSP spacecraft, gather UV data from limb scans of the atmosphere to determine its composition at the time and location of the scan (Ref. 6). This composition data is then applied to the base N00 model to produce a better representation of the atmosphere near where the scan was taken. The present study will provide a baseline for evaluating the effectiveness of the real-time modified MSIS models.

TEST PROCEDURE

The comparison of atmospheric drag models in the context of satellite orbits makes use of the determination of a satellite's orbit using observations, and the propagation to a time later than the range over which observations were taken. Our focus in these studies is not the initial orbit determination process (Ref. 7), but rather the orbit estimation, or differential correction, that starts with some state estimate, and provides a new estimate at some later epoch based on a new set of observations. As such, it inherently involves propagation, because these observations, the original epoch, and the new epoch, will be at varied times. As part of this propagation an accurate force model must be used, and hence the need for good knowledge of atmospheric density.

The tests conducted are designed to compare atmospheric density models by looking at how they affect orbit determination. While we do not have an absolute truth with which to compare our computed result, we can compare one model to the others and identify which is better. For each of the satellite sets used, we have a database of observations, initial elements, and the appropriate solar indices, such as $F_{10.7}$ and the geomagnetic index a_p, for the time period in question. Other necessary data for the functioning of the orbit determination and propagation is present as well. It is important to note that the solar indices used are the retrospective final values. These are usually not available until some time after the time in question, and have been generated by Air Force Weather Agency and NOAA after extensive processing of observational data. An orbit determination on current data is different in that the value of these variables are projected into the future and therefore estimated; the final values may differ somewhat. As a result, these tests would be different from a present-time test because of the use of final values.

Abbreviation	Name
RST	Requested Stop Time
LOBST	Last Observation Before Stop Time
GPCD	General Pert Catalog Date
ESC	Epoch of Satellite in Catalog

Table 1: Important times in for orbit determination.

Before describing the test procedure, some time terminology must be introduced. The orbit determination process requires the use of several different times, and we have given each a name and an abbreviation. These are given in Table 1. First, the user must specify a final time for the orbit determination, the *requested stop time* (RST). This means that no observation after that time will be used for the orbit determination. Together with a fitspan, this determines the complete range of times over which observations may be used. The epoch of the determined vector will not be RST, but rather the time of the last observation before this time, which we call LOBST. Finally, to start the orbit determination we will need to use an element set from the NNSOC catalog; for convenience we give a date to this catalog, *general perturbations catalog date* (GPCD). More importantly, each satellite in this catalog has an epoch, which we call the *epoch of satellite in catalog* (ESC).

There are two sets of data on which the tests have been performed: all unclassified cataloged satellites with perigee below 1000 km drawn from a the set of all unclassified observations in the NNSOC (then Naval Space Command) catalog as of October 1999, which we call *lowsats*, and the HASDM calibration satellites from February 2001. The HASDM satellites were used because of the high number of available observations. The disadvantage is that these orbits, while selected to be representative of all applicable orbits, still are only a small fraction of the possible orbits affected by drag. The full catalog of objects with perigees below 1000 km was selected because these were all known satellites passing through the high drag region and therefore will show how the drag model affects operational analysis.

Both the differential correction and the propagation for this project were conducted using the Naval Research Laboratory's SPeCIAL-K orbit determination software suite (Ref. 8). This software, which is used at NNSOC, uses Special Perturbations (SP) for accurate integration and force modeling. The integration for this study was performed using an 8th-order Gauss-Jackson integrator, a 24×24 geopotential, and lunar and solar perturbations. Standard observational weighting and biasing was used, and, for the differential correction, the fitspan was defined, using standard operational algorithms, to be between 1.5 and 10 days based on the mean motion and the rate of change of mean motion of each object. SPeCIAL-K's automatic parallel processing capabilities allowed the large differential corrections to finish in approximately 4 hours. These runs were completed using 13 workers on a variety of computer platforms, including SGI, AIX and Linux.

There are several ways one might consider comparing the validity of an orbit determination process, including, of course, the atmospheric density model. The purpose in doing an orbit determination is to predict the orbit of a satellite, and as such models can be compared by their predictive power. This process is illustrated in Fig. 1. Additionally, the propagation capabilities of a model can be determined by using observations after the initial RST. With these new observations, one can predict forward and compute residuals, as is shown in Fig. 2. Finally, one can determine a new orbit one day advanced from the first correction and compare the final position with the position

954

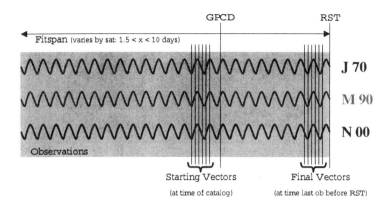

Figure 1: A schematic illustration of differential correction test.

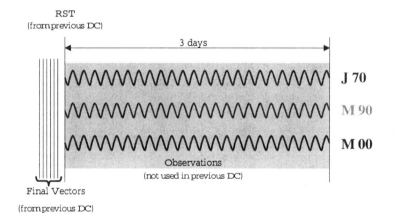

Figure 2: A schematic illustration of orbit propagation test.

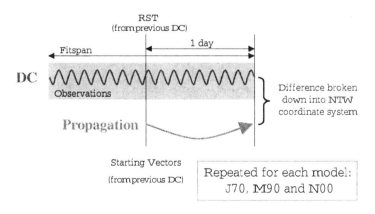

Figure 3: A schematic illustration of day-to-day test.

propagated from the first orbit determination, such as in Fig. 3. Either way, we have a reference truth based on the observations held in reserve.

INITIAL VECTOR QUALITY

When doing orbit determination by performing a differential correction, one needs an initial vector upon which to start the correction process. In the case of the tests described here, that comes from a general perturbations element set containing known orbital parameters of the satellite. It may be felt that such an element set is too inaccurate to provide convergence to the correct orbit directly, and so one may consider a process called *smoothing*. This process consists of the application of orbit determination on the initial state vector derived from the element set. By using a force model more accurate than general perturbations, including drag, it is hoped that the initial conditions will be closer to the true solution.

An important point is that smoothing will use an atmospheric density model as part of the drag force calculation. Prior to performing tests of the atmospheric model, we wish to assess the effect of smoothing. Specifically, we were concerned that we might introduce a bias in favor of the drag model used for smoothing, assuming all tests were smoothed the same way.

In principle a Newton-Raphson process, which is the essence of differential correction, should converge to the same value regardless of the starting vector, provided the starting point is in the basin of convergence (Ref. 9) for the solution. Therefore, one might assume that smoothing is an unnecessary step, and further, would not introduce any bias, presuming the basin boundaries did not change. To test this hypothesis, we performed a test we call "roughing," which is to take the vector converted from GP elements, add a random vector, and then do a differential correction. If the resulting vectors were the same or similar, one could assume this hypothesis true, and assume smoothing unnecessary but also non-biasing. The random vector would be chosen from a Gaussian distribution with a mean of zero and a standard deviation equal to the amount the vector changed in the smoothing calculation.

We chose as the test set the fifty satellites with perigee altitude below 1000 km that had the greatest number of available observations in the database from September–October 1999. Initially, the corresponding fifty element sets distributed by Naval Space Command on September 29, 1999 were converted to state vectors and these vectors were used as the starting points for the differential correction. In this test, GPCD is the day prior to RST; therefore ESC is at least 24 hours before LOBST.

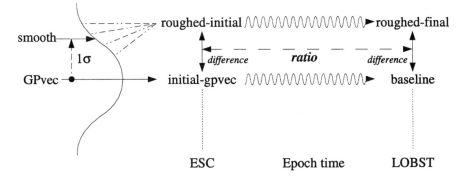

Figure 4: Roughing schematic. Once the smoothed vector has been determined, the difference between it and the GP vector is used as the standard deviation for the random generation of the roughed vector. The orbit is determined for both that and the original vector, and the ratio of the differences of the results formed.

We performed the roughing test on these satellites with the following procedure; see Fig. 4.

First, we retrieved catalog elements with a GPCD of 1999-09-29. These were converted to vectors at ESC. Alternatively, we create the "smoothed" vectors by fitting with observations prior to ESC and using a Jacchia 70 model, propagate the fit vectors to ESC. RST is 1999-09-30 00:00:00.000. Therefore, for each satellite we have two vectors at their ESC, one from the general perturbations propagation of the element set, and one from a differential correction on the observations using the element set as the initial vector, the smoothed vector. The important quantity derived from the smoothed vector is its difference from the general perturbations propagated one. This difference is used as the standard deviation in the generation of a Gaussian noise term to add to the general perturbations vector.

Orbit determination is performed on each of these two sets of vectors, with the final epoch being at LOBST with RST being 1999-10-01 00:00:00. For the position and velocity, the vector magnitude of the difference at this time between these two vectors may be computed, and a ratio formed with the corresponding vector magnitude of the difference of the roughed initial and the initial GP vector. This ratio is a figure of merit for the roughing test.

We have taken measures to make the differential correction consistent between the baseline and perturbed vector cases. First, we performed the differential correction normally on the baseline case. During such a run, there are observations that are rejected for a variety of reasons. These observations were then removed from the database for future runs, and the ability for the program to reject observations disabled. This has the effect of insuring that each baseline and perturbed case has an identical set of reasonable observations. Furthermore, we decreased the convergence tolerance of fractional weighted RMS change from the default 0.01 to 0.001. This means that solving the same orbit with different input vectors will result in more consistent results because the program quits iteration on RMS increase as well as decrease, and a decrease of tolerance makes it more likely to get through a "hump" that is often experienced where the RMS increases and then decreases before converging.

The results of this test are given in Figure 5. The bars indicate the range of ratios, and diamond indicates the mean. The results indicate that generally the ratio is quite small, less than 0.1, indicating that the baseline and the perturbed vectors all converge to essentially the same point. There are a few exceptions; notably satellite 25680, for which the ratios are rather large. This is because for the baseline case, the differential correction iteration was terminated for reasons other than convergence.

Therefore, we conclude that, within reason, perturbations of the initial vector do not affect the result of the differential correction and therefore smoothing is unnecessary. There are two caveats. First, this assumes the differential correction terminates at convergence. Second, in some cases, if observation rejection is not controlled, the set of accepted observations may be different and if so, there could be detectable differences in the resulting fit. There is no reason to believe this biases the result towards a particular drag model, however, it therefore also seems unnecessary to perform a smoothing operation in the first place.

RMS TEST

Comparing atmospheric models is difficult for a variety of reasons: there is usually no direct measurement of density, only the indirect measure from the satellite motion which involves additional unknowns; models vary in their accuracy over regions of the atmosphere and solar conditions, leading to mixed results depending on the orbit studied; and it is difficult to find a good measure of accuracy of orbits.

One way of testing the models' value to orbit propagation is to examine the residuals. Comparisons are done by examining the residual errors of both the differential correction and propagation between the models. A one-day differential correction is performed with each atmospheric model, and the epoch moved forward by the one day. The final root-mean-squared residual errors are then compared to evaluate the relative effectiveness of the particular model in a differential correction. The predictive effectiveness of the drag models is then calculated by predicting forward for three

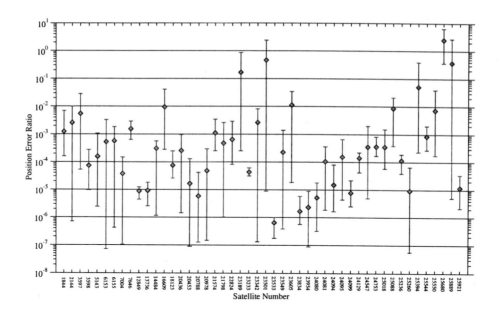

Figure 5: Ratios of final position difference to initial position difference.

days and computing the root-mean-squared residual errors at each of the known observations. In our test, we included only those satellites that updated in all cases. For the lowsats, the total number of such satellites included is 4587; for HASDM, all 60 satellites updated.

	M90	N00
lowsats DC mean	−0.00189	−0.00298
lowsats DC std. dev.	0.0913	0.0848
lowsats Prop. mean	−0.0451	−0.0325
lowsats Prop. std. dev.	0.271	0.328
HASDM DC mean	0.0170	0.0139
HASDM DC std. dev.	0.0754	0.169
HASDM Prop. mean	0.282	0.142
HASDM Prop. std. dev.	0.584	0.456

Table 2: RMS fractional change from J70 to the MSIS models.

In order to see the effect on differential correction and orbit propagation in changing the model from Jacchia 70 to both MSIS models, we have plotted the fractional change in unweighted RMS that is computed for each satellite. In Fig. 6 for differential correction and Fig. 8 for orbit propagation, these fractional changes are plotted by satellite in increasing order. With a scatter plot (Fig. 7 for differential correction and Fig. 9 for propagation) of the RMS with respect to various fit values, one may attempt to isolate certain orbits for which one or the other atmosphere model is superior. No such correlation is apparent, although some trends are discernible. The plots for the HASDM set (Figures 10– 13) are qualitatively similar. The RMS fractional change results are summarized in Table 2.

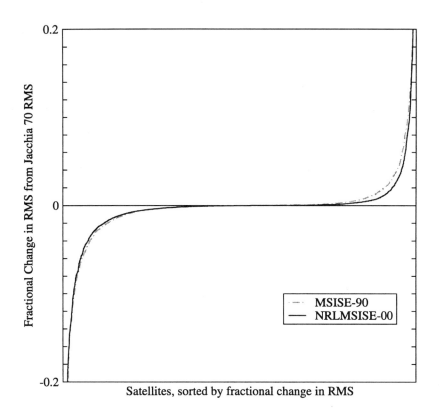

Figure 6: Fractional change in unweighted RMS for differential correction by satellite for the lowsats set, sorted by satellite in order of fractional change.

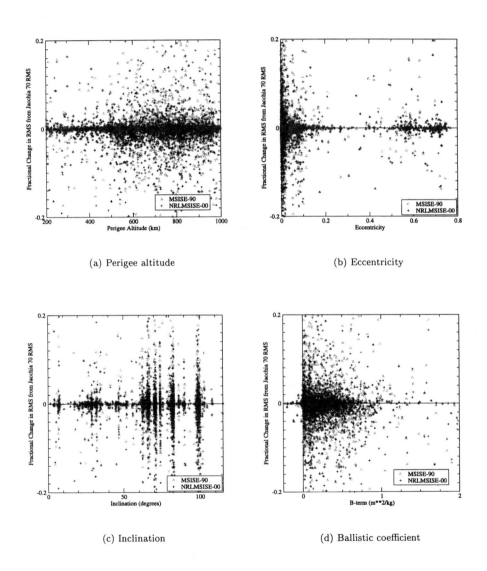

(a) Perigee altitude

(b) Eccentricity

(c) Inclination

(d) Ballistic coefficient

Figure 7: Fractional change in unweighted RMS in lowsats differential correction plotted against various satellite parameters.

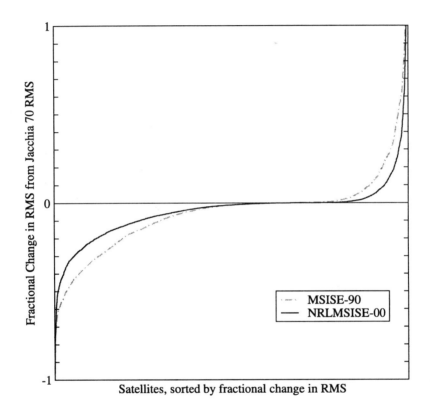

Figure 8: Fractional change in unweighted RMS for orbit propagation by satellite for the lowsats set, sorted by satellite in order of fractional change.

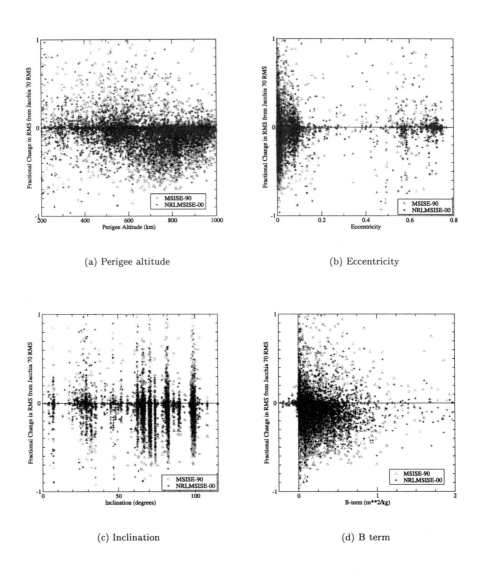

(a) Perigee altitude

(b) Eccentricity

(c) Inclination

(d) B term

Figure 9: Fractional change in unweighted RMS for orbit propagation by satellite for the lowsats set, plotted against various satellite parameters.

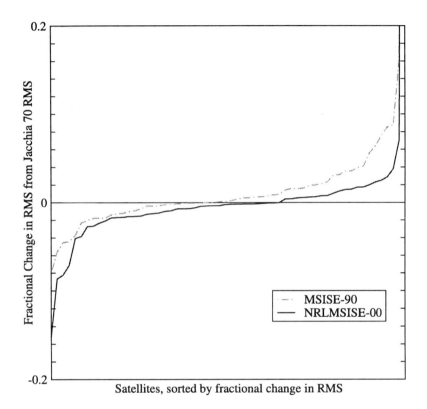

Figure 10: Fractional change in unweighted RMS in differential correction by satellite for the HASDM set, sorted in order of fractional change.

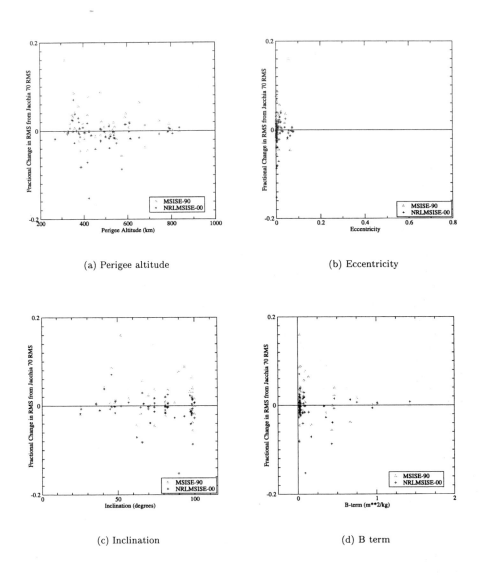

(a) Perigee altitude

(b) Eccentricity

(c) Inclination

(d) B term

Figure 11: Fractional change in unweighted RMS in HASDM differential correction plotted against various satellite parameters.

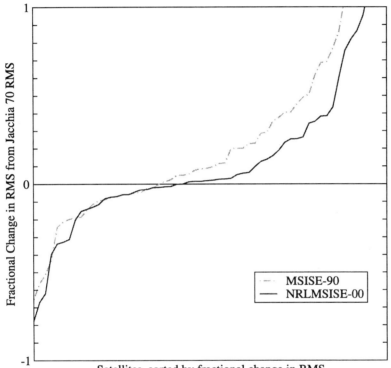

Figure 12: Fractional change in unweighted RMS for orbit propagation by satellite for the HASDM set, sorted in order of fractional change.

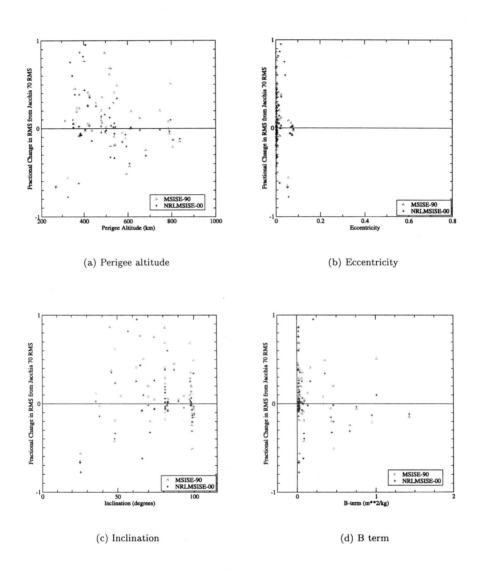

(a) Perigee altitude

(b) Eccentricity

(c) Inclination

(d) B term

Figure 13: Fractional change in unweighted RMS for orbit propagation by satellite for the HASDM set plotted against various satellite parameters.

ONE DAY ORBIT-TO-ORBIT COMPARISON TEST

The atmospheric models may be compared by the relative predictive power of orbits determined with each. Specifically, we determined the orbits using each of three models and observations over a period of time ending (RST) at 1999-10-01 00:00:00 for the lowsats set and at 2001-02-15 00:00:00 for the HASDM set. Then, we redetermined the orbits 24 hours later, including new observations during that additional day. Finally, we propagated each of these determined orbits to the later RST and compared those positions. Thus, we have an orbit-to-orbit comparison. In our test, we included only those satellites that updated in all cases. For the lowsats, the total number of such satellites included is 4423; for HASDM, all 60 satellites updated.

This relative position vector is best presented in the normal, in-track, cross-track coordinate system (NTW) (Ref. 7). As one might suspect, most of the error is to be found in the in-track direction (Fig. 14(a)), though there can be significant error in the other directions (Figs. 14(b), 14(c)) as well. The figures show the distribution of displacement errors for each of the three axes for each of the three density models. The latter two are almost identically distributed for all three density models, and are very symmetric around zero. The distribution is computed by stepping every two meters, finding how many satellites were within a 10 meters of that value (5 meters for cross track and normal), and dividing by the total population times 10 (5) meters. It thus represents an approximate probability distribution.

To summarize, the means and standard deviations of each component for each drag model in the lowsats tests are given in Table 3. The distance between the two predicted values gives a scalar

	Normal	In-track	Cross-track
J70 mean	29.8	493.5	−0.8
J70 std. dev.	1902.4	16592.0	748.8
M90 mean	34.1	166.9	−11.8
M90 std. dev.	1941.1	16588.2	820.0
N00 mean	40.5	172.1	−11.8
N00 std. dev.	1950.5	16534.3	814.4

Table 3: Mean and standard deviation from orbit-to-orbit test for each drag model in the lowsats set. Units are meters.

summary of the error for each satellite in each density model.

Approximately 0.25% of the lowsats set had very high errors, over 10 km, for each of the density models. The most prominent among these, for J70 and M90, was satellite 25228, a Russian rocket body. This was because it decayed in mid-October 1999, shortly after the period that we studied. At the time of study, it was on a reentry trajectory, and its predictability was poor. One could reasonably expect that this would present a severe test of the drag modeling, and indeed, the results obtained showed the one day propagation error to be 2231 km for both J70 and M90. In contrast, the N00 value is 568 km. While too large to be useful, this gives some indication that N00 is better able to handle reentering satellites. In any case, it has been excluded from all the results above.

Other satellites which had large errors were not in obvious difficult trajectories, but still some errors were over 100 km. While these were small in number, the large magnitude significantly skews the mean and standard deviation position error.

The same test was performed on the HASDM set. Because there are so few satellites in this set, it does not lend itself to a probability density or histogram plot, so we simply summarize the results numerically in Table 4.

CONCLUSIONS

In this paper, we examined the relative accuracy for orbit determination and propagation of three atmospheric density models, using two sets of data: all the cataloged satellites with perigee below

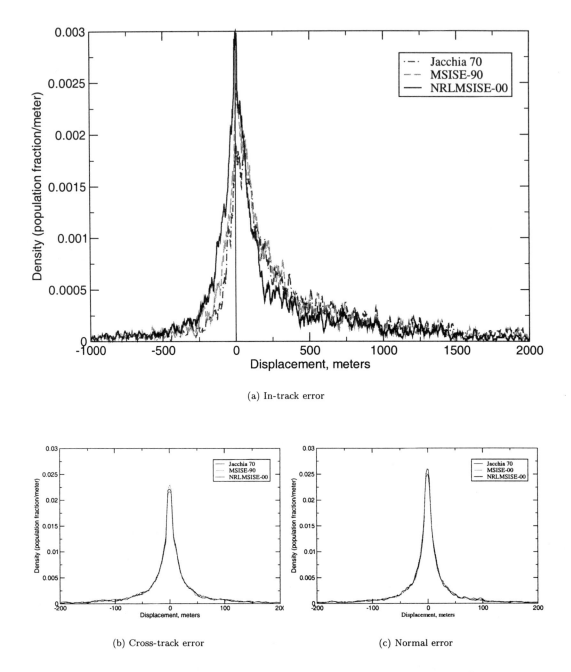

(a) In-track error

(b) Cross-track error

(c) Normal error

Figure 14: Distribution of errors in NTW frame of propagated day-old solution relative to newer solution for lowsats (1999) set.

	Normal	In-track	Cross-track
J70 mean	−16.4	1365.8	−0.2
J70 std. dev.	34.2	2552.1	21.6
M90 mean	−7.4	−7.4	8.3
M90 std. dev.	37.8	2224.4	59.0
N00 mean	−12.6	393.1	3.9
N00 std. dev.	42.4	2128.4	42.0

Table 4: Mean and standard deviation from orbit-to-orbit test for each drag model in the HASDM set. Units are meters.

1000 km, approximately 4500 satellites, and the HASDM calibration set of 60 satellites. In the majority of the tests performed, the MSIS-class models, led by NRLMSISE-00, demonstrate an improvement over the Jacchia model. However, it should be noted that this improvement overall is generally very small in light of the variations of the data and that, for the test where Jacchia proved better, the differences were also very small. In addition, individual satellites can show marked differences, with any of the three models showing the best results. Also, there was no evidence found to show that one model is better for any subset of the orbital regime, though this deserves more investigation. In the end, there is no single model which stands out as demonstrably superior over any other. There is hope, however, that the work currently underway on the implementation of the dynamic NRLMSISE-00 model will change this picture.

ACKNOWLEDGEMENTS

We thank Stefan Thonnard and Andy Nicholas for help with the MSIS model and suggestions on design and interpretation of the tests described. MSIS 90 was previously incorporated into SPeCIAL-K by Alan Segerman and Harold Neal.

REFERENCES

[1] Alan E. Hedin. MSIS-86 thermospheric model. *J. Geophys. Res.*, 92:4649–4662, 1987.

[2] L. G. Jacchia. New static models of the thermosphere and exosphere with empirical temperature models. Technical Report 313, Smithsonian Astrophysical Observatory, 1970.

[3] J. M. Picone, A. E. Hedin, D. P. Drob, and A. C. Aikin. NRLMSISE-00 empirical model of the atmosphere: Statistical comparisons and scientific issues. *J. Geophys. Res.*, 107(A12):1468, 2002.

[4] Alan E. Hedin. Extension of the MSIS thermosphere model into the middle and lower atmosphere. *J. Geophys. Res.*, 96:1159–1172, 1991.

[5] National Technical Information Service, Springfield, Virginia. *U.S. Standard Atmosphere*, 1976. (Product Number: ADA-035-6000); http://nssdc.gsfc.nasa.gov/space/model/atmos/us_standard.html.

[6] S. Knowles, M. Picone, S. Thonnard, A. Nicholas, K. Dymond, and S. Coffey. Applying new and improved atmospheric density determination techniques to resident space obect prosition prediction. In *Advances in Astronautics*, San Diego, CA, August 2001. American Astronautical Society, Univelt, Inc. AAS 01–426.

[7] David A. Vallado. *Fundamentals of Astrodynamics and Applications*. McGraw-Hill, New York, 1997.

[8] Harold L. Neal, Shannon L. Coffey, and Steve Knowles. Maintaining the space object catalog with special perturbations. In F. Hoots, B. Kaufman, P. Cefola, and D. Spencer, editors, *Astrodynamics 1997 Part II*, volume 97 of *Advances in the Astronautical Sciences*, pages 1349–1360, San Diego, CA, August 1997. American Astronautical Society. AAS 97–687.

[9] William H. Press, Brian P. Flannery, Saul A. Teukolsky, and William T. Vetterling. *Numerical Recipes in FORTRAN: The Art of Scientific Computing*. Cambridge University Press, 1992.

IMPLEMENTING THE MSIS ATMOSPHERIC DENSITY MODEL IN OCEAN

Lisa A. Policastri[*] and Joseph M. Simons[†]

The Naval Research Laboratory currently maintains and uses the Orbit Covariance Estimation and ANalysis (OCEAN) software package for orbit determination and propagation. A new Mass Spectrometer-Incoherent Scatter Radar (MSIS) class atmospheric density model, NRLMSISE-00, includes several notable differences from earlier atmosphere models, and has been added as an option in the OCEAN software. In every instance of testing, when comparing the RMS errors of the differential correction, the NRLMSISE-00 model showed improvement over the standard Jacchia-70 model.

INTRODUCTION

Atmospheric drag on a spacecraft is one of the largest sources of error in orbit determination. Accurate modeling of the upper atmosphere has been a topic of great interest for several decades. Thus, the ability to interchange several atmospheric density models in the Orbit Covariance Estimation and ANalysis (OCEAN) software package is a very useful feature. Three types of models have been available in OCEAN for some time: an exponential model, a uniform decay rate estimate, and 'he Jacchia-70 model.

One of the more recently developed classes of atmospheric density models is the Mass Spectrometer-Incoherent Scatter Radar (MSIS) class. In recent years, the Naval Research Laboratory's (NRL) Space Sciences Division has been studying and developing models of this class. That division now has a fairly new and refined MSIS class model, NRLMSISE-00, which has been made available for public use. Since MSIS class models had previously been tested and proven reliable, NRLMSISE-00 was implemented in OCEAN to enhance the software's capabilities.

Background on OCEAN

NRL currently maintains and uses the software package known as OCEAN for orbit determination and propagation. There are five main processing options available in OCEAN: Weighted Least Squares Orbit Determination, Create Ephemeris, Make OCEAN Initial Condition, Modify Database, and Kalman Filter Smoother.

* Spacecraft Development Section, Spacecraft Engineering Department, Naval Research Laboratory, Code 8213, Washington, DC 20375. E-mail: lpolicastri@space.nrl.navy.mil.

† Astrodynamics & Space Applications Office, Spacecraft Engineering Department, Naval Research Laboratory, Code 8103, Washington, DC 20375. E-mail: simons@ssdd.nrl.navy.mil.

The Weighted Least Squares Orbit Determination (WLS-OD) is one of two options used in this analysis. This process estimates spacecraft parameters, such as position, velocity, drag coefficient, and solar pressure coefficient, using a weighted least squares batch algorithm.

Three main files are needed to run the WLS-OD process for a spacecraft: a database command file, an initial condition file, and an observation file. For any WLS-OD process, various parameters can be uniquely defined in the database command file. These parameters include coefficient values, ephemeris-generation information, convergence criteria, propagator types, and force model types. The OCEAN initial condition (OIC) file defines the initial state information of the spacecraft. Finally, the observation file contains actual spacecraft observation data.

The WLS-OD process consists of several steps. First, the user configures the database command file. The OIC file is then used to calculate a nominal trajectory for the spacecraft. The weighted least squares algorithm next estimates the error between this nominal trajectory and the observed trajectory of the spacecraft. This error estimate is then added to the nominal trajectory, and the process is repeated until the computed error estimate meets the specified acceptance criteria.

The Create Ephemeris process, the second of the two OCEAN options used in this analysis, was run after each WLS-OD process. Ephemerides were generated based on the newly computed state. The same model parameters were used in both the WLS-OD and Create Ephemeris processes.

Background on NRLMSISE-00

This NRLMSISE-00 model contains several notable differences from earlier atmospheric density models, such as Jacchia-70. This model incorporates new data sets of orbital drag and satellite accelerometer data. Recent data on temperature and molecular oxygen number density have also been added. Unlike older atmospheric density models, MSIS is dependent on the current time, using data sets that reflect recent states of the atmosphere. In addition, high altitude oxygen ions and hot atomic oxygen components are now included in the total mass density. These species, named anomalous oxygen, were not accounted for in the past.[1]

Because of these added features, implementing the NRLMSISE-00 atmospheric density model into OCEAN seemed a very appropriate solution for enhancing the capabilities of the software.

IMPLEMENTATION

During its implementation, no alterations were made to the NRLMSISE-00 atmospheric density model. The only changes required were to the OCEAN software. Several subroutines were created or updated, and many variables were added to accommodate the model.

Once the model was fully implemented, it was tested with simulated data to ensure it was functioning properly. After analyzing the test results, it was concluded that the new atmospheric density model had been successfully added to OCEAN.

TEST DATA SELECTION

Starlette and GFZ

Once the new model had been integrated into OCEAN, it was necessary to characterize its performance within the software. To do so, actual satellite observation data would be needed. Observation data for two spacecraft were readily available at NRL, namely Starlette and GeoForschungsZentrum-1 (GFZ).

Starlette was chosen as the first test case because of the abundance of CTS (conventional terrestrial system, WGS-84) position data available for it. In addition, a highly accurate ephemeris[2] was available as a reference for ephemeris comparison. However, because Starlette is in a low-to-mid-altitude orbit (see Table 1), it may not show the effects of atmospheric density as clearly as satellites at lower altitudes.

The second satellite tested, GFZ, had a substantial amount of satellite laser ranging (SLR) data available, and, unlike Starlette, its orbit was very low (see Table 1). In fact, GFZ was the lowest geodynamic spacecraft from which SLR data was gathered and whose mission was specifically aimed at improving force modeling.[3] For these reasons, this spacecraft seemed an excellent candidate for this analysis.

HASDM Satellites

In addition to Starlette and GFZ, satellite data sets were chosen from a pool of HASDM data that was made available by NRL colleagues (see Table 1). The HASDM, or High Accuracy Satellite Drag Modeling, group consists of satellites chosen by U.S. Space Command to test their own drag models. Eight of these spacecraft were selected based on their low, nearly-circular orbits. Forty-five days of Space Surveillance Network observation data and their starting vectors were made available for testing purposes.

Table 1 THE TEN SELECTED SATELLITES

Satellite Name	Sat Num	Apo x Per Alt (km)	Inclination (deg)	Avg. # Obs/Day
Starlette	07646	1107 x 805	49.8	1440
GFZ 1	23558	395 x 382	51.6	287
Orbiting Vehicle 1-10	02611	558 x 504	93.4	465
OPS 5712 P/L 160	02826	814 x 801	69.9	404
Agena D Debris	04660	889 x 856	100.1	522
SL-3 Rocket Body	08128	436 x 412	81.2	368
Cosmos 880 Debris	11216	453 x 429	65.8	325
DMSP F8 Debris	18159	653 x 647	98.7	395
NOAA 12 Debris	21298	723 x 716	98.5	365
Pegasus Rocket Body	25647	565 x 521	97.6	499

TEST DESCRIPTION

To characterize the performance of MSIS within OCEAN, WLS-OD runs were performed for each satellite, using the same parameters and varying only the atmospheric density model. The magnitudes of the RMS values from the runs using MSIS would then be compared to those from the runs using Jacchia-70. This comparison would provide a measure of the quality of the fits made using the NRLMSISE-00 model. The parameters used in this analysis may be found in Table 2.

Table 2   PARAMETERS USED BY OCEAN IN THE TEST RUNS

	Measurement Weight	Bad Residual Limit	Statistical Data Editing Limit	RMS Convergence Limit
Azimuth	1×10^{-5} deg	2 σ	2.5σ, 0.7×10^{-3} deg	5×10^{-6} deg
Elevation	1×10^{-5} deg	2 σ	2.5σ, 0.7×10^{-3} deg	5×10^{-6} deg
Range	0.001 m	2 σ	2.5σ, 0.7 m	0.01 m
Range Rate	0.001 m/s	2 σ	2.5σ, 0.7 m/s	0.001 m/s

Force Model		Estimated Parameters	
Geopotential	41x41	Position	
Third Body Sun	ON	Velocity	
Third Body Moon	ON	C_D	
Solar Rad Pressure	ON	Range Bias (pass-by-pass)	
Drag	ON	Range Rate Bias (single-arc)	
Pole Tide	ON		
Ocean Tides	ON	**Max Iterations**	
Indirect Lunar Oblateness	ON	25	
General Relativity	ON		
Solar Earth Solid Tides	ON		
Lunar Earth Solid Tides	ON		

Starlette and GFZ

For the Starlette spacecraft, two MSIS runs were made, using a one-day fit and a seven-day fit. One-day, three-day, and five-day fits were similarly performed for GFZ. These runs were then performed again using Jacchia-70, and the resulting RMS values were compared.

Since a highly accurate reference ephemeris was available for Starlette, the Create Ephemeris option in OCEAN was invoked to generate ephemerides based on the updated states from the orbit determination processes. Ephemerides were generated from both the MSIS and Jacchia states, and these were compared to the reference ephemeris to provide another means of characterizing the performance of MSIS.

HASDM Satellites

Unlike the single runs performed for the Starlette and GFZ spacecraft, many series of linked runs were performed for the eight HASDM satellites. The forty-five-day data sets were divided into one-, three-, or five-day segments, and successive runs were performed for each of these smaller intervals.

For example, the data set for the first satellite, catalog number 02611, was first divided into forty-five one-day segments. The *a priori* initial condition file was used in the first WLS-OD run, which fit an orbit to the data from Jan 15.0 to Jan 16.0, using the NRLMSISE-00 density model and other parameters listed in Table 2. This run produced an estimated final state at Jan 16.0, and this state was used as the initial condition for the next run, Jan 16.0 to Jan 17.0.

This process was continued for all forty-five one-day segments, generating forty-five final RMS values and producing forty-five best-fit ephemerides. These runs were then repeated using the Jacchia-70 atmospheric density model, and the results were compared.

For satellite 02611, series of three-day runs and of five-day runs were also performed. It was determined from their results that three-day runs alone would be sufficient to characterize the performance of MSIS. Thus, for the remaining seven satellites, only series of three-day runs were performed.

After each series of three-day runs, a single-point abutment was performed. Consecutive ephemerides overlap at exactly one data point, and these endpoints should ideally describe the exact same position and velocity, if there are no orbit determination errors. Thus, taking the difference of the endpoints would provide another measure of the quality of the fits, and would either confirm or invalidate the results from the WLS-OD analysis. Figure 1 gives a graphical representation of single-point abutments.

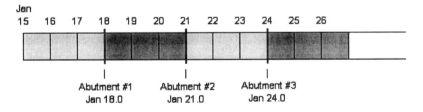

Figure 1 SINGLE-POINT ABUTMENTS

975

RESULTS AND ANALYSIS

Starlette and GFZ

The WLS-OD results for Starlette and GFZ are shown in Table 3. In every instance, the MSIS model produced a lower final RMS value than the Jacchia model. This improvement is more clearly demonstrated by the lower-altitude satellite, GFZ, which experiences significantly more atmospheric drag. In addition, the improvement for GFZ is greater in the longer runs, as the satellite has more time to experience the effects of drag.

Table 3 WLS-OD RESULTS FOR STARLETTE AND GFZ

Satellite	# Days in Test	Density Model	Final RMS (m)	% Improvement with MSIS
Starlette	1	Jacchia-70	0.1954311	1.36%
		NRLMSISE-00	0.1927667	
Starlette	7	Jacchia-70	0.3080477	1.16%
		NRLMSISE-00	0.3044701	
GFZ	1	Jacchia-70	0.0169555	4.05%
		NRLMSISE-00	0.0162673	
GFZ	3	Jacchia-70	2.2663856	38.0%
		NRLMSISE-00	1.4040546	
GFZ	5	Jacchia-70	26.114119	88.7%
		NRLMSISE-00	2.9402079	

Table 4 lists the results of the Starlette ephemeris comparisons. Note that, in both cases, the ephemeris produced by the MSIS run is closer to the reference trajectory than the ephemeris produced by the Jacchia run. Again, the longer run experiences a significantly greater improvement than the shorter run.

Table 4 STARLETTE EPHEMERIS COMPARISONS

Ephemeris Comparison	# Days in Ephemeris	RMS Error (m)	% Improvement with MSIS
Jacchia to Reference	1	0.348	1.72%
MSIS to Reference		0.342	
Jacchia to Reference	7	1.764	38.4%
MSIS to Reference		1.086	

HASDM Satellites

Table 5 shows a series of three-day runs for satellite 02611. The data were broken up into fifteen three-day segments, and the resulting RMS values for each segment are given. Since the data were input into OCEAN as four components, OCEAN computes an RMS value for each component.

Table 5   WLS-OD RESULTS FOR 02611 USING MSIS

Run	Dates	Azimuth (10⁻³ deg)	Elevation (10⁻³ deg)	Range (m)	Range Rate (m/s)
1	Jan 15 - Jan 18	27.5348	26.3871	8.1435	0.4684
2	Jan 18 - Jan 21	28.4048	22.0885	8.2149	0.3989
3	Jan 21 - Jan 24	25.8334	22.3023	10.3678	0.4232
4	Jan 24 - Jan 27	30.1034	24.8949	8.4461	0.5251
5	Jan 27 - Jan 30	30.4447	25.3834	9.6429	0.5383
6	Jan 30 - Feb 2	29.7402	23.7318	9.6959	0.5010
7	Feb 2 - Feb 5	27.1337	22.0743	9.7292	0.3259
8	Feb 5 - Feb 8	24.0638	24.7351	9.5388	0.3587
9	Feb 8 - Feb 11	31.5857	25.2160	8.1326	0.5009
10	Feb 11 - Feb 14	28.6186	25.8804	10.0957	0.5289
11	Feb 14 - Feb 17	29.3189	22.9470	7.8182	0.5092
12	Feb 17 - Feb 20	30.1568	23.6293	7.9125	0.3744
13	Feb 20 - Feb 23	32.6875	24.1508	8.7336	0.4977
14	Feb 23 - Feb 26	27.2877	21.8100	7.9739	0.4691
15	Feb 26 - Mar 1	28.9101	24.8240	11.2559	0.5592
Average RMS:		**28.7883**	**24.0037**	**9.0468**	**0.4653**

Also shown in Table 5 is the average of all fifteen runs for this satellite. These average RMS values were used to compare the effects of the MSIS and Jacchia models. The results for satellite 02611 are given in Table 6 below. In comparing the two models, the results from the NRLMSISE-00 model show improvement over Jacchia-70 in all components. The range RMS value displays the most dramatic improvement of the four measurements.

Table 6   COMPARISON OF AVERAGE THREE-DAY RUNS FOR 02611

	Azimuth (10⁻³ deg)	Elevation (10⁻³ deg)	Range (m)	Range Rate (m/s)
Jacchia	30.4291	24.5467	16.951	0.5165
MSIS	28.7883	24.0037	9.0468	0.4653
% Improvement	**5.39%**	**2.21%**	**46.63%**	**9.91%**

The three-day runs for the remaining seven HASDM objects produced lower RMS values for almost all components when using MSIS. Table 7 shows the percent improvement when using NRLMSISE-00 for all satellites and all components.

977

Table 7 PERCENT IMPROVEMENT IN RMS FOR AVERAGE THREE-DAY RUNS

Satellite	Azimuth	Elevation	Range	Range Rate
02611	5.39	2.21	46.63	9.91
02826	0.55	0.56	21.41	5.09
04660	0.49	-0.15	15.36	7.92
08128	15.20	4.80	43.92	42.55
11216	6.04	4.27	41.42	36.81
18159	12.80	7.85	55.51	52.42
21298	5.81	3.22	50.01	42.46
25647	1.11	-0.16	30.35	17.41
Average %	**6.34**	**2.94**	**39.03**	**32.58**

Azimuth and elevation values showed slight but noticeable improvement when using MSIS, with a 6% improvement in azimuth and a 3% improvement in elevation. As before, the range and range rate RMS values showed the largest improvement, at averages of about 39% and 33%, respectively. Figures 2 and 3 below detail the RMS improvement in range and range rate.

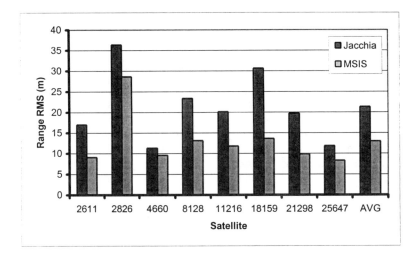

Figure 2 AVERAGED THREE-DAY RANGE RMS VALUES

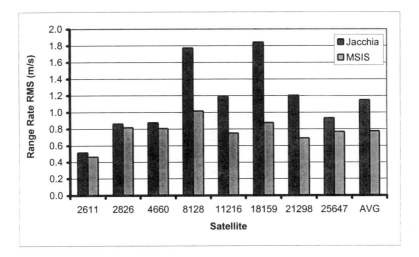

Figure 3 AVERAGED THREE-DAY RANGE RATE RMS VALUES

HASDM Abutment Error Performance

The output from the abutment testing was broken down into in-track, cross-track, and radial position error, as well as total position error and total velocity error. Table 8 shows the abutment error output from satellite 02611, using MSIS. Also shown is the "average abutment error" for that set of runs.

Table 8 ABUTMENT ERRORS FOR 02611 USING MSIS

Abutment	Date	In-Track (m)	Cross-Track (m)	Radial (m)	Position (m)	Velocity (mm/s)
1	Jan 18.0	27.6631	5.1945	19.5749	34.2842	71.4838
2	Jan 21.0	843.7300	11.5363	42.4054	844.8737	878.0304
3	Jan 24.0	97.5045	25.4087	88.1314	133.8652	144.8137
4	Jan 27.0	214.3286	45.4489	19.3332	219.9457	195.9550
5	Jan 30.0	87.7396	68.9616	8.5299	111.9227	91.9718
6	Feb 2.0	192.1170	48.1896	22.3399	199.3245	222.5834
7	Feb 5.0	581.5706	73.3398	32.4040	587.0716	592.0390
8	Feb 8.0	231.1187	22.1896	4.5528	232.2261	255.4208
9	Feb 11.0	195.9024	63.8663	3.7839	206.0849	178.2957
10	Feb 14.0	82.7001	9.8207	27.6610	87.7546	95.5295
11	Feb 17.0	203.1596	28.5346	1.8859	205.1624	163.0607
12	Feb 20.0	254.9306	24.2179	10.4393	256.2911	272.1128
13	Feb 23.0	201.7071	29.3499	7.5069	203.9694	243.9380
14	Feb 26.0	93.8430	9.7610	45.9994	104.9654	114.3281
Avg. Abutment Error		236.2868	33.2728	23.8963	244.8387	251.3973

979

The average abutment errors using Jacchia were computed in the same manner, and these values were compared. A comparison using the output from satellite 02611 is shown in Table 9, where the improvement in the abutment error from using the MSIS model can be clearly seen.

Table 9 COMPARISON OF AVERAGE ABUTMENT ERRORS FOR 02611

	In-Track (m)	Cross-Track (m)	Radial (m)	Position (m)	Velocity (mm/s)
Jacchia	715.7905	41.0578	42.9012	720.6552	799.1107
MSIS	236.2868	33.2728	23.8963	244.8387	251.3973
% Improvement	66.99%	18.96%	44.30%	66.03%	68.54%

The average percent improvement in abutment error for the eight HASDM satellites is shown in Table 10. In nearly every instance, the MSIS model produced a smaller abutment error than the Jacchia model. The average percent improvement for all satellites is also shown.

**Table 10 PERCENT IMPROVEMENT IN ABUTMENT
ERROR FOR AVERAGE THREE-DAY RUNS**

Satellite	In-Track	Cross-Track	Radial	Position	Velocity
02611	66.99	18.96	44.30	66.03	68.54
02826	33.86	-3.89	1.09	27.12	35.09
04660	21.07	-0.42	1.91	17.27	12.21
08128	90.92	40.15	59.33	89.87	90.70
11216	45.57	-24.24	63.86	45.88	47.43
18159	86.63	56.77	47.21	85.93	86.42
21298	85.18	28.96	48.41	83.19	83.73
25647	89.77	32.18	43.33	82.32	85.60
Average %	65.00	18.56	38.68	62.20	63.72

In-track position errors were the largest of the position components, and also showed the greatest improvement. This would be expected, though, as the spacecraft's velocity is in this direction. Cross-track and radial errors also improved significantly. Most noteworthy, however, were the improvements in total position and total velocity errors, at 62% and 64%, respectively. These improvements are detailed in Figures 4 and 5 below.

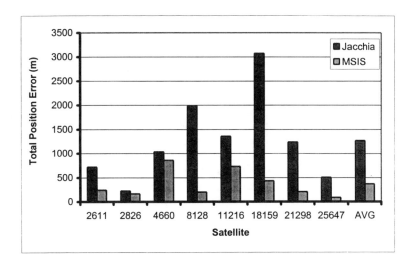

Figure 4   AVERAGED THREE-DAY POSITION ABUTMENT ERROR VALUES

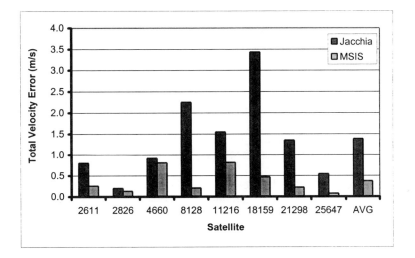

Figure 5   AVERAGED THREE-DAY VELOCITY ABUTMENT ERROR VALUES

CONCLUSIONS

From this analysis, it can be seen that the fit RMS values were consistently lower when using the NRLMSISE-00 atmospheric density model instead of the Jacchia-70 model. Azimuth and elevation values showed slight improvement, with range and range-rate showing much greater improvement.

In-track abutment error showed the largest improvement of the position components, while cross-track and radial errors also decreased substantially. Most importantly, however, total position error and total velocity error decreased by more than half, on average.

NRLMSISE-00 has not yet been extensively tested for orbit determination applications. For those instances to date in which it has been tested, the results have proven inconclusive.[4] When implemented in NRL's OCEAN software package, however, this model has demonstrated considerable improvement over the Jacchia-70 model for satellites in low-earth, nearly-circular orbits.

ACKNOWLEDGEMENTS

Thanks to Dr. J. Michael Picone (Naval Research Laboratory) for providing the NRLMSISE-00 atmospheric density model. Thanks to Mr. Jay Middour (NRL) for supporting and encouraging this analysis. Thanks to Mr. Mark Davis (Honeywell Technical Solutions), Dr. Liam Healy (NRL), and Dr. Alan Segerman (AT&T Government Solutions) for providing invaluable data and ideas. Thanks to Mr. Jacques Fein (Computer Sciences Corporation) for lending his time and sharing his knowledge.

REFERENCES

1. J.M. Picone, A.E. Hedin, D.P. Drob, and J. Lean, "NRLMSISE-00 Empirical Atmospheric Model: Comparisons to Data and Standard Models," AAS 01-394, 2001.

2. J. H. Seago, M. A. Davis, and A. E. Clement, "Precision of a Multi-Satellite Trajectory Database Estimated from Satellite Laser Ranging," AAS 00-180, 2000.

3. "Quicklook: GFZ-1," JPL Mission and Spacecraft Library, September 2001, < http://samadhi.jpl.nasa.gov/msl/QuickLooks/gfz1QL.html >.

4. S. Sheffler, A. Segerman, and D. B. Spencer, "A Comparison of Generated Ephemerides using SpecialK with Various Atmosphere Models," AIAA 2002-4542, 2002.

A VALIDATION OF ATMOSPHERIC DENSITY DETERMINATION FROM LORAAS ULTRAVIOLET SPECTRA BY COMPARISON WITH HASDM SATELLITE DRAG MEASUREMENTS

S. H. Knowles,[*]
A. C. Nicholas,[†] S. E. Thonnard,[†] J. M. Picone,[†] K. F. Dymond[†] and S. McCoy[†]

The extraction of upper atmospheric density from ultraviolet airglow emission measurements shows promise of improving near-real-time knowledge of atmospheric density. This innovative method is in the process of being validated by comparison with other independent information. For this purpose, use of the HASDM data set, consisting of high-tasking-rate Space Surveillance Network observations of about 130 low-earth-orbit satellites with a variety of orbital parameters, provided an excellent opportunity. During the period 15 January 2001 to 21 April 2001, corrections to the modeled atmospheric density were determined by assuming that the satellite inverse ballistic coefficient was constant and then attributing any apparent variation to a change in the mean atmospheric density along the orbit. This validation analysis used the SpecialK (SPK) special perturbations software to fit orbits with a one day fit span for the HASDM data set, with the exception of a few Molniya satellites for which this was not possi-ble. Daily values of the inverse ballistic coefficient were determined for each satellite using SPK with NRLMSISE-00 as an atmospheric reference. The dis-cussion focuses on HASDM objects #22560 and #25648 because of their stable inverse ballistic coefficients and because they represented low (465 km) and high (760 km) altitudes. In particular, these two objects were used for a prelimi-nary comparison of the drag-determined density to LORAAS-derived densities. In addition, the month of July 2000 was examined using MIR observations to confirm the behavior of the ultraviolet density extraction algorithm under storm conditions. The LORAAS density has excellent correlation with the drag de-rived density and a strong dependence on solar and geomagnetic indices. Validation of the behavior of UV density variations as a function of altitude and analysis of the bias between UV and drag-derived density will re-quire further investigation.

---

[*] Raytheon Information Technology and Scientific Services, 4400 Forbes Blvd., Lanham, Maryland 20706.

[†] E.O. Hulburt Center for Space Research, Naval Research Laboratory, 4555 Overlook Avenue, S.W., Washington, DC 20375.

INTRODUCTION

From May 1999 to April 2002 the Naval Research Laboratory (NRL) obtained ultraviolet airglow spectra from the Low Resolution Airglow and Aurora Spectrograph (LORAAS) payload on the Advanced Research and Global Observation Satellite (ARGOS) spacecraft. This instrument provides a passive method to determine neutral atmospheric density in the thermosphere. Although the ARGOS mission ended in April 2002, research continues on the best method of determining atmospheric density from the ultraviolet (UV) data. The launch of the first operational copy of LORAAS is scheduled for the summer of 2003. The operational sensor, called the Special Sensor Ultraviolet Limb Imager (SSULI), will be launched on a new suite of Defense Meteorological Satellite Program (DMSP) satellites. Previous contributions describe the technique used to derive density from UV observations[1,2] and present quantitative results from a ten-day period in December 1999[3]. This contribution covers a more extensive time period, including two major solar-geomagnetic events.

BRIEF DESCRIPTION OF UV DENSITY DETERMINATION METHOD

NRL has developed a method for determining upper atmospheric density from the intensity of dayside ultraviolet airglow emissions obtained from an orbiting platform.[1,2] The LORAAS spectrograph on the ARGOS spacecraft is used to make altitude scans in a sun-synchronous orbit at a local time of about 14^h30^m. The spectral emission profiles are used to adjust an intensity model with four independent parameters. These parameters can then be used with MSIS to estimate the density at other times and locations. In this paper, "MSIS" refers to the NRLMSISE-00 empirical thermospheric model[4]. For this contribution the density at the LORAAS observation location was used. The inversion method requires extensive knowledge of the instrument used, the physics of airglow emissions and the altitude distribution of various atmospheric constituents. The inversion results are sufficiently consistent to justify validation against the extensive geophysical and satellite drag database available at NRL.

The purpose of this contribution is to provide quantitative evaluation of the UV derived density technique against satellite drag derived values. To enable this comparison it is assumed that all geographical variations are quasi-random and will average out over this interval; evidence of this is presented later. Anomalous effects in the auroral zone and the South Atlantic Anomaly have been excised. A more thorough treatment of the latitude, longitude and sun angle effects is expected to result in a reduced variance, but is outside the scope of this contribution.

The LORAAS instrument was active from May 1999 to April 2002, a period during solar maximum when several major solar and geomagnetic events occurred. Two periods during which major solar and geomagnetic activity occurred were chosen for this analysis. The period of January-April 2001 coincided with the start of the HASDM[5] (High Accuracy Satellite Drag Model) effort, and encompassed the high solar and geomagnetic activity period of late March – April 2001. Figure 1 presents solar and geomagnetic indices for this period, including $F_{10.7}$ solar radio flux, a commonly used proxy for the solar extreme ultraviolet (EUV) energy input into the atmosphere, as well as A_p, a daily measure of geomagnetic activity. For solar maximum, this period was relatively quiet until late March, when a major geomagnetic event occurred, followed by lesser activity during April. The month of July 2000, during which the 'Bastille Day' event occurred, was also included to provide a check on the long-term consistency of the measurement method, as well as to include another example of response to major storm events.

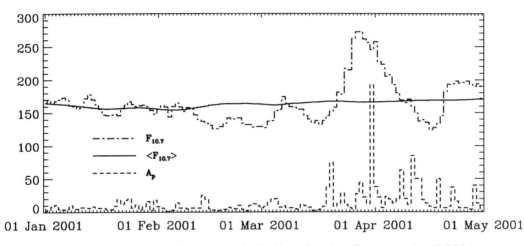

Figure 1 – Solar and Geomagnetic Indices During January-April 2001

During the HASDM effort, a number of objects received high tasking from the Space Surveillance Network in order to evaluate the Dynamic Calibration of the Atmosphere (DCA)[6] self-calibration method. NRL has observation data from all HASDM objects; two of these, at different altitudes, were analyzed for an evaluation of the UV-derived densities.

CALIBRATION ACCURACY

In order to validate the UV density determination method, it is necessary to have an absolute standard against which to calibrate. One method to obtain this would be to rely on a model such as MSIS computed at the location of each LORAAS limb scan. This is subject, however, to the statistical inaccuracy of at least ± 15% commonly observed in satellite drag comparisons[7]. Instead, satellite drag-derived densities were used for comparison. The use of drag determination has two disadvantages: it is an integrated measurement rather than a point measurement, and it requires accurate knowledge of the object's ballistic coefficient to derive an absolute density. An additional factor, which requires validation is the software suite SpecialK[8] (SPK), a special perturbations software suite maintained by NRL.

DETERMINING TOTAL DENSITY FROM SATELLITE ORBITS

For the purpose of this contribution, the mean drag-determined density is defined as

$$\varrho = \frac{B}{\overline{B}}\,\overline{\varrho}\,,\quad (1)$$

where ϱ = the mean drag-determined density
  B = inverse ballistic coefficient determined by SPK
  $\overline{B}$ = the mean ballistic coefficient for that object (proxy for True B)
  $\overline{\varrho}$ = the mean MSIS density over the fitspan

True B's

The calibration of the B terms computed by SPK was performed in two steps. In the first step, the mean value of B determined with SPK was compared with the absolute value computed from physical principles for Starshine 3. Close agreement supports the use of $\overline{B}$ (mean B from SPK) in Equation 1 as a proxy for the true value. Since Starshine 3 was not in orbit during the analyzed periods, mean B terms derived by Bowman ("True Bs")[9] were used to provide an intermediate calibration of the SPK derived $\overline{B}$ values.

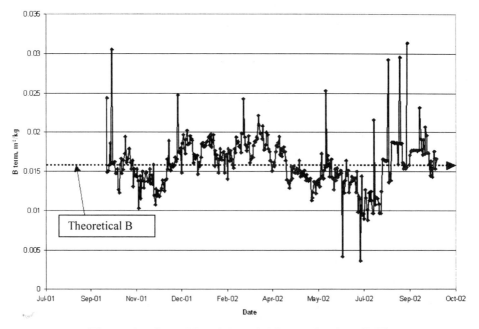

Figure 2 – Starshine 3 Special Perturbations B Terms

One-day special perturbations integrations of Starshine 3 were performed during the period from launch in July 2002 to October 2002 (Figure 2). The theoretical value for B from first principles for Starshine 3 is 0.0156 m^2/kg, based on the published mass and cross-section data, and a coefficient of drag of 2.1(C. Cox private communication 2002).

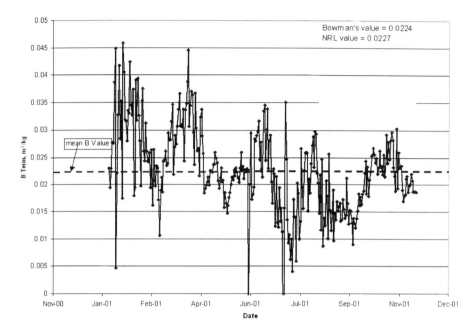

Figure 3 – Sample Variation of B Term For Satellite #60 The Two Negative B Terms are Believed to be Statistical Fluctuations.

The SPK mean B value from September 2001 through September 2002 is 0.0161 m^2/kg. Bowman[9] has discussed issues involved in "True B" determination. These issues include various physical and geometric factors, including the effective coefficient of drag C_D. The estimated uncertainty of 5% in C_D is dominated by the apparent medium and long term fluctuations due to model inadequacies[9]. The observational value of $\overline{B}$ agrees with the theoretical value to well within this uncertainty. Since no Starshine satellite was in orbit during the period of this investigation, several of Bowman's objects were used to provide references for the accuracy of B determinations using SPK. The Bowman objects chosen are shown in Table 1. Figure 3 shows B as determined with SPK for object #60; fluctuations are similar to those presented in Figure 2 for Starshine 3. The two negative B values are consistent with statistical fluctuations. A long-term mean $\overline{B}$ for the ensemble of these objects of 0.0324 m^2/kg $\pm 1\%$ was determined; this compares to the value of 0.0327 m^2/kg obtained by Bowman. Analysis of the residuals between $\overline{B}$(SPK) and True B (Bowman) in Table 1 shows that SPK determined densities are unbiased with an accuracy of better than 16%.

CHOICE OF OBJECTS FOR COMPARISON WITH UV DENSITIES

For comparison with the LORAAS-derived density, two of the HASDM satellites were chosen on the basis of most stable B term over the period of January-November 2001. In order to do this, the orbit of each HASDM object was determined by special perturbations over this period, and it was assumed that the objects with the lowest B variance had the most stable effective frontal area cross-section. There are in fact a number of other HASDM objects with only slightly less stable B terms that can be used for a more comprehensive analysis.

Table 1 – Satellites Chosen From Bowman's True B Calibration Objects

SSN #	Int'l Desig	Shape	Peri-gee ht., km	Apogee ht., km	Incl., Deg.	Bowman 'True B', m²/kg	SPK Mean B, m²/kg	SPK B Error, m²/kg
60	1960-014A	Dbl Cone	386	1145	49.9	0.02237	.0227	0.0005
229	1962-002D	Cylinder	555	600	48.3	0.05157	0.0552	0.0010
2129	1966-026B	Cylinder	507	605	98.2	0.04512	0.0401	0.0006
2611	1966-111B	Cyl+ Boom	504	560	93.4	0.02322	0.0187	0.0003
4221	1969-097A	Cone+ Cyl	375	1930	102.7	0.02111	0.0253	0.0005

Table 2 shows the general characteristics of these objects. Object #22560 has a relatively high mean altitude of 760 km, while also having an easily observable drag effect. Object #25648 has a mean altitude of 465 km, in the lower range of objects that remain in orbit for extended periods. For each object, SPK fits were performed using a one-day fit span.

Table 2 – HASDM Objects Used in UV-Drag Comparison

SSN #	Launch	Type	Perigee, km	Apogee, km	Incl., degrees	Eccen-tricity	SPK Mean B	One-day Allan Variance
22560	1974	Debris	716	803	101.6	0.006	2.032	0.200
25648	1999	Debris	450	479	97.5	0.002	0.0426	0.0028

This fit span, while shorter than the customary LEO fit span, was long enough to give robust convergence. Successive fit spans are non-overlapping; thus each point should be statistically independent. For each object in Table 2 the mean B determined by SPK over the period from mid-January through early November 2001 is shown, together with the Allan variance[10] for lag of one day. The Allan variance is similar to the conventional vari-

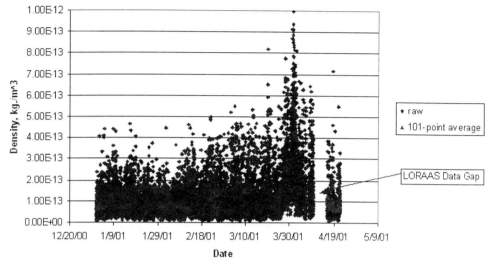

Figure 4 – LORAAS 760 Km UV Density For January-April 2001
Individual Points And 101-Point Means Are Shown.

ance but compares points separated by a certain lag. Using the Allan variance eliminates the effect of long-term modeling errors, which would cause the conventional variance to be 3 to 4 times higher.

UV-DERIVED DENSITY ANALYSIS

The period of 1 January 2001 to 30 April 2001 was chosen for primary analysis. Within this interval limitations were caused by the fact that HASDM data started on 15 January; conversely, during the month of April the spectrograph only obtained limb scans during the periods 1-9 April and 16-21 April. The UV-derived densities were evaluated at the position of the spectrograph scan and the mean altitude of the drag object. Figure 4 shows the UV retrieved density at 760 km altitude for this period. Only retrievals with a goodness of fit measure (χ^2) of less then 5 were used. The individual density retrievals showed significant variability, most likely due to geographical variations as well as geophysical forcing. The gray curve shows a 101-point boxcar average spanning 0.2 days.

Figure 5 shows the daily-averaged UV-derived density for 465 km and 760 km during this period, compared with the SP drag-derived density for #25648 and #22560. Solar and geomagnetic indices are also shown for reference. During the January-February period the drag-derived density showed little change in mean value, but from March 26th to April 5th, high solar and geomagnetic activity caused a major increase in density. The correlation of UV-derived density with drag-derived density can be seen during the storm. The UV-derived density shows an offset in the mean value from the drag-derived density during the quiet period of January and February, 2001; however, short-term variations remain well correlated.

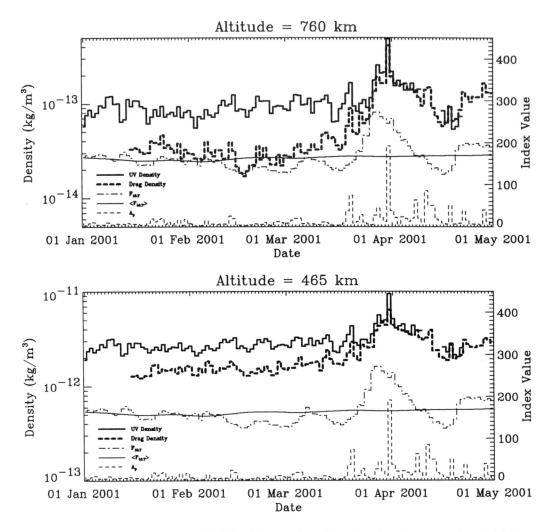

Figure 5 – Comparison of UV And Drag Densities During January-April 2001. Solar And Geomagnetic Indices Are Shown For Reference.

The month of July 2000 (the Bastille Day solar and geomagnetic storm) was also examined to provide a check on the consistency of the UV-derived density for a different mission epoch. Figures 6 and 7 show a comparison of UV-derived density with drag density derived from MIR (#16609). Note that the UV-derived density (90 seconds) shown in Figure 6 has a higher spatial and temporal resolution than the drag-derived density (one day). The UV-derived density in Figure 7 has been averaged over a day for direct comparison to the drag-derived density. MIR had a mean altitude of approximately 325 km

990

during this period. The mean B value for MIR was computed from July 2000 data only. Figure 6 and 7 indicate strong response of both UV-derived and drag-derived density with geomagnetic forcing (A_p). However, the response of the UV-derived density is seen prior to the response of the drag-derived density.

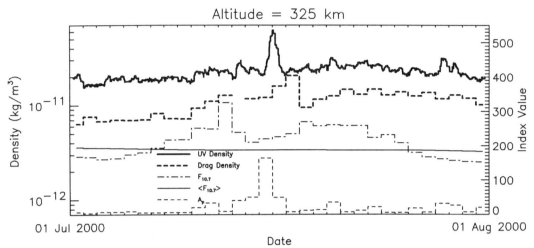

Figure 6 – Comparison of UV Density with MIR Drag Density For July 2000 (51-Point Averages)

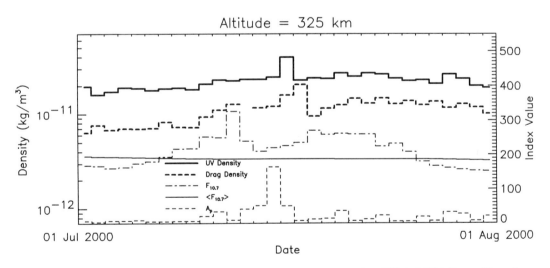

Figure 7 – Comparison of Daily Average UV Density With MIR Drag Density For July 2000

REGRESSION ANALYSIS

In order to quantify the correlation of UV-derived density with drag-derived density or geophysical forcing terms, linear regression coefficients of the form $y = C_1 + C_2 x$ were computed.

Table 3 – Regression Coefficients for UV Density on Solar & Geomagnetic Indices

	$\varrho = (C_1 \pm E_1) + (C_2 \pm E_2) * F_{10.7} + (C_3 \pm E_3) * A_p$								
Alt. and Time	C_1, kg/m³	E_1, kg/m³	E_1/C_1 *100%	C_2	E_2	E_2/C_2 *100%	C_3	E_3	E_3/C_3 *100%
760 km Jan.-Apr 2001	-1.4×10^{-14}	1.2×10^{-14}	86%	5.9×10^{-16}	0.8×10^{-16}	13%	1.7×10^{-15}	0.1×10^{-15}	7%
465 km Jan.-Apr 2001	7.0×10^{-13}	2.1×10^{-13}	29%	1.2×10^{-14}	0.1×10^{-14}	11%	2.9×10^{-14}	0.2×10^{-14}	7%
760 km July 2000	3.2×10^{-14}	0.6×10^{-14}	19%	4.9×10^{-16}	0.3×10^{-16}	6%	8.0×10^{-15}	0.5×10^{-15}	6%
325 km July 2000	1.3×10^{-11}	0.4×10^{-11}	26%	3.8×10^{-14}	1.6×10^{-14}	47%	5.0×10^{-14}	2.0×10^{-14}	40%

An ideal relationship would show a linear term C_2 with a small relative error, and a small bias term C_1 with a large relative error. Regression coefficients for UV-derived density on solar and geomagnetic indices are shown in Table 3, and those for UV-derived density on drag are shown in Table 4.

Daily average densities for all methods were used for the regression analysis. In Table 3 for example, the UV-derived density at 760 km correlates to $F_{10.7}$ with a relative error of 13% and to A_p with an error of 7%, while there is no significant bias term. The other entry in Table 3 also shows strong correlation with $F_{10.7}$ and A_p, but more significant bias values. In Table 4, the UV-derived density at 760 km correlates to drag-derived density with an error of 6%, while there is a residual bias of $5.8*10^{-14}$ kg/m³. For the July 2000 period, the regression coefficients for UV-derived density on solar and geomagnetic indices are shown for 760 km and the mean MIR altitude of 325 km, and the regression coefficient on drag is shown for 325 km.

The existence of a strong functional relationship between the UV-derived density and drag-derived density, and between the UV-derived density and solar and geomagnetic indices, is clear. During periods of high geomagnetic activity the UV-derived density variability agreed well with the drag-derived density. During periods of low geomagnetic activity, an apparent bias was present; however the two were well correlated. Analysis of the bias must account for the lack of spatial coincidence between the LORAAS observations and those of the reference objects.

Table 4 -Regression Coefficients for UV-Derived Density on Drag-Derived Density

	$\varrho = (C_1 \pm E_1) + (C_2 \pm E_2) * \varrho_{drag}$					
Altitude And Time Period	C_1, kg/m^3	E_1, kg/m^3	E_1/C_1 *100%	C_2	E_2	E_2/C_2 *100%
760 km Jan.-Apr. 2001	5.8×10^{-14}	0.03×10^{-14}	1%	0.84	0.05	6%
465 km Jan.-Apr. 2001	1.21×10^{-12}	0.11×10^{-12}	9%	0.88	0.05	6%
325 km July 2000	8.20×10^{-12}	1.36×10^{-12}	16%	1.20	.0.11	9%

CONCLUSION

This article presents a comparison of total thermospheric density derived from LORAAS UV observations and density derived from Space Surveillance Network satellite observations. Two major storms were examined, the Bastille Day event in July 2000 and a storm spanning late March to early April 2001. The results support the use of UV airglow observations to determine total thermospheric density and estimate satellite drag. The regression coefficients of the ultraviolet density on the A_p and $F_{10.7}$ indices, and also on satellite drag-derived density, demonstrate a strong correlation. The UV method has potential to provide density variations for operational use. As such, it naturally complements the DCA orbit self-calibration method. Additionally, the response time of the ultraviolet measurements to solar and geomagnetic forcing occurred on a time scale significantly less than the drag orbit fit span. The UV-derived density correlated well with drag variations during both high and low geomagnetic activity, but a significant bias was present during low activity. The source of this bias requires further investigation.

This work provides a framework for a more comprehensive validation of the UV density determination method. A more accurate analysis requires propagating the LORAAS density to the position of the drag reference object, a larger sample using more reference objects to examine the UV-derived density as a function of altitude, and the use of longer analysis periods.

REFERENCES

1. Meier, R.R. and J.M. Picone, "Retrieval of Absolute Thermospheric Concentrations from the Far UV Dayglow: An Application of Discrete Inverse Theory," J. Geophys. Res. Vol. 99, No. A4, 1.April 1994, pp. 6307-6320.

2. Nicholas, A.C., J.M. Picone, S.E. Thonnard, R.R. Meier, K.F. Dymond, and D.P. Drob, "A Methodology for Using MSIS Parameters Retrieved From SSULI Data to Compute Satellite Drag on LEO Objects", Journal of Atmospheric and Solar-Terrestrial Physics 62 (2000) 1317-1326.

3. Nicholas, A.C., S.E Thonnard, K.F. Dymond and S. Budzien, S. Knowles, and E. Henderlight, "Comparison of Ultraviolet Airglow Derived Density to Satellite Drag Derived Density", presented at the Summer 2002 AIAA/AAS Astrodynamics Specialist Conference, Monterey, CA, 6 August, 2002, AIAA paper # 2002-4735.

4. Picone, J.M., A. E. Hedin, D. P. Drob, and A. C. Aikin, NRLMSISE-00 Empirical Model of the Atmosphere: Statistical Comparisons and Scientific Issues, J. Geophys. Res., Vol. 107, No. A12, 1468, doi:10.1029/2002JA009430, 24 December 2002.

5. Storz, M., B.R. Bowman and J. Branson, "High Accuracy Satellite Drag Model (HASDM)", presented at the Summer 2002 AIAA/AAS Astrodynamics Specialist Conference, Monterey, CA, 7 August, 2002, AIAA paper # 2002-4886.

6. Marcos, F.A., M.J. Kendra, J.M Griffen, J.N. Bass, D.R, Larson and J.F. Liu, "Precision Low Earth Orbit Determination Using Atmospheric Density Calibration", AAS 97-631, In: Hoots, F.R., B. Kaufman, P.J. Cefola, and D.B. Spencer (Eds.), Astrodynamics, Vol. 97(1), AAS, San Diego, pp. 501-513, 1998 (paper presented at the August 1997 AAS/AIAA Astrodynamics Specialist Conference, Sun Valley, Idaho, 7 August, 1997).

7. Knowles, S.H., J.M. Picone, S.E. Thonnard and A.C. Nicholas "The Effect of Atmospheric Drag on Satellite Orbits During the Bastille Day Event", Solar Physics 204, pp.387-397, 2001.

8. Neal, H.L., Coffey, S.L., Knowles, S.H., "Maintaining the Space Object Catalog with Special Perturbations", AAS 97-687, In: Hoots, F.R., B. Kaufman, P.J. Cefola, and D.B. Spencer (Eds.), Astrodynamics, Vol. 97(2), AAS, San Diego, pp. 1349-1360, 1998 (paper presented at the August 1997 AAS/AIAA Astrodynamics Specialist Conference, Sun Valley, Idaho, 7 August, 1997).

9. Bowman, B.R., "True Ballistic Coefficient Determination for HASDM", AIAA paper # 2002-4887, AIAA/AAS Astrodynamics Specialist Conference and Exhibit, Monterey, CA, 5-8 August 2002.

10. see "http://www.allanstime.com/AllanVariance/index.html".

BRIGHTNESS LOSS OF
GPS BLOCK II AND IIA SATELLITES ON ORBIT

Frederick J. Vrba,[*] Henry F. Fliegel[†] and Lori F. Warner[‡]

We present results from an on–going, ground–based, multi–band photometric monitoring program of GPS Block II and IIA satellites, from the U.S. Naval Observatory, Flagstaff Station. Our results show generally well-behaved light curves as a function of Sun–Observer–GPS phase angle, although two occurrences of glints have been observed. Observations of satellites with a range of on orbit time of about ten years show loss of surface reflectivity between 450–790 nm of about 10 percent/year. We discuss implications of how this might relate to solar panel power loss on these spacecraft. We also present early results for Block IIR spacecraft and plans for future work.

INTRODUCTION

Beginning in 1999 the U.S. Naval Observatory (USNO) and the Aerospace Corporation have joined in an effort to obtain optical and infrared ground–based photometry of Global Positioning System (GPS) satellites on orbit. The initial motivations of this work were to understand better brightness versus Sun–Observer–GPS phase angle for use in later USNO ground–based astrometric observations and subsequently the reflectivity of the satellites for refinement of the GPS solar force model. In this paper we review some of the results of our observations, obtained at optical wavelengths, of Block II and IIA satellites which had on–orbit ages at the times of observation of from about one to 12 years.

OBSERVATIONS

Optical observations are obtained with the 1.0–meter aperture telescope at the U.S. Naval Observatory, Flagstaff Station, located in Flagstaff, Arizona. A thinned 1024x1024 Tektronix CCD is used as the detector and is pixelized so as to give an approximate 11x11 arcminute field of view. The telescope is tracked at sidereal rate and pointed toward predicted ephemeridal positions of a given satellite. With telescope pointing and CCD readout overhead, the typical duty cycle is 5 minutes.

---

[*]   U.S. Naval Observatory, Flagstaff Station, P.O. Box 1149, Flagstaff, Arizona 86002-1149.
E-mail: fjv@nofs.navy.mil.

[†]   The Aerospace Corporation, M5-685, 2350 E. El Segundo Blvd., El Segundo, California 90245-4691.
E-mail: henry.f.fliegel@aero.org.

[‡]   The Aerospace Corporation, M5-962, 2350 E. El Segundo Blvd., El Segundo, California 90245-4691.
E-mail: lori.f.warner@aero.org.

The observations are calibrated against astronomical standard stars, resulting in brightness, measured in astronomical magnitudes, accurate to between 1–5%, depending on the sky conditions and the Sun–Observer–GPS phase angle. A description of the observational methods, equipment employed, data reduction techniques, and early results for several satellites have been presented earlier[1].

The standard astronomical filters used are B (blue), V (visual), R (red), and I (infrared) with nominal central wavelengths of 440 nm, 550 nm, 640 nm, and 790 nm, respectively. The flux contained within the V filter, for example in Eq. (1), is

$$F_V = \int_0^\infty \phi_V(\lambda) F_\lambda \, d\lambda \quad (1)$$

where $\phi_V(\lambda)$ is the convolved filter bandpass and detector quantum efficiency and F_λ is the flux distribution of the object observed. Brightness is reported here in terms of the classical astronomical magnitude system. This system is logarithmic with the chosen logarithmic base being the fifth root of 100 = 2.512 and with the sense that the fainter the object, the larger the magnitude which specifies the brightness. The ratio of two flux densities, F1 and F2, is proportional to the difference in apparent astronomical magnitudes via Eq. (2)

$$m_1 - m_2 = -2.5 \log (F_1 / F_2) \quad (2).$$

Thus, an object of magnitude 10 is 100 times brighter than an object of magnitude 15. Magnitudes in each bandpass have been calibrated to absolute flux densities via laboratory measurement by many authors (eg. Ref. 2).

LIGHTCURVES

Between March 1999 and February 2002, the period of time for which we currently have reduced data, we have a total of 15 data sets for the Block II satellites SV13, 15, 17, and 21 and a total of 19 data sets for the Block IIA satellites SV 22, 23, 24, 25, 29, 30, 34, 35, 36, 37, 38, 39, and 40. Additionally, we have a total of 9 data sets for the Block IIR satellites SV 41, 43, 44, 46, 51, and 54. The Block IIR satellites are not the focus of this paper, but results for them are mentioned briefly below. In a few cases above, the data sets cover only a few degrees of the Sun–Observer–SV phase angle, but many cover a significant fraction, if not all, of the observable light curve phase range.

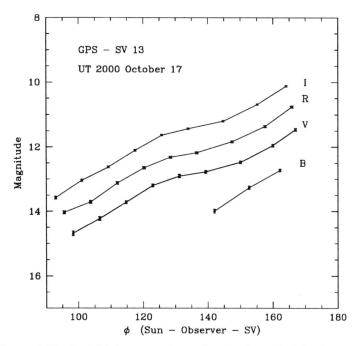

Figure 1 Typical Light curve Behavior During Nightly Overpass

Our data show that Block II and IIA spacecraft show a roughly linear slope of brightness versus phase angle between about 100 degrees phase angle (where the satellites first become front side illuminated) to roughly 160 degrees (where they enter Earth shadow) with brightness slope of about 3–5% per degree of phase angle. A typical set of light curves, in four filters, for one passage over Flagstaff, AZ is shown in Figure 1 for SV 13 on the night of UT 2000 October 17. Repeated observations over several nights near in time show that the light curves are generally quite repeatable.

We have observed two occasions of obvious glints in the 34 light curve data sets for Block II and IIA satellites, all of which have sufficient observed range of Sun–Observer–SV phase range to reveal a glint, if it occurred. Thus, glints are an infrequent, but not an unobserved event. The larger of the two events occurred for SV 23 on 2001 October 23 at about 10 hours UT. This is illustrated in Figure 2 where we have plotted the B–, V–, and I–band magnitudes (with B and I normalized to the V data point nearest in time) as a function of time. The differences in fluxes between the B, V, and I–bands due to the reflected light colors of the satellite are relatively unimportant at this scale. The data show a glint of about 4 astronomical magnitudes (40x), the central part of which lasted about 0.2 hours, but whose wings encompassed about 1.0 hour. In fact, this is a lower limit to the amplitude of the glint since a) the 5 minute duty cycle and 5 sec exposure times of the observations almost certainly did not hit the peak of the glint and b) for the brightest two data points the CCD detector was saturated.

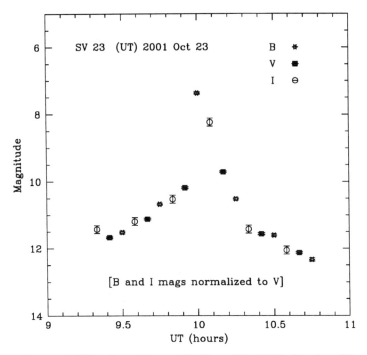

Figure 2 Glint Seen From SV 23 on UT 2001 October 23

The Sun sensors on this particular spacecraft, SV 23, failed early in its mission and its solar panels have subsequently been aligned perpendicularly to the Sun only roughly by frequent commands from the ground. The fact that the solar panels are known to be misaligned, along with the large amplitude of the glint, strongly suggests that the glint was produced almost entirely by nearly specular reflection from the panels and not from the spacecraft body. Since the wings of the glint curve merge smoothly with the light curves of our other observations, this suggests that solar panels provide much of the reflected light, even during normal operation, especially at shorter wavelengths. This is consistent with our suggestion that, although the colors of the spacecraft are much redder than we expect[1], the loss of brightness over many years (see below) is primarily by darkening of the solar panels and not largely of the spacecraft bodies.

BRIGHTNESS LOSS

Once we began our monitoring, even early data began to show indications that the spacecraft must suffer some kind(s) of surface degradation on orbit. It was almost immediately noticed that older spacecraft appeared to be fainter at a given observed phase angle than newer spacecraft. This is illustrated in Figure 3 where V magnitude at two nearby phase angles is plotted for four GPS satellites. Since the data were taken on two consecutive nights, the SV age is roughly coded by SV number. In addition, the optical colors appear much redder than would be expected from surfaces as measured in the factory[1]. These indications led us to investigate further GPS surface degradation with time.

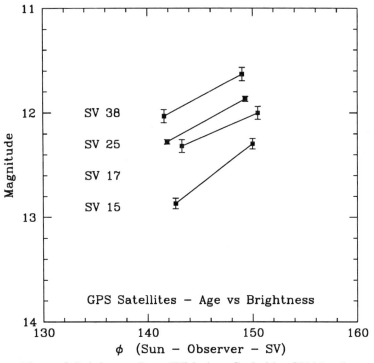

Figure 3 Brightness Loss With Age Coded by SV Number

Based on the behavior of GPS spacecrafts' light curves as a function of Sun–Observer–SV phase angle it was decided to obtain B–, V–, and I–band photometric measurements near phase angle 140 degrees; that is, near the middle of the observable range of phase angle. Observations at lower phase angle would be less accurate due to less reflected light flux, while those at higher phase angle, are subject to spacecraft instability for those orbital passes observed when the spacecraft are coming out of Earth shadow and are reorienting themselves.

In Figure 4 we show the I–band magnitudes at phase angle 140 degrees and normalized to an observer–SV distance of 20K km and at an Earth–Sun distance of 1.0 astronomical unit, plotted as a function of time on orbit for Block II and Block IIA spacecraft, which are shown as solid points. The Block II and IIA vehicles have similar external appearances and form a relatively continuous locus showing declining brightness, consistent with a loss of reflectivity of about 10 percent/year. We note, however, that the scatter in the data is much larger than the error bars of the observations, which are no larger than the plotted points for most observations. We believe that the error bars take into account all known sources of both random and systematic errors, such as photon statistics, CCD read noise, and (most importantly) terrestrial atmospheric extinction variations, which are monitored closely in our observing method. Thus, the scatter appears to be a genuine artifact of the satellites' brightness at the time of observation. Recall, we earlier stated that repeat observations of a given satellite over a time span of a few nights produces essentially identical light curves within the error bars. In fact, two such sets of repeat observations over a time span of 3 nights are hidden in Figure 4 by the fact that their data points overlap in this figure. Aside from confirming the reliability of the observing technique, this result prompts us to ask the question of whether the scatter of the data points is simply due to differences in individual satellites, since the observations shown are an admixture of many different vehicles. The three data points which are connected by solid lines are observations of SV 15 taken at widely scattered times over three years. Clearly the data points encompass the typical scatter seen and illustrate the general situation that repeat observations of satellites taken over time scales of months to years, even when at the same phase angle and normalized to a given range and Earth–Sun distance do not produce the same amount of reflected light.

We suggest that the solar panels may not always be aligned precisely perpendicularly to the Sun and that, from one year to the next, the angle that they present to the observer is slightly different than one would predict from phase angle calculations alone. The possibility that the SV body Z–axis is not properly aligned to the center of the Earth seems to us less likely. SV body misalignment would affect received power, of which we have no reports. Furthermore, the wrinkled Kapton covering of the SV body and its forest of antennas do not provide a smooth specularly reflective surface as do the solar panels, so that small body misalignments would have a smaller effect on observed brightness.

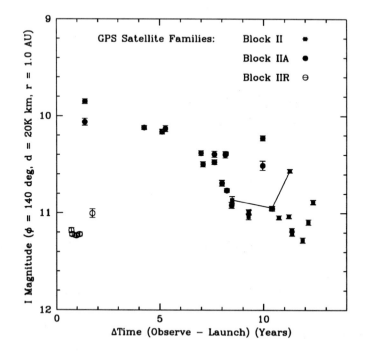

Figure 4 I–Band Brightness versus Time on Orbit

Whatever the ultimate cause(s) of the scatter in the data, it is clear, nonetheless, that the Block II/IIA vehicles have suffered about a 1.2 magnitude reflectance loss (about a factor of 3) over the 11 year time span between the youngest Block IIA vehicle and the oldest Block II vehicle. This amounts to approximately a 10% reflectance loss per year. Similar results are seen in the V– and B– filters which we have used to monitor this phenomenon, although the scatter gets progressively larger toward the shorter wavelength filters. This is probably due to the increasing role of the solar panels' reflectance toward bluer wavelengths and the possible mis–alignment of the panels, as mentioned above.

A few newer Block IIR satellite points (open points) are shown for reference. While the Block IIR data do not have a sufficient on–orbit time spread to determine reflectance loss, their lower brightness, as compared to Block II and IIA vehicles, at an on orbit time of about one year is evident. This is evidence of the total lower reflectivity in the red of the IIR vehicles which have optically–black bodies versus the orange body color for II/IIA satellites.

DISCUSSION

The apparent large reflectance loss of GPS satellites on orbit should be an important input parameter to models of the solar force influence in spacecraft orbit calculations. We note one other potentially significant application of our results. Power loss from from Block II and IIA solar panels has occurred at a rate of about 5 percent/year (Ref 3); a rate about half of that of the reflectance decline of these satellites. That the reflectance and power loss rates might be linked and, hence, indicate a possible reason for the power loss, stems from the fact that photons which are absorbed in the solar panels to produce power are complementary to those photons which are not absorbed but reflected from the solar panels and observed as reflected light at the telescope. Whether the solar panels' cover glasses are slowly being made more opaque by the space environment, such as by impacts from soft protons or micrometeors, or by photochemical self–contamination[4], photons reaching the solar panels will have traversed the affected cover glass perpendicularly and suffer a loss of $e^{-\tau}$ where τ is the glass optical depth. However, although light passes only once through the glass (really quartz) to reach the active solar panel elements to generate power, it passes twice through the glass to reach our telescope, entering perpendicularly and reflected obliquely. The path length for this reflected light is (1 + secant D), where D is the angle the reflected ray makes with the perpendicular. Since our observations are typically made when D is about 40 degrees, the reflected path length is roughly 1.3. Thus, photons reflected off of the solar panel material will have traversed a total path length of 2.3 and suffered a total loss of $e^{-2.3\tau}$. In the linear approximation for small τ, photons reaching the telescope will suffer approximately 2.3 times the loss of those reaching the solar panel material. Thus, the power loss occurring at about half the rate of the reflectance loss is at least consistent with cover glass deterioration. We emphasize that this is merely a hypothesis and depends on several assumptions not yet proven. Much work must be done to characterize the optical properties of the spacecraft surfaces and the way they change in space before our measurements can be compared to a model. Nevertheless our data will help constrain the possibilities in such future models.

FUTURE WORK

The data we have in hand to date suggest several paths for future work, both in observation and interpretation. We first have several more nights of data which we have obtained, but which have not yet been reduced. This will add about 30% more information to plots like Figure 4. We hope to continue our program of optical observations, especially if the GPS community can lend at least vocal support of our work. While we may occasionally observe a few of the oldest vehicles to see if the light declines continue, we will now primarily turn our attention to the Block IIR vehicles. Within about 5 years we should have enough time baseline to see if they suffer the same kind of reflectivity loss as the Block II/IIA vehicles. Since the IIR vehicles have blackened bodies and solar panels which are roughly twice the size of the II/IIA vehicles, these may prove to be useful diagnostic differences in interpreting our II/IIA data. We will also turn to analysis of optical data which we have not discussed in this text, such as color changes as a function of observed phase angle and age on orbit. Finally, we have a high quality near–infrared array camera which we will use to extend our observations to 2250 nm. With observations extending from 440–2250 nm, essentially the entire reflected light spectrum can be analyzed.

REFERENCES

1. H.F. Fliegel, L.F. Warner and F.J. Vrba, "Photometry of Global Positioning System Block II and IIA Satellites on Orbit," *J. of Spacecraft and Rockets*, Vol. 38, No. 4, 2001, pp. 609–616 .
2. M.S. Bessell, "UBVRI Photometry II: The Cousins VRI System, Its Temperature and Absolute Flux Calibration, and Relevance For Two Dimensional Photometry," Pub. Astronomical Society of the Pacific, Vol. 91, No. 543, 1979, pp. 589–607.
3. S. Missirian, private communication, The Aerospace Corp., 2002.
4. A.C. Tribble, "Revised Estimates of Photochemically Deposited Contamination on the Global Positioning System Satellites," *J. of Spacecraft and Rockets*, Vol. 35, No. 1, 1997, pp. 114–116.

COMPARISON OF ACCURACY ASSESSMENT TECHNIQUES FOR NUMERICAL INTEGRATION

Matthew M. Berry[*] and Liam M. Healy[†]

Knowledge of accuracy of numerical integration is important for composing an overall numerical error budget; in orbit determination and propagation for space surveillance, there is frequently a computation time-accuracy tradeoff that must be balanced. There are several techniques to assess the accuracy of a numerical integrator. In this paper we compare some of those techniques: comparison with two-body results, with step-size halving, with a higher-order integrator, using a reverse test, and with a nearby exactly integrable solution (Zadunaisky's technique). Selection of different kinds of orbits for testing is important, and an RMS error ratio may be constructed to condense results into a compact form. Our results show that step-size halving and higher-order testing give consistent results, that the reverse test does not, and that Zadunaisky's technique performs well with a single-step integrator, but that more work is needed to implement it with a multi-step integrator.

INTRODUCTION

Any orbit determination or propagation computation is subject to error from a number of sources: observational error, modeling error, numerical integration error, etc. In building an error budget, it is useful to have a numerical estimate of each. For space surveillance, there is frequently a computation time-accuracy tradeoff that must be balanced. Instead of improving accuracy indefinitely, it may be desirable to accept a degraded accuracy to a certain level in order to improve the computation time or to simplify the integrator in an embedded computer. In this paper, we survey the common techniques of assessing integration error and compare their results.

Although assessment of integration accuracy is a long-standing problem, much of the focus of the available literature is on the three-body problem in astronomy, particularly over long periods of time. Because of the chaotic nature of such systems, accuracy is desired to assess characteristics of chaotic regions in time and space. In astrodynamics, we are more concerned with the problem of orbiting a planet with geopotential, atmospheric drag, solar radiation and other perturbations, and often for a relatively modest number of orbits. Accuracy is desired here for precise knowledge of spacecraft position and orbit over a short period of time. Although the problems are similar, the needs are sufficiently different that the integrator, including order and step-size, may be chosen differently. Nevertheless, as astronomers have developed techniques to assess integrators, we may make use of this work to develop a means of assessing integrators for astrodynamics. Furthermore,

* Graduate Assistant, Department of Aerospace and Ocean Engineering, Virginia Tech, Blacksburg, Virginia 24061; and Naval Research Laboratory, Code 8233, Washington, DC 20375-5355. E-mail: maberry2@vt.edu.

† Research Physicist, Naval Research Laboratory, Code 8233, Washington, DC 20375-5355; and Lecturer, Department of Aerospace Engineering, University of Maryland, College Park, Maryland 20742. E-mail: Liam.Healy@nrl.navy.mil.

we would like a quantitative estimate of error which will assist in choosing a less but still sufficiently accurate method to gain computation speed.

Attempts at characterization of integration error frequently stop with two-body integration because of the ability to determine absolute accuracy. Perturbations have a significant effect on integration accuracy that is not apparent from a two-body study, as is shown below.

An Nth order system of ordinary differential equations, with initial conditions given at $t = t_0$, can be written in the general form (Ref. 1)

$$\dot{x} = f(t, x), \qquad x(t_0) = x_0, \tag{1}$$

where x and f are vectors of N functions and x_0 is a vector of N constants. In our case x consists of three position functions and three velocity functions. If some numerical integration algorithm is used to solve (1), the algorithm generates an approximate solution $\tilde{x}$. After n steps the accumulated error is

$$\xi_n = x(t_n) - \tilde{x}_n, \tag{2}$$

where $x(t)$ is the exact solution, $t_n = t_0 + nh$, and h is the step-size. The error can be written in the form (Ref. 1)

$$\xi_n = [x(t_n) - x_n] + [x_n - \tilde{x}_n], \tag{3}$$

where the first difference is the *truncation error*, and the second difference is the *round-off error* (Ref. 2). Truncation error exists because the numerical integration algorithm has been truncated at some (locally correct) order, p. Truncation error is dependent on the step-size h and the order p, and decreases as the step-size is decreased. Round-off error exists because computers only keep track of numbers to a finite number of significant digits, and some error is introduced during every calculation. Round-off error increases as the step-size is decreased, because more computations are performed.

Various techniques exist to estimate the error, ξ, of a numerical integrator. Each technique has strengths and weaknesses; our purpose in this paper is to describe what these are and to identify a practical procedure testing integrators. These techniques have common features, so at the outset, we consider the choice of a set of test orbits that provide a variety of realistic conditions and thus adequately balance the integrators. Moreover, since all techniques involve the numerical comparison of two prospective orbits, we describe the computation of the error ratio, which provides a figure of relative merit between two integrated orbits. One of the orbits is produced by the integrator being tested; the other, the reference orbit, covers the same time period and initial conditions and may be produced by either integration or analytic computation.

TEST ORBITS AND INTEGRATORS

If the integrator is expected to perform satisfactorily over a wide variety of orbits, some representative sample of these orbits is needed as a test set of initial conditions. In particular, forces that stress an integrator, such as atmospheric re-entry, should be included to give a sense of the worst case. For the case of space surveillance catalog maintenance, circular orbits near the earth and at geosynchronous altitude represent the bulk of the catalog; the addition of a high-eccentricity elliptical orbit with perigee dipping well into the atmosphere stresses the integrator.

Three test cases are considered, a low earth orbit, an elliptical orbit, and a geosynchronous orbit. The initial conditions of these test cases are shown in Table 1. The velocities have been listed to ten significant figures so that the orbital elements have the desired value. The test cases all have an initial epoch of 2001-10-01 00:00:00 UT, and a ballistic coefficient of 0.01 m^2/kg.

Since our purpose here is to compare accuracy assessment techniques, we need not only test orbits, but test integrators on which to try the various techniques. We have chosen two integrators commonly used in astrodynamics. The first is a fourth-order Runge-Kutta, as described in Ref. 2 (pp 320-321). The Runge-Kutta is a single-step, single-integration integrator. The step-sizes used for the Runge-Kutta integrator are 5 seconds for test cases 1 and 2, and 1 minute for test case 3.

Table 1: Test Case Initial Conditions

test #	r (km)	$\dot{r}$ (km/sec)	Perigee Height (km)	Ecc.	Inc. (°)
1	$6678.137\hat{\boldsymbol{I}}$	$5.918276127\hat{\boldsymbol{J}} + 4.966023315\hat{\boldsymbol{K}}$	300	0.0	40
2	$6578.137\hat{\boldsymbol{I}}$	$7.888427772\hat{\boldsymbol{J}} + 6.619176834\hat{\boldsymbol{K}}$	200	0.75	40
3	$42164.172\hat{\boldsymbol{I}}$	$3.074660237\hat{\boldsymbol{J}}$	35786	0.0	0.0

The second integrator is an eighth-order combined Gauss-Jackson and summed Adams, as described in Ref. 3. The Gauss-Jackson is an multi-step, predictor-corrector, double-integration integrator which computes position directly from the accelerations. To get velocity information, the Gauss-Jackson is combined with an eighth-order summed Adams integrator, which is a single-integration, multi-step, predictor-corrector integrator. The corrector is applied only once at each integration step, giving a predict, evaluate, correct cycle. For the Gauss-Jackson integrator, the step-sizes are 30 seconds for test cases 1 and 2, and 20 minutes for test case 3.

ERROR RATIO

In the tests, a metric for integration accuracy is found by defining an error ratio in terms of the RMS error of the integration (Ref. 4). First define position errors as

$$\Delta r = |r_{\text{computed}} - r_{\text{reference}}|. \tag{4}$$

The RMS position error can be calculated,

$$\Delta r_{\text{RMS}} = \sqrt{\frac{1}{N} \sum_{i=1}^{N} (\Delta r_i)^2}. \tag{5}$$

The RMS position error is normalized by the apogee distance and the number of orbits to find the position error ratio,

$$\rho_r = \frac{\Delta r_{\text{RMS}}}{r_A N_{\text{orbits}}}. \tag{6}$$

In Ref. 4, Merson uses a position error ratio to test integrators. A velocity ratio test is added here because velocity and position are often integrated with different numerical integrators, and it may be useful to estimate the error in orbital elements, which depend directly on the velocity. The velocity error ratio is the RMS velocity error normalized by the number of orbits and the perigee speed.

$$\rho_v = \frac{\Delta v_{\text{RMS}}}{v_P N_{\text{orbits}}}. \tag{7}$$

ACCURACY ASSESSMENT TECHNIQUES

Two-body test

If the force is a simple two-body (Kepler) force, an exact solution is available for comparison. The advantage of this technique is that the error is then known exactly, but the disadvantage is that the force may not be realistic. This test does not necessarily indicate how well the integrator handles perturbations. Orbits typically integrated for space surveillance may cover a time period of several days or more. During that time, the integrated effects of the perturbations cause a substantial deviation from the two-body solution. If an integrator handles the two-body force well, but not one or more of the perturbations, the integration error has a great affect on the computed orbit. Typically, drag and solar radiation pressure are the forces that may cause trouble with integrators, drag because of its variability and dependence on velocity, and solar radiation pressure because it is a step function in time, which multi-step integrators have difficulty with because of the back-points.

Fox (Ref. 5) performed extensive tests on integrators using the two-body test. Part of the purpose of his study was to assess accuracy in light of the execution time, so he put "dead weight" into the force evaluation — computations that do nothing but soak up processor time — in an attempt to simulate the evaluation-dominated execution time of integrations with realistic force models. Although we have not addressed the issue of execution time here, the use of a full perturbation model for most tests precludes the need to artificially slow the force evaluation, should that be of interest. Montenbruck (Ref. 6) also used the two body force to assess integrators, even for earth-orbiting artificial satellites.

Table 2 shows position and velocity error ratios with the two-body test for both Runge-Kutta and Gauss-Jackson integrators. The position and velocity error ratios are found using (6) and (7), respectively. Ephemeris generated by the numerical integrators over a three day time span is used for $r_{computed}$ and $v_{computed}$, and $r_{reference}$ and $v_{reference}$ are the values given by the exact analytic solution. Though the cases use different step-sizes, the ephemeris is generated at one minute intervals, so that N in (5) is 4321. For case 3 with the Gauss-Jackson integrator, where the step-size is 20 minutes, the intermediate points are found using a 5th order interpolator. Table 3 gives the maximum position error over three days for each test case. Comparing Table 2 to Table 3 demonstrates how an error ratio corresponds to the maximum position error, which may be of interest.

Table 2: Two-Body Results

test #	Runge-Kutta		Gauss-Jackson	
	pos	vel	pos	vel
1	2.05×10^{-10}	2.05×10^{-10}	7.96×10^{-14}	7.98×10^{-14}
2	2.49×10^{-10}	5.15×10^{-10}	1.03×10^{-11}	2.26×10^{-11}
3	3.27×10^{-11}	3.25×10^{-11}	8.95×10^{-12}	8.57×10^{-11}

Table 3: Two-Body Position Error (mm)

test #	Runge-Kutta	Gauss-Jackson
1	133	0.0494
2	286	14.9
3	7.21	2.60

From Table 2 we see that the Gauss-Jackson integrator is more accurate in every case, except in the velocity for case 3. Gauss-Jackson is roughly three orders of magnitude more accurate than Runge-Kutta for the low earth orbit, roughly one order of magnitude more accurate for the eccentric orbit, and of comparable accuracy for the geosynchronous orbit. Both integrators show the least accuracy in position with the eccentric orbit, though the Gauss-Jackson has the least accuracy in velocity with the geosynchronous orbit.

The two-body test directly applied is a necessary but not sufficient test for assessing integrators, because of the significance to integration of perturbations, but two-body integration used with the other techniques can provide an indicator of uncaptured error in those techniques. The techniques described in the subsequent sections are each tested two ways. In the first test only the two-body force is considered, so that the error measured by the technique can be compared to the known error. In the second test a perturbation model is used. The perturbations forces are 36 × 36 WGS-84 geopotential, the Jacchia 70 drag model (Ref. 7), and lunar and solar forces. Note that all of these perturbations are continuous forces. Solar radiation pressure, a perturbation that has discontinuities at eclipse boundaries, can cause numerical integration errors, especially when using multi-step integrators, because the discontinuity violates the assumption made in the formulation of the integrators that the forces are smooth and continuous (Ref. 8). To simplify our study this perturbation is not considered.

Step-size halving

For the step-size halving test, the reference integration is produced with the same integrator but with the step-size cut in half. Because the truncation error is related to the step-size, this technique can give a good estimate of the former, and even an estimate of the order of the integration method, provided the step-size is large enough that truncation error dominates the total error. If the step-size is too small, round-off error, which will increase as the step-size is decreased, will appear very different than truncation error. However, in the step-size region where truncation error gives way to round-off error, there can be confusion as to what is being measured and to quantify that measurement, so that a further decrease in step-size is necessary to confirm the onset of round-off error. As long as the step-sizes used are much larger than this mixed regime, an error ratio may be computed.

Tables 4 and 5 show error ratios using the step-size halving test, for the two-body force and for the full force model, respectively. The error ratios with the two-body force are the same order of magnitude as the true error ratios in Table 2, and even match to one significant digit, except for case 1 with Gauss-Jackson. This shows that step-size halving gives a reasonable measure of integration error.

Table 4: Two-Body Step-Size Halving Results

test #	Runge-Kutta		Gauss-Jackson	
	pos	vel	pos	vel
1	1.96×10^{-10}	1.96×10^{-10}	2.22×10^{-14}	2.22×10^{-14}
2	2.34×10^{-10}	4.85×10^{-10}	1.03×10^{-11}	2.56×10^{-11}
3	3.07×10^{-11}	3.05×10^{-11}	8.94×10^{-12}	8.62×10^{-11}

Table 5: Perturbed Step-Size Halving Results

test #	Runge-Kutta		Gauss-Jackson	
	pos	vel	pos	vel
1	1.19×10^{-9}	1.19×10^{-9}	4.63×10^{-9}	4.64×10^{-9}
2	1.16×10^{-9}	2.50×10^{-9}	9.93×10^{-9}	2.11×10^{-8}
3	3.07×10^{-11}	3.05×10^{-11}	8.95×10^{-12}	8.62×10^{-11}

Comparing Table 4 to Table 5, we see that the numerical integrators perform worse in the presence of perturbations for the low-earth and eccentric cases, but nearly the same for the geosynchronous case. This may be because the geosynchronous orbit is not subject to drag, and the presence of drag causes an increase in integration error. This shows that the two-body test described above does not capture all of the error of an integrator.

A more formal error analysis is also possible with step-size halving. In theory the global error is on the order of the step size to the power of the order of the numerical integrator (Ref. 9),

$$x - \tilde{x} = \xi \approx Ch^p, \tag{8}$$

where x is the actual solution, $\tilde{x}$ is the numerical solution, and C is some constant that depends on the numerical integrator. An estimate of ξ can be found by comparing results with half the step-size. Considering the numerical solution, $\tilde{x}$, to be a function of the step-size h, we form the equation

$$[x - \tilde{x}(h/2)] \approx Ch^p/2^p \approx [x - \tilde{x}(h)]/2^p. \tag{9}$$

This can be solved for the error,

$$2^p[x - \tilde{x}(h/2)] \approx [x - \tilde{x}(h)]$$
$$(2^p - 1)[x - \tilde{x}(h/2)] \approx \tilde{x}(h/2) - \tilde{x}(h)$$
$$\xi(h/2) \approx \frac{\tilde{x}(h/2) - \tilde{x}(h)}{2^p - 1} \tag{10}$$

In order for (10) to be valid, (8) must hold true. This condition can be checked by forming the quotient

$$\frac{[x - \tilde{x}(h/2)] - [x - \tilde{x}(h/4)]}{[x - \tilde{x}(h)] - [x - \tilde{x}(h/2)]} = \frac{\tilde{x}(h/4) - \tilde{x}(h/2)}{\tilde{x}(h/2) - \tilde{x}(h)} \approx \frac{Ch^p/2^p - Ch^p/4^p}{Ch^p - Ch^p/2^p} = \frac{1}{2^p}. \tag{11}$$

The quotients should approach 2^{-p} as the step size decreases. But when round-off becomes a factor, the quotients drift away from the theoretical value. As long as the quotients indicate that (8) is valid, round-off error is not a major concern and (10) can be used. Equations (10) and (11) have been written in scalar form; in practice they can be applied to each component of the vector state x.

It is possible to improve this process by generalizing the scaling applied to the step size. By using an arbitrary scale factor instead of only one half, more consistent results are possible and it is possible to locate the onset of roundoff error more precisely. Then, the order of the integration method may be computed numerically when one has an exact solution. In fact, one can numerically compute the order of any series expansion this way; the order computation is a special case where the function $g(h)$ is a dependent variable error (like position) after integrating over a fixed time period from fixed initial conditions; it is a function of step-size h only.

If g is an analytic function, it has a Taylor expansion around y_0,

$$g(y) = c_0 + c_1(y - y_0) + \frac{1}{2}c_2(y - y_0)^2 + \dots \tag{12}$$

(In the integration case, $y_0 = 0$, but we will keep it for completeness.) Presuming that the integrator is correct to some order, we may assume the first few coefficients are zero. As above, p is the order (first exponent dropped in the global Taylor expansion), or *error exponent* we wish to solve. As the first nonzero coefficient $c_j = 0$ for $j < p$, and $c_p \neq 0$ it gives the order of the method. The Taylor series may be written starting at order p,

$$|g(y)| = \left| \frac{c_p}{p!}(y - y_0)^p + \dots \right|. \tag{13}$$

By evaluating this function at two different points $y = y_0 + s_1$ and $y = y_0 + s_2$ and taking the ratio, we may determine p,

$$\frac{|g(y_0 + s_1)|}{|g(y_0 + s_2)|} = \left| \frac{\frac{c_p}{p!}s_1^p + \frac{c_{p+1}}{(p+1)!}s_1^{p+1} + \dots}{\frac{c_p}{p!}s_2^p + \frac{c_{p+1}}{(p+1)!}s_2^{p+1} + \dots} \right| \approx \left| \frac{s_1}{s_2} \right|^p, \tag{14}$$

where the approximation presumes that successive terms in both series are insignificant.

The choice of scaling factors s_1 and s_2 are important. In contrast to the "step-size halving" case where the goal is to find the accuracy of the integrator, these will likely not be in the ratio $2 : 1$. In order to make the results of the determination of e more robust, we can evaluate the ratio (14) at a series of different scaling factors $s_1, s_2, \dots$ This series can be constructed from a *geometric step evaluation* series, $\sigma b, \sigma^2 b, \sigma^3 b, \dots$, for some scaling $\sigma > 0$. Say $s_1 = \sigma^n b$ and $s_2 = \sigma^{n-1} b$ so that the ratio (14) becomes

$$\frac{|g(y_0 + \sigma^n b)|}{|g(y_0 + \sigma^{n-1} b)|} \approx \sigma^p. \tag{15}$$

Now, to find p, we take the logarithm

$$p \approx \log \frac{|g(y_0 + \sigma^n b)|}{|g(y_0 + \sigma^{n-1} b)|} \Big/ \log \sigma. \tag{16}$$

The accurate determination of p is hampered by the approximation we have made in (15) and the necessity of dealing with a finite-sized floating point number on a computer. The terms after the first ones in (14) become relatively insignificant if s_1 and s_2 are small, i.e., if b is small. But choosing b too small can cause significant digits to be lost due to finite word size.

Traditionally when "eyeballing" a scaling result, one chooses $\sigma = 10$ (or $\sigma = 0.1$) because it is easy for a human to compute p. For a computer-determined value, $\sigma = 2$ might make more sense (and thus we would live up to the section title), however, we have found that better results are obtained with a σ near 1. Specifically, when $\sigma \approx 1.02 - 1.05$, the error exponent p is often very clearly near an integer. The multiple successive terms in the series, i.e., computing (16) for a series of n, serve to confirm the computed value of p and show at what value of s round-off error sets in; in the transition from truncation to round-off error, a noticeable erratic deviation in the computed truncation error exponent p will be observed. This will help identify the lower bound on step size for the integrator accuracy check, below which round off error is significant.

Comparison with high-order integrator

The comparison integration may be a higher-order - high-accuracy integrator. The advantage of this technique is that perturbations can be tested. However, this technique relies on the assumption that the higher-order integrator is correct, or more correct, than the integrator being tested. This is not necessarily true. Tables 6 and 7 show error ratios comparing the two test integrators to a 14th-order Gauss-Jackson, with the two-body force and full perturbation forces, respectively. The step-size used in the 14th-order Gauss-Jackson is 15 seconds for cases 1 and 2, and 1 minute for case 3.

Table 6: Two-Body High-Order Results

test #	Runge-Kutta		Gauss-Jackson	
	pos	vel	pos	vel
1	2.05×10^{-10}	2.05×10^{-10}	5.34×10^{-14}	5.34×10^{-14}
2	2.49×10^{-10}	5.16×10^{-10}	1.04×10^{-11}	2.30×10^{-11}
3	3.28×10^{-11}	3.25×10^{-11}	9.02×10^{-12}	8.58×10^{-11}

Table 7: Perturbed High-Order Results

test #	Runge-Kutta		Gauss-Jackson	
	pos	vel	pos	vel
1	4.59×10^{-9}	4.61×10^{-9}	4.62×10^{-9}	4.64×10^{-9}
2	7.19×10^{-9}	1.55×10^{-8}	9.94×10^{-9}	2.11×10^{-8}
3	3.27×10^{-11}	3.25×10^{-11}	9.07×10^{-12}	8.58×10^{-11}

Comparing Table 6 to Table 2, we see that the high-order test results are the same order of magnitude as the true results in every case. For the Runge-Kutta integrator, the test matches the true results to at least two significant figures in each case. For the Gauss-Jackson integrator, the test matches the true results to two significant figures for the eccentric case, and one significant figure for the geosynchronous case. Note that the higher the actual error ratio, the closer the high-order results match the true results. This is because the high-order test assumes that the reference integrator is much better than the integrator being tested.

Comparing Table 7 to Table 5, we see that the step-sizing halving results and the high-order results with perturbations are the same order of magnitude in all cases except the velocity of case 2 with Runge-Kutta. The Gauss-Jackson results match to at least one significant figure in each case, while the only case where the Runge-Kutta results match to one significant figure is case 3. The Gauss-Jackson results match closely between the step-size halving test and the high-order test because in both tests, the reference integrator is a Gauss-Jackson integrator with a low step-size.

Reverse test

In this technique the test and reference orbits come from the same integrator. The test orbit is produced by integrating forward from the initial conditions, and the reference orbit is produced by integrating backward to the original starting time using the final state of the first integration as initial conditions. These two integrations should be identical, and any difference between them is due to integration error. Using this technique to measure integration error is advantageous because it is a relatively simple procedure to perform. A disadvantage with this technique is that it does not measure any reversible integration error, that is, any error that is an odd function of the step-size is canceled off when the sign of the step changes on the reverse integration. It has been used extensively for integration accuracy checks, including recently in the N-body problem (Ref. 10).

The two tables 8 and 9 show the reverse test with a two-body force and full perturbation forces, respectively. Again the tests are over a three day interval with ephemeris generated at one minute increments. Comparing Table 8 to Table 2 shows that the reverse test fails to capture a significant portion of the error; in the case of orbit number 3 for the Runge-Kutta integrator, only about a tenth of the error is captured. This may be an example of the weakness of the reverse test pointed out by Zadunaisky (Ref. 11), who demonstrated for the three-body problem that the reverse test will certify that the a second-order Adams-Moulton multi-step method as perfectly accurate because of the symmetry of the equations used under time reversal. Thus two-body integration has proved its value: it can show when another test is deficient. Comparing Table 9 to Tables 5 and 7, we see that the reverse test is also inconsistent with the other techniques in the presence of perturbations.

Table 8: Two-Body Reverse Test Results

test #	Runge-Kutta		Gauss-Jackson	
	pos	vel	pos	vel
1	2.27×10^{-10}	2.27×10^{-10}	4.55×10^{-15}	4.54×10^{-15}
2	5.13×10^{-11}	1.08×10^{-10}	2.21×10^{-11}	4.67×10^{-11}
3	3.53×10^{-12}	3.53×10^{-12}	2.11×10^{-11}	2.12×10^{-11}

Table 9: Perturbed Reverse Test Results

test #	Runge-Kutta		Gauss-Jackson	
	pos	vel	pos	vel
1	2.28×10^{-10}	2.29×10^{-10}	7.79×10^{-10}	7.81×10^{-10}
2	5.18×10^{-11}	1.09×10^{-10}	2.46×10^{-11}	1.11×10^{-10}
3	3.52×10^{-12}	3.52×10^{-12}	1.97×10^{-11}	1.98×10^{-11}

Integral invariants

Another frequently-used technique for integration accuracy assessment is to check the invariance of integral invariants such as energy, which is easy to perform. This technique has been used recently for integrators applied in the N-body problem (Ref. 10). The main drawback is that it does not capture all errors. For example, energy invariance is blind to in-track errors, because an orbit shifted in time has the same energy. In general, the more forces present that break a particular symmetry, the fewer conserved quantities and thus the fewer quantities that can be checked; the presence of drag means that energy is no longer conserved at all. Huang and Innanen (Ref. 12) show that this accuracy check is not exact and reliable and suggest a revised technique. We have not attempted this technique in the present study, because of its limitations.

Zadunaisky's test

Zadunaisky, in Ref. 13, Ref. 1, Ref. 14, and Ref. 11, suggests a technique for measuring numerical integration error. This technique is based on the construction of an analytical function near the solution to the actual problem, then constructing differential equations for which this function is an exact solution. It has two desirable properties: first, we have an exact solution, because we constructed the problem to match the solution we had, and second, this solution is realistic, because it is close in some sense to the solution of the real problem. It thus is like the two-body test in providing an absolute reference for error computation, but has a behavior that mimics real forces.

First a set of polynomials $\boldsymbol{P}(t)$ are determined which represent the components of $\tilde{\boldsymbol{x}}$, the numerical solution of (1). A *pseudo-system* of equations may be constructed from these polynomials,

$$\dot{\boldsymbol{z}} = \boldsymbol{f}(t, \boldsymbol{z}) + \boldsymbol{D}(t), \tag{17}$$

where

$$\boldsymbol{D}(t) = \dot{\boldsymbol{P}}(t) - \boldsymbol{f}(t, \boldsymbol{P}(t)), \tag{18}$$

and where (17) has the same initial conditions as (1),

$$\boldsymbol{z}(t_0) = \boldsymbol{x_0}. \tag{19}$$

The exact solution of the pseudo-system is known, and is $\boldsymbol{P}(t)$. When the pseudo-system is integrated with the same numerical integrator used to create $\tilde{\boldsymbol{x}}$, the errors of the pseudo-system,

$$\boldsymbol{\xi} = \tilde{\boldsymbol{z}} - \boldsymbol{P}(t), \tag{20}$$

may be a good approximation of the error in the original problem.

In Ref. 13, Zadunaisky gives conditions for $\boldsymbol{P}(t)$ under which this technique gives a good approximation of the original error,

$$\left. \begin{array}{r} ||\tilde{\boldsymbol{x}} - \boldsymbol{P}(t)|| \\ ||\tilde{\boldsymbol{x}}^{p+1} - \boldsymbol{P}^{p+1}(t)|| \end{array} \right\} \leq \delta = O(h^2), \tag{21}$$

where p is the order of the numerical integrator, and $\tilde{\boldsymbol{x}}^{p+1}$ are the numerical approximations for the $(p+1)^{\text{th}}$ derivatives of $\boldsymbol{x}$. These conditions are based on error propagation theory, and are meant to ensure that the asymptotic behavior of the errors accumulated in the numerical integration of both the original system and the pseudo-system are the same. In Ref. 11, Zadunaisky gives a different condition, which is that $\boldsymbol{D}(t)$ must not be larger than either the local truncation error or the local round-off error.

In Ref. 14 and Ref. 11, Zadunaisky finds the polynomial, $\boldsymbol{P}(t)$, needed to form the pseudo-system, using Newton's interpolation formula with backward divided differences. Because an $\tilde{N}^{\text{th}}$ degree polynomial is needed to interpolate $\tilde{N} + 1$ points, the original interval is broken into subintervals to avoid the problems involved with interpolating data to a high degree polynomial. After integrating the original problem for $\tilde{N}$ steps, where $\tilde{N} \leq 10$, he applies Newton's formula to obtain polynomials $\boldsymbol{P}(t)$ of $\tilde{N}^{\text{th}}$ degree. Each polynomial $P_i(t)$ interpolates one of the components of position or velocity through the $\tilde{N} + 1$ points spanned by the $\tilde{N}$ steps. These polynomials are used to construct the pseudo-system, (17), and the error estimate is obtained. This process is then repeated using the last point of the previous set as the first point on the next set, and using $\tilde{\boldsymbol{x}}_{\tilde{N}+1}$ and $\tilde{\boldsymbol{z}}_{\tilde{N}+1}$ as initial conditions in the original system and pseudo-system, respectively. Using this method, the solution to the pseudo-system over the entire interval, $z(t)$, is a function of successive $\tilde{N}^{\text{th}}$ degree polynomials that match up the subintervals of $\tilde{N}+1$ points. Therefore $z(t)$ is continuous over the entire interval, but its derivative is discontinuous at the last point of each subinterval. Zadunaisky claims that these discontinuities are irrelevant to the validity of the technique.

To implement Zadunaisky's technique for a given test case, we first generate ephemeris $\tilde{x}(t)$, by numerically integrating the test case. The ephemeris should be generated at time increments equal to the step-size h of the numerical integrator. This ephemeris is then used to find coefficients of the polynomials $P(t)$ at each subinterval. The polynomials are of the form

$$P_i(\tilde{t}) = a_0 + a_1\tilde{t} + a_2\tilde{t}^2 + ... + a_{\tilde{N}}\tilde{t}^{\tilde{N}}, \tag{22}$$

where $\tilde{t}$ is the time since the beginning of the subinterval, and the subscript i refers to the component of the state $\tilde{x} = [r_x r_y r_z v_x v_y v_z]^T$ that the polynomial fits. To make the polynomial exactly match the ephemeris at each of the $\tilde{N}+1$ points on the subinterval, the coefficients $a_0 ... a_{\tilde{N}}$ are found by solving the system

$$\begin{bmatrix} 1 & 0 & 0 & ... & 0 \\ 1 & \tilde{t}_1 & \tilde{t}_1^2 & ... & \tilde{t}_1^{\tilde{N}} \\ 1 & \tilde{t}_2 & \tilde{t}_2^2 & ... & \tilde{t}_2^{\tilde{N}} \\ . & & & & \\ . & & & & \\ 1 & \tilde{t}_{\tilde{N}} & \tilde{t}_{\tilde{N}}^2 & ... & \tilde{t}_{\tilde{N}}^{\tilde{N}} \end{bmatrix} \begin{Bmatrix} a_0 \\ a_1 \\ a_2 \\ . \\ . \\ a_{\tilde{N}} \end{Bmatrix} = \begin{Bmatrix} \tilde{x}_i(t_0) \\ \tilde{x}_i(t_1) \\ \tilde{x}_i(t_2) \\ . \\ . \\ \tilde{x}_i(t_{\tilde{N}}) \end{Bmatrix}, \tag{23}$$

where the subscripts on $\tilde{t}$ and t refer to the points on the subinterval, and $\tilde{t}_0 = 0$. This system can be solved for the coefficients by inverting the first matrix and multiplying it by the vector on the right-hand side. Though this procedure must be repeated for each of the six components and for each subinterval, the matrix inversion only needs to be performed once, because the matrix only depends on the order $\tilde{N}$, and the time step h.

After the coefficients are found, the same numerical integrator is used to generate another set of ephemeris, but with the force model modified so that the pseudo-system (17) is being integrated. Normally, the force model returns an acceleration based on position, velocity, and time, $\ddot{r} = f(t, r, \dot{r})$. Instead, the coefficients for the appropriate subinterval are used to find the values of P and $\dot{P}$, and the force model returns

$$\ddot{r} = f(t, r, \dot{r}) + \dot{P}(t) - f(t, P(t)). \tag{24}$$

Note that P is a six component vector consisting of both position and velocity, though $\ddot{r}$ is only a three component vector. Therefore, the vector $\dot{P}$ used in (24) is a three component vector consisting of the first derivative of the velocity polynomials. This makes (24) have a different form than (17), but this change makes the technique easier to implement.

There are two practical considerations to make when generating and using the coefficients in the modified force model. First, the ephemeris from which the coefficients are generated must be in the same coordinate system in which the integration is performed. Second, conversions may be necessary if the units of the ephemeris, and of the time $\tilde{t}$ used in the polynomial equations, are different from the units used by the integrator.

The ephemeris generated in the second integration, $\tilde{z}$, is compared to the original ephemeris, $\tilde{x}$, to determine an error ratio. For test case 3, with the Runge-Kutta integrator and using only the two-body force, over three days we get a position error ratio of 2.65×10^{-7} with a 9th order polynomial, $\tilde{N} = 9$. From Table 2 we know that the actual error ratio is 3.27×10^{-11}, so the technique is four orders of magnitude too high. If we look at only the first 90 minutes, the technique gives an error ratio that is two orders of magnitude too high. When $\tilde{N} = 6$, the error ratio given by the technique is 1.80×10^{-9}, two orders of magnitude too high. However, the error ratio over 90 minutes is the correct order of magnitude. With $\tilde{N} = 5$, the error ratio is 5.89×10^{-10}, one order of magnitude too high. Over the 90 minute time span, the technique gives an error ratio that is an order of magnitude too low. These results highlight the difficulty in choosing an appropriate degree polynomial.

Because the reliability of Zadunaisky's technique depends on how well the polynomial fits the ephemeris, we suggest another method of determining the polynomials to improve the results. In addition to matching the values of the ephemeris, the derivatives are also matched at the end-points of each subinterval. So for a subinterval consisting of $\tilde{N} - 1$ points, an $\tilde{N}^{\text{th}}$ degree polynomial is found by modifying (23),

$$
\begin{bmatrix}
1 & 0 & 0 & \cdots & 0 \\
1 & \tilde{t}_1 & \tilde{t}_1^2 & \cdots & \tilde{t}_1^{\tilde{N}} \\
\cdot & & & & \\
\cdot & & & & \\
\cdot & & & & \\
1 & \tilde{t}_{\tilde{N}-2} & \tilde{t}_{\tilde{N}-2}^2 & \cdots & \tilde{t}_{\tilde{N}-2}^{\tilde{N}} \\
0 & 1 & 0 & \cdots & 0 \\
0 & 1 & 2\tilde{t}_{\tilde{N}-2} & \cdots & \tilde{N}\tilde{t}_{\tilde{N}-2}^{\tilde{N}-1}
\end{bmatrix}
\left\{
\begin{array}{c}
a_0 \\
a_1 \\
\cdot \\
\cdot \\
\cdot \\
a_{\tilde{N}-2} \\
a_{\tilde{N}-1} \\
a_{\tilde{N}}
\end{array}
\right\}
=
\left\{
\begin{array}{c}
\tilde{x}_i(t_0) \\
\tilde{x}_i(t_1) \\
\cdot \\
\cdot \\
\cdot \\
\tilde{x}_i(t_{\tilde{N}-2}) \\
\dot{\tilde{x}}_i(t_0) \\
\dot{\tilde{x}}_i(t_{\tilde{N}-2})
\end{array}
\right\},
\tag{25}
$$

where $\dot{\tilde{x}}_i(t)$ is the velocity given in the ephemeris when i is a position component, and the acceleration given by the same force model used to generate the ephemeris when i is a velocity component.

Tables 10 and 11 show error ratios given by Zadunaisky's technique with the two-body force and the full force model, respectively. The polynomials used in the method match each point in the subinterval, and the derivatives of the polynomial match the force model at the end points of each subinterval, as described above. For the Runge-Kutta integrator a 5th order polynomial is used, $\tilde{N} = 5$, and for the Gauss-Jackson a 3rd order polynomial is used, $\tilde{N} = 3$. We find these polynomials give the best results. With higher order polynomials, the technique gives error ratios that are too high, compared to the two-body test, and with lower order polynomials the error ratios are too low. In Ref. 14, Zadunaisky uses $\tilde{N} \geq p$, which we have followed for the Runge-Kutta but not for Gauss-Jackson.

Table 10: Two-Body Zadunaisky Results

test #	Runge-Kutta		Gauss-Jackson	
	pos	vel	pos	vel
1	3.08×10^{-10}	3.08×10^{-10}	3.33×10^{-14}	3.33×10^{-14}
2	3.39×10^{-9}	7.30×10^{-9}	6.83×10^{-14}	1.50×10^{-13}
3	3.87×10^{-11}	3.87×10^{-11}	1.86×10^{-14}	1.85×10^{-14}

Table 11: Perturbed Zadunaisky Results

test #	Runge-Kutta		Gauss-Jackson	
	pos	vel	pos	vel
1	1.81×10^{-9}	1.82×10^{-9}	8.06×10^{-8}	8.08×10^{-8}
2	2.11×10^{-9}	4.40×10^{-9}	6.55×10^{-8}	1.44×10^{-7}
3	3.82×10^{-11}	3.82×10^{-11}	1.01×10^{-12}	9.79×10^{-13}

Comparing Table 10 to Table 2, we see that the technique matches the true error ratios in order of magnitude for cases 1 and 3 for Runge-Kutta, and case 1 for Gauss-Jackson. For the eccentric orbit with Runge-Kutta the technique gives an error ratio that is an order of magnitude too high. In Ref. 11 Zadunaisky suggests a variant of his technique that gives improved results for eccentric orbits, but we have not implemented it here. For Gauss-Jackson the technique gives an error ratio that is three orders of magnitude too low for cases 2 and 3. Comparing the perturbed results in

Table 11 to the results from the step-size halving and high-order test in Tables 5 and 7, we see that the Runge-Kutta results with Zadunaisky's technique match the results from the other techniques, at least in order of magnitude. However the Gauss-Jackson results are an order of magnitude higher for Zadunaisky's technique in cases 1 and 2, and an order of magnitude lower in case 3.

Note that for the method of choosing polynomials we have described above, the minimum order polynomial is $\tilde{N} = 3$, because that involves a subinterval of two points. The fact that we found the best results with Gauss-Jackson with this lowest order polynomial, and that this order violates Zadunaisky's criteria of $\tilde{N} \geq p$, indicates that this method of determining polynomials may not be appropriate for Gauss-Jackson, and may explain why the error ratios do not match the error ratios from the other tests.

CONCLUSION

In this paper we demonstrate five techniques for assessing the accuracy of numerical integrators. We test these techniques on a Runge-Kutta integrator and a Gauss-Jackson integrator, with three test orbits. The test orbits are chosen to give a representation of most earth orbits, as well as to stress the integrator with a high drag case.

The two-body test gives an exact measure of error when perturbations are not considered. This exact measure of error is useful to evaluate the other techniques, by testing them without perturbations. However, when these other techniques are used with perturbations, they show a significantly larger error than the two-body test when drag is a factor. The step-size halving test, the higher-order test, and Zadunaisky's test with Runge-Kutta give results that are consistent with one another, and match the two-body error well. The reverse test gives results that are inconsistent, and previous authors have shown it to be unreliable. Zadunaisky's technique with Gauss-Jackson also gives inconsistent results, though this may be due to the method we have chosen to determine the polynomial $P(t)$.

ACKNOWLEDGEMENTS

We thank Prof. Fred Lutze and Prof. Lee Johnson of Virginia Tech for help in understanding numerical interpolation, and its use in implementing Zadunaisky's technique. We also thank Prof. Johnson for notes on step-size halving. We thank Dr. Paul Schumacher of Naval Network and Space Operations Command for introducing us to the literature.

REFERENCES

[1] Zadunaisky, P. E., "On the Accuracy in the Numerical Computation of Orbits," In Giacaglia, G. E. O., editor, *Periodic Orbits, Stability and Resonances*, pp. 216–227, Dordrecht, Holland, 1970. D. Reidel Publishing Company.

[2] Burden, R. L. and Faires, J. D., *Numerical Analysis*, Brooks/Cole Publishing Company, New York, 1997.

[3] Berry, M. and Healy, L., "Implementation of Gauss-Jackson Integration for Orbit Propagation," In *Advances in Astronautics*, San Diego, CA, August 2001. American Astronautical Society AAS 01–426; to appear.

[4] Merson, R. H., *Numerical Integration of the Differential Equations of Celestial Mechanics*, Technical Report TR 74184, Royal Aircraft Establishment, Farnborough, Hants, UK, January 1975. Defense Technical Information Center number AD B004645.

[5] Fox, K., "Numerical integration of the equations of motion of celestial mechanics," *Celestial Mechanics*, Vol. 33, No. 2, 1984, pp. 127–142, June 1984.

[6] Montenbruck, O., "Numerical Integration Methods for Orbital Motion," *Celestial Mechanics and Dynamical Astronomy*, Vol. 53, pp. 59–69, 1992.

[7] Jacchia, L. G., *New Static Models of the Thermosphere and Exosphere with Empirical Temperature Models*, Technical Report 313, Smithsonian Astrophysical Observatory, 1970.

[8] Woodburn, J., "Mitigation of the Effects of Eclipse Boundary Crossings on the Numerical Integration of Orbit Trajectories Using an Encke Type Correction Algorithm," In *AAS/AIAA Space Flight Mechanics Meeting, Santa Barbara, CA, 11–14 February 2001*, AAS Publications Office, P. O. Box 28130, San Diego, CA 92198, 2001. AAS/AIAA Paper AAS 01-223.

[9] Johnson, L. W. and Riess, R. D., *Numerical Analysis*, Addison-Wesley, Reading, Mass., 1982.

[10] Hadjifotinou, K. G. and Gousidou-Koutita, M., "Comparison of Numerical Methods for the Integration of Natural Satellite Systems," *Celestial Mechanics and Dynamical Astronomy*, Vol. 70, No. 2, 1998, pp. 99–113, 1998.

[11] Zadunaisky, P. E., "On the Accuracy in the Numerical Solution of the N-Body Problem," *Celestial Mechanics*, Vol. 20, pp. 209–230, 1979.

[12] Huang, T.-Y. and Innanen, K. A., "The accuracy check in numerical integration of dynamical systems," *Astonomical Journal*, Vol. 88, No. 6, 1983, pp. 870–876, June 1983.

[13] Zadunaisky, P. E., "A Method for the Estimation of Errors Propagated in the Numerical Solution of a System of Ordinary Differential Equations," In Contopoulos, G., editor, *The Theory of Orbits in the Solar System and in Stellar Systems*, pp. 281–287, New York, 1966. International Astronomical Union, Academic Press.

[14] Zadunaisky, P. E., "On the Estimation of Errors Propagated in the Numerical Integration of Ordinary Differential Equations," *Numerische Mathematik*, Vol. 27, No. 1, 1976, pp. 21–39, 1976.

SPACE-BASED INTERFEROMETRY

Chair:

David Hyland
University of Michigan

FORMATION PATH PLANNING FOR OPTIMAL FUEL AND IMAGE QUALITY FOR A CLASS OF INTERFEROMETRIC IMAGING MISSIONS

I. I. Hussein,[*] D. J. Scheeres,[†] D. C. Hyland[‡]

In this paper we state the basic governing relationships between an imaging constellation's motion in physical two dimensional space and imaging objectives. Given this, we introduce a class of spiral motions that satisfies imaging objectives and seek simple solutions to the problem of motion design and control of a formation for imaging applications. Based on the observations drawn from these simple controls, an optimal control problem is formulated for the proposed class of spiral motions to achieve minimum fuel consumption, while satisfying imaging constraints.

1 Introduction

The use of multiple spacecraft formations in interferometric imaging schemes has gained much interest in recent years. The advantages such formations offer include the replacement of large monolithic telescopes and superior angular resolution. Order-of-magnitude advances in optical angular resolution via long baseline interferometry are sought for various NASA missions, such as the Origins program, and high resolution Earth imaging (Ref. [1, 2]).

---

[*]   Ph.D. candidate, Aerospace Engineering, The University of Michigan, Ann Arbor, Michigan 48109-2140.
      E-mail: ihussein@umich.edu.

[†]   Associate Professor of Aerospace Engineering, The University of Michigan, Ann Arbor, Michigan 48109-2140.
      E-mail: scheeres@umich.edu.

[‡]   Professor of Aerospace Engineering, The University of Michigan, Ann Arbor, Michigan 48109-2140.
      E-mail: dhiland@engin.umich.edu.

The dynamics and control of spacecraft formations have been given considerable attention in the past. In these studies, investigators assume certain prescribed motions that satisfy some imaging objective and then seek to achieve these motions via active control. For example, in (Ref. [3]) the authors derive nonlinear and linear spacecraft relative position dynamics and develop a controller design framework for linear control of spacecraft relative position dynamics with guaranteed closed-loop stability. In (Ref. [4]), the authors consider the full nonlinear relative position control problem using a Lyapunov-based, nonlinear, adaptive control law that guarantees global asymptotic convergence of the position tracking error in the presence of unknown, constant, or slowly-varying spacecraft masses, disturbances, and gravity forces.

The problem of rotating a satellite constellation using an adaptive controller with actuator saturation constraints is discussed in (Ref. [5]). An experimental study of synchronized rotation of multiple autonomous spacecraft with rule-based controls is presented in (Ref. [6]). The authors in (Ref. [7]) derive control laws to synchronize formation rotation as well as to control its attitude. In (Ref. [8]), the problem of rotating a constellation using on/off thrusters while maintaining the relative distances within a specified tolerance is addressed. Fuel related problems, such as optimal fuel consumption during a rotation maneuver (Ref. [9]) or constellation reorientation with uniform fuel expenditure and conservation across the constellation (Ref. [10]), have also been addressed.

Natural orbital dynamics have also been exploited for interferometric observatories (Ref. [11, 12]). In (Ref. [13]), the authors design a controller for formation keeping in circular orbit and in (Ref. [14]) for formation keeping during a spiral maneuver while in orbit. In (Ref. [15]) the dynamics of spacecraft formation is considered under the influence of J_2 orbit perturbations, where J_2 invariant orbits are sought for the motion. In (Ref. [16]) the authors design controllers for formation keeping using mean orbit elements, as opposed to spacecraft positions and velocities, as the state of the system. Finally, GPS utilization for spacecraft constellations is discussed in (Ref. [17, 18]) and in (Ref. [19, 20, 21]) the authors propose decentralized controllers for spacecraft formations.

The question of which formation motions yield satisfactory imaging goals has not been discussed in any of the above mentioned research with the exception of (Ref. [12]). In this paper we present a framework such that spacecraft motion planning and controller design meet desired imaging objectives. This paper is organized as follows. In Section 2, we discuss the relationship between the motion of a spacecraft formation in physical two dimensional space and imaging objectives as described by the modulation transfer function (MTF) in the two dimensional plane of spatial frequencies. In Section 3, we derive a class of spiral motions, similar to the ones discussed in (Ref. [14]), that satisfy the imaging objectives. In Section 4, we derive the (constrained) equations of motion based on the class of motions described in Section 3. In section 5, we derive simple controllers that illustrate the concepts discussed in previous sections. In Section 6, we discuss the results obtained when the simple controllers are implemented. Finally, we conclude with a discussion that summarizes our work and our future research directions.

2 Imaging Requirements

Interferometric imaging is performed by measuring the mutual intensity (the two point correlation (Ref. [22])) that results from the collection and subsequent interference of two electric field measurements of a target made at two different observation points. While moving relative to each other, the satellites collect and transmit these measurements, which are later combined at a central node using precise knowledge of their locations and timing of data collection. A least squares error estimate of the image can be reconstructed given the mutual intensity measurements, parameters of the optical system, and the physical configuration of the observatory.

Let $\overline{z}$ be the distance from the image plane to the observation plane. By the term "picture frame" we denote the extent of the intended image on the image plane. The picture frame has a diameter of length $\overline{L}$. Pixelating the image plane into an $m \times m$ grid, the size of each pixel is $L = \overline{L}/m$, and the resulting angular resolution is $\theta_r = L/\overline{z}$. Additionally, the angular extent of the desired picture frame is given by $\theta_p = \overline{L}/\overline{z}$, leading to $\theta_p = m\theta_r$.

To assess the quality of the reconstructed image, the reconstructed image is Fourier transformed into a two dimensional plane of spatial frequencies (the wave number plane.) At any given point on the wave number plane, the modulation transfer function (MTF) is defined as the ratio of the estimated intensity to the true image intensity. For an interferometric imaging constellation, the MTF can be computed given the measurement history and corresponding relative position data between the light collecting spacecraft. In the wave number plane, a point with a zero MTF value implies that the system is "blind" to the corresponding spatial frequency, while a large value of the MTF implies that the image signal can be restored at that wave number via an inverse Fourier transform (Ref. [1, 22, 23]). The MTF, as a measure of the imaging system's performance, is a function of both the optical system and the configuration of the observatory in physical space. The MTF is related to the motion in physical space by (Ref. [1]):

$$M(t, \vec{v}) = \int_0^t d\tau \Sigma_{i=1}^N \Sigma_{j=1}^N \hat{A}_p \left(\vec{v} - \frac{(\vec{\chi}_i(\tau) - \vec{\chi}_j(\tau))}{\lambda} \right), \tag{1}$$

where N is the number of satellites in the formation, $\vec{v}$ is the spatial wave number vector, $\vec{\chi}_i$ is the position vector of spacecraft i relative to a fixed frame, λ is the wavelength of interest and $\hat{A}_p(\vec{\zeta})$ is the Fourier transform of the picture frame function, $A_p(\vec{z})$, given by:

$$A_p(\vec{z}) = \begin{cases} 1 & \text{if } \vec{z} \text{ falls within the desired picture frame} \\ 0 & \text{otherwise} \end{cases}. \tag{2}$$

In this paper we address the issue of motion design and control for an interferometric observatory composed of two satellites that ensures a non-zero value of the MTF within a desired region in the wave number plane. We assume that the system is in free space; the only forces acting on the system are the spacecraft thrusters. Note that this is a reasonable analogue for a spacecraft at the libration point L_2. For example, in (Ref. [24]), the authors discuss the feasibility of nulling natural forces and replacing them with "new" dynamics. One of the spacecraft is stationary and located at the origin of an inertial reference frame. Let $\mathbf{R}_1 \equiv \mathbf{0}$ and $\mathbf{R}_2$ be the position vectors of the first and second satellites, respectively. Thus $\mathbf{R}_2$ also describes the relative position vector between the two satellites. Since we only consider the motion in a plane perpendicular to the line of sight of the observatory, let $\mathbf{r}$ be the projection of $\mathbf{R}_2$ onto such plane.

Dimensions of features in the wave number plane are the reciprocals of the corresponding dimensions in the physical plane. Thus the resolution disc is a disc of diameter $\simeq 1/\theta_r$ and is the region where we desire the MTF to have nonzero values (or, equivalently, wave number plane coverage.) The picture frame region is a circular disc of diameter $\simeq 1/\theta_p$. Therefore, the diameter of the resolution disc is m times the diameter of the picture frame disc in the wave number plane (see Figure (1).) As the relative position vector, $\mathbf{r}$, varies in the physical plane, the picture frame disc moves in the wave number plane, where its center follows the trajectory of the vectors given by $\pm\mathbf{r}/\lambda$, where λ is the imaging wavelength of interest. Let $\tilde{\mathbf{r}} = \mathbf{r}/\lambda$. Each satellite, by itself, will contribute a disc that is centered at the origin with a diameter of $\simeq 1/\theta_p$, and each pair of satellites will contribute two discs of diameter $\simeq 1/\theta_p$ located 180 degrees apart with a radius of $\tilde{r} = \frac{r}{\lambda}$ from the center, where $r = |\mathbf{r}|$. In polar coordinates, r and $\tilde{r}$ define the radial position of the second spacecraft in the physical plane and the picture frame disc centers in the wave number plane, respectively, and θ is the angular position of the spacecraft in the physical space and the first picture frame disc center, while the second picture frame disc center has an angular position of $\theta + \pi$.

3 Motion Design for Wave Number Plane Coverage

Here we only consider the motion of the second spacecraft (the first is fixed at the origin) and only one of the picture frame discs (while the second will have an identical motion that is symmetric about the origin.) Thus, by $(\tilde{r}, \theta)$ we imply the polar coordinates of one picture frame disc center. One way to ensure full coverage of the resolution disc is to initialize the second spacecraft such that at $t = 0$ we have $(\tilde{r} = \frac{1}{\theta_p}, \theta = 0)$, make it follow a linear spiral as a function of θ, and to impose the terminal condition that at $t = t_f$ we have $(\tilde{r} = \frac{(m+1)}{2\theta_p}, \theta = \frac{(m-1)\pi}{2})$, where t_f is the terminal time. This motion implies that the two picture frame discs are initialized such that they lie outside the picture frame disc whose center is fixed at the origin,

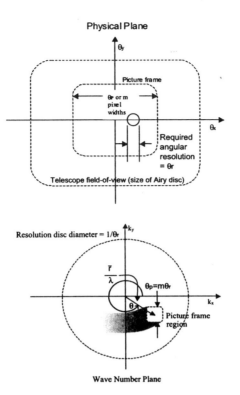

Figure 1: Physical and Wave Number Plane Variables

and moves spirally outwards till they lie outside resolution disc.

Note that $\tilde{r} = 0$ is a singularity, which produces poor numerical results for the motion near the origin, and thus we do not initialize the system at this value. However, the discussion in Section 2 implies that initializing the system with $\tilde{r} = 1/\theta_p$ results in some regions close to the origin of the wave number plane that will not be covered (the shaded region in Figure (2).) It will be shown in simulation that this is not necessarily true, and that the initial condition we impose suffices for full coverage, especially near the origin. Thus $\tilde{r}$ and θ are constrained to satisfy

$$\tilde{r}(\theta(t)) = \frac{1}{\pi\theta_p}\left(\pi + \theta\right), \ \theta \in \left[0, \frac{(m-1)\pi}{2}\right]. \tag{3}$$

This implies that

$$r(\theta(t)) = \frac{\lambda}{\pi\theta_p}\left(\pi + \theta\right), \ \theta \in \left[0, \frac{(m-1)\pi}{2}\right]. \tag{4}$$

The first and second time derivatives of the constraint (4) also need to be satisfied:

$$\dot{r} = \frac{\lambda}{\pi\theta_p}\dot{\theta}, \ \forall t \in [0, t_f], \tag{5}$$

and

$$\ddot{r} = \frac{\lambda}{\pi\theta_p}\ddot{\theta}, \ \forall t \in [0, t_f], \tag{6}$$

where $\dot{r}$ and $\ddot{r}$ are the first and second time derivatives of r with respect to time.

Figure (2) shows an example of the trajectory in the physical and wave number planes for an object that is located at $\overline{z} = 15$pc (1pc$= 3.085 \times 10^{13}$km), with a picture frame that is $\overline{L} = 12,760$km wide, with $m = 17$ pixels and, thus, a pixel size of $L = 750.6$km (i.e. this constellation maneuver is capable of detecting any object whose size is greater than L.) These are the values that are used throughout this paper. These values correspond to applications such as JPL's Terrestrial Planet Finder (TPF) and they could also be adjusted for Earth imaging applications. In the latter case, the maneuver spans a few meters only, as opposed to few thousands of kilometers as in the example we treat in this paper.

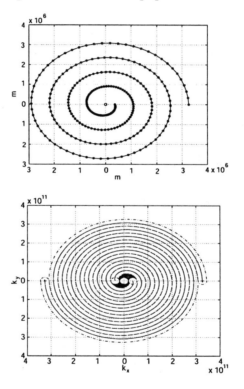

Figure 2: Motion In The Physical (In Meters, Top) and Wave Number (Dimensionless, Bottom) Planes

4   Equations of Motion with a Spiral Constraint

In this section we derive the constrained equations of motion of the second spacecraft relative to the origin. After dividing throughout by the mass of the second spacecraft, the two degree of freedom equations of motion for the system are described in polar coordinates by

$$\ddot{r} - r\dot{\theta}^2 = a_r, \tag{7}$$

and

$$r\ddot{\theta} + 2\dot{r}\dot{\theta} = a_t, \tag{8}$$

where a_r is the radial component of the thrust vector and a_t is the tangential component of the thrust vector, both divided by the mass of the spacecraft. Applying the configuration constraint in (4) to the system of equations (7)-(8) results in the following two *dependent* equations that describe the motion

$$\ddot{\theta} - f(\theta)\dot{\theta}^2 = \tilde{a}_r, \tag{9}$$

and

$$f(\theta)\ddot{\theta} + 2\dot{\theta}^2 = \tilde{a}_t, \tag{10}$$

where $f(\theta) = \pi + \theta$, $\tilde{a}_r = \frac{\pi\theta_p}{\lambda}a_r$ and $\tilde{a}_t = \frac{\pi\theta_p}{\lambda}a_t$. Note that these two equations are equivalent. By solving for $\ddot{\theta}$ in (9) and (10) and equating the resulting expressions, one finds that the tangential and radial components of the force-per-mass vector are related such that

$$\tilde{a}_r = -\left[f(\theta) + \frac{2}{f(\theta)} \right] \dot{\theta}^2 + \frac{1}{f(\theta)}\tilde{a}_t. \tag{11}$$

By solving either (9) or (10) such that θ goes from 0 to $\frac{(m-1)\pi}{2}$ we are guaranteed that the spacecraft will have gone through one entire "image" maneuver and covered the wave number plane.

5 Simple Controllers to Achieve Imaging Objectives

In this section we investigate the performance of five controllers that are designed to satisfy the spiral motion constraint. The two main performance considerations are image quality and fuel consumption. Image quality requires full coverage of the resolution disc and the attainment of a desired signal-to-noise ratio. An image quality performance measure can be expressed as:

$$\mathcal{I} = \int_0^{t_f} dt \int d\vec{v} \left(1 - \Gamma^R \left(\vec{v}, t \right) \right), \tag{12}$$

where R is a "risk factor" and $\Gamma(\vec{v}, t)$ is the estimated signal-to-noise ratio of the interferometric measurement divided by the desired signal-to-noise ratio. Using Poisson statistics in a semi-classical photon arrival fluctuation calculation (Section 4.4.3

1025

in (Ref. [25])), this can be shown to be proportional to the square root of the MTF magnitude. Hence, Γ is given by:

$$\Gamma(\vec{v}, t) = \begin{cases} 1 & \text{if } \alpha M^{1/2} \geq 1 \\ \alpha M^{1/2} & \text{otherwise} \end{cases}, \tag{13}$$

where α is proportional to the desired signal-to-noise ratio and M is computed from Eq. (1). The larger the risk factor, R, the more conservative the imaging performance measure becomes. Regions in the wave number plane where Γ is less than unity correspond to spatial frequencies of the signal that do not satisfy the desired signal-to-noise ratio. On the other hand, $\Gamma = 1$ implies both coverage and achievement of the desired signal-to-noise ratio at the corresponding spatial frequency. $\Gamma = 1$ can be achieved if the relative speeds between the spacecraft in the constellation are less than or equal to a certain threshold value, which we denote by V_{SNR}. If the relative speed between any two spacecraft in the constellation is larger than V_{SNR} in some region of the wave number plane, the desired signal-to-noise ratio is not attained there.

In Section 2, we have assumed that the area covered by the picture frame disc is being covered in a uniform fashion. However, the coverage it achieves has a nonuniform distribution that resembles a Gaussian. It can be shown that the width of this distribution is determined by the relative speed between the spacecraft. For low speeds the picture frame disc could even have a diameter that is slightly larger than $1/\theta_p$. For large speeds the picture frame disc will have a distribution with an effective width that is smaller than $1/\theta_p$. Thus, speed primarily affects signal-to-noise ratio and the distribution of the signal within the picture frame disc as it moves in the wave number plane.

The performance measure used for fuel consumption is

$$\mathcal{U} = \int_0^{t_f} \left(a_r^2(t) + a_t^2(t) \right) dt. \tag{14}$$

The two cost functions in (12) and (14) are the basis for evaluating the imaging performance and fuel consumption. The sum $\mathcal{J} = w_1 \mathcal{I} + w_2 \mathcal{U}$, where w_i $(i = 1, 2)$ are weighting coefficients, could be used as the objective function for a nonlinear optimal controller that optimizes both fuel consumption and image quality, which our group is currently investigating. Next we consider benchmark problems for controller design.

Maneuver 1: Constant Speed Motion As mentioned above, it is desirable to have a uniform signal-to-noise ratio over the entire resolution disc region in the wave number plane. This can be achieved by setting the magnitude of the velocity to be:

$$V = \sqrt{\left(r\dot{\theta} \right)^2 + \dot{r}^2} = V_{SNR}, \tag{15}$$

which is constant. Taking the derivative of this condition,

$$r\dot{\theta} \left(\dot{r}\dot{\theta} + r\ddot{\theta} \right) + \dot{r}\ddot{r} = 0, \tag{16}$$

substituting r, $\dot{r}$ and $\ddot{r}$ using (4)-(6) and rearranging one gets the closed loop differential equation

$$\ddot{\theta} = -\frac{f(\theta)}{1 + f^2(\theta)}\dot{\theta}^2, \tag{17}$$

which will achieve full coverage of the wave number plane if it satisfies the correct boundary conditions. From Eq. (9), the radial component of the control thrust is given by

$$\tilde{a}_r = -f(\theta)\dot{\theta}^2 \left[1 + \frac{1}{1 + f^2(\theta)}\right]. \tag{18}$$

Using this and Eq. (11), the tangential component, $\tilde{a}_t$ is thus

$$\tilde{a}_t = \dot{\theta}^2 \left(2 - \frac{f^2(\theta)}{1 + f^2(\theta)}\right). \tag{19}$$

Maneuver 2: Constant Tangential Velocity A simplified version of the situation considered in the previous paragraph is to assume that only the tangential component of the velocity vector is constant; $V_t = r\dot{\theta}$ is constant. Taking the the derivative of this condition,

$$\dot{r}\dot{\theta} + r\ddot{\theta} = 0, \tag{20}$$

substituting r, $\dot{r}$ and $\ddot{r}$ using (4)-(6) and rearranging one gets the closed loop differential equation

$$\ddot{\theta} = -\frac{1}{f(\theta)}\dot{\theta}^2. \tag{21}$$

From Eq. (9), the radial component of the control thrust is given by

$$\tilde{a}_r = -\dot{\theta}^2 \left[f(\theta) + \frac{1}{f(\theta)}\right]. \tag{22}$$

Using this and Eq. (11), the tangential component, $\tilde{a}_t$, is thus

$$\tilde{a}_t = \dot{\theta}^2. \tag{23}$$

Maneuver 3: Constant Angular Rate Next, consider the situation where $\dot{\theta}$ is a constant. Thus $\ddot{\theta} = 0$, which implies that θ is a linear function of time. After applying the boundary conditions, one obtains an explicit solution for θ

$$\theta(t) = \frac{(m - 1)\pi}{2t_f}t. \tag{24}$$

Since $\dot{r} = \frac{\lambda}{\pi\theta_p}\dot{\theta}$, it is also constant and $\ddot{r} = 0$. From Eq. (9), the radial component of the thrust vector is given by

$$\tilde{a}_r = -f(\theta)\dot{\theta}^2. \tag{25}$$

And from Equation (10), the tangential component of the thrust vector is given by

$$\tilde{a}_t = 2\dot{\theta}^2. \tag{26}$$

1027

Maneuver 4: Zero Tangential Acceleration Consider not applying any tangential thrust; then $\tilde{a}_t = 0$ and Eq. (11) implies that

$$\tilde{a}_r = -\left[f(\theta) + \frac{2}{f(\theta)}\right]\dot{\theta}^2. \tag{27}$$

Finally, the closed loop differential equation is

$$\ddot{\theta} = -\frac{2}{f(\theta)}\dot{\theta}^2. \tag{28}$$

5.1 Results

In the results we present here, Maneuver (1) represents the benchmark solution to the problem. It is the solution that achieves the two imaging objectives since (1) it follows a linear spiral that completely covers the desired resolution disc and (2) it moves at a constant speed that is sufficient to attain the desired signal-to-noise ratio. First we determined the value of V_{SNR} and set the initial condition for Maneuver (1) such that the spacecraft has an initial speed equal to V_{SNR}. This speed is guaranteed to be maintained by the additional speed constraint (15). We simulated the system and determined the terminal time t_f. Thus, for all other cases, we set the initial conditions such that each individual maneuver is completed at $t = t_f$. Finally, we use Eqs. (12) and (14) to evaluate the performance of all four motions. A reference speed, V_{SNR}, that achieves the desired signal-to-noise ratio is assumed to be 30m/s and t_f was found to be 17 days and 17 hours.

Figures (3)-(7) show the time evolution of r, θ, V, a_r and a_t for all four maneuvers. Figures (8) and (9) show how the Γ function accumulates and how the image forms with time for Maneuvers (1) and (4). These represent the two interesting cases: Maneuver (1) is the benchmark solution and Maneuver (4) shows that Γ by itself evaluated at the end of the maneuver is not sufficient to judge the quality of the final image. Maneuver (2) produces results almost identical to Maneuver (1) and we will only comment on the results for Maneuver (3).

The image formation algorithm was developed by our group and has been used in this section to demonstrate the concepts presented in this paper. The algorithm assumes a given target planet, which includes planet surface details. It also assumes a statistical model for noise. It simulates the motion of the spacecraft, computes the resulting Γ function, and estimates the image.

For Maneuver (1), note that the speed is constant throughout the maneuver (upper left chart in Figure (5).) Consequently, we expect a uniform Γ function throughout the wave number plane (Figure (8)); the desired signal-to-noise ratio is attained everywhere in the plane. However, for Maneuver (4), we note that we start off with a speed of 110m/s, which is larger than V_{SNR}. That is why near the center of the resolution disc (i.e. by the fifth day) we can see some partial coverage, resulting in a signal-to-noise ratio that is lower than the desired. We also expect that for Maneuver (4), the picture frame disc to be significantly narrower than the nominal value, $1/\theta_p$.

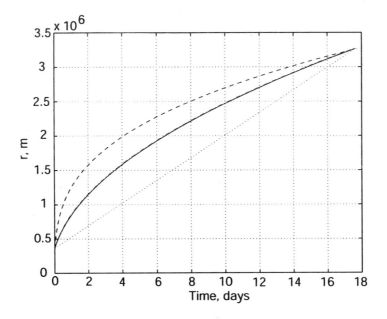

Figure 3: $r(t)$ For Maneuver (1) (Solid), Maneuver (2) (Dash-Dotted), Maneuver (3) (Dotted) and Maneuver (4) (Dashed)

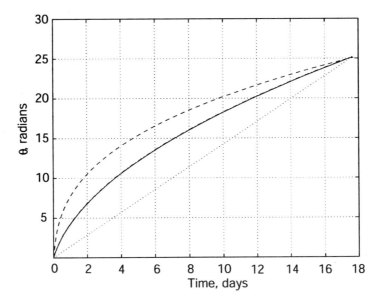

Figure 4: $\theta(t)$ For Maneuver (1) (Solid), Maneuver (2) (Dash-Dotted), Maneuver (3) (Dotted) and Maneuver (4) (Dashed)

Thus, at low frequencies the formation in Maneuver (1) performs better than the formation in Maneuver (4).

On the other hand, as the the spacecraft in Maneuver (4) proceeds beyond the

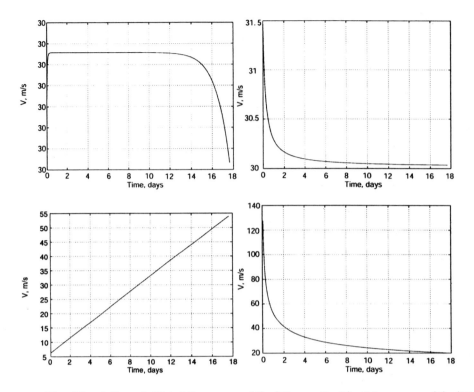

Figure 5: The Total Speed For Maneuver (1) (Upper Left), Maneuver (2) (Upper Right), Maneuver (3) (Lower Left) and Maneuver (4) (Lower Right)

fifth day, its speed drops below V_{SNR}, thus, achieving the desired signal-to-noise ratio. Moreover, one should expect a wider picture frame disc signal distribution. Beyond the fifth day of the maneuvers, both cases will achieve equal imaging performance for all but the last 2π radians of the maneuvers. During the last spiral revolution, the spacecraft in Maneuver (4) will have a wider picture frame disc than that of Maneuver (1) and will, thus, furnish bonus high frequency coverage in the wave number plane.

Therefore, during one phase of the maneuvers the formation in Maneuver (1) has superior performance than the formation in Maneuver (4). During the second phase the reverse is true. Figure (10) is a surface plot of the difference between the Γ functions of Maneuvers (1) and (4). Computing the imaging performance measure (12) one finds that Maneuver (4) has an overall smaller imaging performance measure than Maneuver (1) (Table (1).) This implies that the bonus coverage the formation in Maneuver (4) achieves at high frequencies outweighs the deficient coverage at low frequencies. If the imaging performance is restricted to the resolution disc only, then Maneuver (1) will achieve better imaging performance than Maneuver (4) because the latter leaves gaps at low frequencies.

On the other hand, because the spacecraft in Maneuver (4) is forced to move at a speed lower than that in Maneuver (1), more fuel is required to slow it down than that in Maneuver (1). Indeed, Maneuver (4) results in the least efficient maneuver

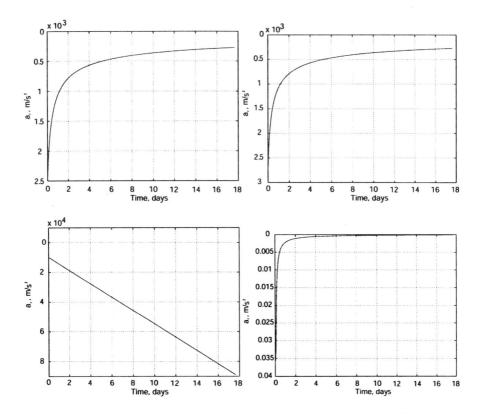

Figure 6: Radial Acceleration, a_r, For Maneuver (1) (Upper Left), Maneuver (2) (Upper Right), Maneuver (3) (Lower Left) And Maneuver (4) (Lower Right)

as far as fuel consumption is concerned but it is the most efficient in terms of image quality.

Table (1) shows a summary of the performance of all four maneuvers. We note that, as one might expect, Maneuvers (1) and (2) perform almost equally. Both result in intermediate image quality and fuel expenditure. Maneuver (3) requires the least amount of fuel but with the poorest performance. The reverse statement is true for Maneuver (4). Therefore, this result suggests that there exists a tradeoff between image quality and fuel expenditure.

	$\mathcal{I}, \times 10^9$	$\mathcal{U}$, m/s
Maneuver (1)	5.14	0.53
Maneuver (2)	5.13	0.55
Maneuver (3)	6.27	0.46
Maneuver (4)	4.62	15.02

Table 1: Performance Measures For Maneuvers (1)-(4)

Note that the imaging performance index (12) is a function of the final time, t_f.

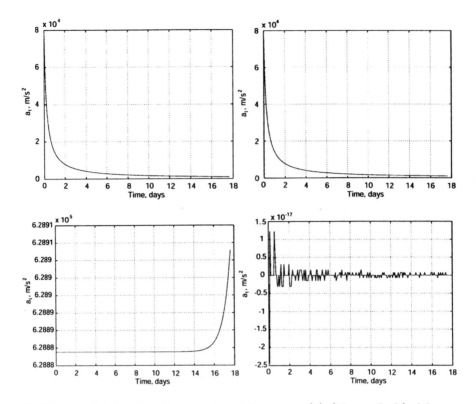

Figure 7: Tangential Acceleration, a_t, For Maneuver (1) (Upper Left), Maneuver (2) (Upper Right), Maneuver (3) (Lower Left) And Maneuver (4) (Lower right)

As t_f is increased sufficiently the above result will change. For instance, in the limit as t_f approaches infinity, the partial coverage ($\Gamma < 1$) that Maneuver (4) leaves behind at low frequencies represents a nonzero constant quantity which is being integrated over an infinite time horizon. Thus, the resulting performance index will be infinite, which means that Maneuver (4) is infeasible in the limit. For Maneuver (1), on the other hand, satisfying the critical speed threshold, V_{SNR}, ensures that for all past coverage we have $\Gamma = 1$. However, note that the integral computed up to any finite point in time will be nonzero due to the remaining uncovered region of the resolution disc. Moreover, any maneuver that results in $\Gamma = 1$ everywhere in the resolution disc with a finite imaging performance index in finite time, such as Maneuver (1), is also expected to do so in the limit as t_f approaches infinity. Thus we see that in the limit as t_f approaches infinity Maneuver (1) will be feasible whereas maneuver (4) will be considered infeasible.

Finally, one may construct maneuvers by splitting the boundary value problem in θ into multiple boundary value problems with different controls over each segment. For example one can use the controller of Maneuver (3) and then switch to the controller of Maneuver (4) when the spacecraft's speed crosses 30m/s. This combined maneuver's performance is expected to have a lower fuel cost than Maneuver (4) but better imaging performance since $\mathcal{I}$ will be accumulated over low speeds at both low and high frequencies.

6 Optimal Control Problem

Based on the conclusions drawn from the above results, the main objectives are: (1) achieve minimum fuel consumption during the maneuver, (2) attain the desired signal to noise ratio and (3) cover the whole wave number plane at the end of the maneuver. The third objective is achieved by the spiral constraint in Eq. (4). The second objective is achieved if we constrain the speed at which the spacecraft moves to a value not larger than V_{SNR}. Working with the constrained equations only, all three objectives can be achieved if one finds a solution to the following optimal control problem: Minimize

$$J = \int_0^{t_f} (\tilde{a}_r^2 + \tilde{a}_t^2)dt, \tag{29}$$

subject to the constraint:

$$\dot{x}_1 = x_2, \tag{30}$$

$$\dot{x}_2 = f(x_1)x_2^2 + \tilde{a}_r,$$

and initial conditions $x_1(0) = 0$ and $x_2(0) = x_{2i}$, where $x_1 = \theta$, $x_2 = \dot{\theta}$ and x_{2i} is fixed. There is a terminal condition $x_1(t_f) = \frac{(m-1)\pi}{2}$, while $x_2(t_f)$ is kept free. To satisfy the imaging objective, the speed of the spacecraft, V, should be constrained such that $V \leq V_{SNR}$ throughout the maneuver. This and Eq. (4) imply that the solution to the optimal control problem should satisfy the inequality constraint:

$$S(x_1, x_2) = x_2^2 \left(f^2(x_1) + 1\right) - V_{SNR}^2 \left(\frac{\pi \theta_p}{\lambda}\right)^2 \leq 0. \tag{31}$$

Note that the constraint in Eq. (11) must be satisfied and, thus, a_t is not a free variable. If a solution exists, then that solution should achieve all three objects of the mission.

For this problem the Hamiltonian is given by:

$$H = \tilde{a}_r^2 + \tilde{a}_t^2 + P_1 x_2 + P_2 \left(f(x_2)x_2^2 + \tilde{a}_r\right) + P_3 \left[2f(x_1)x_2^3 \left(f^2(x_1) + 2\right) + 2x_2 \left(f^2(x_1) + 1\right)\tilde{a}_r\right], \tag{32}$$

where P_1, P_2 and P_3 are multipliers to be determined. For the above state-constrained optimal control problem, the first order necessary conditions entail (Ref. [26]):

$$\tilde{a}_r = -\frac{1}{2\left(1 + f^2(x_1)\right)} \left[2f(x_1)\left(f^2(x_1) + 2\right)x_2^2 + P_2 + 2P_3 x_2 \left(f^2(x_1) + 1\right)\right], \tag{33}$$

$$\dot{P}_1 = -2\left(f(x_1)\tilde{a}_r + \left(f^2(x_1) + 2\right)x_2^2\right)\left(\tilde{a}_r + 2f(x_1)x_2^2\right) - P_2 x_2^2 \\ -4f(x_1)x_2 P_3 \left(f(x_1)x_2^2 + \tilde{a}_r\right), \tag{34}$$

1033

and

$$\dot{P}_2 = -4\left(f^2(x_1)+2\right)x_2\left(f(x_1)\tilde{a}_r+\left(f^2(x_1)+2\right)x_2^2\right)-P_1-2P_2f(x_1)x_2$$
$$-P_3\left[6f(x_1)x_2^2\left(f^2(x_1)+2\right)+2\left(f^2(x_1)+1\right)\tilde{a}_r\right], \tag{35}$$

Note that $P_3 = 0$ whenever $S(x_1, x_2) < 0$. For $S(x_1, x_2) = 0$, Eq. (31) is differentiated as many times as required until the control $\tilde{a}_r$ appears in the equation. Say that $\tilde{a}_r$ appears in the qth derivative: $S^{(q)} = 0$. Then, $S^{(q)}(x, \tilde{a}_r, t) = 0$ and Eq. (33) are used to solve for P_3. Thus $P_3(t)$ is determined if the state $x(t)$ and co-states $P_i(t)$ $(i = 1, 2)$ are known. From Eq. (33) and Eq. (11), $\tilde{a}_t$ must satisfy

$$\tilde{a}_t = \frac{1}{2\left(1+f^2(x_1)\right)}\left[2\left(f^2(x_1)-f(x_1)+1\right)\left(f^2(x_1)+2\right)x_2^2+P_2-2P_3x_2\left(f^2(x_1)+1\right)\right].$$
$$\tag{36}$$

The transversality conditions imply that $P_1(t_f) = p_1$ is unknown and $P_2(t_f) = 0$. This is a nonlinear optimal control problem with mixed boundary values for the state and costate variables and a state inequality constraint. Numerical methods exist in the literature to obtain the optimal path, if it exists (see for example (Ref. [26])).

7   Conclusions

In this paper we have introduced some basic notions that relate the imaging problem to the motion and controller design problem. Results show evidence of a tradeoff between image quality and fuel expenditure. Based on the observations drawn from solutions to the problem, an optimal control problem was formulated. Currently, our research is focused on computing the solution to this optimal control problem.

References

[1] D. C. Hyland. Interferometric Imaging Concepts With Reduced Formation-Keeping Constraints. *AIAA Space 2001 Conference*, Albuquerque, NM, August, 2001.

[2] S. E. Gano et. al. A Baseline Study of a Low-Cost, High-Resolution, Imaging System Using Wavefront Reconstruction. *AIAA Space 2001 Conference*, Albuquerque, NM, August, 2001.

[3] V. Kapila, A. G. Sparks, J. M. Buffington and Q. Yan. Spacecraft Formation Flying: Dynamics and Control. *Proceedings of the American Control Conference*, June 1999, pp. 4137-4141.

[4] S. Marcio, V. Kapila, Q. Yan. Adaptive Nonlinear Control of Multiple Spacecraft Formation Flying. *Journal of Guidance, Control, and Dynamics*, Vol. 23, No. 3, May-June 2000, pp. 385-390.

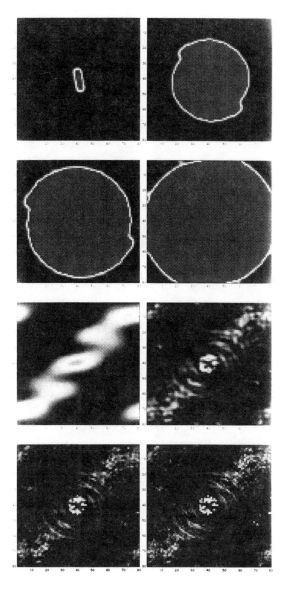

Figure 8: Γ (Top Four Figures) And Estimated Image (Bottom Four Figures) At $t = 0$, At $t = 5.9$ Days, At $t = 11.8$ Days And At $t = 17.7$ Days For Maneuver(1)

[2] S. E. Gano et. al. A Baseline Study of a Low-Cost, High-Resolution, Imaging System Using Wavefront Reconstruction. *AIAA Space 2001 Conference*, Albuquerque, NM, August, 2001.

[3] V. Kapila, A. G. Sparks, J. M. Buffington and Q. Yan. Spacecraft Formation Flying: Dynamics and Control. *Proceedings of the American Control Conference*, June 1999, pp. 4137-4141.

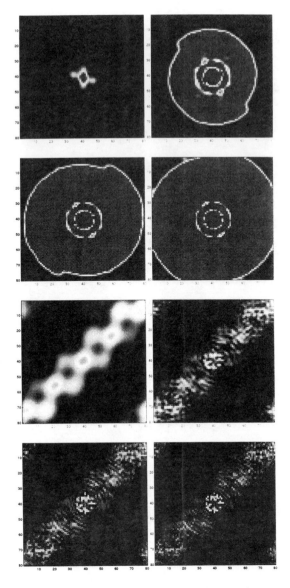

Figure 9: Γ (Top Four Figures) And Estimated Image (Bottom Four Figures) At $t = 0$, At $t = 5.9$ Days, At $t = 11.8$ Days And At $t = 17.7$ Days For Maneuver(4)

[4] S. Marcio, V. Kapila, Q. Yan. Adaptive Nonlinear Control of Multiple Spacecraft Formation Flying. *Journal of Guidance, Control, and Dynamics*, Vol. 23, No. 3, May-June 2000, pp. 385-390.

[5] J. Lawton, R. W. Beard, F. Y. Hadaegh. An Adaptive Approach to Satellite Formation Flying With Relative Distance Constraints. *Proceedings of the American Control Conference*, June 1999, pp. 1545-1549.

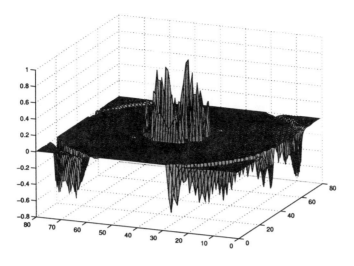

Figure 10: $\Gamma_1 - \Gamma_4$

[9] R. W. Beard and F. Y. Hadaegh. Fuel Optimized Rotation for Satellite Formations in Free Space. *Proceedings of the American Control Conference*, June 1999, pp. 2975-2979.

[10] R. W. Beard, T. W. McLain and F. Y. Hadaegh. Fuel Equalizated Retargeting for Separated Spacecraft Interferometry. *Proceedings of the American Control Conference*, pp. 1580-1584, June 1998.

[11] E. M. C. Kong, D. W. Miller and R. J. Sedwick. Exploiting Orbital Dynamics for Interstellar Separated Spacecraft Interferometry. *Proceedings of the American Control Conference*, pp. 4153-4157, 1999.

[12] I. I. Hussein, D. J. Scheeres and D. C. Hyland. Interferometric Observatories in Earth Orbit, *2003 AAS/AIAA Space Flight Mechanics Meeting*, Ponce, Puerto Rico, February 9-13, 2003.

[13] R. H. Vassar and R. B. Sherwood. Formationkeeping for a Pair of Satellites in a Circular Orbit. *Journal of Guidance*, Vol. 8, No. 2, pp. 235-242, 1985.

[14] A. B. DeCou. Orbital Station-Keeping for Multiple Spacecraft Interferometry. *The Journal of the Astronautical Sciences*, Vol. 39, No. 3, July-September 1991, pp. 283-297.

[15] H. Schaub and K. T. Alfriend. J_2 Invariant Relative Orbits for spacecraft Formations. *Celestial Mechanics and Dynamical Astronomy*, Vol. 79, pp. 77-95, 2001.

[16] H. Schaub, S. R. Vadali, J. L. Junkins and K. T. Alfriend. Spacecraft Formation Flying Control Using Mean Orbital Elements. *The Journal of the Astronautical Sciences*, Vol. 48, No. 1, pp. 69-87, 2000.

[17] A. Robertson, T. Corazzini and J. P. How. Formation Sensing and Control Techniques for a Separated Spacecraft Interferometer. *Prceedings of the American Control Conference*, June 1998, pp. 1574-1579.

[18] K. Lau, S. Lichten, L. Young and B. Haines. An Innovative Deep Space Application of GPS Technology for Formation Flying Spacecraft. *Guidance, Navigation and Control Conference*, July 1996, pp. 3819-3828.

[19] J. R. Carpenter. A Preliminary Investigation of Decentrralized Control for Satellite Formations. *IEEE Aerospace Conference Proceedings*, Vol. 7, 2000, pp. 63-74.

[20] J. R. T. Lawton, B. J. Young and R. W. Beard. A Decentralized Approach to Elementary Formation Maneuvers. *Proceedings - IEEE International Conference on Robotics and Automation*, Vol. 3, 2000, pp. 2728-2733.

[21] W. Kang, A. Sparks and S. Banda. Coordinated Control of Multisatellite Systems. *Journal of Guidance, Control and Dynamics*, Vol. 24, No. 2, March-April 2001, pp. 360-368.

[22] M. Born and E. Wolf. *Principles of optics*, Pergamon Press, New York, 1964.

[23] D. C. Hyland. The Inverse Huygens-Fresnel Principle and Its Implications for Interferometric Imaging. *Journal of the Astronautical Sciences*, to appear.

[24] F. Y. Hsiao and D. J. Scheeres. Design of Spacecraft Formation Orbits Relative to a Stabilized Trajectory. *2003 AAS/AIAA Space Flight Mechanics Meeting*, Ponce, Puerto Rico, February 9-13, 2003.

[25] L. Mandel and E. Wolf. *Optical Coherence and Quantum Optics*. Cambridge University Press, 1995.

[26] A. E. Bryson and Y. Ho. *Applied Optimal Control*. Hemisphere Publishing Corporation, New York, 1975.

WORST CASE AND MEAN SQUARED PERFORMANCE OF IMAGING SYSTEMS: A FEATURE-BASED APPROACH[*]

Suman Chakravorty,[†] Pierre T. Kabamba[‡] and David C. Hyland[**]

In this paper, we quantify the effect of random noise on the probability of misclassification of images. We consider two metrics for the noise corrupting the image: the mean squared error (MSE) and the worst case error (WCE). We show that these are consistent with the goal of image classification in that, as the MSE or WCE tends to zero, the probability of misclassifying an image also tends to zero. Given a feature map, i.e., a real-valued function of an image variable, we assume that classification is done by applying a threshold to the feature map. In this feature-based classification, we find bounds on the MSE and the WCE such that the probability of misclassifying the image is guaranteed to be less than some pre-specified value. We illustrate the theory through an example where the banded appearance of the image of a planet is detected. We also show that, in the special case of a linear feature map, finding an estimate that minimizes the probability of misclassification reduces to a problem of finding a minimum weighted-MSE estimate. The results of this paper could be used for the reliable characterization of exo-solar planets and similar astronomical studies.

Keywords: imaging, mean squared error, worst case error, feature-based classification.

1 Introduction

This paper is motivated by the prospect of taking high resolution images of exo-solar planets at distances of up to 15 parsecs [1]. The images obtained would be used to classify the planet based on some feature of the image, a feature being defined as a real-valued function of an image variable. Some of the features that may be considered are the percentage of oxygen in the atmosphere of the planet, the ratio of land to water on the surface or the percentage of carbon dioxide in the atmosphere . We are given a threshold on the feature, such that if the value of the feature is greater than the threshold, the image is classified in class $\mathcal{I}_1$, and otherwise, it is classified into class $\mathcal{I}_2$. This paper examines the effect of noise on such a feature-based classification problem. We consider two error metrics between the image and its noisy estimates : the mean squared error (MSE) and the worst case error (WCE). We first show the consistency of these metrics, namely, that the probability of misclassifying an image tends to zero as each of these error metrics converges to zero. This yields bounds on the MSE and the WCE such that the probability of misclassifying an image can be guaranteed lower than a pre-specified value. We show that if the feature being considered is a linear map, then an optimal imaging problem can be formulated based upon an appropriately weighted MSE criterion. We show the existence of a solution to this optimization problem under certain assumptions and give necessary conditions of optimality. The work presented in this paper is related to the fields of signal detection/pattern recognition, medical imaging and image processing, and the relationships are discussed hereafter.

*   This work was supported by NASA under grant NAG5-10336.

†   Graduate Student Research Assistant, Department of Aerospace Engineering, The University of Michigan, Ann Arbor, Michigan 48109-2140. E-mail: schakrav@engin.umich.edu.

‡   Corresponding Author: Professor, Department of Aerospace Engineering, The University of Michigan, Ann Arbor, Michigan 48109-2140. E-mail: kabamba@engin.umich.edu.

**   Professor and Chairman, Department of Aerospace Engineering, The University of Michigan, Ann Arbor, Michigan 48109-2140. E-mail: dhiland@engin.umich.edu.

In pattern recognition and signal detection [2, 3] the problem of designing optimal classifiers has been studied in detail. It reduces to a problem of statistical hypothesis testing [2]. Given the statistical characteristics of the noise corrupting the signal and the partition of the set of signals into two disjoint classes, the optimal classifier is designed using the Bayesian likelihood ratio test (BLRT) or the Most Powerful Likelihood Ratio Test (MP-LRT) [2]. In robust signal detection the problem of optimal detection of signals when there is uncertainty in the signal or the noise involved in the process has been studied [4, 5, 6]. This is done by obtaining the detector that optimizes the worst case combination of signal and noise , given that the uncertainty classes for these have been specified. The problem considered in this paper is similar to the robust detection problem in the sense that we are also dealing with classification of an image. However, the purpose of this paper is not the design of an optimal classifier. In fact, we assume that a classifier is given (the feature map and a threshold on it) and we presume no model on the noise. The objective of the paper is, given a set of images, to characterize in precise quantitative terms (misclassification error), the effect of the noise corrupting the system so that some worst case performance, in terms of the misclassification error, is achieved.

In medical imaging research , the Mean Squared Error (MSE) has been used as the criterion for posing a variety of static and sequential observation selection problems [7]-[10]. In this paper, we also perform an MSE minimization. However, we motivate the use of MSE as an optimization criterion by showing that it is consistent with the goal of image classification. In particular, we show that in the case of a linear feature, minimizing a weighted MSE criterion (weighted by the feature), results in an estimate that minimizes the misclassification probability of the image. We also note that the problem in our case is underedetermined, in that the number of observations is smaller than the number of pixels in the image. The references cited above consider overdetermined problems.

The image formed by an optical instrument is obtained by a spatial convolution of the object field with the point-spread function (psf) of the optical instrument [11]-[13]. The psf of an optical instrument is a low-pass filter because any optical instrument records the field due to the object over a finite area. As a consequence, the problem of image reconstruction from the observations, viewed as a deconvolution, is ill-posed [13]. This problem of deconvolution in the presence of noise has been widely studied in the image processing community and is solved using regularization methods [13]-[15]. In some applications, the point-spread function of the optical instrument might not be known perfectly. These problems are treated as blind-deconvolution problems where the psf of the instrument and the image are estimated simultaneously [16, 17].

If the field is reconstructed over the whole infinite plane, the problem of image reconstruction is no longer ill-posed. Under conditions of bandlimitedness or analyticity on the image, respectively, the image can be reconstructed perfectly over the entire infinite plane by taking measurements on a countable set or a set of arbitrarily small measure, respectively [11, 18]. If we have further apriori information that the image belongs to a certain finite dimensional subspace, then the image can be reconstructed perfectly, in the absence of noise, by taking a finite number of observations. In this paper we consider the problem of obtaining the optimal estimate of an image by taking a finite number of observations, given that we know that it is constrained in such a finite dimensional subspace, given the statistical characteristics of the noise corrupting the measurements, and given that the feature map is linear . The image reconstruction algorithm proposed results in an unbiased estimate of the image. In fact, the estimate resulting from the algorithm is the minimum variance unbiased estimate. Finally note that the problem of optimal imaging is similar to the optimal sensor placement problem in various other applications such as large flexible structures, chemical processes and power systems [19]-[22].

The original contributions of this paper are as follows. We show the consistency of the MSE and the WCE criteria with the goal of image classification in that the probability of misclassifying an image goes to zero as the MSE or WCE of its estimates tends to zero. We obtain error bounds on the noise corrupting an imaging system such that a pre-specified level of performance in terms of the probability of misclassifying an image is guaranteed. The error bounds are obtained in terms of the MSE and the WCE for the case of classification based upon a feature. We show that, in the special case of a linear feature, finding an optimal estimate of an image from a family of its estimates, in the sense that the misclassification error is minimized, reduces to a problem of minimizing the weighted-MSE between the image and its estimates. We consider the case when we know that the image is constrained in a known finite dimensional subspace. Then, the optimal imaging problem is one of choosing the finite number of vantage points from which to take measurements, given the statistical characteristics of the noise corrupting the measurements, such that the MSE is minimized. We show that, under certain assumptions, an optimal solution exists and also give necessary conditions for optimality.

The rest of the paper is organized as follows. Section 2 shows the consistency of the MSE and WCE criteria with respect to the goal of classification. In section 3 we obtain error bounds on the MSE and WCE such that some pre-specified level of performance, in terms of the probability of correctly classifying an image, is guaranteed. In section 4 we present an example in which the banded appearance of an image is detected. In section 5 we treat the special case of linear feature maps and show that, under this assumption, an optimal imaging problem can be formulated in terms of a weighted-MSE criterion such that the solution minimizes the probability of misclassifying an image. In section 6 we formulate the problem of choosing optimal vantage points to form images, based on the MSE criterion. In section 7 we present a numerical example illustrating the results of section 6. In section 8 we draw conclusions on the work presented in this paper, and discuss various avenues of future research.

2 Classification : Consistency of Error Metrics

In this section, we introduce the notion of Misclassification Error (ME). We also show that two different measures of image error, the Mean Squared Error (MSE) and the Worst Case Error (WCE), are consistent with the ME, in that the ME tends to zero as the MSE or the WCE tends to zero. This will allow us to find bounds on the MSE or the WCE such that if these bounds are satisfied by the estimates, the ME is guaranteed to be below a certain pre-specified level.

Let $\mathcal{I}$ denote the space of images of interest, $\mathcal{I} \subseteq L_2(D)$ where D denotes the support of the images and is equipped with the standard L_2 norm, denoted by $||.||$. Let $\mathcal{I}$ be partitioned into two non-empty classes $\mathcal{I}_1$ and $\mathcal{I}_2$ with $\mathcal{I} = \mathcal{I}_1 \cup \mathcal{I}_2$ and $\mathcal{I}_1 \cap \mathcal{I}_2 = \phi$. Let $P : \mathcal{I} \to \{0, 1\}$ denote the mapping such that if $i \in \mathcal{I}_1$, then $P(i) = 0$ and $P(i) = 1$ otherwise.

Definition 2.1 *The Misclassification Error (ME) of an estimate $\hat{i}$ of an image i is defined as*

$$\epsilon(\hat{i}, i) = prob(P(\hat{i}) \neq P(i)).$$

Definition 2.2 *The Mean Squared Error (MSE) between an image i and its estimate $\hat{i}$ is defined as*

$$||i - \hat{i}||_2 = (E||i - \hat{i}||^2)^{1/2},$$

whenever $E||i - \hat{i}||^2$ exists and where $E(.)$ denotes the expectation operator.

Definition 2.3 *The Worst Case Error (WCE) between an image i and its estimate $\hat{i}$ is defined as*

$$||i - \hat{i}||_\infty = (sup_{x \in D} E|i(x) - \hat{i}(x)|^2)^{1/2}.$$

Definition 2.4 *Given an image i, we say that a metric $||.||_q$ is consistent with classification if, for any sequence of estimates $\{\hat{i}_n\}$ of i, $||i - \hat{i}_n||_q \to 0$ implies $\epsilon(\hat{i}_n, i) \to 0$.*

In the following propositions, we show that the MSE and the WCE are consistent with classification.

Proposition 2.1 *The MSE is consistent with classification whenever the image does not lie on the boundary between the two partition classes.*

Proof: Let $\{\hat{i}_n\}$ be an arbitrary sequence of random estimates of the given image i. Let $\{\hat{i}_n\}$ be convergent, with limit i, in the MSE sense, i.e., $\lim_{n\to\infty}||i - \hat{i}_n||_2 = 0$. Let $i \in \mathcal{I}_1$ without loss of generality and let $d(i, \mathcal{I}_2) = \eta$ where $d(x, S) = inf_{y \in S}||x - y||$. We have $\eta > 0$, since the image i does not lie on the boundary separating the two classes. Hence, for estimate $\hat{i}$ of i,

$$\epsilon(\hat{i}, i) = prob(||i - \hat{i}|| > \eta).$$

Therefore, for sequence $\{\hat{i}_n\}$ we have that

$$0 \leq \epsilon(\hat{i}_n, i) = prob(||i - \hat{i}_n|| > \eta) \leq \frac{E||i - \hat{i}_n||^2}{\eta^2},$$

where the last inequality follows due to the Tchebyshev inequality [23]. Noting that $\eta > 0$, it follows that $\epsilon(\hat{i}_n, i) \to 0$ as $n \to \infty$. Hence, we have shown that the MSE is consistent with classification. **Q.E.D**

Proposition 2.2 *The WCE is consistent with classification whenever the image does not lie on the boundary between the two partition classes.*

Proof: Let $\{\hat{i}_n\}$ be an arbitrary sequence of random estimates of the given image i. Let $\{\hat{i}_n\}$ be convergent, with limit i, in the WCE sense, i.e., $\lim_{n\to\infty}||i - \hat{i}_n||_\infty = 0$. Note that

$$E||i - \hat{i}_n||^2 = E \int_D |i(x) - \hat{i}(x)|^2 dx.$$

By Fubini's theorem [24], we have that

$$0 \leq E||i - \hat{i}_n||^2 = \int_D E|i(x) - \hat{i}(x)|^2 dx.$$

Since $sup_{x \in D} E|i(x) - \hat{i}_n(x)|^2 \to 0$ as $n \to \infty$, we can conclude, from the Lebesgue Dominated Convergence Theorem [24], that $E||i - \hat{i}_n||^2 \to 0$. From Proposition 2.1, it follows that $\epsilon(\hat{i}_n, i) \to 0$. Hence, the WCE is consistent with classification. **Q.E.D**

In the preceding development, we have shown that the MSE and the WCE are consistent with the goal of image classification. Suppose now that we are interested in keeping the ME below a certain level, say p_{min}. Finding the ME of an estimate of the image may not be feasible. However, the preceding development allows us to find bounds on the WCE or the MSE such that if the image estimates satisfy these bounds, then the ME of these estimates cannot be above the level p_{min}. Note that the MSE or the WCE of an estimate can be easily calculated from the second-order statistical characteristics of the noise. In the next section we find such bounds for the case of feature-based classification with thresholding.

3   Guaranteed Performance of Feature-Based Classification

In this section we find bounds on the MSE and the WCE such that some pre-specified level of performance, in terms of the ME, is guaranteed. Let $i \in L_2(D)$ be an image that is unknown but fixed. Assume that the support D is compact. Let $\hat{i}$ be a noisy estimate of i such that

(1)
$$\hat{i} = i + e,$$

where e is a random process on D.

We make the following assumptions on the noise corrupting the image and the classification scheme.

$\mathcal{A}$ 3.1 *We are given a feature map $F : L_2(D) \to \Re$ and a threshold $x_0 \in R$ such that*

$$F(i) \leq x_0 \iff i \in \mathcal{I}_1,$$
(2)
$$F(i) > x_0 \iff i \in \mathcal{I}_2.$$

$\mathcal{A}$ 3.2 *The set of images, $\mathcal{I} \subseteq L_2(D)$, is convex.*

$\mathcal{A}$ 3.3 *The feature map F is continuously differentiable on $\mathcal{I}$ and its gradient is bounded uniformly by a constant $L < \infty$.*

It follows from the Riesz representation theorem [25] that the gradient can be uniquely represented by $f_i \in L_2(D)$, i.e.,

(3)
$$\forall g \in L_2(D), \nabla F_i g = (f_i | g),$$

where $(.|.)$ denotes the innner product on L_2.

We now obtain uniform bounds for the WCE and the MSE such that a pre-specified minimum acceptable performance level is attained by the measurement system. Suppose we require that for all $i \in \mathcal{W}_k$, $\epsilon_p(\hat{i}, i) \leq p_{min}$, where for a given $k > 0$,

$$\mathcal{W}_k = \{i \in \mathcal{I} : |F(i) - x_0| > k\},$$

represents the set of unambiguous images with level k. Then, the problem of finding a uniform bound on the MSE such that a minimum acceptable performance is achieved can be posed as :
find $\delta_2 > 0$ such that if $\forall i \in \mathcal{W}_k$, $||i - \hat{i}||_2 < \delta_2$, then $\forall i \in \mathcal{W}_k$, $\epsilon_p(\hat{i}, i) \leq p_{min}$.

We have the following results:

Lemma 3.1 *Under assumption $\mathcal{A}3.3$, the feature map F is globally lipschitz, with Lipschitz constant L, i.e., for all $i_1, i_2 \in \mathcal{I}$,*

$$|F(i_1) - F(i_2)| \leq L||i_1 - i_2||.$$

Proof: See Lemma 2.2 in [26].
Q.E.D.

Proposition 3.1 *Under assumptions $\mathcal{A}$ 3.1 - $\mathcal{A}$ 3.3, if $\forall i \in \mathcal{W}_k$, $||i - \hat{i}||_2 \leq \delta_2$, where $\delta_2 = (p_{min}B^2)^{1/2}$, where $B = \frac{k}{L}$, then $\forall i \in \mathcal{W}_k$, $\epsilon(\hat{i}, i) \leq p_{min}$.*

Proof: We have that
$$E|F(i) - F(\hat{i})|^2 \leq E\{L^2||i - \hat{i}||^2\} = L^2||e||_2^2.$$
The first inequality follows from the fact that $|F(i) - F(\hat{i})|^2$ and $||i - \hat{i}||_2$ are non-negative random variables and Lemma 3.1. Also,

$$\epsilon(\hat{i}, i) \leq prob\{|F(i) - F(\hat{i})| \geq |F(i) - x_0|\} \leq \frac{E|F(i) - F(\hat{i})|^2}{|F(i) - x_0|^2} \leq \frac{L^2}{k^2}||e||_2^2.$$

Hence, if we set $\delta_2 = (p_{min}B^2)^{1/2}$, it follows that $\epsilon(\hat{i}, i) \leq p_{min}$.
Q.E.D.

Proposition 3.2 *Under assumptions $\mathcal{A}$ 3.1 -$\mathcal{A}3.3$, if $\forall i \in \mathcal{W}_k$, $||i - \hat{i}||_\infty \leq \delta_\infty$, where $\delta_\infty = (\frac{p_{min}B^2}{\mu(D)})^{1/2}$, then $\forall i \in \mathcal{W}_k$, $\epsilon(\hat{i}, i) \leq p_{min}$.*

Proof: Note that $\|e\|_2^2 \leq \|e\|_\infty^2 \mu(D)$, where $\mu(D)$ represents the measure of the support D. Since it is compact, it follows that $\mu(D) < \infty$. Hence, if we set $\delta_\infty \leq \left(\frac{p_{min}B^2}{\mu(D)}\right)^{1/2}$, Proposition 3.1 implies that $\epsilon(\hat{i}, i) \leq p_{min}$.

Q.E.D

At this point we would like to note that the key assumption required to derive the above Propositions is $A3.3$. A feature with an unbounded gradient would not be useful in reliable characterization of images, since a small change in the image may result in an arbitrarily large change in the feature value. Hence, any feature that is used in a practical system must have a bounded gradient, which then guarantees that the probability of misclassifying an image can never be greater than p_{min}.

In the above results we have used the Tchebyshev inequality to find bounds on the errors that corrupt the measurements. However, the bounds that were obtained can be very conservative since the Tchebyshev bound is generally very conservative. If the noise is small enough, then the resultant randomness in the feature could be approximated linearly. Thus, if the noise was zero mean Gaussian, the resulting randomness in the feature would also be Gaussian. This allows us to find much tighter bounds. This is formally stated in the following assumptions and corollaries.

A **3.4** *The noise process corrupting the measurements is zero mean Gaussian.*

A **3.5** *The random variable $F(\hat{i}) - F(i)$ is zero-mean gaussian.*

Then, under the above assumption, we have the following corollary.

Corollary 3.1 *Let assumptions A3.1- A3.5 hold. If the estimates $\hat{i}$ of any $i \in W_k$ satisfy $\|i - \hat{i}\|_2 \leq \delta_2$ where δ_2 is such that*

$$prob\{N(0,1) \geq \frac{B}{\delta_2}\} \leq p_{min},$$

where $B = \frac{k}{L}$, then $\forall i \in W_k, \epsilon(\hat{i}, i) \leq p_{min}$.

Similarly, we have the following for the WCE.

Corollary 3.2 *Let assumptions A3.1- A3.5 hold. If the estimates $\hat{i}$ of any $i \in W_k$ satisfy $\|i - \hat{i}\|_\infty \leq \delta_\infty$ where δ_∞ is such that*

$$prob\{N(0,1) \geq \frac{B}{(\mu(D)\delta_\infty)^{1/2}}\} \leq p_{min},$$

where $B = \frac{k}{L}$, then $\forall i \in W_k, \epsilon(\hat{i}, i) \leq p_{min}$.

The proofs of the above corollaries follow from Propositions 3.1 and 3.2 and the fact that the random variable $F(\hat{i}) - F(i)$ is zero-mean Gaussian. Hence, given a feature F with threshold x_0 and performance indices p_{min} and k, the above results could be used to find a noise level, represented here by δ_2 or δ_∞, such that if the measurement noise intensity is below this level, then some minimum acceptable performance is achieved. We illustrate this in the next section through an example.

4 Example: Banded Appearance of an Image

In this section, we present an example that illustrates the results of the previous section. In this example we classify an image as banded or otherwise. The example is motivated by the prospect of finding "Jupiter like" gas-giants in other solar systems. The bandedness criterion is a natural way of identifying such gas giants.

Let an image be denoted in the spatial domain by $i(x)$, $x \in D$ where D denotes the compact support of the image. Let $\bar{i}$ denote the Fourier transform of the image, with support denoted by Ω. A banded image would have most of its energy concentrated in a few frequencies. Hence, let $d\Omega$ denote a subset of Ω that is small compared to Ω. We classify an image as being banded in $d\Omega$ if the proportion of the total energy in it is above a given threshold. Thus, the bandedness feature can be represented as

$$(4) \qquad F(i) = \frac{\int_\Omega Q(\omega)|\bar{i}(\omega)|^2 d\omega}{||\bar{i}||^2},$$

where

$$Q(\omega) = \begin{cases} 1 & if \quad \omega \in d\Omega \\ 0 & if \quad \omega \notin d\Omega \end{cases}.$$

Assume we have a threshold x_0 such that if $F(i) > x_0$, then the image is classified as banded; otherwise the image is classified as not banded in $d\Omega$.

For simplicity of treatment let us consider the discrete version of the same problem. Note that any two-dimensional image is a matrix of intensity values and can be represented as a vector if we stack the rows of the matrix in lexicographic order. Hence, from now on, the image shall be represented as a vector. In this case the image is represented by an n-vector and we are interested in finding if the energy of the signal is concentrated in a maximum of m elements of the vector. Thus the feature in this case is

$$(5) \qquad F(i) = \frac{(\bar{i})^* Q(\bar{i})}{\bar{i}^* i},$$

where $\bar{i}^*$ denotes the complex conjugate transpose of the vector $\bar{i}$, Q is a diagonal matrix with ones on the diagonal corresponding to a particular choice of the m-elements of the n-vector and zeroes everywhere else, $\bar{i} = Hi$ where H is a unitary matrix that represents the discrete fourier transform operator. Let x_0 be a threshold on the feature. Let $k > 0, p_{min} > 0$ be given. Then our design (in the case of MSE) reduces to finding δ_2 such that if $i \in \mathcal{W}_k$ and estimate $\hat{i}$ of i satisfies $||i - \hat{i}||_2 \leq \delta_2$, then $\epsilon(\hat{i}, i) \leq p_{min}$.

In order to use the results of the previous section, we need to find an upper bound on $||f_i||$. Taking the derivative of $F(i)$ with respect to i we have that

$$||f_i||^2 = 4\frac{\bar{i}^* Q\bar{i}}{(\bar{i}^* i)^2}[1 - \frac{\bar{i}^* Q\bar{i}}{\bar{i}^* i}],$$
$$= 4\frac{F(i)(1 - F(i))}{||\bar{i}||^2}.$$

Note that since H is unitary, $||i|| = ||\bar{i}||$. Let us assume that it is known that the image cannot have an energy content lower than a given lower bound M, i.e., for all i, $||i||^2 \geq M$. This assumption is not restrictive since we might have prior knowledge of the energy content of the image of the planet that we are imaging. Also, photodetectors generally cannot detect intensity below a certain threshold and that would in turn result in a lower limit on the energy of the planets that can be imaged. Then, it can be shown that $L = \frac{1}{M^{1/2}}$. Hence, if we assume normality of the error process, δ_2 has to be chosen such that

$$prob\{N(0,1) \geq \frac{B}{\delta_2}\} \leq p_{min},$$

where $B = kM^{1/2}$. In case the error process is not normal, we have

$$\delta_2 \leq (p_{min}B^2)^{1/2}.$$

In Fig.1 we show plots of p_{min} as a function of the normalized MSE for various values of k (the MSE is normalized by the lower limit on the 2-norm of the signal L). The threshold x_0 is assumed to be 0.9 in this case. It can be seen that the maximum allowable value of the MSE would critically depend on the values

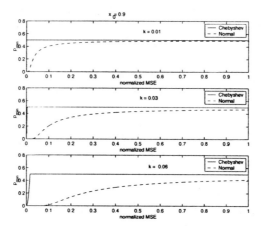

Figure 1: Plot of p_{min} as a function of the normalized MSE bounds.

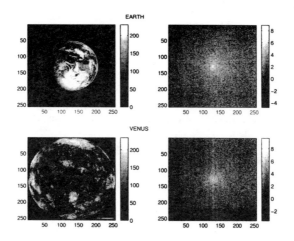

Figure 2: Image of planets and the corresponding Fourier transforms

of k and p_{min}. Too high a value of p_{min} with a correspondingly low value of k would lead to δ_2 that might be prohibitively low, in that it might be too expensive to achieve it in practice. Another factor that weighs in heavily is the lower bound B. If it is too conservative, then that could also lead to values of δ_2 that are very low (In this example the lower bound obtained is tight). Hence, we can conclude that a good design of δ_2 would require carefully chosen values of k and p_{min} and calculation of the tightest possible bound B for the feature given. The case of WCE, where we obtain δ_∞, is similarly treated.

In Fig.2, Fig.3 and Fig.4 we show the images of the planets in the solar system and their corresponding Fourier transforms. The bandedness feature values of the planets are shown in Fig.5. We chose the threshold for bandedness of the images at 0.9, i.e., an image is classified as banded if more than 90% of its energy is concentrated in less than 1% of the frequencies, i.e., Q was a diagonal matrix with ones on $0.1N$ of the diagonal elements and zeroes everywhere else, for an N-pixellated image. Only Jupiter and Neptune were classified as banded through this criterion. Also we note that the prominent banded appearance of Jupiter reflects in the feature value of 0.93. This was concentrated in just 11 spatial frequencies.

We also performed Monte Carlo simulations in order to validate the theory presented in the previous sec-

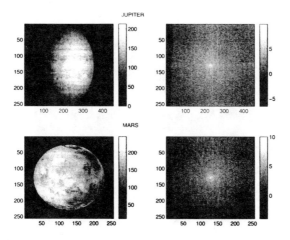

Figure 3: Image of planets and the corresponding Fourier transforms

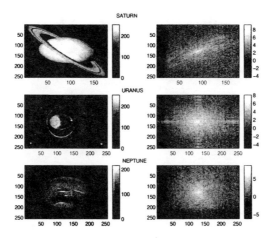

Figure 4: Image of planets and the corresponding Fourier transforms

tion. We assumed that the noise corrupting the image was zero mean uncorrelated Gaussian. The MSE value prescribed by the theory was 2% of the energy of the image (the squared 2-norm of the image) for a maximum of 10% misclassification error. The simulations were done for the case of Jupiter and we found that the misclassification error was between 4.85% and 4.88% with a confidence level of 99% [27, 28]. However, it was noticed that the noisy feature is biased. The nonlinearity of the feature is responsible for this, along with the fact that $A3.5$ is not satisfied. We also note that the MSE level that is allowable using the Tchebyshev inequality is around 0.5%, which is more conservative than the level afforded by the normal approximation.

In the above example, the bandedness of the image of a planet was detected. Also we found bounds on the MSE such that the misclassification error was below 10%. Monte Carlo simulations confirmed that if the bound on the MSE was satisfied by an estimate, then the probability of misclassification was indeed less than 10%.

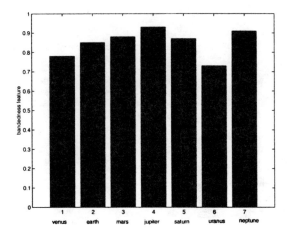

Figure 5: Values of the bandedness feature for the solar planets.

5   Optimal Imaging with Linear Feature Maps

In the development so far, we have found worst case bounds on the MSE and the WCE of the estimate of an image such that performance in terms of the ME is guaranteed. Now we turn our attention to the case of optimal imaging.

The problem of optimal imaging is stated as follows : Let i denote a fixed but unknown image. Let $\hat{i}(\theta)$ denote a family of estimates of i parametrized by $\theta \in S$. Find

$$\theta^* = arg \min_{\theta \in S} \epsilon(\hat{i}(\theta), i).$$

In the following, we show that if the feature is a linear map, minimizing a weighted mean squared error criterion (weighted by the feature), results in an optimal estimate.

We have the following result.

Proposition 5.1 *Let F be a linear map. Suppose assumptions $\mathcal{A}3.1$ and $\mathcal{A}3.4$ hold. Let $\hat{i}_1$ and $\hat{i}_2$ be noisy estimates of i such that $E|F(i) - F(\hat{i}_1)|^2 \leq E|F(i) - F(\hat{i}_2)|^2$. Then $\epsilon(\hat{i}_1, i) \leq \epsilon(\hat{i}_2, i)$.*

Proof:
By Assumption $\mathcal{A}3.1$, there exists $x_0 \in \Re$ such that $F(i) < x_0 \Rightarrow i \in \mathcal{I}_1$ and $F(i) \geq x_0 \Rightarrow i \in \mathcal{I}_1$. Let $\delta i_1 = i - \hat{i}_1$ and $\delta i_2 = i - \hat{i}_2$. Since F is a linear map and $\delta i_1, \delta i_2$ are Gaussian random processes, $(F(i) - F(i_{1,2}))$ are gaussian random variables. Note that by Assumption $\mathcal{A}3.4$, the estimates $\hat{i}_1$ and $\hat{i}_2$ are unbiased and hence, we can set

(6) $$F(\hat{i}_1) \equiv N(F(i), \sigma_1),$$

(7) $$F(\hat{i}_2) \equiv N(F(i), \sigma_2),$$

where, by hypothesis, we have that $\sigma_1 \leq \sigma_2$. Let $i \in \mathcal{I}_1$.

$$\epsilon(\hat{i}_1, i) = prob(\hat{i}_1 \notin \mathcal{I}_2) = prob(F(\hat{i}_1) > x_0)$$

(8) $$= prob(\frac{F(\hat{i}_1) - F(i)}{\sigma_1} > \frac{x_0 - F(i)}{\sigma_1}).$$

Noting that $\frac{F(\hat{i}_1)-F(i)}{\sigma_1} \equiv N(0,1)$,

(9)
$$\epsilon(\hat{i}_1, i) = prob(N(0,1) > \frac{x_0 - F(i)}{\sigma_1}).$$

(10)

Similarly,

(11)
$$\epsilon(\hat{i}_1, i) = prob(N(0,1) > \frac{x_0 - F(i)}{\sigma_2}).$$

Noting that $x_0 > F(i)$ and $\sigma_1 \leq \sigma_2$, it follows from above that $\epsilon(\hat{i}_1, i) \leq \epsilon(\hat{i}_2, i)$. The case when $i \in \mathcal{I}_2$ can be treated similarly.

Q.E.D

From the above proposition, we can conclude that if $\mathcal{A}3.1$ and $\mathcal{A}3.4$ are satisfied by the estimates, then the optimal imaging problem reduces to finding θ^* where

$$\theta^* = arg \min_{\theta \in S} E|F(i) - F(\hat{i}(\theta))|^2.$$

Hence, to choose the optimal estimate we need to evaluate the quantity $E|F(i) - F(\hat{i}(\theta))|^2$ for the various estimates. It is possible to evaluate the above for the case of a linear map. In fact,

(12)
$$E|F(i) - F(\hat{i}(\theta))|^2 = \int_D \int_D f(t) R_{\delta i(\theta)}(t,s) \bar{f}(s) dt ds,$$

where $R_{\delta i(\theta)}(t,s)$ denotes the covariance function of the error process $\delta i(\theta)$. Thus, in the case of a linear feature map, evaluating

$$\theta^* = arg \min_{\theta \in S} \int_D \int_D f(t) R_{\delta i(\theta)}(t,s) \bar{f}(s) dt ds$$

results in an optimal estimate in the sense that the misclassification error is minimized.

Remark : A sufficient condition for the MSE to be optimal with respect to the ME is that the feature map in consideration be "isotropic", i.e., have a gradient whose norm is independent of direction. However, the only isotropic map in any finite dimensional or separable space is the trivial zero map. Hence, if ME based upon a feature is the optimality criterion, then the MSE is not a good criterion to use.

6 Optimal Image Acquisition : Finite Number of Observations

In this section we formulate the problem of optimal selection of vantage points for imaging, given that only a finite number of observations can be made. The estimates in this case are parametrized by the spatial coordinates of the vantage points. We utilize the results of the previous section in order to show optimality of the vantage points. In this section we shall be assuming that the feature involved is a linear map. We require that the image be perfectly determined through a finite number of observations in the absence of noise. In the presence of noise, the optimal choice of observations would be those that minimize the effect of noise on the synthesized image.

Definition 6.1 *We define an N-sampling domain as a set* $\bar{x} = \{x_1, x_2, ..., x_N\}$ *where* $\forall i, x_i \in \Re^2$.

We denote by $i/\bar{x}$ the set of the sample values of the function i on the set $\bar{x}$. Let H denote the convolution operator that relates the electromagnetic field at the object plane to the electromagnetic field at the measurement plane. It denotes mathematically, the physical process of light propagation from the object to the measurement plane.

We make the following assumption.

$\mathcal{A}$ 6.1 *Let* $i \in span[\phi_1, .., \phi_N]$ *where* $\{\phi_1, .., \phi_N\}$ *is a known orthonormal set of vectors.*

Suppose we take noise corrupted measurements of Hi at $\bar{x} = \{x_1, .., x_N\}$. We need to estimate i from these observations. By the linearity of H and $\mathcal{A}6.1$, it follows that $Hi \in span[H\phi_1, .., H\phi_N]$. Let $H\phi_k = \psi_k$. We obtain the estimate of the image, $\hat{i}$, in the following steps:

Step 1 First we form an estimate of Hi, $\hat{H}i_{\bar{x}}$, as:

(13)
$$\hat{H}i_{\bar{x}}(x) = [\psi_1(x), \cdots, \psi_N(x)]\Psi_{\bar{x}}^{-1}(Hi/\bar{x} + \epsilon),$$

where

$$\Psi_{\bar{x}} \equiv \begin{bmatrix} \psi_1(x_1) & \psi_2(x_1) & \cdots & \psi_n(x_1) \\ \vdots & \vdots & \ddots & \vdots \\ \psi_1(x_N) & \psi_2(x_N) & \cdots & \psi_n(x_N) \end{bmatrix},$$

assuming that Ψ_S is invertible and

$$\epsilon = \begin{bmatrix} \epsilon(x_1) \\ \vdots \\ \epsilon(x_N) \end{bmatrix},$$

$\epsilon(x_i)$ being the noise corrupting the observation at x_i.

Step 2 Then we obtain the estimate of i, $\hat{i}$, by setting $\hat{i} = H^{-1}\hat{H}i_{\bar{x}}$.

The above algorithm results in an unbiased estimate of the image. In fact, the algorithm results in a minimum variance unbiased estimate and the proof is a straightforward application of the Gauss-Markov Theorem [25].

It can be shown that the value of $E|F(i) - F(\hat{i})|^2$ is given by

(14)
$$e_f(S) = trace(\Psi_S^{-1} R_\epsilon^S \Psi_S^{-*} G),$$

where

$$G = FF^*,$$

and

$$F = [<f, \phi_1>, .., <f, \phi_N>].$$

Let S be constrained to remain in some $D \subset \Re^N$. Then the problem of optimal imaging can be stated as: Find
(15)
$$S^* = arg \min_{S \in D} e_f(S).$$

In the following proposition we show the existence of a solution to the optimization problem (15) for the case of a linear feature.

Proposition 6.1 *If the basis functions* $\{\psi_1, .., \psi_N\}$ *are continuous and the constraint set* D *is compact then the functional* e_f *attains its minimum on* D.

Proof:
By the assumption that the functions $\{\psi_1, .., \psi_n\}$ are orthonormal, it follows that there exists some $S_0 \in D$ such that $e_f(S_0) < \infty$. Consider the set $\mathcal{G} = \{S | e_f(S) \le e_f(S_0)\}$. The function e_f is continuous at all points in $\mathcal{G}$ since the only points of discontinuity of the functional e_f is where it is unbounded. Hence, by the definition of continuous functions, it follows that $\mathcal{G}$ is closed and also compact (since it is a subset of D). Thus e_f attains its minimum on $\mathcal{G}$ and hence in D.

Q.E.D.

It can be shown that the above optimization problem can be framed as a constrained optimization problem with inequality constraints (the set constraint is treated as the inequality constraint). The following are first order necesssary conditions for a relative minimum [29]:

$$\bigtriangledown e_f(S^*) + \mu^t \bigtriangledown \bar{g}(S^*) = 0,$$
$$\mu^t \bar{g}(S^*) = 0, \mu \geq 0,$$
$$where$$

(16)
$$\frac{\partial e_f}{\partial t_i} = -2\sigma^2 trace(\Psi^{-1} \frac{\partial \Psi}{\partial t_i} \Psi^{-1} \Psi^{-*} G)),$$

$$\bar{g}(S) = \begin{bmatrix} a_1 - x_1 \\ \vdots \\ a_n - x_N \end{bmatrix},$$

assuming that the noise is independent identically distributed, i.e., $R_\epsilon^S = \sigma^2 I$ and $t_i \geq a_i \, \forall i = 1, .., N$.

Remark 1 : It can be shown that for a given subspace, the choice of the optimal vantage points is independent of the choice of the basis vectors.

Remark 2 : If we consider the case when the basis consists of only one vector, the optimization problem is to choose one point such that the MSE is minimized. It turns out that this optimal vantage point is precisely that point where the "signal to noise ratio" of the basis function to the noise is maximum. To see this, note that for the case of a single basis vector, (14) reduces to $e_2(S) = \frac{\sigma^2(x)}{\psi^2(x)}$, which is the inverse of the signal-to-noise ratio of the basis function to the noise corrupting the measurements. Hence, the optimal vantage point to take the measurement from is the point at which the signal-noise-ratio is maximum.

Remark 3 : The error criterion (14) looks like a generalized signal to noise ratio .

Thus, if we have apriori knowledge about the objects that we are imaging, the results presented above could be used to design optimal observatories. For example , the above results could be used in the design of multi-spacecraft imaging systems, given that we have apriori knowledge about the physical characteristics of the heavenly bodies that we are imaging, and are interested in obtaining higher resolution images of the same.

7 Example of Optimal Image Acquisition

In this section we present a simple numerical example that illustrates the problem formulation that was presented in the previous section. The feature-map considered here is linear and the example is finite-dimensional. In the example, we find the least weighted-MSE estimate and the minimum ME estimates. We show that these match, as predicted by the theory.

In Fig.6, the topmost plot represents the image at the object plane. It is a discretized image with four pixels on each side. The bottom plot in the same figure represents the image at the measurement plane. The feature considered in this example is the sum of the pixel values of the image. It is assumed that the image lies in class $\mathcal{I}_1$ if the sum is less than a threshold value of 4.1 and it belongs to class $\mathcal{I}_2$ otherwise. The sum of the pixel values of the object as shown in Fig.6 is 4, and hence, the object belongs to class $\mathcal{I}_1$.

We assume that it is known that the support of the object is constrained to the central four pixels of the image. Note that the image can be represented as a vector if we stack the rows of the image lexicographically.

In doing this, the image is represented as an element in $\Re^{16}$. It then follows that the knowledge that the support is constrained to the central 4 pixels of the picture translates into saying that the basis of the image is $\{e_6, e_7, e_{10}, e_{11}\}$, where e_i denotes the ith co-ordinate vector in $\Re^{16}$. Hence, the basis for the image at the measurement plane is $\{He_6, He_7, He_{10}, He_{11}\}$. These basis images are shown in Fig.7.

The problem of optimal imaging with respect to the weighted-MSE is to choose the four pixels of the image at the measurement plane at which to make measurements so that the weighted-MSE in estimating the image is minimized, assuming that the statistics of the noise corrupting the measurements are known. We assume that the noise corrupting the measurements is independent identically distributed with Gaussian statistics. The variance of the noise is assumed to be 0.05. In Fig.8 we show the object and the optimal estimated image. The optimal locations of the measurements on the measurement plane are represented by the '@' in Fig.6.

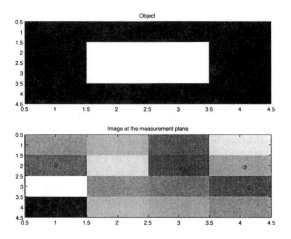

Figure 6: Image at object and measurement planes.

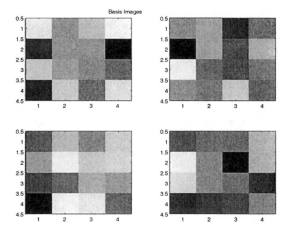

Figure 7: Basis images.

In Fig.9, we show a comparison plot of the weighted-MSE and the misclassification error. Every point on the X-axes in the plots represents a particular choice for the measurement points on the measurement

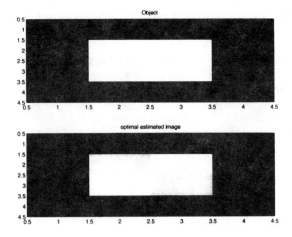

Figure 8: Actual and the optimal reconstructed image.

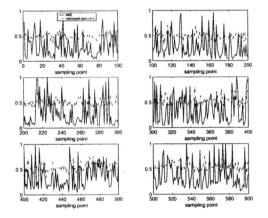

Figure 9: Plot of weighted-MSE and ME.

plane. We observed that the optimal weighted-MSE estimate and the optimal ME estimate were the same. However there were cases for which the above did not hold true in the simulations. (The different cases were obtained by different choices of the transformation between the image and the measurement plane). The misclassification errors for this example were obtained through Monte-Carlo simulations and would account for the discrepancy between the minimum weighted-MSE estimate and the minimum ME estimate in these cases. Note that in our experience, the minimum weighted-MSE estimate often exactly matches the minimum ME estimate.

8 Conclusions

In this paper, we have considered the effects of noise on image classification. We have considered two measures of the noise corrupting the image, namely the MSE and the WCE, and have shown that they are consistent with the goal of image classification, in that the ME tends to zero as the WCE or MSE tends to zero. In the case of image classification based upon a feature, we have obtained upper bounds on the noise corrupting the image such that the misclassification error for unambiguous images is guaranteed to be less than some pre-specified value. We have also argued that if the feature is a linear map, a weighted-MSE criterion (weighted by the feature itself) results in an optimal ME estimate.

The Tchebyshev inequality was used to find upper bounds on the MSE or WCE such that satisfactory performance was guaranteed. However, the Tchebyshev inequality is very conservative and to obtain less conservative estimates, we used the normality of the noise corrupting the image, and concluded that the feature of the noisy estimates would be normal with the true mean as the actual feature value. However, this assumption is violated in the example that was presented to detect the bandedness of an image. The disparity can be attributed to the fact that the noise corrupting the image is not small enough and hence, the nonlinearity of the feature map results in a bias in the noisy feature values. However, it can be seen that the feature maps are homogeneous and convex. Hence, the problem has structure that could be exploited to obtain tighter bounds. This shall be a direction of our future research.

References

[1] *http://planetquest.jpl.nasa.gov/TPF/*

[2] E. L. Lehmann, *Testing Statistical Hypotheses*, Wiley, New York, 1959

[3] S. Theodoridis, *Pattern Recognition*, Academic Press, San Diego, 1999

[4] S. Kassam and H. Poor, "Robust techniques for signal processing : a survey", *Proceedings of the IEEE*, vol.73, n.3 (Mar.1985), pp.433-481

[5] R. Martin and S. Schwartz, "Robust detection of a known signal in nearly gaussian noise", *IEEE Transactions on Information Theory*, IT-17(1971), pp.50-56

[6] P. Willett and B. Chen, "Robust detection of small stochastic signals", *IEEE Transactions on Aerospace and Electronic systems*, vol.35, n.1, Jan.1999, pp.15-30

[7] Y. Cao, D. N. Levin, "Using an Image Database to Constrain the Acquisition and Reconstruction of MR Images of the Human Head", *IEEE Transactions on Medical Imaging*, vol.14, n.2, pp.350-361, 1995

[8] S. J. Reeves, "Sequential Algorithms for Observation Selection", *IEEE Transactions on Signal Processing*, vol.47, n.1, pp.123-131, 1999

[9] Y. Gao, S. J. Reeves, "Optimal K-Space Sampling in MRSI for Images with a Limited Region of Support", *IEEE Transactions on Medical Imaging*, vol.19, n.12, pp.1168-1178, 2000

[10] S. K. Nagle, D. N. Levin, "Multiple Region MRI", *Magnetic Resonance in Medicine*, vol.41, pp.774-786, 1999

[11] J. W. Goodman, *Introduction to Fourier Optics*, Boston, MA, Mcgraw Hill, 1996

[12] J. W. Goodman, *Statistical Optics*, Wiley Classics Library, 2000

[13] G. Demoment, "Image Reconstruction and Restoration: Overview of Common Estimation Structures and Problems", *IEEE Transactions on Acoustics, Speech and Signal Processing*, vol.37, pp2024-2036, 1989

[14] A. K. Katsagellos, "Iterative Image Restoration Algorithms", *Optical Engineering*, vol.28, n.7, pp735-748, 1989

[15] A. K. Katsagellos(ed.), *Digital Image Restoration*, Springer-Verlag, New York, 1991

[16] M. Cannon, "Blind Deconvolution of Spatially Invariant Image Blurs with Phase", *IEEE Transactions on Acoustics, Speech and Signal Processing*, vol.24, pp58-63, 1976

[17] D. Kundur, D. Hatzinakos, "A Novel Blind Deconvolution Scheme for Image Restoration using Recursive Filtering", *IEEE Transactions Signal Processing*, vol.46, n.2, 1998

[18] R. J. Marks, *Introduction to Shannon Sampling and Interpolation*, Springer Verlag, New York, 1991

[19] R. E. Skelton, D. Chiu, "Optimal Selection of Inputs and Outputs in Linear Stochastic Systems", *The Journal of Astronautical Sciences*, vol.XXXI, n.3, pp.399-414, 1983

[20] P. G. Maghami, S. M. Joshi, "Sensor/Actuator Placement for Flexible Space Structures", *IEEE Transactions on Aerospace and Electronic Systems*, vol.29, n.2, pp.345-351,1993

[21] W. H. Chen, J. H. Seinfeld, "Optimal Location of Process Measurements", *Int. J. Control*, vol.21, n.6, pp.1003-1014, 1975

[22] J. E. Farach, W. M. Grady, A. Arapostathis, "An Optimal Procedure for Placing Sensors and Estimating the Locations of Harmonic Sources in Power Systems", *IEEE Transactions on Power Delivery*, vol.8, n.3, pp.1303-1310, 1993

[23] H. Stark and J. W. Woods, *Probability, random processes and estimation theory for engineers*, Prentice Hall, Englewood Cliffs, NJ, 1986

[24] M. Loeve, *Probability Theory*, Third edition, D. Van Nostrand and Company, Princeton, NJ.

[25] D. G. Luenberger, *Optimization by Vector Space Methods*, Wiley, New York, 1968

[26] H. K. Khalil, *Nonlinear Systems*, Second Edition, Prentice Hall, NJ 1996

[27] T. W. Anderson, H. Burstein, "Approximating the Upper Binomial Confidence Limit", *Journal of the American Statistical Association*, vol.62, issue 319, 1967, pp 857-861.

[28] T. W. Anderson, H. Burstein, "Approximating the Lower Binomial Confidence Limit", *Journal of the American Statistical Association*, vol.63, issue 324, 1968, pp 1413-1415.

[29] D. G. Luenberger, *Linear and Nonlinear Programming*, Addison-Wesley, Reading, MA, 1984

INTERFEROMETRIC OBSERVATORIES IN EARTH ORBIT

I. I. Hussein,[*] D. J. Scheeres,[†] D. C. Hyland[‡]

We propose a class of satellite constellations that can act as interferometric observatories in Low Earth Orbit (LEO), capable of forming high resolution images in time scales of a few hours without the need for active control. First we discuss the requirements to achieve these imaging goals. Next we define a class of constellations that can achieve these goals in LEO. An optimization procedure is also defined that supplies m pixels of resolution with a minimum number of satellites. For the example considered, this procedure results in an observatory that is within 0-2 satellites from a lower bound of $\sqrt{m}$ satellites. We introduce a linear imaging constellation and formulate a concise 0-1 mathematical program, the solution of which is the solution to optimal aperture configuration for full coverage of the wave number plane. Next, we extend the LEO observatory to a general Earth orbit and relate the solution of the (linear array) 0-1 program to that of the Earth orbiting constellation. We discuss how the zonal J_2 effect can be utilized to scan the observatory across the celestial sphere and, finally, we discuss some practical implementation issues.

1 Introduction to the Imaging Requirements

Interferometric imaging is performed by measuring the mutual intensity (the two point correlation[1]) that results from the collection and subsequent interference of two electric field measurements of a target made at two different observation points. While

[*] Ph.D. candidate, Aerospace Engineering, The University of Michigan, Ann Arbor, Michigan 48109-2140.
E-mail: ihussein@umich.edu.

[†] Associate Professor of Aerospace Engineering, The University of Michigan, Ann Arbor, Michigan 48109-2140.
E-mail: scheeres@umich.edu.

[‡] Professor of Aerospace Engineering, The University of Michigan, Ann Arbor, Michigan 48109-2140.
E-mail: dhiland@engin.umich.edu.

moving relative to each other, the satellites collect and transmit these measurements, which are later combined at a central node using precise knowledge of their locations and timing of data collection. A least squares error estimate of the image can be reconstructed given the mutual intensity measurements, parameters of the optical system, and the physical configuration of the observatory. To assess the quality of the reconstructed image, the reconstructed image is Fourier transformed into a two dimensional plane of spatial frequencies (the wave number plane). At any given point on the wave number plane, the modulation transfer function (MTF) is defined as the ratio of the estimated intensity to the true image intensity. For an interferometric imaging constellation the MTF can be computed given the measurement history and corresponding relative position data among the light collecting spacecraft. In the wave number plane, a point with a zero MTF value implies that the system is "blind" to the corresponding sinusoidal pattern, while a large value of the MTF implies that the image signal can be restored at that wave number via an inverse Fourier transform[1,2]. The MTF, as a measure of the imaging system's performance, is a function of both the optical system and the configuration of the observatory in physical space. In this note we address the issue of designing the configuration of an interferometric observatory that ensures a non-zero value of the MTF within a desired region in the wave number plane.

For a general satellite constellation, denote the position vector of satellites i and j by $\mathbf{R}_i$ and $\mathbf{R}_j$, respectively, $i, j = 0, 1, \ldots, N-1$, where N is the number of satellites. Let $\mathbf{R}_{ij} = \mathbf{R}_j - \mathbf{R}_i$ be the relative position vector between satellites i and j and $\mathbf{r}_{ij}$ be the projection of $\mathbf{R}_{ij}$ onto a plane perpendicular to the line of sight of the observatory. Let $\bar{z}$ be the distance from the image plane to the observation plane. Denote by the term "picture frame" the angular extent of the intended image on the image plane. The picture frame has a diameter of length $\overline{L}$. Pixelating the image plane into an $m \times m$ grid, the size of each pixel is $L = \overline{L}/m$, and the resulting angular resolution is $\theta_r = L/\bar{z}$. Additionally, the angular extent of the desired picture frame is given by $\theta_p = \overline{L}/\bar{z}$, leading to $\theta_p = m\theta_r$.

Dimensions of features in the wave number plane are the reciprocals of the corresponding dimensions in the physical plane. Thus the resolution disc is a disc of diameter $1/\theta_r$ and is the region where we desire the MTF to have nonzero values (henceforth, simply denoted by wave number plane coverage). The picture frame region is a circular disc of diameter $1/\theta_p$. Therefore, the diameter of the resolution disc is m times the diameter of the picture frame disc in the wave number plane (see Figure (1)). As the relative position vector of two spacecraft varies in the physical plane, the picture frame disc moves in the wave number plane, where its center follows the trajectory of the vector given by $\pm\mathbf{r}_{ij}/\lambda$, where λ is the imaging wavelength of interest. Each satellite, by itself, will contribute a disc that is centered at the origin with a diameter of $1/\theta_p$, and each pair of satellites will contribute two discs of diameter $1/\theta_p$ located 180 degrees apart with a radius of $\frac{r_{ij}}{\lambda}$ from the center, where $r_{ij} = |\mathbf{r}_{ij}|$. Define the minimum relative distance between satellites to be $d_{\min} = \lambda/\theta_p$. To completely cover the resolution disc in the wave number plane it is sufficient to have satellites distributed such that there exist pairs with relative distances $d_{\min}$, $2d_{\min}, \cdots, \frac{1}{2}(m-1)d_{\min}$. Let $d_{\max} = \frac{1}{2}(m-1)d_{\min}$.

2 Circular Orbit Constellations

We propose a class of very long baseline constellations that achieves the requirement that the wave number plane be completely covered. The satellite constellation is placed on a circular arc that is a segment of a low Earth orbit and whose center is located at the center of the Earth (see Figure (1)). The satellites are distributed such that the second satellite is located at a distance of $d_{\min}$ from the first satellite, the third at $2d_{\min}$ from the first, the fourth at $3d_{\min}$ from the first, and so on. Thus, a constellation of N_f satellites will have the Nth satellite located at a distance of $(N-1)d_{\min}$ from the first. This distribution, defined as the "fundamental" constellation, implies that there are $m = 2N_f - 1$ pixels and ensures the complete coverage of the wave number plane, once the constellation is rotated $180°$ (i.e. after half an orbit period). Figure (1) shows the geometry of this configuration for $N_f = 3$ satellites ($m = 5$ pixels). We nominally assume that the orbit plane is perpendicular to the line of sight to the target.

To compute the precise locations of the satellites in the constellation, specify wavelength of interest, λ, and the desired angular extent of the picture frame $\theta_p = \overline{L}/\overline{z}$. Given a number of satellites N_f, or the number of pixels m, one then obtains the corresponding angular resolution, θ_r, and knowledge of θ_p enables us to compute $d_{\min}$ and $d_{\max}$. Throughout this note we use the following values: $\lambda = 10\mu$m, $\overline{z} = 7.408 \times 10^{14}$km ($\sim 24$ parsec from the Earth), $\overline{L} = 13 \times 10^3$km $r_o = 7,200$km and $d_{\min} = 569.52$km.

Let $\hat{i}$ and $\hat{j}$ be two orthogonal unit vectors in the orbit plane, the position vector of the kth satellite, $k = 0, \cdots, N_f - 1$, is

$$
\mathbf{r}_k(t) = r_o \Bigg\{ \left[\cos(\omega t)\left(1 - \frac{k^2}{2}\left(\frac{d_{\min}}{r_o}\right)^2\right) - \sin(\omega t)k\frac{d_{\min}}{r_o}\sqrt{1 - \left(\frac{k}{2}\right)^2\left(\frac{d_{\min}}{r_o}\right)^2} \right]\hat{i}
$$
$$
+ \left[\sin(\omega t)\left(1 - \frac{k^2}{2}\left(\frac{d_{\min}}{r_o}\right)^2\right) + \cos(\omega t)k\frac{d_{\min}}{r_o}\sqrt{1 - \left(\frac{k}{2}\right)^2\left(\frac{d_{\min}}{r_o}\right)^2} \right]\hat{j} \Bigg\} \tag{1}
$$

where ω is the orbit angular velocity of the nominal circular orbit

$$
\omega = \sqrt{\frac{\mu}{r_o^3}}, \tag{2}
$$

and r_o is the orbit radius. The relative position vector from satellite l to satellite k

is given by

$$\mathbf{r}_{lk}(t) = d_{\min}\left\{\left[\cos(\omega t)(l^2 - k^2)\frac{d_{\min}}{2r_o} + \sin(\omega t)\left(-k\sqrt{1 - k^2\left(\frac{d_{\min}}{2r_o}\right)^2} + l\sqrt{1 - l^2\left(\frac{d_{\min}}{2r_o}\right)^2}\right)\right]\hat{\imath}\right.$$
$$\left. + \left[\sin(\omega t)(l^2 - k^2)\frac{d_{\min}}{2r_o} + \cos(\omega t)\left(k\sqrt{1 - k^2\left(\frac{d_{\min}}{2r_o}\right)^2} - l\sqrt{1 - l^2\left(\frac{d_{\min}}{2r_o}\right)^2}\right)\right]\hat{\jmath}\right\}. \tag{3}$$

In the wave number plane the relative position vector is $\bar{r}_{lk} = r_{lk}/\lambda$, a vector emanating from the origin with its tip at the center of the picture frame disc. Ignoring orbit perturbations, the above satellite arrangement guarantees that each $\bar{r}_{lk}$ has a constant magnitude (since they are distributed along the same circular orbit), which is given by

$$\bar{r}_{lk} = \frac{2r_o\sigma}{\lambda}\sqrt{\left((\hat{l}^2 - \hat{k}^2)\sigma\right)^2 + \left(\hat{k}\sqrt{1 - \left(\hat{k}\sigma\right)^2} - \hat{\imath}\sqrt{1 - \left(\hat{l}\sigma\right)^2}\right)^2}, \tag{4}$$

where $\hat{l} = l/(N_f - 1)$, $\hat{k} = k/(N_f - 1)$ ($l, k = 0, 1, 2, \ldots, N_f - 1$) and $\sigma = \frac{d_{\max}}{2r_o}$. Note that $0 < \sigma \le 1$, where $\sigma \to 0$ as either $d_{\max} \to 0$ or $r_o \to \infty$. The latter case arises if the constellation is placed on an orbit with small curvature. As $\sigma \to 0$ we have $\bar{r}_{lk} \to d_{\min}|k - l|/\lambda$. $\sigma = 1$ only when $d_{\max} = 2r_o$ (i.e. when the constellation spans 180°). On the other hand, note that for all N_f we have $0 \le \hat{l}, \hat{k} \le 1$ and that variations in N_f do not induce variations on $\bar{r}_{lk}$.

Note that all $\bar{r}_{lk}$'s rotate at the same (constant) rate, ω, and that this constellation will sweep out the resolution disc in the wave number plane over half an orbit. If the line of sight is tilted away from the orbit normal by an angle ϵ, coverage of the wave number plane will range from full resolution θ_r to a minimum resolution of $\theta_r \cdot \cos\epsilon$. Figure (1) shows the wave number plane coverage for $N_f = 3$. Note that imaging in the opposite direction is possible by rotating the spacecraft 180° about the radius vector.

For the above parameters, the observatory is performing 1.9461×10^{-6} milliarcsec imaging at 10μm. Formation keeping and spacecraft pointing of a formation having a maximum baseline of about 14,000km, with a 24pc target and under the influence of J_2, drag and other perturbations is expected to be a difficult control problem. This may require much tighter pointing requirements than the Hubble Space Telescope. However, note that each aperture in our constellation will probably not be as large as and will not involve as many flexible structures as the Hubble Space Telescope. Moreover, note that since all the spacecraft lie on the same circular orbit, they will all be subject to the same differential perturbations, whose short-term effects are small. Still, these short-term effects can be accounted for by performing accurate relative position measurements between the spacecraft -an issue that is not specific to our observatory, but that is common to a typical interferometric formation.

In summary, Section (1) describes the mapping of wave number plane filling into the motion of N spacecraft in space. Section (2) proposes one way of achieving this mapping in a satisfactory manner. Given this mapping one can seek other motions that achieve the imaging requirements in the wave number plane. Kong et. al.[4] propose one such motion, where the linearized motion about a circular orbit is considered (thus, Hill's equations). In their work, the spacecraft are relatively close to each other, allowing for interstellar imaging applications only. Moreover, the imaging region of their "interferometer is limited to only objects that are located in the positive z and positive x directions in the Hill frame"[4]. In contrast, the observatory we propose here provides high resolution imaging of targets that are several parsecs away. More significantly, our observatory does not require any active thrusting to keep the constellation in a near rigid formation.

3 Minimizing the number of satellites for a given resolution

In the fundamental constellation, we define the "fundamental" baselines by $\bar{r}_{0,k}$ and the "bonus" baselines by $\bar{r}_{l,k}$, $l \neq 0$. By themselves, the fundamental baselines guarantee complete coverage of the wave number plane over half an orbit period, and the bonus baselines provide redundant coverage. For large N_f, there will be an excessive number of multiple coverage areas, implying that the number of satellites can be reduced with the resolution disc still being completely covered.

To carry out this minimization it is not necessary to consider the two dimensional wave number plane, and is sufficient to consider the one dimensional wave number space. Define a ray in the wave number plane parameterized by the radius $\bar{k}_r \in [0, \bar{k}_{\max}]$, where $\bar{k}_{\max} = 1/(2\theta_r)$. Let the contribution of each pair of satellites (l, k) to the image coverage be given by

$$f_{lk}(\bar{k}_r) = \begin{cases} 1 & \text{if } \bar{k}_r \in [\bar{r}_{lk} - \frac{\bar{d}_{\min}}{2}, \bar{r}_{lk} + \frac{\bar{d}_{\min}}{2}] \\ 0 & \text{otherwise} \end{cases}, \tag{5}$$

for $l = 0, \ldots, N-1$ and $k = l, \ldots, N-1$. Next, define the function $Q(\bar{k}_r) = \frac{1}{N} \sum_{l=0}^{N-1} \sum_{k=l}^{N-1} f_{lk}(\bar{k}_r)$ which is the superposition of all contributions. Figure (1) shows Q for $N = N_f = 3$ ($m = 5$ pixels) and Figure (2) shows Q for $N_f = 16$ ($m = 31$). For the $N_f = 1, 2, 3$ and 4 cases removing any satellite will immediately cause a portion of the resolution disc to not be covered, thus the minimum number of satellites for these cases is $N_{\min}(m) = \frac{1}{2}(m+1)$. For larger numbers of satellites (i.e., larger number of pixels, m) this is not true.

Our current minimization problem is stated as: "Starting from a fundamental constellation, with a corresponding fixed number of pixels m, maximize the number of satellites that can be removed from the constellation under the constraint that $Q(\bar{k}_r) > 0$ on the interval $\bar{k}_r \in [0, \bar{k}_{\max}]$." The constraint ensures complete coverage of the wave number line, meaning that each point on the line is covered by at least one

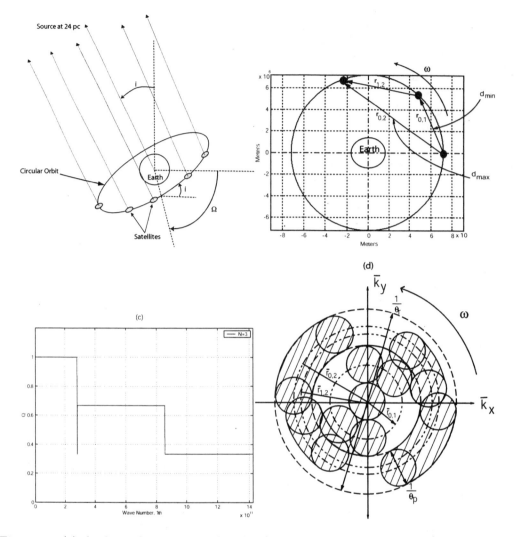

Figure 1: (a) A three dimensional sketch of the imaging observatory (not to scale). (b) Physical distribution in the orbit for $N_f = 3$ (not to scale). (c) Q curve for $N_f = 3$ in the fundamental constellation. (d) Physical distribution in the wave number plane for $N_f = 3$, $m = 5$ (not to scale).

satellite pair. Satellite arrangements that violate the lower bound are immediately discarded as they will have "gaps" in the wave number line, which lead to spatial frequencies that will not be covered.

To solve this problem, an algorithm was implemented that computes the Q function for the fundamental constellation and all its subsets, found by removing one satellite at a time, two at a time, and so forth. Satellite combinations that violate the lower threshold are discarded and the remaining solutions with a minimum number of satellites, $N_{\min}(m)$, constitute the minimal set. Note that for a given m there may be several different constellations with the same, minimum, number of satellites.

In a fundamental constellation of N_f satellites, there are up to

$$\sum_{k=1}^{N_f} \binom{N_f}{k} = 2^{N_f} - 1 \qquad (6)$$

trials that this algorithm may need to make, for large N_f this is unreasonably large. There are, however, numerous ways to speed up the computation by restricting the space of trials considered, some of which have been used in our computations. This algorithm has been implemented for $m = 3, 5, 7, \ldots, 39$, the results summarized in Table (1). Figure (2) shows Q for a fundamental constellation of $N_f = 16$ satellites ($m = 31$ pixels) and a minimum of $N_{\min}(31) = 8$ satellites. The $N_{\min}$ curve shown is the one that maximizes the area under the Q curve over all the 28 possible constellations with 8 satellites, and is comprised of satellites $0, 1, 2, 3, 4, 5, 10$ and 15. It is important to note that the minimal sets may change with the factor σ in Eq. (4).

A lower bound on the size of a constellation can be determined as follows. For a constellation of N satellites, there are exactly

$$\binom{N}{2} = \frac{1}{2}N(N-1)$$

baselines. Each baseline provides 2 pixels, plus one for the self pixels giving a total of $m = N(N-1) + 1$. Thus a lower bound on the number of satellites to cover m pixels is given by:

$$N_{lb} = \text{int}^+ \left[\frac{1}{2} \left(1 + \sqrt{4m - 3} \right) \right].$$

where $\text{int}^+[x]$ is the smallest integer larger than or equal to x. A solution can have no fewer than this number of satellites in the constellation without having gaps in the wave number plane. Moreover, there may not exist solutions with $N_{\min} = N_{lb}$. For example, for $m = 15$ the minimal solution has $N_{\min} = 5$, which is equal to the lower bound. For $m = 29$ the minimal solution has $N_{\min} = 7$, which has one more satellite than the lower bound of 6 (see Table(1)). For large m, the lower bound is approximately $\text{int}^+[\sqrt{m}]$.

4 The Linear Array and its Relation to Earth-Orbiting Constellations

Assume now that we distribute the spacecraft on a linear segment instead of a circular arc (see Figure (3)). There are two ways in which such a situation may rise. The first situation is exemplified by a multi-aperture single spacecraft mission. With the apertures arranged on a line, as opposed to a circular arc, we could still achieve full wave number coverage by a simple 180° rotation of the spacecraft about an axis that is along the line of sight. Since the solution we propose for the LEO observatory hinges

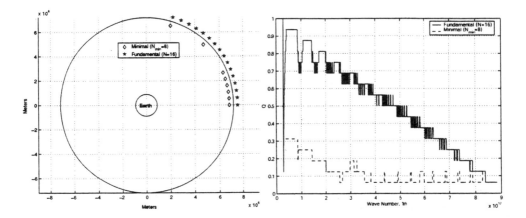

Figure 2: Fundamental and minimal distributions (not to scale) and Q curves for $m = 31$.

Fundamental number of satellites (N_f)	Number of Pixels $m = 2N_f - 1$	Minimum Number of Satellites N_{min}	Number of Solutions	Lower Bound N_{lb}
1	1	1	1	1
2	3	2	1	2
3	5	3	1	3
4	7	4	1	3
5	9	4	2	4
6	11	5	3	4
7	13	5	3	4
8	15	5	1	5
9	17	6	10	5
10	19	6	3	5
11	21	6	2	5
12	23	7	18	6
13	25	7	12	6
14	27	7	4	6
15	29	7	1	6
16	31	8	28	6
17	33	8	19	7
18	35	8	3	7
19	37	9	142	7
20	39	9	91	7

Table 1: Summary Of Results For A 7200 km Orbit, With $\frac{d_{min}}{r_o} = 0.0791$ And $\theta_p = 1.75 \times 10^{-11}$

on the assumption that the apertures are rigidly connected and the whole constellation performs a simple rotation about an axis passing through the Earth center, then the concept design proposed above should apply for the linear aperture spacecraft as well in a space-based short baseline interferometric mission. We may then want to address the same question posed above: how to achieve maximum resolution with the minimum number of satellites and without gaps on the wave number line? For example, University of Michigan's proposed EV3M[5] imaging spacecraft is one where all apertures are positioned linearly to maximize the achievable resolution of the spacecraft with three apertures only ($N_f = 4$ spacecraft, $N_{min} = 3$ spacecraft and $m = 7$ pixels), while fully covering the wave number plane.

Second, note that in the limit as $\sigma \to 0$, an Earth-orbiting constellation may be approximated, to first order, as a linear array constellation. Thus a solution of the linear constellation will be the same as that for the Earth-orbiting constellation for σ sufficiently small (i.e. either d_{max} sufficiently small or r_o sufficiently large). As will be shown below, the optimization problem for the linear array can be expressed as a 0-1 mathematical program that can be solved using existing techniques. Techniques, such as evolutionary programming, furnish solutions for high dimensional problems with small computational time, as opposed to exhaustive search algorithms as the one discussed above. This is an advantage in constellation design especially in the case where the constellation contains a very large number of small-sized satellites, where an exhaustive algorithm may take weeks of computation time even on the fastest available computers. Below, the linear array constellation problem will be formulated, solution techniques will be discussed and, in the following section, an application of this solution for an Earth-orbiting array will be discussed.

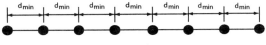

Linear Array of Apertures

Figure 3: Linear Array Layout

In the "fundamental" arrangement of a linear array of apertures, the relative distance between satellites l and k on the wave number line is given by:

$$\bar{r}_{lk} = |k - l| \frac{d_{min}}{\lambda}. \tag{7}$$

In Section (2), we parameterized the wave number line by the actual wave number, $\bar{k}$. Instead, suppose we parameterize it by $\frac{\bar{k}\lambda}{d_{min}}$. Thus, for the linear constellation, each satellite pair will contribute to a "wave number bin" $[\bar{\chi} - 0.5, \bar{\chi} + 0.5]$ centered at $\bar{\chi}$, where $\bar{\chi} = \frac{\bar{r}_{lk}\lambda}{d_{min}}$.

Let x_i denote the state of each aperture: x_i is 1 if it is selected as a member of the constellation or 0 if not. So, $x \in \mathbb{B}^{N_f}$, where $\mathbb{B} = \{0, 1\}$ and N_f is the number of satellites in the fundamental set. Note that for a particular choice of apertures, $c^T x$ represents the total number of satellites for that particular choice of apertures, where c is an N_f vector of 1's.

Next, it can be shown that $b_k(x) = x^T \mathbb{I}_k x$, $k = 1, \ldots, N_f$, is equal to the number of contributions to interval number k, where the first interval is centered at the origin and the N_fth interval is the outermost one and where $\mathbb{I}_k$ is a matrix of zeros except for the $(k-1)$st super diagonal. For example, if $N_f = 5$ and satellites $1, 2, 4$ are only selected, then the total number of satellites is equal to $[1\ 1\ 1\ 1\ 1] \cdot [1\ 1\ 0\ 1\ 0]^T = 3$ satellites. Also, for this case $b_1 = 3$ contributions due to satellites 1, 2 and 4 each paired with itself, $b_2 = 1$ contributions due to the pairing of satellites 1 and 2, $b_3 = 1$ contributions due to pairing of satellites 2 and 4, $b_4 = 1$ contributions due to pairing of satellites 1 and 4 and finally $b_5 = 0$ due to no pairing of satellites which are $5d_{min}$ apart from each other:

$$b_1 = [1\ 1\ 0\ 1\ 0] \begin{bmatrix} 1 & 0 & 0 & 0 & 0 \\ 0 & 1 & 0 & 0 & 0 \\ 0 & 0 & 1 & 0 & 0 \\ 0 & 0 & 0 & 1 & 0 \\ 0 & 0 & 0 & 0 & 1 \end{bmatrix} \begin{bmatrix} 1 \\ 1 \\ 0 \\ 1 \\ 0 \end{bmatrix} = 3,$$

$$b_2 = [1\ 1\ 0\ 1\ 0] \begin{bmatrix} 0 & 1 & 0 & 0 & 0 \\ 0 & 0 & 1 & 0 & 0 \\ 0 & 0 & 0 & 1 & 0 \\ 0 & 0 & 0 & 0 & 1 \\ 0 & 0 & 0 & 0 & 0 \end{bmatrix} \begin{bmatrix} 1 \\ 1 \\ 0 \\ 1 \\ 0 \end{bmatrix} = 1,$$

$$b_3 = [1\ 1\ 0\ 1\ 0] \begin{bmatrix} 0 & 0 & 1 & 0 & 0 \\ 0 & 0 & 0 & 1 & 0 \\ 0 & 0 & 0 & 0 & 1 \\ 0 & 0 & 0 & 0 & 0 \\ 0 & 0 & 0 & 0 & 0 \end{bmatrix} \begin{bmatrix} 1 \\ 1 \\ 0 \\ 1 \\ 0 \end{bmatrix} = 1,$$

$$b_4 = [1\ 1\ 0\ 1\ 0] \begin{bmatrix} 0 & 0 & 0 & 1 & 0 \\ 0 & 0 & 0 & 0 & 1 \\ 0 & 0 & 0 & 0 & 0 \\ 0 & 0 & 0 & 0 & 0 \\ 0 & 0 & 0 & 0 & 0 \end{bmatrix} \begin{bmatrix} 1 \\ 1 \\ 0 \\ 1 \\ 0 \end{bmatrix} = 1, \text{ and}$$

$$b_5 = [1\ 1\ 0\ 1\ 0] \begin{bmatrix} 0 & 0 & 0 & 0 & 1 \\ 0 & 0 & 0 & 0 & 0 \\ 0 & 0 & 0 & 0 & 0 \\ 0 & 0 & 0 & 0 & 0 \\ 0 & 0 & 0 & 0 & 0 \end{bmatrix} \begin{bmatrix} 1 \\ 1 \\ 0 \\ 1 \\ 0 \end{bmatrix} = 0.$$

Because $b_5 = 0$, $x = [1\ 1\ 0\ 1\ 0]^T$ is not a solution to the problem.

Thus the set of designs with the minimum number of satellites that completely cover the wave number line are all *global* solutions to the following minimization problem:

$$\min_x \ c^T x$$
$$\text{s.t. } b_k(x) \geq 1, \ k = 1, \ldots, N_f \tag{8}$$
$$x \in \mathbb{B}^{N_f}.$$

· This problem is generally known in the literature as a combinatoric/integer 0-1 programme, usually with linear cost function and linear constraints[6]. The solution of this problem requires the minimization of a linear cost function subject to a quadratic constraint. General 0-1 programming techniques exist to solve the program in Eq. (8).

4.1 Numerical Results

We first attempt to solve this problem by applying a thorough search algorithm as discussed in previous Section (3). Figure (4) shows the number of feasible solutions (top left), the number of minimal solutions (top right), $N_{\min}$ (bottom left) and the CPU time (using a 1.5GHz IBM platform) in hours (bottom right). The advantage of this algorithm is that it gives complete information on all possible minimal solutions (e.g. the configuration of spacecraft in each solution). The main drawback is the computational time involved to obtain the results. For example, for $N_f = 21$ spacecraft there are 2.5×10^5 feasible solutions that all require evaluation of their Q functions, consuming about 7 hours using the exact search algorithm.

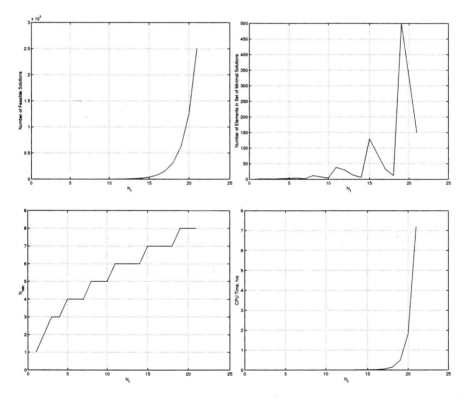

Figure 4: Exact Search Algorithm Results

To decrease the amount of CPU time, a statistical approach to solving the above 0-1 program is to utilize an evolutionary programming (EP) method. This is schematically summarized in Figure (5). The results of applying this algorithm are shown in Figure (6). We notice a tremendous amount of computation time savings. Using this algorithm we can arrive at solutions in about 11 seconds for $N_f = 21$ spacecraft. Due to the random nature of the search algorithm of the EP method, we note that we do not have full information regarding the total number of solutions available, the size of the feasible set or the exact design of the constellation (i.e. we may only know what $N_{\min}$ is, but not the full set of solutions that achieves $N_{\min}$ spacecraft).

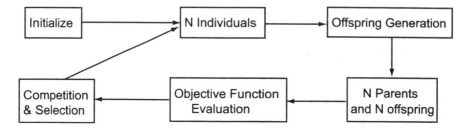

Figure 5: An Evolutionary Programming Method

A closer look at the right hand plot in Figure (6) shows that we seem to be able to get the correct solution up to $N_f = 32$. However, for $N_f = 37$ we notice a drop in $N_{\min}$. Since it is not possible for a larger fundamental constellation (that achieves higher resolution) to achieve a smaller $N_{\min}$ than a smaller fundamental constellation, then we know that the $N_{\min}$ obtained using the EP algorithm for $N_f = 37$ is not correct. This result casts larger doubt that the results obtained for $N_f > 37$ will be correct either, though we could be confident that results for values of $N_f \leq 36$ seem to be plausible.

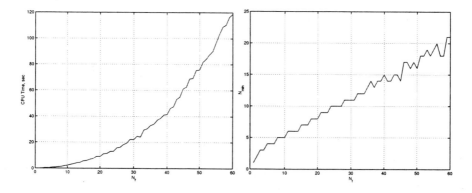

Figure 6: EP Algorithm Results

We observe that the algorithm performs poorly for higher values of N_f, though it is much faster than the exact search algorithm. In general one would trade off accuracy of solutions for speed in an EP algorithm. One may wish to improve the results, though at the expense of longer computational time, by constricting the offspring generated to be ones that are visible in the first place. The results are shown in Figure (7), where we observe that consistent results are obtained for values of $N_f \leq 23$. For $N_f = 23$, the computation time is 1.8 hours versus 18 seconds for the general EP algorithm and more than 7 hours for the thorough search algorithm.

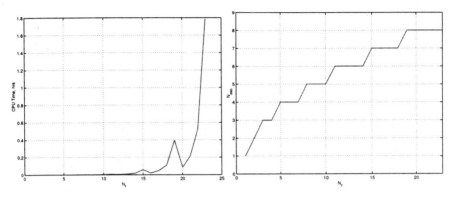

Figure 7: Restricted EP Algorithm Results

4.2 The Linear Array and the Earth-Orbiting Observatory

Assume that we can find a distance preserving isometry, ϕ, between the (curved) Earth-orbiting and the linear array geometries discussed above. As mentioned above, the significance of the linear problem is that we can readily solve this problem (as stated in Eq. (8)) and then compute the corresponding solution on the curved one-dimensional space through the inverse mapping ϕ^{-1} (see Figure (9)). The main benefit is that we can utilize techniques available in the literature for solving the 0-1 program in Eq. (8) to compute the solution to an Earth-orbiting configuration. First, recall the definition for an isometry and isometric spaces[7]:

Definition 4.1 (Isometry and Isometric Spaces) *Let M_1 and M_2 be two topological spaces. An isometry $\phi : M_1 \rightarrow M_2$ is a one-to-one correspondence such that $d_2(\phi(x), \phi(y)) = d_1(x, y)$ for all $x, y \in M_1$, where $d_1(\cdot, \cdot)$ and $d_2(\cdot, \cdot)$ are distance functions on M_1 and M_2, respectively. If there exists an isometry $\phi : M_1 \rightarrow M_2$, then M_1 and M_2 are called isometric.*

S^1 being the circle in $\mathbb{R}^2$ with radius r_o, let $M_C \subset S^1$ denote the one-dimensional space for a curved Earth-orbiting constellation and $M_L \subset \mathbb{R}^1$ be the one-dimensional space, which is simply a line segment on $\mathbb{R}^1$, for the linear aperture constellation.

Let $d_C(s_l, s_k)$ be the Euclidean distance in $\mathbb{R}^2$ between satellites k and l on M_C and let $d_L(s_l, s_k)$ be the Euclidean distance in $\mathbb{R}^1$ between satellites k and l on M_L. d_C and d_L are given by

$$d_C(s_l, s_k) = \frac{2r_o\sigma}{\lambda}\sqrt{\left((\hat{l}^2 - \hat{k}^2)\sigma\right)^2 + \left(\hat{k}\sqrt{1 - \left(\hat{k}\sigma\right)^2} - \hat{l}\sqrt{1 - \left(\hat{l}\sigma\right)^2}\right)^2} \quad (9)$$

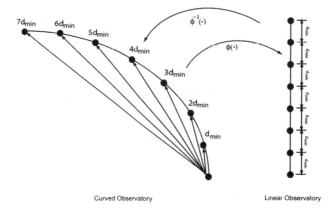

Figure 8: Relation between the Linear And Earth-Orbiting Constellations

where $\hat{l}$ and $\hat{k}$ are as defined in Section (2), and

$$d_L(s_l, s_k) = |k - l| \, d_{\min}/\lambda \qquad (10)$$

Due to the way spacecraft are arranged in the fundamental configuration, M_C and M_L may not have an isometry. That is because if M_C is simply unfolded onto M_L, spacecraft nodes on M_C do not map onto spacecraft nodes on M_L. Though it is true that the real line $\mathbb{R}^1$ and the circle S^1 have the isometry $\phi(t) = (r_o \sin t, r_o \cos t)$: $\mathbb{R}^1 \to S^1 \subset \mathbb{R}^2$ using a metric that measures distance *along* the curves, this is not true in our case because the metric we must use for imaging is the direct shortest distance between points in $\mathbb{R}^2$ as opposed to a metric along S^1.

For $\sigma \neq 0$, points on $M_C \subset S^1$ are shifted when M_C is unfolded onto M_L. Despite the fact that the total distance from the zeroth spacecraft to the $(N_f - 1)^{\text{st}}$ spacecraft is preserved and is equal to $d_{\max}$, distances between intermediate spacecraft are not.

It can be shown that:

$$d_C(s_l, s_k) = \frac{2r_o\sigma}{\lambda} \left|\hat{k} - \hat{l}\right| + O\left(\sigma^3\right) = d_L(s_l, s_k) + O\left(\sigma^3\right).$$

Hence, we have:

$$\lim_{\sigma \to 0} d_C(s_l, s_k) = d_L(s_l, s_k). \qquad (11)$$

In other words, $d_C(s_l, s_k) \to d_L(s_l, s_k)$ as the curvature of M_C or $d_{\max}$ approach zero. Thus, to a first order approximation, distance is conserved under the isometry $\phi(t) = (r_o \sin t, r_o \cos t)$ and the solution to the linear array should be identical to that of the curved array. In the next section we show results that indicate that for sufficiently small σ, the solution to the mathematical program in Eq. (8) is also a solution to the Earth-orbiting constellation configuration. On the other hand, note that since $0 \leq \hat{l}, \hat{k} \leq 1$, N_f will not have variational effects on the solution. Thus, one only needs to take into consideration variations in σ when seeking solutions to the Earth-orbiting constellation that are based on solutions to the linear array.

4.3   Numerical Results

In this section, we solve (8) using an exact search algorithm and compare the result with those obtained for a curved constellation. The results show that the two solutions indeed match, as predicted, for all values of σ small and up to a critical value σ^*. For $\sigma > \sigma^*$, the solution to the Earth-orbiting constellation deviates from that obtained for the linear array. For instance, for $N_f = 9$, the algorithm finds 8 minimal solutions to the linear array, whereas for the curved constellation it finds only 6 for $\sigma = \sigma^* = 0.019$. Figure (9) shows a plot for σ^* as a function of N_f for $5 \leq N_f \leq 10$. Note that for $1 \leq N_f \leq 4$, there exists just a single minimum and, thus, σ^* is undefined there.

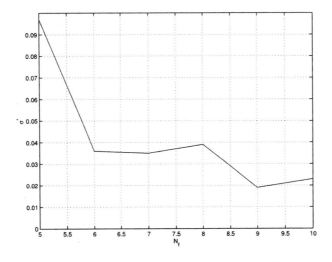

Figure 9: σ^* For $5 \leq N_f \leq 10$

5   The Interferometric Observatory

The Earth-orbiting constellation configurations proposed above will completely cover the wave number plane in half an orbit period, while imaging for several orbital periods will result in improved image quality. Thus, over a short period (days at most) an image can be formed. If we place the constellation in an inclined orbit, the orbit plane will precess relative to inertial space due to J_2 and the constellation will scan across the celestial sphere at a constant rate, effectively repeating its coverage after half a nodal period. The precession rate of the orbit plane is given by[8]

$$\dot{\overline{\Omega}} = -\frac{3}{2}\sqrt{\frac{\mu}{r_o^3}}\,\frac{R_o^2 J_2}{r_o^2}\cos(\imath),$$

where $R_o = 6378.14$km is the Earth's radius, $J_2 = 0.00108263$ is the second zonal harmonic of the Earth, $\mu = 3.986005 \times 10^5$km^3/s^2 is the Earth's gravitational constant,

ι is the inclination, r_o is the orbital radius, and the precession period of the node is $T = \frac{2\pi}{\dot{\Omega}}$. For an 800km altitude orbit inclined at 45° to the equator, the precession period is 77 days. For a constellation in a 45° or 135° inclination orbit, every point on the celestial sphere can be imaged with a resolution ranging from θ_r to $\theta_r/\sqrt{2}$ within half a nodal period.

An important design consideration is the speed at which the picture frame disc moves in the wave number plane, as this affects the image quality. The larger this speed is, the poorer the image quality becomes. Given an upper bound on the wave-plane velocity, $\bar{v}$, and a desired angular resolution, θ_r, this constrains the angular rate at which the picture frame disc moves in the wave number plane, equal to the mean motion of the orbit, $\omega \leq \frac{1}{2}\bar{v}\theta_r$. This bounds the desired orbit radius, $r_o \geq \sqrt[3]{\frac{4\mu}{(\bar{v}\theta_r)^2}}$. Thus, the choice of orbit radius does not depend only on the desired baselines (determined from the desired angular resolution), but also on the desired image quality. Note that it is possible to trade a higher speed in the wave number plane (shorter period) with additional observations, striking a balance between the two.

Other issues of concern are the signal detection, transmission and interference. There are certain optical technologies that are assumed to exist for this proposed very long baseline LEO observatory to be feasible. We assume that either a heterodyne or a direct detection method is used. Heterodyne detection has the advantage of selecting and detecting only the components of the wavefront of the source that are in phase with the wavefront of a local laser oscillator[9]. Thus, heterodyne detection furnishes phase information. On the other hand, direct detection, though less efficient as far as signal to noise is concerned, is still feasible via spatial filtering with a glass fiber to obtain a single geometric mode[10]. Local heterodyne detection, however, has a major advantage over direct detection Michelson interferometry. Direct detection requires that the detected signal be divided into $N-1$ equal parts, where N is the number of satellites in the constellation, corresponding to $N-1$ baselines. This results in the reduction of the signal by a factor of $N-1$. Each of the $N-1$ signals will possess a reduced SNR. This is exacerbated due to the presence of large background noise and the long distances over which the signals are transmitted from each spacecraft to a combiner spacecraft. For wavelengths above $\sim 4\mu$m, heterodyne detection is likely to be superior because of these problems with direct detection. It could be shown, however, that below $\sim 4\mu$m, direct detection will have better SNR properties than heterodyne detection. For a 10μm mission, such as the one proposed in this paper, heterodyne detection is advantageous.

A technique different from heterodyne detection is also under investigation by our group. This is Fourier Transform Spectral Interferometry for electric field reconstruction in a separated spacecraft interferometric mission. Novel optical techniques exist, such as Dual-Quadrature Spectral Interferometry (DQSI) and Spectral Phase Interferometry for Direct Electric Field Reconstruction (SPIDER), that aim at the full characterization of an electric field, both temporally and spatially[11,12]. The goal of such research, which has so far been developed for highly coherent sources and is currently being developed for non-coherent sources, is to extract necessary information in digital form that allow for performing the interference process digitally on a microchip. Once such digital information is available, these can be sent via communication links such as radio frequency signals to a central processing unit located on one of the spacecraft for the mutual intensity computations and metrology measurements. In light of technologies such as heterodyne detection or electric field reconstruction, a very long baseline mission such as the one we propose in this paper should be feasible as far as the optics are concerned.

A final remark is that JPL's Terrestrial Planet Finder (TPF) technology is an IR interferometer that currently does not involve baselines longer than 100m between apertures. TPF may provide an angular resolution that is as small as 0.75 milli-arcsec at $3\mu m$ and 1000m baseline. This corresponds to single pixel detection of a planet located 0.5AU from a sun-like star[13]. The main aim of TPF is to detect an earth-like planet by separating the planet from its parent star and capturing its light on a single pixel. In contrast, the underlying aim of the proposed observatory is to form a multi-pixel image of the disk of the planet. In consequence, the very long baseline constellation, such as the example we use here, offers 1.9461×10^{-6} milli-arcsec resolution at $10\mu m$ and a longest baseline of over 14,000km.

6 Conclusion

In this paper, the imaging objectives are stated and a class of constellations that can achieve high resolution images in LEO was discussed. An optimization procedure is also defined that supplies m pixels of resolution with a minimum number of satellites. We introduced a linear imaging constellation and formulated a 0-1 mathematical program, the solution of which is the solution to the optimal aperture configuration for full coverage of the wave number plane. This, in turn, helps to numerically solve the constellation design problem for a general Earth-orbiting constellation. We discussed how the zonal J_2 effect can be utilized to scan the observatory across the celestial sphere. Finally, we discussed some practical implementation issues. Future research will study the behavior of similar constellations subject to more general gravitational fields.

References

[1] M. Born and E. Wolf. *Principles of optics*, Pergamon Press, New York, 1964.

[2] D. C. Hyland. Interferometric Imaging Concepts With Reduced Formation-Keeping Constraints. *AIAA Space 2001 Conference*, Albuquerque, NM, August, 2001.

[3] D. C. Hyland. Formation Control for Optimal Image Reconstruction in Space Imaging Systems. *International Symposium on Formation Flying*, October 29-31, Centre National d'Etudes Spatiales, Toulouse, France.

[4] E. M. C. Kong, D. W. Miller and R. J. Sedwick. Exploiting Orbital Dynamics for Interstellar Separated Spacecraft Interferometry. *Proceedings of the American Control Conference*, pp. 4153-4157, 1999.

[5] S. E. Gano et. al. A Baseline Study of a Low-Cost, High-Resolution, Imaging System Using Wavefront Reconstruction. *AIAA Space 2001 Conference*, Albuquerque, NM, August, 2001.

[6] Y. J. Cao, L. Jiang and Q. H. Wu. An Evolutionary Programming Approach to Mixed-Variable Optimization Prooblems. *Applied Mathematical Modelling*, Vol. 24, pp. 931-942, 2000.

[7] W. A. Sutherland. *Introduction to Metric and Topological Spaces*, Oxford Science Publications, Oxford, 1975.

[8] A. E. Roy. *Orbital Motion*, 3rd ed., Institute of Physics Publishing, Philadelphia, 1998.

[9] H. Townes. Noise and Sensitivity in Interferometry. Chapter 4 of *Principles of Long Baseline Interferometry*. Course Notes from the 1999 Michelson Summer School, August 15-19, 1999, P. Lawson, Ed. JPL Publication 00-009, July 2000.

[10] R. H. Kingston. Detection of Optical and Infrared Radiation. *J. Oppt. Soc. Am. B.*, Vol. 4, 1450-1741, 1978.

[11] L. Lepetit, G. Chriaux, and M. Joffre. Linear Techniques of Phase Measurement by Femtosecond Spectral Interferometry for Applications in Spectroscopy. *J. Opt. Soc. Am. B*, Vol. 12, No. 12, December 1995.

[12] C. Iaconis and I. Walmsley. Self-Referencing Spectral Interferometry for Measuring Ultrashort Optical Pulses. *IEEE J. of Quantum Mechanics*, Vol. 35, No. 4, April 1999.

[13] *The Terrestrial Planet Finder: A NASA Origins Program to Search for Habitable Planets*. JPL Publication 99-003, May 1995.

DESIGN OF SPACECRAFT FORMATION ORBITS RELATIVE TO A STABILIZED TRAJECTORY

F. Y. Hsiao[*] and D. J. Scheeres[†]

This paper investigates the design of spacecraft formation orbits traveling relative to a general trajectory. To describe the motion we approximate the time-varying linear dynamics about the trajectory with a locally time-invariant system and use linear orbit elements to describe the relative trajectories. We find sufficient conditions under which the relative motion is stable and a controller can be designed. We consider the problem of specifying the orientation of the fundamental orbits, which is equivalent to eigenvector placement, and show how these modes can be combined to force the formation to fly in a range of orientations. Applications of our approach to relative motion in rotating and non-rotating systems are given.

INTRODUCTION

A general control law to stabilize the relative motion about a general trajectory in a Hamiltonian system and related methods of relative trajectory design are studied in this paper. Among the possible applications of this work is formation flight about a halo orbit in the Hill three body problem, an unstable periodic orbit in the vicinity of libration points in the Sun-Earth system. This work is motivated by potential future applications of spacecraft formation flight for interferometric imaging of distant stellar systems. Flying formations of spacecraft in the vicinity of L_2 halo orbits has been recognized as a feasible environment for this application. Results of this study will contribute to the possible application of formation flight techniques in this unstable environment.

The idea for using the center manifold of a halo orbit for formation flight was proposed in Ref. [1]. However, the utilization of the center manifold of an unstable orbit for formation flight is constrained by its long frequency of relative motion and the difficulty in control and computation of the orbits. The current state of the art for the computation and control of motion in an unstable halo orbit center manifold is reviewed in Ref. [2, 3] and indicates that highly precise computation must be made in order to keep the trajectory from diverging. The technique we propose here eliminates both of these concerns, as we are able to generate relative trajectories with shorter periods, and by stabilizing the relative motion relieve many of the concerns about controlling and computing the relative motions over longer spans of time. Our current approach to the design and control of formation flight trajectories should be distinguished from previous work, such as that reported in Refs. [4–6], as those efforts focus on the control and design of formation flight in near-Earth orbits.

---

[*]   Ph.D. candidate, Aerospace Engineering, The University of Michigan, Ann Arbor, Michigan 48109-2140. E-mail: fhsiao@umich.edu.

[†]   Associate Professor of Aerospace Engineering, The University of Michigan, Ann Arbor, Michigan 48109-2140. E-mail: scheeres@umich.edu. Senior Member AIAA.

The current work in this paper is an extension of the work reported in Ref. [10] that investigated the feasibility of the "local time" algorithm for the description of relative motion about a halo orbit. Here we investigate the general rule from which different controllers are designed to stabilize the relative motion. We also investigate different approaches to design relative trajectories for application to various space missions. The goal of this analysis is to come up with a general control law which allows formation flight dynamics to be arbitrarily assigned. The ultimate goal of this work is to develop efficient approaches to the implementation of distributed interferometric imaging by relying on the motion of spacecraft relative to each other to cover the necessary elements of the "image plane" that the imaging system is trying to reconstruct [7]. Our approach will be used in conjunction with innovative interferometric imaging approaches that can relieve the tight relative position control constraints that previous system concepts have had to deal with. Thus, by defining the preferred planes of motion for our stabilized system we define the preferred imaging surfaces as a function of time.

In detail, our paper covers the following items. First we find a condition under which the relative motion about a general trajectory in a celestial system is stabilized. Then we apply our algorithm to the relative motions about a halo orbit in the Hill three body problem. The control law creates a linear, time varying system about the periodic orbit, where all motion winds on tori about the periodic orbit. It can be shown that these orbits are stable over both short and long periods of time, and that their winding number about the periodic orbit can be controlled by the gain of the feedback law and can be made arbitrarily large. Then, in order to better study the resulting dynamics of our stabilized system, we introduce a "short time" approximation, which allows us to characterize relative motion with a simpler, time-invariant linear system. With this approximation, we also evaluate the control cost over a whole period and show that it is feasible. Finally, based on the concept of "linear orbit elements" that relates the geometric description of relative motion along a periodic orbit to the eigenstructures of the dynamics matrix, we come up with several algorithms for trajectory design. By studying the techniques of eigenstructure assignment we can better understand how our system can be used for interferometric imaging applications.

MODEL OF SPACECRAFT MOTION

Equations of Motion

To deal with the formation flight of spacecraft, we can view it as a constellation of spacecrafts flying in the vicinity of a spacecraft in the nominal trajectory in a specific celestial circumstance. Assume the triad (x, y, z) are the coordinates in the body fixed frame of the first spacecraft, and this frame is rotating about the z axis with rotation rate ω. The force potential in this circumstance is defined as $\mathbf{U}(\mathbf{r})$. According to dynamics we can write the equations of motion as:

$$\ddot{\mathbf{r}}_I = \frac{\partial U(\mathbf{r})}{\partial \mathbf{r}}$$

(1)
$$= \ddot{\mathbf{r}}_b + \dot{\omega} \times \mathbf{r} + 2\omega \times \dot{\mathbf{r}} + \omega \times (\omega \times \mathbf{r})$$

where $\mathbf{r} = \sqrt{x^2 + y^2 + z^2}\hat{r}$. Assume ω is constant. Eq. (1) can be expressed as:

(2)
$$\ddot{x} - 2\omega\dot{y} = \frac{\partial V}{\partial x}$$

(3)
$$\ddot{y} + 2\omega\dot{x} = \frac{\partial V}{\partial y}$$

(4)
$$\ddot{z} = \frac{\partial V}{\partial z}$$

(5)
$$V = U(\mathbf{r}) + \frac{1}{2}\omega^2(x^2 + y^2)$$

Although Eqs. (2)–(4) are derived from dynamics in a body fixed frame in a specific circumstance, we can view ω as a free parameter and V as an arbitrary force potential so that all space missions satisfying these equations of motion can be analyzed. In fact, if ω is taken to be zero, the above equations of motion degenerate to those in an inertial frame.

The Hill Problem

A spacecraft's trajectory in the vicinity of a Halo orbit in Sun-Earth system is highly unstable and non-Keplerian. To compute the non-linear equations of motion, we use an approximation to the restricted three-body problem, known as the Hill problem[8]. The three dimensional motion is governed by the equations of motion which have the same structure as Eqs. (2)–(4), where the x axis points towards Earth from the Sun, the z axis points normal to the Earth orbital plane and the y axis completes the triad. In this problem, ω is the rotation rate of the Earth's orbit about the Sun and the force potential is taken as

$$(6) \qquad V = \frac{\mu}{r} + \frac{1}{2}\omega^2(3x^2 - z^2)$$

where μ is the gravitational parameter of the Earth. The Lagrange equilibrium points have the coordinates $(\pm(\frac{\mu}{3\omega^2})^{1/3}, 0, 0)$[8] in this system.

The Hill Equations have been numerically integrated with proper initial conditions to find a periodic halo orbit, similar to the Genesis halo orbit during its main mission phase[2]. The period of this orbit is 179 days. It has two pairs of stable oscillation modes, one with the same period as that of the rotating frame leading to unity eigenvalues, with the other slightly longer leading to a pair of eigenvalues on the unit circle, and one pair of characteristic exponents $\sigma = \pm4.757 \times 10^{-7}/s$ (a characteristic time of 24.3 days), one of which causes a hyperbolic instability.

Linearized Dynamics

To derive equations for relative motion between the spacecraft, we assume the second spacecraft is on a non-periodic orbit in the vicinity of the first spacecraft. Assume the first spacecraft has a trajectory defined as $\mathbf{R}(t; \mathbf{R}_o, \mathbf{V}_o)$ with the property, $\mathbf{R}(t+T) = \mathbf{R}(t)$, where T is the period. Naturally, the velocity, $\mathbf{V}(t; \mathbf{R}_o, \mathbf{V}_o)$, is also periodic. Similarly, we can define the position and velocity vectors for the second spacecraft by $\mathbf{r}(t; \mathbf{r}_o, \mathbf{v}_o)$ and $\mathbf{v}(t; \mathbf{r}_o, \mathbf{v}_o)$. Since the second spacecraft is initially in the vicinity of the first spacecraft, linear systems theory can be applied to describe their relative dynamics. We define $\mathbf{X} = [\mathbf{R}; \mathbf{V}]$, $\mathbf{x} = [\mathbf{r}; \mathbf{v}]$, and $\delta\mathbf{x}$ as the difference between them, then the dynamics of $\delta\mathbf{x}$ can be approximated by:

$$(7) \qquad \delta\mathbf{x} = \mathbf{x} - \mathbf{X}$$

$$(8) \qquad \delta\dot{\mathbf{x}} = A(t)\delta\mathbf{x}$$

$$(9) \qquad A(t) = \frac{\partial \mathbf{F}}{\partial \mathbf{X}}$$

$$(10) \qquad A(t+T) = A(t)$$

where $\mathbf{F}$ is the dynamics function corresponding to Eqs. (2)-(4). The solution for relative motion can be expressed as:

$$(11) \qquad \delta\mathbf{x} = \Phi(t, t_o)\delta\mathbf{x}_o$$

$$(12) \qquad \Phi(t, t_o) = \frac{\partial \mathbf{X}(t; \mathbf{X}_o)}{\partial \mathbf{X}_o}$$

$$(13) \qquad \dot{\Phi}(t, t_o) = A(t)\Phi(t, t_o)$$

$$(14) \qquad A(t) = \left[\begin{array}{c|c} 0 & I_{3\times3} \\ \hline V_{\mathbf{rr}} & 2\omega J \end{array}\right]$$

$$(15) \qquad J = \begin{bmatrix} 0 & 1 & 0 \\ -1 & 0 & 0 \\ 0 & 0 & 0 \end{bmatrix}$$

$$(16) \qquad \Phi(t_o, t_o) = I_{6\times6}$$

where Φ is the state transition matrix (STM), and $\delta\mathbf{x}_0$ is the initial offset from the first spacecraft, i.e., the periodic orbit.

STABILIZING THE RELATIVE MOTION

In Ref. [9], the dynamics of long and short time spans for this relative motion are discussed more fully. In Ref. [10], we see that the relative motion of the spacecraft over a short time span centered at t_i can be understood by analyzing the eigenvalues and eigenvectors of the matrix $A(t_i)$. Therefore, in deriving our control law we will use these "short-term" dynamics to guide our thinking.

The goal of our investigation is to stabilize relative motions in the sense of Lyapunov (not asymptotic stability). According to Ref. [11], Lyapunov stability is defined as following:

An equilibrium point x_e of the system $\dot{x} = A(t)x$ is stable in the sense of Lyapunov if for any $\epsilon > 0$, there exists a value $\delta(t_0, \epsilon) > 0$ such that if $||x(t_0) - x_e|| < \delta$, then $||x(t) - x_e|| < \epsilon$, regardless of t, where $t > t_0$

In celestial mechanics, the stabilized trajectory will consist of oscillatory motions about the nominal trajectory, which in this context can be interpreted as motions in the center manifold. Now consider Eq. (8) over a short time span centered at $t = t_i$ and expand it.

$$\text{(17)} \qquad \delta\mathbf{r}'' - 2\omega J \delta\mathbf{r}' - V_{\mathbf{rr}}(t_i)\delta\mathbf{r} \;=\; 0$$

A feedback controller can be designed as

$$\text{(18)} \qquad \begin{aligned} \mathcal{T}_c &= -V_{cv}\delta\mathbf{r}' - V_{cr}\delta\mathbf{r} \\ &= \mathcal{T}_{cv}\delta\mathbf{r}' + \mathcal{T}_{cr}\delta\mathbf{r} \end{aligned}$$

Then the equations of motion in the feedback system are

$$\text{(19)} \qquad \delta\mathbf{r}'' - (2\omega J + V_{cv})\delta\mathbf{r}' - (V_{\mathbf{rr}} + V_{cr})\delta\mathbf{r} \;=\; 0$$

$$\text{(20)} \qquad \delta\mathbf{r}'' - S\delta\mathbf{r}' - V\delta\mathbf{r} \;=\; 0$$

If V_{cv} and V_{vr} are chosen skew symmetric and symmetric respectively, the Hamiltonian structure is maintained. Namely, S and V in Eq. (20) are skew symmetric and symmetric respectively. Moreover, if we can find a condition under which relative motions are kept in the center manifolds, the controller can then be obtained by simply subtracting the original Coriolis and force potential matrices from the desired ones. This would increase our choice in future orbit design. Therefore, we are concentrating on finding the condition for stability rather than proposing a specific control law.

Assume V in Eq. (20) is negative definite and consider a Lyapunov function $\mathcal{V}(\delta\mathbf{r}, \delta\dot{\mathbf{r}})$,

$$\text{(21)} \qquad \mathcal{V}(\delta\mathbf{r}, \delta\dot{\mathbf{r}}) \;=\; -\frac{1}{2}\delta\mathbf{r}^T V \delta\mathbf{r} + \frac{1}{2}\delta\dot{\mathbf{r}}^T \delta\dot{\mathbf{r}}$$

Then $\mathcal{V}(0) = 0$ and $\mathcal{V}(\mathbf{x}) > 0$ if $\mathbf{x} \neq 0$, where $\mathbf{x} = (\delta\mathbf{r}, \delta\dot{\mathbf{r}})$.

$$\begin{aligned} \dot{\mathcal{V}}(\delta\mathbf{r}, \delta\dot{\mathbf{r}}) &= -\delta\dot{\mathbf{r}}^T V \delta\mathbf{r} + \delta\dot{\mathbf{r}}^T \delta\ddot{\mathbf{r}} \\ &= -\delta\dot{\mathbf{r}}^T V \delta\mathbf{r} + \delta\dot{\mathbf{r}}^T S \delta\dot{\mathbf{r}} + \delta\dot{\mathbf{r}}^T V \delta\mathbf{r} \\ &= \delta\dot{\mathbf{r}}^T S \delta\dot{\mathbf{r}} \\ \text{(22)} \qquad &= 0 \end{aligned}$$

Since the requirement for Lyapunov stability is that the time derivative of the Lyapunov function is less than or equal to zero, our linear system is proved stable in the sense of Lyapunov. Also, due to a Hamiltonian structure, this result implies that our local time invariant system has all poles on the imaginary axis. Figures 1 and 2 provide two examples of the robustness of the controller developed from the sufficient condition. The controlled trajectory remains stable after two halo orbit periods (about 1 year), while the uncontrolled trajectory diverges after 66 days.

Another important feature in the realizability of our control law is its cost. Here we define cost as the integral of control acceleration over time. Assume the greatest oscillation frequency in the system is ω_{max}, the total cost would be bounded as

$$
\begin{aligned}
|\Delta \mathbf{V}| &= \int_0^t |\mathbf{a}| d\tau \\
&= \int_0^t |\mathcal{T}_{cv}\delta \mathbf{r}' + \mathcal{T}_{cr}\delta \mathbf{r}| d\tau \\
&\leq \int_0^t (|\mathcal{T}_{cv}\delta \mathbf{r}'| + |\mathcal{T}_{cr}\delta \mathbf{r}|) d\tau \\
&\leq \int_0^t (\omega_{max}||\mathcal{T}_{cv}||_2 + ||\mathcal{T}_{cr}||_2)|\delta \mathbf{r}| d\tau \\
&\leq \int_0^t (\omega_{max}||\mathcal{T}_{cv}||_2 + ||\mathcal{T}_{cr}||_2)(\omega_{max}||\Phi_{\mathbf{rv}}||_2 + ||\Phi_{\mathbf{rr}}||_2)|\delta \mathbf{r_0}| d\tau \\
&\leq |\delta \mathbf{r_0}| \left[\int_0^t (\omega_{max}||\mathcal{T}_{cv}||_2 + ||\mathcal{T}_{cr}||_2)(\omega_{max}||\Phi_{\mathbf{rv}}||_2 + ||\Phi_{\mathbf{rr}}||_2) d\tau \right]
\end{aligned}
$$

(23)

where $|| \cdot ||_2$ is the 2-norm of a matrix which is usually defined as $\sqrt{\lambda_{max}(AA^T)}$, where λ_{max} denotes the maximum eigenvalue of AA^T. Φ is the state transition matrix, and

(24)
$$
\Phi(t,0) = \left[\begin{array}{c|c} \Phi_{\mathbf{rr}} & \Phi_{\mathbf{rv}} \\ \hline \Phi_{\mathbf{vr}} & \Phi_{\mathbf{vv}} \end{array} \right]
$$

Systems Without Coriolis Force

To simplify our analysis, we first consider a system where the Coriolis force is nulled out. To discuss the stability of this system we take the free parameter, S, as zero, and, obviously, the same result is obtained. That is, an inertial frame has the same sufficient condition for stability. Furthermore, this also implies that we can make an "artificial inertial environment" by nulling out the natural Coriolis force and utilize properties of inertial frame for formation flight. For example, if the force potential is controlled diagonal and negative definite, the relative motions are just harmonic oscillations in three dimensional space. We are able to design a general trajectory by combining these oscillations with particular initial conditions.

In addition to force potential control, we can compute the upper bound of cost in velocity term in Eq. (18) to estimate the extra cost of this control.

$$
\begin{aligned}
\mathcal{T}_{vc} &= -V_{cv}\delta \mathbf{r}' \\
&= 2\omega J \delta \mathbf{r}' \\
&\leq 2\omega ||\delta \mathbf{r}'|| \\
&\sim 2k\omega^2 ||\delta \mathbf{r}|| \\
&\sim 0.079k||\delta \mathbf{r}|| \quad nm/s^2
\end{aligned}
$$

(25)

where k is the ratio of highest mode frequency to the frame rotation rate. The cost is low enough to implement using a low-thrust engine. The upper limit of the total cost is estimated in Eq. (23) and numerically simulated in Figure 3.

Systems With Coriolis Force

Given V_{cv} zero or skew symmetric and $V_{\mathbf{rr}} + V_{cr}$ negative definite, Eq. (19) is proved stable. One example is the feedback control law proposed in Ref. [9]. To implement this control at a given time t_i, we evaluate the

1079

local eigenstructure of the matrix $A(t_i)$, find the characteristic exponents of the hyperbolic motion $\pm\sigma(t_i)$, and find the eigenvectors that define the stable and unstable manifolds of this motion $\mathbf{u}_\pm(t_i)$, where the "+" denotes the unstable manifold and "−" denotes the stable one. Then the applied control acceleration is:

$$(26) \qquad \mathcal{T}_c = -\sigma(t_i)^2 G \left[\mathbf{u}(t_i)_+ \mathbf{u}(t_i)_+^T + \mathbf{u}(t_i)_- \mathbf{u}(t_i)_-^T \right] \delta\mathbf{r}$$

where G is the gain parameter and $\delta\mathbf{r}$ is the measured relative position vector, i.e., the offset between the periodic orbit and the spacecraft at time t_i. This control law maintains the structure of the problem, and provides local and global stability if the gain constant G is chosen sufficiently large. In fact, as shown in Figure 4, after the control law in Eq. (26) was applied, the force potential becomes negative definite. The upper limit of total cost is also estimated in Eq. (23) and numerically simulated in Figure 5. Figure 6 and Figure 7 give two more examples about the stability and cost of this control law.

LINEAR ORBITAL ELEMENTS

Local dynamics formulation

Having found a sufficient condition under which the closed-loop system is stable, we now concern ourselves with the dynamics of the controlled system. In previous sections, we approached the problem by cutting the orbit into several pieces and treating each of segment as time invariant instead of considering the original linearized time varying system over a whole period. If the time interval is sufficiently small, we can understand our system by investigating the eigen-structure of each local $\bar{A}(t_i)$ (stabilized) matrix. For the case of a completely stable map, there are three pairs of imaginary eigenvalues, known as the modes of the system. The trajectory described by each mode forms an elliptical orbit relative to the nominal trajectory with the origin of the frame at its the center. The actual trajectory is the linear combination of these three modes. Since $\bar{A}(t_i)$ is not constant over the whole orbit period, we cannot define one set of some constants describing the relative motion over the entire orbit period. Instead, we define a secular set of "Linear Orbital Elements" to describe the geometry of the trajectory.

Classical Orbital Elements

Among the six classical orbital elements defined in Astrodynamics, five of them are sufficient to completely describe the size, shape and orientation of an orbit. The sixth element pinpoints the position of the satellite along the orbit at a particular time. These elements are defined as follows[12]: *the semi-major axis* a, *the eccentricity* e, *the inclination* i, *longitude of the ascending node* Ω, *the argument of periapsis* ω, *and the the mean anomaly* M. In traditional astrodynamics the utilization of classical orbital elements provides us with a set of coordinates that allow us to immediately understand the geometry of motion. The same idea can be applied to our controlled trajectory as well, defining a set of geometrically meaningful elements that can be used in the design of relative motion, as illustrated in Figure 8. Thus we would like to develop a set of elements, called "Linear Orbital Elements", to describe the geometry of the controlled trajectory relative to the periodic halo orbit, as shown in Figure 9. Provided with proper assumptions, each mode generates a slowly varying elliptical orbit relative to the halo orbit. However, there are some fundamental differences between our case and elliptical orbits in two body problem:

1. The length of semi-major axis is not needed. Since these are linear motions, we don't concern ourselves with their amplitude. Hence, the semi-major axis a, and the eccentricity e are replaced by *the ratio of major-axis to minor-axis* (a/b) to describe the in-plane orbital shape.

2. The origin of this elliptic motion is located at the center of the ellipse instead of at its focus. As a result, the argument of periapsis is replaced by the *in-plane orientation of the major axis* denoted by ψ (since ω represents frequency in this paper).

3. The mean anomaly is not defined in the current application, but could be in future work.

Linear Orbital Elements

Assume the eigenvalue and eigenvector of $A(t_i)$ are $\pm j\omega_k$ and $\mathbf{u}_k = \vec{\alpha}_k \pm \vec{\beta}_k$, where the subscription k denotes each mode. The mathematic descriptions can be obtained by following usual procedures used in Astrodynamics, and are listed as follows [10].

$$\hat{h}_k = \frac{\vec{\beta}_k \times \vec{\alpha}_k}{|\vec{\beta}_k \times \vec{\alpha}_k|} \tag{27}$$

$$\hat{n}_k = \hat{z} \times \hat{h}_k \tag{28}$$

$$(\frac{a}{b})_k^2 = \frac{1 + \sqrt{(|\vec{\alpha}_k|^2 - |\vec{\beta}_k|^2)^2 + 4(\vec{\alpha}_k \cdot \vec{\beta}_k)^2}}{1 - \sqrt{(|\vec{\alpha}_k|^2 - |\vec{\beta}_k|^2)^2 + 4(\vec{\alpha}_k \cdot \vec{\beta}_k)^2}} \tag{29}$$

$$i_k = \arccos \hat{z} \cdot \hat{h}_k \tag{30}$$

$$\Omega_k = \arctan \frac{n_{y_k}}{n_{x_k}} \tag{31}$$

$$\psi_k = \arctan \frac{\mathbf{r}_k(\theta_\perp) \cdot \hat{t}_k}{\mathbf{r}_k(\theta_\perp) \cdot \hat{n}_k} \tag{32}$$

where $\hat{n}_k$ is the ascending node vector and $\hat{t}_k$ is its unit transverse vector, which can be obtained by the cross product of $\hat{h}_k$ and $\hat{n}_k$.

Note that the above results are obtained for each mode $k = 1, 2, 3$. The true trajectory is the linear combination of all three modes. By understanding the behavior of each mode over an interval of time, we can sketch out the whole picture of the trajectory. These elements are osculating elements in that they vary slowly as a function of time. As we move through the halo orbit we are interested in observing the change in these elements, as this defines how the relative motion of the formation will be modified over time. Moreover, the mathematical formulae link the eigenvectors of the dynamic matrix with linear orbital elements, which converts the geometric distribution into the problem of eigenstructure properties.

TRAJECTORY DESIGN

In the previous section we defined sequences of linear orbit elements to understand the geometry of motion. Naturally, we are able to change the relative trajectories by adjusting those elements which are specified by the eigenstructures of $\bar{A}(t_i)$ matrix. Thus, the design of relative motion is controlled by the matrix $\bar{A}(t_i)$.

It is well known that for a controllable system we can arbitrarily place its poles [13]. However, the desired closed-loop eigenvalues and eigenvectors may not be attainable at the same time since this is an overconstrained problem. This implies that closed-loop right eigenvectors cannot be assigned arbitrarily, and only those that belong to the span of the corresponding allowable subspace can be assigned precisely [14]. Most people deal with this sort of problem with the technique of optimization, such as Ref. [14] and Ref. [15], either output feedback or state feedback. This algorithm tries to "rotate" the subspace of eigenstructures such that the weighting is minimum. With certain constraints the system must also satisfy, stability can then be guaranteed.

It is indeed a very useful tool for a regular control problem which requires poles on negative real plane. In our problem, however, these approaches are not applicable. Formation flight requires all closed-loop poles on the imaginary axis. Optimization of eigenvector placement might move those poles onto the real axis which would destabilized the formation dynamics. Therefore, we need to investigate other approaches to solve our unique problem.

Non-rotational System

To simplify the analysis we first consider a non-rotational system. By applying a controller to null out the Coriolis force and to stabilize system, we obtain the following equation of motion.

$$(33) \qquad \delta\mathbf{r}'' - V(t)\delta\mathbf{r} = 0$$

Assume $V(t)$ is constant over a short time interval to derive the "instantaneous" motion. The characteristic equation is

$$
\begin{aligned}
det(\omega^2 I - V) &= det(M)det(\omega^2 I - \Lambda_V)det(M^{-1}) \\
&= det(\omega^2 I - \Lambda_V) \\
&= (\omega^2 - \lambda_{v1})(\omega^2 - \lambda_{v2})(\omega^2 - \lambda_{v3}) \\
(34) \qquad &= 0
\end{aligned}
$$

where M is the eigenvector matrix and Λ_V is the diagonalized force potential. From linear theory, the force potential can be written as $V = M\Lambda_V M^{-1}$ and the eigenvectors satisfy the equation:

$$(35) \qquad [\omega_i^2 I - V]\mathbf{v}_i = 0$$

Here we should note that the controlled V is negative definite so that $\lambda_{vi} < 0$. Then we can solve for this differential equation.

$$(36) \qquad \delta\mathbf{r}(t) = \sum_{i=1}^{3} \rho_i \mathbf{v}_i \cos(\sqrt{|\lambda_{vi}|}t + \phi_i)$$

where $\mathbf{v}_i$ is the eigenvector for each mode; ρ_i and ϕ_i are determined by initial conditions. This equation tells us that, for each mode, the relative trajectory performs linear harmonic oscillations along the direction of eigenvectors with their associated eigenvalues as their frequencies. The real trajectory is just the linear combination of all three modes. Moreover, all the eigenvectors are orthonormal to each other because V is symmetric. That is, the eigenvector matrix M can be viewed as a rotation matrix. To generate a desired trajectory, we can initially assemble diagonal V matrix. Then choose a proper frequency ratio and initial conditions to generate a desired trajectory. Finally, apply frame rotation by pre- and post-multiplication by a rotation matrix to obtain a specific orientation.

Figures 10 and 11 show an example of mode combinations where a circular and an "8"-shaped trajectory are created. Having shown previously that the extra cost to eliminate Coriolis force is not very large, we have considerable freedom to perform trajectory design in this way.

Rotational System – Algebraic Approach

To understand the relationship between eigenvalues and eigenvectors, we need to look back at Eq. (20) in a 3DOF system. Assume $\pm jk_i$ are a pair of eigenvalue and $\vec{\alpha}_i \pm j\vec{\beta}_i$ are their associating eigenvectors. If we only apply a controller to the force potential, Eq. (20) can be expanded as

$$(37) \qquad \begin{bmatrix} -k_i^2 - \bar{V}_{xx} & \mp 2k_i j - \bar{V}_{xy} & -\bar{V}_{xz} \\ \pm 2k_i j - \bar{V}_{xy} & -k_i^2 - \bar{V}_{yy} & -\bar{V}_{yz} \\ -\bar{V}_{xz} & -\bar{V}_{yz} & -k_i^2 - \bar{V}_{zz} \end{bmatrix} \begin{bmatrix} a_{1i} \pm jb_{1i} \\ a_{2i} \pm jb_{2i} \\ a_{3i} \pm jb_{3i} \end{bmatrix} = 0$$

And the characteristic equation would be

$$(38) \qquad det(-k_i^2 I - 2k_i j J - \bar{V}) = 0$$

where $\vec{\alpha}_i = (a_{1i}, a_{2i}, a_{3i})$, $\vec{\beta}_i = (b_{1i}, b_{2i}, b_{3i})$, $i = 1, 2, 3$ denote the ith mode, and $\bar{V}$ denotes the normalized controlled force potential with respect to the rotation rate ω of the frame. Now we only consider one mode and rewrite Eq. (37) as a function of $\bar{V}_{ij}$.

$$
(39) \quad
\begin{bmatrix}
a_1 & a_2 & a_3 & 0 & 0 & 0 \\
b_1 & b_2 & b_3 & 0 & 0 & 0 \\
0 & a_1 & 0 & a_2 & a_3 & 0 \\
0 & b_1 & 0 & b_2 & b_3 & 0 \\
0 & 0 & a_1 & 0 & a_2 & a_3 \\
0 & 0 & b_1 & 0 & b_2 & b_3
\end{bmatrix}
\begin{bmatrix}
\bar{V}_{xx} \\
\bar{V}_{xy} \\
\bar{V}_{xz} \\
\bar{V}_{yy} \\
\bar{V}_{yz} \\
\bar{V}_{zz}
\end{bmatrix}
= -k^2
\begin{bmatrix}
a_1 \\
b_1 \\
a_2 \\
b_2 \\
a_3 \\
b_3
\end{bmatrix}
- 2k
\begin{bmatrix}
-b_2 \\
a_2 \\
b_1 \\
-a_1 \\
0 \\
0
\end{bmatrix}
$$

Or in the form of $\mathbf{P_i}\bar{\mathbf{V}} = \mathbf{Q_i}(\bar{\mathbf{V}})$, $i = 1, 2, 3$. Since all the three modes have to satisfy the same force potential, Eq. (39) are valid for all the three modes. Let

$$
(40) \quad \mathbf{P} = \begin{bmatrix} \mathbf{P_1} \\ \mathbf{P_2} \\ \mathbf{P_3} \end{bmatrix}
$$

$$
(41) \quad \mathbf{Q} = \begin{bmatrix} \mathbf{Q_1} \\ \mathbf{Q_2} \\ \mathbf{Q_3} \end{bmatrix}
$$

Then we can combine those three modes and write as $\mathbf{P}\bar{\mathbf{V}} = \mathbf{Q}(\bar{\mathbf{V}})$, where $\mathbf{P} \in \mathbb{R}^{18 \times 6}$ and $\mathbf{Q} \in \mathbb{R}^{18}$. It is an overconstraint simultaneous nonlinear equations with 6 unknowns and 18 equations. The solution exists *if and only if* $\mathbf{Q}(\bar{\mathbf{V}}) \in \mathcal{RA}\{\mathbf{P}\}$. We should note that both $\mathbf{P}$ and $\mathbf{Q}$ are functions of k, $\vec{\alpha}$ and $\vec{\beta}$, and k in turn is also a function of $\bar{\mathbf{V}}$. Generally speaking, this problem is difficult to solve.

In a practical issue, we can pick a set of candidate eigenvectors and eigenvalues, and plug them into Eqs. (38) and (39) to see if they satisfy both equations. Since we have more constraints if we consider all three modes at the same time, one way to simplify the situation is to only specify one mode and let the other two modes be free. Upon finding some candidate force potentials, we can then pick the negative definite ones so that the stability is guaranteed.

Application to A Specific Case

The most specific case is to design a circular relative trajectory. As our design method above, we only consider one mode here and pick the negative force potential at the end. From Eqs.(27) and (29), we can conclude that $\vec{\alpha} \perp \vec{\beta} \perp \hat{h}$ and $|\vec{\alpha}| = |\vec{\beta}| = 1/\sqrt{2}$ for a circular trajectory mode. Assume $\vec{\alpha} = (\hat{a}_1, \hat{a}_2, \hat{a}_3)/\sqrt{2}$, $\vec{\beta} = (\hat{b}_1, \hat{b}_2, \hat{b}_3)/\sqrt{2}$ and $\hat{h} = (h_1, h_2, h_3)$, where $\hat{a}_i$ and $\hat{b}_i$ denotes the unit vector components of $\vec{\alpha}$ and $\vec{\beta}$. Then $\hat{h} = 2(\vec{\beta} \times \vec{\alpha})$. Accordingly,

$$
\begin{aligned}
\hat{\alpha} &= \hat{h} \times \hat{\beta} \\
&= \begin{bmatrix} h_2\hat{b}_3 - h_3\hat{b}_2 \\ h_3\hat{b}_1 - h_1\hat{b}_3 \\ h_1\hat{b}_2 - h_2\hat{b}_1 \end{bmatrix} \\
\hat{\beta} &= \hat{a} \times \hat{h} \\
&= \begin{bmatrix} -h_2\hat{a}_3 + h_3\hat{a}_2 \\ -h_3\hat{a}_1 + h_1\hat{a}_3 \\ -h_1\hat{a}_2 + h_2\hat{a}_1 \end{bmatrix}
\end{aligned}
$$

Plug it into Eq. (39) and perform some algebraic manipulation to find two relations.

$$
(42) \quad
\begin{bmatrix}
k^2 + 2kh_3 + \bar{V}_{xx} & \bar{V}_{xy} & -2kh_1 + \bar{V}_{xz} \\
\bar{V}_{xy} & k^2 + 2kh_3 + \bar{V}_{yy} & -2kh_2 + \bar{V}_{yz} \\
\bar{V}_{xz} & \bar{V}_{yz} & k^2 + \bar{V}_{zz}
\end{bmatrix}
\vec{\alpha} = 0
$$

$$
(43) \quad N\vec{\alpha} = 0
$$

where

$$
\begin{aligned}
N_{11} &= -\bar{V}_{xy}h_3 + \bar{V}_{xz}h_2 \\
N_{12} &= (k^2 + \bar{V}_{xx})h_3 - \bar{V}_{xz}h_1 + 2k \\
N_{13} &= -(k^2 + \bar{V}_{xx})h_2 + \bar{V}_{xy}h_1 \\
N_{21} &= -(k^2 + \bar{V}_{yy})h_3 + \bar{V}_{yz}h_2 - 2k \\
N_{22} &= \bar{V}_{xy}h_3 - \bar{V}_{yz}h_1 \\
N_{23} &= -\bar{V}_{xy}h_2 + (k^2 + \bar{V}_{yy})h_1 \\
N_{31} &= -\bar{V}_{yz}h_3 + (k^2 + \bar{V}_{zz})h_2 \\
N_{32} &= \bar{V}_{xz}h_3 - (k^2 + \bar{V}_{zz})h_1 \\
N_{33} &= -\bar{V}_{xz}h_2 + \bar{V}_{yz}h_1
\end{aligned}
$$

Note that $\vec{\beta}$ also has to satisfy Eqs. (42) and (43). Since $\vec{\alpha}$ is perpendicular to $\vec{\beta}$, the rank of Eqs. (42) and (43) cannot be greater than one, and $\vec{\alpha}$ and $\vec{\beta}$ must be located in the null space of both matrices.

Consider a specific case in which the off diagonal terms in $\bar{V}$ are all zero. Then Eqs. (42) and (43) simplify to

(44)
$$
\begin{vmatrix}
k^2 + 2kh_3 + \bar{V}_{xx} & 0 & -2kh_1 \\
0 & k^2 + 2kh_3 + \bar{V}_{yy} & -2kh_2 \\
0 & 0 & k^2 + \bar{V}_{zz}
\end{vmatrix} = 0
$$

(45)
$$
\begin{vmatrix}
0 & (k^2 + \bar{V}_{xx})h_3 + 2k & -(k^2 + \bar{V}_{xx})h_2 \\
-(k^2 + \bar{V}_{yy})h_3 - 2k & 0 & (k^2 + \bar{V}_{yy})h_1 \\
(k^2 + \bar{V}_{zz})h_2 & -(k^2 + \bar{V}_{zz})h_1 & 0
\end{vmatrix} = 0
$$

The expansion of Eq. (45) is

(46)
$$
2h_1 h_2 k(k^2 + \bar{V}_{zz})(-\bar{V}_{xx} + \bar{V}_{xx}) = 0
$$

one of the possible solutions in Eq. (46) is $k^2 + \bar{V}_{zz} = 0$, which is consistent with Eq. (44). Since we know the rank of Eqs. (44) and (45) are not greater than one, this is not a qualified solution. The other one solution is $\bar{V}_{xx} = \bar{V}_{yy}$, and, from Eq. (44), we conclude that $\bar{V}_{xx} = \bar{V}_{yy} = -(k^2 + 2kh_3)$. Therefore, $\hat{\beta} = (1, 0, 0)$, $\hat{\alpha} = (0, 1, 0)$ and $\hat{h} = (0, 0, 1)$. On the other hand, $-k^2 = \bar{V}_{xx} + 2k = \bar{V}_{yy} + 2k$. We find that the characteristic equation, Eq. (38), is also satisfied. Moreover, Eqs. (44) and (45) both degenerate to $rank \leq 1$ with these results. That shows the consistency in our derivation.

Rotational System – Geometric Approach

Augmented Matrix

Another approach to dealing with complex matrices is to form an augmented matrix in which the real part and imaginary part are included. The matrix in Eq. (37) is complex and this approach can be applied. Assume the eigenvectors are in the form $\mathbf{u} = \vec{\alpha} \pm j\vec{\beta}$. Without loss of generality, we only take the "plus sign" into consideration. Re-write Eq. (37) as

(47)
$$
\begin{aligned}
(\mathcal{A} + j\mathcal{B})(\vec{\alpha} + j\vec{\beta}) &= 0 \\
\mathcal{A} &= -k^2 I - \bar{V} \\
\mathcal{B} &= -2kJ
\end{aligned}
$$

where $\mathcal{A}$ is symmetric and $\mathcal{B}$ is skew symmetric. Thus, a linear equation of augmented real block matrices is established.

(48)
$$
\mathcal{A}\vec{\alpha} - \mathcal{B}\vec{\beta} = 0
$$
(49)
$$
\mathcal{A}\vec{\beta} + \mathcal{B}\vec{\alpha} = 0
$$

Taking inner product of Eq. (48) with $\vec{\alpha}$ and Eq. (49) with $\vec{\beta}$, we obtain,

$$\vec{\alpha}^T \mathcal{A} \vec{\alpha} = \vec{\alpha}^T \mathcal{B} \vec{\beta} \tag{50}$$

$$\vec{\beta}^T \mathcal{A} \vec{\beta} = -\vec{\beta}^T \mathcal{B} \vec{\alpha} \tag{51}$$

However, Eq. (50) and Eq. (51) can be expanded as $\vec{\alpha}^T \mathcal{B} \vec{\beta} = -\vec{\beta}^T \mathcal{B} \vec{\alpha} = 2kH_3$, where H_3 is the third component of $\vec{\beta} \times \vec{\alpha}$, the "linear angular momentum", and is a scalar. If $H_3 \neq 0$, Eqs. (50) and (51) can be reformed as

$$\vec{\alpha}^T \left(\frac{1}{2kH_3} \mathcal{A} \right) \vec{\alpha} = 1 \tag{52}$$

$$\vec{\beta}^T \left(\frac{1}{2kH_3} \mathcal{A} \right) \vec{\beta} = 1 \tag{53}$$

Geometric Interpretation

The geometric interpretation of an equation of the form $X^T A X = 1$ is an conoid in three dimensional space, where X is a vector and A is an 3 by 3 matrix. The principle axes of this conoid lie along with the eigenvectors of matrix A, and the size as well as the shape of this conoid is determined by the eigenvalues of matrix A. The expansion of the quadratic equation $X^T A X = 1$ usually contains all combination of x, y and z. However, with proper choice of bases, or coordinate rotation in the geometric viewpoint, we can get a very neat expression. Assume λ_1, λ_2 and λ_3 are eigenvalues of A. Under a specific set of bases we obtain,

$$
\begin{aligned}
X^T A X &= \lambda_1 x^2 + \lambda_2 y^2 + \lambda_3 z^2 \\
&= \pm \frac{x^2}{a^2} \pm \frac{y^2}{b^2} \pm \frac{z^2}{c^2} \\
&= 1
\end{aligned}
\tag{54}
$$

where the choice of sign depends on the signs of λ's. The geometric shapes of Eq. (54) are classified into seventeen groups [16]. As a result, in our case, $a^2 = 2kH_3/(k_i^2 - \lambda_j)$, where i denotes the ith mode of whole system and j denotes the jth eigenvalue of $\bar{V}$. Similarly, b^2 and c^2 has the same form but different values.

It's obvious that Eq. (52) and Eq. (53) have similar structure. Moreover, since the force potential $\bar{V}$ is symmetric, its eigenvector matrix can be chosen orthonormal. Assume M is the orthonormal eigenvector matrix, so that $M^{-1} = M^T$ and $\bar{V} = M \Lambda_{\bar{V}} M^T$, where $\Lambda_{\bar{V}}$ is the diagonalized force potential. Then,

$$
\begin{aligned}
\frac{1}{2kH_3} \mathcal{A} &= \frac{1}{2kH_3} (-k^2 I - \bar{V}) \\
&= M \left[\frac{1}{2kH_3} (-k^2 I - \Lambda_{\bar{V}}) \right] M^T
\end{aligned}
\tag{55}
$$

Eq. (55) shows that matrix $\mathcal{A}$ has the same eigenvector as $\bar{V}$. That is, the conoid formed by matrix $\mathcal{A}$ has the same orientation as that formed by $\bar{V}$. Nevertheless, it has a different size and shape and is either an ellipsoid or a hyperboloid. The shape and size are determined by the relationship between k and eigenvalues of $\bar{V}$.

The above analysis tells us that both $\vec{\alpha}$ and $\vec{\beta}$ must lie on the surface of the conoid. On the other hand, $\vec{\alpha}$ and $\vec{\beta}$ also form the orbital plane. Consequently, $\vec{\alpha}$ and $\vec{\beta}$ must lie on the intersecting ellipse or hyperbola, as depicted in Figure (12). This instead limits the choice of orientation of an orbital plane, which is consistent with our previous conclusion. This idea may help us to analyze or design a trajectory from the viewpoint of geometry.

Application to A Specific Case

To design a trajectory by choosing eigenvectors, we must constrain $\vec{\alpha}$ and $\vec{\beta}$ to lie on the intersecting curve of a conoid and the desired orbital plane. One of the specific cases is to place the orbit plane on the one formed by e_1 and e_2, any two eigenvectors of the force potential. By choosing these eigenvectors, it may be possible for us to orient the orbit plane in a specific attitude. Specifically, let

(56)
$$\begin{aligned}
\vec{\alpha} &= \alpha_1 e_1 + \alpha_2 e_2 \\
\vec{\beta} &= \beta_1 e_1 + \beta_2 e_2 \\
M &= \begin{bmatrix} e_1 & e_2 & e_3 \end{bmatrix} \\
&= \begin{bmatrix} e_{11} & e_{12} & e_{13} \\ e_{21} & e_{22} & e_{23} \\ e_{31} & e_{32} & e_{33} \end{bmatrix}
\end{aligned}$$

where M is the orthonormal eigenvector matrix of $\bar{\mathbf{V}}$. Substitute into Eq. (48) and Eq. (49) and after a few algebraic manipulations (Appendix I), we can show that the plane normal to the Z-axis is the only solution. Actually, this result can also be obtained by directly solving for the equations of motion, which gives us a more dynamical interpretation of this phenomenon.

First we can express the force potential in terms of its eigenvalues and eigenvectors.

(57)
$$\mathbf{V} = [\lambda_1 e_1 e_1^T + \lambda_2 e_2 e_2^T + \lambda_3 e_3 e_3^T]$$

Assume the orbit plane lies on the e_1-e_2 plane. The trajectory, $r(t)$, can be written as the linear combination of these two eigenvectors

$$\begin{aligned}
\mathbf{r}(t) &= r_1(t)e_1 + r_2(t)e_2 \\
\dot{\mathbf{r}}(t) &= \dot{r}_1(t)e_1 + \dot{r}_2(t)e_2 \\
\ddot{\mathbf{r}}(t) &= \ddot{r}_1(t)e_1 + \ddot{r}_2(t)e_2
\end{aligned}$$

Plug into the equations of motion

$$\ddot{\mathbf{r}}(t) = 2\omega J \dot{\mathbf{r}}(t) + V\mathbf{r}(t)$$

Consider the Z-component to this equation,

$$\begin{aligned}
\ddot{r}_1(t)e_{31} &= \lambda_1 e_{31} r_1(t) \\
\ddot{r}_2(t)e_{32} &= \lambda_2 e_{32} r_2(t)
\end{aligned}$$

Since $\lambda_1 < 0$ and $\lambda_2 < 0$, if e_{31} and e_{32} are not zero, the solutions would be

$$\begin{aligned}
r_1(t) &= \rho_1 \cos(\sqrt{|\lambda_1|}t) \\
r_2(t) &= \rho_2 \cos(\sqrt{|\lambda_2|}t)
\end{aligned}$$

However, these don't satisfy the differential equations in the X and Y components because the Coriolis force will slightly change the system frequency. So, we can't find an oscillating frequency which makes the trajectory remain on the e_1-e_2 plane if we have Z component. Namely, the only possibility is that the orbit remains on the XY plane.

CONCLUSIONS

In this paper we analyze the dynamics of motion relative to an unstable periodic orbit. We first consider general equations of motion and try to find a sufficient condition under which the relative motions are stable. Then, based on previous work on linear orbital elements, we develop different algorithms for trajectory design. This work is being developed for future application to spacecraft-based interferometric imaging located in the vicinity of the Earth-Sun L_2 libration point.

ACKNOWLEDGEMENTS

The work described here was funded in part by NASA's Office of Space Science and by the Interplanetary Network Technology Program by a grant from the Jet Propulsion Laboratory, California Institute of Technology which is under contract with the National Aeronautics and Space Administration.

APPENDIX I

Let e_1, e_2, e_3 be the orthonormal eigenvectors of $\bar{V}$ matrix and $\lambda_1, \lambda_2, \lambda_3$ be the corresponding eigenvalues. Assume $\vec{\alpha}$ and $\vec{\beta}$ are located on the plane formed by any two eigenvectors. Without loss of generality, the e_1-e_2 plane is chosen. Then

$$(58) \qquad \vec{\alpha} = \alpha_1 e_1 + \alpha_2 e_2$$
$$(59) \qquad \vec{\beta} = \beta_1 e_1 + \beta_2 e_2$$
$$\mathcal{A} = -k^2 I - \bar{V}$$
$$= M(-k^2 I - \Lambda_{\bar{V}})M^T$$
$$= \left[(-k^2 - \lambda_1)e_1 e_1^T + (-k^2 - \lambda_2)e_2 e_2^T + (-k^2 - \lambda_3)e_3 e_3^T \right]$$

Plug Eqs.(58) and (59) into Eqs.(48) and (49) to obtain

$$(60) \qquad \alpha_1(-k^2 - \lambda_1)e_1 + \alpha_2(-k^2 - \lambda_2)e_2 = 2J(\beta_1 e_1 + \beta_2 e_2)$$
$$(61) \qquad \beta_1(-k^2 - \lambda_1)e_1 + \beta_2(-k^2 - \lambda_2)e_2 = -2J(\alpha_1 e_1 + \alpha_2 e_2)$$

Consider the third components of Eqs. (60) and (61), we can conclude that

$$\alpha_1(-k^2 - \lambda_1)e_{31} + \alpha_2(-k^2 - \lambda_2)e_{32} = 0$$
$$\beta_1(-k^2 - \lambda_1)e_{31} + \beta_2(-k^2 - \lambda_2)e_{32} = 0$$

Or in matrix form

$$(62) \qquad \begin{bmatrix} \alpha_1 & \alpha_2 \\ \beta_1 & \beta_2 \end{bmatrix} \begin{bmatrix} p \\ q \end{bmatrix} = 0$$

where $p = (-k^2 - \lambda_1)e_{31}$ and $q = (-k^2 - \lambda_2)e_{32}$. Since $\vec{\alpha}$ is not parallel to $\vec{\beta}$, the matrix is full rank. Therefore, $[p \quad q]^T = 0$. Moreover, due to the Coriolis term in the characteristic equation, the system eigenvalues, $-k_i^2$, are not equal to the eigenvalues of force potential, λ_j. Hence, the only solution to Eq. (62) is $e_{31} = e_{32} = 0$, i.e., $e_3 = (0, 0, 1)$.

REFERENCES

1   Barden, B.T. and K.C. Howell, "Fundamental Motions Near Collinear Libration Points and Their Transitions", *Journal of the Astronautical Sciences*, Vol. 46, 1998, pp. 361–378

2. Howell, K.C., B.T. Barden, M.W. Lo, "Application of Dynamical Systems Theory to Trajectory Design for a Libration Point Mission", *J. of the Astronautical Sciences*, Vol. 45, 161–178, 1997

3   Gómez, G., J. Masdemont, and C. Simo, "Quasihalo Orbits Associated with Libration Points", *Journal of the Astronautical Sciences*, Vol. 46, 1998, pp. 135–176

4   Schaub, H. and K.T. Alfrend, "Impulsive Feedback Control to Establish Specific Mean Orbit Elements of Spacecraft Formations", *Journal of Guidance, Control and Dynamics*, Vol. 24, No. 4, 2001, pp. 739–745

5   Mesbahi, M. and F.Y. Hadaegh, "Formation Flying Control of Multiple Spacecraft via Graphs, Matrix Inequalities, and Switching", *Journal of Guidance, Control and Dynamics*, Vol. 24, No. 2, 2001, pp. 369–377

6   Sabol, C., R. Burns, and C.A. Mclaughlin, "Satellite Formation Flying Design and Evolution", *Journal of Spacecraft and Rockets*, Vol. 38, No. 2, 2001, pp. 270–278

7   Hussen, I., D.J. Scheeres and D.C. Hyland, "Interferometric Observatories in Low Earth Orbit", submitted to *Journal of Guidance, Control and Dynamics*

8. Marchal. C., *The Three-Body Problem, Elsevier*, 1990, pp. 64

9   Scheeres, D.J., F.Y. Hsiao and N.X. Vinh, "Stabilizing Motion Relative to An Unstable Orbit: Applications to Spacecraft Formation Flight", *Journal of Guidance, Control and Dynamics* accepted

10. Hsiao, F.Y. and D.J. Scheeres, "The Dynamics of Formation Flight About a Stable Trajectory", *AAS Paper* 02-189

11. Nemytskii V. V, V.V. Stepanov, *Qualitative Theory of Differential Equations*, Dover, 1989, pp. 152

12. Bate, R.R, Mueller D.D., White, J.E., *Fundamentals of Astrodynamics*, Dover, New York, 1971, pp. 58–63

13. Bay, J.S., *Fundamentals of Linear State Space systems*, 1999, pp. 419

14. Liu, G.P., R.J. Patton, *Eigenstructure Assignment for Control System Design*, 1998, pp. 20

15. Subrahmanyan, P., D. Trumper, "Eigenvector Assignment", 1999 AACC

16. Zwillinger, D., *CRC Standard Mathematical Tables And Formulae*, 30th Edition, pp. 316 – 319

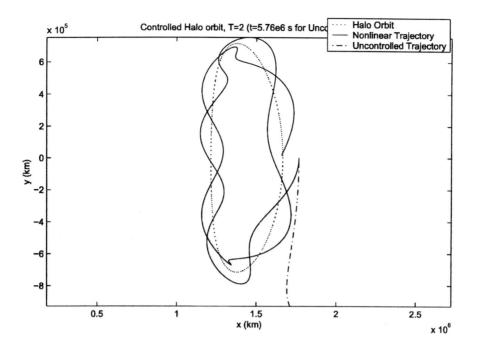

Figure 1: A control law $\mathcal{T}_c = -\sigma^2 G[u_+ u_+^T + u_- u_-^T]$ with $G = 10$ is applied (Eq. (26)) Initial deviation is 1×10^5 km in x-direction. This plot shows the stability in a nonlinear version.

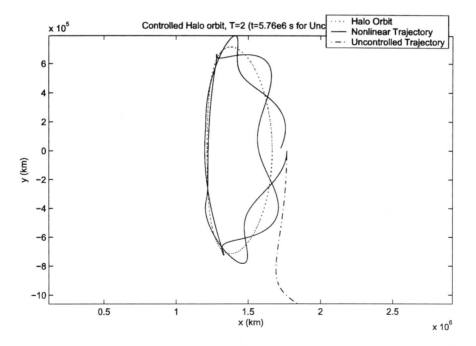

Figure 2: A desired force potential $\bar{V}_{rr} = -2.7 \times 10^{-12} I_{3\times3}$ is chosen and the controller is $\mathcal{T}_c = -(\bar{V}_{rr} - V_{rr})\delta\mathbf{r}$. Initial deviation is 1×10^5 km in x-direction. This plot shows the stability in a nonlinear version.

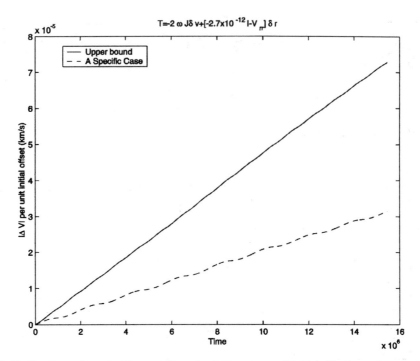

Figure 3: The first figure shows that the upper bound on total cost per unit initial offset to create a non-Coriolis system, and the cost of a specific example with $r_0 = 1$ km.

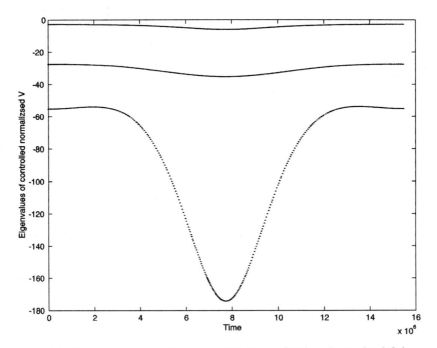

Figure 4: Eigenvalues of control law in Eq. (26). We see that the V is negative definite

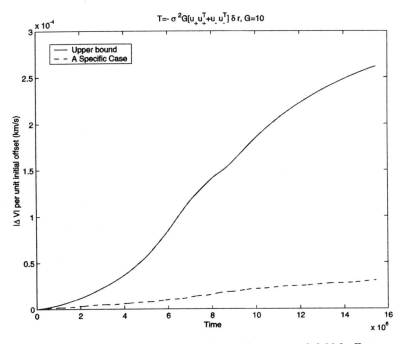

Figure 5: The first figure shows that the upper bound on total cost per unit initial offset as we apply the controller $T = -\sigma^2 G[\mathbf{u}_+ \mathbf{u}_+^T + \mathbf{u}_- \mathbf{u}_-^T]$ and $G = 10$, and the cost of a specific example with $r_0 = 1$ km.

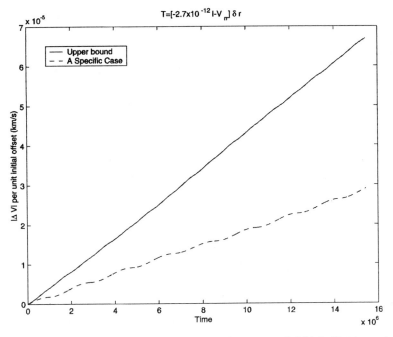

Figure 6: The first figure shows that the upper bound on total cost per unit initial offset to create circular trajectory with constant frequency over an entire period, and the cost of a specific example with $r_0 = 1$ km.

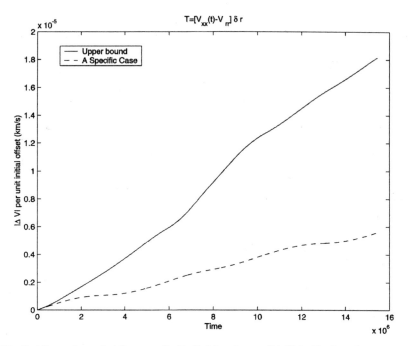

Figure 7: The first figure shows that the upper limit of total cost per unit initial offset in order to create circular trajectory with varying frequency over a whole period. The second figure gives the actual cost when $r_0 = 1$ km.

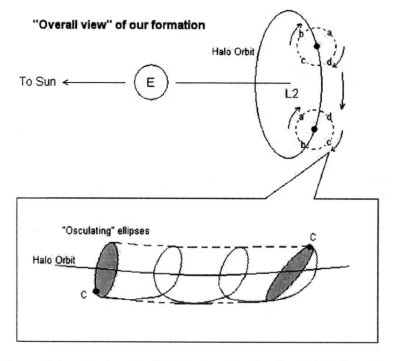

Figure 8: An overall view of the formation flight of spacecraft and how "osculating" ellipses describe relative linear trajectories.

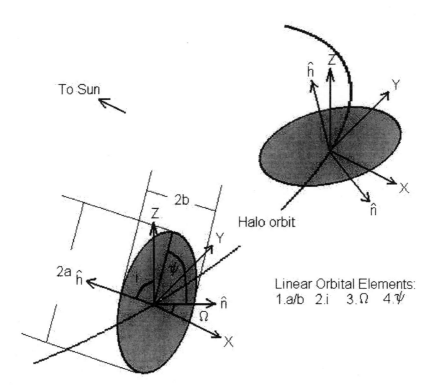

Figure 9: The definition of the "Linear Orbital Elements".

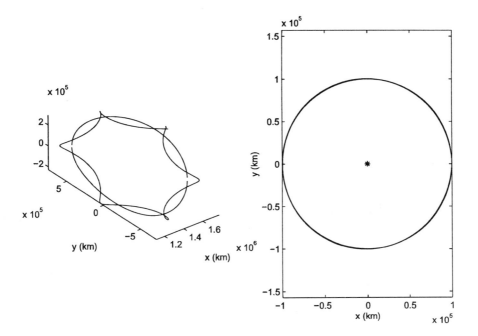

Figure 10: A relative circular trajectory is created in a non-Coriolis system. (a) Integrated in the original non-linear equations about a halo orbit. (b) Same Orbit relative to the halo orbit in (a)

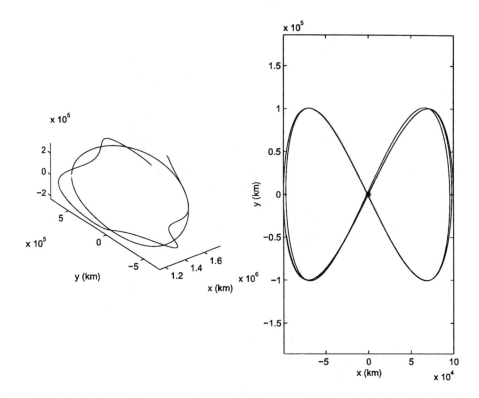

Figure 11: A relative figure-"8" trajectory in a non-Coriolis system. (a) Integrated in the original non-linear equations about a halo orbit. (b) Same Orbit relative to the halo orbit in (a)

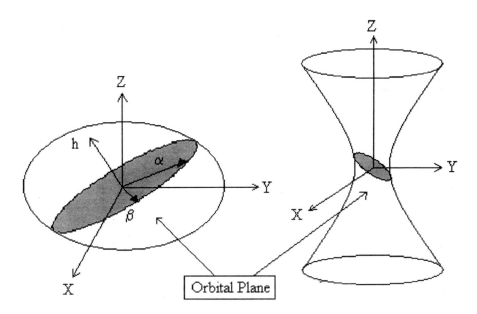

Figure 12: The Geometric Approach and the relationship between eigenvectors and conoids. $\vec{\alpha}$ and $\vec{\beta}$ denote the real and imaginary part of an eigenvector in our controlled system, respectively.

OPTIMAL COORDINATION OF MOBILE AGENTS: APPLICATION TO SPACE-BASED INTERFEROMETERS[*]

Venkatesh G. Rao[†] and Pierre T. Kabamba[‡]

This paper poses and provides solutions for an abstract class of coordinated-motion optimization problems, along with proofs of optimality. An exhaustive characterization of the behavior of the optimal solutions is provided. The results are applied to a scheduling problem for a two-spacecraft interferometric observatory.

INTRODUCTION

In this paper, optimal solutions are derived for a class of one-dimensional, two-agent, coordinated-motion problems with a variety of inequality constraints on the velocities. The original motivation for this analysis is the problem of time-optimal coordination in one-dimensional formation travel, but the results are applicable to a variety of mobile agent coordination scenarios. In this paper, we apply the results to a problem of interpolating secondary imaging goals in the observation schedule of a pair of interferometric telescopes.

Space-based interferometric observatories are multiagent systems comprising several spacecraft. Currently planned missions in NASA's Origins program, such as the Terrestrial Planet Finder (TPF) fall in this class. The motivating application in this case is a mission concept based on the exact inverse of the Huygens-Fresnel principle,[1] that is being studied at the University of Michigan and JPL. Scheduling observations for a multi-spacecraft observatory involves challenges beyond those posed by single-spacecraft observatories such as Hubble.[2, 3] In particular, since the observation capacities of the system depend on the relative positions of independent spacecraft, scheduling an observation can involve translational motion planning as well as attitude planning. It is likely that any feasible system will require the use of passive orbital motions to achieve imaging goals,[4] given the high cost of achieving arbitrary translations for spacecraft, but depending on the details of the mission goals, some relative translation, accomplished by active thrusting, may be necessary. In this paper we consider the conceptually simple case of an observatory in gravity-free space, where the spacecraft are capable of active motion to achieve reconfiguration.

The rest of the paper is organized as follows. We begin by developing an abstract motion-planning problem formulation and summarizing the results qualitatively. Next, we present the main results in the form of theorems and lemmas, and discuss their interpretations. We conclude with an application involving interferometric imaging, and some conclusions. Detailed proofs are provided in the appendix.

[*] This research was funded by NASA grant number NRA-99-05-OSS-0077.

[†] Department of Aerospace Engineering, The University of Michigan, 1320 Beal Ave., Ann Arbor, Michigan 48109-2140. E-mail: raov@engin.umich.edu.

[‡] Department of Aerospace Engineering, The University of Michigan, 1320 Beal Ave., Ann Arbor, Michigan 48109-2140. E-mail: kabamba@engin.umich.edu.

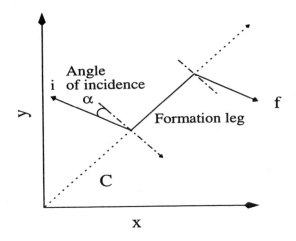

Figure 1: Position plane and geometry of connection for $p = 2$

PROBLEM STATEMENT AND SUMMARY OF RESULTS

In this section, we develop the problem statement in terms of a simple, one-dimensional formation travel application, for clarity of exposition. In a later section, it will be shown that a particular problem of coordinated motion in two dimensions, for a simplified model of an interferometric observatory, can be cast in this form.

Consider the problem of finding time-optimal trajectories for two agents traveling in a one dimensional space, such as a railroad track, in the presence of joint velocity constraints. Let the positions of the agents be given by $x(t)$ and $y(t)$ in some coordinate system in the space.

Definition 1: For this system, we refer to the x, y plane as the *position plane* (Fig. 1).

Let the velocities of the agents be given by $v_x(t)$ and $v_y(t)$ respectively, let the velocities be commandable subject to one of the following p-norm constraints:

$$\left|\frac{v_x(t)}{V_1}\right|^p + \left|\frac{v_y(t)}{V_1}\right|^p \leq 1, \ (x(t), y(t)) \in C'$$

$$\left|\frac{v_x(t)}{V_2}\right|^p + \left|\frac{v_y(t)}{V_2}\right|^p \leq 1, \ (x(t), y(t)) \in C, \tag{1}$$

when $p = 1$ or $p = 2$ and

$$|v_x| \leq V_1 \text{ and } |v_y| \leq V_1, \ (x(t), y(t))(t) \in C'$$

$$|v_x| \leq V_2 \text{ and } |v_y| \leq V_2, (x(t), y(t)) \in C. \tag{2}$$

for the $p = \infty$ norm. C is the *coordination line* in the position plane, given by the line $x = y$, and C' is its complement. Let $V_1/V_2 \triangleq \gamma < 1$. The velocities are therefore constrained to lie inside a p-norm ball ($p = 1, 2, \infty$) about the origin in the (v_x, v_y) plane, whose size is determined by location of the system in position space.

Problem: Let the initial and final positions of the agents on the position plane be given by be $X(t_i) \triangleq (x_i, y_i)$ and $X(t_f) \triangleq (x_f, y_f)$. Let the cost functional J be given by $J(t, x(t), y(t), v_x(t), v_y(t)) = t_f - t_i$, where t_f and t_i are the times at the beginning and end of a given trajectory. Given the system and cost function as described, we want to find the optimal trajectory from i to f in the position plane, and find any locations where the optimal trajectory connects and disconnects from C, which are refered to as rendezvous and breakup points respectively.

Due to symmetry in the velocity constraint, time-optimal paths are clearly reversible, so we assume, without loss of generality, that $x_i \leq x_f$.

Remark: The problem statement describes the simplest case of formation travel, where there is some mutual benefit when two agents travel together. We assume the agents are much smaller than the distances being traversed, so we approximate proximity by collocation. In this simple formulation, the cost function is time, and the agents travel faster (at V_2) when traveling together. We can interpret the velocity constraint model in several ways[5] as joint power consumption limits. As we will show in a later section, all the three norm cases have interesting interpretations in terms of particular models of interferometric observatories.

Summary of Results

The main results of the paper are presented in three theorems, one for each of the three p-norm cases. The theorems define the conditions under which coordinated motion is beneficial, and the locations of rendezvous and breakup points. Qualitatively, the major results are as follows:

- For each of the three p-norm cases, for a given initial point, the position plane may be partitioned into two sets. For destination points in the first set S_p, the optimal trajectory does not involve rendezvous, travel on C, and breakup. For destination points in the second set, S'_p, it does. The sets are sketched in Fig. 2 for the three norm cases. The broad structure of the partition, it can be seen, is similar in each case.

- While the behavior of the optimal trajectories in C' is varied, the behavior on C is simple: the optimal trajectory connects and disconnects from the coordination line at most once. The angle of incidence onto C is uniquely defined for each norm, being $\pi/4$, 0 and $\arcsin V_1/V_2$ for the 1, ∞ and 2 norm cases respectively. The angle of incidence, α, in the position plane determines the location of the rendezvous and breakup points for the agents in the physical, one-dimensional space.

- For the case $p = 2$, the allowable trajectories are either straight or polygonal lines, and are always unique. For $p = 1$ or ∞, the time-optimal trajectory is usually not unique when the destination point is in S_p. In both these cases, any continuous curve inside a bounding box that satisfies certain conditions on its slope at all points is time-optimal. The bounding boxes, B_p, and representative allowable trajectories for $p = 1, \infty$ are sketched in Fig. 3.

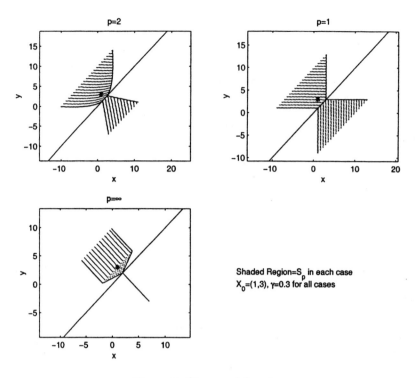

Figure 2: Shapes of S_p sets

MAIN RESULTS

Preliminaries

In this section, we provide some preliminary definitions, and state three lemmas that are relevant to all three p-norm cases. We then state the results for each p-norm case individually. Proofs and notational conventions are at the end of the paper.

Definition 2: We define six points, k, v and h, k', v', h' on C as follows.

$$k \triangleq \left(\frac{x_i + y_i}{2}, \frac{x_i + y_i}{2} \right)$$

$$v \triangleq (x_i, x_i)$$

$$h \triangleq (y_i, y_i), \tag{3}$$

with the k', v', h' being defined similarly with respect to (x_f, y_f). k, v and h are, respectively, the foot of the perpendicular from i to C, and the intersections of C with the vertical and horizontal lines through i. k', v' and h' are similarly defined with respect to f (Fig. 4).

Definition 3: We define the set of allowable trajectories, Φ to be the set of all continuous, piecewise differentiable curves with end points i and f in the position plane, that have at most a finite number of corners.

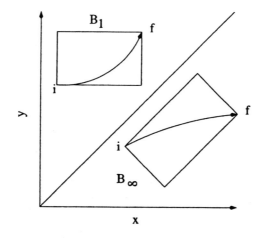

Figure 3: Bounding boxes $B_p(i, f)$ for $p = 1$, $p = \infty$

Lemma 1: For a given curve $\phi \in \Phi$, and a given p-norm, the minimum time of traversal occurs when the velocities of the agents at every point (x, y) on ϕ that is not a corner, are such that the velocity vector is on the boundary of the p-norm ball.

Lemma 2: A time-optimal arc ϕ^* cannot intersect itself.

Lemma 3: Let p and q be the endpoints of any subarc of a time-optimal trajectory ϕ^* that does not intersect C, such that p and q lie on a straight line parallel to C. Let the straight line segment pq also be time-optimal. Let the time of traversal of any straight line segment of length d and slope 1 in C' be $k_1 d$ and let the time of traversal of any segment of length d on C be $k_2 d$. Further, let $k_2 < k_1$. Let there exist an upper bound V on the magnitude of the velocity in any direction in the position plane away from C. If all the preceding conditions above are satisfied, then ϕ^* can connect and disconnect from C at most once.

We now consider the problem for the three cases $p = 1, 2, \infty$ separately.

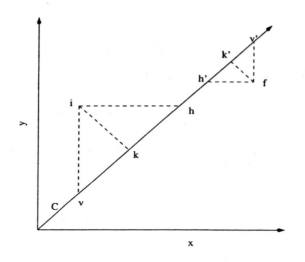

Figure 4: Points k, v, h, k', v' and h'

Case $p = 2$

Definition 4: For a given initial point X_i we define the set $S_2(X_i)$ to be the set of points X_f in the position plane such that:

$$\|X_f - X_i\|_2 < \sqrt{\frac{1-\gamma^2}{2}}(|x_i - y_i| + |x_f - y_f|) + \frac{\gamma}{\sqrt{2}}(|(x_i + y_i) - (x_f + y_f)|). \tag{4}$$

We define $S'_2(X_i)$ to be the complement of $S_2(X_i)$ in the position plane (Fig. 2).

Lemma 4: Consider the case $p = 2$. Every subarc of a time-optimal trajectory ϕ^* that does not contain points of C must be a straight-line segment.

Corollary 1: The minimum time of traversal between points m and n, for $p = 2$, along an arc that does not intersect C, is given by $l_s(m, n)/V_1$. (The proof is trivial and is omitted).

Theorem 1: For the case $p = 2$, given X_i, the following hold:

1. If $X_f \in S_2(X_i)$, then the time-optimal trajectory is the straight line if.

2. If $X_f \in S'_2(X_i)$, then the unique time-optimal trajectory is a polygonal line, with at most two corners. The corners, if they exist, must lie on C and form vertex angles of $\pi/2 + \alpha$, where $\sin(\alpha) = V_1/V_2 = \gamma$.

Case $p = \infty$

Definition 5: For a given initial point i we define the set $S_\infty(X_i)$ to be the set of points X_f in the position plane such that:

$$\max(|x_i - x_f|, |y_i - y_f|) < \frac{|x_i - y_i| + |x_f - y_f|}{2} + \gamma \left(\frac{|(x_i + y_i) - (x_f + y_f)|}{2} \right). \qquad (5)$$

We define $S'_\infty(X_i)$ to be the complement of $S_\infty(X_i)$ in the position plane (Fig. 2).

Definition 6: Given 2 points p and q in the position plane, we define $B_\infty(p, q)$ to be the closed set of points contained within the unique rectangle whose diagonal is the straight line pq and whose sides are segments of lines of slope 1 and -1, passing through p and q (Fig. 3).

Lemma 5: Consider the case $p = \infty$. Let (m, n) be any subarc of a time-optimal trajectory ϕ^*, such that $(m, n) \cap C$ is empty. Then the following statements are true:

1. (m, n) must be one of two types of curves. The first type has slope less than or equal to 1 in magnitude at all points where it is defined, and the second type has slope greater than or equal to 1 in magnitude at all points where it is defined.

2. (m, n) must remain within the rectangle $B_\infty(m, n)$.

3. Any other curve between m and n of the same type as (m, n) is also time-optimal.

Corollary 2: The minimum-time of traversal between points m and n, for $p = \infty$, along an arc that does not intersect C is given by

$$t = \max(|x_m - x_n|/V_1, |y_m - y_n|/V_1). \qquad (6)$$

(The proof is trivial and is omitted).

Theorem 2: For the case $p = \infty$, given X_i, the following hold:

1. If $X_f \in S_\infty(X_i)$, all trajectories between i and f, satisfying condition 1 of Lemma 5 are time-optimal.

2. If $X_f \in S'_\infty(X_i)$, then the unique time-optimal trajectory is a polygonal line with at most two corners. The corners, if they exist must lie on C, and form vertex angles of $\pi/2$.

Case $p = 1$

Definition 7: For a given initial point i we define the set $S_1(X_i)$ to be the set of points X_f in the position plane such that:

$$\|X_f - X_i\|_1 < |x_i - y_i| + |x_f - y_f| + \gamma(|(x_i + y_i) - (x_f - y_f| - |x_i - y_i| - |x_f - y_f|). \qquad (7)$$

We define $S'_1(X_i)$ to be the complement of $S_1(X_i)$ in the position plane (Fig. 2).

Definition 8: Given two points p and q in the position plane, we define $B_1(p,q)$ to be the closed set of points contained within the unique rectangle whose diagonal is the straight line pq and whose sides are segments of lines of slope 0 and ∞, passing through p and q (Fig. 3).

Lemma 6: Consider the case $p = 1$. Let (m,n) be a subarc of any time-optimal trajectory ϕ^*, such that $(m,n) \cap C$ is empty. Then the following statements are true:

1. (m,n) belongs to one of two classes of curves. The first class consists of curves on which $y(t)$ and $x(t)$ are nondecreasing in time, and the second consists of curves such that $y(t)$ it is nonincreasing and $x(t)$ is nondecreasing in time.

2. (m,n) remains confined within the rectangle $B_1(m,n)$.

3. Any other curve between m and n in the same class as (m,n) is also time-optimal.

Corollary 2: The minimum-time of traversal between points m and n, for $p = 1$, along an arc that does not intersect C is given by

$$t = |x_m - x_n|/V_1 + |y_m - y_n|/V_1. \tag{8}$$

(The proof is trivial and is omitted).

Theorem 3: For the case $p = 1$, given X_i, the following hold:

1. If If $X_f \in S_1(X_i)$, then all trajectories between i and f satisfying condition 1 of Lemma 6 are time-optimal.

2. If $X_f \in S'_1(X_i)$, then the unique time-optimal trajectory is a polygonal line, with at most two corners. The corners, if they exist, must lie on C and form vertex angles of $3\pi/4$.

Discussion

In this section we characterize the results geometrically in terms of behavior of optimal trajectories in the position plane.

The optimal solutions can be succinctly described in terms of partitions of the position plane into the sets S_p and S'_p (Fig. 2). The sets as sketched are for a representative initial condition where the agents are initially not collocated. Each point in the shaded area represents a destination to which uncoordinated travel is the optimum solution, while each point outside the shaded areas represents a destination for which coordinated motion, including rendezvous, a period of formation travel, and breakup, is optimal.

Broadly, it can be seen that the three sets, for the three norm cases are roughly the same in structure (Fig. 2). In the area on the 'opposite' side of C (representing an 'overtaking' case, where C must be crossed by any trajectory) coordinated optimal solutions are relatively predominant, while on the same side as the initial point, non-coordinated optimal solutions are predominant. The former observation makes intuitive sense, because crossing C involves a rendezvous, and the benefits of coordination are easier to realize. The latter observation is also expected since the agents have to make an added effort to get to C, and therefore there is less likelihood of the benefits of coordination outweighing the cost of attaining it.

Notice that all three sets S_p (Fig. 2) are symmetric about an axis inclined at $-\pi/4$ (the line ik in Fig. 4). It can be seen from the equations that the left and right hand sides remain unchanged under reflection across ik. Geometrically, this is due to the fact that mirror images of candidate optimal trajectories to a destination point f, across ik, are the candidate optimal trajectories to f', the mirror image of f. The actual traversal times of these mirrored trajectories are also equal, due to the symmetry of the velocity norm-balls about an axis inclined at $-\pi/4$. Since the sets are defined in terms of the traversal times of candidate optimal trajectories, they are symmetric about ik.

The sets S_2 and S_∞ increase in size as γ increases. This again is as expected, since the benefit of coordination reduces with a smaller velocity advantage on C. S_1, however, is invariant with respect to γ. This follows from solving the inequality (7) for the set boundary, in which case γ drops out.

A detailed interpretation of the results in terms of time and fuel-optimal formation travel is given in,[5] along with examples and graph-theoretic extensions to the formation travel case. In this paper, we focus on formulating an optimal imaging problem within this framework and applying the results to the problem of optimally interpolating scan observations between dwell observations for an interferometric space observatory.

APPLICATION: INTERFEROMETRIC IMAGING

Consider two light-collector spacecraft oriented in a target heading h. Let each spacecraft be equipped with an impulsive thruster that can accelerate/decellerate it to/from a maximum allowable velocity of $V/2$ in any direction, with respect to an inertial frame in a plane containing the two spacecraft, called the *observation plane*. This plane is assumed perpendicular to the target heading h initially. The spacecraft are assumed to be very distant from large masses, so that the observation plane stays normal to h over time periods much longer than the duration of imaging activities. The relative position velocity vector does not change during imaging, except due to impulses generated by the spacecraft themselves. We may approximate motion on this plane as two-dimensional free body motion due to impulsive, in-plane thrusts. Let $\bar{r}$ be the relative position vector of the two spacecraft in the observation plane.

As defined, the system has four position states. For interferometric imaging of very distant targets, only the relative position matters, and the relevant motion can be described in terms of a *wave number* plane,[1] which is a representation of the spatial frequencies of the (two-dimensional) image of the target. A point on the wave number plane represents a particular spatial frequency (proportional to the distance from the origin). The abscissa and ordinate of a point on this plane are also equal to the components of $\bar{r}$ in the observation plane divided by the wavelength of observation. Points in this plane map to an equivalence class of configurations in the observation plane. A configuration corresponding to a point on the wave number plane is capable of imaging features on the target at the spatial frequency corresponding to that point.

In the wave number plane, the maximum velocity of the system in any direction is V/λ. Now assume that the system functions primarily in *dwell* mode, where the location of the system on the wave number plane is constant for an epoch of time, following which the system moves to another dwell point. These dwell epochs, when the relative position vector of the spacecraft does not change, permit the highest-quality measurements to be made. We are interested in interpolating a scan observation in between two dwell epochs. Specifically, consider scan observations comprising traversal of a specified straight line through the origin, which we take, without loss of generality, to be the line $x = y$ in the wave number plane. Such observations might be used, for instance, to quickly check for banded structure on a planetary target along the scan line. Assume that idle time between dwell points costs c_1 units, and that any time spent scanning any portion of a scan line accrues a lower cost of c_2 units (scanning still has a cost with respect to high-value dwell observations). We can now state the problem as follows: find the trajectory that minimizes the cost of transit between dwell points.

This problem, it is easy to see, is isomorphic to our one-dimensional formation travel problem with $p = 2$. The cost function is now a positive definite function of time, and the 'velocities' on and off the coordination (scan) line are given by $V_1 = V/c_1\lambda$ and $V_2 = V/c_2\lambda$. The results of the previous sections may therefore be used to compute the coordinated and non-coordinated trajectories (which correspond to scanning and non-scanning solutions) and the associated costs, to determine whether it is worth interpolating the scan observation. All of the discussion presented previously applies, *mutatis mutandis*, to this case. In the next section, we present a numerical example.

Numerical Example

We now consider a numerical example. Let the initial and final configurations of the spacecraft be given by the relative position vectors [1000 km, 2000 km] and [6000 km, 4000 km]. Let the scan line for the observation to be interpolated be the line $x = y$ on the wave number plane. Let the observation wavelength be $\lambda = 10^{-6}$. Let the costs of idle time and scanning time be $c_1 = 1$ and $c_2 = 0.5$ respectively, and let $V = 10$ km/s. With this data, we can compute that the cost of directly moving from the initial to the final dwell point is 538.5, and can be achieved in 538.5 seconds. An alternate path, which performs a 372.5 second scan along the direction of interest, achieves a cost of 431.2, but takes slightly longer overall, 617.5 seconds. The trajectories of the spacecraft in the wave number plane and in the observation plane are shown in Fig. 5 and 6. Fig. 5 also shows, for reference, a disc corresponding to the highest resolution (spatial frequency) at which the target is being imaged, as well as smaller circles representing the patch of frequencies around a point in the wave number plane, that can be imaged at a given time. The radius of these smaller circles depends on the individual resolving powers of the spacecraft.

The trajectory in the wave number plane can be traversed at any velocity. We assume it is traversed at the maximum velocity, which yields the unique observation plane trajectories in Fig. 6

CONCLUSION

In this paper, the solution to the problem of coordinated motion for two agents in one dimension, for a variety of velocity constraints, was presented. The low-dimensionality allowed the exhaustive characterization of the behavior of optimal solutions, including conditions governing starting and ending of 'coordination' which can model various physical situations such as formation travel or scan observations in interferometric imaging.

Potential extensions include incorporation of orbital motion, more realistic cost functions that

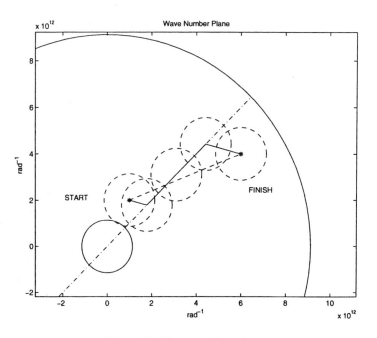

Figure 5: Wave number plane

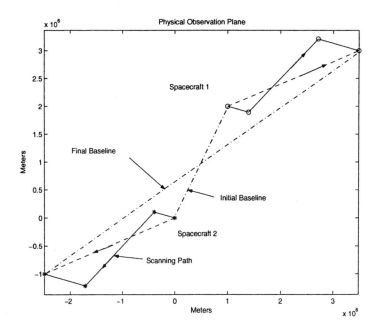

Figure 6: Observation plane

include the real costs of thrusting in various directions, more general 'coordination regions' in the position plane, a wider class of velocity constraints, and incorporation of scheduling constraints. These will be subject of future work.

ACKNOWLEDGEMENTS

The authors thank Professor Jason Speyer for a useful discussion.

NOTATION

1. We denote trajectories by lower-case greek letters (ϕ, ψ, ...). We denote time-optimal trajectories with star superscripts, ϕ^*, ψ^*, ...

2. We denote points on trajectories in the position plane by lower case letters (a, b, ...). In particular, the initial and final points are denoted i and f respectively. The associated coordinates of a point a are denoted $X_a \triangleq (x_a, y_a)$.

3. We denote portions of trajectories with set notation using boundary points. Thus, (p, q) and $[p, q]$ are the open and closed arcs between points p and q on a trajectory.

4. We denote straight lines between points p and q by pq.

5. We denote the arc-length of a curve between points p and q by $l(p, q)$, and the straight line distance between them by $l_s(p, q)$.

6. We denote the minimum time for traversing an arc between p and q by $t(p, q)$.

References

[1] Hyland, D. C., "Interferometric Imaging Concepts with Reduced Formation-Keeping Constraints,," *AIAA Space 2001 Conference*, Albuquerque, NM, August 2001.

[2] Muscettola, N., "HSTS: Integrating Planning and Scheduling," *Intelligent Scheduling*, edited by M. Zweben and M. Fox, Morgan Kaufmann, 1994.

[3] Johnston, M. D. and Miller, G. E., "Intelligent Scheduling of Hubble Space Telescope Observations," *Intelligent Scheduling*, edited by M. Zweben and M. S. Fox.

[4] Hussain, I. I., Scheeres, D. J., and Hyland, D. C., "Interferometric Observatories in Earth Orbit," *Proc. AAS/AIAA Space Flight Mechanics Meeting*, Ponce, PR, February 2003.

[5] Rao, V. G. and Kabamba, P. T., "Time-Optimal Graph Traversal for Two Agents," *Technical Report NRA2002R1, Aero. Engg., U. Michigan.*, September 2002, http://www-personal.umich.edu/~ raov/nra2002r1.pdf.

[6] Gelfand, I. M. and Fomin, S. V., *Calculus of Variations*, Dover, Mineola, NY, 2000.

APPENDIX: DETAILED PROOFS

Proof of Lemma 1: Let ds be an infinitesimal portion of an allowable curve ϕ, about a point $q = (x_q, y_q)$ that is not a corner. Let m be the slope of the tangent to the curve at q. Then dt, the time of traversal of ds is given by

$$dt = ds / \sqrt{v_x(q)^2 + v_y(q)^2} = ds / |V(q)|, \qquad (9)$$

where $v_x(q)$ and $v_y(q)$ are the instantaneous velocities at q. Since the curve is predetermined, the velocity $V(q)$ must satisfy both the governing p-norm constraint (1 or 2) and the slope constraint, $v_y(q)/v_x(q) = m$. If $V(q)$ is not on the boundary of the p-norm ball, then we can find a scalar $k > 1$ such that $kV(q)$ is on the boundary of the p-norm ball, and the slope constraint is still satisfied. The time for traversing ds with velocity $kV(q)$ is obviously less than dt in (9). Therefore, for the minimum time of traversal of ϕ, the velocity must be on the boundary of the p-norm ball at all non-corner points. Also note that (1) must hold with equality for $p = 1, 2$ and at least one of the relations of (2) must hold with equality for $p = \infty$. $\square$

Proof of Lemma 2: Let, if possible, ϕ^* be an optimal curve that intersects itself at p. Let the curve be parameterized by time t, and let $X(t_1) = X(t_2)$, $t_2 > t_1$. Now, the curve constructed by deleting the portion corresponding to the interval (t_1, t_2) is also an admissible curve, with traversal time less by $t_2 - t_1$ than the original curve. ϕ^*, therefore, is suboptimal, a contradiction. Therefore, a time-optimal curve cannot intersect itself. $\square$

Proof of Lemma 3: Let $\phi_c^* = \phi^* \cap C$. Since $\phi^* \triangleq [i, f]$ and C are closed, so is ϕ_c^*. We will first show that ϕ_c^* is connected. Now let, if possible, ϕ_c^* be disconnected. Since ϕ_c^* is closed, it contains its boundary points. Since it is disconnected, ϕ^* must disconnect and reconnect at least once. Therefore there exists at least one pair of boundary points m_1, $m_2 \in \phi_c^*$, $m_1 \neq m_2$, such that (m_1, m_2) lies entirely outside C. Let $d = l_s(m_1, m_2)$ on C. By continuity, we can pick points $\bar{m}_1$ and $\bar{m}_2$ on $m_1, m_2)$ such that

$$x_{\bar{m}_2} - x_{\bar{m}_1} = y_{\bar{m}_2} - y_{\bar{m}_1}, \qquad (10)$$

and such that for sufficiently small $\delta_1 > 0$, $\delta_2 > 0$,

$$l(m_1, \bar{m}_1) < \delta_1$$
$$l(m_2, \bar{m}_2) < \delta_1$$
$$|d - \bar{d}| < \delta_2, \qquad (11)$$

where $\bar{d} = l_s(\bar{m}_1, \bar{m}_2)$. By the second assumption, the $t(\bar{m}_1, \bar{m}_2)$ is the same as the time of traversal of the straight line $\bar{m}_1\bar{m}_2$. $t(m_1, m_2)$ is therefore given by:

$$t = \lim_{\delta_1, \delta_2 \to 0} k_1 (d \pm \delta_2) + f(\delta_1) = k_1 d, \qquad (12)$$

where $f(\delta_1)$ represents terms of the order of δ_1, the time to traverse the arcs $(m_1, \bar{m}_1)$ and $(\bar{m}_2, m_2)$. Clearly $f(\delta_1) \to 0$ as $\delta_1 \to 0$, since velocity is bounded. By assumption the time of traversal of the straight-line $m_1 m_2$ on C is given by $\bar{t} = k_2 d$. Taking the difference,

$$\bar{t} - t = (k_2 - k_1)d > 0 \qquad (13)$$

Therefore ϕ^* is not time-optimal, a contradiction. Therefore ϕ^* is connected. By Lemma 2, ϕ^* cannot intersect itself, so the connectedness of ϕ_c^* implies that ϕ^* cannot disconnect and reconnect at the same point. Therefore, ϕ^* can connect and disconnect to C only once $\square$

Proof of Lemma 4:   Let $[m, n]$ be any closed subarc or ϕ^* that does not contain points of C. $[m, n]$ must also be time-optimal since every subtrajectory of an optimal path is also optimal. Let $[p, q] \subset [m, n]$ be the end points of a smooth subarc of $[m, n]$. Since there are at most a finite number of corners on ϕ^*, $[m, n]$ must be composed of smooth segments such as $[p, q]$.

Let the $[p, q]$ be partitioned into n segments with end points $p = m_1, \ldots, m_{n+1} = q$. Since $[p, q]$ is smooth, we can always pick a sufficiently high n such that for each segment, ϕ^* can be locally described by either a continuously differentiable function $y = f(x)$, or a continuously differentiable function $x = f(y)$.

Now consider a segment $[m_i, m_{i+1}]$, and assume, without loss of generality, that it is described by a continuously differentiable function $y = f(x)$. Since ϕ^* is time-optimal, $y = f(x)$ must be an extremal of the functional

$$J(y) = \int_{x_{m_i}}^{x_{m_{i+1}}} \frac{\sqrt{1 + y'^2}}{V_1} dx. \tag{14}$$

V_1, the tangential velocity, is constant. The Euler equation[6] for the functional is

$$F_y + \frac{d}{dx} F_{y'} = 0, \tag{15}$$

which reduces to

$$\frac{d}{dx} \left(\frac{y'}{\sqrt{1 + y'^2}} \right) = 0, \tag{16}$$

which is a second order ODE, with the general solution:

$$y = ax + b, \tag{17}$$

which is a straight line. The segment $[m_i, m_{i+1}]$ of $[p, q]$ is therefore a straight line. Since, by assumption, there are no corners in $[p, q]$, it must be a straight line segment. Therefore $[m, n]$, which is composed of arcs such as $[p, q]$, can at most be a polygonal line. Next we prove that there cannot exist any corners on $[m, n]$, and it must be a single straight line.

Let, if possible, there be a corner point $r \in [m, n]$. Pick p and q such that the only corner on the arc $[p, q]$ is r, which is possible since there are at most a finite number of corners on ϕ^*. From (17) the arcs $[p, r]$ and $[r, q]$ must be straight lines. Pick a coordinate system (u, w) such that the axes are not parallel to either pr or rq, and such that the arc prq is described by a continuous function $w = f(u)$. We seek then, to minimize the functional

$$J(w) = \int_{u_p}^{u_r} \frac{\sqrt{1 + w'^2}}{V_1} dx + \int_{u_r}^{u_q} \frac{\sqrt{1 + w'^2}}{V_1} dx = J(pr) + J(rq), \tag{18}$$

where u_p, u_r and u_q are the abscissae of p, q and r respectively. The time-optimal curve, prq satisfies the Euler equations for the two integrands above, and in addition satisfies the Weirstrass-Erdmann (W-E) corner conditions at r. Since the Euler equations must be satisfied for the pr and rq for any arbitrary corner point, the variation of the functional is simply given by terms depending on the corner:[6]

$$\begin{aligned} \delta J(pq) &= F_{w'}|_{u=u_r-0} \delta w + (F - w' F_{w'})|_{u=u_r-0} \delta u \\ \delta J(qr) &= -F_{w'}|_{u=u_r+0} \delta w - (F - w' F_{w'})|_{u=u_r+0} \delta u, \end{aligned} \tag{19}$$

which lead to the Weierstrass-Erdmann corner conditions:

$$\begin{aligned} (F_{w'})|_{u=u_r-0}^{u=u_r+0} &= 0 \\ (F - w' F_{w'})|_{u=u_r-0}^{u_r=c+0} &= 0. \end{aligned} \tag{20}$$

Let the two straight lines determined from the Euler equations for pr and rq be $w = a_1u + b_1$ and $w = a_2u + b_2$ respectively. The W-E conditions then reduce to:

$$\frac{a_1}{\sqrt{1 + a_1^2}} = \frac{a_2}{\sqrt{1 + a_2^2}}$$
$$\frac{1}{\sqrt{1 + a_1^2}} = \frac{1}{\sqrt{1 + a_2^2}}. \tag{21}$$

It is clear that $a_1 = a_2$ and $b_1 = b_2$ is the unique solution of the above equations. This implies that pq is a single straight line segment, and $[m, n]$ cannot have a corner point on it. Therefore we have proved that any closed subarc of ϕ^* that does not intersect C is a straight line. We now generalize this to any subarc. Any open or half-open subarc which is a subset of a closed subarc that does not intersect C, is clearly also a straight line segment. The only non-trivial case occurs when an open subarc (m, n) that does not intersect C has a point on C as one of its limit points (say m). Now any closed subset of (m, n) is a straight line. We can pick $m' \notin C$ arbitrarily close to m such that $[m', n)$ is a straight line. Taking the limit as $m' \to m$, it is clear that (m, n) is also a straight line. $\square$

Proof of Theorem 1: Assume that ϕ^* traverses C. Since ϕ_c^* is a connected, closed set by Lemma 3, and portions of ϕ^* outside C are straight lines by Lemmas 4, the time-optimal trajectory must be a polygonal line with at most three segments, with any corners lying on C. We now show the conditions that govern any corners that may exist. We derive the condition for connection, and the condition for disconnection follows immediately by time-reversibility of optimal paths.

Consider an arc (i, q), such that $i \notin C$ and $q \in C$. Let $r \in (i, q)$ be the connection point. (i, r) and $[r, q)$ are straight line segments lying in C' and C respectively, by Lemmas 3 and 4. We consider two exhaustive cases. We will show that the conclusions agree when the cases overlap:

a. $x_q \geq x_k$

b. $x_q \leq x_k$

Consider case a. We claim that $x_r \geq x_k$. Let, if possible, $x_r < x_k$. Then the straight line segments ir and ir', where r' is the point on C such that $x_{r'} - x_k = x_k - x_r'$, are of equal length. Therefore, $t(i, r) = t(i, r')$ by Corollary 1. Since $l(r', q) < l(r, q)$, $t(i, r) + t(r, q) > t(i, r') + t(r', q)$. Therefore the arc (i, q) cannot be time-optimal. Therefore $x_r \geq x_k$. Now we derive the location of r.

The W-E conditions (20) govern the corner as in Lemma 4. The conditions must, however, be modified, since the velocities on (p, r) and (r, q) are not the same, and because the corner is constrained to lie on C. As before, the equations for the two segments from the Euler equations are:

$$y = a_1x + b_1, x \in (x_p, x_r) \tag{22}$$
$$y = a_2y + b_2, x \in (x_r, x_q). \tag{23}$$

For this case, the variations δx and δy in (20) are not independent, but are related by the constraint $r \in C$. C is given by the equation $y = x \stackrel{\triangle}{=} \Psi(x)$, this is equivalent to the condition $\delta y = \Psi' \delta x$. The single modified Weirstrass-Erdmann condition for this case is:

$$(F_y \Psi' + F - y' F_{y'})|_{x = x_r - 0}^{x = x_r + 0} = 0. \tag{24}$$

This reduces to

$$\frac{1 + y'}{\sqrt{1 + y'^2}}\bigg|_{x_r - 0} = \gamma \frac{1 + y'}{\sqrt{1 + y'^2}}\bigg|_{x_r + 0}. \tag{25}$$

Now $(r, q) \in C$ implies $a_2 = 1$ and $b_2 = 0$. Therefore we may write (25) as

$$\frac{1 + a_1}{\sqrt{1 + a_1^2}} = \sqrt{2}\gamma. \tag{26}$$

Now, from geometry $a_1 = \tan(\beta)$ in Fig. (1) (measured positive counterclockwise), therefore:

$$\sin(\beta) + \cos(\beta) = \sqrt{2}\gamma. \tag{27}$$

From the geometry of the situation, it is clear that $\alpha = \beta + \frac{\pi}{4}$, which leads to:

$$\sin(\alpha) = \sin(\beta + \frac{\pi}{4}) = \frac{\sin(\beta) + \cos(\beta)}{\sqrt{2}} = \gamma, \tag{28}$$

or

$$\sin(\alpha) = V_1/V_2. \tag{29}$$

Similarly, it can be shown that case b. also yields (29), with $x_r \leq x_k$. The overlapping case, when $q = k$ occurs when $a_1 = -1$, yielding the limiting case $\gamma = 0$. If $q \neq k$, then r, the corner, is on the same side of k as q. For every nonzero γ, it follows that ϕ^* cannot contain k and k'. The time-optimal trajectory is therefore either the straight line between i and f or the unique polygonal line with at most two corners, and vertex angles of $\pi/2 + \alpha$ on C, that does not include k and k'. The regions $S_2(X_i)$ and $S'_2(X_i)$ are defined by explicitly comparing the traversal times for the alternate candidate optimal paths. $\square$

Proof of Lemma 5: For the $p = \infty$ case, we consider two exhaustive cases, we will show that the conclusions agree when they overlap.

$$
\begin{aligned}
&\text{a. } |x_m - x_n| \geq |y_m - y_n| \\
&\text{b. } |x_m - x_n| \leq |y_m - y_n|.
\end{aligned}
\tag{30}
$$

Let case a hold. Since v_x and v_y are independently constrained to a magnitude of V_1 in C', we must have $t(m, n) \geq |x_m - x_n|/V_1$. Let the *canonical trajectory* $\psi^*(m, n)$ be defined by the velocity profile

$$
\begin{aligned}
(v_x, v_y) &= (V_1, V_1),\, t < |y_m - y_n|/V_1 \\
&= (V_1, 0),\, |y_m - y_n|/V_1 < t \leq |x_m - x_n|/V_1.
\end{aligned}
\tag{31}
$$

It is clear that $\psi^*(m, n)$ is a minimum-time path, since it achieves the minimum possible time of $|x_m - x_n|/V_1$. Therefore any other time-optimal path, including subarc (m, n) of ϕ^*, must also have $v_x = V_1$ throughout. Since $x(t)$ is monotonically increasing on (m, n), it can be described by a continuous function $y = f(x)$. Since $v_x = V_1$, $v_y \leq V_1$, and $|y'(x)| = |v_y|/|v_x|$ wherever it is defined, we must have $|y'(x)| \leq 1$. Similarly, it follows that for case b, $|y'(x)| \geq 1$. Applying both conditions, $|y'(x)| = 1$ identically when $|x_m - x_n| = |y_m - y_n|$. Since the two cases exhaustively cover all situations, claim 1 is proved.

To prove the second claim, we again consider case a. first. Let, if possible, there exist a point q on an optimal trajectory outside the rectangle $B_\infty(m, n)$. Since $x(t)$ is monotonically increasing,

$$x_m \leq x_q \leq x_n. \tag{32}$$

Geometrically, it is clear that for a point outside $B_\infty(m, n)$ satisfying this condition, either

$$|y_q - y_m| > |x_q - x_m|, \tag{33}$$

or

$$|y_q - y_n| > |x_q - x_n|, \tag{34}$$

1110

Let, if possible, $|y_q - y_m| > |x_q - x_m|$. There are at most a finite number of corners in (m, n). Since $x(t)$ is monotonically increasing and the time-optimal path is described by a function $y = f(x)$, we can write $y_q - y_m$ as an integral computed piecewise on the differentiable portions of ϕ^*.

$$y_q - y_m \;=\; \int_{x_m}^{x_q} y'(x)dx$$

$$|y_q - y_m| \;\leq\; \int_{x_m}^{x_q} |y'(x)|dx \leq \int_{x_m}^{x_q} 1 dx$$

$$\leq\; x_q - x_m = |x_q - x_m|, \tag{35}$$

a contradiction. Therefore $y_q - y_m > x_q - x_m$ cannot hold. Similarly, $|y_q - y_n| > |x_q - x_n|$ also leads to a contradiction. Therefore q must lie within $B_\infty(m, n)$. The proof for case b. is similar. In the case when $|x_m - x_n| = |y_m - y_n|$, the rectangle $B_\infty(m, n)$ degenerates to a straight line of slope 1 or -1. The second claim is therefore proved.

To prove the third claim for case a, let ϕ be any curve between m and n such that its slope at no point exceeds 1 in magnitude. Therefore $x(t)$ on ϕ must be monotonically increasing, and $v_x = V_1$ at all points by Lemma 1. Since the vertical line through n does not intersect $B_\infty(m, n)$ except at n itself, $y(t) = y_n$ at the same time as $x(t) = x_n$. Therefore ϕ is also time-optimal. Again the proof for case b. is similar. $\square$

Proof of Theorem 2: First note that the conditions of Lemma 3 apply, and therefore $\phi_c^* \stackrel{\triangle}{=} C \cap \phi^*$ is closed and connected, and ϕ^* connects and disconnects from C at most once. Since ϕ^* intersects C, we can pick $p \in \phi_c^*$. We consider two exhaustive cases. We will show that the conclusions agree where the two cases overlap.

a. $x_p \geq x_k$

b. $x_p \leq x_k$.

Consider case a. We consider two possible sub-cases:

a.i) $\phi_c^* = p$, an isolated point

a.ii) $\phi_c^* = [m, n]$, a closed and connected interval on C, such that $p \in [m, n]$

First, we show that $\phi_c^* = p \Rightarrow p = k$. Let, if possible, $p \neq k$, therefore $x_p > x_k$. By hypothesis, $[i, p) \cap C$ is empty, since p is the only point in ϕ_c^*. Pick $\delta > 0$. Let $p_\delta \in [i, p)$ be such that $l(p, p_\delta) < \delta$. Then, since $[i, p_\delta]$ is also time-optimal for any δ, we have

$$t(i, p) = \lim_{\delta \to 0} t(i, p_\delta). \tag{36}$$

Let k_δ be the point of intersection of ik and the straight line with slope 1 passing through p_δ. Therefore we also have $l(k, k_\delta) < \delta$. Since $x_p > x_k$, $|x_p - x_i| > |y_i - y_p|$, applying Lemma 5, the time $\bar{t} = t(i, p_\delta)$ is the same as the time to traverse the path $(i, k_\delta, p_\delta) \in B_\infty(i, p_\delta)$. Thus

$$t(i, p) \;=\; \lim_{\delta \to 0} \bar{t} = \lim_{\delta \to 0} t(i, k_\delta) + t(k_\delta, p_\delta)$$

$$=\; \frac{|x_k - x_i|}{V_1} + \frac{|x_p - x_k|}{V_1}. \tag{37}$$

But the path (i, k, p), which traverses C over $[k, p]$ has a traversal time of

$$t = \frac{|x_k - x_i|}{V_1} + \frac{|x_p - x_k|}{V_2}, \tag{38}$$

which is less than $t(i, p)$ on ϕ^*, since $V_2 > V_1$, which implies that ϕ^* is not optimal, a contradiction. Therefore $p = k \Rightarrow \phi_c^* = p$. Applying time-reversibility, it is clear that we also have $p = k'$ in this case, therefore it also follows that ϕ_c^* can be an isolated point only when k and k' are coincident. Next, we consider the case a.ii when $p \in [m, n]$.

Since $x_p \geq x_k$. Assume without loss of generality m precedes p in the forward direction of traversal of ϕ^*. It is clear that $m = k$, by the same argument used to show that $p = k$ in the previous case, since if $m \neq k$, the time of traversal along (i, k, m) will be less than the time of traversal along any path (i, m) that does not intersect C. Applying the same argument to n, the point where the ϕ_c^* leaves C, it is clear that $n = k'$.

For case b, it can be similarly shown that the first subcase yields $p = k = k'$ and the second subcase yields $m = k$, $n = k'$ as here. The conclusions are clearly in agreement when $p = k$.

Therefore, if ϕ_c^* is non-empty, it must connect at k and disconnect at k'. Finally, we show that ϕ^* is unique. Consider $B_\infty(i, k_\delta)$ and $B_\infty(f, k'_\delta)$, where $l(k_\delta, k) < \delta$ and $l(k'_\delta, k') < \delta$ as $\delta \to 0$. Since ik and fk' have slope -1, by Lemma 5, in the limit, both rectangles must degenerate to straight lines. Therefore ϕ^* is unique.

ϕ^* is therefore either a curve that does not intersect C and obeys the conditions of Lemma 5, or it is a polygonal line with at most two corners, with vertex angles of $\pi/2$ on C. The sets S_∞ and S'_∞ are defined by explicitly comparing the traversal times for the alternate candidate optimal paths. $\square$

Proof of Lemma 6: Consider the coordinate transformation

$$\begin{bmatrix} u \\ v \end{bmatrix} = \begin{bmatrix} 1/\sqrt{2} & 1/\sqrt{2} \\ -1/\sqrt{2} & 1/\sqrt{2} \end{bmatrix} \begin{bmatrix} x \\ y \end{bmatrix}, \tag{39}$$

which represents a rotation of the axes by $\pi/4$ of the position plane. The velocities v_x and v_y transform similarly to v_u and v_v. Now writing out the first equation of constraint (1) with $p = 1$, and substituting v_u and v_v for v_x and v_y, we get

$$\begin{aligned} v_u &< V_1/\sqrt{2} \\ -v_u &< V_1\sqrt{2} \\ v_v &< V_1\sqrt{2} \\ -v_v &< V_1\sqrt{2}, \end{aligned} \tag{40}$$

which reduces to

$$\begin{aligned} |v_u| &< V_1/\sqrt{2} \\ |v_v| &< V_1/\sqrt{2}, \end{aligned} \tag{41}$$

which we recognize as the constraint for the ∞ case with $V_1/\sqrt{2}$ instead of V_1. (It is clear that this is due to the fact that both the $p = 1$ and $p = \infty$ norm balls are squares). We need conditions in the (x, y) coordinates for $p = 1$ that will be equivalent to the conditions (30) in Lemma 5. Let (m, n) be a subarc of ϕ^* that does not intersect C. Equation (30) for the (u, v) coordinate system yields the two exhaustive cases case a. $|u_n - u_m| \geq |v_n - v_m|$

case b. $|u_n - u_m| \leq |v_n - v_m|$

Assume, without loss of generality, that $x_n > x_m$. The two corresponding cases in the (x, y) coordinate system are

case a. $y_n \geq y_m$

case b. $y_n \leq y_m$

Let case a. hold. To derive the slope condition, we transform the corresponding slope condition from Lemma 5, case a,

$$v_u(t) = V_1/\sqrt{2} \ \forall t$$

$$|v_v(t)|/|v_u(t)| \overset{\triangle}{=} |\lambda(t)| \leq 1 \ \forall t, \tag{42}$$

where $\lambda(t)$ is the slope of ϕ^* in the (u, v) frame wherever it is defined. Transforming back to the original coordinates, we arrive at:

$$v_x(t) = (v_u(t) - v_v(t))/\sqrt{2} \ = \ V_1(1 \pm \lambda(t))/2 \geq 0$$
$$v_y(t) = (v_u(t) + v_v(t))/\sqrt{2} \ = \ V_1(1 \mp \lambda(t))/2 \geq 0, \tag{43}$$

which is equivalent to saying that y and x must be nondecreasing functions of time.

The result for case b corresponds to case b in Lemma 5, and y on (m, n) in this case must be a non-decreasing function of time.

Similarly, we prove claim 2 by transforming $B_\infty(m, n)$ into the (x, y) coordinates and obtain the condition that (m, n) must stay within the rectangle $B_1(m, n)$. The proof of claim 3 again carries over directly from Lemma 5. $\square$

Proof of Theorem 3: First note that the conditions of Lemma 3 apply, and therefore $\phi_c^* \overset{\triangle}{=} C \cap \phi^*$ is closed and connected, and ϕ^* connects and disconnects from C at most once. Since ϕ^* intersects C, we can pick $p \in \phi_c^*$. We consider two exhausitive cases. We will show that the conclusions agree where the two cases overlap.

a. $x_p \geq x_k$

b. $x_p \leq x_k$.

Consider case a. We consider two possible sub-cases:

a.i) $\phi_c^* = p$

a.ii) $\phi_c^* = [m, n]$, a closed and connected interval on C, containing p.

We now derive conditions under which each case occurs.

We claim that $p \in (v, h) \subset \Rightarrow C\phi_c^* = p$. To prove this, let $p \in (v, h) \subset C$. Let if possible ϕ_c^* contain a point q other than p. Then it must contain the interval $[p, q]$ by Lemma 3. Since $p \in (v, h)$, this means there exists a point $c \in [p, q] \cap (v, h)$, $p \neq c$. Now note that for any point $c \in [m, n]$, the time of traversal $t(i, c)$ is the same, since

$$t(i, c) = |x_i - x_c|/V_1 + |y_i - y_c|/V_1 = (y_i - x_i)/V_1 = t(i, p). \qquad (44)$$

Therefore if the optimal path traverses C via an intermediate point c, the time of traversal would be $t(i, c) + t(c, p)$, which is obviously greater than the time $t(i, p)$ of traversal for a path from i to p that does not intersect C. Thus ϕ^* is suboptimal, a contradiction. Thus, p is the only point in ϕ_c^*.

Now consider the case when $p \notin (v, h)$. We claim that in this case, $p \in [m, n]$, such that $m = v$ or $m = h$, and $n = v'$ or h' (this includes the possibility that $m = n$, which is the other situation where p is an isolated point). Without loss of generality, assume $x_p \geq x_h$, since the conclusions for $x_p \leq x_v$ follow from symmetry.

First we show that $m = h$. Let, if possible $m \neq h$. Therefore (i, m) does not intersect C. Pick $\delta > 0$. Let m_δ be such that $l(m, m_\delta) < \delta$. Define h_δ to be the point of intersection of the straight line ih and the line through m_δ with slope 1. By Lemma 6 and Corollary 2, $\bar{t} = t(i, m)$ is

$$
\begin{aligned}
\bar{t} &= \lim_{\delta \to 0} t(i, h_\delta) + t(h_\delta, m_\delta) \\
&= \lim_{\delta \to 0} (|x_i - x_{h_\delta}| + |y_i - y_{h_\delta}|)/V_1 + (|x_{m_\delta} - x_{h_\delta}| + |y_{m_\delta} - y_{h_\delta}|)/V_1 \\
&= (|x_i - x_h| + |y_i - y_h|)/V_1 + (|x_m - x_h| + |y_m - y_h|)/V_1. \qquad (45)
\end{aligned}
$$

The time of traversal through (i, h, m), traversing C is

$$t = (|x_i - x_h| + |y_i - y_h|)/V_1 + (|x_m - x_h| + |y_m - y_h|)/V_2, \qquad (46)$$

which is clearly less than $\bar{t}$, since $V_2 > V_1$. Therefore ϕ^* is suboptimal, a contradiction. Therefore $m = h$.

Therefore, if ϕ_c^* contains points of C outside (v, h), it must connect at either v or h. By time-reversibility of ϕ^*, ϕ_c^* must disconnect at v' or h'.

Therefore, if ϕ_C^* is non-empty, it must be a point $p \in [v, h] \cap [v', h']$ or traverse C connecting at v or h and disconnecting at v' or h'. It is clear that in the latter case, since $(v, h) \notin \phi_c^*$ and $(v', h') \notin \phi_c^*$, and since $B_1(i, h)$, $B_1(i, v)$, $B_1(f, v')$ and $B_1(f, h')$ all degenerate to straight lines, that ϕ^* is unique.

ϕ^* is therefore either a curve that does not intersect C and obeys the conditions of Lemma 5, or it is the unique polygonal line with at most two corners and vertex angles of $3\pi/4$ on C, which does not contain (v, h) and (v', h'). The sets S_1 and S'_1 are defined by explicitly comparing the traversal times for the alternate candidate optimal paths. $\square$

PROBABILISTIC CONTROLLER ANALYSIS AND SYNTHESIS: THE METHOD OF HPD INSCRIPTION[*]

Hiroaki Fukuzawa[†] and Pierre T. Kabamba[‡]

Probabilistic control of LTI and linear periodic systems is considered. Here, the plant is subject to constant parametric uncertainty, where the uncertain parameters are assumed to be random variables with Gaussian distribution. The method is based on the fact that quadratic performance imposes linear constraints on the eigenvalues of the solution of a Lyapunov equation. The probability of performance is approximated by the content of the largest Gaussian HPD ellipsoid inscribed in the performance region. The probability of performance is optimized by inscribing the HPD ellipsoids into the performance region. An example is given.

INTRODUCTION

Probabilistic control is an approach to robust control where we assume that the plant is subject to parametric uncertainty and the uncertainty is described in probabilistic terms. This probabilistic description of the plant naturally arises when knowledge of plant parameters is acquired through experimental measurements.[1] Within this framework, system properties such as stability and performance are considered probabilistic. This leads to the following analysis and synthesis questions, respectively: For a given uncertain plant and a given controller, what is the probability of stability and performance? And for a given uncertain plant, how does one obtain a controller that maximizes the probability of stability and performance? These are the questions treated in this paper.

The computational complexity of exact robustness analysis has long been recognized,[2] and methods to overcome this difficulty belong mostly to two broad classes: worst-case bounds, and approximations, the latter of which encompasses probabilistic control. While the literature on worst-case bounds is very extensive (see, e.g. Refs.[3-6] and the references therein), probabilistic control has only recently received increasing attention. Representative articles in this area include Refs.[7-9], where Monte Carlo-based techniques are used to evaluate the probability of stability and performance based on the distribution of closed-loop eigenvalues. References[10,11] give the minimum

---

*   This research was funded by NASA grant number NRA-99-05-OSS-0077.

†   Graduate Student Research Assistant, Department of Aerospace Engineering, The University of Michigan, 1320 Beal Ave., Ann Arbor, Michigan 48109-2140. E-mail: hfukuzaw@engin.umich.edu.

‡   Professor, Department of Aerospace Engineering, The University of Michigan, 1320 Beal Ave., Ann Arbor, Michigan 48109-2140. E-mail: kabamba@engin.umich.edu.

number of samples required to ensure a prescribed level of confidence in probability estimates, while Refs.[12,13] suggest using uniform distributions to generate parameter samples when there is no a priori knowledge of the parameter distribution. Sensitivity of controller design with respect to uncertain parameters has been studied before the emergence of probabilistic control. For instance, Refs.[14–16] present methods to reduce or optimize the sensitivity norms in controller design. Other relevant research includes covariance averaging[17] and upper-bounding of the variance of uncertain parameters for stability robustness[18]. Probabilistic control is also closely related to linear eigenvalue sensitivity analysis, which has been applied in Refs.[19,20] to derive covariance matrices of eigenvalues.

In Ref.[21] the authors consider probabilistic control of linear time-invariant systems with quadratic performance indices and introduce the HPD inscription method to give approximate solutions to the analysis and synthesis problems. This paper describes the method of HPD inscription in a more general context since we have realized that the method is quite general in nature and have potential capacities to be applied to broader control problems. In addition to linear time-invariant systems we also consider probabilistic control of linear periodic systems in this paper.

The remainder of the paper is as follows. The next section describes the HPD inscription method in a general context. The following sections treats probabilistic control of LTI and linear periodic systems followed by an illustrative example with a second order LTI system. We also discuss possible application to orbital controls. The paper concludes with some remarks.

HPD INSCRIPTION METHOD

For a system, let θ be the vector of random parameters and K be the design variable. Define λ as a function of θ and K: $\lambda(\theta, K)$. Then, λ is a random vector. We call λ a *performance vector* if there exists a set Λ such that the system has acceptable performance if and only if $\lambda \in \Lambda$. We call Λ the *performance set*. Let the probability density function of λ be $f(\lambda)$. Then,

$$\Pr[\text{performance}] = \int_\Lambda f(\lambda)d\lambda. \tag{1}$$

However, this integral is in general difficult to evaluate. Our analysis and synthesis methods give approximate evaluation and maximization of this integral, respectively, under the following two assumptions: 1) λ is Gaussian, 2) Λ is a set bounded by linear inequalities.

Let λ be an n dimensional Gaussian random vector with mean $\bar{\lambda}$ and covariance R_λ, i.e., with probability density function

$$f(\lambda) = \frac{1}{(2\pi)^{n/2}|R_\lambda|^{1/2}} \exp\{-\frac{1}{2}(\lambda - \bar{\lambda})^T R_\lambda^{-1}(\lambda - \bar{\lambda})\}. \tag{2}$$

The *highest posterior density (HPD) region of content* ρ is defined as

$$H^\rho = \{\lambda \in \mathbb{R}^n : (\lambda - \bar{\lambda})^T R_\lambda^{-1}(\lambda - \bar{\lambda}) \leq F^{-1}(\rho|n)\}, \tag{3}$$

where $F^{-1}(\rho|n)$ is the inverse of the χ^2 *cumulative density function* for probability ρ with n degree of freedom.[22,23] Hence, the HPD regions are concentric coaxial similar ellipsoidal regions satisfying

$$\Pr[\lambda \in H^\rho] = \int_{H^\rho} f(\lambda)d\lambda = \rho. \tag{4}$$

For analysis, we find the largest HPD region included in the performance set, then approximate the probability of performance by its content, i.e.,

$$\Pr[\text{performance}] \approx \max_{H^\rho \subseteq \Lambda} \rho. \tag{5}$$

For synthesis, we approximately maximize the probability of performance in terms of K by maximizing the approximate probability obtained in the analysis part, i.e.,

$$\max_K \Pr[\text{performance}] \approx \max_K \max_{H^\rho \subseteq \Lambda} \rho. \tag{6}$$

Visually this process inscribes the HPD regions into the performance set.

Let the performance set Λ have the following structure

$$\Lambda = \bigcap_{i=1}^{m} \Lambda_i, \quad \Lambda_i = \{\lambda \in \mathbb{R}^n : c_{0i} + c_i^T \lambda \leq 0\}. \tag{7}$$

Then, it can be shown that if $\bar{\lambda} \in \Lambda$,

$$\max_{H^\rho \subseteq \Lambda} \rho = F(\mu|n), \tag{8}$$

where

$$\mu = \min_i \frac{(c_{0i} + c_i^T \bar{\lambda})^2}{c_i^T R_\lambda c_i}, \quad 1 \leq i \leq m, \tag{9}$$

and $F(\mu|n)$ is the χ^2 cumulative density function for value μ with n degrees of freedom, i.e.,

$$F(\mu|n) = \int_0^\mu \frac{t^{(n-2)/2} e^{-t/2}}{2^{n/2} \Gamma(n/2)} dt. \tag{10}$$

It follows that with $\bar{\lambda} \in \Lambda$ our approximations are

$$\Pr[\text{performance}] \approx F(\mu|n), \tag{11}$$

and

$$K_{\text{optimal}} = \arg \max_K \Pr[\text{performance}] \tag{12}$$

$$\approx \arg \max_K \mu. \tag{13}$$

The quantity μ in (9) represents minimum covariance-weighted distance between $\bar{\lambda}$ and the performance set boundary.[23] We have so far assumed that λ is Gaussian and reached the analysis and synthesis scheme (11) and (12). But we can rather define the analysis scheme (11) and synthesis scheme (12) as such independently of the distribution of λ so as to allow us to use the schemes for non-Gaussian performance vectors. After all it is quite reasonable to maximize minimum covariance-weighted distance between $\bar{\lambda}$ and the performance set boundary to increase the probability of performance. One way to obtain mean and covariance of λ is by linearization, that is to approximate using mean and covariance of θ, $\bar{\theta}$ and R_θ, as

$$\bar{\lambda} \approx \lambda(\bar{\theta}), \tag{14}$$

$$R_\lambda \approx \left(\frac{\partial \lambda}{\partial \theta}\right)_{\bar{\theta}} R_\theta \left(\frac{\partial \lambda}{\partial \theta}\right)_{\bar{\theta}}^T. \tag{15}$$

LTI SYSTEMS

The symbols $\xi : N(\bar{\xi}, R_\xi)$ mean that ξ is a Gaussian random vector with mean $\bar{\xi}$ and covariance R_ξ. The notation vec(A) represents a column vector produced by stacking the columns of matrix A, trace(A) means trace of A. The matrices I and O are the identity and zero matrices, respectively, the sizes of which are occasionally indicated in their subscripts. The symbols $\otimes$ and $\oplus$ denote the Kronecker product and sum operators, respectively (See e.g. Ref.[24]).

Consider the continuous-time, linear, time-invariant state-space system

$$\dot{x} = A(\theta)x + B(\theta)u, \quad x \in \mathbb{R}^n, u \in \mathbb{R}^p, \tag{16}$$

where θ is a vector of random parameters such that $\theta : N(\bar{\theta}, R_\theta)$. The initial state, $x(0) = x_0$, is assumed to be random with mean 0 and covariance X_0. Define the quadratic performance index as

$$J = \int_0^\infty E_{x_0}[x^T Q x + u^T R u]\, dt, \tag{17}$$

where $E_{x_0}[\cdot]$ is the expectation operator with respect to x_0, and for some threshold τ let $J \leq \tau$ be acceptable performance. We use the constant state feedback control $u = Kx$, where the gain K is design variable.

With

$$x(t) = e^{(A+BK)t}x_0, \tag{18}$$

it is straightforward to see that

$$J = \text{trace}(X X_0), \tag{19}$$

where

$$X \triangleq \int_0^\infty e^{(A+BK)^T t}(Q + K^T R K)e^{(A+BK)t}\, dt. \tag{20}$$

If $A + BK$ is stable, then X is the symmetric solution of the Lyapunov equation

$$(A + BK)^T X + X(A + BK) + Q + K^T R K = O. \tag{21}$$

Let S be Cholesky decomposition factor of X_0 such that $X_0 = S^T S$, then

$$J = \text{trace}(X S^T S) = \text{trace}(S X S^T) = \text{trace}(P), \tag{22}$$

where $P \triangleq SXS^T$. Pre-multiplied by S and post-multiplied by S^T, Eq. (21) becomes

$$S(A + BK)^T S^{-1} S X S^T + S X S^T S^{-T}(A + BK)S^T$$
$$+ S(Q + K^T R K)S^T = O, \tag{23}$$

which can be written as

$$FP + PF^T + V = O, \tag{24}$$

where

$$F \triangleq S(A + BK)^T S^{-1}, \quad V \triangleq S(Q + K^T R K)S^T. \tag{25}$$

Thus, P is the symmetric solution of the Lyapunov equation (24). Conditions for stability and performance can be completely expressed in terms of eigenvalues of P, $\{\lambda_i, 1 \leq i \leq n\}$, as

$$\lambda_1 > 0, \lambda_2 > 0, \cdots, \lambda_n > 0, \sum_{i=1}^n \lambda_i \leq \tau. \tag{26}$$

Consequently, we can take $\lambda \triangleq [\lambda_1, \lambda_2, \ldots, \lambda_n]^T$ as a performance vector with the performance set

$$\Lambda = \{\lambda \in \mathbb{R}^n : \lambda_1 > 0, \lambda_2 > 0, \cdots, \lambda_n > 0, \sum_{i=1}^{n} \lambda_i \leq \tau\}. \tag{27}$$

Mean and covariance of λ can be approximated by linearization. In particular

$$\frac{\partial \lambda}{\partial \begin{bmatrix} \text{vec}(A) \\ \text{vec}(B) \end{bmatrix}} = LT, \tag{28}$$

$$L = \begin{bmatrix} L_1 & L_2 & \cdots & L_n \end{bmatrix}^T, \quad L_i = \text{vec}(v_i v_i^T), \tag{29}$$

$$T = -(F \oplus F)^{-1}\{(PS^{-T} \otimes S) + (S \otimes PS^{-T})M\}\begin{bmatrix} I_{n^2} & K^T \otimes I_n \end{bmatrix}, \tag{30}$$

where M is the permutation matrix such that $\text{vec}(A^T) = M\text{vec}(A)$, and v_i is the normalized eigenvector of P associated with λ_i.

EXAMPLE

Consider the second order system

$$\dot{x} = Ax + Bu, \quad x \in \mathbb{R}^2, u \in \mathbb{R}^1, \quad X_0 = \begin{bmatrix} 1 & 0 \\ 0 & 1 \end{bmatrix},$$

where the elements of A and B matrices are jointly Gaussian such that

$$\begin{bmatrix} \text{vec}(A) \\ \text{vec}(B) \end{bmatrix} : N\left(\begin{bmatrix} \text{vec}(\bar{A}) \\ \text{vec}(\bar{B}) \end{bmatrix}, R_{AB}\right), \quad \bar{A} = \begin{bmatrix} 1 & 2 \\ 3 & 4 \end{bmatrix}, \quad \bar{B} = \begin{bmatrix} 1 \\ 1 \end{bmatrix},$$

$$R_{AB} = \begin{bmatrix} 0.0912 & 0.0987 & 0.0444 & 0.0336 & 0.0564 & 0.0651 \\ 0.0987 & 0.1473 & 0.0762 & 0.0396 & 0.0633 & 0.1017 \\ 0.0444 & 0.0762 & 0.0597 & 0.0243 & 0.0405 & 0.0702 \\ 0.0336 & 0.0396 & 0.0243 & 0.0195 & 0.0324 & 0.0339 \\ 0.0564 & 0.0633 & 0.0405 & 0.0324 & 0.0600 & 0.0489 \\ 0.0651 & 0.1017 & 0.0702 & 0.0339 & 0.0489 & 0.0969 \end{bmatrix}.$$

Let the performance specification be

$$J = \int_0^\infty E_{x_0}[x^T \begin{bmatrix} 1 & 0 \\ 0 & 1 \end{bmatrix} x + (0.5)u^2] \, dt \leq 10.$$

The performance set is in this case

$$\Lambda = \{\lambda \in \mathbb{R}^2 : \lambda_1 > 0, \lambda_2 > 0, \lambda_1 + \lambda_2 \leq 10\}.$$

Let

$$\hat{p} = \text{approx. of Pr[performance] by HPD inscription method,}$$

$$\hat{p}_{MC} = \text{approx. of Pr[performance] by 10,000 Monte Carlo simulation.}$$

We obtained the following results. With initial non-optimal gain $K_{ini} = [-2, -5]$,

$$\hat{p} = 0.2214, \qquad \hat{p}_{MC} = 0.6676.$$

With optimal gain $K_{opt} = [-7.8367, -16.4565]$,

$$\hat{p} = 0.9925, \qquad \hat{p}_{MC} = 0.9323.$$

Figures 1–4 compare HPD regions and distributions of Monte Carlo samples of λ.

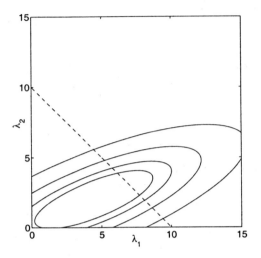

Figure 1: Gaussian HPD regions of contents 0.5, 0.7, 0.9, and 0.99 with non-optimal gain K_{ini}

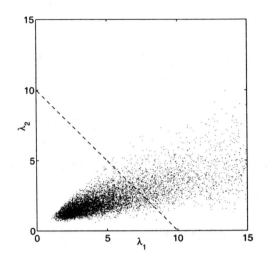

Figure 2: Distribution of 10,000 Monte Carlo sample with non-optimal gain K_{ini}

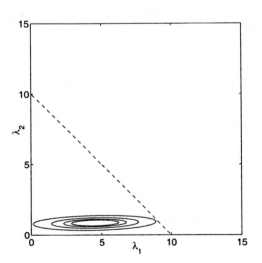

Figure 3: Gaussian HPD regions of contents 0.5, 0.7, 0.9, and 0.99 with optimal gain K_{opt}

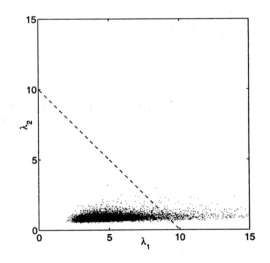

Figure 4: Distribution of 10,000 Monte Carlo sample with optimal gain K_{opt}

LINEAR PERIODIC SYSTEMS

Consider the continuous-time, linear, periodic state-space system

$$\dot{x} = A(t;\theta)x + B(t;\theta)u, \quad x \in \mathbb{R}^n, u \in \mathbb{R}^p, \tag{31}$$

where θ is a vector of random parameters such that $\theta : N(\bar{\theta}, R_\theta)$. The initial state, $x(0) = x_0$, is assumed to be random with mean 0 and covariance X_0. The time-varying matrices A and B are periodic with period T, i.e., $A(t+T;\theta) = A(t;\theta)$, $B(t+T;\theta) = B(t;\theta)$ for all t and θ. Let $\Phi(t_2, t_1;\theta)$ be the state transition matrix of $A(t;\theta)$. We use the *Sampled State Periodic Hold* (SSPH) control of the form[25, 26]

$$u(t) = B(t;\bar{\theta})^T \Phi^T((k+1)T, t;\bar{\theta})Kx(kT), \quad t \in [kT, (k+1)T], \tag{32}$$

where the constant gain matrix K is design variable. Define the quadratic performance index as

$$J = \sum_{k=0}^{\infty} E_{x_0}[x^T(kT)(Q + K^T RK)x(kT)], \tag{33}$$

where $E_{x_0}[\cdot]$ is the expectation operator with respect to x_0, and for some threshold τ let $J \leq \tau$ be acceptable performance.

Apply the variation of constants formula to (31) to obtain

$$x((k+1)T) = \Phi((k+1)T, kT;\theta)x(kT) + \int_{kT}^{(k+1)T} \Phi((k+1)T, \tau;\theta)B(\tau;\theta)u(\tau)\,d\tau. \tag{34}$$

From (32), (34), and the identity $\Phi(t+T, t_0+T) = \Phi(t, t_0)$, for all t, t_0, we obtain

$$x((k+1)T) = [\Phi(T, 0;\theta) + W(T, 0;\theta)K]x(kT), \tag{35}$$

where

$$W(t_2, t_1;\theta) \triangleq \int_{t_1}^{t_2} \Phi(t_2, \tau;\theta)B(\tau;\theta)B^T(\tau;\bar{\theta})\Phi^T(t_2, \tau;\bar{\theta})\,d\tau. \tag{36}$$

Eq. (35) represents a discrete-time, linear, time-invariant system. The matrix $\Psi \triangleq \Phi(T, 0;\theta) + W(T, 0;\theta)K$ is the closed-loop monodromy matrix of (31) with the control (32). Stability of the closed loop system is determined by the eigenvalues of Ψ. Note that SSPH control is a fairly general control method in the sense that it can arbitrarily assign the eigenvalues of the monodromy matrix, if and only if the pair $(\Phi(T, 0;\theta), W(T, 0;\theta))$ is controllable.[26]

In the following, let $\Phi(T, 0;\theta)$ and $W(T, 0;\theta)$ be simply denoted by $\Phi(\theta)$ and $W(\theta)$, respectively. With

$$x(kT) = (\Phi + WK)^k x_0, \tag{37}$$

it is straightforward to see that

$$J = \text{trace}(XX_0), \tag{38}$$

where

$$X \triangleq \sum_{k=0}^{\infty} ((\Phi + WK)^k)^T (Q + K^T RK)(\Phi + WK)^k. \tag{39}$$

If $\Phi + WK$ is stable, then X is the symmetric solution of the discrete-time Lyapunov equation

$$(\Phi + WK)^T X(\Phi + WK) - X + (Q + K^T RK) = O. \tag{40}$$

Let S be Cholesky decomposition factor of X_0 such that $X_0 = S^T S$, then,

$$J = \text{trace}(X S^T S) = \text{trace}(S X S^T) = \text{trace}(P), \tag{41}$$

where $P \triangleq S X S^T$. Pre-multiplied by S and post-multiplied by S^T, Eq. (40) becomes

$$S(\Phi + WK)^T S^{-1} S X S^T S^{-T} (\Phi + WK) S^T - S X S^T$$
$$+ S(Q + K^T RK) S^T = O, \tag{42}$$

which can be written as

$$FPF^T - P + V = O, \tag{43}$$

where

$$F \triangleq S(\Phi + WK)^T S^{-1}, \quad V \triangleq S(Q + K^T RK) S^T. \tag{44}$$

Thus, P is the symmetric solution of the Lyapunov equation (43). Conditions for stability and performance can be completely expressed in terms of eigenvalues of P, $\{\lambda_i, 1 \le i \le n\}$, as

$$\lambda_1 > 0, \lambda_2 > 0, \cdots, \lambda_n > 0, \sum_{i=1}^{n} \lambda_i \le \tau. \tag{45}$$

Consequently, we can take $\lambda \triangleq [\lambda_1, \lambda_2, \ldots, \lambda_n]^T$ as a performance vector with the performance set

$$\Lambda = \{\lambda \in \mathbb{R}^n : \lambda_1 > 0, \lambda_2 > 0, \cdots, \lambda_n > 0, \sum_{i=1}^{n} \lambda_i \le \tau\}. \tag{46}$$

Mean and covariance of λ can be approximated by linearization. In particular

$$\frac{\partial \lambda}{\partial \begin{bmatrix} \text{vec}(\Phi) \\ \text{vec}(W) \end{bmatrix}} = LT, \tag{47}$$

where

$$L = \begin{bmatrix} L_1 & L_2 & \cdots & L_n \end{bmatrix}^T, \quad L_i = \text{vec}(v_i v_i^T), \tag{48}$$

$$T = -((F \otimes F) - I_{n^2})^{-1}((FPS^{-T} \otimes S) + (S \otimes FPS^{-T})M) \begin{bmatrix} I_{n^2} & K^T \otimes I_n \end{bmatrix}, \tag{49}$$

where M is the permutation matrix such that $\text{vec}(\Phi^T) = M\text{vec}(\Phi)$, and v_i is the normalized eigenvector of P associated with λ_i.

APPLICATION TO ORBITAL CONTROL

Potential applications include orbital controls where there are uncertain parameters. Orbital motion is generally described by a set of nonlinear differential equations

$$\dot{x} = g(x, u; \theta), \tag{50}$$

where θ is the vector of uncertain parameters. If (50) is linearized around a nominal periodic solution $(x_0(t), u_0(t))$ such that

$$\dot{x}_0 = g(x_0, u_0; \bar{\theta}), \tag{51}$$

then, resulting system is a linear periodic system with uncertain parameters

$$\delta\dot{x} = \frac{\partial g(x_0, u_0; \theta)}{\partial x} \delta x + \frac{\partial g(x_0, u_0; \theta)}{\partial u} \delta u + \frac{\partial g(x_0, u_0, ; \bar{\theta})}{\partial \theta} \delta\theta. \tag{52}$$

CONCLUSIONS AND REMARKS

A Method has been proposed for probabilistic controller analysis and synthesis. It was shown that the method can be applied to control of LTI systems and linear periodic systems with quadratic performance indices. There, the probability of performance was approximately maximized. The example clearly showed improvement of the probability of performance as a result of this approximate optimization. An application to orbital controls was also discussed. Note that one way to obtain mean and covariance of the performance vector is by linearization. However, this way of approximating mean and covariance can become problematic during optimization process in some cases. As an alternative there is a way that uses the statistics of a sample set. Validity and possibility of this alternative way especially in connection with the optimization process should be studied more deeply in the future.

References

[1] J. R. Taylor, *An Introduction to Error Analysis: The Study of Uncertainties in Pysical Measurements, Second Edition,* University Science Books, 1997.

[2] R. P. Braatz, P. M. Young, J. C. Doyle, and M. Morari, "Computational Complexity of μ Calculation, "*IEEE Transactions on Automatic Control,* IEEE, vol. 39, no. 5, pp. 1000-1002, May 1994.

[3] P. Dorato and R. K. Yedavalli, ed., *Recent Advances in Robust Control,* IEEE Press, 1990.

[4] J. Ackermann, *Robust Control: Systems with Uncertain Physical Parameters,* Springer-Verlag, 1993.

[5] B. R. Barmish, *New Tools for Robustness of Linear Systems,* Macmillan, 1994.

[6] K. Zhou with J. Doyle and K. Glover, *Robust and Optimal Control,* Prentice Hall, 1995.

[7] R. F. Stengel and L. R. Ray, "Stochastic Robustness of Linear Time-Invariant Control Systems, "*IEEE Transactions on Automatic Control,* vol. 36, no. 1, pp. 82-87, January 1991.

[8] L. R. Ray and R. F. Stengel, "A Monte Carlo Approach to the Analysis of Control System Robustness, "*Automatica,* vol. 29, no. 1, pp. 229-236, 1993.

[9] R. F. Stengel, L. R. Ray, and C. I. Marrison, "Probabilistic Evaluation of Control System Robustness, "*Int. J. Systems Sci.,* vol. 26, no. 7, pp. 1363-1382, 1995.

[10] P. Khargonekar and A. Tikku, "Randomized Algorithms for Robust Control Analysis and Synthesis Have Polynomial Complexity, "*Proceedings of the 35th Conference on Decision and Control,* IEEE, pp. 3470-3475, December 1996.

[11] R. Tempo, E. W. Bai and F. Dabbene, "Probabilistic Robustness Analysis: Explicit Bounds for the Minimum Number of Samples, "*Proceedings of the 35th Conference on Decision and Control,* IEEE, pp. 3424-3428, December 1996.

[12] E. Bai, R. Tempo, and M. Fu, "Worst-Case Properties of the Uniform Distribution and Randomized Algorithms for Robustness Analysis, "*Proceedings of the American Control Conference,* IEEE, vol. 1, pp. 861-865, 1997.

[13] B. R. Barmish and C. M. Lagoa, "The Uniform Distribution: A Rigorous Justification For its Use in Robustness Analysis, "*Proceedings of the 35th Conference on Decision and Control,* IEEE, pp. 3418-3423, December 1996.

[14] R. K. Yedavalli and R. E. Skelton, "Controller Design for Parameter Sensitivity Reduction in Linear Regulators," *Optimal Control Applications & Methods,* vol. 3, pp. 221-240, 1982.

[15] E. G. Gilbert, "Conditions for Minimizing the Norm Sensitivity of Characteristic Roots, "*IEEE Transactions on Automatic Control,* vol. AC-29, no. 7, pp. 658-661, July 1984.

[16] R.E. Skelton and D.A. Wagie, "Minimal Root Sensitivity in Linear Systems, "*J. Guidance Control Dynam.,* vol. 7, no. 5, pp. 570-574, 1984.

[17] S. R. Hall, D. G. MacMartin, and D. S. Bernstein, "Covariance Averaging in the Analysis of Uncertain Systems, "*IEEE Transactions on Automatic Control,* vol. 38, no.12, December 1993.

[18] E. Yaz, "Deterministic and Stochastic Robustness Measures for Discrete Systems, "*IEEE Transactions on Automatic Control,* vol. 33, no. 10, pp. 952-955, 1988.

[19] K. B. Lim and J. L. Junkins, "Probability of Stability: New Measures of Stability Robustness for Linear Dynamical Systems, "*The Journal of the Astronautical Sciences,* vol. 35, no. 4, pp. 383-397, October-December 1987.

[20] C. Pierre, "Root Sensitivity to Parameter Uncertainties: A Statistical Approach, "*Int. J. Control,* vol. 49, no. 2, pp. 521-532, 1989.

[21] H. Fukuzawa, P. T. Kabamba, "Probabilistic Controller Analysis and Synthesis for Quadratic Performance: The Method of HPD Inscription, "*Preceedings of 2002 American Control Conference,* May 2002.

[22] G. E. P. Box and G. C. Tiao, "Multiparameter Problems from a Bayesian Point of View, "*Annals of Mathematical Statistics,* vol. 36, pp. 1468-1482, 1965.

[23] E. S. Hamby, "A Bayesian Approach to Modeling and Control, "Ph. D Dissertation, University of Michigan, Ann Arbor, 1998.

[24] A. Graham, *Kronecker Products and Matrix Calculus With Applications,* Ellis Horwood, 1981.

[25] S. Bittanti, G. Guardabassi, C. Maffezzoni, and L. Silverman, "Periodic systems: Controllability and the Matrix Riccati Equation, "*SIAM J. Contr. and Optimiz.,* vol. 16, no. 1, pp. 37-40, 1978.

[26] P. T. Kabamba, "Monodromy Eigenvalue Assignment in Linear Periodic Systems, "Technical Notes and Correspondence, *IEEE Transactions on Automatic Control,* vol. AC-31, no. 10, pp. 950-952, October 1986.

CONSTELLATIONS

SESSION 11

Chair:

Michael Zedd
Naval Research Laboratory

The following papers were not available for publication:

AAS 03-181
"Analysis of Small Constellations of Satellites for Mid-Course Tracking," by Irene A. Budianto, Daniel Gerencser and Patrick J. McDaniel, USAF Research Laboratory (Paper Withdrawn)

AAS 03-183
"A New Satellite Constellation Augment System-Space Station Utilisation," by Qun Fang, Lvping Li, Zhiwei Du and Dou Xiaomu, Northwestern Polytechnical University, Xi'an, China (Paper Withdrawn)

WALKER CONSTELLATIONS TO MINIMIZE REVISIT TIME IN LOW EARTH ORBIT

Thomas J. Lang[*]

In the design of satellite constellations, continuous coverage of the region of interest is not always necessary. By allowing viewing gaps or revisit times, it may still be possible to meet mission objectives while substantially reducing the required number of satellites. In this study, a brute force technique is employed to find the best symmetric, Walker type constellation (including inclination) that minimizes the global maximum revisit time (MRT) for numbers of satellites between 3 and 40. A parametric approach is used to cover the LEO regime from 700 to 1500 km in altitude and for minimum elevation angles from 0 to 20 degrees.

INTRODUCTION

Much research has been done on optimizing constellations (minimizing the number of satellites to do the job) for continuous coverage of the earth (References 1-8). It was long ago noted that a single parameter, the radius of the coverage circle of a single satellite, as shown in Figure 1, could be used to decide how many such satellites would be required for continuous global coverage. It was not necessary to know the altitude of the satellite and minimum elevation angle required for viewing, just the coverage circle size that resulted from these parameters. This means that tables of optimal constellations could be created based only on the coverage circle size and then applied to any appropriate combination of altitude and elevation angle. This convention of optimizing constellations based on the value of coverage circle size (often called theta) has been adopted by many of the earlier referenced researchers who offer tables of optimized constellations for up to 100 or more satellites. A user can quickly find the best constellation for his continuous coverage needs simply by knowing the coverage circle

† Director, Astrodynamics Department, The Aerospace Corporation, Mail Stop M4/947, P.O. Box 92957, Los Angeles, California 90009-2957. E-mail: Thomas.j.lang@aero.org.

size of a single satellite. In fact, for continuous coverage, the rotation of the earth (or even the fact that it does rotate) is immaterial. Even the regression of the line of nodes of the orbits is not important since all orbits are assumed to be at the same altitude and inclination.

The same simplicity is not available for optimizing constellations to minimize revisit time (the time during which a point on the ground is not in view of a satellite). When a gap opens up on the surface of the earth, it is necessary to know the rotation rate of the earth, the satellite, and the line of nodes in order to determine when the point will be viewed again. Yet, analysis of constellations that minimize revisit time is significant because in many instances continuous coverage is not required or is not cost effective. This is particularly true if the sensors are to be located in low Earth orbit (LEO). As an example, for a constellation of satellites at 700 km altitude, at least 44 satellites are required to provide continuous global coverage (assuming zero degree minimum elevation angle for viewing). For monitoring slowly changing conditions such as weather or earth resources this is clearly excessive. Even for monitoring more quickly changing events, allowing viewing gaps (revisit times) of as little as twelve minutes would cut the required constellation size in half. Clearly, this would represent a significant cost saving.

The objective of this paper is to present, in parametric form, optimized constellations of LEO satellites that minimize the maximum revisit time to any point on the Earth's surface. The tables and plots presented should allow the mission planner to estimate the number of LEO satellites that would be required to obtain a certain value of maximum revisit time given the altitude and elevation angle constraints of the system.

METHOD

In optimizing the constellations for revisit time, the first problem is to decide what revisit time to use. Some popular metrics are the average revisit time, maximum revisit time, and the nth percentile revisit time. Contrary to intuition, optimizing for the best average revisit time may yield a poor maximum revisit time, and vice versa (see references 9 and 10). For the purposes of this study, the maximum revisit time (i.e., the largest coverage gap) or MRT to any point on the earth's surface has been used as the metric. This has been done to remain conservative in showing the benefit that can be obtained by reducing the constellation size from the continuous coverage case.

The constellation optimization has been limited to Walker type constellations. Walker constellations encompass all symmetric arrangements of satellites and can be described by the traditional parameters T, P, and F. Here T is the total number of satellites and P is the number of orbit planes at the same inclination and spread evenly around the equator. There are T/P satellites evenly distributed in each plane. The parameter F describes the phasing of satellites in one plane relative to those in another. The satellites in a plane are shifted in phase by F*360/T relative to the plane to the west.

To generate the results, a two step process was used. The first step was a coarse search mode utilizing a computer tool (CHCKCVR) developed by Eric George. This tool was utilized in brute force mode to evaluate all possible symmetric Walker arrangements, varying each one to obtain the inclination that gave the minimum MRT to a set of 50 points spread over the globe. This series of runs was performed for three LEO altitudes; 700 km, 1100 km and 1500 km. For each of these three altitudes, elevation angle constraints of 0, 10, and 20 degrees were analyzed. This search was conducted separately for both posigrade and retrograde orbits. The second step was to run these optimized Walker arrangements through a higher fidelity simulation tool (COVERIT) with more ground points (every one degree in latitude and two degrees in longitude) and smaller time steps (one minute) to obtain more accurate values of MRT. In most cases the two tools agreed to within a minute or two. In some cases the fine search revealed high MRT values and the offending constellation was so noted.

RESULTS

Figures 2 through 4 show the maximum revisit times (MRT) as a function of the number of satellites in the constellation for altitudes of 700, 1100, and 1500 km, respectively. On each plot the curves for minimum elevation angles of 0, 10, and 20 degrees are depicted. In Figures 5 through 7, the same data is plotted but with curves for each altitude on the same plot. The data is also shown in Tables 1 through 3 including the optimal T/P/F arrangement, the inclination, and the resulting value of MRT (in minutes). This has been done for both posigrade and retrograde constellations. The figures display the lower of the two MRT values.

General Observations

On initial inspection the following generalizations, not surprisingly, can be made
1. The MRT usually decreases as the number of satellites increases.
2. The MRT decreases as the altitude increases.
3. The MRT decreases as the elevation angle decreases.
4. There is generally good agreement between the MRT for the posigrade and retrograde cases. Exceptions are marked with an "X" in the tables.

There are notable exceptions to the first generalization, particularly at low numbers of satellites. In numerous cases raising the number of satellites may actually increase the MRT. This is nothing new and has been seen often in optimizing for continuous global coverage in earlier work. Here the trend is quite pronounced at the lower number of satellites that odd numbers of satellites in the constellation are better than even numbers. The reason for this can be seen by comparing constellations of three and four polar planes evenly spaced in right ascension of ascending node as shown in Figure 8. It can be seen that an even number of planes cause overlap, whereas an odd number of planes create a more dispersed arrangement.

Other than this, the curves are fairly well behaved, indicating that the method has produced consistent results that tend to support the claim of optimality. If the method were haphazard and the results non-optimal, we should expect to see more variability in the data. A researcher should be able to use these figures to estimate the MRT for a given number of LEO satellites at a specified altitude and with a specified minimum elevation angle. Even if the instantaneous coverage footprint of the satellite is not a circle (see Figure 9), the figures should enable one to estimate the MRT. For a circular footprint with a nadir hole (center footprint in Figure 9), as long as the time for a ground point to pass through the nadir hole is less than the MRT, the constellations given in the tables and figures of this paper will still be optimal. For the butterfly pattern sensor on the right of Figure 9, the constellations in this paper can still be used as an estimate. In this case the actual MRT values will be slightly higher since a ground point may have to wait until the satellite is nearly overhead in order to be viewed.

Comparison to Published Constellation Data

The constellations of Tables 1 through 3 can be compared to some that have been previously optimized at both ends of the MRT spectrum. We already have optimal constellations for continuous global coverage, or MRT=0, in references 1, 3, 7, and 8. For high values of MRT, we have optimal constellations of three to six satellites, obtained using a Genetic Algorithm approach, in references 9, 10, and 11.

Optimal constellations for continuous global coverage for the cases of interest are shown below using data taken from reference 8.

Altitude (km)	Elevation (deg)	Optimal Constellation T	P	F	Inclination
700	0	44	22	17	67
700	10	100	50	43	75
700	20	205			
1100	0	30	10	6	61
1100	10	59	59	27	69
1100	20	109	109	51	74
1500	0	22	11	8	58
1500	10	41	41	18	66
1500	20	74	74	68	80

The cases above with 41 or fewer satellites agree exactly with the current data in Figures 2 through 7 for MRT values of zero. The cases with higher numbers of satellites appear to lie comfortably on the current curves extended to MRT values of zero. Thus, in the limit as MRT approaches zero, we achieve the constellations that were optimized for continuous global coverage.

Moreover, it is interesting to note that, for low values of MRT (5 to 10 minutes or less), the best MRT constellation (in terms of T/P/F) is quite often a constellation that performed well as a continuous global coverage constellation in studies such as Reference 8. In these cases the MRT optimization leads to a slightly higher inclination (enabling it to see the poles with a slightly smaller coverage circle). To point out this trend, the posigrade constellations in Tables 1 through 3 have been annotated with a "#" sign if their T/P/F values match those of optimal or very nearly optimal continuous global coverage constellations from the tables of Reference 8. Notice that most of the posigrade constellations with MRT<10 fall into this category. The T/P/F arrangements that gave good performance for continuous global coverage also tend to be good MRT constellations as long as the resulting MRT value is less than 10 minutes. The implication is that for low values of MRT we already have a good first guess constellation (from the tables of optimal continuous coverage constellations) with an inclination that allows coverage of the poles. In other words, in cases where the desired MRT is 10 minutes or less, a researcher could look in the continuous global coverage tables of Reference 8 for constellations requiring slightly larger values of theta (coverage circle radius). These constellations, with inclinations to reach the poles, would typically offer good MRT performance.

Detailed Observations and Trends

Looking closer at the data in Tables 1 through 3 yields additional insights. The most obvious trend is that for most of the cases there is a preferred number of planes (P). This can be recognized by noting that, whenever the number of satellites is a multiple of the preferred P, the constellation has P planes. The data below shows these preferred values of P for each case.

Altitude (km)	Elevation (deg)	Theta (deg)	Preferred P
700	0	25.70	5
700	10	17.45	7
700	20	12.14	9
1100	0	31.47	3
1100	10	22.87	5
1100	20	16.73	7
1500	0	35.94	3
1500	10	27.13	none
1500	20	20.47	5

This means that the optimal constellations tend to favor arrangements with a specified number of planes. For the case of 700 km altitude and zero elevation angle, for example, note in Table 1 that whenever the total number of satellites is a multiple of five, the optimal arrangement consists of five polar planes (except for T = 40). Plotting this data in Figure 10 as P versus theta shows a significant trend. As the coverage circle radius, theta, increases in value, there is a trend toward the next lower odd integer value of preferred planes P. These constellations are polar such as those depicted in Figure 8. We have already discussed the significance of using an odd number of polar planes. As theta is made smaller, the number of such planes must be increased in order to allow the coverage circle to reach the regions of Earth between the planes. A simple approach would be to say that for P polar planes (where P is an odd integer), the angle α between the planes is $\alpha = 360/(2*P)$. The regions of Earth between the planes can be reached as long as theta is greater than α. This is a first order conclusion, of course, since the Earth is rotating beneath the planes. In reality, theta must be somewhat larger than α to accommodate this effect. The equation theta = α = 360/(2*P) has been plotted on Figure 10 as a curve. Note that it predicts fairly well the behavior of the preferred number of planes as a function of coverage circle radius, theta. When the curve drops to the next lower odd integer, so does the preferred number of planes P for the cases studied. This relationship can be used to design MRT optimal constellations with a low number of polar planes. A low number of planes is often desired to promote multiple satellites per launch or to simplify sparing (one spare per plane).

Another trend can be noted in Tables 1 through 3. For the most part, the optimal constellations tend to consist of one satellite per plane or T=P (other than the cases of preferred number of planes noted above). In many cases there are sequences where T=P and F=M or F=T-N (where M and N are even integers) even as T is varied. For example, note the sequences below.

Altitude (km)	Elevation (deg)	Sequence
700	0	For posigrade T= 11 to 28, F=T-4
700	10	For posigrade T= 12 to 40, F=T-6
		For retrograde T= 11 to 17, F=2
		For retrograde T= 18 to 34, F=T-8
700	20	For posigrade T= 12 to 40, F=T-10
		For retrograde T= 19 to 40, F=T-10
1100	0	For posigrade T= 13 to 30, F=T-4
		For retrograde T= 23 to 31, F=4
1100	10	For posigrade T= 11 to 31, F=T-6
1100	20	For posigrade T= 10 to 40, F=T-6
		For retrograde T= 17 to 39, F=T-8
		For retrograde T= 10 to 16, F=2
1500	0	For retrograde T= 5 to 12, F=2
1500	10	For posigrade T= 11 to 30, F=T-4
		For retrograde T= 6 to 12, F=2
		For retrograde T= 21 to 31, F=4
1500	20	For posigrade T= 11 to 39, F=T-6
		For retrograde T= 4 to 14, F=2
		For retrograde T= 16 to 29, F=T-6
		For retrograde T= 31 to 39, F=6

While these sequences are sometimes interrupted, there is still a strong trend present in the data. Understanding these trends would likely improve our knowledge of optimal constellations.

A final trend is the mirror image constellation. Posigrade and retrograde constellations are mirror images of each other if they have the same T and P, the sum of their inclinations is 180, and the sum of their F's is equal to P. This is the case for about a quarter of the constellations listed in Tables 1 through 3. That is, the optimization process found both the posigrade and retrograde twins to be optimal in about one fourth of the cases run. It should be noted that from a revisit time standpoint the constellations are not truly mirror images, because of Earth rotation.

CONCLUSIONS

In situations where continuous coverage is not strictly required, the number of satellites required for coverage can be reduced significantly by allowing a maximum viewing gap or maximum revisit time (MRT) to points on the ground. The included Figures and Tables of optimized Walker constellations are intended to enable a researcher to quickly estimate how many LEO satellites are required to achieve a specified MRT using satellites at a given altitude and with specified elevation angle limits. As part of this parametric examination a number of conclusions have been reached concerning the Walker constellations optimized for MRT.

1. The optimized constellations found agree well with those published earlier for continuous global coverage and for sparse constellation coverage.
2. For low values of MRT (5 to 10 minutes or less), the optimal Walker constellations are usually ones that performed well for continuous global coverage. That is, the satellite arrangements that performed well at MRT=0 are also good performers for MRT as large as 10 minutes. Many tables of these are available (e.g. References 7 and 8) and a user can search through them seeking minimal constellations of satellites that require slightly larger values of theta (coverage circle radius) than offered by his choice of altitude and elevation angle.
3. For low values of satellites (typically 3 to 6), the non-symmetric constellations generated by the Genetic Algorithm (References 9 and 10) usually offer better values of MRT. A good rule of thumb is that if the MRT is on the order of one revolution or more, the non-symmetric GA constellation will usually be better.
4. Posigrade and retrograde Walker constellations offer similar MRT performance.
5. There is generally a preferred number of polar planes (ranging from 3 to 9 for the cases studied) for these Walker constellations, that depends on the value of the coverage circle size, theta. The preferred number of polar planes can be predicted from the altitude and elevation angle employed. Using the preferred number of planes allows advantages in launching and maintaining the constellation.
6. There appears to be a pattern in many of the optimal constellations that is not yet understood. More study in this area could lead to a better knowledge of Walker constellation optimization.

ACKNOWLEDGEMENTS

This study could not have been performed without the use of a constellation optimization tool (CHCKCVR) developed in 1997 by Eric George also from The Aerospace Corporation. Much of the motivation for the current study is derived from his work during that period.

REFERENCES

1. Walker, J.G., "Continuous Whole Earth Coverage by Circular Orbit Satellite Patterns," Royal Aircraft Establishment Technical Report 77044, Mar. 1977.
2. Beste, D.C., "Design of Satellite Constellations for Optimal Continuous Coverage," *IEEE Transactions on Aerospace and Electronics Systems*, May 1978.
3. Ballard, A.H., "Rosette Constellations of Earth Satellites," *IEEE Transactions on Aerospace and Electronic Systems*, Vol. AES-16, No. 5, Sep. 1980, pp. 656-673.
4. Rider, L., "Optimized Polar Orbit Constellations for Redundant Earth Coverage," *The Journal of the Astronautical Sciences*, Vol. 33, Apr. - Jun. 1985, pp. 147-161.
5. Draim, J.E., "A Common Period Four-Satellite Continuous Global Coverage Constellation," AIAA preprint 86-2066-CP, AIAA/AAS Astrodynamics Conference, Williamsburg, VA, Aug. 18-20, 1986.
6. Adams, W.S. and Rider, L., "Circular Polar Constellations Providing Continuous Single or Multiple Coverage Above a Specified Latitude," *The Journal of the Astronautical Sciences*, Vol. 35, No. 2, Apr.-Jun. 1987, pp. 155-192.
7. Lang, T.J., "Optimal Low Earth Orbit Constellations for Continuous Global Coverage," paper AAS 93-597, AAS/AIAA Astrodynamics Specialist Conference, Victoria, B.C., Canada, Aug. 16-19, 1993.
8. Lang, T.J., and Adams W.S., "A Comparison of Satellite Constellations for Continuous Global Coverage", paper 1.4, IAF International Workshop on Mission Design and Implementation of Satellite Constellations, Toulouse, France, Nov 17-19, 1997.
9. George, E. "Optimization of Satellite Constellations for Discontinuous Global Coverage Via Genetic Algorithms," AAS 97-621, AAS/AIAA Astrodynamics Soecialist Conference, Sun Valley, ID, August4-7, 1997.
10. Williams, E.A., Crossley, W.A., and Lang, T.J., "Average and Maximum Revisit Time Trade Studies for Satellite Constellations using a Multiobjective Genetic Algorithm", AAS 00-139, AAS/AIAA Space Flight Mechanics Meeting, Clearwater, Florida, January 23-26, 2000.
11. Lang, T.J.,"A Parametric Examination of Satellite Constellations to Minimize Revisit Time for Low Earth Orbits Using a Genetic Algorithm", AAS 01-345, AAS/AIAA Astrodynamics Specialist Conference, Quebec City, Quebec, Canada, July 30 – Aug 2, 2001.

Figures

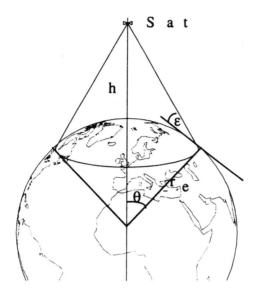

Figure 1. Definition of the Satellite Coverage Circle

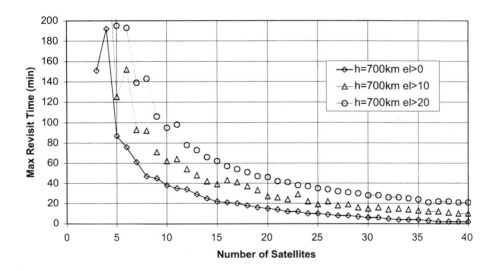

Figure 2. Maximum Revisit Time for 700 km Altitude

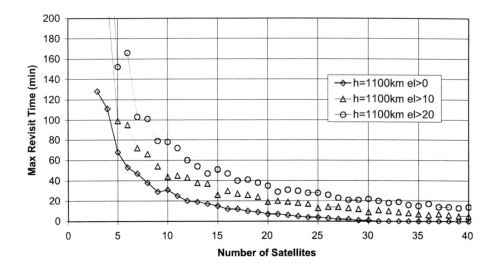

Figure 3. Maximum Revisit Time for 1100 km Altitude

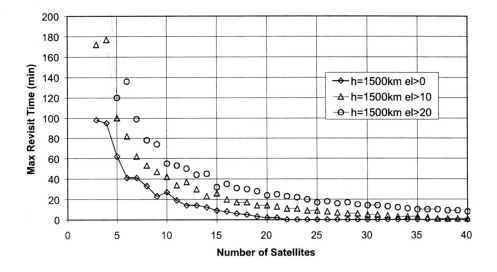

Figure 4. Maximum Revisit Time for 1500 km Altitude

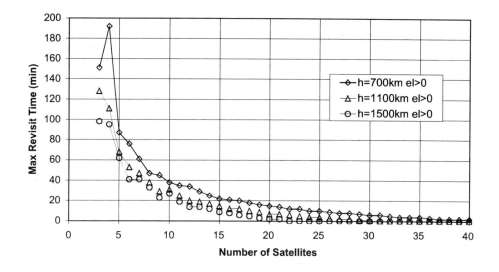

Figure 5. Maximum Revisit Time for 0 deg Elevation Angle

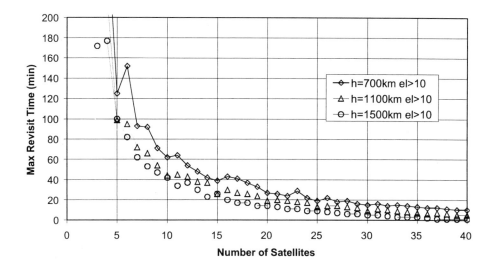

Figure 6. Maximum Revisit Time for 10 deg Elevation Angle

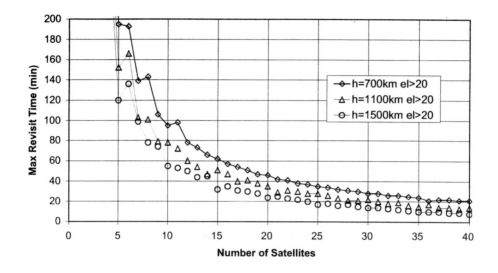

Figure 7. Maximum Revisit Time for 20 deg Elevation Angle

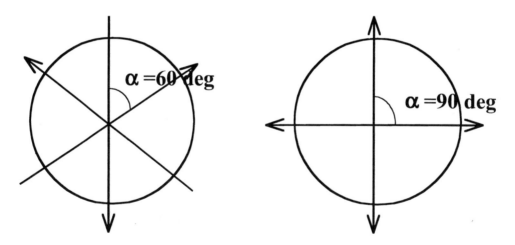

Three Polar Planes   Four Polar Planes

Figure 8. Odd vs Even Polar Planes (Top View)

Figure 9. Various Sensor Footprints

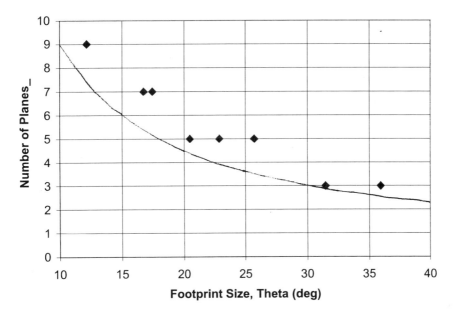

Figure 10. Preferred Number of Orbit Planes

TABLE 1. OPTIMAL WALKER CONSTELLATIONS AT 700 KM ALTITUDE

ELEVATION = 0 DEG

POSIGRADE					RETROGRADE				
T	P	F	Inc	MRT	T	P	F	Inc	MRT
3	3	0	90	156	3	3	0	96	151
4	4	2	66	192	4	4	2	114	200
5	5	3	90	87	5	5	3	90	87
6	6	4	66	76	6	6	4	114	85 X
7	7	5	66	61	7	7	2	114	83 X
8	8	4	74	51	8	8	2	114	47
9	9	7	66	51	9	9	2	110	45
10	5	1	84	38	10	5	2	96	38
11	11	7	78	35	11	11	2	114	35
12	6	3	66	42 X	12	12	2	112	34
13	13	9	72	29	13	13	2	112	32
14	14	10	80	25	14	7	3	114	31
15	5	0	88	22	15	5	0	90	22
16	16	12	74	21	16	8	3	114	22
17	17	13	82	20	17	17	10	114	20
18	18	14	72	19 #	18	9	3	114	18
19	19	15	76	16	19	19	27	90	31 X
20	5	0	86	15	20	5	0	90	15
21	21	17	72	14	21	21	12	114	15
22	22	18	76	12	22	11	3	112	15
23	23	9	66	12	23	23	13	114	13
24	24	20	74	10 #	24	12	5	114	13
25	5	0	88	10	25	5	0	90	10
26	13	8	66	9 #	26	13	5	112	11
27	27	11	66	8	27	27	16	114	9
28	28	24	80	8 #	28	28	4	102	10
29	29	12	66	7	29	29	17	110	7
30	5	2	90	6	30	5	1	92	7
31	31	13	72	6	31	31	4	106	7
32	16	11	70	5 #	32	16	5	110	5
33	33	14	68	4	33	33	19	110	5
34	17	12	66	4 #	34	17	5	112	4
35	5	0	90	4 #	35	5	0	90	4
36	18	13	68	3 #	36	18	5	112	3
37	37	16	66	2 #	37	37	21	106	4
38	19	14	70	2 #	38	19	5	110	3
39	39	17	66	2 #	39	39	22	112	2
40	20	15	74	2 #	40	20	5	106	2

ELEVATION = 10 DEG

POSIGRADE					RETROGRADE				
T	P	F	Inc	MRT	T	P	F	Inc	MRT
3	3	2	90	228.	3	3	2	94	225
4	4	0	78	289	4	4	0	90	289
5	5	4	90	125	5	5	4	92	193 X
6	6	0	74	152	6	6	0	106	157
7	7	3	90	93	7	7	3	90	93
8	8	0	84	92	8	8	0	90	93
9	9	7	74	71	9	9	7	90	84 X
10	10	4	82	62	10	5	1	92	73 X
11	11	9	74	64	11	11	2	106	90
12	12	6	78	54	12	12	2	104	54
13	13	5	86	54	13	13	2	106	48
14	7	6	88	42	14	7	4	92	42
15	15	9	82	39	15	15	2	104	46
16	16	8	78	43	16	16	2	104	44
17	17	11	78	46	17	17	2	106	41
18	18	10	74	37	18	18	10	90	37
19	19	13	78	33	19	19	11	90	36
20	20	14	82	27	20	20	12	90	32
21	7	0	88	26	21	7	0	90	26
22	22	16	80	24	22	22	14	94	29
23	23	17	76	29	23	23	13	106	30
24	24	18	78	22	24	24	16	90	27
25	25	19	82	19	25	25	17	90	25
26	26	20	78	22	26	13	3	106	27
27	27	21	80	18	27	27	19	90	22
28	7	0	88	19	28	7	0	90	19
29	29	23	78	16	29	29	21	90	21
30	30	24	82	15	30	30	8	98	22 X
31	31	25	78	16	31	31	23	90	19
32	32	26	80	14	32	32	24	90	18
33	33	27	80	15	33	33	8	96	18
34	34	28	78	14	34	34	26	90	17
35	35	29	80	13	35	7	0	90	13
36	36	30	80	12	36	36	6	98	17
37	37	31	78	12	37	37	6	100	17
38	38	32	78	11	38	38	6	100	15
39	39	33	80	10	39	39	8	98	14
40	40	34	82	10	40	40	6	100	14

ELEVATION = 20 DEG

POSIGRADE					RETROGRADE				
T	P	F	Inc	MRT	T	P	F	Inc	MRT
3	3	0	78	539 X	3	3	1	96	456
4	4	2	82	340	4	4	2	90	340
5	5	0	84	195	5	5	3	94	233 X
6	6	0	86	193	6	6	0	90	194
7	7	6	84	211 X	7	7	4	92	139
8	8	4	86	143	8	8	4	90	143
9	9	3	86	106	9	9	1	92	112
10	10	0	84	95	10	10	8	90	113 X
11	11	7	86	105	11	11	5	90	98
12	12	2	84	78	12	12	2	90	78
13	13	11	78	73	13	13	11	90	86 X
14	14	4	82	66	14	14	4	90	69
15	15	5	86	62	15	15	3	90	74 X
16	16	6	82	57	16	16	6	90	66
17	17	7	84	54	17	17	2	102	54
18	18	8	82	51	18	9	2	92	53
19	19	9	86	47	19	19	9	90	48
20	20	10	82	46	20	20	4	100	52
21	21	11	84	42	21	21	11	90	44
22	22	12	80	41	22	22	12	90	42
23	23	13	84	38	23	23	13	90	41
24	24	14	82	37	24	24	14	90	37
25	25	15	82	35	25	25	15	90	40
26	26	16	84	34	26	26	16	90	34
27	27	17	82	32 #	27	9	5	92	44 X
28	28	18	86	31	28	28	18	90	32
29	29	19	82	30	29	29	19	90	31
30	30	20	84	28	30	30	20	90	30
31	31	21	82	28	31	31	21	90	28
32	32	22	84	26	32	32	22	90	28
33	33	23	82	26	33	33	23	90	26
34	34	24	84	25	34	34	24	90	26
35	35	25	86	24	35	35	25	90	25
36	9	6	90	21	36	9	6	90	21
37	37	27	84	22	37	37	27	90	24
38	38	28	82	22	38	38	28	90	22
39	39	29	86	21	39	39	29	90	22
40	40	30	82	21	40	40	30	90	21

TABLE 2. OPTIMAL WALKER CONSTELLATIONS AT 1100 KM ALTITUDE

ELEVATION = 0 DEG

POSIGRADE					RETROGRADE				
T	P	F	Inc	MRT	T	P	F	Inc	MRT
3	3	2	90	128	3	3	1	100	178 X
4	4	0	60	142 X	4	4	2	120	111
5	5	3	62	68	5	5	3	120	77
6	3	0	90	54	6	3	0	92	53
7	7	5	60	52	7	7	2	118	47
8	8	4	66	42	8	8	2	120	38
9	3	1	90	30	9	3	1	92	29
10	5	2	60	34	10	10	2	120	31
11	11	4	60	26	11	11	7	120	25
12	3	2	88	20	12	3	2	92	31 X
13	13	9	60	19	13	13	4	120	20
14	14	10	76	17	14	7	3	120	17
15	3	0	86	15	15	3	1	92	15
16	16	12	68	14	16	8	3	120	12
17	17	13	80	12	17	17	10	116	12
18	18	14	60	11 #	18	9	3	118	10
19	19	8	60	9	19	19	11	116	9
20	10	7	60	8 #	20	10	3	120	7
21	21	17	76	7	21	21	12	116	7
22	22	18	72	6	22	22	16	120	6
23	23	19	68	6	23	23	4	114	5
24	24	20	80	5 #	24	24	4	102	4
25	25	21	74	4	25	25	4	110	4
26	26	16	60	3 #	26	26	10	118	3
27	27	23	74	2	27	27	4	104	2
28	28	24	74	2 #	28	28	4	104	2
29	29	25	76	1 #	29	29	4	102	1
30	30	26	70	1 #	30	30	4	102	1
31	31	9	62	1 #	31	31	4	100	0
32									
33									
34									
35									
36									
37									
38									
39									
40									

ELEVATION = 10 DEG

POSIGRADE					RETROGRADE				
T	P	F	Inc	MRT	T	P	F	Inc	MRT
3	3	0	86	261	3	3	1	100	273
4	4	0	68	254 X	4	4	2	110	220
5	5	3	86	99	5	5	1	94	102
6	6	2	68	118 X	6	6	4	112	95
7	7	5	68	72	7	7	5	100	89 X
8	8	4	82	66	8	8	2	90	71
9	9	7	68	62	9	9	2	112	54
10	5	4	88	44	10	5	2	92	45
11	11	5	80	49	11	11	2	110	45
12	12	6	74	44	12	12	2	110	43
13	13	7	84	38	13	13	2	108	41
14	14	8	68	37	14	14	2	112	37
15	5	0	88	26	15	5	0	90	27
16	16	10	86	30	16	8	3	112	30
17	17	11	74	27	17	17	10	112	28
18	18	12	80	26	18	9	3	112	27
19	19	13	72	24	19	19	11	110	26
20	5	0	90	19	20	5	0	90	19
21	21	15	86	20	21	21	15	90	21
22	22	16	78	19	22	22	6	100	21
23	23	17	76	18	23	23	13	112	19
24	24	18	80	17	24	12	3	112	18
25	5	3	88	13 #	25	5	3	90	13
26	26	20	74	14	26	26	6	102	16
27	27	11	68	15	27	27	6	98	14
28	28	22	72	13	28	14	5	110	15
29	29	12	68	12	29	29	6	100	12
30	5	0	88	10	30	5	4	90	9
31	31	25	80	11	31	31	18	110	12
32	16	11	68	10 #	32	32	6	98	10
33	33	14	68	9	33	33	19	108	11
34	17	12	68	8 #	34	17	5	112	9
35	5	0	88	7 #	35	5	0	90	7
36	18	13	68	6 #	36	18	5	108	9
37	37	16	70	7 #	37	37	6	98	7
38	19	14	68	6 #	38	19	5	108	7
39	39	17	68	5 #	39	39	28	102	14 X
40	5	0	88	5 #	40	5	2	94	5

ELEVATION = 20 DEG

POSIGRADE					RETROGRADE				
T	P	F	Inc	MRT	T	P	F	Inc	MRT
3	3	2	90	249	3	3	2	94	245
4	4	0	78	313	4	4	0	90	313
5	5	4	84	222 X	5	5	2	94	152
6	6	0	74	166	6	6	0	106	172
7	7	3	90	103	7	7	3	90	103
8	8	0	82	101	8	8	0	90	101
9	9	7	74	79	9	9	5	90	99
10	10	4	84	106 X	10	10	2	90	78
11	11	9	74	72	11	11	9	90	88
12	12	6	80	62	12	12	2	104	60
13	13	5	84	59	13	13	2	106	54
14	7	6	88	47	14	7	2	94	48
15	15	9	84	63 X	15	15	2	104	51
16	16	8	80	47	16	16	2	106	47
17	17	11	80	40	17	17	9	90	44
18	18	10	82	41	18	18	10	90	41
19	19	13	78	38	19	19	11	90	40
20	20	14	84	43 X	20	20	12	90	35
21	7	0	88	29	21	7	5	90	29
22	22	14	86	31	22	22	14	90	32
23	23	15	78	30	23	23	13	106	34
24	24	18	80	28	24	24	16	90	30
25	25	19	84	30	25	25	17	90	28
26	26	18	84	26	26	26	18	90	27
27	27	21	82	23	27	27	19	90	25
28	7	1	88	21	28	7	0	92	21
29	29	23	80	21	29	29	21	90	23
30	30	24	84	22	30	30	22	90	22
31	31	25	78	20	31	31	23	90	22
32	32	26	82	18	32	32	24	90	20
33	33	27	78	19	33	33	25	90	20
34	34	28	80	16	34	34	26	90	19
35	7	0	88	16	35	7	3	92	15
36	36	30	78	17	36	36	8	100	21
37	37	31	82	14	37	37	8	96	18
38	38	32	78	14	38	38	30	90	17
39	39	33	80	13	39	39	31	90	15
40	40	34	82	14	40	40	8	96	15

TABLE 3. OPTIMAL WALKER CONSTELLATIONS AT 1500 KM ALTITUDE

| ELEVATION = 0 DEG | | | | | | | | | | ELEVATION = 10 DEG | | | | | | | | | | ELEVATION = 20 DEG | | | | | | | | | |
| POSIGRADE | | | | | RETROGRADE | | | | | POSIGRADE | | | | | RETROGRADE | | | | | POSIGRADE | | | | | RETROGRADE | | | | |
T	P	F	Inc	MRT	T	P	F	Inc	MRT	T	P	F	Inc	MRT	T	P	F	Inc	MRT	T	P	F	Inc	MRT	T	P	F	Inc	MRT
3	3	1	90	99	3	3	1	92	98	3	3	0	90	180	3	3	0	100	172	3	3	1	90	227	3	3	1	98	298 X
4	4	2	80	95	4	4	0	120	111 X	4	4	2	64	215 X	4	4	0	116	177	4	4	2	70	280	4	4	2	90	279
5	5	3	60	64	5	5	2	120	62	5	5	1	64	100	5	5	1	98	102	5	5	1	90	120	5	5	1	90	120
6	3	2	90	41	6	3	2	90	41	6	6	4	64	82	6	3	2	96	94 X	6	6	2	70	136	6	6	2	110	142
7	7	5	60	49 X	7	7	2	120	41	7	7	5	64	67	7	7	2	116	62	7	7	3	74	99	7	7	3	92	105
8	8	4	68	39	8	8	2	120	33	8	8	4	72	56	8	8	2	112	53	8	8	2	80	78	8	8	2	90	80
9	3	0	88	23	9	3	1	96	24	9	9	7	64	56 X	9	9	2	114	47	9	9	7	70	74	9	9	7	90	93 X
10	5	2	60	28	10	10	2	110	27	10	5	4	90	43	10	10	2	116	42	10	5	2	90	55	10	5	2	90	55
11	11	7	60	19	11	11	7	120	20	11	11	7	74	34	11	11	2	116	38	11	11	5	84	55	11	11	2	110	53
12	3	0	90	16	12	3	2	92	14	12	12	8	64	40	12	12	2	110	37	12	12	6	74	50	12	12	2	110	51
13	13	5	62	14	13	13	8	118	14	13	13	9	70	30	13	13	8	116	37 X	13	13	7	82	44	13	13	2	108	49
14	7	4	60	12	14	7	3	118	12	14	7	3	76	23	14	7	3	116	31 X	14	14	2	80	46	14	14	2	110	45
15	3	0	88	9	15	15	9	118	10	15	15	11	76	28	15	15	9	116	26	15	5	3	88	34	15	5	3	90	32
16	8	5	60	8 #	16	8	3	118	8	16	16	12	72	20	16	8	3	116	22	16	16	10	84	35	16	16	10	90	35
17	17	7	60	6 #	17	17	10	118	6	17	17	13	78	17	17	17	10	114	21	17	17	11	76	31	17	17	4	104	41 X
18	3	0	86	6	18	9	3	118	5	18	18	14	72	18 #	18	9	3	116	17	18	18	12	82	30	18	18	12	90	32
19	19	5	60	3 #	19	19	11	120	3	19	19	15	76	14	19	19	11	116	16	19	19	13	74	28	19	19	13	90	29
20	10	7	60	2 #	20	10	3	118	2	20	20	16	76	14 #	20	10	3	116	15	20	5	2	90	24	20	5	2	90	24
21	7	3	60	2 #	21	7	4	116	2	21	21	17	72	13	21	21	4	106	15	21	21	15	80	25	21	21	15	90	25
22	22	6	60	0 #	22	11	3	118	1	22	22	18	76	11	22	11	3	116	13	22	22	16	78	23	22	11	3	110	27
23										23	23	19	76	11	23	23	4	106	13	23	23	17	78	22	23	23	17	90	24
24										24	24	20	74	9 #	24	24	4	106	11	24	24	18	80	20	24	24	18	90	22
25										25	25	21	78	9	25	5	0	90	10	25	5	3	90	17 #	25	5	3	90	17
26										26	26	22	76	8	26	26	4	108	9	26	26	20	76	18	26	26	20	90	19
27										27	27	23	78	7	27	27	4	106	8	27	27	21	80	16	27	27	21	90	19
28										28	28	24	78	6 #	28	28	4	100	7	28	28	22	74	17	28	7	0	90	19
29										29	29	12	66	6	29	29	4	102	7	29	29	23	78	15	29	29	23	90	17
30										30	30	26	76	5 #	30	30	4	104	6	30	5	0	90	14	30	5	0	90	14
31										31	31	13	64	5	31	31	4	100	6	31	31	25	78	14	31	31	6	102	16
32										32	16	11	66	4 #	32	16	5	108	5	32	32	26	80	13	32	32	6	96	15
33										33	33	14	68	3	33	33	19	112	3	33	33	27	78	12	33	33	6	100	14
34										34	17	12	68	3 #	34	17	5	110	3	34	34	28	80	11	34	34	6	96	14
35										35	5	0	90	3 #	35	5	0	90	3	35	5	3	90	10	35	5	1	92	10
36										36	18	13	66	2 #	36	18	15	114	20 X	36	36	30	78	10	36	36	6	100	11
37										37	37	16	68	1 #	37	37	21	112	1	37	37	31	78	10	37	37	6	96	11
38										38	19	14	68	1 #	38	19	5	108	1	38	38	32	80	9	38	19	5	110	12
39										39	39	17	68	1 #	39	39	22	110	1	39	39	33	78	9	39	39	6	100	9
40										40	20	15	74	1 #	40	20	5	106	1	40	5	1	90	8	40	5	1	90	8

In these tables an "X" denotes a constellation for which there is significant disagreement in MRT value with its posigrade or retrograde counterpart. An "#" denotes a T/P/F arrangement that was a good performer for continuous global coverage (MRT=0).

AAS 03-179

DAILY REPEAT-GROUNDTRACK MARS ORBITS

Gary Noreen,[*] **Stuart Kerridge,**[*] **Roger Diehl,**[*]
Joseph Neelon,[*] **Todd Ely**[*] **and Andrew E. Turner**[†]

This paper derives orbits at Mars with groundtracks that repeat at the same times every solar day (sol). A relay orbiter in such an orbit would pass over in-situ probes at the same times every sol, ensuring consistent coverage and simplifying mission design and operations. 42 orbits in five classes are characterized:

- 14 circular equatorial prograde orbits
- 14 circular equatorial retrograde orbits
- 11 circular sun synchronous orbits
- 2 eccentric equatorial orbits
- 1 eccentric critically inclined orbit

The paper reports on the performance of a relay orbiter in some of these orbits.

INTRODUCTION

NASA has been building a Mars network to relay communications between Marscraft[‡] and Earth and to provide navigation and timing services. This network currently includes UHF relays on two NASA orbiters – Mars Global Surveyor (MGS) and Mars Odyssey[1] – both in low orbits selected to optimize the collection of remote science rather than relay performance. The network will be augmented by UHF relays on the ESA Mars Express orbiter in 2004 and by the NASA Mars Reconnaissance Orbiter (MRO) in 2006.

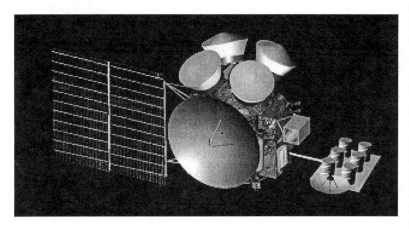

Figure 1 Artist Conception of Mars Telesat

[*]   Jet Propulsion Laboratory, California Institute of Technology, 4800 Oak Grove Drive, Pasadena, California 91109-8099.

[†]   Space Systems Loral, 3825 Fabian Way, Palo Alto, California 94303.

[‡]   A Marscraft is a lander, rover, aerobot or small orbiter in the vicinity of Mars.

NASA plans to send a telecommunications satellite to Mars in 2009. This Mars Telesat (Figure 1) will be placed into an orbit optimized for Marscraft relay support. While the Telesat may carry 4 CNES NetLanders to Mars and may also support a deep space optical communications experiment and some scientific instruments, its primary mission will be to relay communications to and from Marscraft. It thus will be placed into an orbit optimized for its relay communications mission.

Marscraft operators typically desire repeated contacts at the same times every day. This enables them to design their missions and operations based on an invariant pattern of relay contacts. To identify desirable Mars Telesat orbits, we have characterized orbits with groundtracks that repeat every sol. This paper describes the Mars orbits we characterized and reports on the performance of the Mars Telesat operated in several of these orbits.

Similar collections of orbits with daily repeating groundtracks could be easily assembled for other planetary bodies.

CIRCULAR EQUATORIAL ORBITS

An object in a circular equatorial orbit repeats the same groundtrack every orbit. We wish to find orbits that have daily repeating groundtracks, i.e. that pass over the same locations at the same time each solar day.

In order to pass over an object on Mars at the same time each sol, the time it takes for a circular prograde equatorial orbiter to complete Q periods P_Q (where Q, the Trace Repetition Parameter, is an integer), plus the amount of time it takes for the orbiter to cover the angular distance traversed by an object on Mars between a Martian sidereal day (P_{sid}) and a Mars sol (P_{sol}), must add up to a sol. In equation form,

$$P_{sol} = QP_Q + \frac{P_Q(P_{sol} - P_{sid})}{P_{sid}} \tag{1}$$

Solving for P_Q, for a prograde orbit we have

$$P_Q = \frac{P_{sol}P_{sid}}{P_{sid}(Q-1) + P_{sol}}. \tag{2}$$

Similarly, for a retrograde orbit

$$P_Q = \frac{P_{sol}P_{sid}}{P_{sid}(Q+1) - P_{sol}}. \tag{3}$$

The Keplerian orbit period is equal to

$$P_K = 2\pi\sqrt{\frac{a^3}{\mu}} = \frac{2\pi}{n} \tag{4}$$

where

a = semi-major axis of the orbit,

μ = gravitational constant of the planet (GM), and

n = mean motion.

Values for μ and other Mars constants are given in Table 1.

Table 1
MARS PLANETARY CONSTANTS

Gravitational constant μ	42,828.376645 km^3/s^2
Equatorial radius R_M	3,396.2 km
Constant of oblateness J_2	0.001958705252674
Sidereal rotation period P_{sid}	24.6230 hours
Solar rotation period P_{sol}	24.6598 hours
Orbit period around sun (Mars year)	668.6 sols
Mean rotation rate of Mars around the sun n_{sol}	1.059 x 10^{-7} rad/s

Eq. 4 does not account for the effects of the oblateness of Mars on the orbit, which affects the node, argument of periapsis, and mean anomaly. In the degenerate case of equatorial orbits, all motion from oblateness effects contribute to the in-plane motion of the spacecraft. When summing the motion from oblateness effects, the effects on argument of periapsis and mean anomaly are projected into the equatorial plane, which can be accounted for by using a factor of $\cos i$. When summing the oblateness effects with the mean motion of equatorial orbits, the direction of the orbit is important; for prograde orbits, the oblateness effects add to the magnitude of the total motion, and the reverse is true for retrograde orbits. To account for this behavior, an additional factor of $\cos i$ must be applied to all oblateness effects so that the equation for total motion is

$$n_t = n + \left[\dot{\Omega} + (\dot{\omega} + \dot{M}_0)\cos i\right]\cos i \tag{5}$$

where

$$\dot{\Omega} = -\frac{3\sqrt{\mu}J_2 R_M^2}{2a^{7/2}(1-e^2)^2}\cos i \tag{6}$$

1145

is the node regression rate,

$$\dot{\omega} = -\frac{3\sqrt{\mu}J_2 R_M{}^2}{4a^{7/2}\left(1-e^2\right)^2}(5\sin^2 i - 4) \tag{7}$$

is the precession rate of the argument of periapsis, and

$$\dot{M}_0 = -\frac{3\sqrt{\mu}J_2 R_M{}^2 \sqrt{1-e^2}}{4a^{7/2}\left(1-e^2\right)^2}(3\sin^2 i - 2) \tag{8}$$

is the correction to the rate of change of the mean anomaly, and where

J_2 = constant of planet oblateness,

e = eccentricity of the orbit, and

i = orbit inclination.

In the case of circular equatorial orbits, we can combine Eqs. 5 through 8 and set $e = 0$ and $i = 0°$ or $180°$ to get

$$n_t = \sqrt{\frac{\mu}{a^3}} + \frac{3\sqrt{\mu}J_2 R_M{}^2}{a^{7/2}}. \tag{9}$$

We need the orbiter to go completely around in its orbit once each period P_Q, so

$$P_Q = \frac{2\pi}{n_t}. \tag{10}$$

Eqs. 9 and 10 can be solved iteratively for a for each period P_Q in Eqs. 2 and 3. The results of these calculations are in Table 2.

Table 2
CIRCULAR EQUATORIAL ORBITS

	Prograde			Retrograde		
Q	P_Q, Hours	Radius, km	Altitude, km	P_Q, Hours	Radius, km	Altitude, km
1	24.6230	20,429.9	17,033.7	24.6967	20,470.7	17,074.5
2	12.3207	12,878.6	9,482.4	12.3391	12,891.4	9,495.2
3	8.2158	9,831.7	6,435.5	8.2240	9,838.3	6,442.1
4	6.1626	8,118.4	4,722.2	6.1673	8,122.4	4,726.2
5	4.9305	6,998.2	3,602.0	4.9334	7,001.0	3,604.8
6	4.1089	6,199.0	2,802.8	4.1110	6,201.1	2,804.9
7	3.5221	5,595.2	2,199.0	3.5236	5,596.8	2,200.6
8	3.0819	5,120.2	1,724.0	3.0831	5,121.5	1,725.3
9	2.7395	4,735.0	1,338.8	2.7404	4,736.0	1,339.8
10	2.4656	4,415.2	1,019.0	2.4663	4,416.0	1,019.8
11	2.2415	4,144.7	748.5	2.2421	4,145.4	749.2
12	2.0547	3,912.4	516.2	2.0552	3,913.0	516.8
13	1.8967	3,710.3	314.1	1.8971	3,710.9	314.7
14	1.7612	3,532.7	136.5	1.7616	3,533.2	137.0

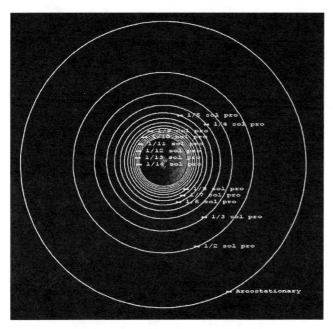

The prograde orbits in Table 2 are illustrated in Figure 2. The equatorial circular prograde orbit with $Q=1$ is an areostationary orbit, akin to geostationary orbits on Earth – an orbiter in this orbit will always be over the same location on the equator of Mars.

Figure 2 Equatorial Circular Orbits

EQUATORIAL ECCENTRIC ORBITS

To produce a daily repeat-groundtrack in an equatorial eccentric orbit, not only must the total motion of the orbit provide the desired period, but the line of apsides must maintain the same orientation with respect to the Sun. The rotation rate $\dot{\varpi}_a$ of the line of apsides projected onto the equatorial plane is

$$\dot{\varpi}_a = \dot{\varpi}\cos i + \dot{\Omega}. \tag{11}$$

For a prograde equatorial orbit,[†] we set $i = 0°$ and combine Eqs. 6, 7 and 11 to get

$$\dot{\varpi}_a = \frac{3J_2\sqrt{\mu}R_M{}^2}{2a^{7/2}(1-e^2)^2}. \tag{12}$$

Setting $\dot{\varpi}_a = n_{sol}$ so that the line of apsides rotates in inertial space at a sun synchronous rate and solving for e, we have

$$e = \sqrt{1 - \sqrt{\frac{3J_2\sqrt{\mu}R_M{}^2}{2n_{sol}a^{7/2}}}}. \tag{13}$$

Combining Eqs. 5 through 8 and setting $i = 0°$, we have for total motion

$$n_t = \sqrt{\frac{\mu}{a^3}}\left(1 + \frac{3J_2R_M{}^2}{2a^2(1-e^2)^2}\left(1 + \sqrt{1-e^2}\right)\right). \tag{14}$$

Eqs. 13 and 14 can be solved iteratively to find eccentric equatorial orbits with periods equal to submultiples of a sol (Eq. 10). There are just two such orbits, both prograde. Their characteristics are shown in Table 3. Turner[2] denotes such orbits at Earth Apogee at Constant time-of-day Equatorial (ACE) orbits; at Mars, we use the same acronym to denote Apoapsis at Constant time-of-day Equatorial.

[†] $\dot{\varpi}_a$ is negative for a retrograde equatorial orbit ($i = 180°$). n_{sol} must be positive, so there are no viable retrograde equatorial eccentric orbits with daily repeating ground traces.

Table 3

MARS ACE ORBIT CHARACTERISTICS

Orbit period, sols	½ Sol	⅓ Sol
Orbit period, hours	12.321 hours	8.216 hours
Semi Major Axis a	12,886 km	9,833 km
Eccentricity e	0.691	0.402
Periapsis Altitude	583 km	2,482 km
Apoapsis Altitude	18,397 km	10,392 km

CIRCULAR SUN SYNCHRONOUS ORBITS

For a circular inclined orbit to have a daily repeating groundtrack, the node regression rate $\dot{\Omega}$ must be equal to the rotation rate n_{sol} of Mars about the sun. Such orbits are generally denoted sun synchronous because they always maintain the same orientation with respect to the sun. Setting Eq. 6 equal to n_{sol} and solving for i, we have

$$i = \cos^{-1}\left(-\frac{2a^{7/2}n_{sol}}{3J_2\sqrt{\mu}R_M^2}\right) \tag{15}$$

We need the nodal period P_Ω (the period between crossings of the equatorial plane) to be a submultiple of the solar rotation period of Mars, i.e.

$$P_\Omega = \frac{P_{sol}}{Q} \tag{16}$$

From Vallado[3], the nodal period of an orbit P_Ω is related to the Keplerian period P_K by

$$P_\Omega = \frac{P_K}{1 + \dfrac{3R_M^2 J_2\left(4 - 5\sin^2 i + \sqrt{1-e^2}\left(2 - 3\sin^2 i\right)\right)}{4a^2(1-e^2)^2}} \tag{17}$$

For a circular orbit, $e = 0$ and we have

$$P_\Omega = \frac{P_K}{1 + \dfrac{3R_M^2 J_2\left(3 - 4\sin^2 i\right)}{2a^2}} \tag{18}$$

Eqs. 15 and 18 can be solved iteratively to find i and a for each sun synchronous circular orbit with a nodal period equal to a submultiple of a sol (Eq. 16). Table 4 characterizes these orbits, which are illustrated in Figures 3 and 4.

Table 4

CIRCULAR SUN SYNCHRONOUS ORBITS

Q	Nodal Period, Hours	Inclination	Radius, km	Altitude, km
4	6.165	136.683°	8118	4722
5	4.932	115.563°	6992	3596
6	4.110	106.362°	6190	2794
7	3.523	101.330°	5585	2188
8	3.082	98.266°	5108	1712
9	2.740	96.267°	4721	1325
10	2.466	94.894°	4400	1004
11	2.242	93.914°	4129	733
12	2.055	93.192°	3895	499
13	1.897	92.647°	3692	296
14	1.761	92.225°	3514	118

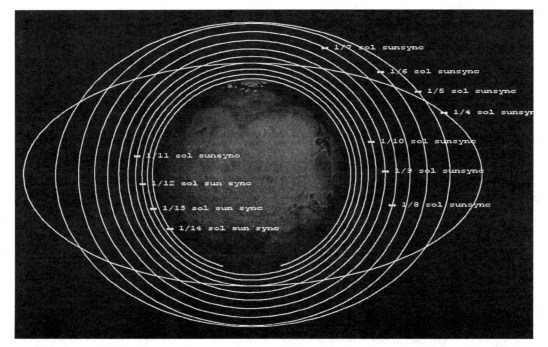

Figure 3 Circular Sun Synchronous Orbits Viewed Perpendicular to Nodes

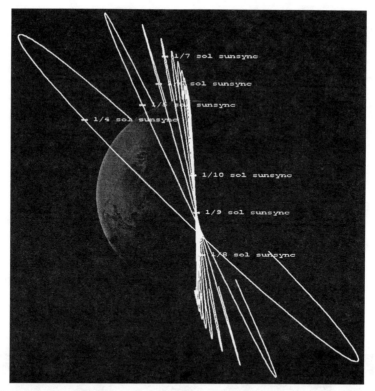

Figure 4 Circular Sun Synchronous Orbits Viewed Along Nodes

INCLINED ECCENTRIC ORBITS

Consider critically inclined eccentric orbits, in which there is no apsidal rotation with respect to the line of nodes. The node regression rate must then be maintained at the sun synchronous rate to ensure a daily repeating groundtrack.

$\dot{\varpi} = 0$ at the critical inclinations, i=116.565 and i=63.435. Setting $\dot{\Omega} = n_{sol}$ and $\cos i = -\sqrt{1/5}$ ‡, we can solve eq. 7 for e:

$$e = \sqrt{1 - \sqrt[4]{\mu/5}\sqrt{\frac{3J_2R_M^2}{2a^{7/2}n_{sol}}}} \tag{19}$$

Setting $\sin^2 i = 4/5$ in eq. 17, the nodal period is

---

‡ $\cos i$ must be negative for positive regression, so only 116.565° is viable.

$$P_\Omega = \cfrac{P_K}{1 - \cfrac{3R_M{}^2 J_2}{10a^2(1-e^2)^{3/2}}} \qquad (20)$$

We can now find the values of a and e which result in nodal periods that are sub-multiples of a sol as before through iteration. There is one such orbit at Mars (see Table 5). We denote this orbit Apoapsis at Constant time-of-day Critically Inclined, or ACCI. In the past, a similar orbit at Earth has been called "triply synchronous."[4]

Table 5
MARS ACCI ORBIT CHARACTERISTICS

Nodal period in sols	¼ sol
Nodal period in hours	6.165 hours
Semi-major Axis a	8,114 km
Eccentricity e	0.464
Periapsis Altitude	953 km
Apoapsis Altitude	8,483 km

Figure 5 illustrates the Mars ACCI orbit and both Mars ACE orbits.

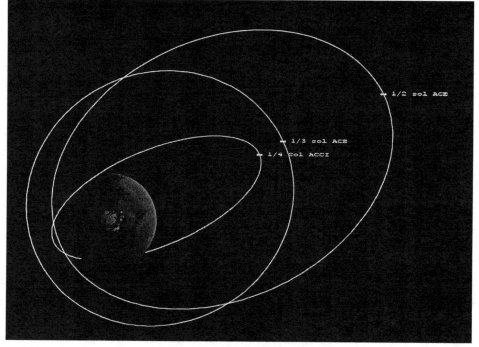

Figure 5 Elliptical Orbits

RELAY PERFORMANCE

Figure 6 illustrates the connectivity to Mars Science Laboratory (MSL) from various orbits and the sun (i.e. the times at which MSL would be in view). Note that connectivity through the Mars Telesat is typically much longer than through MRO, which is representative of the other low polar science orbiters now at Mars.

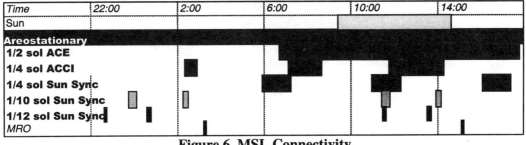

Figure 6 MSL Connectivity

The ½ sol ACE orbit has some interesting attributes that make it attractive for mid-latitude Mars rovers. Its highly eccentric orbit requires relatively low ΔV for insertion, and if appropriately set up, the orbiter can be in view every day for essentially all the time that the rover is in sunlight. This could enable continuous rover monitoring and control.

Figure 7 shows the data volume that could be relayed from MSL through various orbiters. The Mars Telesat will increase the amount of data that can be relayed to Earth from MSL by more than an order of magnitude.

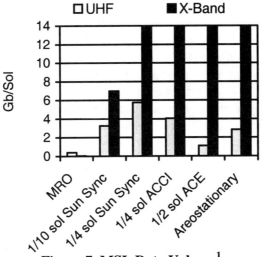

Figure 7 MSL Data Volume[1]

Figures 8 and 9 show mean pass lengths for links between landers and an orbiter in the ½ sol ACE and ¼ sol ACCI orbits, respectively.

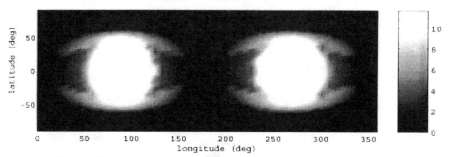

Figure 8 Mean Pass Length of ½ Sol ACE Orbit, Hours

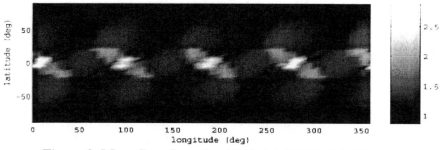

Figure 9 Mean Pass Length of ¼ Sol ACCI Orbit, Hours

While daily repeat ground track orbits are generally desirable for landers for operational reasons, they have certain limitations that must be considered. Coverage tends to be granular in latitude and longitude – i.e. it is not uniform over the entire planet. Also, because the orbit passes over the same areas each day, secular perturbations caused by inhomogeneities in the gravity field of Mars can build up over time (Ely[5]). However, exact repeat ground tracks are not important operationally for relay users, and this effect can be mitigated with East-West stationkeeping. Since the orbiter passes over landers on the surface of Mars with the same geometry each day, there is a reduction in the diversity of radiometric data that can be collected by the orbiter and a consequent reduction in the accuracy of radiometric-based navigation.

FUTURE WORK

MSL will be JPL's flagship mission of this decade and will be the dominant initial user of the Mars Telesat. The Mars Telesat and MSL teams at JPL are jointly constructing computer simulations of MSL operations using the Mars Telesat in various candidate orbits to characterize the science products that MSL could return through the Mars Telesat in each orbit and to help assess other operational implications of the candidate orbits. The Mars Telesat team is also simulating data returned from the CNES NetLanders through the Mars Telesat. The results of these simulations will be used to help select a final orbit for the Mars Telesat.

Mars Telesat mission designers are also determining the ΔV required to place the Telesat into various candidate orbits. This depends on arrival geometry and many other factors.

The Mars Telesat will be just one part of the Mars network, albeit a key part. NASA requires a relay radio on every orbiter sent to Mars with a design life in orbit of more than one year. NASA science orbiters are generally put into low polar orbits, which provide excellent relay coverage in polar regions (up to 13 contacts per day per orbiter) but limited coverage of mid-latitude regions (1 contact per day or less). If put into an equatorial orbit, the Mars Telesat could fill in the gaps left by the other orbiters. On the other hand, the Mars Telesat is to be a long-life orbiter and might be the only orbiter available to support important polar missions, so it may require global coverage – precluding an equatorial orbit.

Final selection of the Mars Telesat orbit will await the results of the simulation studies, assessments of the longevity of the other Mars orbiters, and consideration of how the Mars Telesat can best fit into the mix of future orbiter and lander missions to Mars.

ACKNOWLEDGEMENTS

The work described in this paper was carried out at the Jet Propulsion Laboratory, California Institute of Technology, under contract with NASA.

[1] Noreen, G., et. al., G. Marconi Orbiter and the Mars Relay Network, AIAA Communications Satellite Conference, Montreal, May 13-15, 2002.

[2] Turner, Andrew E., New Non-Geosynchronous Orbits for Communications Satellites to Off-Load Daily Peaks in Geostationary Traffic, AAS/AIAA Astrodynamics Specialist Conference, Kalispell, Montana, August 10-13, 1987.

[3] Vallado, David A., Fundamentals of Astrodynamics and Applications, 2nd Edition, Microcosm Press, 2001.

[4] Bachtell, Edward E., et. al., Satellite Voice Broadcast System Study, Martin Marietta Aerospace, NASA Contract NAS3-24233, November, 1985.

[5] Ely, Todd A. and Kathleen C. Howell, Long-Term Evolution of Artificial Satellite Orbits Due to Resonant Tesseral Harmonics, Journal of the Astronautical Sciences, Vol. 44, No. 2, April-June 1996, pp. 167-190.

SINGLE SATELLITE ORBITAL FIGURE-OF-MERIT[*]

John L. Young III,[†] David W. Carter,[‡]
John E. Draim[**] and Paul J. Cefola[††]

This paper investigates the properties of a new figure-of-merit for single-satellite orbits proposed by Draim: the coverage provided per unit of launch ΔV. This study limits orbits to critically inclined repeat ground track orbits. The coverage aspect of the figure-of-merit was investigated and found to exhibit complex behavior for certain cases. Coverage defined in this study is the time over a given satellite repeat period that the satellite is in view of a specified point on the ground. Constraints are placed on the elevation angle required for viewing the ground station. Through coverage maximization, several known orbits were reproduced and shown to be well designed in terms of coverage. The launch ΔV is calculated using analytic formulas, where the ΔV is used as a measure of launch cost.

This figure-of-merit has potential for use by satellite system designers to compare the cost-effectiveness of different orbits to determine the optimal orbit for a single satellite and by extension, a constellation of satellites. Trends in the figure-of-merit were investigated with respect to repeat ground track pattern, ground station location, orbit eccentricity, and minimum elevation angle. Using the proposed figure-of-merit, various repeat ground track orbits were examined and compared to draw conclusions on the cost-effectiveness of each orbit.

---

[*] This work was sponsored by Draper Laboratory under Independent Research and Development Funding and at the Lincoln Laboratory by the Air Force under Air Force Contract F19628-00-C-0002. "Opinions, interpretations, conclusions, and recommendations are those of the authors and are not necessarily endorsed by the United States Government."

[†] Ensign John L. Young III, USNR, and Draper Laboratory Fellow, The Charles Stark Draper Laboratory, 555 Technology Square, Cambridge, Massachusetts 02139-3563. E-mail: jlyoung@draper.com.

[‡] Dr. David W. Carter, Technical Staff, The Charles Stark Draper Laboratory, 555 Technology Square, Cambridge, Massachusetts 02139-3563. E-mail: dcarter@draper.com.

[**] Captain John E. Draim, USN (retired), and Aerospace Consultant, 9310 Telfer Court, Vienna, Virginia 2212-3438. E-mail: jdraim@aol.com.

[††] Dr. Paul J. Cefola, Technical Staff, MIT/Lincoln Laboratory, 244 Wood Street, Lexington, Massachusetts 02420-9108. Also Lecturer in Aeronautics and Astronautics, MIT. E-mail: cefola@ll.mit.edu.

INTRODUCTION

A metric that can be used by satellite designers to compare the cost-effectiveness of different orbits would be a valuable way to determine the optimal orbit for a single satellite. This paper investigates the properties of the new figure-of-merit for single-satellite orbits proposed by Draim. The figure-of-merit examined is primarily for analyzing communications satellites, but it can be tailored to different applications.

The orbital figure-of-merit, J, examines the coverage time versus the launch cost in ΔV and is defined by:

$$J = \max_{\Omega,\omega,M} \frac{T(a,e,i,\Omega,\omega,M)}{\Delta V(a,e,i)} g \tag{1}$$

where T is a measure of the coverage time in seconds over the interval $[t_0, t]$ where $t - t_0$ is the repeat period. The orbit elements a, e, i, Ω, ω, and M correspond to time t_0. The quantity ΔV is the velocity increment in m/s required to attain the satellite's mission orbit. The constant g is the acceleration due to gravity on the Earth's surface (9.81 m/s^2) and is used to non-dimensionalize the figure-of-merit.

Coverage T, which we show as a function of orbit elements, will generally depend on additional parameters. For this study, we focus on coverage provided to a given location on Earth; thus T depends implicitly on latitude, longitude, and on a minimum elevation parameter.

Note that to first order, ΔV depends only on orbit semimajor axis, eccentricity and inclination. The coverage metric is maximized over the orbit elements Ω, ω, and M. Thus, the figure-of-merit is a function of only three orbit elements: a, e, and i. In this study we restrict consideration to repeat ground track orbits, inclined at either 63.4° or 116.6°. Figure-of-merit is therefore a function of just two of the orbit elements – a and e.

The ΔV required to establish the mission orbit can be estimated using simple analytic formulas, can be computed using software packages such as DAB Ascent [1], or can be obtained from data in launch vehicle users guides. In this study, we rely on analytic formulas.

We make use of graphics visualization and nonlinear optimization tools provided with Matlab to show in detail how the figure-of-merit J depends on user location and on the relevant orbit elements. We draw general conclusions as to the most cost-effective single-satellite repeat ground track orbital parameters for observing ground points at various latitudes, particularly with respect to mean motion, eccentricity, and argument of perigee, for elliptic orbits.

Previous studies have been conducted to compare the cost-effectiveness of different satellite systems. Extensive work has been done in comparing different satellite systems by Shaw [2], Violet [3] and Gumbert [4, 5]. Violet and Gumbert focused on analyzing mobile phone satellite systems by employing a cost per billable minute metric to compare the different systems. Another metric that has been used for satellite system analysis is the cost per billable T1 minute metric for broadband satellite applications [6, 7]. In terms of coverage, Hanson [8] analyzed satellite constellations to determine the optimum constellation design, where he defined optimum as the "minimum number of satellites at the minimum possible inclination with the smallest possible maximum time gap." His work only considered circular orbits, where he showed that repeat ground track orbits provided better coverage than non-repeat ground track orbits.

This paper aims to examine the cost-effectiveness of the satellite's orbit, applied from the standpoint of not only coverage, but also the launch cost of the satellite. The launch ΔV can be used as a measure of launch cost by examining the rocket equation. The rocket equation,

$$\Delta V = g \cdot I_{sp} \cdot \ln\left(\frac{m_i}{m_f}\right) \tag{2}$$

shows that ΔV is a function of the acceleration due to gravity, g, the specific impulse, I_{sp}, and the ratio of the initial mass (m_i) to the final mass (m_f) of the launch vehicle. In this equation, g and I_{sp} are constants. Although the function is logarithmic, in the ranges we are examining the ΔV can be assumed to vary linearly with the mass ratio. In particular, this assumption is true in cases of multi-staged vehicles. Thus, a higher ΔV results in a higher mass ratio, yielding a larger launch vehicle. The size of the launch vehicle can be assumed to be directly related to the dollar cost for a satellite launch.

METHODOLOGY

In calculating the figure-of-merit, the orbit which provides the maximum coverage time to a point and the velocity required to attain the optimized orbit were necessary.

Coverage Calculation

A coverage function was written to calculate the coverage time over a ground station as a function of the órbit elements, where the semimajor axis, eccentricity and inclination are held fixed. Only repeat ground track orbits are examined in this study, therefore a given commensurability constraint, eccentricity and inclination correspond to a specific value for the semimajor axis.

Due to the nature of repeat ground track orbits, the location of the ascending node is independent of the epoch. By using the Earth-fixed longitude of ascending node, rather than the right ascension of the ascending node, the dependence of that orbit element on the epoch can be eliminated.

Since three of the orbit elements are held fixed (a, e, and i), coverage for a particular repeat period was determined to be a function of the three remaining orbit elements, the longitude of ascending node (LAN), the argument of perigee, and the mean anomaly.

When calculating coverage for an N:1 repeat ground track orbit, it is necessary to consider the coverage over a full repeat cycle. In an N:1 orbit, the satellite orbits the Earth N times per day, while the ground track repeats each day. Since the entire ground track is covered, it is acceptable to begin the integration at an equator crossing. By placing the satellite at an equator crossing the dependence on the mean anomaly can be eliminated and coverage is reduced to a function of two variables, thus simplifying the optimization.

To further simplify the optimization, constraints were placed on the range of the longitude of ascending node. The argument of perigee varies between −180° to 180°, but the longitude of ascending node could be constrained due to the fact that repeat ground track patterns were the only orbits examined in this study. As a result, the ascending nodes repeat at longitude intervals of 360°/N. Therefore, the range chosen for the longitude of ascending node was ± 180°/N about the longitude of the ground station.

The constrained nonlinear optimization function "fmincon," found in the Matlab Optimization Toolbox, was used to optimize the coverage function. Constraints were placed on the bounds of the optimization variables. "Fmincon" uses sequential quadratic programming to find a local minimum which satisfies user constraints, starting from the user's initial guess. Should there be multiple maxima, it is not guaranteed that the optimal orbit would be found unless the area where the global maxima is known.

It was discovered that some orbit / ground station combinations exhibit strange coverage behavior with multiple local maxima. To determine the global maximum coverage, a two step process was devised. Coverage was evaluated over a coarse interval over the range of longitudes of ascending node and arguments of perigee. These data were then formed into a contour plot to show the variation of coverage with respect to the optimization variables, as well as to allow the optimizer to obtain a suitable starting point close to the global maximum of the function. This starting point was input into the optimization function to determine the exact maximum coverage time.

Delta-V (ΔV) Calculation

There are several possible methods to compute the cost in ΔV required to attain the optimized orbit. Two options considered were an analytic approach and the use of a launch trajectory software tool. Since this study attempts to observe trends in the figure-of-merit, an exact solution to the required ΔV is not necessary. Therefore the analytic approach was chosen in that it provides an good estimate of the necessary ΔV.

Boltz [9] provides a complex method where one can calculate the ΔV from launch to final orbit by using the equations of motions. However, this study took a simplified technique to determine the ΔV, similar to the approach presented by Loftus, Teixeira, and Kirkpatrick [10] where the estimated Delta-V required for a launch vehicle (ΔV_{design}) is given by:

$$\Delta V_{design} = \Delta V_{burnout} + \Delta V_{gravity} + \Delta V_{drag} \tag{3}$$

In Equation 3, $\Delta V_{burnout}$ is the orbital velocity at injection, $\Delta V_{gravity}$ are velocity losses due to gravity and ΔV_{drag} are velocity losses due to atmospheric drag. In this paper, the total ΔV to orbit was found by:

$$\Delta V = \Delta V_{Hohmann} + \Delta V_{losses} \tag{4}$$

where $\Delta V_{Hohmann}$ was the change in velocity for a Hohmann transfer from the Earth's surface to the final orbit. The ΔV_{losses} term accounts for losses due to gravity and drag. Typical values range from 4000 - 6000 ft/s [11] and in this study, the velocity losses due to gravity and drag was set to 5000 ft/s.

To verify the validity of this method, ΔV values for launch into low-Earth orbit published by Humble, Henry and Larson [12] were compared with the ΔV values calculated using the analytic approach. Deviations from the published ΔV values were on the order of 3 - 5%. One comparison done was for an Ariane 44L launch vehicle. The ΔV for the Ariane 44L was estimated to be 9138 m/s to a circular low-Earth orbit with a semimajor axis of 6548 km and an inclination of 7.0°, while the analytic approach yielded a ΔV of 8766 m/s.

To facilitate a launch to any inclination, the launch site was placed on the equator. With the knowledge of the three orbit elements, semimajor axis, eccentricity, and inclination, the velocity change for the Hohmann transfer can be calculated, and thus the total ΔV to attain orbit is determined.

COVERAGE RESULTS

Coverage / Orbit Variables

This study examined six different N:1 repeat ground track patterns where N included 1, 2, 3, 4 and 8 revolutions. All of the orbits considered were critically inclined at 63.4°, except the 8:1 repeat period where both 63.4° and 116.6° inclinations were analyzed. The 8:1 case at an inclination of 116.6° allows for a sun-synchronous repeat ground track orbit. Three different eccentricities were chosen for each repeat period. Eccentricities of 0, 0.4 and the maximum eccentricity possible for an 800 km perigee altitude were selected. In the case of the 8:1 orbits, the maximum eccentricity was approximately 0.32. Because of this, an eccentricity of 0.2 replaced the 0.4 eccentricity point. The maximum eccentricity can be calculated as a function of the repeat cycle, inclination, and a minimum perigee radius. Ground station locations were varied from the equator to the North Pole in 15° latitude increments. Unless otherwise noted, all ground stations were located on the Prime Meridian.

The final variable that was introduced in the coverage calculation was the minimum elevation angle. Minimum elevation angles of 10° and 30° were chosen in the figure-of-merit analysis. The 10° elevation angle constraint is suited for commercial satellite telephony systems such as Iridium, Ellipso, and Globalstar. The 30° elevation angle constraint is relevant to higher frequency systems that are currently being considered.

Coverage Contour Plots

The variation of coverage with respect to the optimizing variables was examined through the use of coverage contour plots. These plots give the coverage versus the Earth-fixed longitude of ascending node and the argument of perigee. When examining the coverage behavior of the orbits, it was found that ground station locations at low latitudes often resulted in coverage contour plots with unusual features. An example of this unusual behavior is shown in Figure 1, a contour plot of a 3:1 repeat ground track orbit with an eccentricity of 0.4. The ground station for this case is located on the equator and the minimum elevation angle is 10°. The contour plots show the coverage as a percentage of time in view of the ground station to the total time for one repeat cycle. An examination of Figure 1 shows that there are multiple maxima. Due to the ground station's location at 0° latitude, there is symmetry in the contour plot and two global maxima result. These are shown in Figure 1, where one global maximum occurs in a small region at approximately a longitude of ascending node $\Omega = -30°$ and an argument of perigee $\omega = -90°$, while the other occurs at $\Omega = 30°$, $\omega = 90°$. The global maximum, while optimal, is not robust in that small deviations in either the argument of perigee or longitude of ascending node results in a large change in the coverage.

While low latitude ground stations exhibited complex behavior, as the ground station's latitude was increased, the coverage contours became straightforward and the behavior was as expected. Figure 2 illustrates the coverage variation for the same orbit as in Figure 1, however the ground station is located at 75° N latitude. There is little dependence on the longitude of ascending node and the best argument of perigee is $\omega = -90°$.

Coverage Percentage vs. Optimization Variables
3:1 Repeat Ground Track Orbit | Ground Station Latitude: 0°
i = 63.4° | e = 0.4 | Min. Elevation Angle = 10°

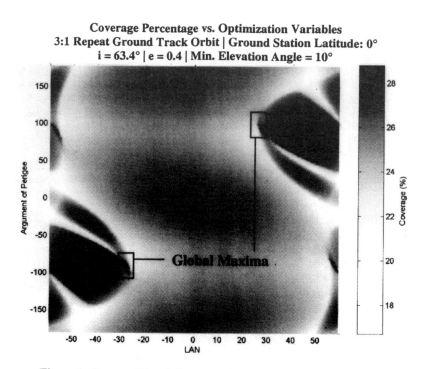

Figure 1: Contour Plot of Coverage, Low Latitude Ground Station

Coverage Percentage vs. Optimization Variables
3:1 Repeat Ground Track Orbit | Ground Station Latitude: 75° N
i=63.4° | e = 0.4 | Min. Elevation Angle = 10°

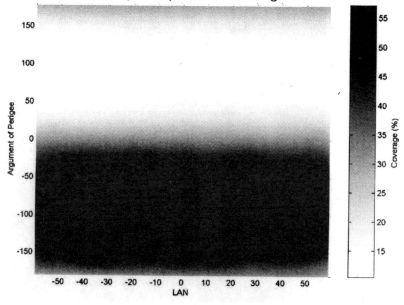

Figure 2: Contour Plot of Coverage- High Latitude Ground Station

In optimizing the coverage for the orbits, previously developed orbits resulted as the optimal solution for some repeat ground track cases that were examined. Orbits that were reproduced with remarkable similarity in the optimization routine were the Tundra [13, 14, 15, 16, 17, 18], Molniya [19], COBRA [20], and Ellipso-Borealis [21, 22] orbits.

Tundra Orbit

The Sirius satellite system is a constellation of three satellites designed to provide 24 hour commercial digital-quality radio coverage to the United States. It is the first system to use a critically inclined 1:1 repeat ground track orbit, or Tundra orbit. Two coverage optimizations were performed using Tundra orbits. The first optimization was for a 1:1 repeat ground track orbit where the eccentricity was set to that of the Sirius-1 orbit, while the second was for a 1:1 repeat ground track orbit with the maximum eccentricity possible for an 800 km perigee altitude. Since the Sirius constellation was designed to provide coverage to the U.S., the ground station for the optimizer was located at 36.5° N and 96.5° W, the approximate center of the U.S.

Table 1 lists the orbit elements for three orbits: the actual Sirius-1 satellite, the optimized Sirius orbit, and the overall optimal 1:1 repeat ground track orbit. The arguments of perigee and longitudes of ascending node for the Sirius-1 orbit and the optimized Sirius orbit are nearly identical. When the ground tracks are plotted in Satellite Tool Kit (shown in Figure 3), there are only minor differences in the two ground tracks. The Sirius-1 orbit is shown in blue, while the optimized orbit is shown in black. Figure 3 also plots the ground track for the overall optimal orbit for the given ground station. The optimal orbit exhibits the best coverage over the ground station, but other constraints are not taken into account. Due to design constraints, the Sirius orbit has a lower eccentricity, accounting for the differences between the two orbits. All three orbits exhibit similarities in that the arguments of perigee are approximately 270° while the longitudes of ascending node are chosen to place the apogee over the target.

Although the Sirius orbit does not provide the maximum coverage possible for a 1:1 repeat ground track orbit, the orbit is optimal in terms of coverage for its given eccentricity. An orbit with a higher eccentricity can provide greater coverage than the Sirius system, but due to design constraints an eccentricity of approximately 0.26 was chosen.

Table 1: Orbit Elements for Sirius-1 and Optimized Orbits for a Ground Station at (36.5° N, 96.5° W, 0 km)

	Sirius-1 [23]	Optimized Sirius Orbit	Optimal Orbit
a (km)	42163.79	42163.40	42158.17
e	0.2635	0.2635	0.8297
i (degrees)	63.24	63.40	63.40
ω (degrees)	269.56	269.98	270.04
Ω (degrees)	294.00	293.33	346.15

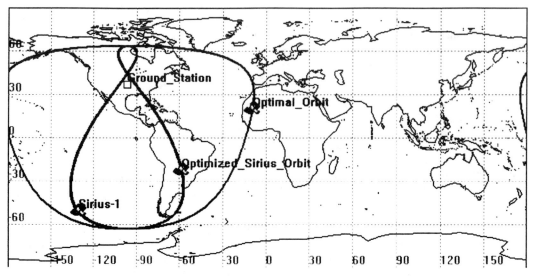

Figure 3: STK Ground Track Comparison for Sirius-1 Orbit and Optimized Orbit

Molniya Orbit

Another well known orbit that was generated through coverage optimization was the Molniya orbit. The Molniya orbit is a 2:1 repeat ground track orbit that has been used since 1965 [19, 24, 25]. A Molniya orbit generated by Satellite Tool Kit was compared to that of the optimized 2:1 repeat ground track orbit set for a ground station at 45° N latitude. Inputs for the STK generated orbit were the apogee longitude and the perigee altitude. Table 2 shows the element sets for the two orbits, while Figure 4 shows the ground tracks corresponding to the two orbits. In the STK generated orbit, the apogee was placed over the target's longitude. Contrary to intuition, this is not the optimal coverage point. In fact, the optimum coverage occurs when the ground station is located halfway between the two apogees, as shown in the red ground track. By examining the coverage contour for the 2:1 repeat ground track case at maximum eccentricity as shown in Figure 5, we can see how the coverage varies as a function of the longitude of ascending node. Although Figure 5 shows the optimal coverage for a ground station located at 45° N latitude, the contour plots for both 60° N and 75° N latitudes also show that the optimal coverage point is identical to the 45° N location. When the minimum elevation angle is increased to 30°, the orbit changes for the 45° N ground station location, however for the 60° N and 75° N case, the orbit remains as shown in Figure 4.

Table 2: Orbit Elements for an STK Generated Molniya Orbit and Optimized Orbit for a Ground Station at (45° N, 0° W, 0 km)

	STK Generated Molniya	Optimized Orbit
a (km)	26553.94	26553.94
e	0.7297	0.7297
i (degrees)	63.40	63.40
ω (degrees)	270.00	270.00
Ω (degrees)	100.67	82.73

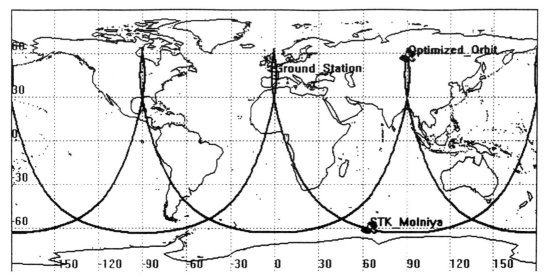

Figure 4: STK Ground Track Comparison for a Typical Molniya Orbit and Optimized Orbit

Coverage Percentage vs. Optimization Variables
2:1 Repeat Ground Track Orbit | Ground Station Latitude: 45° N
i = 63.4° | e = 0.7297 | Min. Elevation Angle = 10°

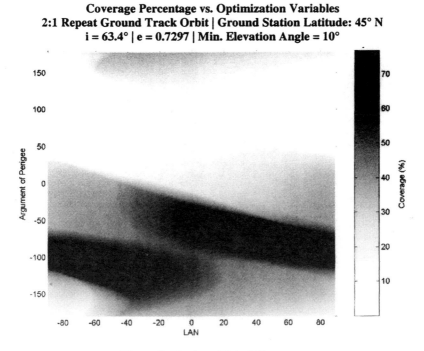

Figure 5: Contour Plot of Coverage

COBRA Orbit

The "Communications Orbiting Broadband Repeating Array" or COBRA orbit proposed by Draim [20], is a leaning 3:1 repeat ground track orbit with an orbit period of 8 hours. The "lean" in the orbit is caused by the argument of perigee. These orbits can be either right leaning or left leaning where right leaning orbits correspond to those that have arguments of perigee greater than 270°, while the left leaning orbits have arguments of perigee less than 270°. Table 3 shows the orbit elements for both Draim's COBRA Orbit and the optimized orbit. The ground station in this optimization run was set to 30° N latitude. The coverage optimization produced a COBRA-like orbit in that both the COBRA and the optimized orbits are leaning. Draim's orbit leans further to the left with an argument of perigee ω = 226.45°, while the optimized orbit has an argument of perigee of 245.03°, but the optimized orbit still shows a significant lean. The coverage contour plot for this orbit / ground station combination is shown in Figure 7, and it indicates that there are two optimal orbits that exist. These two orbits correspond to a right leaning and a left leaning orbit and both have the same amount of coverage. The difference in the arguments of perigee can be attributed to additional constraints that were applied to the COBRA orbit such as the requirement to remain clear of the GEO region.

Table 3: Orbit Elements for a Typical COBRA Orbit and Optimized Orbit for a Ground Station at (30° N, 0° W, 0 km)

	COBRA Orbit [26]	Optimized Orbit
a (km)	20260.85	20260.86
e	0.6458	0.6457
i (degrees)	63.40	63.40
ω (degrees)	226.45	245.03
Ω (degrees)	60.00	51.13

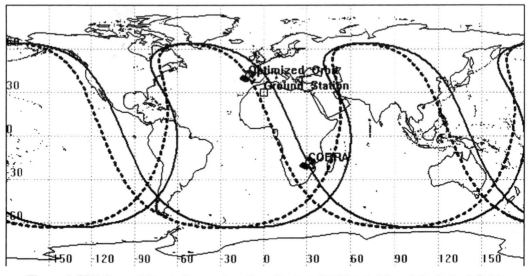

Figure 6: STK Ground Track Comparison for a Typical COBRA Orbit and Optimized Orbit

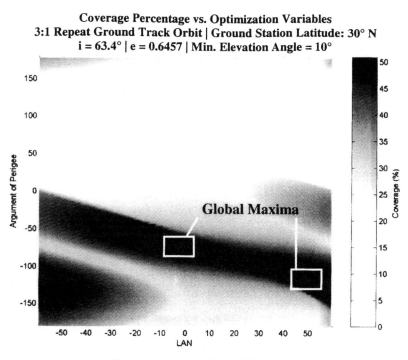

Coverage Percentage vs. Optimization Variables
3:1 Repeat Ground Track Orbit | Ground Station Latitude: 30° N
i = 63.4° | e = 0.6457 | Min. Elevation Angle = 10°

Figure 7: Contour Plot of Coverage

Ellipso-Borealis

The final orbit that appeared through the optimization routine was similar to the Ellipso-Borealis orbit. Ellipso aims to provide low cost global communication coverage through two orbit constellations. One of these constellations is the Ellipso-Borealis which is a series of elliptical 8:1 repeat ground track orbits at an inclination of 116.6°. The Ellipso-Borealis constellation is designed to provide coverage to northern latitudes.

An optimization of a retrograde 8:1 orbit for a ground station located at 30° N was performed. Table 4 shows orbit element sets of both the proposed Ellipso-Borealis constellation and that of the optimized orbit. As with the previous two cases, the orbit was optimized for a ground station located at 0° longitude. Therefore, the arguments of perigee are similar, however the longitudes of ascending nodes differ because the Borealis orbit was optimized for coverage over the U.S. and Canada. Figure 9 illustrates the variation in coverage vs. the longitude of ascending node and the argument of perigee for this orbit. This contour plot shows that there are two areas of maximum coverage, both of which have an argument of perigee near $\omega = -90°$, but the optimal point is located at a longitude of ascending node near 16°.

Table 4: Orbit Elements for a Typical Ellipso-Borealis Orbit and Optimized Orbit For a Ground Station at (30° N, 0° W, 0 km)

	Ellipso-Borealis [26]	Optimized Orbit
a (km)	10559.25	10558.58
e	0.3453	0.3202
i (degrees)	116.57	116.6
ω (degrees)	270.00	269.96
Ω (degrees)	10.00	15.75

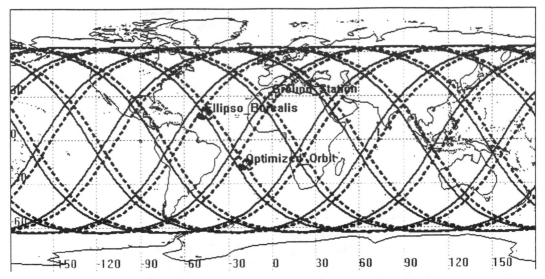

Figure 8: STK Ground Track Comparison for a Typical Ellipso-Borealis Orbit and Optimized Orbit

Coverage Percentage vs. Optimization Variables
8:1 Repeat Ground Track Orbit | Ground Station Latitude: 30° N
i = 116.6° | e = 0.3202 | Min. Elevation Angle = 10°

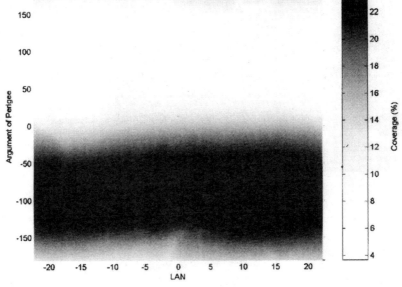

Figure 9: Contour Plot of Coverage

With a specified eccentricity, inclination, orbit repeat period, and ground station location, the optimization over the longitude of ascending node and argument of perigee was shown to yield known orbits. The optimization has also demonstrated that some of the known orbits presented are either optimal, or near optimal in terms of coverage. In addition, the Sirius case shows that the optimum coverage over a single point provides a good approximation for coverage over a region. Although the coverage times and thus the figure-of-merit will not be the same for coverage to a region, similar orbits will likely result.

Optimized coverage times for each orbit examined are shown in Table 5. Aside from the 1:1 case, as the eccentricity increases and the ground station latitude increases, the coverage time increases. The other trend to note in this table is that as the repeat period increases, the coverage time decreases.

Table 5: Coverage Time (s) vs. Ground Station Latitude

(10° Minimum Elevation Angle)

Repeat Period	Latitude / Eccentricity	0°	15° N	30° N	45° N	60° N	75° N	90° N
1:1	e = 0	86163	74179	64365	56674	49111	41181	33111
	e = 0.4	86162	75691	71389	68023	64665	60718	55361
	e = 0.8297	58064	67059	72644	76218	78059	78291	78359
2:1	e = 0	35488	31177	27728	28427	29357	30068	30329
	e = 0.4	39625	38149	36515	44959	49772	51962	52590
	e = 0.7297	41382	40981	58585	66173	69469	70795	71167
3:1	e = 0	22598	22566	23186	25181	26642	27573	27920
	e = 0.4	24819	33773	34445	40853	46460	49294	50194
	e = 0.6457	27361	42591	43861	50820	59728	63367	64408
4:1	e = 0	19934	21523	21790	21409	23472	25189	25686
	e = 0.4	26665	29577	32791	35246	43407	46945	47983
	e = 0.5707	28396	31355	36900	41457	52685	56825	58001
8:1 (i=63.4°)	e = 0	9413	10893	12695	13360	13715	15667	17094
	e = 0.2	11428	14076	17515	19873	21863	26799	28129
	e = 0.3178	11932	15681	20013	23697	27406	33491	34944
8:1 (i=116.6°)	e = 0	9692	11008	12635	13439	13613	15836	17243
	e = 0.2	10988	14140	17496	19677	22287	27008	28335
	e = 0.3202	11447	15918	20321	23842	27524	33900	35355

DELTA-V (ΔV) RESULTS

Once the optimized orbit was determined, the launch ΔV was computed to complete the figure-of-merit calculation. The results for the ΔV values for each orbit are summarized in Table 6. The table shows that as expected, the circular 1:1 case has the highest ΔV. As the eccentricity increases, the ΔV decreases for all repeat periods, even though the semimajor axis remains almost constant for each repeat period.

Table 6: ΔV Calculations (m/s)

Repeat Period	e = 0	e = 0.4	e = max
1:1	13242.2	12810.1	12118
2:1	12826.9	12417.7	11877.2
3:1	12441.5	12058.8	11666
4:1	12092	11732.1	11470
8:1	10955.7	10860.9	10750.1
8:1 (i=116.6°)	11380	11285.7	11173.5

FIGURE-OF-MERIT RESULTS

Given the optimal orbit in terms of coverage, the ΔV was calculated to determine the figure-of-merit values for each of the cases in the study. Trends within each class of repeat ground track orbits were examined. The figure-of-merit values were then compared with each other to determine which provided the most cost-effective solution.

Figure-of-Merit Contour Plots

Contour plots of figure-of-merit vs. the orbit's eccentricity and ground station latitude were created to examine the behavior of each repeat ground track pattern. Aside from the 1:1 repeat ground track orbits, the general pattern seen is that the figure-of-merit increases as the eccentricity increases. This is not surprising in that eccentric orbits allows for concentrated coverage in the appropriate hemisphere. In addition, as the eccentricity increases, the launch ΔV decreases. A typical figure-of-merit plot is shown in Figure 10, a contour plot of the 3:1 repeat ground track orbit with a minimum elevation angle of 10°. The best figure-of-merit occurs around 90° N latitude at the maximum eccentricity. As the eccentricity increases, the figure-of-merit increases for all latitudes. The other trend that can be noted from Figure 10 is that for a given eccentricity, the figure-of-merit increases as the latitude increases.

As the minimum elevation angle increases from 10° to 30°, these patterns remain consistent, but the figure-of-merit values decrease and the region with the best figure-of-merit is smaller. This can be seen by comparing Figures 10 and 11, where Figure 11 is the 3:1 case with a minimum elevation angle of 30°.

The 1:1 repeat ground track case is the exception to the previously noted trends. Figure 12 shows the figure-of-merit contour of the 1:1 case with a 10° minimum elevation angle. This plot indicates that there are two regions of high figure-of-merit values. For the lower latitudes, the figure-of-merit is the best between eccentricities of 0.1 and 0.5, but for the mid to upper latitudes, the figure-of-merit is the best at the maximum eccentricity.

When the minimum elevation angle is increased to 30° in the 1:1 case, the figure-of-merit behavior at the lower latitudes changes significantly. As seen in Figure 13, for ground stations located below 10° N latitude, the optimal orbit has a low eccentricity, but for latitudes between 10° and 30°, the optimal orbit has an eccentricity around 0.4. If the region of interest is in the mid to upper latitudes, then the optimal orbit is still at the maximum eccentricity.

1170

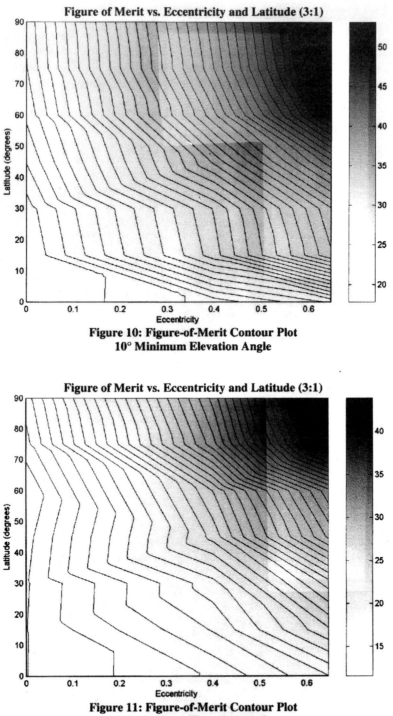

Figure 10: Figure-of-Merit Contour Plot
10° Minimum Elevation Angle

Figure 11: Figure-of-Merit Contour Plot
30° Minimum Elevation Angle

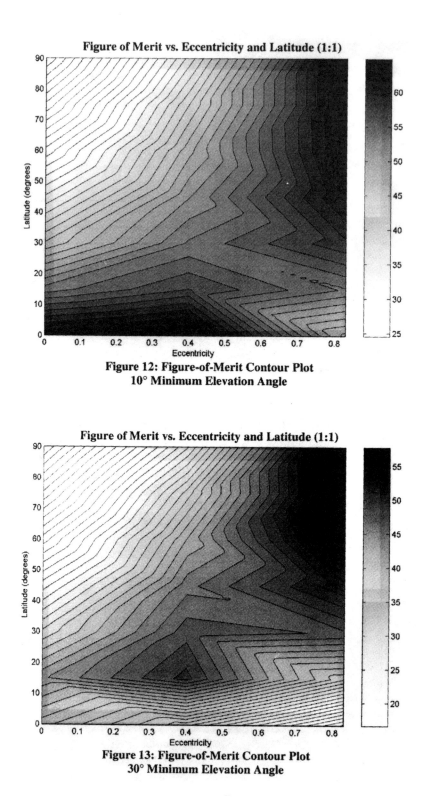

**Figure 12: Figure-of-Merit Contour Plot
10° Minimum Elevation Angle**

**Figure 13: Figure-of-Merit Contour Plot
30° Minimum Elevation Angle**

The figure-of-merit plots above were created using 21 total data points. A test case using 35 data points was created to determine if the contour plots changed significantly with additional data. It was shown that the plots did change slightly, with certain regions having different values for figure-of-merit, however the general trends remained consistent.

Repeat Ground Track Comparison

Ultimately, the different repeat ground track orbits were compared with each other based on their eccentricity to determine which was the most cost-efficient in terms of the figure-of-merit.

In each of the figures below, the 1:1 repeat ground track orbit has a higher figure-of-merit than any of the other repeat ground track orbits examined in this paper. Figures 14, 15, and 16 show plots of figure-of-merit values vs. ground station latitudes for each repeat pattern. The minimum elevation angle for these plots is 10°. These three figures correspond to the 0, 0.4 and maximum eccentricity cases. For the 8:1 case, the maximum eccentricity is approximately 0.32, so the 0.2 eccentricity case was plotted in Figure 15.

Not shown in these figures is the calculated figure-of-merit for a geostationary satellite. The figure-of-merit for geostationary satellites is a step function where the figure-of-merit for a point is approximately 66.9 for latitudes ranging from 71.4° S to 71.4° N, and zero for latitudes north and south of 71.4° in the 10° minimum elevation case. In the 30° case minimum elevation angle case, the figure-of-merit is 66.9, but the latitude region for coverage is reduced to areas between 52.5° S and 52.5° N.

In both Figure 14 and Figure 15, the figure-of-merit values tend to increase as the ground station latitude increases. The exception to this trend is the 1:1 repeat ground track orbit. Another trend that can be seen in these plots is that as the ground station locations increase in latitude, the figure-of-merit values for the different ground track patterns move towards each other. Although the 1:1 repeat ground track is shown to be the best orbit, the 2, 3, and 4 to 1 repeat patterns have figure-of-merit values that are similar to each other, especially for the 0 and 0.4 eccentricity cases.

In most of the cases, the figure-of-merit for the 2:1 repeat ground track orbit was better than the 3:1 case, however, a close examination of Figure 16 reveals that at the ground station with a latitude of 15° N, the 3:1 case is in fact better. Even though this is a single point, it illustrates that selecting a 3:1 repeat ground track orbit could be a better choice, depending on the application.

When the minimum elevation angle was increased to 30°, there were slight changes in the plots. All of the figure-of-merit values decreased, but the general trends remained the same as in the cases with a minimum elevation angle of 10°.

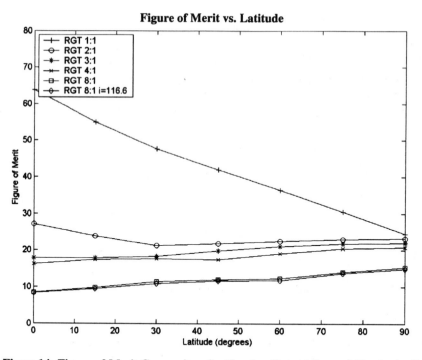

Figure 14: Figure-of-Merit Comparison for Varying Repeat Ground Tracks (e=0)

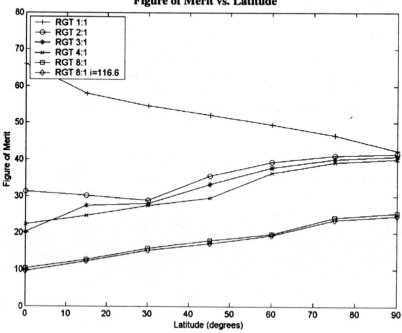

Figure 15: Figure-of-Merit Comparison for Varying Repeat Ground Tracks (e=0.4)

1174

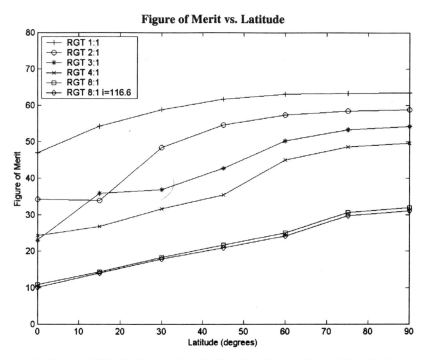

Figure 16: Figure-of-Merit Comparison for Varying Repeat Ground Tracks (e = max)

CONCLUSIONS

This paper shows that the proposed figure-of-merit has interesting implications regarding orbit design. In addition, the coverage for repeat ground track orbits for a single point is not necessarily straightforward. Finally this study has reproduced several known orbits and has shown them to be optimal for their repeat period.

The results of this study show that the 1:1 repeat ground track orbit is optimal in terms of the new figure-of-merit calculation. However, it is also important to note that the 2:1, 3:1, and 4:1 cases are roughly equivalent. Although the 2:1 and 3:1 repeat patterns do not provide as much coverage to one point on the Earth as the 1:1 case, when these satellites are not in view of the ground station, there are other points on the Earth receiving equal coverage. For example, the 3:1 repeat pattern provides the same percentage of coverage over three regions in the world, while the 1:1 pattern only provides coverage to one region. When considering global coverage applications, the figure-of-merit could be modified to take this into account.

The use of elliptical orbits allows satellite system designers to concentrate coverage over regions of interest. The eccentricity of the orbit also significantly affects the figure-of-merit. In general, better coverage and figure-of-merit values result when the orbit approaches the maximum eccentricity, especially for the northern latitudes. The minimum elevation angle has little impact on the trends of the figure-of-merit, but it does lower the figure-of-merit due to the reduced coverage.

1175

FUTURE WORK

A potential area for improvement in the figure-of-merit would be to include coverage to a region, rather than to a single point. Since the single point coverage exhibited interesting behavior, it is likely that the coverage over a region would provide different, but just as unusual coverage behavior. The final area for improvement in the figure-of-merit calculation is the ΔV calculation. The current method employed provides a good approximation for the actual ΔV required to attain orbit, however a more accurate value would allow the figure-of-merit to possibly change.

There are many considerations that must be taken into account while designing orbits. This figure-of-merit provides valuable insight into the potential coverage issues for a single satellite or orbit constellation. While the figure-of-merit developed and studied in this paper cannot be used solely to determine the best orbit for satellite systems, it provides a useful tool in the orbit design process.

ACKNOWLEDGEMENTS

The authors would like to acknowledge the assistance of our many co-workers at the Charles Stark Draper Laboratory and at the MIT Lincoln Laboratory for their support during the preparation of this paper. The first author would like to thank the Charles Stark Draper Laboratory for support of his graduate work as a Draper Fellow. In addition, we would like to acknowledge the utility of The Mathworks MATLAB and Analytical Graphics Inc.'s Satellite Tool Kit (STK) software in preparing the graphics for this paper.

REFERENCES

[1] Baker, David A. and Earley, Sidney M., "Capabilities, Methods, and Accuracy of DAB Ascent," AAS 93-637, Paper Presented at AAS/AIAA Astrodynamics Conference, Victoria, British Columbia, Canada. August 16-19 1993.

[2] Shaw, Graeme B., "The Generalized Information Network Analysis Methodology for Distributed Satellite Systems," Sc.D. Thesis, Massachusetts Institute of Technology, Cambridge, Massachusetts, 1999.

[3] Violet, Michael David, "the Development and Application of A Cost Per Minute Metric for the Evaluation of Mobile Satellite Systems in a Limited-Growth Voice Communications Market," Master's Thesis, Massachusetts Institute of Technology, Cambridge, Massachusetts, 1995.

[4] Gumbert, Cary C., "Assessing Future Growth Potential of Mobile Satellite Sytstems Using a Cost Per Billable Minute Metric," Master's Thesis, Massachusetts Institute of Technology, Cambridge, Massachusetts, 1996.

[5] Gumbert, Cary C., Violet, Michael D. et al., "Cost Per Billable Minute Metric For Comparing Satellite Systems," Journal of Spacecraft and Rockets, Vol. 34, No. 6, November - December 1997.

[6] Kelic, Andjelka, "Assessing the Technical and Financial Viability of Broadband Satellite Systems Using a Cost Per T1 Minute Metric," Master's Thesis, Massachusetts Institute of Technology, Cambridge, Massachusetts, 1998.

[7] Kashitani, Tatsuki, "Development and Application of an Analysis Methodology for Satellite Broadband Network Architectures," Masters Thesis, Massachusetts Institute of Technology, Cambridge, Massachusetts, 1999.

[8] Hanson, John M., Evans, Maria J., and Turner, Ronald E., "Designing Good Partial Coverage Satellite Constellations." Journal of Astronautical Sciences, Vol. 40, No. 2, April-June 1992.

[9] Boltz, F.W., "Analytic Solution for Vertical Launch Into Orbit," Journal of the Astronautical Sciences, Vol. 37, No.4, October-December 1989, pp 491-511.

[10] Wertz, James, R., and Wiley J. Larson. Space Mission Analysis and Design. Third Edition. El Segundo, California: Microcosm Press, 1999.

[11] Space Handbook, AU-18 Air University Press, Maxwell Air Force Base, Alabama. 1985. (p.3-9)

[12] Humble, Ronald W., Henry, Gary N., and Wiley J. Larsen. Space Propulsion and Analysis and Design. New York, New York: McGraw Hill, 1995.

[13] Barker, Lee, and Stoen, Jeffrey. "Sirius Satellite Design: The Challenges of the Tundra Orbit in Commercial Spacecraft Design." 24th Annual AAS Guidance and Control Conference, AAS 01-071, 2001.

[14] Bruno, Michael J., and Pernicka, Henry J. "Mission Design Considerations for the Tundra Constellations." AIAA/AAS Astrodynamics Specialist Conference and Exhibit, Monterey, California, August 5 – 8, 2002.

[15] Turner, Andrew E. "Molniya/Tundra Orbit Constellation Considerations for Commercial Applications," AAS 01-215, 2001.

[16] Efficient High Latitude Service Arc Satellite Mobile Broadcasting Systems; US Patent and Trademark Office; 6223019B1; April 24, 2001.

[17] Briskman, R., and Prevaux, J. "S-DARS Broadcast from Inclined, Elliptical Orbits" IAF-01-M.5.05. Paper Presented at the 52nd International Astronautical Congress, Toulouse , France, 1 - 5 Oct 2001.

[18] Briskman, Robert D., and Sharma, Surinder P., "DARS Satellite Constellation Performance," Preprint AIAA 2002-1967, Presented at the 20th AIAA International Communications Satellite Systems Conference, Montreal, Quebec, Canada, 12-15 May 2002.

[19] Johnson, Nicholas L., Soviet Space Programs 1980 – 1985, San Diego, California: AAS Publications, 1987.

[20] Draim, J.E. Inciardi, R., Cefola, P. Proulx, R., Carter, D., and Larsen, D., "Beyond GEO- Using Elliptical Orbit Constellations to Multiply the Space Real Estate", IAF-01-A.1.03. Paper Presented at the 52nd International Astronautical Congress, Toulouse, France, October 1-5, 2001.

[21] Draim, John E. Cefola, Paul J., Proulx, Ronald J., Larsen, Duane, and Granholm, George. "Elliptical Sun-synchronous Orbits With Line of Apsides Lying In or Near the Equatorial Plane." AAS 99-311, Presented at the 1999 AAS/AIAA Astrodynamics Specialist Conference, Girdwood, Alaska, August 15 – 19, 1999.

[22] Draim, John E., and Castiel, David. "Optimization of the Borealis and Concordia Sub-constellations of the Ellipso Personal Communications System," IAF-96-A.1.01. Paper presented at the 47th International Astronautical Congress, Beijing, China, Oct. 7-11, 1996.

[23] Sirius-1 Two-Line Elements <u>Celestrak WWW</u> TS Kelso,
<http://www.celestrak.com/NORAD/elements/other-comm.txt>

[24] Fieger, Marty E., "Evaluation of Semianalytical Satellite Theory Against Long Arcs of Real Data for Highly Eccentric Orbits," Master's Thesis, Massachusetts Institute of Technology, Cambridge, Massachusetts, 1987.

[25] Fischer, Jack D., "The Evolution of Highly Eccentric Orbits. Master of Science Thesis," Master's Thesis, Massachusetts Institute of Technology, Cambridge, Massachusetts, 1998.

[26] Sabol, Chris, "Application of Sun-Synchronous, Critically Inclined Orbits to Global Personal Communications Systems," Master's Thesis, Massachusetts Institute of Technology, Cambridge, Massachusetts, 1994.

DRAG AND STABILITY OF A LOW PERIGEE SATELLITE

Joseph R. Schultz[*] and Mark J. Lewis[†]

Knowledge of how a rarefied gas interacts with a spacecraft surface is important in determining the drag and stability of flexible spacecraft. This paper reviews stability of flexible spacecraft. This paper reviews stability parameters for several boomed spacecraft configurations and examines how those stability parameters change with changing gas-surface interaction parameters. It is found that lack of knowledge of the gas-surface interaction makes it difficult to predict stability for all but the most robust spacecraft configurations. By making reasonable assumptions on the gas-surface interaction limits, though, the region of stability uncertainty can decrease dramatically and allow for more flexibility in the design.

INTRODUCTION

Most satellites and spacecraft fly in high enough orbits so that they will not experience significant aerodynamic drag in an orbital time period. However, there are some missions where a spacecraft may wish to enter the upper regions of the Earth's atmosphere for a short period of time. These missions may be operational or scientific in nature such as to conduct an aeroassist for an orbital plane change or to take magnetosphere readings in the upper atmosphere.

In some cases, those spacecraft may have large flexible booms. Those booms may be used for solar panels, antenna, or to hold scientific instruments away from the body of the spacecraft. An important engineering and design question is how those large, flexible booms affect the drag and overall stability of the spacecraft during the atmospheric portion of flight. Much of the literature on aeroassist maneuvers, however, focuses on spacecraft with a space-shuttle-like configurations where the spacecraft is treated like a point-mass model.[1,2,3] Because of the rigid-body nature of these spacecraft, internal flexibility modes are usually not of concern.

Spacecraft with long, flexible booms, however, present an extra layer of complexity for the dynamics. Even in free-space, instabilities may arise that would not normally be predicted by a point-mass model. Although the literature covers many aspects of the dynamics and control of these flexible spacecraft in free-space[4,5,6], there is little analysis for when those types of spacecraft undergo the aerodynamic effects of an atmospheric passage.

Schultz, Pines[7] have made an initial investigation of the dynamics of flexible, boomed spacecraft during an aeroassist, but did not go into detail of how the dynamics change with variations in the gas-surface model. It is presumed that the details of how the rarefied gas interacts with the spacecraft body and boom surfaces will play an important role in determining the stability of a particular spacecraft design.

Lewis[8] has examined gas-surface interaction models and developed encompassing equations to determine lift and drag forces on spacecraft surfaces. Bowman[9] has applied those equations to optimal spacecraft body design. Sentman[10] used similar techniques for predicting lift and drag coefficients for

---

*   Graduate Student, Department of Aerospace Engineering, University of Maryland, College Park, Maryland 20742-3015. E-mail: jrs@wam.umd.edu. Student Member AIAA.

†   Professor, Department of Aerospace Engineering, University of Maryland, College Park, Maryland 20742-3015. E-mail: lewis@eng.umd.edu. Associate Fellow AIAA.

several spacecraft shapes. In that body of work, however, there was not application of those techniques to spacecraft with flexible booms.

This paper will combine the research on gas-interactions with several boomed spacecraft models to determine what role those interactions have on the stability of the entire spacecraft during its flight through the upper atmosphere.

BACKGROUND

Rarefied Gas-Surface Aerodynamic Models

Rarefied conditions can be considered to prevail if the vehicle dimensions are substantially larger than ambient mean-free-path. The atmospheric mean-free-path is 0.1 m at 98 km altitude, 1.0 m at 112 km, and 10 m at 132 km[11], so the validity of using a rarefied gas-surface model will depend on the size of the spacecraft and its altitude at perigee. For larger vehicles or lower altitudes, a continuum or transitional aerodynamic model will need to be used. It is assumed in this paper, however, that the rarefied conditions are satisfied and would be applicable for most satellites during a shallow atmospheric pass.

As outlined by Lewis[8], a simplified gas-surface interaction model can be used to determine reasonable values of aerodynamic coefficients for a spacecraft in rarefied flow. In rarefied flow at orbital speeds, where the freestream velocity is much greater than the molecules' mean thermal velocity, surface pressure is estimated directly from the momentum exchange of molecules colliding at velocity, V_o, with a surface at angle α, and rebounding with a fraction of their velocity ε at rebound angle $\delta\alpha$. The pressure force, p, due to impingement of molecules on a surface of unit area A is:

$$\frac{p}{A} = m_o I_m V_o \left(\sin\alpha + \varepsilon\sin(\delta\alpha)\right) \tag{1}$$

The impingement rate, I_m, is the number of particles of mass m_o striking a surface of unit area at given angle α per unit time. The impingement rate can be derived directly from a Maxwellian distribution with net velocity equal to orbital velocity. Goodman and Wachman[12] subdivided the impingement rate into a convected contribution and a thermal contribution. The convected term is:

$$I_{convect} = nV_o \sin\alpha \tag{2a}$$

where n is the molecule number density.

The thermal term is due to random molecular motion at mean thermal speed, $\bar{c}$, which at zero freestream velocity would be:

$$I_{thermal} = n\bar{c}/4 \tag{2b}$$

Combining Eqs. (2a) and (2b),

$$I_m = nV_o\left(\sin\alpha + 1/(4s)\right) \tag{2c}$$

where $s \equiv V_o/\bar{c}$, and s is typically between 10 and 25 at the altitudes and conditions of interest.

The thermal contribution is thus generally much smaller than the convected impingement except at very small angles, (i.e. $\alpha \leq 1°$). Note also that the rarefied flow is "shadowed" at negative values of α.

A traditional view of the impingement process is that it yields a distribution of molecules that fit somewhere between the range of diffusely scattered to specularly reflected, with some ratio of reflected velocity to collision velocity. Thus, lift and drag can each be written as the sum of a fraction ξ of specular reflections and $(1-\xi)$ diffusive reflections. The results of Hurlbut and Sherman[13], based on the Nocilla reflection model, are very closely matched by this simpler formulation. For purposes of establishing reasonable performance limits, the outflow has been modeled as a linear combination of the specular and diffuse scattering limits. As Bird[14] points out, there is no firm justification of this model, but it brackets the expected limits. In fact, the actual velocity distribution is lobular, with a dominant direction near the specular rebound angle, and may have secondary nodes at other directions.

The diffuse-specular model has the advantage for this study in that it yields simplified predictions for lift and drag as:

$$\frac{L}{A} = m_o I_m V_o \varepsilon \begin{cases} \frac{2}{3}\cos\alpha & \text{diffuse } (\xi = 0) \\ \sin 2\alpha & \text{specular } (\xi = 1) \end{cases} \tag{3a}$$

$$\frac{D}{A} = m_o I_m V_o \left[1 + \varepsilon \begin{cases} \frac{2}{3}\sin\alpha & \text{diffuse } (\xi = 0) \\ -\cos 2\alpha & \text{specular } (\xi = 1) \end{cases} \right] \tag{3b}$$

where the coefficient ε represents an average fraction of velocity *magnitude* that is reflected from a collision ($\varepsilon = 1$ corresponds to a perfectly elastic specular collision, while $\varepsilon = 0$ is a hit-and-stick inelastic collision) with molecules rebounding either specularly, or diffusely, via a cosine scattering law[12]. Note that $\varepsilon = \sqrt{1-\zeta}$, where ζ is the energy accommodation coefficient, which is generally a weak function of the surface angle.

With the standard notation of:

$$L = \frac{1}{2}\rho V^2 S C_L \tag{4a}$$

$$D = \frac{1}{2}\rho V^2 S C_D \tag{4b}$$

the Eqs. (3a) and (3b) can be rearranged to provide equations for the lift and drag coefficients for a flat surface in a rarefied flow:

$$C_{Dspec} = 2\left(\sin\alpha + \frac{\bar{c}}{4V_o}\right)(1 - \varepsilon\cos 2\alpha) \tag{5a}$$

$$C_{Lspec} = 2\left(\sin\alpha + \frac{\bar{c}}{4V_o}\right)\varepsilon\sin 2\alpha \tag{5b}$$

$$C_{Ddiff} = 2\left(\sin\alpha + \frac{\bar{c}}{4V_o}\right)\left(1 + \varepsilon\frac{2}{3}\sin\alpha\right) \tag{5c}$$

$$C_{Ldiff} = 2\left(\sin\alpha + \frac{\bar{c}}{4V_o}\right)\varepsilon\frac{2}{3}\cos\alpha \tag{5d}$$

$$C_D = \xi C_{Dspec} + (1-\xi)C_{Ddiff} \tag{6a}$$

$$C_L = \xi C_{Lspec} + (1-\xi)C_{Ldiff} \tag{6b}$$

Thus, the lift and drag coefficients depend only on the ratio of thermal to freestream velocities, the gas-surface interaction parameters ε and ξ, and the angle of the surface to the flow. Graphs of these lift and drag coefficients for various surface angles and gas-surface parameters are shown in Figures 1 and 2. For

1181

those graphs, the value of $\bar{c}$ was set to 640 m/s and Vo to 8300 m/s which would be typical for an aeroassist down to 130 km above the Earth's surface. Also note that when $\varepsilon = 0$, the lift and drag coefficients are the same for any value of ξ.

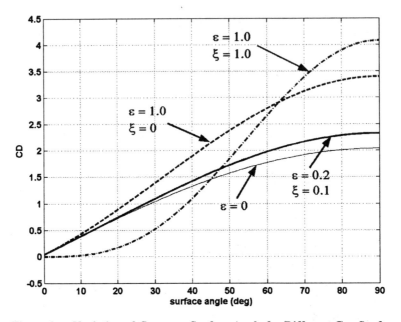

Figure 1   Variation of C_D versus Surface Angle for Different Gas-Surface
Interaction Properties

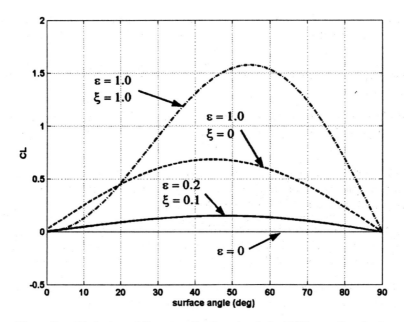

Figure 2   Variation of C_L versus Surface Angle for Different Gas-Surface
Interaction Properties

Gas-Surface Interaction Parameters

Potter and Rockaway[15] have explored the effect of rarefied conditions on the L/D ratio of a half-cone at 120 km. The L/D ratio varies between 0.2 for total accommodation ($\varepsilon = 0$), to 5 for specular reflection. This analysis includes transitional bridging effects to represent the intermediate behavior between continuum and pure rarefied flow. Other data have been used to point out the strong dependence of lift and drag coefficients on Knudsen number[16]. Data from Seidl and Steinheil and others, assembled by Herrero[17], for helium impacting at 2 km/sec on sapphire is well matched by values of $\varepsilon = 0.25$, $\xi = 0.53$. These values corresponds to a maximum $L/D = 0.19$ at 40° surface angle. Measurements of nitrogen (N_2 and N_2^+) on realistic spacecraft surfaces are well-approximated by $\varepsilon = 0.2$, $\xi = 0.4$. This yields a maximum $L/D = 0.135$ at an oblique surface angle of 31°. Harvey and Lord also report energy accommodation coefficients that yield values of ε between 0 and 0.5.[16]

Aerodynamic forces will be a function of the specific gas composition at altitude because of the dependence of gas-surface interactions on both surface material and gas properties. Highly reactive atomic oxygen is the dominant specie at orbital altitudes, with a local density that is approximately an order of magnitude greater than nitrogen beginning at approximately 230 km.[18] The specifics of atomic oxygen gas/surface interactions will dictate the flight values of accommodation and reflection pattern at atmospheric entry. For instance, the measurements of Cross and Blais[19] using high-energy atomic oxygen beams have shown substantially different reflection behavior depending on whether or not surface reactions occur. They report that atomic oxygen exhibits complete accommodation with near-diffuse scattering on Kapton and Teflon surfaces. Atomic oxygen on non-reactive nickel oxide exhibits nearly specular reflection with 50% translational energy accommodation ($\varepsilon = 0.7$).

The specific gas composition encountered by a spacecraft will depend on its altitude. At the altitudes up to 100 km, molecular nitrogen is the dominant species. However, in the range of minimum altitudes for low-perigee satellite orbits, molecular nitrogen, molecular oxygen, and atomic oxygen all have comparable density. At 120 km, the models of Hedin, et. al. [18], predict equatorial number densities of N_2, O, and O_2 as 2.4×10^{17} m^{-3}, 6.1×10^{16} m^{-3}, and 2.8×10^{16} m^{-3} respectively; at 130 km, predicted number densities of N_2, O, and O_2 are 8.8×10^{16} m^{-3}, 3.1×10^{16} m^{-3}, and 7.9×10^{15} m^{-3} respectively, and at 140 km N_2, O, and O_2, number densities are 4.3×10^{16} m^{-3}, 1.9×10^{16} m^{-3}, and 3.3×10^{15} m^{-3} respectively. Thus, at 120 km gas composition is approximately 73% nitrogen and 19% atomic oxygen, while at 140 km nitrogen is 66% and atomic oxygen is 29% of the gas. The number density of N_2 drops to 48% of the total at 180 km altitude (where O is also 48%) and is only 35% at 210 km, at which height the number density of O is 60% of the total.

Predictions of gas-surface interaction parameters, ε and ξ, are difficult to make. As seen, they are highly dependant on atmospheric composition and surface material. There is little experimental data available to help in predictions except for those specific cases noted. It would be prudent, therefore, for a spacecraft designer to take into account a range of parameter values when making lift and drag predictions.

Stability of Different Spacecraft Configurations

Knowledge of the gas-surface interaction is not only important for lift and drag predictions of spacecraft traveling in the upper reaches of the atmosphere, but also for predicting the static and dynamic stability of those spacecraft during their atmospheric flight. A low drag spacecraft may be desirable to limit lifetime fuel consumption, but if the low drag configuration is unstable, the design will be of little practical use.

The general symmetric 2-boomed and 4-boomed spacecraft that will be investigated are shown in Figures 3 and 4.

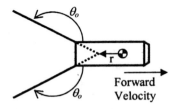

Figure 3 Symmetric 2-Boom Spacecraft

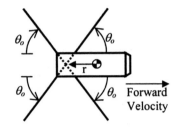

Figure 4 Symmetric 4-Boom Spacecraft

Note that θ_o is the non-deflected boom angle, and r is the distance from the spacecraft's center of mass to the connection point of the booms (normalized by the spacecraft's body length, L^*). It is assumed that the booms are connected to the spacecraft body at the same point, and the connection point lies on the centerline of a symmetric spacecraft.

Additional spacecraft parameters that will be referred to include:

m = mass of one boom (normalized by the total mass of the spacecraft, m_{tot}),
S = effective surface area of one boom (normalized by the spacecraft body's effective surface area, S^*),
L = length of one boom (normalized by the spacecraft body's length, L^*),
k = effective spring constant of the boom (normalized by $1/2\,\rho\,V_o^2\,S^*\,L^*$ where ρ is the atmospheric density and V_o is the spacecraft's forward velocity),
x_{AC} = location of aerodynamic center (normalized by L^*),
C_{Dbo} = drag coefficient of the spacecraft body,
$C_{Lb\alpha}$ = slope of the spacecraft body's lift coefficient,
$C_{Mb\alpha}$ = slope of the spacecraft body's moment coefficient.

Schultz, Pines have developed analytic solutions from Ref. 7 that map out the stable and unstable designs for the two configurations. These analytic solutions are derived from linear assumptions. The assumptions are that the deflections are small, the mass of the booms is small, and the effective aerodynamic surface area of the booms is reasonably large, i.e.:

$$\alpha' \ll 1 \tag{7a}$$
$$m \ll 1 \tag{7b}$$
$$\frac{S}{mC_{Dbo}} \gg 1 \tag{7c}$$

Note: α' is defined as the deflection angle of the boom from its original position, θ_o.

With these assumptions, it can be shown that for a 2-boom spacecraft, the spacecraft will be stable if the following criterion is met:

$$
\left(\begin{array}{l}
S^2 \dfrac{2r}{L} C_{Do}\left(\left(C_{Da}-C_{Lo}\right)s_{\theta_o}+\left(C_{La}+C_{Do}\right)c_{\theta_o}\right) \\[6pt]
+S\left[\begin{array}{l}
\left(-\dfrac{2k}{L}+\dfrac{C_{Mba}}{L}\right)\left(\left(C_{Da}-C_{Lo}\right)s_{\theta_o}+\left(C_{La}+C_{Do}\right)c_{\theta_o}\right) \\[6pt]
-\dfrac{2r}{L}\dfrac{2k}{L}\left(C_{La}+C_{Do}\right)
\end{array} \right] \\[12pt]
-\dfrac{C_{Mba}}{L}\dfrac{2k}{L}
\end{array} \right) > 0
\tag{8a}
$$

Note: s_{θ_o} and c_{θ_o} are abbreviations for $\sin\theta_o$ and $\cos\theta_o$, respectively.

For a 4-boom spacecraft, the spacecraft will be stable if the following criterion is met:

$$
\left(\begin{array}{l}
S^3 \dfrac{2r}{L} C_{Do}\left(\left(C_{Da}-C_{Lo}\right)s_{\theta_o}+\left(C_{La}+C_{Do}\right)c_{\theta_o}\right)^2 \\[6pt]
+S^2\left(-\dfrac{2k}{L}+\dfrac{C_{Mba}}{2L}\right)\left(\left(C_{Da}-C_{Lo}\right)s_{\theta_o}+\left(C_{La}+C_{Do}\right)c_{\theta_o}\right)^2 \\[6pt]
-S\dfrac{2r}{L}\left(\dfrac{2k}{L}\right)^2\left(C_{La}+C_{Do}\right) \\[6pt]
-\dfrac{C_{Mba}}{2L}\left(\dfrac{2k}{L}\right)^2
\end{array} \right) > 0
\tag{8b}
$$

The variables C_{Lo}, C_{Do}, C_{La}, and C_{Da} are the lift and drag coefficients of the booms and the slope of those coefficients as the angle changes. Or, in other words, the total lift and drag coefficient can be written suitably with a linear expression:

$$
C_L = C_{Lo} + C_{La}\alpha'
\tag{9a}
$$

$$
C_D = C_{Do} + C_{Da}\alpha'
\tag{9b}
$$

(Note that the coefficients used for Eqs. (8a) and (8b) are the coefficients that correspond to the top-most boom in the 2-boom configuration, and the top-forward-most boom in the 4-boom configuration.)

These lift and drag coefficients are completely dependant on the aerodynamic model used. Since these terms appear in the stability criteria, the choice of gas-surface interaction model may have a significant impact on determining if a particular spacecraft design is stable or not.

Selection of the Gas-Surface Interaction Model

Eqs. (5a) thru (6b) will be used for determining the lift and drag coefficients. Evaluating C_L and C_D at $\alpha = \theta_o$ provides the equation for C_{Lo} and C_{Do} for a flat surface. By differentiating and evaluating at $\alpha = \theta_o$, equations for C_{La} and C_{Da} are determined.

Although there are more detailed gas-surface models available[13,14], the above formulas provide an excellent means to bound the predicted behavior by simply changing the variables, ε and ξ. The formulas also easily provide the aerodynamic coefficients that will be needed in the spacecraft stability analysis with Eqs. (8a) and (8b).

Modeling of the Booms

The shape of the booms will play a role in their aerodynamic behavior. The aerodynamic force on a cylindrical boom will be different than that on a square boom. The force on a flat boom will be different than a boom with a complex lattice structure.

These differences, however, become less prevalent under certain flight regime parameters. As the thermal speed ratio, $\bar{c}/V_o$, becomes small, the flow will see less of the back side of a surface. Thus, any complex structure on the back side of a surface can be ignored as the thermal speed ratio goes to zero. As the momentum rebound coefficient, ε, drops, the surface absorbs more of the incoming flow and rebounds less into its directional nature. Therefore surface details in the flow become less important. As the gas-surface interaction becomes more diffuse, the reflections become less directional. Here, too, the surface details become less important.

Thus, under certain gas-surface parameters, the booms can be approximated as flat plates with the same projected surface area for predicting the lift and drag of the booms. The flat plate is chosen because it is the simplest model for the boom.

Fortunately, the gas-surface interactions that are expected for an actual spacecraft and its booms come very close to meeting those parameter assumptions. As mentioned earlier, for orbital flight above 120 km above Earth, the thermal speed ratio will typically be less than 0.1, and a non-shiny or rough spacecraft surface is expected to be highly diffuse in nature.

As an example, if a designer uses cylindrical booms with likely values of $\xi = 0.1$ and $\varepsilon = 0.2$, the predicted lift and drag on a cylinder is, at most, only about 4% lower than that predicted by a flat plate.

If desired, one can account for this drag-relieving effect of a non-flat surface and still use the flat plate model. The flat plate model is used, but the calculated surface area is decreased to match the drag predicted by a cylindrical or other shaped model. This technique is especially useful for configurations with fixed boom positions.

ANALYSIS

Stability of 2-boom and 4-boom Spacecraft with Different Gas-Surface Interaction Parameters

· By substituting different values of ξ and ε into the coefficient equations, Eqs. (5a) thru (6b), and substituting the coefficients into Eqs. (8a) and (8b), one can see how the stability of a particular spacecraft changes with the choice of the boom's gas-surface interaction model.

Rewriting Eqs. (5a) and (5b) with the fixed spacecraft parameters, one can see how $C_{Mb\alpha}$ changes with boom position, θ_o.

For the 2-boom spacecraft:

$$\frac{C_{Mb\alpha}}{SL} < \frac{-\left(\left(\frac{2k}{SL} - \frac{2r}{L}C_{Do}\right)\left((C_{D\alpha} - C_{Lo})s_{\theta o} + (C_{L\alpha} + C_{Do})c_{\theta o}\right) + \frac{2r}{L}\frac{2k}{SL}(C_{L\alpha} + C_{Do})\right)}{\left(\frac{2k}{SL} - (C_{D\alpha} - C_{Lo})s_{\theta o} + (C_{L\alpha} + C_{Do})c_{\theta o}\right)} \tag{10a}$$

For a 4-boom spacecraft:

$$\frac{C_{Mba}}{SL} < \frac{-2\left(\left(\frac{2k}{SL}-\frac{2r}{L}C_{Do}\right)\left((C_{Da}-C_{Lo})s_{\theta_o}+(C_{La}+C_{Do})c_{\theta_o}\right)^2+\frac{2r}{L}\left(\frac{2k}{SL}\right)^2(C_{La}+C_{Do})\right)}{\left(\left(\frac{2k}{SL}\right)^2-\left((C_{Da}-C_{Lo})s_{\theta_o}+(C_{La}+C_{Do})c_{\theta_o}\right)^2\right)} \qquad (10b)$$

The term C_{Mba} is an important spacecraft parameter because for a symmetric spacecraft body it can be shown that the aerodynamic center, x_{AC}, and C_{Mba} are related by the equation:

$$C_{Mba} = \left(C_{Dbo} + C_{Lba}\right)x_{AC} \qquad (11)$$

Thus, C_{Mba}, is important in determining the aerodynamic center of the spacecraft if the location of the center of gravity is known.

To simplify the stability study, several of the spacecraft parameters will be fixed according to Table 1. These parameters are representative of a planned NASA spacecraft that is discussed in the next section.

Table 1

NOMINAL SPACECRAFT PARAMETERS

Variable	Value
L^{*}	5 m
L	2
S^{*}	1 m
S	0.2
k	4
r	−0.5
C_{Dbo}	3
C_{Lba}	0.5
$\dfrac{\bar{c}}{V_o}$	0.1

Note: For the ensuing analysis, it is assumed that the spacecraft body's lift and drag is completely known so that C_{Dbo} and C_{Lba} are fixed by the values in Table 1, and the parameters ξ and ε are unknown only for the booms. This will allow one to isolate the effect of the uncertainty of the boom's gas-surface interaction parameters on stability. Other investigations can also include varying ξ_{body} and ε_{body} for the spacecraft body as well as the booms so that C_{Dbo} and C_{Lba} also change. (Bowman, in Ref. 9, provides computational techniques for determining C_{Dbo} and C_{Lba}.) Eq. (11) would still simply apply but with the more detailed values.

Eqs (10a) and (10b) can be plotted for C_{Mba} vs. θ_o, for various values of ξ and ε. These are shown in Figures 5 and 6. The figures show the combinations of C_{Mba} and θ_o that produce either a stable spacecraft configuration or an unstable one. The division between the stable and unstable regions is represented by the various lines depending on the choice of the gas-surface interaction parameters shown. The thick, solid line represents the division between these two regions for a nominal gas-surface interaction parameter set of $\xi = 0.1$ and $\varepsilon = 0.2$. However, these parameters may not be actually known. Since both ξ and ε can vary from 0 to 1, the division line could theoretically be any of those shown. However, those lines are the extremes of the parameters. Any combination will fall somewhere inside the light-grey region.

1187

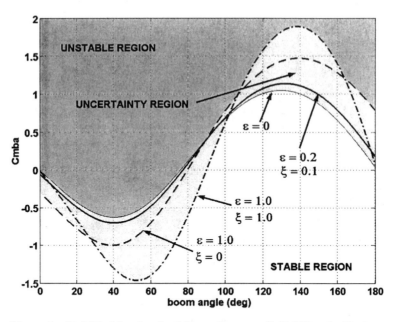

Figure 5   Stability Map for the 2-Boom Spacecraft, Full ξ and ε Variance

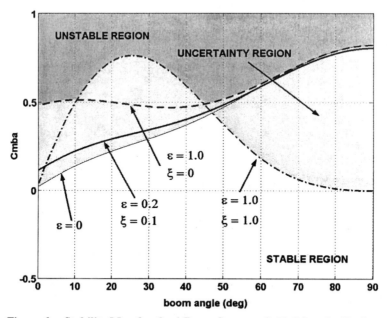

Figure 6   Stability Map for the 4-Boom Spacecraft, Full ξ and ε Variance

As a design tool, the spacecraft is guaranteed to be stable if the variable spacecraft parameters are in the clear area and unstable if they are in the dark-grey region even if knowledge of the gas-surface

1188

interaction is unknown. However, if the parameters are located in the light-grey region, stability or instability cannot be guaranteed.

One can see that regions of stability vary tremendously depending on the range of ξ and ε. (Note: For legibility, different scales were used for the 2-boom and 4-boom spacecraft, but in either case, the size of the uncertainty region is of the same order.)

In all probability, however, the range of ξ and ε is expected to be much smaller. It is very likely that both ξ and ε will only be between 0 and 0.4 for most spacecraft surfaces.

Redrawing Figures 5 and 6 for these new ranges produces the following graphs (Figures 7 and 8):

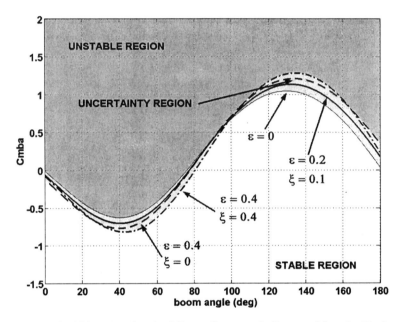

Figure 7   Stability Map for the 2-Boom Spacecraft, Expected ξ and ε Variance

One sees that the light-grey uncertainty region has shrunk considerably when the range of the boom's likely gas-surface interaction parameters is limited. Thus, even some improved knowledge of the gas-surface interaction expectations will be of great aid to the spacecraft designer.

One of the greater benefits of improved gas-surface interaction parameters is that the known stable region becomes larger. This will allow for a greater range of spacecraft designs while still being confident of their stability. It gives flexibility to the designer.

Note, however, that depending on the boom position, estimates for desired stable $C_{Mb\alpha}$ values still can differ by a large percentage difference, so it is still a design challenge when knowledge of the gas-surface interaction is not fully known.

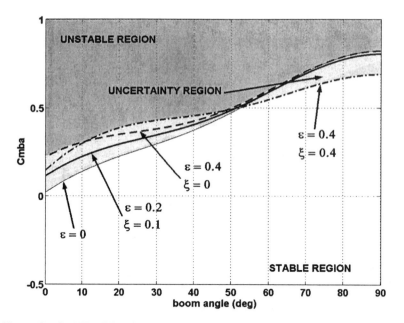

Figure 8   Stability Map for the 4-Boom Spacecraft, Expected ξ and ε Variance

Applications to an Orthogonally Positioned Six-Boomed Spacecraft

The proposed NASA/Goddard Geospace Electrodynamic Connections Mission (GEC)[20] is an example of a six-boomed spacecraft that will be conducting "dipping" campaigns into the upper regions of the Earth's atmosphere for scientific measurements. For this spacecraft, the boom positions are fixed in an orthogonal arrangement (see Figure 9) with 4 booms at 45° angles in the pitching plane and 2 booms at 90° angles in the yawing plane.

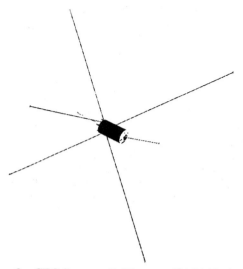

Figure 9   GEC Spacecraft (Courtesy NASA/Goddard)

A couple of interesting design questions are how the placement of the booms and location of the aerodynamic center of the spacecraft affects the stability results, and how the incomplete knowledge of the gas-surface interaction influences the prediction.

Because the design parameter θ_o is already set for the GEC spacecraft, the design variables x_{AC} and r will be examined.

Since the booms of the 6-boom satellite are in different and perpendicular planes, the same equations, Eqs. (10a) and (10b), can be used as there are two booms in one plane and four in the other plane. The value of C_{Dbo} in Eq. (11), however, needs to be modified to include the drag of those booms not in the plane being examined. In the pitching plane, there are 2 booms that are non-influential "dead weight" as far as that plane is concerned and Eq. (12a) applies. In the yawing plane, there are 4 booms that are non-influential "dead weight", and Eq. (12b) applies.

$$C_{Mb\alpha} = \left(C_{Dbo} + 2SC_{Do} + C_{Lb\alpha}\right)x_{AC} \tag{12a}$$

$$C_{Mb\alpha} = \left(C_{Dbo} + 4SC_{Do} + C_{Lb\alpha}\right)x_{AC} \tag{12b}$$

After taking this into account, the aerodynamic center location, x_{AC}, was plotted vs. boom location, r, to determine stability in the pitch and yaw planes. (Note that because of the assumed symmetric nature of the spacecraft, x_{AC} is the same for both the pitching and yawing plane.) See Figure 10.

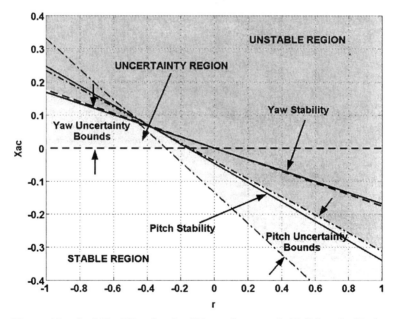

Figure 10   Stability Map for the 6-Boom Spacecraft, Full ξ and ε Variance

The two solid lines represent the border between the stable and unstable regions for the gas-surface interaction parameters of $\xi = 0.1$ and $\varepsilon = 0.2$. One line denotes the stability in the pitching plane and the other line, the yawing plane as illustrated.

These lines are flanked by the uncertainty bounds where all combinations of ξ and ε for each plane will lie. The dashed lines show the border of all the combinations in the yaw plane, and the dot-dashed lines show the border of all the combinations in the pitch plane. Those lines bound all combinations of ξ and ε from 0 to 1.

By taking the intersection of the pitch and yaw sets of bounds, Figure 10 shows which combinations of x_{AC} and r are guaranteed to be stable (clear region) or unstable (dark-grey region). No guarantees can be made if the combination lies in the light-grey region because of the uncertainty of the boom's gas-surface interaction.

As before, the expected range of ξ and ε will likely be much smaller. Figure 10 is redrawn to illustrate the uncertainty when the values of ξ and ε range from only 0 to 0.4 (See Figure 11).

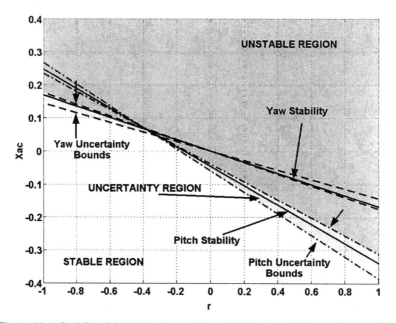

Figure 11   Stability Map for the 6-Boom Spacecraft, Expected ξ and ε Variance

With improved assumptions of the gas-surface interaction parameters, the design space becomes more manageable. A designer has a larger stable region which gives more design flexibility.

Additionally, even if the design parameters of the spacecraft are deemed stable at the beginning of a spacecraft's life, it is possible that they may change over its lifetime. During the spacecraft's mission life, the values of x_{AC} and r could change, say, as fuel is consumed.

Using Figure 10 as an example, if the fuel was stowed near the front of the spacecraft, as it is consumed, the center of gravity will shift rearward, and the values of x_{AC} and r will become more positive. From any arbitrary initial position in Figure 9, that point will move up and to the right toward the unstable region. If the uncertainty bounds are large, it may not be known whether the spacecraft will drift into an unstable region or not during its lifetime. Additionally, since it may be unclear where the true yaw and pitch stability lines intersect, it may not be known if the spacecraft becomes unstable first in the pitching plane or the yawing plane.

Fortunately, for expected values of ξ and ε, the stability lines don't vary as drastically, and the question seems to be easily answered at most points except where the two lines intersect. At this point, the

1192

answer is still fairly uncertain. A designer may wish to avoid a configuration that is near this intersection point.

CONCLUSION

The gas-surface interaction plays an important role in determining the stability of a flexible spacecraft. When knowledge of the gas-surface interaction is unknown, it is very difficult to determine whether a spacecraft configuration is stable or unstable. Fortunately, when one makes reasonable bounds on the gas-surface interaction parameters, the stability results are better predicted. As knowledge of the gas-surface interactions improves, the number of guaranteed stable spacecraft configuration increases. This gives the spacecraft designer more flexibility in his or her design.

In practice, it is easier to predict the gas-surface interaction behavior of a very rough, absorptive material. In such a case, both ξ and ε will be very close to 0. For a very shiny and reflective material, the higher values of ξ and ε are generally more difficult to determine. Additionally, ξ and ε tend to decrease as the spacecraft ages due to atmospheric wear and oxidation.

Because of the variability of the parameters for a specular, reflective material, the region of stability uncertainty will be larger than that for a diffuse, absorptive material. Therefore, if stability is of great concern to the designer, he or she should choose very diffusive and absorptive materials for the boom design. With such materials, the stability predictions are more accurate. The trade-off, though, is that such materials could produce a larger amount of drag.

If drag is of a larger concern, then with using shiny, reflective boom materials, the designer will either have to use a spacecraft configuration that is well within the spacecraft's stable region under the large range of interaction parameters, or undertake a large number of tests and experiments to better pin down the exact gas-surface interaction parameter values.

FUTURE WORK

This paper has focused on how the uncertainty in the gas-surface interaction parameters affects the booms for spacecraft stability. Uncertainty in the gas-surface interaction parameters of the spacecraft body was not included. Since the spacecraft booms and body would generally be made of the different materials, further work could include how the gas-surface interaction parameters change the spacecraft body's lift and drag, and how that change affects the overall spacecraft stability.

ACKNOWLEDGMENTS

The authors would like to thank the NASA/Goddard Graduate Student Research Program, technical advisors, Mr. Thomas Stengle and Mr. Michael Mesarch, and GEC program manager, Mr. Paul Buchanan, for their research opportunities and support.

REFERENCES

[1]London, H. S., "Change of Satellite Orbit Plane by Aerodynamic Maneuvering", *Journal of the Aerospace Sciences*, Vol. 29, No. 2, 1962, pp.323-332.

[2]Citron, S. J., Meir, T. C., "An Analytic Solution for Entry into Planetary Atmospheres", *AIAA Journal*, Vol 3, No. 3, March 1965.

[3]Vinh, N. X., *Optimal Trajectories in Atmospheric Flight*, Elsevier Scientific Publishing Co., New York, NY, 1981.

[4]Turner, J. D., Chun, H. M., "Optimal Distributed Control of a Flexible Spacecraft during a Large-Angle Maneuver", *Journal of Guidance and Control*, Vol. 7, No. 3, May-June 1984.

[5]Park, S.-Y., "Thermally Induced Attitude Disturbance Control for Spacecraft with a Flexible Boom", *Journal of Spacecraft and Rockets*, Vol. 39, No. 2, Mar-Apr 2002.

[6]Budynas, R., Poli, C., "Three-Dimensional Motion of a Large Flexible Satellite", *Automatica*, Vol. 8, Pergamon Press, Great Britain, 1972, pp. 275-286.

[7]Schultz, J. R., Pines, D. J., "Stability and Control of Flexible Spacecraft during a Shallow Aeroassist", *Proceedings of the AIAA/AAS Astrodynamics Specialist Conference*, AIAA-2002-4521

[8]Lewis, M. J., "Aerodynamic Maneuvering for Stability and Control of Low-Perigee Satellites," *Proceedings of the AAS/AIAA Space Flight Mechanics Meeting*, AAS Paper 01-239, Feb 2001.

[9]Bowman, D. S., "Numerical Optimization of Low-Perigee Spacecraft Shapes," Masters Thesis, University of Maryland, College Park, MD, UM-AERO 01-03, 2001.

[10]Sentman, L. H., "Free Molecule Flow Theory and Its Application to the Determination of Aerodynamic Forces", Lockheed Missiles & Space Company, Tech Report LMSC-448514, 1 October 1961.

[11]*U.S. Standard Atmosphere, 1976*, National oceanic and Atmospheric Administration/National Aeronautics and Space Administration/United States Air Force, Washington, D.C. 1976.

[12]Goodman, F. O., and Wachman, H. Y., *Dynamics of Gas-Surface Scattering*, Academic Press, New York, 1976, pp. 307-308.

[13]Hurlbut, F.C., and Sherman, F.S., "Application of the Nocilla Wall Reflection Model to Free-Molecule Kinetic Theory," *Physics of Fluids*, Vol. 11, No. 3, March 1968, pp, 486-496.

[14]Bird, G.A. *Molecular Gas Dynamics and the Direct Simulation of Gas Flows*, Oxford Engineering Science, Clarendon Press, Oxford, England 1995, pp. 118-122.

[15]Potter, J. L., and Rockaway, J. K., "Aerodynamic Optimization for Hypersonic Flight at Very High Altitudes," *Progress in Astronautics and Aeronautics*, Vol. 160,, ed. by Shizgal, B, and Weaver, D., American Institute of Aeronautics and Astronautics, Washington, D.C., 1994.

[16]Harvey, J., and Lord, G., *Rarefied Gas Dynamics*, Oxford University Press, Oxford, England 1995.

[17]Herrero, F. A., "The Lateral Surface Drag Coefficient of Cylindrical Spacecraft in a Rarefied Finite Temperature Atmosphere," *AIAA Journal*, Vol 23, No. 6, June 1984. pp. 862-867.

[18]Hedin, A. E. "MSIS-86 Thermospheric Model," *Journal of Geophysical Research*, Vol. 92, 1987, pp. 4649-4662.

[19]Cross, J. B., and Blais, N. C., "High-Energy/Intensity CW Atomic Oxygen Beam Source," in *Rarefied Gas Dynamics: Space Related Studies, Progress in Astronautics and Aeronautics*, American Institute of Aeronautics and Astronautics, Washington, D.C., Volume 116, 1989, pp. 143-155.

[20]"Understanding Plasma Interactions with the Atmosphere: The Geospace Electrodynamic Connections (GEC) Mission", NASA/TM-2001-209980.

ORBITAL DEBRIS

Chair: David Spencer
                                    Pennsylvania State University

The following paper was not available for publication:

AAS 03-189
   "Kalman Filter Approach Using Constant Gains for Predicting the Re-Entry of
   Space Debris Objects," by A.K. Anilkumar, M.R. Ananthasayanam, P.V. Subba
   Rao (Paper Withdrawn)

IMPROVED ANALYTICAL EXPRESSIONS
FOR COMPUTING SPACECRAFT COLLISION PROBABILITIES

Ken Chan[*]

This paper presents an improved analytical expression for computing the collision probability between two orbiting objects. In the encounter plane, a scale transformation converts the bivariate Gaussian probability density function (pdf) to an isotropic Gaussian pdf and the circular cross-section to an elliptical one. By approximating this ellipse with an equivalent circle, the two-dimensional Gaussian distribution is replaced by a one-dimensional Rician, the integral of which is known in the form of a convergent infinite series. This study shows that it is permissible to truncate after the first or second term and still obtain results to within 0.4% error for a wide range of collision parameters considered. This method of computing the collision probability is then extended to include the case of more general cross-sections such as the International Space Station (ISS). The model presently used by NASA is based on a two-dimensional numerical integral evaluation and takes ten thousand times longer than the analytical formulation obtained in this study. Moreover, the NASA model assumes that the collision cross-section is circular and is not applicable to a complex spacecraft structure such as the ISS. Comparisons have been made with other methods involving numerical evaluation of the collision integral. Agreement is observed for two or three significant figures for extremely close encounters with collision probabilities of 10^{-2} to distant encounters with probabilities of 10^{-28}.

INTRODUCTION

An earlier formulation[1,2,3] was derived primarily for computing the collision probability for encounters with numerous satellites over an extended period of time. For each pair-wise conjunction, the probability density function (pdf) of the relative position error is given by a three dimensional Gaussian form having in general three unequal standard deviations. In the encounter plane which is perpendicular to the relative velocity vector at the point of closest approach, this pdf is bivariate Gaussian. In the previous derivation, because the relative position at the point of closest approach in the encounter plane was not taken into account, the ellipses of constant pdf were replaced by equivalent circles of constant pdf of the same area. The consequence was that the two dimensional Gaussian distribution was replaced by a one-dimensional Rician distribution which involved the modified Bessel function of the first kind of zero order. This integral was obtained in the form of an infinite series convergent for all values of the relevant parameters: the miss distance, the geometric mean standard deviation and the

*   The Aerospace Corporation, 15049 Conference Center Drive, Chantilly, Virginia 20151.
E-mail: kenneth.f.chan@aero.org.

radius of the effective cross-section of collision. Thus, for each pair-wise conjunction, the angle of the relative position in the encounter plane was not taken into account. In effect, it was averaged out by replacing the ellipses with circles. The result was that the long-term cumulative collision probability was amenable to analytical evaluation.

The present analysis undertakes to include the angle of the relative position in the encounter plane. Thus, we have the problem of integrating a bivariate Gaussian pdf over a circular cross-section displaced from the center of the ellipses. By performing a scale transformation, we convert this into the problem of integrating an isotropic Gaussian pdf over an elliptical cross-section. Up to this step, the analysis is still exact. If we now approximate the elliptical cross-section with an equivalent circular cross-section of the same area, we have thus reduced the problem again into the same form as the previous. Thus, we now once again consider the Rician integral which involves one variable and whose analytical form we have already obtained. It is noted that the two problems are not tautologically the same. In the previous one, the elliptical pdf contours were replaced by equivalent circular pdf contours. In the present one, the elliptical cross-section is replaced by an equivalent circular cross-section. However, both are now handled by the same mathematics from this point onward. The only difference is that our formulation now involves two kinds of standard deviations: the mean standard deviation and an additional "auxiliary" standard deviation. This improved formulation has been derived and reported by the present author earlier[4].

At a conjunction involving a fly-by (which holds for most encounters except possibly for some infrequent ones involving geosynchronous satellites), an encounter coordinate system is defined as follows: Let $\mathbf{v}_{ps}$ denote the relative velocity of the primary spacecraft with respect to the secondary object which may be another active spacecraft, a rogue satellite or space debris. In the encounter coordinate system, let the y-axis be defined along this vector. Then, the (x, z)-plane is normal to this vector. At the instant of closest approach, let the x-axis be defined such that the primary spacecraft is at $(x_e, 0, 0)$. Thus, we have defined the encounter coordinate system whose unit basis vectors are $\hat{x}$, $\hat{y}$, and $\hat{z}$. This is illustrated in Figure 1.

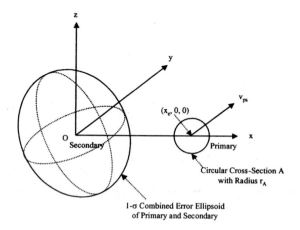

Figure 1 - Combined error ellipsoid and collision cross-section in encounter coordinate system

Let C be the combined covariance matrix in the encounter system, defined to be the sum of the individual covariance matrices for the case of independent or (less stringently) uncorrelated random variables. This covariance matrix gives the pdf of the relative position of the satellites. Let r_p denote the radius of the sphere, which just circumscribes the primary spacecraft, the center of the sphere being located at the center of mass. Similarly, let r_s denote the radius of the circumscribing sphere of the secondary object. Let us now assume that both objects are spherical, the reasons being that we simplify the problem and also we do not account for their orientation (attitude). Then, if the secondary comes within a sphere with radius $r_A = r_p + r_s$ centered at the primary, there will be a collision between the two.

In the vicinity of the encounter, an assumption is usually made that the trajectories of the two objects are straight lines. From a qualitative view, one argues that the orbiting velocities are of the order of several kilometers per second and the time spent in the encounter region is only a fraction of a second or at most a few seconds so that the effects of gravitational force are negligible. This results in essentially rectilinear motion over a large region of tens or hundreds of standard deviations in extent. Then, the volume swept out by the sphere of radius r_A is a long cylinder essentially extending along the y-direction from $-\infty$ to $+\infty$. Thus, instead of having to deal with a three-dimensional pdf, a little consideration reveals that we need only consider the marginal two-dimensional pdf. In the case of the marginal pdf of a joint Gaussian distribution of random variables, no tedious explicit integral evaluation need be performed. We merely set $y = 0$ and appropriately change the multiplicative factor in the three-dimensional pdf to obtain the desired result[5]. Hence, at the instant of closest approach when the primary crosses the (x, z)-plane, its relative position uncertainty is described by the following bivariate Gaussian pdf

$$f\left(x, z\right) = \frac{1}{2\pi \, \sigma_x \sigma_z \sqrt{1-\rho_{xz}^2}} e^{-\left[\left(\frac{x}{\sigma_x}\right)^2 - 2\rho_{xz}\left(\frac{x}{\sigma_x}\right)\left(\frac{z}{\sigma_z}\right) + \left(\frac{z}{\sigma_z}\right)^2\right]/2\left(1-\rho_{xz}^2\right)} . \tag{1}$$

This expression can also be interpreted as the conditional pdf for collision given that the primary crosses the encounter plane. Hence, the probability of collision is given by the two-dimensional integral

$$P = \iint\limits_A f(x,z) \, dx \, dz \tag{2}$$

where A is the collision cross-section area which is a circle with radius r_A centered at the primary.

TRANSFORMATION TO PRINCIPAL AXES

In general, the form of the pdf in equation (1) contains an undesirable diagonal term when the correlation coefficient ρ is not zero. In order to simplify analysis, we rotate the coordinate system (x, z) to a new coordinate system (x', z') such that the off-diagonal terms no longer appear in the pdf. In this new coordinate system, the major and minor axes of the ellipse associated with the covariance matrix C will be along the coordinate axes, which are the principal directions. To accomplish this, we proceed as follows: The covariance matrix C is given by

$$C = \begin{bmatrix} \sigma_x^2 & \rho_{xz}\sigma_x\sigma_z \\ \rho_{xz}\sigma_x\sigma_z & \sigma_z^2 \end{bmatrix}. \tag{3}$$

Rotation of coordinates to principal directions yields the covariance matrix C' given by

$$C' = \begin{bmatrix} \sigma_{x'}^2 & 0 \\ 0 & \sigma_{z'}^2 \end{bmatrix}. \tag{4}$$

Then, we may show that the requisite rotational angle θ is given by

$$\theta = \frac{1}{2}\tan^{-1}\left[\frac{2\rho_{xz}\sigma_x\sigma_z}{\left(\sigma_x^2 - \sigma_z^2\right)}\right]. \tag{5}$$

For the case of $\sigma_x \neq \sigma_z$, equation (5) yields two values of the rotational angle θ from the x-axis to the x'-axis corresponding to the two non-collinear eigenvectors. We may take one of these to be in the first quadrant and the other in the fourth quadrant. If we wish to assign $\sigma_{x'} \geq \sigma_{z'}$, then we choose θ to be in the first or fourth quadrant according to the correlation coefficient $\rho > 0$ or < 0. This is illustrated in Figure 2.

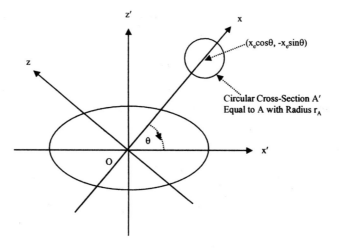

Figure 2 - Rotation to Principal Axes

For the case of $\sigma_x = \sigma_z$, the Right Hand Side (RHS) of equation (5) is undefined. A little consideration reveals that we choose θ to be $\pm \pi/4$ depending on the sign of ρ. However, in general, equation (5) applies to rotation to either one of the two principal axes x' or z' without requiring that $\sigma_{x'} \geq \sigma_{z'}$.

Since the determinant remains invariant under a rotation, therefore from equations (3) and (4)

$$\sigma_{x'}\sigma_{z'} = \sqrt{1-\rho_{xz}^2}\,\sigma_x\sigma_z \ . \tag{6}$$

In the (x', z') system, the pdf (1) becomes

$$f(x',z') = \frac{1}{2\pi\sigma_{x'}\sigma_{z'}}\,e^{-\frac{1}{2}\left[\left(\frac{x'}{\sigma_{x'}}\right)^2 + \left(\frac{z'}{\sigma_{z'}}\right)^2\right]} \tag{7}$$

and the collision probability (2) is given by

$$P = \frac{1}{2\pi\,\sigma_{x'}\sigma_{z'}}\iint_{A'} e^{-\frac{1}{2}\left[\left(\frac{x'}{\sigma_{x'}}\right)^2 + \left(\frac{z'}{\sigma_{z'}}\right)^2\right]}\,dx'dz' \tag{8}$$

where A' is the circular cross-section with radius $r_{A'}$ centered at the primary. Let (x_p', z_p') denote the coordinates of the center of mass of the primary in the (x', z') system. Therefore, we have

$$A' = A, \qquad r_{A'} = r_A, \qquad x_p' = x_e\cos\theta, \qquad z_p' = -x_e\sin\theta \ . \tag{9}$$

TRANSFORMATION TO ISOTROPIC DISTRIBUTION

We next wish to map the ellipses of constant pdf

$$\frac{x'^2}{\sigma_{x'}^2} + \frac{z'^2}{\sigma_{z'}^2} = k^2 \tag{10}$$

into circles of constant pdf having the equation

$$\frac{x''^2}{\sigma_{z'}^2} + \frac{z''^2}{\sigma_{z'}^2} = k^2 \ . \tag{11}$$

The constant k is a label for the size of the ellipse or circle. This mapping is accomplished by the following transformation

$$x'' = \frac{\sigma_{z'}}{\sigma_{x'}}x', \qquad z'' = z' \ . \tag{12}$$

In the (x'', z'') system, the pdf (7) becomes isotropic, viz,

$$f(x',z') = \frac{1}{2\pi\,\sigma_{z'}^2}\,e^{-\left(x''^2 + z''^2\right)/2\sigma_{z'}^2} \tag{13}$$

and the collision probability (8) is equivalently given by

$$P = \frac{1}{2\pi \, \sigma_{z'}^{2}} \iint_{A''} e^{-\left(x''^{2} + z''^{2}\right)/2\sigma_{z'}^{2}} \, dx'' \, dz'' \tag{14}$$

where A″ is the elliptical cross-sectional area with semi-major axis a and semi-minor axis b centered at the primary. Let (x_p'', z_p'') denote the coordinates of the center of mass of the primary in the (x″, z″) system. Therefore, we have

$$A'' = \frac{\sigma_{z'}}{\sigma_{x'}} A', \qquad a = r_{A'}, \qquad b = \frac{\sigma_{z'}}{\sigma_{x'}} r_{A'}, \qquad x_p'' = \frac{\sigma_{z'}}{\sigma_{x'}} x_p', \qquad z_p'' = z_p' \ . \tag{15}$$

APPROXIMATION OF CROSS-SECTION BY CIRCLE

Up to this step, we have not made any approximations. Suppose we now approximate the elliptical cross-section A″ with a circular cross-section A‴ of the same area with radius $r_{A'''}$. This is illustrated in Figure 3.

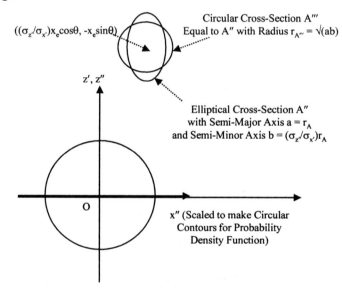

Figure 3 - Approximation of Cross-Section by Circle

Let r_p'' denote the distance of the primary from the origin in the (x″, z″) system. Then, we have

$$A''' = \pi r_{A''}^{2}, \qquad r_{A''} = \sqrt{ab} = \sqrt{\frac{\sigma_{z'}}{\sigma_{x'}}} r_{A'}, \qquad r_p''^{\,2} = x_p''^{\,2} + z_p''^{\,2} \ . \tag{16}$$

1202

Thus, the collision probability (14) is approximated by

$$P = \frac{1}{2\pi\,\sigma_{z'}^2} \iint\limits_{A''} e^{-\left(x''^2 + z''^2\right)/2\sigma_{z'}^2}\, dx''\, dz'' \tag{17}$$

By using equations (9), (15) and (16), we may easily show that the distance r_p'' of the primary from the origin in the (x'', z'') system is given by the following two equivalent forms

$$r_p''^{\,2} = x_e^2 \left\{ 1 + \left[\left(\frac{\sigma_{z'}}{\sigma_{x'}}\right)^2 - 1 \right] \cos^2\theta \right\} \tag{18}$$

$$r_p''^{\,2} = x_e^2 \left\{ 1 + \left[\left(\frac{\sigma_{z'}}{\sigma_{x'}}\right)^2 - 1 \right] \left(\frac{x_p'^{\,2}}{x_p'^{\,2} + z_p'^{\,2}}\right) \right\} . \tag{19}$$

IMPROVED ANALYTICAL EXPRESSIONS

There is a benefit from the previous approximation: The two-dimensional isotropic Gaussian pdf with standard deviation $\sigma_{z'}$ in equation (17) integrated over a circle of radius $r_{A'''}$ displaced a distance r_p'' from the origin can be transformed[6] into the following one-dimensional Rician pdf integrated from 0 to $r_{A'''}$

$$P = \int_0^{r_{A'''}} \frac{r}{\sigma_{z'}^2} e^{-\left(r^2 + r_p''^{\,2}\right)/2\sigma_{z'}^2} I_0\left(\frac{r\,r_p''}{\sigma_{z'}^2}\right) dr . \tag{20}$$

In this equation, $I_0(\cdot)$ denotes the modified Bessel function of the first kind of order zero. This pdf arises frequently in the detection of signals in the presence of noise. The transformation was originally obtained by Rice[7] in 1944 who studied the case of a sine wave plus random noise and obtained asymptotic expressions for the integral under some simplifying assumptions. The Rician pdf also arises in target detection by pulsed radar, non-coherent detection of band-passed binary signals, artillery effectiveness of fragmentation bombs, and presently in collision of spacecraft with space debris. This pdf has the following properties: When the cross-section displacement is equal to zero, it is precisely the Rayleigh pdf. If it is large, the pdf tends to be symmetrical in the interval of several units of standard deviation containing the maximum point and resembles the Gaussian pdf in this interval (keeping in mind that it is defined from 0 to ∞ whereas the latter is defined from $-\infty$ to $+\infty$). Computer programs are presently used for the numerical evaluation of this integral. However, an analytical solution in the form of an infinite series which is convergent for all values of the relevant parameters has been reported[8] by the author in addition to the expressions with error bounds obtained previously[1,2,3]. This solution is

$$P = e^{-v/2} \sum_{m=0}^{\infty} \frac{v^m}{2^m\, m!} \left[1 - e^{-u/2} \sum_{k=0}^{m} \frac{u^k}{2^k\, k!} \right] \tag{21}$$

where, by using equations (16) through (20), it is convenient to introduce the dimensionless variables u and v (similar to those in [1], [2] and [3]) but which are now defined in terms of the mean standard deviation σ, the auxiliary standard deviation σ^*, miss distance x_e, and the radius r_A of collision cross-section integration by

$$u \equiv \left(\frac{r_{A''}}{\sigma_{z'}}\right)^2 = \left(\frac{r_A^2}{\sigma_{x'}\sigma_{z'}}\right) = \left(\frac{r_A}{\sigma}\right)^2 > 0 \tag{22}$$

$$v \equiv \left(\frac{r_p''}{\sigma_{z'}}\right)^2 = x_e^2\left\{1+\left[\left(\frac{\sigma_{z'}}{\sigma_{x'}}\right)^2 - 1\right]\left(\frac{x_p'^2}{x_p'^2 + z_p'^2}\right)\right\} \Big/ \sigma_{z'}^2 = \left(\frac{x_e}{\sigma^*}\right)^2 > 0 \tag{23}$$

$$\sigma^2 \equiv \sigma_{x'}\sigma_{z'} \tag{24}$$

$$\sigma^{*2} \equiv \sigma_{z'}^2\left\{1+\left[\left(\frac{\sigma_{z'}}{\sigma_{x'}}\right)^2 - 1\right]\left(\frac{x_p'^2}{x_p'^2 + z_p'^2}\right)\right\}^{-1} . \tag{25}$$

If only the first term in the series is retained ($m = 0$ in equation (21)), it is proved that this collision integral may be expressed analytically as

$$P = e^{-v/2}(1 - e^{-u/2}) + S_1 \tag{26}$$

where S_1 satisfies the inequality

$$S_1 < \frac{1}{8}u^2ve^{-v/2}e^{uv/4} . \tag{27}$$

We may refer to S_1 as the truncation error after the first term. Unlike signal detection with different parameter ranges in its applications, it turns out that for the spacecraft encounters considered in this study, this truncation after the first term yields an error which appears in the fourth significant digit. Thus, it suffices to retain only the explicit terms in equation (21). If not, then we would have to truncate after the second term ($m = 1$ in equation (21)) and use the explicit terms in the following equation:

$$P = e^{-v/2}\left\{\left(1 - e^{-u/2}\right) + \frac{v}{2}\left[1 - \left(1 + \frac{u}{2}\right)e^{-u/2}\right]\right\} + S_2 \tag{28}$$

where S_2 satisfies the inequality

$$S_2 < \frac{1}{128}u^3v^2e^{-v/2}e^{uv/4} . \tag{29}$$

It suffices to use equation (28) because any additional terms beyond this will contribute to insignificant accuracy in the numerical results. Moreover, it must be borne in mind that we have already made an approximation in representing the elliptical cross-section by a circular cross-section in the transition from equation (14) to (17).

COMPARISION WITH OTHER MODELS

As mentioned before, the analytical expressions (26) and (28) yield results that agree extremely well with those of other formulations. In 1992, Foster derived the two-dimensional NASA model[9] which obtains the collision probability by numerically integrating the following integral

$$P = \frac{1}{2\pi\sigma_{x'}\sigma_{z'}} e^{-\frac{1}{2}x_e^2\left[\left(\frac{\sin\varphi}{\sigma_{x'}}\right)^2+\left(\frac{\cos\varphi}{\sigma_{z'}}\right)^2\right]} \int_0^{r_A} r \int_0^{2\pi} e^{-\frac{1}{2}r^2\left[\left(\frac{\sin\theta}{\sigma_{x'}}\right)^2+\left(\frac{\cos\theta}{\sigma_{z'}}\right)^2\right]+rx_e\left[\left(\frac{\sin\theta\sin\varphi}{\sigma_{x'}^2}\right)+\left(\frac{\cos\theta\cos\varphi}{\sigma_{z'}^2}\right)\right]} d\theta dr \quad (30)$$

where ϕ is the angle of the primary from the z'-axis. In the implementation, the angle stepsize is $0.5°$ and the radius stepsize is $r_A/12$.

In 1995, Chan[1] reduced the two-dimensional integral to the following one-dimensional integral in the (x,z)-encounter plane

$$P = \frac{1}{\sqrt{2\pi}\sigma_x} \int_{x_e-r_A}^{x_e+r_A} e^{-x^2/2\sigma_x^2} \frac{1}{2}\left[\text{erf}\left(\frac{\lambda+m}{\sqrt{2}\sigma'}\right)+\text{erf}\left(\frac{\lambda-m}{\sqrt{2}\sigma'}\right)\right] dx \quad (31)$$

where

$$\frac{\lambda \pm m}{\sqrt{2}\sigma'} \equiv \frac{\sqrt{r_A-(x-x_e)^2} \pm \left(\frac{\rho_{xz}\sigma_z}{\sqrt{2}\sigma'}\right)}{\sqrt{2}\sigma_z\sqrt{1-\rho_{xz}^2}} . \quad (32)$$

The integrand involves the error function which is difficult to deal with analytically, especially if the argument is large. Thus, the collision probability has to be computed by numerically evaluating the error function using standard algorithms. Because of this reason, this formulation was not pursued further.

In 2000, Patera[10] derived the following mathematically equivalent model which is a one-dimensional "line" integral

$$P = -\frac{1}{2\pi} \oint_{\text{Ellipse}} e^{-\alpha r^2} d\theta \quad \text{if origin of encounter lies outside elliptical cross-section} \quad (33)$$

$$P = 1-\frac{1}{2\pi} \oint_{\text{Ellipse}} e^{-\alpha r^2} d\theta \quad \text{if origin of encounter lies inside elliptical cross-section.} \quad (34)$$

The expression for α appearing in the exponent of the integrand is extremely complicated and depends on expressions involving elements of the transformation matrix from the three-dimensional encounter system to the three-dimensional principal axes system and also on the standard deviations in the diagonal covariance matrix. The integration is performed numerically around a circuit in the counter-clockwise direction.

In 2002, Alfano[11] obtained the integral (31) but expressed it in the (x′,z′)-encounter plane in which the pdf has only diagonal terms $\sigma_{x'}$ and $\sigma_{z'}$

$$P = \frac{1}{2\sqrt{2\pi}\sigma_{x'}} \int_{x'_p - r_A}^{x'_p + r_A} e^{-\frac{x'^2}{2\sigma_{x'}^2}} \left\{ \text{erf}\left[\frac{z'_p + \left(r_A^2 - (x' - x'_p)^2\right)^{1/2}}{\sqrt{2}\sigma_{z'}} \right] + \text{erf}\left[\frac{-z'_p + \left(r_A^2 - (x' - x'_p)^2\right)^{1/2}}{\sqrt{2}\sigma_{z'}} \right] \right\} dx' \quad (35)$$

where (x'_p, z'_p) denotes the center of mass of the primary. Even though the integrand is simpler, the evaluation of this integral has still to be performed numerically. Again, when the argument exceeds approximately 6 (approximately 8.5 standard deviations), it is difficult to accurately evaluate the error function numerically even by the use of recursion relations because of truncation errors. The value of 6 corresponds to the limits of computing the error function using double precision. All the above formulations rely strictly on numerical integration for implementation. However, Alfano[12] also used the algebraic series processing algorithms in Mathcad to expand the integrand and then integrate termwise the RHS of equation (8) to obtain the following expression

$$P = \left(1 + P_1 + P_2 + P_3 + P_4 + P_5 + P_6\right) \frac{1}{2\sigma_{x'}\sigma_{z'}} e^{-\frac{1}{2}\left[\left(\frac{x'_p}{\sigma_{x'}}\right)^2 + \left(\frac{z'_p}{\sigma_{z'}}\right)^2\right]} . \quad (36)$$

The polynomials P_i are very complicated expressions involving the diagonal standard deviations, the coordinates of the primary in the (x′,z′)-encounter plane and the radius of the circular cross-section. This expression yields results which agree to at least four significant digits in comparison with the more accurate numerical integration models. Thus, in essence, equations (26), (28) and (36) are the only analytical models which are applicable over a fairly large range of parameters for spacecraft collision probability computations.

The above four formulations (NASA, Chan, Patera and Alfano) described by equations (26), (30), (33) and (35) were coded and investigations were performed to study their comparative accuracies. The following tables show results comparing equations (26) and (28) with the NASA, Patera and Alfano formulations. Hereafter, for convenience, we shall refer to them as "models" even though they are not truly different models but are all derived from the same original three-dimensional Gaussian pdf. (In common usage, a model usually involves mathematical assumptions entering into the description of a physical problem.) The NASA model is used as a basis of comparison for all the other models. To replicate the exact NASA model, we retain the same number of integration steps (12 x 720 = 8640). Since there are two trigonometric evaluations and one exponential evaluation in the inner loop, there are approximately 25921 evaluations (abbreviated as evals) in the integration. For the Patera model, 1000 steps were used. Since each step involves one exponential evaluation and one inverse sine function evaluation (for dθ), therefore there are 2000 evals in the integration. For the Alfano model, 1000 steps were used. Since each step involves one exponential and two error function evaluations, therefore there are 3000 evals in the integration. Finally, there are only two exponential evaluations for the one-term Chan model and three evals for the two-term Chan model required to compute the collision probability.

Table 1 - Comparison for Extremely Close Encounters

This table compares the probability of collision for the NASA, Patera, Alfano and Chan Models using the NASA Model as the basis of comparison. The collision cross-section is a circle.	Extremely Close Encounters with Small Covariance	Extremely Close Encounters with Small Covariance	Extremely Close Encounters with Small Covariance	Extremely Close Encounters with Small Covariance
INPUTS ARE IN BOLD NUMBERS				
SigmaX' (km)	**0.05**	**0.05**	**0.075**	**0.075**
SigmaZ' (km)	**0.025**	**0.025**	**0.025**	**0.025**
Collision Cross-section Radius RA (km)	**0.005**	**0.005**	**0.005**	**0.005**
X'p (km)	**0.01**	**0**	**0.01**	**0**
Z'p (km)	**0**	**0.01**	**0**	**0.01**
OUTPUTS ARE IN REGULAR NUMBERS				
Collision Probability (NASA)	9.742E-03	9.181E-03	6.571E-03	6.125E-03
Collision Probability (Patera)	9.741E-03	9.181E-03	6.571E-03	6.125E-03
Percent Error	0.00%	0.00%	0.00%	0.00%
Collision Probability (Alfano)	9.741E-03	9.181E-03	6.571E-03	6.125E-03
Percent Error	0.00%	0.00%	0.00%	0.00%
Collision Probability (Chan One Term)	9.753E-03	9.185E-03	6.586E-03	6.134E-03
Percent Error	0.12%	0.04%	0.22%	0.14%
Collision Probability (Chan Two Terms)	9.754E-03	9.189E-03	6.586E-03	6.135E-03
Percent Error	0.13%	0.08%	0.22%	0.17%

Table 1 shows comparisons for the case of extremely close encounters between a small spacecraft and space debris with combined cross-sectional radius of 5 meters. The nominal miss distance is 10 meters. In the first two columns, their standard deviations in the encounter plane are described by ellipses with aspect ratio of 2:1 and are 50 and 25 meters respectively. Note that this does not necessarily mean that their three-dimensional covariances are small because they could be approaching head-on with a much larger (maybe 10 times) standard deviation in the intrack direction. The first column is for the case of the primary on the major axis of the error ellipse while the second column is for the primary on the minor axis. In the first column, it is seen that the collision probability is 9.74 E-3 computed by the numerical NASA, Patera and Alfano models. The analytical Chan model yields 9.75 E-3. In the last two columns, their standard deviations in the encounter plane are described by ellipses with aspect ratio of 3:1 and are 75 and 25 meters respectively. The third column is for the case of the primary on the major axis of the error ellipse while the fourth column is for the primary on the minor axis. It is seen that the above general observations hold for the numerical results. The only difference is that the collision probability has decreased by one-third when the aspect ratio of the error ellipses changes from 2:1 to 3:1.

Table 2 - Comparison for Close and Distant Encounters

This table compares the probability of collision for the NASA, Patera, Alfano and Chan Models using the NASA Model as the basis of comparison. The collision cross-section is a circle.	Close Encounters with Moderate Covariance	Close Encounters with Moderate Covariance	Distant Encounters with Moderate Covariance	Distant Encounters with Moderate Covariance
INPUTS ARE IN BOLD NUMBERS				
SigmaX' (km)	**3**	**3**	**3**	**3**
SigmaZ' (km)	**1**	**1**	**1**	**1**
Collision Cross-section Radius RA (km)	**0.01**	**0.01**	**0.01**	**0.01**
X'p (km)	**1**	**0**	**10**	**0**
Z'p (km)	**0**	**1**	**0**	**10**
OUTPUTS ARE IN REGULAR NUMBERS				
Collision Probability (NASA)	1.577E-05	1.011E-05	6.443E-08	3.219E-27
Collision Probability (Patera)	1.577E-05	1.011E-05	6.443E-08	3.219E-27
Percent Error	0.00%	0.00%	0.00%	0.00%
Collision Probability (Alfano)	1.577E-05	1.011E-05	6.443E-08	*0.000E+00*
Percent Error	0.00%	0.00%	0.00%	*-100.00%*
Collision Probability (Chan One Term)	1.577E-05	1.011E-05	6.443E-08	3.215E-27
Percent Error	0.00%	0.00%	0.00%	-0.12%
Collision Probability (Chan Two Terms)	1.577E-05	1.011E-05	6.443E-08	3.216E-27
Percent Error	0.00%	0.00%	0.00%	-0.08%

The first two columns of Table 2 show comparisons for the case of close encounters between two small spacecraft or a slightly larger spacecraft and space debris with combined cross-sectional radius of 10 meters. The nominal miss distance is 1 km. In the first two columns, their standard deviations in the encounter plane are described by ellipses with aspect ratio of 3:1 and are 3 and 1 km respectively. In the first two columns, it is seen that the collision probability is 1.577 E-5 and 1.011 E-5 respectively for the cases of the primary on the major and the minor axes. These numbers are obtained using the numerical NASA, Patera and Alfano models as well as for the two analytical Chan models. The last two columns are for distant encounters with the nominal miss distance being 10 km and everything else being unchanged. It is seen that the above general observations hold for the collision probabilities which are 6.443 E-8 and 3.22 E-27 (the latter being essentially zero). The Alfano model was implemented on Mathcad which gives an output of 0.00 E0 for very small numbers. This is the reason for the misleading 100% error as recorded for that inconsequentially insignificant case. Numerous cases were run by keeping all parameters the same but increasing the aspect ratio of the error ellipse in the encounter plane from 3:1 progressively to 10:1. These results are not shown here because the general trends of consistency among the four models were observed, i.e., they agree to three or four significant figures.

Table 3 - Comparison for Distant Encounters for Large Spacecraft

This table compares the probability of collision for the NASA, Patera, Alfano and Chan Models using the NASA Model as the basis of comparison. The collision cross-section is a circle.	Distant Encounters with Large Covariance	Distant Encounters with Large Covariance	Distant Encounters with Moderate Covariance for Large Collision Cross-Section	Distant Encounters with Moderate Covariance for Large Collision Cross-Section
INPUTS ARE IN BOLD NUMBERS				
SigmaX' (km)	**10**	**10**	**3**	**3**
SigmaZ' (km)	**1**	**1**	**1**	**1**
Collision Cross-section Radius RA (km)	**0.01**	**0.01**	**0.05**	**0.05**
X'p (km)	**10**	**0**	**5**	**0**
Z'p (km)	**0**	**10**	**0**	**5**
OUTPUTS ARE IN REGULAR NUMBERS				
Collision Probability (NASA)	3.033E-06	9.656E-28	1.039E-04	1.564E-09
Collision Probability (Patera)	3.033E-06	9.656E-28	1.039E-04	1.564E-09
Percent Error	0.00%	0.00%	0.00%	0.00%
Collision Probability (Alfano)	3.033E-06	*0.000E+00*	1.039E-04	1.564E-09
Percent Error	0.00%	*-100.00%*	0.00%	0.00%
Collision Probability (Chan One Term)	3.033E-06	9.644E-28	1.039E-04	1.552E-09
Percent Error	0.00%	-0.12%	0.00%	-0.76%
Collision Probability (Chan Two Terms)	3.033E-06	9.645E-28	1.039E-04	1.556E-09
Percent Error	0.00%	-0.11%	0.03%	-0.50%

The first two columns of Table 3 show comparisons for the case of distant encounters between two space-orbiting objects with combined cross-sectional radius of 10 meters. The nominal miss distance is 10 km. In the first two columns, their standard deviations in the encounter plane are described by ellipses with aspect ratio of 10:1 and are 10 and 1 km respectively. It is seen that the four models yield collision probabilities of 3.03 E-6 and 9.65 E-28 respectively for the cases of the primary on the major and the minor axes. In the third column, we consider encounters of debris with an extremely large spacecraft such as the International Space Station (ISS) of dimensions of 2 x 50 = 100 meters. The standard deviations in the encounter plane are described by ellipses with aspect ratio of 3:1 and are 3 and 1 km respectively. The nominal miss distance of the debris is 5 km. The third column is for the case of the ISS on the major axis of the error ellipse. It is seen that all the four models yield collision probability of 1.04 E-4. The fourth column is for the case of the primary on the minor axis of the error ellipse. It is seen that the three numerical models yield collision probability of 1.564 E-9. The Chan results are respectively off by 0.76% and 0.5% for the one-term and the two-term model. From the above three tables, we observe that there is not much significant difference between all these models for the wide range of cases studied.

From the standpoint of the number of trigonometric, exponential and error function evaluations, either Chan model is still much faster because it requires only one-thousandth of the time taken by the Patera (2000 evals) and the Alfano (3000 evals) models and approximately one-ten thousandth the time taken by the NASA model (26,000 evals). If collision probability is used in screening thousands of pairs of orbiting objects which are candidates for potential collision, this difference in computation speed is an important factor to be considered.

GENERALIZATION TO MORE COMPLEX CROSS-SECTIONS

Up to this point, the analysis is based on modeling the primary and secondary spacecraft by spheres of radii r_p and r_s centered at their respective centers of mass. The approximation of the secondary by a sphere is usually justified because the attitude is not known in general. However, the approximation of the primary by a sphere does not yield accurate results because the spacecraft comprises many components which may be modeled as spheres, cylinders, cones, or circular, rectangular and triangular plates, etc. These components may also eclipse one another depending on the projection onto the encounter plane. Thus, the collision cross-section can be very complicated to describe. Even if the outline of the cross-section is available, the method of integrating using radial and angular variables as in the NASA model is not applicable because that formulation is strictly based on a circular collision cross-section. One would then have to perform an extremely time-consuming two-dimensional integration most likely in Cartesian coordinates. The methods of Patera and Alfano may be extended to spacecraft structures more complicated than a sphere. The collision probability of a tethered satellite colliding with a secondary has been formulated by Patera[12]. For this case, the collision cross-section is a ribbon of width equal to $2r_s$ and center-line coinciding with the tether. To be precisely correct, the ends of this ribbon should be two semi-circles, but Patera implicitly simplified these to be straight lines. For practical reasons, this simplification does not introduce substantial errors. If we consider very complex spacecraft structures, even this method of "line" integration does not prove feasible because it relies on knowing the analytical description of the various components and this knowledge explicitly depends on the detailed projection on the encounter plane even for the case of no self-shadowing. In investigative studies, it has been found that the following **Method of Equivalent Cross-Section Areas** (MECSA) may be fruitfully applied to compute the collision probability for the case of complex spacecraft structures. Basically, it simply involves the cosine of the angle between the unit normal vector $\mathbf{n}_{comp}$ of the particular component and the relative velocity vector $\mathbf{v}_{ps}$ so that the projected area A on the encounter plane is given in terms of the area A_{comp} by the relation

$$A = \left| \mathbf{n}_{comp} \cdot \mathbf{v}_{ps} / v_{ps} \right| A_{comp} . \tag{37}$$

This is substituted into equations (9) and (15) to obtain in succession the cross-sections A' and A". Finally, A" is converted into an equivalent circular cross-section A''' of radius $r_{A'''}$ having the same area as A". The center of A''' is chosen to be at the centroid of A" which also corresponds to the centroids of A' and A because the transformations between the various coordinates are linear. (Linear transformations include rotation, dilation, contraction, shear and translation.) This is another salient feature of MECSA. The justification (but not proof) of this method lies in the fact that if the cross-section dimensions are small compared to the standard deviations or the miss distance, then the pdf does not vary much within the cross-section area. Under these conditions, the collision integral (14) is essentially given by the value of the integrand at the centroid multiplied by the area.. What we do here is not a simple straight-forward multiplication of a function value by the area to obtain the integral. We still invoke the Rician integral (20) with a circular area A''' equal to the non-circular area A". The question that remains to be answered is the merit of this method if the cross-section dimensions are not small compared to the standard deviations or the miss distance. The subject of this applicability is taken up in the following case studies. To illustrate the accuracy of this method, we first consider the case of A being a square.

Table 4 - Comparison for Extremely Close Encounters for Squares

This table compares the probability of collision for the Patera, Alfano and Chan Models using the Patera Model as the basis of comparison. The collision cross-section is a square.	Extremely Close Encounters with Small Covariance	Extremely Close Encounters with Small Covariance	Extremely Close Encounters with Small Covariance	Extremely Close Encounters with Small Covariance
INPUTS ARE IN BOLD NUMBERS				
SigmaX' (km)	**0.05**	**0.05**	**0.075**	**0.075**
SigmaZ' (km)	**0.025**	**0.025**	**0.025**	**0.025**
Base of Rectangle (km)	**0.01**	**0.01**	**0.01**	**0.01**
Height of Rectangle (km)	**0.01**	**0.01**	**0.01**	**0.01**
X'p (km)	**0.01**	**0**	**0.01**	**0**
Z'p (km)	**0**	**0.01**	**0**	**0.01**
OUTPUTS ARE IN REGULAR NUMBERS				
Collision Probability (Patera)	1.238E-02	1.167E-02	8.351E-03	7.786E-03
Collision Probability (Alfano)	1.238E-02	1.167E-02	8.351E-03	7.786E-03
Percent Error	0.00%	0.00%	0.00%	0.00%
Collision Probability (Chan One Term)	1.240E-02	1.168E-02	8.378E-03	7.802E-03
Percent Error	0.19%	0.09%	0.32%	0.21%
Collision Probability (Chan Two Terms)	1.240E-02	1.168E-02	8.378E-03	7.805E-03
Percent Error	0.20%	0.14%	0.32%	0.24%

Table 4 shows the comparison between the numerical integration and MECSA for extremely close encounters with small covariance similar to the parameters in Table 1. The numerical integration is performed using equation (33) with 400 integration steps. It is observed that the approximation of a square by an equivalent circle gives very accurate collision probabilities, being less than 0.4% for the cases studied. Other cases involving rectangles with aspect ratios up to 10:1 have been studied with comparably small errors introduced. (See Table 8.)

Table 5 - Comparison for Close and Distant Encounters for Squares

This table compares the probability of collision for the Patera, Alfano and Chan Models using the Patera Model as the basis of comparison. The collision cross-section is a square.	Close Encounters with Moderate Covariance	Close Encounters with Moderate Covariance	Distant Encounters with Moderate Covariance	Distant Encounters with Moderate Covariance
INPUTS ARE IN BOLD NUMBERS				
SigmaX' (km)	**3**	**3**	**3**	**3**
SigmaZ' (km)	**1**	**1**	**1**	**1**
Base of Rectangle (km)	**0.02**	**0.02**	**0.02**	**0.02**
Height of Rectangle (km)	**0.02**	**0.02**	**0.02**	**0.02**
X'p (km)	**1**	**0**	**10**	**0**
Z'p (km)	**0**	**1**	**0**	**10**
OUTPUTS ARE IN REGULAR NUMBERS				
Collision Probability (Patera)	2.007E-05	1.287E-05	8.204E-08	4.100E-27
Collision Probability (Alfano)	2.007E-05	1.287E-05	8.204E-08	*0.000E+00*
Percent Error	0.00%	0.00%	0.00%	*-100.00%*
Collision Probability (Chan One Term)	2.007E-05	1.287E-05	8.204E-08	4.093E-27
Percent Error	0.00%	0.00%	0.00%	-0.17%
Collision Probability (Chan Two Terms)	2.007E-05	1.287E-05	8.204E-08	4.095E-27
Percent Error	0.00%	0.00%	0.00%	-0.11%

Table 5 shows the comparison between the numerical integration and MECSA for close encounters and distant encounters with moderate covariance similar to the parameters in Table 2. Again, the numerical integration is performed using equation (33) with 400 integration steps. It is observed that the approximation of a square by an equivalent circle gives very accurate collision probabilities, being less than 0.2% for the cases studied. Other cases involving rectangles with aspect ratios up to 10:1 have been studied with comparable resulting accuracies. (See Table 8.)

Table 6 - Comparison for Extremely Close Encounters for Triangles

This table compares the probability of collision for the Patera, Alfano and Chan Models using the Patera Model as the basis of comparison. The collision cross-section is an isosceles triangle.	Extremely Close Encounters with Small Covariance	Extremely Close Encounters with Small Covariance	Extremely Close Encounters with Small Covariance	Extremely Close Encounters with Small Covariance
INPUTS ARE IN BOLD NUMBERS				
SigmaX' (km)	**0.05**	**0.05**	**0.075**	**0.075**
SigmaZ' (km)	**0.025**	**0.025**	**0.025**	**0.025**
Base of Triangle (km)	**0.02**	**0.02**	**0.02**	**0.02**
Height of Triangle (km)	**0.01**	**0.01**	**0.01**	**0.01**
X'p (km)	**0.01**	**0**	**0.01**	**0**
Z'p (km)	**0**	**0.01**	**0**	**0.01**
OUTPUTS ARE IN REGULAR NUMBERS				
Collision Probability (Patera)	1.239E-02	1.167E-02	8.364E-03	7.796E-03
Collision Probability (Alfano)	1.239E-02	1.167E-02	8.364E-03	7.795E-03
Percent Error	0.00%	0.00%	0.00%	0.00%
Collision Probability (Chan One Term)	1.240E-02	1.168E-02	8.378E-03	7.802E-03
Percent Error	0.13%	0.07%	0.17%	0.09%
Collision Probability (Chan Two Terms)	1.240E-02	1.168E-02	8.378E-03	7.805E-03
Percent Error	0.14%	0.12%	0.17%	0.12%

To illustrate the accuracy of MECSA, we next consider the case of A being an isosceles triangle. Table 6 shows the comparison between the numerical integration and MECSA for extremely close encounters with small covariance similar to the parameters in Tables 1 and 4. It is observed that the approximation of an isosceles triangle by an equivalent circle gives very accurate collision probabilities, being less than 0.2% for the cases studied. Other cases involving general triangles with aspect ratios up to 10:1 have been studied with comparable results obtained. (See Table 8.)

Table 7 - Comparison for Close and Distant Encounters for Triangles

This table compares the probability of collision for the Patera, Alfano and Chan Models using the Patera Model as the basis of comparison. The collision cross-section is an isosceles triangle.	Close Encounters with Moderate Covariance	Close Encounters with Moderate Covariance	Distant Encounters with Moderate Covariance	Distant Encounters with Moderate Covariance
INPUTS ARE IN BOLD NUMBERS				
SigmaX' (km)	**3**	**3**	**3**	**3**
SigmaZ' (km)	**1**	**1**	**1**	**1**
Base of Triangle (km)	**0.04**	**0.04**	**0.04**	**0.04**
Height of Triangle (km)	**0.02**	**0.02**	**0.02**	**0.02**
X'p (km)	**1**	**0**	**10**	**0**
Z'p (km)	**0**	**1**	**0**	**10**
OUTPUTS ARE IN REGULAR NUMBERS				
Collision Probability (Patera)	2.007E-05	1.287E-05	8.204E-08	4.097E-27
Collision Probability (Alfano)	2.007E-05	1.287E-05	8.204E-08	*0.000E+00*
Percent Error	0.00%	0.00%	0.00%	*-100.00%*
Collision Probability (Chan One Term)	2.007E-05	1.287E-05	8.204E-08	4.093E-27
Percent Error	0.00%	0.00%	0.00%	-0.11%
Collision Probability (Chan Two Terms)	2.007E-05	1.287E-05	8.204E-08	4.095E-27
Percent Error	0.00%	0.00%	0.00%	-0.06%

Table 7 shows the comparison between the numerical integration and MECSA for close encounters and distant encounters with moderate covariance similar to the parameters in Tables 2 and 5. It is observed that the approximation of an isosceles triangle by an equivalent circle gives very accurate collision probabilities, being less than 0.2% for the cases studied. Other cases involving general triangles with aspect ratios up to 10:1 have been studied, yielding results of comparable accuracies. (See Table 8.)

Table 8 - Comparison for Close Encounters for Elongated Rectangles and Triangles

This table compares the probability of collision for the Patera and Chan Models using the Patera Model as the basis of comparison. The collision cross-section is a rectangle or a triangle as indicated.	Close Encounters with Moderate Covariance for Rectangles with 10:1 Aspect Ratio	Close Encounters with Moderate Covariance for Rectangles with 10:1 Aspect Ratio	Close Encounters with Moderate Covariance for Triangles with 10:1 Aspect Ratio	Close Encounters with Moderate Covariance for Triangles with 10:1 Aspect Ratio
INPUTS ARE IN BOLD NUMBERS				
SigmaX' (km)	**1**	**1**	**1**	**1**
SigmaZ' (km)	**1**	**1**	**1**	**1**
Base of Rectangle or Triangle (km)	**0.2**	**0.2**	**0.2**	**0.2**
Height of Rectangle or Triangle (km)	**0.02**	**0.02**	**0.02**	**0.02**
X'p (km)	**1**	**0**	**1**	**0**
Z'p (km)	**0**	**1**	**0**	**1**
OUTPUTS ARE IN REGULAR NUMBERS				
Collision Probability (Patera)	3.861E-04	3.855E-04	1.931E-04	1.929E-04
Collision Probability (Chan One Term)	3.860E-04	3.860E-04	1.930E-04	1.930E-04
Percent Error	-0.03%	0.13%	-0.01%	0.07%
Collision Probability (Chan Two Terms)	3.861E-04	3.861E-04	1.930E-04	1.930E-04
Percent Error	-0.01%	0.15%	-0.01%	0.08%

Table 8 shows the comparison between the numerical integration and MECSA for close encounters with moderate covariance similar to the parameters in Tables 2, 5 and 7. For convenience, the pdf is chosen to be isotropic. The first two columns are for the case of elongated rectangles having aspect ratio of 10:1 oriented parallel or perpendicular to the axes respectively. The last two columns are similarly for the case of elongated triangles. It is observed that the corresponding approximations by an equivalent circle give very accurate collision probabilities, being less than 0.2% for the cases studied even for large linear dimensions of 200 meters. The cases for which the pdf is bivariate Gaussian may be reduced to the case of isotropic pdf by changing the aspect ratio of the rectangle or triangle. It is quite remarkable that the MECSA computations have such high fidelity.

DISCUSSION AND CONCLUSION

The above results show that for an extremely wide range of encounter parameters, the four models (NASA, Patera, Alfano and Chan) yield collision probabilities which agree to three or four significant figures for the case of circular collision cross-section. The NASA model is strictly applicable to the case of circular cross-section whereas the other three may be extended in principle to non-circular areas. However, the Patera and the Alfano models are not easily extendable to complex shapes because they rely on the detailed analytical description of the boundary so that the numerical integration may be performed. The Chan model is easily generalized to include these general shapes because it is based only on areas.

Tables 1, 4 and 6 show cases of extremely good agreement for which the collision cross-section dimensions are of the order of the standard deviations and the miss distance. We have noted that a sufficient (but not necessary) condition for the Chan model to apply is that the collision cross-section dimensions are small compared to the standard deviations or the miss distance. This sufficient condition holds for most problems because the standard deviations are hundreds of meters to a few kilometers whereas the collision cross-section dimensions are of the order of tens of meters. It is possible to assign values to these parameters such that this model yields poor approximations. For these cases, it is permissible to break up the cross-section into a number of smaller pieces such that the sufficient condition holds and the Chan model will then apply.

It remains to discuss how various components in a complex spacecraft structure will project onto the encounter plane. For simplicity, we shall ignore eclipsing of one spacecraft component by another. We know that a sphere will project as a circle. A cylinder will project as a parallelogram plus two semi-ellipses. A cone will generally project as a triangle plus a segment of an ellipse or in some cases as an ellipse. A circular plate will project as an ellipse. A rectangle will project as a parallelogram. A triangle will project as a triangle. For each of these components, we know the centroid in the spacecraft body frame. We may easily obtain the centroid of the projection of each component in the encounter plane. The area of the projected cross-section is determined using equation (37). Thus, we may now use equations (22) through (29) to compute accurate collision probabilities with approximately 0.4% error for the cases considered.

Finally, in MECSA, we recall that we convert A'' into an equal-area circular cross-section A''' of radius $r_{A'''}$ for use in equation (22). Because of the linear transformations between A, A', A'' and A''', a little consideration of equations (9), (15) and (16) reveals that we can also convert the area A into an equal-area circle, thus simplifying the computations. That is, we may immediately obtain the radius r_A for use in equation (22) by simply setting at the start

$$A = \pi r_A{}^2 \, . \tag{38}$$

When there is no substantial self-shadowing, MECSA may be applied to rapidly and accurately compute the collision probability for the case of complex spacecraft structures. It does not rely on knowing the detailed analytical description of the various components on the encounter plane. Rather, it depends only on the angle between the unit normal spacecraft component vector and the relative velocity vector. The key step is the conversion of the component into an equivalent circle of the same area centered at the centroid. It is hoped that MECSA will be used for efficiently computing the collision probabilities of the International Space Station when it encounters other space-orbiting objects.

ACKNOWLEDGMENT

The author wishes to thank Drs. Salvatore Alfano, Glenn Peterson and Mr. Alan Jenkin of The Aerospace Corporation for their painstaking efforts in reading this manuscript.

REFERENCES

1. F. K. Chan, "Collision Probability Analyses for Earth-Orbiting Satellites," Proceedings of the 7th International Space Conference of Pacific Basin Societies, Nagasaki, Japan, July 1997 (Abridged). Aerospace Corporation Report No. ATR-2000(8456)-1, September 2000 (Unabridged).

2. F. K. Chan, "Analytical Expressions for Computing Spacecraft Collision Probabilities," AAS/AIAA Space Flight Mechanics Meeting, Santa Barbara, California, February 2001.

3. F. K. Chan, "Close Encounters with Multiple Satellites," AAS/AIAA Space Flight Mechanics Meeting, San Antonio, Texas, January 2002.

4. F. K. Chan, "Spacecraft Maneuvers to Mitigate Potential Collision Threats," AIAA/AAS Astrodynamics Specialist Conference, Monterey, California, August 2002.

5. A. Papoulis, *Probability, Random Variables, and Stochastic Processes*, 3rd Edition, McGraw-Hill, New York, 127–128 (1991).

6. A. Papoulis, *Probability, Random Variables, and Stochastic Processes*, McGraw-Hill, New York, 195–196 (1965).

7. S. O. Rice, "Mathematical Analysis of Random Noise," Bell System Tech. J., **23**, 282-332 (1944) and **24**, 46-156 (1945).

8. F. K. Chan, "Signal Fading in the Ionosphere for a Variety of Disturbance Sources," 10th International Ionospheric Effects Symposium, Alexandria, Virginia, May 2002.

9. J. L. Foster and H. S. Estes, "A Parametric Analysis of Orbital Debris Collision Probability and Maneuver Rate for Space Vehicles," NASA/JSC-25898, August 1992.

10. R. P. Patera, "A General Method for Calculating Satellite Collision Probability," AAS/AIAA Space Flight Mechanics Meeting, Clearwater, Florida, January 2000.

11. S. Alfano, "Aerospace Support to Space Situational Awareness," MIT Lincoln Laboratory Satellite Operations and Safety Workshop, Haystack Observatory, Chelmsford, Massachusetts, October 2002.

12. R. P. Patera, "A Method for Calculating Collision Probability between a Satellite and a Space Tether," AAS/AIAA Space Flight Mechanics Meeting, Santa Barbara, California, February 2001.

RISK OF COLLISION FOR
THE NAVIGATION CONSTELLATIONS:
THE CASE OF THE FORTHCOMING GALILEO

A. Rossi,[*] G. B. Valsecchi[†] and E. Perozzi[‡]

The satellite global positioning systems presently in space are the American NAVSTAR-GPS and the Russian GLONASS. Within this decade the European system, Galileo, should be operational in the same altitude range, dubbed MEO, Medium Earth Orbit. In this paper the fragmentation of a spacecraft related to one of these three constellations has been simulated and the collision risk faced by the operational satellites has been analyzed. Both the intra-constellation and the inter-constellation risk have been studied. An improvement in the collision risk calculation method developed in (Valsecchi et al., *Space Debris*, 2000) is described in the paper. The new method overcomes the limitation in the application of Öpik's theory of planetary encounters, dictated by the assumption of random orientation of the argument of perigee and the longitude of node of the projectiles, and allows its application to the Medium Earth orbital regime. In general terms it has been observed that the flux following a generic fragmentation is by far larger than the low background flux in MEO. The strong interrelation of the three constellations has been shown by analyzing the inter-constellation effects of the fragmentations. In particular the GPS and the GLONASS are strongly interacting, while the Galileo constellation, orbiting at somewhat higher altitude, is less affected by any unfortunate event happening in the two lower constellations.

INTRODUCTION

Satellite global positioning systems have been deployed in the early 90's as military support systems by the US and the former Soviet Union. Later on, the global positioning systems first became available for private use in 1995, providing an extremely accurate and valuable tool, nowadays used by a huge number of people for many different applications. The satellite positioning systems presently in space are the American NAVSTAR-GPS (Navigation System with Time and Ranging – Global Positioning System) and the Russian GLONASS (Globaluaya Navigatsionnaya Sputnikovaya Sistema).

---

*   ISTI-CNR, Via Moruzzi, 1, I-56124 Pisa, Italy.

†   IASFC-CNR, Via Fosso del Cavaliere 100, I-00133 Roma, Italy.

‡   Telespazio S.p.A., Via Tiburtina 965, I-00156 Roma, Italy.

	a [km]	e	i [deg]	Number of satellites	Number of planes
Galileo	29994	0	56	30	3
GPS II	26559	0	55	28	6
GLONASS	25478	0	64.8	24	3

Table 1: Characteristics of the present and foreseen navigation constellations. Note that the number of satellites include in-orbit spares.

On March 26, 2002 the European Union Transport Ministers gave the final go-ahead on the Galileo project. Galileo, developed by the European Space Agency in collaboration with the European Union, is a civil system, designed to be operational from 2008. Galileo is a 27 satellite Walker constellation (3 planes with 9 satellites each) plus 3 active in-orbit spares, at an altitude of about 23 600 km, with an inclination of 56°. The goal of the Galileo project is to provide Europe, and in general the world, with an accurate, secure and certified satellite positioning system. Galileo should be inter-operable with the existing satellite navigation systems and particularly with the GPS. One of the most ambitious goals of the future satellite systems is, for example, the automatic guidance and control of commercial aircraft. Such an application would of course require an extreme level of reliability.

As seen from Table 1, the Galileo constellation will orbit a region of space (dubbed MEO, Medium Earth Orbit) close to the other two navigation constellations presently deployed. Although the nominal orbits of the different constellations are well separated in altitude, there is the possibility that old uncontrolled spacecraft could intersect the operational orbits. Moreover, although not nearly so crowded as the Low Earth Orbit (LEO), also the MEO region is populated by a large number of space debris. The European Space Agency (ESA) debris environment model[1], MASTER 2001, includes about 60 000 objects larger than 1 cm with orbital elements that are possibly crossing the orbits of the navigation constellations. Actually most of the objects in MEO are clustered about the Molnyia orbits (i.e. with $a \approx 26\,500$, $e \approx 0.7$ and $i \approx 63°$). As a matter of fact these objects have a minimal interaction with the navigation constellations. But, even if we exclude the objects close to Molnyia orbits, about 16 000 objects with diameter larger than 1 cm have orbits potentially crossing the navigation constellations. Figure 1 shows the orbital distribution of this population of objects. Note how, in the lowest panel in Figure 1, the GPS orbit appears within reach of several thousand objects, due to the non-zero eccentricity of most of the debris in the MEO zone. Although lower than in LEO, the average collision velocity at the GPS/GLONASS altitude is still about 5 km/s. At this velocity, an impacting particle of about 1 cm in diameter (corresponding to about 1.5 g) delivers an energy of the order of 10^4 J and is capable of producing severe damages to a spacecraft.

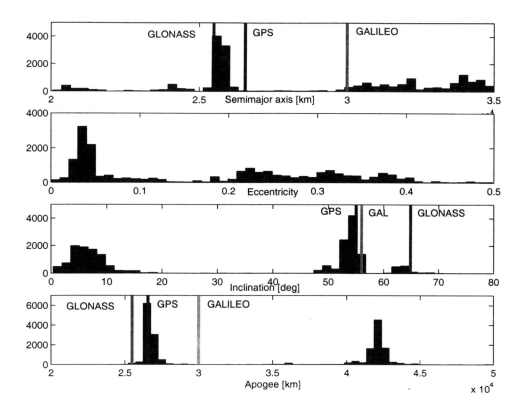

Figure 1: Distribution of the objects with diameters larger than 1 cm in MEO, from the ESA's debris model MASTER 2001. The semimajor axis, eccentricity, inclination and apogee of the objects is shown. The values of the semimajor axis and of the inclination of the GPS, GLONASS and GALILEO orbits is also shown for reference. Note that a cutoff at an eccentricity of 0.5 has been used to exclude all the objects close to Molnyia type orbits.

The sensible applications of the navigation constellations and the absence of any natural sink mechanism (such as the atmospheric drag in LEO) led the GPS operators to the adoption of a debris prevention policy. In particular, as proposed for the geostationary ring, the GPS satellites are moved to a disposal region, about 500 km above the operational orbit, at the end-of-life. Spent upper stages of evolved expandable launch vehicles (EELVs) may also be placed in the disposal region after the completion of their mission. This procedure should prevent any accidental collision between operational satellites and old spacecraft. The disposal zone is in principle well separated both from the GPS operational orbit and from the Galileo planned orbit. Unfortunately the picture is more complicated. In a number of recent papers[2,3] the instability of the GPS disposal orbit, with an increase of the eccentricity that could lead to dangerous crossings of the operational orbits, has been clearly shown. The same is true for the GLONASS related spacecraft.

It is therefore apparent that in a debris mitigation policy the MEO region must be viewed as a whole and that any action undertaken in a constellation should take into account the presence of the other analogous systems in the vicinity. To further explore the concept of the interdependence of the navigation constellations, in this paper a particularly dangerous situation will be analyzed. The fragmentation of a spacecraft related to one of the navigation constellation (following an explosion or an accidental collision) is simulated. Then the interaction of the debris produced by the fragmentation event with the three constellations is studied. In the next Section the method developed for this analysis will be outlined. Then, in the following Section, the results of the collision risk analysis will be showed and discussed.

COLLISION RISK CALCULATION

Öpik's theory of planetary encounters[4] can be used to analytically calculate the magnitude and direction of the relative velocity vector at impact of a projectile on a given target on circular orbit. Making use of this theory, we developed a method to assess the collision risk for an Earth orbiting satellite. The method is described in Refs. 5 and 6. Here the fundamentals of the method will be briefly recalled.

Let a and e be the semimajor axis and eccentricity of the orbit of the projectile, and I its inclination with respect to the plane of the orbit of the target; the latter, in turn, is on a circular orbit of radius a_0. The velocity U at which the projectile encounters the target is

$$U = \sqrt{\frac{Gm_\oplus}{a_0}} \cdot \sqrt{3 - \frac{a_0}{a} - 2\sqrt{\frac{a(1-e^2)}{a_0}} \cos I} \,,$$

and its component U_r along the direction from the center of the Earth to the target is

$$U_r = \pm\sqrt{\frac{Gm_\oplus}{a_0}} \cdot \sqrt{2 - \frac{a_0}{a} - \frac{a(1-e^2)}{a_0}} \,,$$

where G is the constant of gravity and $m_\oplus$ is the mass of the Earth.

Then, Öpik's expression for the intrinsic collision probability p per revolution of the projectile[4] is given by:

$$p_{rev} = \frac{U}{\pi |U_r| \sin I} \,, \tag{1}$$

while the intrinsic probability per unit time is obtained by dividing for the orbital period of the projectile:

$$p = \frac{\sqrt{Gm_\oplus}}{2\pi a^{1.5}} \cdot \frac{U}{\pi |U_r| \sin I} \,.$$

Thus, given a target orbit, p is only a function of the orbital elements (a, e and I) of the projectile.

Öpik's theory makes basic assumptions that pose some caveats for its practical application. In particular, it assumes that the argument of perigee ω and the longitude of node Ω of the projectile orbit, evaluated using as reference plane the orbital plane of the target, are randomly distributed between 0 and 2π. This means, for instance, that the theory is not applicable to situations in which a resonance is constraining the distribution of w, as is the case, for example, of the Molnyia orbits. These orbits lie at the critical *equatorial* inclination of $i_{eq} = 63°$, and have the *equatorial* argument of perigee constrained in the Southern hemisphere, in order to be high above the horizon for the users located in the former Soviet Union countries; thus, for Molnyia projectiles aimed at a target on circular orbit in the equatorial plane, Öpik's theory would clearly not be applicable. Nonetheless, in LEO, the randomization induced by the drift of w_{eq} and Ω_{eq} due to the Earth's quadrupole J_2 is so effective that Öpik's theory can be easily applied without significant loss of accuracy[5,6].

On the other hand, the precession rates $\dot{w}_{eq}$ and $\dot{\Omega}_{eq}$ are about two orders of magnitude smaller at the GPS altitude than in LEO; in particular, we have $\dot{\Omega}_{eq} \simeq -0.042$ deg/day and $\dot{w}_{eq} \simeq -0.02$ deg/day at the GPS altitude. This slower evolution prevents the direct application of our original method to MEOs; we therefore devised an extension to the method presented in Refs. 5 and 6 to take into account also orbital regimes where the randomization of the angular elements cannot be granted, along a line of reasoning similar to that of Ref. 7.

Let us discuss the reasoning that is at the basis of the assumption of a flat distribution in Ω and w in the derivation of Eq. (1). Necessary conditions for a collision to occur are:

- that the perigee and apogee of the projectile orbit are such that $q = a(1-e) < a_0 < Q = a(1+e)$;

- that the geocentric distance of the projectile, at its crossings of the orbital plane of the target, be equal to a_0.

These crossings take place at the ascending and the descending nodes of the projectile orbit on the target orbital plane, where we have $w + f = 0$ (at the ascending node, f being the true anomaly of the projectile), and $w + f = 180°$ (at the descending node); the geocentric distances of the nodal points are given by

$$r_a = \frac{a(1-e^2)}{1 + e\cos(-w)} = \frac{a(1-e^2)}{1 + e\cos w}$$

$$r_d = \frac{a(1-e^2)}{1 + e\cos(180° - w)} = \frac{a(1-e^2)}{1 - e\cos w}.$$

When both r_a and r_d are sufficiently different from a_0, collisions are impossible; this happens most of the time during the perturbation-induced rotation w, but sooner or later, unless a specific dynamical mechanism limits the range of values attainable by w, it will happen that, for a particular value w_c of w,

$$a_0 = \frac{a(1-e^2)}{1 + e\cos w_c}.$$

This collision condition is, in general, not alone; in fact a collision is also possible for $w_c + 180°$, i.e. at the other node, as discussed above. Moreover, taking into account

1221

that, for any angle α, $\cos\alpha = \cos(360° - \alpha)$, collisions will also be possible for $360 - \omega_c$ and for $180 - \omega_c$. It is easy to check that these four values of ω are all in different quadrants: let us call ω_{c1} the one in the first quadrant, then $\omega_{c4} = 360° - \omega_{c1}$ is in the fourth, while $\omega_{c3} = \omega_{c1} + 180°$ and $\omega_{c2} = 180 - \omega_{c1}$ are, respectively, in the third and in the second quadrant.

The derivation of Öpik's expression (Eq.(1)) consists in a particle-in-the-box evaluation of the probability of presence of the projectile in a solid ring of radius a_0, with very small horizontal and vertical thickness; in this context, the assumption of the randomness of ω allows the explicit computation of the probability of presence of the projectile in the ring. If the probability distribution of the argument of perigee of the projectile orbits is not flat, we need a way to take this into account by appropriately rescaling the probability of presence of the projectile in the above described ring. Let us introduce the quantity

$$\omega^\star = \min(\omega - \omega_{c1}, \omega - \omega_{c2}, \omega - \omega_{c3}, \omega - \omega_{c4});$$

it is a simple function that expresses the difference between the current value of ω and the nearest collision solution.

The ω-randomness assumption of Öpik's theory can be considered equivalent to the assumption that the distribution of $\omega^\star$ is flat between $-180°$ and $180°$; in particular, this means that, given a population consisting of N projectiles, Öpik's assumption implies that there should be

$$y_f = \frac{N\delta\omega^\star}{2\pi}$$

projectiles in a small interval of width $\delta\omega^\star$ centered on $\omega^\star = 0$.

On the other hand, it is easy to compute the true distribution of $\omega^\star$ for all the projectiles (from the results of the numerical integration of the orbit of the debris cloud). We can then compare at any given instant in time, as in Figure 2, the value of the true distribution of the $\omega^\star$ at $0°$, y_t, with the value of the flat distribution, y_f. Then, the ratio y_t/y_f gives the correction factor by which the probability of collision, given by Eq. (1), calculated assuming the flat distribution of perigee arguments, has to be multiplied. In this way, for each projectile, the true collision probability, taking into account the slow diffusion of the orbital elements in MEO, is obtained.

The remaining open question is a procedure to compute ω, the argument of perigee of the projectile computed with respect to the orbital plane of the target; this is dealt with in the following.

Given an orbit with semimajor axis a, eccentricity e, equatorial inclination i_{eq}, equatorial longitude of node Ω_{eq}, and equatorial argument of perigee ω_{eq}, the magnitude of its angular momentum vector $\vec{h}$ is given by[8]:

$$h = \sqrt{Gm_\oplus a(1 - e^2)}.$$

The vector $\vec{h}$ can be written as $\vec{h} = \vec{i}h_x + \vec{j}h_y + \vec{k}h_z$, where $\vec{i}$, $\vec{j}$ and $\vec{k}$ are orthogonal unit vectors oriented along the usual x-y-z axes (the x and y-axes in the equatorial

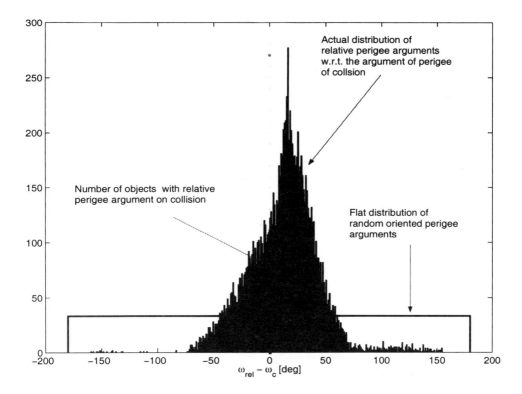

Figure 2: Example of the distribution of the true $\omega^\star$ for a debris cloud generated by an explosion, with respect to a flat distribution of relative arguments of perigee.

plane, the z-axis normal to it, with the x-axis directed toward the γ-point). We have:

$$h_{xy} = \sqrt{h_x^2 + h_y^2} = h \sin i_{eq}$$

$$h_x = h_{xy} \sin \Omega_{eq} = h \sin i_{eq} \sin \Omega_{eq}$$

$$h_y = -h_{xy} \cos \Omega_{eq} = -h \sin i_{eq} \cos \Omega_{eq}$$

$$h_z = h \cos i_{eq}, \tag{2}$$

Then, the vector $\vec{\varepsilon} = \vec{i}\varepsilon_x + \vec{j}\varepsilon_y + \vec{k}\varepsilon_z$, is defined as $\vec{\varepsilon} = GM\vec{e}$ where $\vec{e}$ is a vector with the magnitude of the osculating eccentricity drawn from the center of the Earth toward the perigee. The components of $\vec{\varepsilon}$ can be written as:

$$\varepsilon_x = \frac{h_x \varepsilon_y - \varepsilon h_{xy} \cos \omega_{eq}}{h_y}$$

$$\varepsilon_y = \frac{\varepsilon h_x \cos \omega_{eq} - \varepsilon_z h_z h_y / h_{xy}}{h_{xy}}$$

$$\varepsilon_z = \frac{h_{xy} \varepsilon \sin \omega_{eq}}{h} \tag{3}$$

so that, by using Eqs. (2) and (3), we can write the unit vectors $\hat{h}$ and $\hat{\varepsilon}$ as:

$$\hat{h} = \begin{pmatrix} \sin i_{eq} \sin \Omega_{eq} \\ -\sin i_{eq} \cos \Omega_{eq} \\ \cos i_{eq} \end{pmatrix}$$

$$\hat{\varepsilon} = \begin{pmatrix} \cos \omega_{eq} \cos \Omega_{eq} - \cos i_{eq} \sin \omega_{eq} \sin \Omega_{eq} \\ \cos \omega_{eq} \sin \Omega_{eq} + \cos i_{eq} \sin \omega_{eq} \cos \Omega_{eq} \\ \sin i_{eq} \sin \omega_{eq} \end{pmatrix}. \tag{4}$$

Suppose we now have two sets of orbital elements, one relative to the target $(a_0, e_0, i_{0eq}, \omega_{0eq}, \Omega_{0eq})$, and the other relative to the projectile $(a, e, i, \omega_{eq}, \Omega_{eq})$.
We want to compute the inclination, I and the argument of perigee, ω of the projectile in a reference frame X-Y-Z, in which the orbit of the target has inclination equal to zero, and the X-axis is directed along the nodal line, on the equator, of the target orbit. To obtain I, we just have to compute:

$$\cos I = \hat{h} \cdot \hat{h}_0$$
$$= \sin i_{eq} \sin i_{0eq} (\sin \Omega_{eq} \sin \Omega_{0eq} + \cos \Omega_{eq} \cos \Omega_{0eq}) + \cos i_{eq} \cos i_{0eq}. \tag{5}$$

To compute ω, we apply a rotation of $-\Omega_{0eq}$ about the z-axis (given by the rotation matrix $R_{-\Omega_0}$); after this rotation the old x-axis has been transformed into the new X-axis, directed along the nodal line, on the equator, of the target orbit. We then apply a rotation of $-i_{0eq}$ about the X-axis (given by the rotation matrix R_{-i_0}). The rotation of the unit vectors of the target orbit, from Eq. (4), is given by:

$$\hat{h}_{0r} = R_{-i_0} R_{-\Omega_0} \hat{h}_0 = \begin{pmatrix} 0 \\ 0 \\ 1 \end{pmatrix}$$

$$\hat{\varepsilon}_{0r} = R_{-i_0} R_{-\Omega_0} \hat{\varepsilon}_0 = \begin{pmatrix} \cos \omega_{0eq} \\ \sin \omega_{0eq} \\ 0 \end{pmatrix}.$$

For the projectile orbit we have analogously:

$$\hat{h}_r = R_{-i_0} R_{-\Omega_0} \hat{h}$$
$$= \begin{pmatrix} \sin i_{eq} (\sin \Omega_{eq} \cos \Omega_{0eq} - \cos \Omega_{eq} \sin \Omega_{0eq}) \\ -\sin i_{eq} \cos i_{0eq} (\sin \Omega_{eq} \sin \Omega_{0eq} + \cos \Omega_{eq} \cos \Omega_{0eq}) + \cos i_{eq} \sin i_{0eq} \\ \sin i_{eq} \sin i_{0eq} (\sin \Omega_{eq} \sin \Omega_{0eq} + \cos \Omega_{eq} \cos \Omega_{0eq}) + \cos i_{eq} \cos i_{0eq} \end{pmatrix}$$

and

$$\hat{\varepsilon}_r = R_{-i_0} R_{-\Omega_0} \hat{\varepsilon}.$$

By substituting Eq. (4), and developing, we obtain the rotated unit vectors of $\hat{\varepsilon}_r$ as:

$$\hat{\varepsilon}_{xr} = \cos \omega_{eq} (\cos \Omega_{eq} \cos \Omega_{0eq} + \sin \Omega_{eq} \sin \Omega_{0eq})$$

$$+ \cos i_{eq} \sin \omega_{eq} (\cos \Omega_{eq} \sin \Omega_{0eq} - \sin \Omega_{eq} \cos \Omega_{0eq})$$

$$\hat{\varepsilon}_{yr} = \sin \omega_{eq} [\sin i_{eq} \sin i_{0eq} + \cos i_{eq} \cos i_{0eq} (\cos \Omega_{eq} \cos \Omega_{0eq} + \sin \Omega_{eq} \sin \Omega_{0eq})]$$
$$- \cos \omega_{eq} \cos i_{0eq} (\cos \Omega_{eq} \sin \Omega_{0eq} - \sin \Omega_{eq} \cos \Omega_{0eq})$$

$$\hat{\varepsilon}_{zr} = \sin \omega_{eq} [\sin i_{eq} \cos i_{0eq} - \cos i_{eq} \sin i_{0eq} (\cos \Omega_{eq} \cos \Omega_{0eq} + \sin \Omega_{eq} \sin \Omega_{0eq})]$$
$$+ \cos \omega_{eq} \sin i_{0eq} (\cos \Omega_{eq} \sin \Omega_{0eq} - \sin \Omega_{eq} \cos \Omega_{0eq}) .$$

Finally, from Eqs. (3), we have:

$$\cos \omega = \frac{\hat{\epsilon}_{yr} \hat{h}_{xr} - \hat{\epsilon}_{xr} \hat{h}_{yr}}{\hat{h}_{xyr}} \tag{6}$$

$$\sin \omega = \frac{\hat{\epsilon}_{zr}}{\hat{h}_{xyr}} . \tag{7}$$

With Eqs. (5), (6), and (7) it is possible to completely describe the dynamics of the debris cloud with respect to the target orbit. In the next Section the results of the application of this improved method to the Navigation Constellations are shown.

DESCRIPTION OF THE SIMULATIONS

In Refs. 6, 9 and 10 the hazard posed to a LEO multi-plane constellation by the fragmentation of a spacecraft inside the constellation itself has been analyzed. It has been shown that the interaction of the evolving debris orbits with the global dynamics of the system gives way to a long lasting collision flux exceeding the background debris flux for long interval of times after the event. It has also been shown how the effects are different on the various constellation planes, according to the plane in which the fragmentation takes place.

Exploiting the advancements in the collision probability calculation method described in the previous Section, in this paper a similar study is performed for the MEO navigation constellations. In particular a number of tests have been performed, simulating the fragmentation either of a Galileo or of a GPS or of a GLONASS related spacecraft. Both an explosion induced or a collision induced break-up have been studied. In Table 2 the type of event, the orbital elements and the mass of the parent spacecraft and the number of debris larger than 1 cm produced by the fragmentation, are given. In the simulation process the fragments down to 1 mg are produced, but only those with diameters larger than 1 cm (corresponding to about 1.5 g) are considered. In Table 2 it can be noted that the first two simulated fragmentations pertain to spacecraft in Galileo-like orbits, the next two events are related to spacecraft in GPS-like orbits and the last two fragmentations relate to spacecraft in GLONASS-like orbits. The planes of the fragmentation events are taken as "reference" planes (i.e., plane number 1) for the corresponding constellations, in the following analysis. Note that in all the simulated collision events the mass and the velocity of the projectile were set to 10 kg and 5 km/s, respectively. With these values the specific energy for catastrophic break-up, i.e. the ratio between the projectile energy and the target mass, is larger

Type of event	a [km]	e	i [deg]	Ω [deg]	ω [deg]	Mass of the spacecraft [kg]	Number of debris larger than 1 cm
Explosion	29994	1×10^{-3}	56.0	262.7	56.9	2000	72 100
Collision	29994	1×10^{-3}	56.0	262.7	56.9	2000	18 300
Explosion	26559.74	0	55.0	274.0	50.0	922	33 200
Collision	26559.74	0	55.0	274.0	50.0	2339	14 000
Explosion	25478	1×10^{-3}	64.7	262.7	56.9	2121	76 200
Collision	25478	1×10^{-3}	64.7	262.7	56.9	2121	16 800

Table 2: **Characteristics of the simulated fragmentation events.**

than the threshold discriminating between a localized, crater-like, target damage and a total fragmentation. Therefore the targets are fragmented. The mass distribution of these fragments is a power law with an energy-dependent exponent[11] and the velocity increment ΔV of the fragments as a function of size is obtained extracting from a triangular distribution with the peak value given by the "intermediate" model proposed in Ref. 12. In the explosion cases, a high intensity explosion according to the model described in Ref.11 is simulated. Then the orbit of the fragments larger than 100 g is individually propagated, for 20 years, taking into account all the main gravitational and non-gravitational perturbations (Earth geopotential harmonics, luni-solar perturbations, air drag and solar radiation pressure). The smaller fragments are sampled, with a sampling factor of 5. The orbit of the constellation satellites is propagated, again for 20 years, by taking into account only the J_2 perturbations, to simulate the station keeping of the controlled operative satellites.

Considering all the fragments produced by the simulated fragmentation event, the total projectile flux as a function of altitude and mass is computed, with the method described in the previous Section, by adding up the contributions of all the fragments for which an impact becomes possible. Then the flux of the simulated collision fragments is compared, for any given range of projectile energies, to the reference value corresponding to the background flux resulting from the entire debris population currently present in space. The background flux is estimated by the SDM[11] code, using the MASTER 2001 population of orbital debris. The difference between these two fluxes will give a clear estimate of the extra collision risk faced by the constellation satellites due to the consequences of the simulated break-up.

RESULTS

The effect of the fragmentations described in Table 2 have been analyzed in terms of flux of particles on the different planes of a constellation. Both the intra-constellation

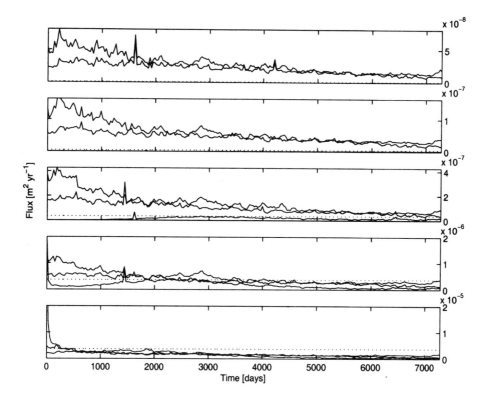

Figure 3: Flux of fragments vs. time, on the three Galileo constellation planes, produced by a simulated explosion of a spacecraft on a constellation plane. Each curve represents the flux affecting one of the three constellation planes. The five panels show, from bottom to top, the fluxes at increasing impact energies E: $10^3 < 2E < 10^4$ J; $10^4 < 2E < 10^5$ J; $10^5 < 2E < 10^6$ J; $10^6 < 2E < 10^7$ J; $10^7 < 2E < 10^8$ J. The horizontal dotted lines represent the background fluxes in the same energy ranges, computed from the overall space debris population included in MASTER 2001.

and the inter-constellation effects have been studied. This means that, e.g., an explosion of a spacecraft in a Galileo-like orbit has been studied with respect both to the Galileo constellation itself and with respect to the GPS constellation. Figure 3 shows the flux of debris, in m^{-2} yr^{-1}, coming from the first explosion of Table 2, on the three planes of the Galileo constellation (i.e., including the plane of the fragmentation itself). In every panel, each one of the three curves shows the flux with respect to one of the planes; in particular, looking at the bottom panel, the curve starting from the highest flux value and then rapidly decreasing, represents the flux on the same explosion plane. This flux of low energetic particles is composed by a large number of debris that, just after the fragmentation event, remain clustered around the parent body orbit. The small relative inclination between the target plane and the debris orbits account for low impact velocities, hence for the low energy of the impact. In the same panel the two lower, almost overlapping, curves represent the flux on the planes with nodes shifted by 120° (lowest curve) and 240° (intermediate curve) from the explosion plane. In the second panel from bottom, corresponding to the interval $10^4 < 2E < 10^5$ J in impact energy, in the very first days the curve relative to the flux on the same explosion plane is the highest one, although rapidly declining. After this short transient, the top curve becomes the one related to the plane at 240° from the explosion. The curve of the flux on the plane at 120° becomes the intermediate one. In the upper panels, the flux of more energetic particles on the same plane of the explosion is negligible (again due to the low relative inclination, very slowly evolving due to the slow relative nodal regression). The curves of the flux on the other two planes are again in the same order an in the previous panel. The small number of well separated planes and, especially, the slower dynamics of MEO account for a global behavior different from the one shown in Refs. 6, 9 and 10 for the LEO constellations. As expected, the fluxes of debris are about two orders of magnitude lower than the fluxes registered in the case of a LEO IRIDIUM-like constellation (see, e.g., Figure 6 in Ref. 6), mainly due to the lower orbital velocity and relative inclinations. It has however to be noted how, in the top three panels, the fluxes of explosion debris on the planes shifted from the explosion one, are several times larger than the background flux. It is worth stressing that the higher panels correspond to impact energies that can severely damage a spacecraft and that, in the top panel, the range of energy between 10^7 and 10^8 J, is of the order of the fragmentation threshold for a Galileo satellite. Nonetheless it must also be remembered that the flux levels are quite low and account for a hazard of damaging impacts of the order of $10^{-3} - 10^{-4}$ over a decade.

In Figure 4 the flux, on the three planes of the Galileo constellation, of the debris produced by a collisional break-up of a spacecraft in a Galileo-like orbit is shown. The five lowest panels correspond to the five panels of Figure 3. These panels display a behavior similar to those of the explosion case. The main difference with respect to Figure 3 is the presence of the two additional panels covering the highest ranges of energy. This is due to the presence, in the mass distribution of the collision events, of a small number of large fragments capable of delivering these higher impact energy to the target. Nonetheless, the resulting fluxes are very low due to the low number of fragments involved.

The situation becomes more involved in the case of the GPS constellation (Figure 5). The presence of six orbital planes, spaced by only 60°, makes the picture closer to the one observed in the LEO constellations. As in Figure 3, in the two lowermost panels of Figure 5, after a short interval following the fragmentation, the flux of debris on the same plane of the explosion falls abruptly. In the same panels, the flux on the other four planes is comparable, with the planes from 60° to 180° apart experiencing the highest values several years after the explosion. In the three upper panels the effect of the differential precession of the orbital nodes becomes more apparent (as it was in LEO) and the fluxes rise at successive times, according to the nodal separation with respect to the event plane. In particular, in the second panel from top, it can

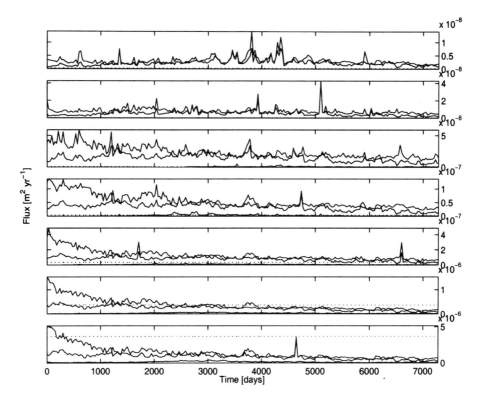

Figure 4: The same as Figure 3 for the case of a collisional break-up of a spacecraft in a Galileo-like orbit. The seven panels show, from bottom to top, the fluxes at increasing impact energies E: $10^3 < 2E < 10^4$ J; $10^4 < 2E < 10^5$ J; $10^5 < 2E < 10^6$ J; $10^6 < 2E < 10^7$ J; $10^7 < 2E < 10^8$ J, $10^8 < 2E < 10^9$ J; $2E > 10^9$ J.

be noted, starting from about 1500 days after the explosion, a first rise of the line related to the plane at 60°, followed by all the others until the one at 240° apart, that start to rise at about 4500 days. Only the flux on the plane 300° apart does not display a significant enhancement during the simulation time span. By the time this plane should be reached by the drifting projectile nodes, the cloud is already quite dispersed. The values of the fluxes are similar to those observed in Figure 3. The background values are instead higher at the GPS altitude. This is due to the fact that presently no constellation is actually deployed at the Galileo altitude. Nonetheless, also for the GPS case, the explosion debris flux is, for several years, a few times higher than the background in all the energy ranges.

In the case of a collision on a GPS-like orbit the situation is similar to the explosion one. The main difference is again that also the highest energy intervals become populated, as in Figure 4. The cases related to events in a GLONASS-like orbit display behaviors comparable to the Galileo constellation and are therefore not shown.

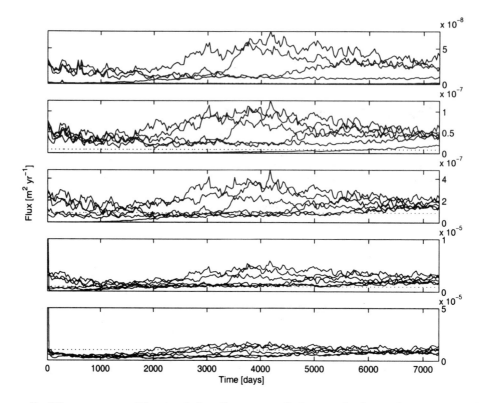

Figure 5: The same as Figure 3 for the case of the explosion of a spacecraft in a GPS-like orbit.

The next step was to analyze the effect of a fragmentation happening in a given constellation with respect to the other ones. As pointed out before, the proximity of the orbits of the three constellation makes them interdependent. Figure 6 shows the flux, on the three Galileo planes, of the debris coming from the explosion of a spacecraft in a GPS-like orbit. Note that the plane of the exploding spacecraft has $\Omega = 274°$, while the Galileo plane number 1 in Figure 6 is supposed to be at $\Omega = 262.7°$ (the following two planes are, as usual, 120° and 240° apart). In the three lower panels the fluxes on the Galileo planes are comparable, only a few times lower, with those observed in Figure 3. The curve that has a long peak, after about 1500 days from the explosion, represents the flux on the plane shifted by 120°. After this interval of time, due to the evolution of the mutual inclinations, this plane faces almost head-on collisions with the debris that gained enough eccentricity, from the breakup ΔV, to reach Galileo near apogee. In the two upper panels the flux is about one order of magnitude lower than in Figure 3, also for the most exposed plane at 120°. This should be due to the fact that heavier debris, capable of delivering the higher impact energies displayed in these panels, do not get enough ΔV to reach the Galileo altitude.

In Figure 7 the opposite situation is shown. The effects, on the GPS constellation, of

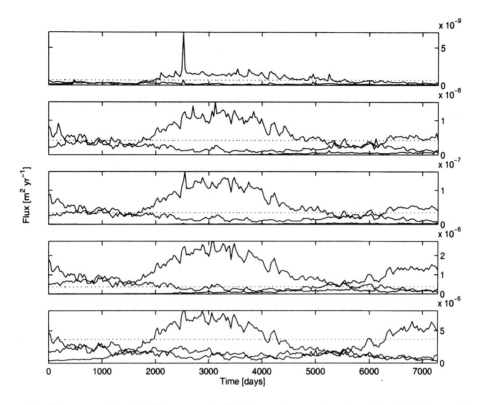

Figure 6: Flux on the Galileo constellation planes due to the debris produced by the explosion of a spacecraft in a GPS-like orbit. The panels are the same as in Figure 3.

the explosion of a spacecraft in a Galileo-like orbit are analyzed. The most noticeable feature is that the flux in the energy range $10^5 < 2E < 10^6$ J (third panel) is at the same level of the case of the explosion of a spacecraft in the GPS orbit (Figure 5). In the other panels the fluxes are between one order of magnitude and one half times lower than in Figure 5. In all the panels the highest flux is observed on the plane at 180° from the reference plane. This plane is nearly counter rotating with respect to the plane of the exploding Galileo spacecraft and is therefore subject to high velocity collisions. The second highest flux is then observed, starting from about 900 days after the event, on the plane at 240°, as the mutual inclination evolution brings the plane to interact with the cloud of debris.

Finally the different effect of an explosion of a spacecraft in a GLONASS-like orbit on the Galileo (Figure 8) and the GPS (Figure 9) constellations is analyzed. In Figure 8 it can be noted how the effects on the Galileo constellation are mitigated by the large separation in altitude between this constellation and the GLONASS. Only the flux on the plane shifted by 120° is noticeable. In the four lowest panels this flux is actually comparable to the fluxes of Figure 3. For the upper panel the same considerations as in Figure 6 hold. The other planes appears to be almost not affected

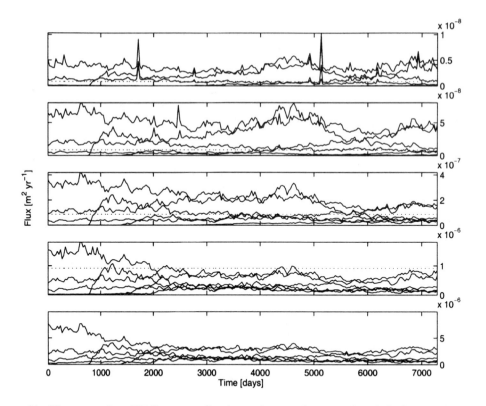

Figure 7: Flux on the GPS constellation planes due to the debris produced by the explosion of a spacecraft in a Galileo-like orbit. The panels are the same as in Figure 3.

at all by the fragmentation debris. The picture becomes quite different if the effects of the GLONASS explosion on the GPS constellation are displayed. In Figure 9 the fluxes on all the GPS planes are generally at least at the same level as in Figure 5. In particular, in the two uppermost panels the fluxes are even about one order of magnitude larger than the corresponding ones in Figure 5. The most exposed planes are again those at 180°, 120° and 240° (in the order in which the curves appear, from top to bottom, in the second panel from top). Due to the comparatively small separation in altitude of the two constellations an increase of eccentricity of less than 5 % makes a debris coming from the GLONASS altitude a possible projectile for a GPS satellite. This increment in eccentricity is easily obtained with the ΔVs of a few hundred m/s due to the explosion.

CONCLUSIONS

Thanks to the improvement of the method described in this paper with respect to the one developed in Refs. 5 and 6, the collision risk for the navigation constellations, following a fragmentation of a spacecraft, has been analyzed. The slower dynamics (in

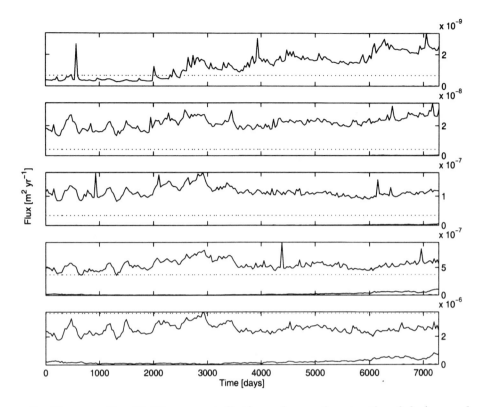

Figure 8: Flux on the Galileo constellation planes due to the debris produced by the explosion of a spacecraft in a GLONASS-like orbit. The panels are the same as in Figure 3.

terms of precession of the angular arguments of an orbit) of the MEO region prevent the appearance of the strong global effects observed for the LEO constellations[6,10]. Nonetheless a selection of the incident flux according to the initial location of the fragmentation event with respect to a given plane can still be observed. The planes that have a node displaced by about 180° from the one of the event are generally more exposed to strong incident fluxes. Hence they face a higher risk of a damaging collision. In general terms it has been observed the the flux following a generic fragmentation is by far larger than the low background flux in MEO. The values, spanning the range $10^{-6} \div 10^{-8}$ m^{-2} yr^{-1} according to the different impact energy levels, still account for low risks in terms of impacts per year. However, the very sensible applications of the navigation constellation call for a high level of reliability that could be seriously endangered by such prolonged levels of debris fluxes. Moreover the strong interdependence of the three constellations has been highlighted, by displaying the inter-constellations effects of an explosion event. In some cases the geometry of the systems is such that a larger flux is experienced if the fragmentation event happens on a different constellation. This effect is apparent comparing Figure 5 with Figure 9. On the other hand the results shown by Figure 8 would suggest that

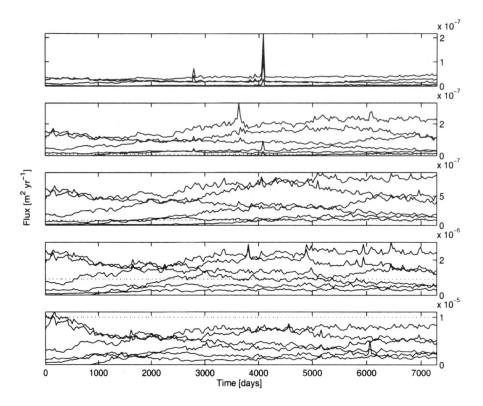

Figure 9: Flux on the GPS constellation planes due to the debris produced by the explosion of a spacecraft in a GLONASS-like orbit. The panels are the same as in Figure 3.

an increased separation in altitude could protect the forthcoming Galileo from any unfortunate event happening in the two lower constellations.

Therefore the mitigation policies of the different constellations should be harmonized, viewing the navigation constellation orbital regime as a whole region to be protected against a possible future debris growth.

BIBLIOGRAPHY

1. H. Klinkrad, J. Bendisch, K.D. Bunte, H. Krag, H. Sdunnus and P. Wegener, "The MASTER-99 space debris and meteoroid environment model", *Advances in Space Research*, **28**, 1355-1366, 2001.
2. C.C. Chao and Gick R.A., "Long-term evolution of navigation satellite orbits: GPS/Glonass/Galileo", paper COSPAR 02-A-02858, *Advances in Space Research*, in press, 2002.
3. A.B. Jenkin and Gick R.A., "Collision risk posed to the Global Positioning System by disposal orbit instability", *Journal of Spacecraft and Rockets*, **39**, No. 4, 532–539, 2002.

4. E.J. Öpik, *Interplanetary Encounters*, Elsevier, New York, USA, 1976.

5. G.B. Valsecchi, A. Rossi and P. Farinella, "Visualizing impact probabilities of space debris", *Space Debris*, **1** (2), pp. 143 – 158, 2000.

6. A. Rossi, G.B. Valsecchi and P. Farinella, "Collision risk for high inclination satellite constellations", *Planetary and Space Science*, **48**, 319–330, 2000.

7. A. Dell'Oro and Paolicchi P., "Statistical properties of encounters among asteroids: a new, general purpose, formalism", *Icarus*, **136**, 328–339, 1998.

8. A.E. Roy, *Orbital motion*, Adam Hilger Ltd., Bristol, UK, 1982.

9. A. Rossi, G.B. Valsecchi and P. Farinella, "Risk of collision for constellation satellites", *Nature*, **399**, 743–744, 1999.

10. A. Rossi and G.B. Valsecchi, "Self generated debris hazard for satellite constellations", proceedings of the *Second International Workshop on Satellite Constellations and Formation Flying*, Feb. 19–20, 2001, Haifa, Israel, 123–133, 2000.

11. A. Rossi, L. Anselmo, A. Cordelli, P. Farinella P. and C. Pardini, "Modelling the evolution of the space debris population", *Planetary and Space Science*, **46**, 1583–1596, 1998.

12. Su, S.-Y., "The velocity distribution of the collisional fragments and its effect on the future space debris environment", *Adv. Space Res.*, **10**, 389–392, 1990.

THE 2002 ITALIAN OPTICAL OBSERVATIONS OF THE GEOSYNCHRONOUS REGION

Manfredi Porfilio,[*] Fabrizio Piergentili[†] and Filippo Graziani[‡]

In the April 2002 the first dedicated observations of space debris in Italy have been performed by the Group of Astrodynamics of the University of Rome "La Sapienza" (GAUSS). Such test campaign, devoted to the Geostationary Earth Orbit (GEO) region objects detection, was accomplished using one of the Campo Catino Astronomical Observatory telescopes and was successful. As a consequence, GAUSS decided to continue the GEO observation activity from Campo Catino, also using other facilities. Since November 2002 the team has also joined the Inter-Agency Space Debris Co-ordination Committee (IADC) GEO campaign even if only a few clear-sky nights could be exploited.

The paper reports the results from all Italian GEO optical campaigns performed since April 2002 (test campaign) until January 8th 2003.

INTRODUCTION

Since many years the space debris concern is faced mostly through space environment observation and mitigation; also in Italy the debris issue has been dealt with, although only in April 2002 the first dedicated campaign has been performed. In that month the Group of Astrodynamics of the University of Rome (Gruppo di Astrodinamica dell'Università degli Studi di Roma "La Sapienza", GAUSS) has carried out a test campaign for Geostationary Earth Orbit (GEO) debris detection.

The GEO ring has been observed for a number of years by Russia, Japan and especially U.S.A.[1] and European Space Agency (ESA)[2]; since a few years the Inter-Agency Space Debris Coordination Committee (IADC) organises systematic joint optical searches of the geosynchronous region. From the Italian side, as the April 2002 test campaign was a success[3], GAUSS has scheduled and performed other campaigns since September, while since November 2002 the Group tried to join the IADC campaigns themselves. Unfortunately only a small number of clear-sky nights occurred in the IADC scheduled observation days throughout the period November 2002-January 2003, as it will be detailed in the relevant section.

GAUSS searches have been carried on exploiting the facilities of "Associazione Astronomica Frusinate" (Frosinone Astronomical Society); the society manages two observatories in the central part of Italy: the Campo Catino Astronomical Observatory (one of the best amateur astronomers' facilities in Italy) and the Collepardo Automatic Telescope (CAT).

---

[*] University of Rome "La Sapienza", Scuola di Ingegneria Aerospaziale, via Eudossiana 16, I-00174 Roma, Italy. E-mail: manfredi_porfilio@hotmail.com.

[†] University of Rome "La Sapienza", Scuola di Ingegneria Aerospaziale, via Eudossiana 16, I-00174 Roma, Italy. E-mail: fapierge@tin.it.

[‡] University of Rome "La Sapienza", Scuola di Ingegneria Aerospaziale, via Eudossiana 16, I-00174 Roma, Italy. E-mail: gauss@caspur.it.

CAMPO CATINO AND COLLEPARDO OBSERVATORIES

Both "Associazione Astronomica Frusinate" observatories are some tens of kilometres far from Rome: the geodetic coordinates of Campo Catino Astronomical Observatory are: latitude 41° 49' 16", longitude 13° 19' 47" altitude 1500 m; the coordinates of Collepardo Automated Telescope are: latitude 41° 45' 54", longitude 13° 22' 29" altitude 555 m.

Campo Catino Observatory main dome (Figure 1) hosts an equatorial mount usually carrying three optical tubes. The largest one is a 80 cm aperture Ritchey-Chrétien (the main structure in Figure 2). On the other hand CAT is based on a much lighter, quicker and more automated mount (Figure 3).

Figure 1 Campo Catino Astronomical Observatory

Apart from the cited 80 cm device, other 3 lighter optical tubes are available, which can be mounted on both Campo Catino main mount and CAT: a 25 cm Baker-Schmidt (in Figure 2 almost completely hidden by the Ritchey-Chrétien device, but well visible in Figure 3), a 15 cm apochromatic refractor practically always mounted on Campo Catino main mount (narrow tube in Figure 2) and a 40 cm Ritchey-Chrétien. Two Charged Couple Device (CCD) sensors are available for such telescopes: an AP-8 (Figure 4) and a ST-9.

For the April test campaign, performed from Campo Catino, the 25 cm Baker-Schmidt device was chosen; as its focal length is 800 mm, such device allows to have a quite wide field of view (FOV): the telescope was coupled with the AP-8 1024×1024 back-illuminated CCD (Figure 4), resulting in a 1°, 52' FOV. The pixel size is 24 μm, so the pixel FOV is about 6.6 seconds of arc. The system can detect 18th magnitude stars in a 20 seconds integration time.

Figure 2 Campo Catino's main dome telescopes

Figure 3 Collepardo Automated Telescope (CAT)

Figure 4 The AP-8 1024×1024 back-illuminated CCD

In Table 1 are detailed the number of clear-sky nights, the observatory, the optical tube, the focal length, the sensor used and the resulting FOV for each campaign.

Table 1

Equipment used during each campaign

Campaign	Clear-sky nights	Observatory	Optical tube	Focal length (mm)	CCD	FOV
April 2002 test campaign	2	Campo Catino	Baker-Schmidt 25 cm	800	AP-8	1°, 46'
September 2002	3	Campo Catino	Baker-Schmidt 25 cm	800	AP-8	1°, 46'
October 2002	1	CAT	Ritchey-Chrétien 40 cm with focal reducer	1550	AP-8	55'
IADC Nov-Dec 2002	2	CAT	Ritchey-Chrétien 40 cm	2950	AP-8	29'
IADC Dec02-Jan03	1	CAT	Ritchey-Chrétien 40 cm	2950	AP-8	29'

THE GEO CAMPAIGNS OVERVIEW

All the observations were devoted to surveying the GEO ring, although a number of Low Earth Orbit (LEO) objects were recognised. Throughout the 9 clear-sky nights (up to January 8th 2003), the total number of observing hours was 28, during which 1840 frames were collected and 180 objects detected. In Table 2 such results are detailed for each campaign, showing the weather impact in autumn and winter time, not only in terms of number of clear-sky nights per campaign but even in terms of number of observation hours per night.

Table 2

Observation hours and detected objects

Campaign	Clear-sky nights	Obs. hours	N. of frames	Detected objects				US
				CT	USCT	OF-USCT	Total	
April 2002 test campaign	2	7	430	99	6	5	110	1
September 2002	3	13	915	48	12	4	64	3
October 2002	1	0.5	34	0	0	2	2	0
IADC Nov-Dec 2002	2	4.5	285	0	1	0	1	0
IADC Dec02-Jan03	1	3	176	2	0	1	3	0
TOTAL	**9**	**28**	**1840**	**149**	**19**	**12**	**180**	**4**

The involved objects have been grouped in the following categories:

CTs   Catalogued Targets: present in NASA's "Geosynchronous Catalog Report";

USCTs   UnSuccessfully Correlated Targets: not present in the above cited catalogue; note that these bodies can not referred to as "uncatalogued": they might simply be non geosynchronous. Unfortunately in Italy we can not access the whole NASA catalogue, so we can not discriminate between catalogued and not catalogued bodies. We deliberately do not employ the usual notation "UCT" (UnCorrelated Targets) since our frames processed by someone with full access to the catalogue would produce different results in terms of CTs and UCTs[3];

OF-USCTs   One Frame UnSuccessfully Correlated Targets: while we refer to High Earth Orbit (HEO) not correlated bodies simply as USCT, we add "OF" when an object appears in one only frame. The detected body may be a LEO satellite, quickly passing through the telescope field of view (FOV), or it may appear in one frame only for any other reason. The fundamental difference is the possibility to perform orbit determination for the objects which at least appear in two images.

USs UnSeen: objects from NASA's "Geosynchronous Catalog Report" that should have appeared in one or more frames but result undetectable.

The CTs count one per night, that is, if a CT appears on more than one frame, it is counted only once all through the night. The USCTs count one per night as long as they are easily identified.

With regard to Table 2 results, we have to notice that the number of detected object per observation hour mostly depends on which zone of the sky we are pointing at and on the FOV. For instance, the April campaign was apparently most fruitful because, as a test, we just pointed at the geostationary declination (about −6 degrees from Campo Catino and Collepardo). Furthermore the FOV was the greatest among the campaigns performed so far. Comparing the April results with those of September (performed with the same equipment), we see a similar number of (HEO) USCT per frame, while the same thing does not happen for CTs, just because in April the weight of controlled (equatorial) satellites is significant.

The number of USs is relatively small. This is why we are proceeding with our observations. Moreover, some of these bodies should have been present on bad quality frames.

Unfortunately, we began the 2002 IADC campaigns late in November, when the weather conditions were quite unfair. The best period for searching from Campo Catino and Collepardo starts in April to end in October.

THE SEARCH STRATEGY

All the observations were mainly devoted to the geosynchronous region analysis. However, some differences in the search mode have been implemented in the various campaigns. We will address the observation features by campaign.

April 2002 test campaign

As the purpose of the first campaign was collecting as many tracks as possible, we decided to point an arc of the actual geostationary ring. The telescope was operated in star-tracking mode: the observed star field was kept constant along some hours of observation, letting the GEO ring cross such field.

A 20 seconds exposure time was selected. Such a relatively long time allows to improve the ratio of integration time to gap time between frames (the CCD read-out time is about 36 seconds), thus allowing to more easily detect GEO objects passing near the field boundary and to observe LEO objects possibly crossing the FOV. Furthermore, the GEO satellites are easily detected as stripes (longer and longer as the exposure time increases) on a dot-shaped background of stars. It is important to have rather long stripes if a visual detection of the targets must be performed on each image.

With the described observing strategy, for each detected body, two points of the track (leading and ending edges) per image are identified and their positions are determined in terms of topocentric right ascension (RA) and declination.

Some tests were also performed with the telescope fixed (rather than star-tracking). In this case the GEO satellites appear as points while the stars as streaks. With this search method we can not derive precise information about the satellites positions using off-the-shelf astrometry software, but it is possible to integrate on the same pixel a much greater amount of light from the GEO objects, thus allowing to reveal fainter debris[1].

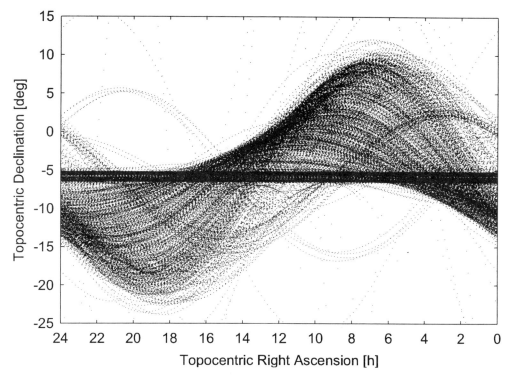

Figure 5 NASA's "Geosynchronous Catalog Report" objects as seen from Collepardo

September and October 2002 campaigns

After the successful April campaign, we decided to better organise our observation in order to systematically scan some interesting sections of the geosynchronous region. The Figure 5 shows NASA's "Geosynchronous Catalog Report" objects as seen from Collepardo. Their orbits have been propagated throughout 24 hours in order to achieve such diagram. It is apparent that the GEO ring declination relative to Collepardo (and Campo Catino) is about $-6°$. Having in mind this diagram, we decided to scan some regions characterised by:

- a good Sun phase angle;

- a declination as distant as possible from the GEO ring one (obviously remaining 15 degrees apart from the ring itself);

- a right ascension such as to look at a region "crowded" by satellites according to the catalogue. We remind that the Figure 5 diagram shape depends on physical reasons (the precession of the satellites orbital planes due to the coupling of luni-solar and J_2 perturbations), so that a similar graph must apply to non-catalogued bodies potentially observable.

The star-tracking mode was used again, but from this campaign on, the exposure time was reduced to 15 seconds, thus allowing to get more frames of each HEO object and, at the same time, improving the signal to noise ratio. The stripes are still enough long to allow an easy visual detection: further exposure time reductions seem to be possible.

IADC campaigns

Since late November 2002, the observation from Collepardo were managed in the frame of IADC GEO campaigns: the campaigns co-ordinators decide for each scheduled day which part (in terms of absolute right ascension and declination) of the GEO region has to be observed. Of course the general criterion is to scan, within a few years, the whole ring or, at least, the more interesting zones. The single observers are free to choose the observation details:

- star-tracking or GEO tracking mode (we chose the first);

- how to take account of the variable topocentric right ascension and declination corresponding to the fixed absolute ones. As the variation of relative RA is not so great throughout a few hours for the GEO ring distance (of course the declination change is even smaller), the following approaches can be used: taking account of a mean value of the parallax during the observation period or changing the observed star field a certain number of time during the night. We chose the second approach changing the star field every 30 minutes.

The IADC "main strategy" just described, must be pursued when the resulting elevation is greater than 30°; when this is not the case (to say, before the search field rising and after its setting), a "secondary strategy" is prescribed to be followed: the assigned RA is a function of the local sidereal angle (in such a way great elevations are guaranteed to hold).

Comparison of the search strategies

The frame in Figure 6 (from the April campaign, thus the exposure time is 20 s and the FOV 1° 46') clearly shows how easy it is to detect the Earth satellites on the fixed star background: the two horizontal stripes are GEO bodies, while the diagonal long streak is an OF-USCT orbiting at lower altitude. It is also apparent that three hot lines (vertical stripes in the picture) are present in the sensor.

In Figure 7, a photograph with the same FOV is presented, but with an exposure time of 15 s (September campaign). It is evident that the satellites streaks (GEO objects) are still quite easily identifiable, letting us think that it is possible to proceed with further reductions of the integration time without introducing any difficulties in the visual detection process. The exposure time can be reduced much more in case of longer focal lengths. In Figure 8, the track of a GEO object is clearly visible. The exposure time is again 15 seconds, but the FOV is now 29' (December 02 – January 03 campaign).

**Figure 6 19-4-2002, 01:15:05 Universal Time (UT), 20 seconds exposure, 1° 46' FOV.
3 objects in the FOV**

Figure 7 18-9-2002, 23:35:25 UT, 15 seconds exposure, 1° 46' FOV. 4 objects in the FOV

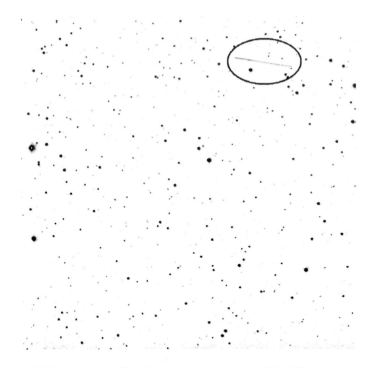

Figure 8 1-1-2003, 21:04:51 UT, 15 seconds exposure, 29' FOV. 1 object in the FOV

Finally, the Figures 9 and 10 (from the April campaign) show some tests performed with the telescope in fixed pointing. The GEO satellites are now points. In Figure 9 the exposure time was 20 s, resulting in long star streaks, while in Figure 10 it was 1 s. In the latter case both the stars and the satellite have a point-like shape. Of course the objects are fainter than in the case of the 20 s integration, but it is still possible to see them. The horizontal FOV is 1° 46'.

Some considerations about benefits and drawbacks of using the different search modes here presented, are reported in [3].

RESULTS

In the section "The GEO campaigns overview", we have already summarised the results of the GEO campaigns, detailing for each campaign the number of CTs, USCTs and USs (see Table 2). We report here, other interesting outcomes.

The Figure 11 shows one of the most "crowded" frames: 8 objects are present in the 1° 46' FOV, 7 of which turned out to be CTs. The 8th object (the one in the upper left corner) is a USCT, even though it is very bright (so, probably catalogued). As a matter of fact the orbit determination performed with the methods described in the section "Data handling", showed that this USCT is probably a GTO object launched from Kurou. For this reason it is clearly not included in NASA's "Geosynchronous Catalog Report".

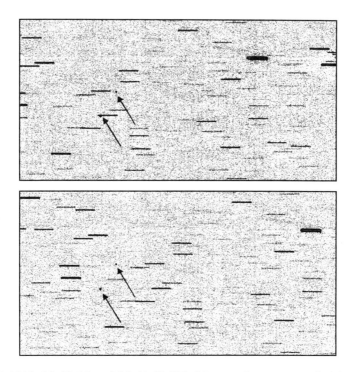

Figure 9 23-4-2002, 22:40:25 and 22:41:22 UT, 20 second exposures. 2 objects in the FOV

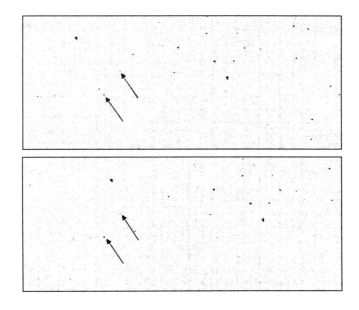

Figure 10 23-4-2002, 22:50:21 and 22:50:59 UT, 1 second exposures. 2 objects in the FOV

The CT in the upper part of Figure 11 (close to the USCT) is a rocket body (catalogue number 26101). Due to the 20 seconds star-tracking exposure, it is possible to determine for this debris a "flash period" of about 5 seconds, showing a precession of the body.

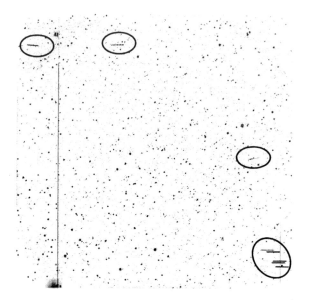

Figure 11 23-4-2002, 23:53:55 UT, 20 seconds exposure. Eight objects in the FOV

Figure 12 shows the position of the CTs in terms of right ascension and declination. Each object is represented by a single (great) point; as a background (small points) we have plotted the data already seen in Figure 5. This diagram also gives an idea of the scanned fields.

The inclination versus the right ascension of ascending node (RAAN) for the observed objects is reported in Figure 13. As it is well known (see for instance [4]) the evident structure in Figure 13 depends on the precession of the orbital planes of the uncontrolled GEO satellites due to the luni-solar perturbation coupled with the Earth equatorial bulge (J_2). The equatorial (probably controlled) and uncontrolled (subdued to precession) satellites are plainly identifiable in the diagram.

Figure 14 shows the apparent (i.e. not corrected for the solar phase angle and the distance) visual magnitudes of the detected objects. We have to remark that the magnitudes here reported are not evaluated in the standard mode, which is very useful if you want to have an idea of the target dimension (of course in this case the absolute magnitude must be taken into account[2]). To evaluate the magnitude (according to the standard definition), it is necessary to take account of the light collected all along the exposure time. In the case of a trailing target, this implies to comprise the whole streak. The problem with trailing targets, is that a magnitude evaluated in this way, is definitely meaningless from the signal to noise ratio (thus from the detection probability) standpoint.

For instance, let us consider the case of two targets, one in LEO the other in HEO, having the

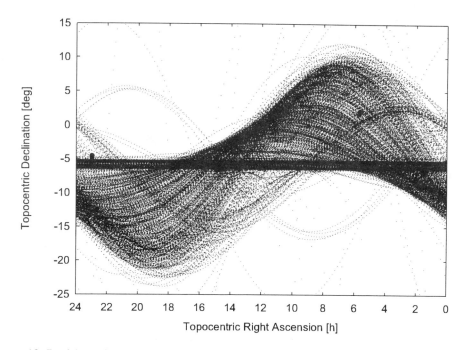

Figure 12 Position of the detected CTs with respect to NASA's "Geosynchronous Catalog Report" objects

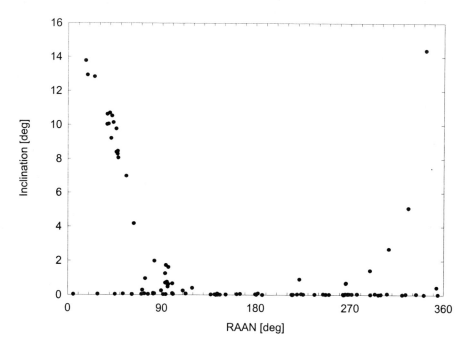

Figure 13 The detected objects orbital planes

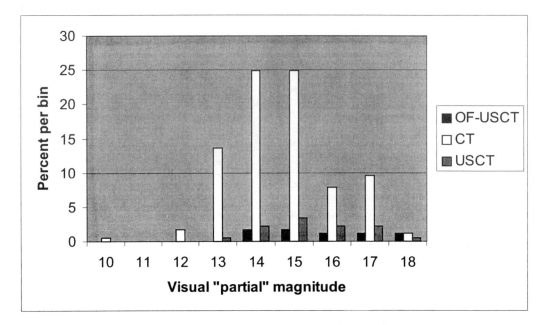

Figure 14 Visual "partial" magnitude of the detected objects

same apparent magnitude (standard definition). As their trails on the frames would be completely dissimilar in length (due to their motion relative to the observer), it is evident that their signal to noise ratios (thus their detection probabilities) would be very different (the light from the faster object is spread on a greater number of pixels).

As we are mostly interested in a S/N-oriented parameter (in order to compare the results from a detection probability point of view) we defined a non-standard or "partial" magnitude, evaluated taking into account only a part of the streak as long as its thickness. With this parameter, we can evaluate the detection limit of our equipment independently on the target relative velocity.

Of course, the "partial" magnitude of a given object decreases (the object looks brighter) with the exposure time. This is exactly what is desirable for a S/N-related parameter: as the integration time diminishes, the background noise as well as the stars signals decrease, while the average pixel light level from the target remains the same (we remind that the targets trail). As a result, the signal to noise ratio increases. As the exposure time goes to zero, the "partial" magnitude tends to the apparent standard one (the streak tends to a point). This implies that the limit "partial" magnitudes give a precise idea of the limit (standard) magnitudes we would have in case of short exposures.

Therefore, the "partial" magnitudes are not comparable (their numerical values are greater) with those obtained (with the same exposure time) in a target-tracking mode (that is telescope-parked mode in case of GEO satellites): for instance, comparing the results presented by Jarvis et al.[1] with those achieved by GAUSS and reported in [3] (which refer only to 20 second exposures, that is the same integration interval as in [1]), a two magnitudes difference can be pointed out.

DATA HANDLING

The data processing, which is presently carried out off-line, comprises the following tasks: the star field identification, the objects classification by comparison with the results of the propagation of NASA's "Geosynchronous Catalog Report" TLEs, the visual search for bodies not present in this catalogue and the orbit determination.

The orbits of the USCTs are estimated by several methods: the Laplace method, the Gauss method and the double r-iteration method. To take into account more than 3 observations (i.e. 3 declination and 3 right ascension measures), the least squares method is used as well as a method based on the determination of the "most probable" orbital element set, performed on the basis of a joint probability distribution function of some "critical" elements (typically the semi-major axis and the semi-latus rectum[5]). This probability distribution is computed starting with the elements determined from each possible combination of 3 observations of the considered object. The latter method can be conveniently joined with the least squares one (thus using combinations of more than 3 observations) in order to improve the results. The orbit determination work is still on going and will be presented in the future.

CONCLUSIONS AND FUTURE DEVELOPMENTS

In 2002 Italy has finally started the space debris detection activity. The Group of Astrodynamics of the University of Rome (GAUSS) is carrying on this activity using the facilities of "Associazione Astronomica Frusinate" and particularly the Collepardo Automated Telescope, which can be programmed in order to keep on a whole night of observations.

The orbit determination task has been faced and the methods will be improved. Furthermore, much effort is presently being spent on image processing, in order to automate in the future the object identification process.

In order to understand if uncatalogued object may be detected with the facilities presently employed, it would be extremely important to access the whole NASA's catalogue. Such a political issue will be brought to the attention of the IADC members, at the next IADC meeting (March 2003).

The observation activity will be carried on during the next months, also in the frame of IADC GEO campaigns.

ACKNOWLEDGMENTS

The authors would like to thank Dr. S. Isobe, of Japanese National Astronomical Observatory, who provided the "Asteroid Catcher" software by Japan Space Guard Association, Dr. M. Di Sora, President of Frosinone Astronomical Society, who is making possible the Italian activity on space debris detection, Mr. F. Mallia for the reliable "on line" technical support, Mr. U. Tagliaferri, who carried out in Collepardo the greatest part of the observations and Mr. F. Galante who analysed part of the collected data.

REFERENCES

1. Jarvis K. S., T. L. Thumm, K. Jorgensen, J. L. Africano, P. F. Sydney, M. J. Matney, E. G. Stansbery, and M. K. Mulrooney, *CCD Debris Telescope Observations of the Geosynchronous Orbital Debris Environment, Observing Year: 1998*, JSC-29537, National Aeronautics and Space Administration, Houston, Texas, August 2001.

2. T. Schildknecht, R. Musci, M. Ploner, S. Preisig, J. de Leon Cruz, H. Krag, "Optical Observation of Space Debris in the Geostationary Ring", *Proceedings of the Third European Conference on Space Debris*, Darmstadt, Germany, 19-21 March 2001.

3. M. Porfilio, F. Piergentili, F. Graziani, "First optical space debris detection campaign in Italy", presented at the 34th COSPAR scientific Assembly, The World Space Congress 2002, Houston, USA, 10-19 October 2002, COSPAR paper n. PEDAS1-B1.4-0063-02.

4. Graziani F., *Appunti delle lezioni di Astrodinamica applicata*, edited by Emanuela D'Aversa, Aerospace Department, University of Rome "La Sapienza", Rome, 1992. (In Italian)

5. Piergentili F., *Sperimentazione di sistemi ottici per la determinazione orbitale*, Laurea Thesis in Aerospace Engineering, University of Rome "La Sapienza", Department of Aerospace and Astronautical Engineering, Rome, 2001. (In Italian)

A GEOMETRICAL APPROACH
TO DETERMINE BLACKOUT WINDOWS AT LAUNCH

V. Rabaud[*] and B. Deguine[†]

Determining the blackout launch windows is a combinatory problem using a large amount of data. As there are many possible cases, we have been looking for a new and general continuous method which can give results in a few minutes. To solve this problem, we have to confront a primary object trajectory and several debris trajectories obtained thanks to a given extrapolation model. In fact, it will be shown that we first have to solve it by assuming that debris evolve in a keplerian way. The method uses several basic geometrical filters to determine when and where a debris gets into a volume based upon a one sheeted hyperboloid. To be more accurate, some perturbations are then applied to the debris trajectories and new kinds of filters are run. This paper will address this original approach and some interesting results of its efficiency and benchmarks. Moreover, some ideas to improve the method will also be presented at the end.

Nomenclature

CNES	Centre National d'Etudes Spatiales
LEOP	Launch and Early Orbit Phase
AVOIDANCE	Program developed by CNES using the presented method (acronym for "close Approach detection Via one-sheeted hyperbolOID for lAuNCh Event")
d_{CA}	Secure threshold distance
$d_{perturbation}$	Maximum distance between the Keplerian trajectory and the SGP4/SDP4 extrapolated one, for the same debris
$d_{CA\,Kepler}$	Security threshold distance when the debris is Keplerian. Equal to $d_{CA} + d_{perturbation}$
$a, e, i, \omega, \Omega, M$	Keplerian elements of the debris
T	Period of the debris orbit
t	Universal time: t indicates a precise moment and describes the following intervals
$[tinf_{Kepler}; tsup_{Kepler}]$	Temporal interval concerning the Keplerian debris
$[tinf_{NORAD}; tsup_{NORAD}]$	Temporal interval concerning the NORAD extrapolated debris
$[tinf_{PO}; tsup_{PO}]$	Temporal interval concerning the primary object trajectory segment we study
$[METinf_{PO}; METsup_{PO}]$	MET interval concerning the primary object trajectory segment we study. It is defined relatively to an initial date (which can be the launch date)
Segment	Geometrical linear object defined by two consecutive points of a trajectory
Debris	All orbiting object tracked by US Space Command
NORAD debris	Debris whose trajectory is based upon two-lines and NORAD propagators.
Corresponding Keplerian debris	For a NORAD debris, it is the closest object evolving in a Keplerian way that can be found

[*] Ecole Nationale et Supérieure de l'Aéronautique et de l'Espace 10, avenue E. Belin, F-31400 Toulouse, France. E-mail: Vicent.Rabaud@m4x.org.

[†] Centre National d'Etudes Spatiales, Space Mechanics Division, 18, avenue E. Belin, F-31401 Toulouse Cedex, France. E-mail: Beatrice.Deguine@cnes.fr.

INTRODUCTION

As sensors used to observe debris around Earth are becoming more accurate and more numerous, our knowledge of debris environment increases. Thus, we become more aware of the danger they may represent towards our satellites or launchers.

Our study focuses on the possible close approaches during the LEOP phase of a primary object trajectory, which includes the launching and the early orbit phase. Therefore, the primary object can be either a launcher or a satellite.

Usually, primary object and debris trajectories are discretized to observe the distances between them. In order to have a faster method than the usual time consuming iterative ones, a completely geometrical approach was developed to solve the blackout launch window problem. Its efficiency is based upon the solvability of simple geometrical equations.

We are first going to present a close approach determination method between a Keplerian debris and a primary object during LEOP. We will then focus on a more precise method that takes into account perturbations acting on debris.

A benchmark of AVOIDANCE (the program we have developed at CNES using this method) will be given at the end of this article.

GENERAL CONSIDERATIONS

Our input data are the discretized trajectory of the launched object (launcher or satellite), the initial launch window, and two-lines of all debris we want to confront.

The launch trajectory of the primary object is given at a precise time. It is assumed that between two consecutive points of its ephemeris, the primary object is moving linearly: we have then a trajectory segment.

Concerning the debris, as the initial data are two-lines coming from US Space Command catalog, the SGP4/SDP4 NORAD extrapolation model is used to describe their evolution. Depending on the debris initial data sources, any other associated propagation model could be used with this method.

GENERAL PROBLEM

Our aim is to determine the time windows over which we can launch our primary object safely so that it stays above a given close-approach threshold distance d_{CA} from every debris.

As the primary object trajectory is given as an ephemeris and as linear movement between two consecutive ephemeris points is assumed, the trajectory can be considered as a set of segments. Thus, the problem is reduced to the determination of the time intervals over which one debris may be dangerous with one segment of the primary object trajectory. The method used to solve this problem is then reproduced for every other debris and segment of the primary object trajectory.

Moreover, as we are looking for a geometrical approach, it would be easier if the debris were Keplerian. Let us assume that the debris behaves in a keplerian way that is "rather close" to the NORAD debris we confront to the primary object segment. It means that the two debris positions obtained by a Keplerian and a NORAD extrapolation are not more than $d_{perturbation}$ far over the study time interval (defined by the launch interval and the ephemeris).

Our problem is to find the time interval $[tinf_{NORAD}; tsup_{NORAD}]$ during which the NORAD debris is less than d_{CA} close from our primary object segment. Solving the problem "find the time interval during which the keplerian debris is less than $d_{CA} + d_{perturbation}$ close from our primary object segment" allows to solve our problem when NORAD debris are considered. This statement can be demonstrated by a reductio ad absurdum.

By definition of $d_{perturbation}$, over this time interval, the Keplerian debris is less than $d_{CA} + d_{perturbation}$ close from our primary object segment. Let us define $d_{CA\ Kepler} = d_{CA} + d_{perturbation}$ and also the time interval over which the Keplerian debris is $d_{CA\ Kepler}$ close from our primary object segment: $[tinf_{Kepler}; tsup_{Kepler}]$.

What is the relation between $[tinf_{NORAD}; tsup_{NORAD}]$ and $[tinf_{Kepler}; tsup_{Kepler}]$?

If $[tinf_{NORAD}; tsup_{NORAD}]$ is not completely included in $[tinf_{Kepler}; tsup_{Kepler}]$, it means it exists a time at which the NORAD debris is d_{CA} close from our primary object segment and the corresponding Keplerian debris is not $d_{CA\ Kepler}$ close from our primary object segment. This statement leads to a contradiction according to the definition of $d_{perturbation}$ and by reminding that $d_{CA\ Kepler} = d_{CA} + d_{perturbation}$.

Therefore :
$$[tinf_{NORAD}; tsup_{NORAD}] \subset [tinf_{Kepler}; tsup_{Kepler}]$$

Our geometrical approach will first consider the corresponding Keplerian debris. It will solve the problem "find the time interval during which the keplerian debris is less than $d_{CA} + d_{perturbation}$ close from our primary object segment". For that, some equations will be solved in a completely continuous problem: there won't be any discretization. If a potential danger is found on a $[tinf_{Kepler}; tsup_{Kepler}]$ interval with a threshold distance of $d_{CA\ Kepler}$, we will then focus on the NORAD debris whose behavior is described by NORAD propagators, by discretizing its trajectory and using the security threshold distance d_{CA}. It is important to mention that the necessary discretization only happens at the last step of the method, and thus only for a few cases.

LEADING IDEAS

Let's consider a primary object trajectory segment: it is a physical segment defined during a given time interval. The launcher is assumed to be everywhere on it during the corresponding time interval. Therefore, to have accurate results, an ephemeris time step of a second or less has to be used. In the following, the expression "trajectory segment" will refer to this primary object trajectory segment.

Depending on the exact launch time, our trajectory segment is going to be at a different place, due to the Earth rotation: all its different positions are in fact going to form a surface.

Figure 1 : Surface described by the rotated segments

By extending our different rotated segments in lines, we recognize the surface: it's an one-sheeted hyperboloid.

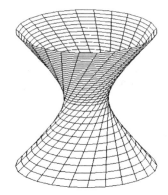

Figure 2 : The one-sheeted hyperboloid is a doubly ruled surface

Due to some symmetries, its equation is here: $x^2 + y^2 - \dfrac{(z - z0)^2}{b^2} = c^2$

But we're only interested in a slice of it. Moreover, as our possible launch window isn't 24-hour long (only few hours are usually considered), only a portion of this slice has to be taken into account.

Now, let's consider the debris. As it has already been mentioned, the geometrical approach would be easier to implement if the debris behaved according to a keplerian propagation model. This has consequently led to the following method composed of two steps. The first step consists of a simplified approach by considering that the debris evolves in a Keplerian way. It reduces the number and the length of the analyzed time intervals. Next, the second step implements the NORAD model of perturbations.

KEPLERIAN APPROACH

Let's wrap our trajectory segment in a volume made with the security threshold distance $d_{CA\ Kepler} = d_{CA} + d_{perturbation}$.

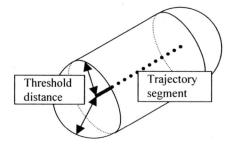

Figure 3 : A trajectory segment in its security zone

Depending on the launch time, this volume is going to rotate around the Earth rotation axis, forming a security volume V.

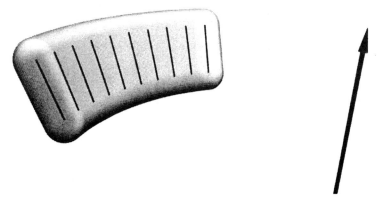

Figure 4 : The security volume V (the axis represents the Earth rotation axis)

If the debris comes $d_{CA\ Kepler}$ close to the trajectory segment, it must penetrate this security volume V. But as its equation is rather complex, instead of considering it, it is wrapped in a bigger but simpler volume: a portion of a rectangular torus.

Figure 5 : The portion of a rectangular torus

Thus, as a simplified approach, we are now meant to know if an ellipse (corresponding to the debris trajectory) gets through a portion of a rectangular torus. To solve this geometrical problem, three successive simple geometrical filters are going to be applied. If a dangerous zone is found, a fourth filter, called "time filter", will check whether the debris and the primary object can be inside it at the same time.

The following graph shows the logic of the keplerian approach:

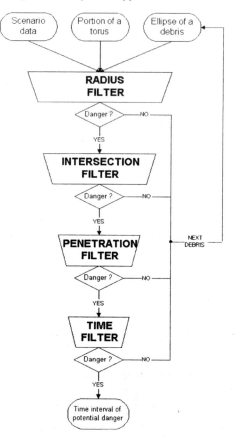

Figure 6 : The geometrical approach

FIRST FILTER: RADIUS FILTER

If the debris gets into the portion of torus, then its distance r to the center of the Earth must verify:

$$R_{\min} \le r \le R_{\max}$$

where $R_{\min}$ and $R_{\max}$ are determined by the dimensions of the torus.

However, as the debris is considered evolving in a Keplerian way, r must also verify:

$$a \cdot (1-e) \le r \le a \cdot (1+e)$$

Therefore, if there is no intersection between $[a \cdot (1-e), a \cdot (1+e)]$ and $[R_{\min}, R_{\max}]$, the debris cannot be dangerous. Then, the trajectory segment is confronted to another debris. Otherwise, the next filter is applied.

SECOND FILTER: INTERSECTION FILTER

To have the debris going through the portion of the torus, the debris orbital plane must go through it. A test is applied to check this situation.

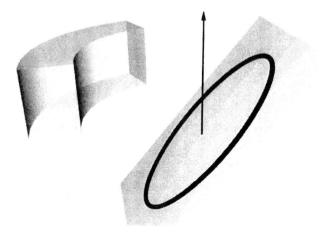

Figure 7 : Intersection Filter

We first define the torus with cylindrical coordinates:

$$r_{min} \leq r \leq r_{max}$$
$$\theta_{min} \leq \theta \leq \theta_{max}$$
$$z_{min} \leq z \leq z_{max}$$

Then, we take a point in the debris orbital plane which coordinates are (α, β). It is assumed to be in the torus and thus α and β have to verify some properties. If there is no contradiction between these properties, it means the debris orbital plane goes through the torus.

Moreover, we find that α and β have to verify equations as:

$$\alpha \in [\alpha_{min}, \alpha_{max}]$$
$$\beta \in [\beta_{min}, \beta_{max}]$$

which means that the intersection surface of the ellipse plane with the torus by a rectangle has been surrounded.

THIRD FILTER: PENETRATION FILTER

Now, the next step aims at determining whether the debris ellipse goes through the rectangle obtained at the end of the previous filter. It consists in solving four second degree equations (one for each side). If solutions exist, the intersection points can be defined by a true anomaly interval $[vinf_{deb}; vsup_{deb}]$ on the debris ellipse. This anomaly interval is then converted in a temporal one: $[tinf_{deb}; tsup_{deb}]$.

Figure 8 : Penetration Filter

TIME FILTER

If this step is reached, it means the debris goes through the rectangular torus during the time interval $[tinf_{Kepler}; tsup_{Kepler}]$, modulo its Keplerian period T. As the entering and exiting points are known, a sub-portion of the original rectangular torus, in which the launcher trajectory mustn't be at the same time, can be defined (**Figure 8 : Penetration Filter**).

The primary object enters this sub-portion thanks to Earth rotation at $tinf_{PO}$. And it gets out at a time $tsup_{PO}$. This defines a forbidden temporal interval:

$$[tinf_{PO}; tsup_{PO}]$$

To determine it, the extreme angles of the security volume are determined as shown on the **Figure 9** hereafter, and then converted in temporal values knowing the Earth rotation speed.

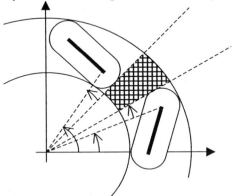

Figure 9 : Rotation of the security zone determining $[tinf_{PO}; tsup_{PO}]$, seen from above

If there is no intersection between one of the intervals $[tinf_{Kepler}; tsup_{Kepler}]$ modulo the period of the debris and $[tinf_{PO}; tsup_{PO}]$, it means the primary object and the debris aren't in the danger zone at the same time.

PERTURBATION FILTER

As it has been described in the General Problem section, the time interval $[\text{tinf}_{\text{NORAD}};\text{tsup}_{\text{NORAD}}]$, over which the NORAD debris is dangerous, is included in $[\text{tinf}_{\text{Kepler}};\text{tsup}_{\text{Kepler}}]$.

Therefore, if after running the previous filters with a $d_{CA\,Kepler} = d_{CA} + d_{perturbation}$ threshold, an interval of potential danger $[\text{tinf}_{\text{Kepler}};\text{tsup}_{\text{Kepler}}]$ is found, we have to check if the NORAD debris is less than d_{CA} close to the primary object trajectory segment over this interval. The following method is applied:

- The interval $[\text{tinf}_{\text{Kepler}};\text{tsup}_{\text{Kepler}}]$ is discretized in shorter time intervals depending on the precision required (0.1 second or less is usually used) : $[t_0;t_1]$, $[t_1;t_2]$...

- For each $[t_i;t_{i+1}]$, the debris trajectory segment is computed by using NORAD propagators and the debris is assumed to be everywhere on it over the given time interval. Then, the method verifies whether this interval can be dangerous by applying two filters:

  o It first checks if the debris trajectory segment has a common part with the d_{CA} torus thanks to a segment-torus intersection filter.

  o Then, the same time filter as before is applied between the primary object segment and the debris segment.

- If the debris segment is still dangerous, its distance to the portion of hyperboloid slice is then computed.

The following figure illustrates this process for one primary object trajectory segment and one debris segment.

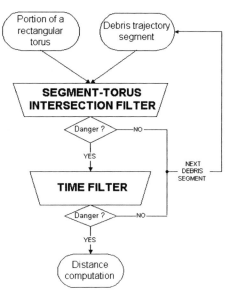

Figure 10 : The perturbation filter

At the end of the process, if the debris segment is dangerous, a time interval $[\text{tinf}_{\text{NORAD}};\text{tsup}_{\text{NORAD}}]$ is determined.

HOW TO DETERMINE THE DISTANCE BETWEEN A SEGMENT AND A PORTION OF A SLICE OF A ONE SHEETED HYPERBOLOID

This part of the method is the most complex. To determine the minimum distance, the corresponding minimum point on the segment and the corresponding minimum point on the hyperboloid are going to be determined. It can be shown that this problem corresponds to find the minima of a function defined by

$$\Delta(z,k)=(\sqrt{A\cdot z^2 + B\cdot z+C} - \sqrt{D\cdot k^2 + E\cdot k + F})^2 + (G\cdot z + H\cdot k + I)^2$$

where z is the spot height of the point on the one-sheeted hyperboloid and k the position of the other minimum point on the segment.

The minima of this function are calculated by solving the zeros of its derivatives, which can lead to solve a fourth degree polynomial. However, as only a segment and a slice of hyperboloid are of interest, z and k check some relations like:

$$z_{min} \leq z \leq z_{max}$$
$$0 \leq k \leq 1$$

Therefore, after studying some function variations, we are most of the time not bound to solve this polynomial, which increases the computational speed.

DEDUCING THE BLACKOUT LAUNCH WINDOW

Finally, the NORAD debris will be d_{CA} close to the primary object trajectory segment over $[t\text{inf}_{NORAD}; t\text{sup}_{NORAD}]$. The primary object trajectory segment studied corresponds to the ephemeris interval $[\text{METinf}_{PO}; \text{METsup}_{PO}]$. It means the primary object mustn't be launched on the following time interval:

$$[t\text{inf}_{NORAD} - \text{METinf}_{PO}; t\text{sup}_{NORAD} - \text{METsup}_{PO}]$$

DETERMINING $d_{perturbation}$

This distance has to be initially determined through a discretization of the studied time interval (determined by the original launch window and the primary object ephemeris time interval). Between each step, the evolution is going to be considered as linear. Therefore, a high time step should not be used. But a small one would drastically increase the computational time. With a one-minute time step, the committed errors are very slight.

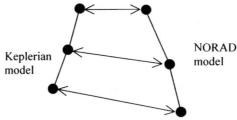

Keplerian model

NORAD model

Figure 11 : Computation of the distance between the NORAD and the Keplerian models

At each time step, the distance between the positions given by the two propagation models is computed and the biggest one named $d_{perturbation}$ is kept.

BENCHMARK OF *AVOIDANCE*

The method presented above has been implemented at CNES in a FORTRAN 90 program called AVOIDANCE. It must be noticed that no real work has been done to improve its speed.

We have run it on several launcher ephemeris lasting from 10 to 30 minutes and reaching an altitude varying between 300 and 1500 km. Their time step was one second long. And we also decided to compute the NORAD extrapolation with a 0.1 second discretization. The launch window is two hours long.

case	Length (min)	Altitude (km)
1	10	300
2	20	1000
3	10	1000
4	10	800
5	10	600
6	20	900
7	30	400
8	10	1200
9	30	1500
10	20	400

Table 1: Description of the different ephemeris

To know the efficiency of every filter (the three geometrical ones, the temporal and the perturbation ones), we have run AVOIDANCE on the described ephemeris with a 5km security distance, and done the mean of the efficiencies for each filter. A potentially dangerous case associates a debris and a trajectory segment. The efficiency written here above for a filter is the percentage of eliminated non-dangerous cases on all the cases it has to manage (which means after the previous filters have been applied).

	Geometrical filters			Temporal filter	Perturbation filter
	1	2	3		
% of elimination	77%	85%	57%	93%	99.99%

Table 2: Efficiency of the different filters

The computing time on a 900 MHz PC is shown below for several cases. The peaks correspond to high altitude trajectories with a long duration.

Moreover, the following graph shows the execution time needed to compute the dangers with a larger security distance (10km or 20 km). This graph clearly shows that the impact of the security threshold from 5 km to 20 km on the execution time is rather small.

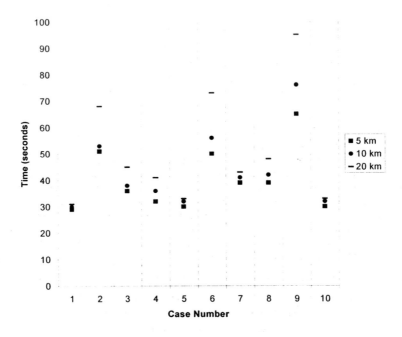

Figure 12 : A benchmark for some results

It must be mentioned, that at least 20 to 30 seconds are used to pre-compute the distance $d_{perturbation}$.

Moreover, to be sure that the results given by AVOIDANCE were right, we compared them with the ones given by AGI's Satellite Tool Kit® version 5.0 Beta.

STK and AVOIDANCE found that the same debris could be dangerous but AVOIDANCE found larger blackout launch windows (from 3 to 6 seconds long) that included STK's ones (which were only 1 to 2 seconds long).

Therefore, we decided to use more precise ephemeris, defined with a 0.1 second step (and no longer 1 second). Then, the intervals found were less than half of a second larger. Of course, the execution time increases by 20 to 40 seconds, but this is not the best way to proceed: in fact, the 0.1 second step should be used only during the perturbation filter, which is the last step of the method.

LIMITATIONS OF THE METHOD

This method uses two approximations which intensity is given by the user:

- The primary object is assumed to evolve linearly between two consecutive ephemeris points.

- For each segment considered, the primary object is assumed to be everywhere at the same time.

- The same two hypotheses are made for the NORAD debris at the very end of the method.

In both cases, the user can choose the threshold of the time discretization. A time step of 0.1 second has been used successfully. This only corresponds to 1km if a velocity of 10 km/s is considered (which is already fast for a launcher or a satellite).

STRENGTHS OF THE METHOD

Nonetheless, this method is interesting for the following reasons:

- It first uses a continuous approach, which is much faster and provides exact solutions. It also enables the computing time to grow slowly with the security distance (there is not much difference between 5km or 20 km).

- Every primary object trajectory segment is analyzed independently from the others. Therefore, we can imagine an approach considering differently these segments depending on their length or duration. That is to say, a trajectory segment corresponding to a small primary object velocity could be longer than the one related to a high velocity. The execution time would obviously be improved.

- Any propagation model associated to the debris orbital parameter sources can be used to take into account the perturbations acting on debris.

THE FUTURE OF *AVOIDANCE*

Two alternatives can be envisaged:

- To use adjustable time steps for the primary object ephemeris and the NORAD extrapolation: a large step like 1 second should first be used and then, a smaller one should be considered when concentrating on the potential dangers.

- To remove the hypothesis of omnipresence on a trajectory segment for both the launcher and the debris. To do so, we are going to use a simple evolution model between two points (still a linear one but with a constant acceleration). The problem will therefore remain continuous and the equations will only have two parameters: the universal time and the launch time.

Use one or maybe both of these implementations, will allow to reduce the blackout window lengths and to be more accurate on the minimum close approach distance.

Moreover, our efforts have only been put to the heart of the algorithm and not on the calculus of $d_{perturbation}$, which might simply be improved.

The presented theory is still under development but seems to be strong, fast and reliable.

ORBITAL DEBRIS ANALYSIS OF
TIMED SPACECRAFT MISSION

S. S. Badesha,[*] S. K. Dion[†] and R. E. O'Hara[‡]

Collision with orbital debris is a hazard of growing concern as past practices and procedures have allowed man-made objects to accumulate in orbit. To control further generation of orbital debris NASA has issued safety standard NSS1740.14. The Johns Hopkins University Applied Physics Laboratory (JHU/APL) developed the Thermosphere, Ionosphere, Mesosphere, Energetics and Dynamics (TIMED) spacecraft for an atmospheric remote sensing mission sponsored by the NASA Office of Space Science. The TIMED spacecraft was placed in a circular orbit, 625 km altitude, and 74.1 degrees inclination for a two year mission to study the influences of the sun and human activities on the Earth's thermosphere, ionosphere, and mesosphere which are the least explored regions of the atmosphere. This paper demonstrates compliance of the TIMED spacecraft with all requirements of NSS 1740.14.

This paper describes the TIMED spacecraft hardware, operational scenarios, and orbital debris analysis, along with the assumptions and results. The analysis was done using NASA's Debris Assessment Software (DAS) program. This program has five main modules corresponding to the five main pertinent operational scenarios. The DAS predicted that the area-time product exceeds the requirement limit of 0.1 m2/yr close to the potential 650 km altitude. However, an orbital de-cay program that uses real time solar flux values in the density model, and the drag calculation using a 50-percentile solar flux, predicted a significantly lower orbital lifetime in compliance with NSS 1740.14. Additionally, the total debris casualty area of 9.2 m2 slightly exceeded the guideline requirement value of 8 m2. However, reanalysis of the reentry survivability, using NASA's enhanced DAS code Object Reentry Survival Analysis Tool (ORSAT), resulted in a sig-nificantly lower value of 5.2 m2 in compliance with NSS 1740.14.

Keywords: TIMED, Orbital Debris.

INTRODUCTION

Collision with orbital debris is a hazard of growing concern as past practices and procedures have allowed man-made objects to accumulate in orbit. If unchecked, the number of man-made objects in low Earth orbits (LEO) will reach a critical level[1]. Collisions could then become the main source of further orbital debris. In approximately 100 years, there will be over 50,000 objects larger than 10cm in LEO. Collisions will have caused over 40,000 of them[1]. Even an object as small as 0.1 mm could cause damage to sensitive components such as sensor lenses. Ob-

[*]   The Johns Hopkins University, Applied Physics Laboratory, 11100 Johns Hopkins Road, Laurel, Maryland 20723-6099. E-mail: surjit.badesha@jhuapl.edu.

[†]   SRS Information Services, NASA Goddard Space Flight Center, Greenbelt, Maryland 20771.

[‡]   Lockheed Martin Space Operations, Houston, Texas 77058.

jects larger than 1 cm cannot be shielded against. Objects in the 1 mm to 1 cm range could be shielded against, but there would be a cost and weight penalty.

As evidenced by the number of recently proposed or deployed satellite-based information transmission systems such as Globalstar, Iridium, and Teledesc, the commercialization of space is becoming compelling. Orbital debris could, however, make the emerging satellite revolution difficult. Furthermore, scientific exploration of space would be greatly compromised. To control further generation of orbital debris, NASA has issued safety standard NSS 1740.14, Guidelines and Assessment Procedures for Limiting Orbital Debris[2]. NASA requires all of its future programs to be in compliance with the guidelines set forth in this document. It is hoped that other countries and organizations will follow NASA's lead in the very near future.

In compliance with NASA guidelines, the following five issues were analyzed:

1. Debris generated during normal operations
2. Debris generated by explosive and intentional breakups
3. Post-mission orbital lifetime and disposal procedure
4. Debris generated by on-orbit collisions
5. Reentry survivability of debris

This paper describes the analysis, along with the assumptions and results. The analysis was done using NASA's Debris Assessment Software (DAS) program[3]. This program has five main modules corresponding to the five issues mentioned above. In NSS 1740.14, detailed guidelines and evaluation methods for each of the five issues is documented. These evaluation methods are the basis of the DAS program, version X.09.

This spacecraft orbital debris assessment considers only the spacecraft and associated instruments and does not include analysis of the launch vehicle or any payload attach fittings. Furthermore, this assessment is not an integrated assessment of the entire launch system.

DESCRIPTION OF DESIGN AND OPERATIONS FACTORS

Spacecraft Hardware

The TIMED Spacecraft shown in Figure 1 is a three axis stabilized nadir pointing, low earth orbit scientific satellite. The spacecraft consists of four instruments mounted on a spacecraft bus with solar arrays, battery cells, antennas, ordnance, electrical system, and a guidance and control system.

Payload (Instruments)

The TIMED Payload is comprised of four instruments, the Global Ultra-Violet Imager (GUVI); the TIMED Doppler Interferometer Telescope (TIDI); the Solar EUV Experiment (SEE); and, the Sounding of the Atmosphere using Broadband Emission Radiometry (SABER). A detailed description of each of the instruments is given in Reference 4.

Spacecraft Bus

The spacecraft primary structure consists of four side panels and two decks constructed of aluminum honeycomb core decks with 0.02-inch thick aluminum face sheets on the inside and outside of the spacecraft. The panels are bolted to the spacecraft frame. The spacecraft antennas and star cameras are mounted to the upper deck in addition to the SEE instrument, TIDI instrument, and the TIDI telescopes. The battery, reaction wheels, SABER pulse cooler and torque rods are mounted internal to the spacecraft. Most of the instruments are mounted external to the spacecraft. Figure 1 shows external views of the spacecraft that include the instruments.

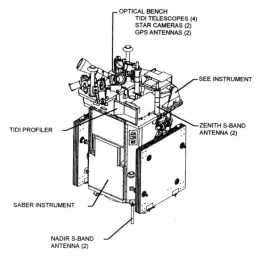

Figure 1. Spacecraft External View (Top).

Description of Surfaces/Materials Exposed to Space

The instrument and the spacecraft are covered in multi-layer insulation (MLI) blankets manufactured to industry standards and materials. The outer layer is Kapton. Radiators are covered with silver Teflon. Exposed spacecraft structural components are primarily aluminum honeycomb panels, several small titanium fittings, an aluminum payload adapter, and graphite epoxy composite components. Surface finishes of exposed areas (other than MLI and radiators) are typically Alodine (aluminum) and industry standard flight approved thermal control paint. Other exposed surfaces are primarily solar cell cover glass.

Description of Spacecraft Components Most Sensitive to Debris Impact

TIMED has no particular sensitivities to debris impacts. As with all satellites, thermal and optical surfaces will be slowly degraded by long-term exposure to atomic oxygen and micrometeor impacts. Shielding of any components from debris impacts is not needed, nor is it practical to do so, to retain functional requirements.

Description of TIMED On-Board Stored Energy

TIMED has six sources of stored energy. The nickel-hydrogen battery, the SABER Instrument cooler, the SEE Instrument heat pipes, the spacecraft heat pipes, the SEE EUV Grating Spectrograph (EGS) component, and the reaction wheels. The spacecraft battery, the SABER cooler, SEE EGS component, and the SEE and spacecraft heat pipes are pressure systems.

The nickel-hydrogen battery consists of 22 pressure vessel cells. Battery halves are internal to the spacecraft bus. A half section of the battery is shown in Figure 2.

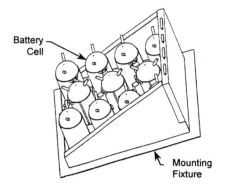

Battery Cell

Mounting Fixture

Figure 2. Battery Half.

Mission Parameters

The TIMED mission consists of a spacecraft launched from Vandenburg Air Force Base. The TIMED spacecraft was launched with the Jason-1 spacecraft. The Jason-1 spacecraft was located in the upper section of the Duel Payload Attached Fitting (DPAF) and the TIMED spacecraft was in the lower section. After many delays the launch took place on December 7, 2001. Orbital parameters are shown below:

Apogee:	625 km ± 25 km
Perigee:	625 km ± 25 km
Inclination:	74.1 ± 0.1 Degrees

ASSESSMENT OF DEBRIS GENERATED DURING NORMAL OPERATIONS

Debris Potential During Launch Phase

TIMED was launched on a DELTA II 7920-10 ELV. The DELTA II second stage placed TIMED in the desired orbit. The second stage is believed to produce no on-orbit debris. The second stage was preprogrammed to perform a retrograde fuel depletion burn after payload separation to reduce the probability of accidental debris generation.

Debris Released After Orbital Insertion

During normal operation the TIMED Spacecraft will generate only one piece of debris; the SABER cover. The SABER instrument cover was released during initial contact after orbital insertion via a paraffin wax actuator (STARSYS Research Model EP-5025).

Capturing the cover was considered impractical for two reasons. First, a captured cover would impede the instrument scanning field-of-view given the instrument's location on the deck. Second, a hinged cover mechanism would cover the radiators and have a negative effect on the performance of the instrument. Since there is no need to close the aperture again, the instrument cover was designed as a less complicated, lighter, throw-away cover.

The TIMED Spacecraft has an initial (insertion) circular orbit of 625 ± 25 km. 25 km is a conservative 3-sigma initial orbit insertion error. The SABER cover was to be ejected at relatively low speed anti-ramward. The ejection speed is assumed to have a negligible effect on the SABER cover motion, as far as the orbital debris analysis is concerned.

The SABER cover is shown in Figure 3. The elliptical SABER cover mass is 0.331 kg, and front, top, and side areas are 0.033200 m^2, 0.006345 m^2, and 0.005168 m^2, respectively. Since the uncontrolled SABER cover is assumed be tumbling during its descent, the average cross-sectional area is used for the drag calculation (per NSS 1740.14). The average cross-sectional area of 0.0223 m^2 is smaller than the frontal (maximum) area, therefore, uncontrolled flight will have a longer orbital lifetime. The average area-to-mass ratio is 0.0672.

Since the SABER cover mean diameter is larger than 1 mm (small debris threshold), the following analysis was completed in accordance with NSS 1740.14, Chapter 3.

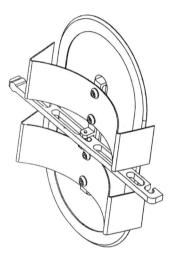

Figure 3. SABER Cover.

The SABER cover has a large frontal area to mass ratio; therefore, the drag effect can be very large. The cover was released 45 degrees from the anti-ramward direction. Releasing the cover in this direction eliminated the potential for the spacecraft to collide with the cover in the future. A long north to south pass was used to maximize the inclination change, with the cover being released near the end of the pass.

The probability of debris colliding with the spacecraft is dependent on the number and size of the debris and on the length of time it remains in orbit. According to NSS 1740.14 guidelines, the total area-time product should be no longer than 0.1 m^2-yr and the total object-time product should be no larger than 100 object-yr. The area-time product is the sum of the cross-sectional area multiplied by the total time the debris spends at an altitude below 2000 km during the orbit lifetime of the debris. The total object-time product is the sum of the total time spent below 2000 km during the orbit lifetime of each debris object.

The DAS program was used to compute the orbital lifetime and area-time product, assuming a mean solar activity of 130 sfu. Figures 4 and 5 show the variation of the orbital lifetime and area-time product as a function of initial orbital altitude. Since there is only one piece of debris associated with the TIMED Spacecraft, the object-time product value required is the same as the orbital lifetime value.

Figure 4 shows that over the initial orbital altitude range of 600 to 650 km, the object-time product values are well within the requirement of 100 object-year by an order of magnitude. However, the area-time product exceeds the requirement limit of 0.1 m²-yr near the 650 km initial orbital altitude (see Figure 5). Since 650 km initial orbital altitude represents a 3 sigma value, the chance of SABER cover debris releasing from this altitude is very, very small; less than 0.14 percent. Nevertheless, we chose to further analyze the problem.

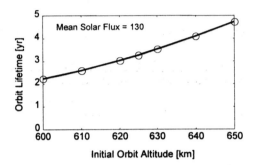

Figure 4. SABER Cover Orbit Lifetime vs. Initial Orbit Altitude.

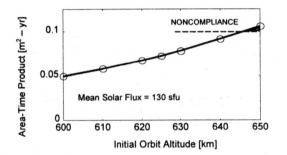

Figure 5. SABER Cover Area-Time Product vs. Initial Orbit Altitude.

An orbital decay prediction program, written by M. Packard of JHU/APL, which uses real time solar flux values in the density model, and 50 percentile solar flux values in the drag calculation, predicts a significantly lower orbital lifetime as shown in Figure 6. Using these values, the area-time product turns out to be much lower, as shown in Figure 7, in compliance with the NASA requirements.

Figure 8 shows the 10.7-cm 50 percentile solar flux variation, beginning with the year 2000, the original TIMED Spacecraft launch. Since the year 2000 coincides with the maximum solar flux cycle of about 11 years, a higher than average density value in early years results in higher drag, leading to a lower orbital lifetime.

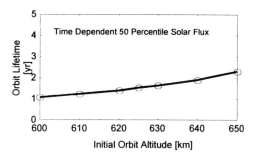

Figure 6. SABER Cover Orbit Lifetime vs. Initial Orbit Altitude (M.P.).

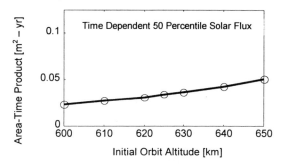

Figure 7. SABER Cover Area-Time Product vs. Initial Orbit Altitude (M.P.).

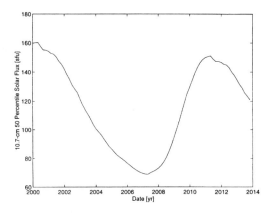

Figure 8. 10.7-cm 50 Percentile Solar Flux vs. Date.

The TIMED Spacecraft meets the criteria in Chapter 3 of the NASA Safety Standard Guidelines and Assessment Procedures for Limiting Orbital Debris, NSS 1740.14 for debris released during normal operations using the appropriate solar flux values for the launch window.

ASSESSMENT OF DEBRIS GENERATED BY EXPLOSIONS AND INTENTIONAL BREAKUPS

Explosions from On-Orbit Stored Energy

TIMED contains no sources of stored energy capable of causing significant orbital debris. Only the nickel-hydrogen batteries, SABER cooler, SEE and spacecraft heat pipes, and the reaction wheels contain stored energy. The assessments for explosion are covered in this section.

Spacecraft Bus Nickel Hydrogen Battery Assessment

An analysis was conducted by the nickel-hydrogen battery manufacturer, Eagle-Picher Technologies, to assess the explosion potential of these cells. The manufacturer concluded that these cells are incapable of exploding from internal pressure and elaborate precautions are taken to minimize the possibility of electrolyte leakage.

The nickel-hydrogen battery cells employed safety controls to prevent explosion as a possible event. The batteries have a 3 to 1 safety factor for design optimization. The manufacturer, Eagle-Picher, has qualification tested the design. Each of the battery halves is mounted internal to the spacecraft housing. This allows a layer of containment for the battery.

Eagle-Picher nickel-hydrogen cells have 200,000,000 cell-hours in actual orbital operation (collectively) with no prior history of on-orbit failure, or debris generation.

There is a low probability of the battery generating orbital debris due to design and testing (MIL- STD-1522) and cell heritage.

SABER Pulse Tube Cooler Assessment

The cooler has two layers of containment: the cooler housing (3/8" thick aluminum) and the spacecraft panels and decks. The cooler and cold fingers have been assessed to insure that they provide adequate containment. The cooler is designed to MIL-STD-1222A and the ASME Boiler Code and has a 4:1 safety factor. The cooler has flown on the LEWIS Mission.

The most likely failure mode of the pulse tube would be a slow leak rather than an explosive event. Even a rapid decompression of this gas should not cause any damage outside of the SABER housing. There are layers of MLI blankets and aluminum radiation shields that would absorb the released energy and contain the small mass of the pulse tube debris. The expanded gas inside the radiation shield would have a pressure of 15 psig or less, which can vent through existing gaps.

The probability that the pulse tube cooler would generate orbital debris is low. The cooler is designed to MIL STD 1522 and the ASME Boiler Code, has two layers of containment (instrument and spacecraft), has limited pressurant (35.5 ml of helium), no heat sources near it, and has heritage.

SEE and Spacecraft Heat Pipe Assessment

The two SEE heat pipes will not be a source of orbital debris since both ammonia heat pipes are leak-before-burst design. These heat pipes are designed to handle pressures 2.5 times the maximum operating pressure (MOP). The MOP is 334 psig for each heat pipe.

The four spacecraft heat pipes are imbedded in the lower deck. These heat pipes are designed to MIL-STD-1522A and are capable of handling 2.5 times the Maximum Design Pressure (MDP). The MOP is 210 psig at 40° C and the MDP is 1000 psia at 105° C. Since these heat pipes are contained in the spacecraft structure and are designed with safety margin, they are at low risk to produce orbital debris.

SEE EGS Component Assessment

The SEE EGS component is pressurized during ground operations, including launch, to 1.2 atm (17.6 psia) with nitrogen to maintain a clean environment. After the spacecraft is on-orbit, the SEE cover is opened and the pressure is released through a small orifice. The SEE components have been structurally analyzed and tested for their space application.

There is a low probability that the SEE EGS component will generate orbital debris because of the low differential pressure, large structural design safety factors for analyses and tests, and planned release of pressure at the beginning of the mission.

Spacecraft Reaction Wheels Assessment

The TIMED reaction wheels, manufactured by ITHACO Space Systems, Inc., are flight proven with no prior history of on-orbit failure, much less debris generation. TIMED's reaction wheels are identical to the ones that were approved to fly on the Far UV Spectroscopic Explorer (FUSE) spacecraft.
Intentional Breakups

There will be no intentional breakup.

ASSESSMENT OF DEBRIS GENERATED BY ON-ORBIT COLLISIONS

Assessment of Collisions with Large Objects During Mission Operations

Large debris is defined as an object with a diameter greater than 10 cm in LEO. It is known that a collision with a 10 cm object in LEO could result in catastrophic damage to a spacecraft.

The TIMED spacecraft has the average cross-section area of 10.275 m^2 and an operational lifetime of two years. At a 625 km operational altitude, the probability of collision with large debris or meteoroids is predicted to be 8×10^{-5} and 4.5×10^{-9}, respectively. The probability of collision with man-made debris is about a four order of magnitude greater than with the meteoroids. The overall probability of 8×10^{-5} is much smaller than the 10^{-3} value recommended by the NASA guidelines. Decaying of the orbital altitude over the two-year period has a very small effect on the overall probability number.

Assessment of Collisions with Small Debris During Mission Operations

The TIMED Spacecraft disposal procedure has it tumbling in order to exit the LEO within 25 years. Since the spacecraft Center of Pressure (CP) is off-set from the Center of Gravity (CG) position, it does not require any special system in order to go into the tumbling motion. Therefore, the damage due to small debris during operational lifetime will have no effect on the spacecraft's ability to exit the orbit in less than 25 years in compliance with NSS 1740.14, Chapter 5.0.

The general characteristics of the impact probability and impact frequency are presented in Figures 9 and 10, respectively. In each case three curves are presented representing the effect of the meteoroid, man-made debris, and combined environment. It should be noted that the chance of the spacecraft colliding with 1 cm debris is about 0.22 percent, and the probability goes up to about 50 percent for 1 mm debris.

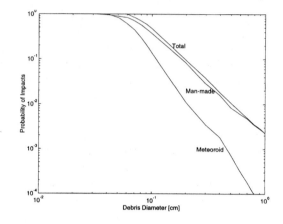

Figure 9. TIMED Spacecraft Probability of Impacts vs. Debris Diameter.

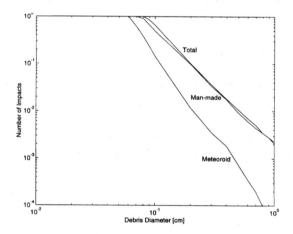

Figure 10. Number of Impacts vs. Debris Diameter.

ASSESSMENT OF POSTMISSION DISPOSAL PROCEDURE AND ORBITAL LIFETIME

Ideally, at the end of the mission, the spacecraft will be orientated into the maximum drag configuration. Because the spacecraft CP is offset by about 3 to 10 cm from the CG in all three directions, it is predicted that without attitude control, the spacecraft will go into a tumbling motion. This does not require the spacecraft to be reoriented. Since aerodynamic damping in space is very low, the possibility of the spacecraft tumbling motion being damped out about a statically stable orientation is negligible. Therefore, a conservative drag value during this phase of the flight is obtained using the average cross-sectional area of 10.275 m^2.

The average cross-sectional area is significantly greater than the operational frontal area of 3.378 m^2, due mainly to the solar arrays. This reduces the orbital lifetime within the required value of 25 years, over the initial orbital altitude range as shown in Figure 11.

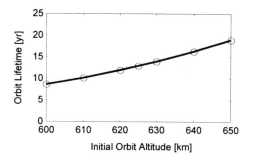

Figure 11. TIMED Spacecraft Orbit Lifetime vs. Initial Orbit Attitude.

At end of life, the battery will be discharged and the reaction wheels will be spun-down to 0 rpm and turned off. The batteries will be discharged by turning all the subsystem loads on while the solar arrays are oriented away from the sun. It is not possible to disconnect the battery from the solar arrays; however, if the solar arrays were to be exposed to sun during tumbling, the spacecraft loads will continue to discharge the battery. No significant residual pressure is expected. At end of life, the pulse tube cooler and the heat pipes will continue to contain pressure. There is no way to release the pressure from these sealed units. The cooler has minimal pressure (35.5 ml of helium) and can be contained by the instrument housing as well as the spacecraft bus housing. The heat pipes are designed with large safety factors and the spacecraft heat pipes are contained.

ASSESSMENT OF DEBRIS SURVIVABILITY FROM POSTMISSION DISPOSAL ATMOSPHERIC REENTRY

As stated in the previous Section, an atmospheric reentry approach will be used to remove the TIMED spacecraft from LEO in less than 25 years. This may result in some spacecraft components surviving reentry into the atmosphere. An analysis was done to assess the reentry survivability of the spacecraft components (debris).

In accordance with the NSS 1740.14, the TIMED spacecraft "parent object" was best represented as a cylinder with dimensions approximately 2.8 m diameter and 11.3 m length. The DAS program uses the parent object data to compute the spacecraft trajectory from reentry interface altitude to the break up altitude, at which point the trajectory of the spacecraft parent object then becomes the trajectory of the individual spacecraft components upon The reentry interface altitude and the break up altitude are assumed to be 122 km and 78 km, respectively.

Most of the spacecraft components are modeled as a sphere or cylinder, as shown in Table 1. Since survival of a component depends on the material properties, size, shape, and mass, all likely components that would survive were modeled. For example, it is expected that the aluminum solar arrays with a very large surface area will not survive, however, a small titanium bracket may survive reentry. Since the DAS (version x.09) program does not have iron properties in its database (however, new materials can be added to the database), the iron torque rods were modeled as stainless steel. Their specific heat capacity and melting point temperature are very close; Cp_I = 452.0 J/kg/°K vs. Cp_{SS} = 439.2 J/kg/°K, Tm_I = 1812 °K vs. Tm_{SS} = 1728 °K.

1277

The DAS program predicts that 25 objects will survive to the ground, as shown in Table 2. The total debris casualty area is 19.364 m^2, which is over the guideline requirement value of 8 m^2.

According to D. Persons of JHU/APL, who was responsible for the structural analysis of the TIMED Spacecraft solar arrays, when 0.7585 N/m^2 pressure is applied, the solar arrays will fold at the yoke. At the reentry interface orbital altitude of 122 km, the solar arrays would be subjected to the maximum aerodynamic pressure of 1.1 N/m^2, thus the arrays will already be folded. Folding of the solar arrays reduces the spacecraft's average cross-sectional area significantly. Consequently, the velocity at the break up altitude is greater, increasing burn up through the atmosphere. For the folded solar array configuration, the DAS program predicts that significantly fewer objects will survive the atmosphere, as shown in Table 3, with a total debris casualty area of 9.182 m^2.

Table 1. Reentry Object Data.

Object Description	Model Type	Diameter (m)	Length (m)	Mass (kg)	Material Type	Qty
TIMED Spacecraft	Cylinder (C)	2.804	11.276	600	Aluminum	1
TIDI Telescope	C	0.1016	0.4826	4.6	Aluminum	4
TIDI Profiler	C	0.2032	0.3048	14.1	Aluminum	1
TIDI Electronics	C	0.2032	0.2794	5.5	Aluminum	1
SEE	C	0.508	0.6858	21.4	Aluminum	1
GUVI	C	0.2032	0.2540	12.8	Aluminum	1
GUVI ECU	C	0.2540	0.3048	6.41	Aluminum	1
SABER	C	0.762	0.9144	61.7	Aluminum	1
Solar Array Drive Motor	C	0.0889	0.127	1.72	Titanium	2
Battery	C	0.4572	0.5588	27.66	Aluminum	2
PSE/DPU	C	0.3048	0.4572	20.63	Aluminum	1
Star Camera	C	0.1778	0.508	4.9	Aluminum	2
Reaction Wheel	C	0.2794	0.2032	5.92	Aluminum	4
Torque Rod	C	0.0762	0.9144	4.13	Stainless Steel	3
IRU (gyros)	C	0.2286	0.2032	7.71	Aluminum	1
AIU	C	0.2032	0.3556	6.43	Aluminum	1
IEM (9 cards & chassis)	C	0.3302	0.2032	12.24	Aluminum	2
Top Deck & Framing	C	1.3716	0.0508	12.99	Aluminum	1
Bottom Honeycomb Deck	C	1.016	0.0508	5.62	Aluminum	1
Bottom Aluminum Deck	C	1.1684	0.0508	12.44	Aluminum	1
+x Panel	C	1.524	0.0381	6.12	Aluminum	3
+y Panel	C	0.762	0.0381	3.45	Aluminum	1
+x/+y Panel & Long.	C	1.524	0.0381	4.08	Aluminum	2
-x/+y Panel & Long (SABER)	C	1.524	0.0381	5.58	Aluminum	2
+x/+y Corner Brackets	Sphere (S)	1.524	-----	1.36	Aluminum	4
Payload Adapter	C	0.9652	0.1397	6.32	Aluminum	1
Instrument Bench	C	1.1684	0.0508	6.58	Aluminum	1
Star Camera Bracket	S	0.127	-----	0.68	Titanium	2
RW Bracket	C	0.4572	0.2.32	6.58	Aluminum	4
Solar P. Motion Bracket	C	0.2032	0.1016	1.0	Aluminum	2
Solar P. Hold-Down Bracket	C	0.1016	0.1778	1.36	Aluminum	2
TIDI T. Bracket	C	0.1143	0.1753	0.865	Titanium	4

Table 2. Objects Survived to the Ground (Operational Configuration).

Object Description	Area (m^2)	Quantity	Total Area (m^2)
TIDI Profiler	0.7267	1	0.7267
SEE	1.4247	1	1.4247
GUVI	0.6859	1	0.6859
SABER	2.0628	1	2.0628
Solar Array Drive Motor (SADM)	0.5008	2	1.0016
Battery	1.2251	2	2.4502
PSE/DPU	0.9566	1	0.9566
Torque Rod	1.0240	3	3.0720
AIU	0.6655	1	0.6655
IEM	0.7471	2	1.4942
x/y Panel Bracket	0.4446	4	1.7784
Star Camera Bracket	0.4151	2	0.8302
TIDI Telescope Bracket	0.5538	4	2.2152
Total Debris Casualty Area =			**19.364 m^2**

**Table 3. Objects Survived to the Ground (Folded Solar Array Configuration)
ORSAT Reentry Analysis of the TIMED Spacecraft.**

Object Description	Area (m²)	Quantity	Total Area (m²)
SABER	2.0628	1	2.0628
Solar Array Drive Motor	0.5008	2	1.0016
Torque Rod	1.0240	3	3.0720
Star Camera Brackets	0.4151	2	0.8302
TIDI Telescope Bracket	0.5538	4	2.2152
		Total Debris Casualty Area =	9.182 m²

The DAS tool provides a more conservative reentry analysis and is typically used for a first cut reentry analysis of space hardware. DAS's inability to perform heat conduction in cylinders and spheres, lack of oxidation heating, and the assumption of an emissivity of 1.0 for all material types results in it being a more conservative reentry tool.

With the DAS predicted total debris casualty area of the TIMED spacecraft exceeding the NASA Safety Standard 1740.14 guideline, an analysis using the higher fidelity Object Reentry Survival Analysis Tool (ORSAT) model was warranted. An analysis of the reentry survivability of the TIMED spacecraft was performed by the Johnson Space Center (JSC) Orbital Debris Program Office at the request of JHU/APL. However, only the objects predicted to survive in the DAS analysis (operational configuration) were evaluated with ORSAT. Furthermore, the Star Camera Bracket and the TIDI Telescope Bracket did not need to be analyzed in the original DAS analysis since, per the NSS 1740.14, only objects which have at least one dimension greater than 0.25 m must be evaluated. All of the mass must be accounted for in the analysis, but objects of that size do not have to be evaluated individually. It was decided at the time that these two components would not be included in the ORSAT study, though typically all components are evaluated in the higher fidelity ORSAT reentry analyses. Dimensions of some of the surviving objects, re-measured, were slightly different (larger or smaller) from those given in the Table 1.

ORSAT was developed and documented in 1993 by a joint effort between NASA JSC and Lockheed Martin Space Operations[5]. The latest published version of the code, ORSAT 5.0, is described in the User's Manual[6]. ORSAT allows for breakup and survivability evaluation of space hardware reentering the Earth's atmosphere and the resulting risk to the ground population from the surviving debris. The ORSAT program is comprised of six key physics modules: trajectory, atmosphere, aerodynamics, aero thermodynamics, thermal and debris casualty expectation.

ORSAT accounts for several heating factors in the thermal analysis of the reentering object. These include the hot wall heating rate, q_{hw}, the oxidation heating rate, q_{ox}, the gas cap radiative heating rate, q_{rad}, and the re-radiation heating rate, q_{rr}, where the resulting net heating rate is calculated as:

$$q_{net} = q_{hw} + q_{ox} + q_{rad} - q_{rr} \qquad (1)$$

In the TIMED ORSAT analysis, the gas cap radiative heating rate is left out as it is only applicable to high velocity entries (greater than 10 km/sec). Also, for each component the oxidation heating was initially left out (chemical heating efficiency factor, τ, is set to zero), providing a more conservative analysis. If a component was determined to survive, the analysis was repeated with τ set to 0.5, implying 50% heating efficiency. The result for this case is usually considered the final result since τ is an experimental value which tends to vary by material and by altitude. However, a τ of 1.0 (100% heating efficiency) was also considered when doing the analysis and with the exception of the SABER and PSE/DPU components, did not alter the final result. Table 4 illustrates the affect of varying τ for the individual components.

All but the batteries and the torque rods were modeled as solid objects, per the provided data from JHU/APL. For the cylindrical objects, the heat conduction capability was implemented and a nodal analysis was performed. Only a lumped mass approach can be used for boxes at this time. The use of heat conduction did not have a significant impact on the results, however.

The battery was modeled using several material layers. For the initial analysis of the battery, it was necessary to calculate the heat of ablation, h_{abl}, for the aluminum casing. The heat of ablation is simply the thermal mass, m, multiplied by the sum of the heat of fusion, h_f, and the specific heat capacity, Cp, times the difference between the melting temperature, T_m, and the initial temperature, T_i, as illustrated in the following equation:

$$h_{abl} = m \, [h_f + Cp \, (T_m - T_i)] \qquad (2)$$

The battery was then modeled as a solid aluminum box using the given dimensions. The results were then evaluated for the point at which the heat of ablation of the box reached the calculated heat of ablation for the aluminum casing thermal mass. At this point, the aluminum battery casing was considered demised. The trajectory at the demise point was then used for the initial conditions of the next portion of the analysis, which looked at the individual battery cells. It was only necessary to look at one battery cell since each of the battery cells was identical (See Figure 2). The battery cell was modeled as a cylinder with an outer layer of magnesium ZK60A-T5 and an inner layer of Inconel. Both layers were determined to demise, as shown in the results Table 5.

The torquer rod was modeled in ORSAT as a three material cylindrical object. The outer shell of the rod was modeled as fiberglass. The second layer was modeled using a copper material to represent the copper wiring coiled around the final layer, a solid stainless steel core. Each of these layers was determined to demise.

Table 5 presents the final results of the TIMED ORSAT analysis. As seen in the table, four object types were determined to survive reentry: the SEE, SABER, SADM, and the PSE/DPU. The final total debris casualty area predicted for the TIMED spacecraft is approximately 5.249 m^2.

NSS 1740.14 proposes a debris casualty area upper limit of 8 m^2 as the guideline for uncontrolled reentry. This general number was derived from the guiding principle to limit the risk of human casualty to 0.0001 per reentry event. Since the casualty area limit for meeting the 1 in 10,000 guideline is actually dependent on the specific orbital inclination of a spacecraft, the NASA Office of Safety and Mission Assurance is now evaluating the risk numbers, or casualty expectation, for specific orbits instead of employing a set casualty area which approximates 1 in 10,000. The casualty expectation is calculated for each reentry event using the affected population density (based on the spacecraft orbital inclination) and the total debris casualty area for a particular spacecraft. Thus, for the TIMED orbital inclination of 74.1 degrees, the 2001 population density under the spacecraft is approximately 9.477e-06 /m^2. Based on the 5.249 m^2 total debris casualty area, the resulting casualty expectation for the TIMED spacecraft is approximately 1 in 20,100, which is well within the 1 in 10,000 NASA guideline.

Table 4: Oxidation Heating Affect on Object Demise.

Component	$\tau = 0.0$	$\tau = 0.5$	$\tau = 1.0$
TIDI Profiler	Survived	Demised	N/A
SEE	Survived	Survived	Survived
GUVI	Survived	Demised	N/A
SABER	Survived	Survived	Demised
SADM	Survived	Survived	Survived
Battery	Demised	N/A	N/A
PSE/DPU	Survived	Survived	Demised
Torque Rod	Demised	N/A	N/A
AIU	Demised	N/A	N/A
IEM	Demised	N/A	N/A
X/Y Panel Brackets	Demised	N/A	N/A

Table 5: Final Results of ORSAT TIMED Reentry Analysis.

Object	Qnty.	Debris Casualty Area (x Qnty.)
TIDI Profiler	1	0.0
SEE	1	1.44
GUVI	1	0.0
SABER	1	1.947
SADM	2	1.002
Battery	2	0.0
PSE/DPU	1	0.86
Torque Rod	3	0.0
AIU	1	0.0
IEM	2	0.0
X/Y Panel Brackets	4	0.0
TOTAL DEBRIS CASUALTY AREA		**5.249 m^2**

CONCLUSION

The TIMED Spacecraft is in compliance with the guideline requirements of the NSS 1740.14.

ACKNOWLEDGEMENTS

Funding for this program was provided by NASA. We would like to thank TIMED project manager, Mr. D. Grant, mission system engineer, Mr. G. Cameron, and integration and test director, Mr. S. Kozuch of JHU/APL for their support and direction.

REFERENCES

1. Opiela, J. N.,and P. H. Krisko, "Evaluation of Orbital Debris Mitigation Practices Using EVOLVE 4.1," IAA-01-IAA.6.6.04, in *Space Debris 2001*, J. Bendisch, ed., AAS *Science and Technology Series*, Vol. 105, pp.209-219, 2002, (paper presented at the IAF, Toulouse, France, Oct. 1-5, 2001).

2. NASA Safety Standard, 1995. Guidelines and Assessment Procedures for Limiting Orbital Debris. NSS 1740.14. Office of Safety and Mission Assurance, Washington, D.C. 20546.

3. NASA, 1996, Orbital Debris Assessment Software. Lyndon B. Johnson Space Center, Houston, TX 77058.

4. Dion, S. K, etal, " TIMED Mission Orbital Debris Assessment Report ," 1999.

5. Bouslog, S. A., Ross, B. P., and Madden, C. B., "Space Debris Reentry Risk Analysis," AIAA 94-0591, Jan. 10-13, 1994.

6. Rochelle, W. C., Kirk, B. S., and Ting, B. C., "User's Guide for Object Reentry Survival Analysis Tool (ORSAT) – Version 5.0, Vols. I and II," JSC-28742, July 1999.

ATTITUDE DETERMINATION
AND DYNAMICS

SESSION 13

Chair:

Beny Neta
Naval Postgraduate School

The following papers were not available for publication:

AAS 03-193

"Attitude Determination Algorithm for Gyroless Spacecraft Using Disturbance Accommodation Technique," by Injung Kim, Seoul National University; Jinho Kim, Swales Aerospace; Youdan Kim, Seoul National University (Paper Withdrawn)

AAS 03-195

"Practically Oriented Attitude Stabilization of an Underactuated Satellite Using Two Wheels," by N.M. Horri, Stephen Hodgart (Paper Withdrawn)

AAS 03-196

"Stability Analysis of Coupled Slosh-Vehicle Model Using Dynamical Systems Tools," by C. Nichkawde, N. Ananthkrishnan (Paper Withdrawn)

CONFORMAL MAPPING AMONG ORTHOGONAL, SYMMETRIC, AND SKEW-SYMMETRIC MATRICES

Daniele Mortari[*]

This paper shows that Cayley Transforms, which map Orthogonal and Skew-Symmetric matrices, may be considered the extension to matrix field of the complex conformal mapping function $f_1(z) = \dfrac{1-z}{1+x}$. Then, by using a set of real matrices which are, simultaneously, Orthogonal and Symmetric (the *Ortho−Sym* matrices), it similarly shows how to extend two complex conformal mapping functions (namely, the $f_2(z) = \dfrac{i-z}{i+z}$, and the $f_3(z) = \dfrac{1-z}{1+z} i$ - here called the *clockwise*, and the *counter clockwise* functions), to matrix field. This extension consists of some new one-to-one mapping relationships between Orthogonal and Symmetric, and between Symmetric and Skew-Symmetric matrices. This new relationships complete the picture of the one-to-one matrix mapping among Orthogonal, Symmetric, and Skew-Symmetric matrices. Finally, this paper shows how to map among Orthogonal, Symmetric, and Skew-Symmetric matrices, by means of a direct product.

Introduction

Cayley Transforms are a beautiful and useful tool to perform one-to-one mapping between Orthogonal C and Skew-Symmetric Q real matrices. Cayley Transforms consists of two formally identical relationships that can be written as

$$\begin{cases} C = (I - Q)(I + Q)^{-1} = (I + Q)^{-1}(I - Q) & \text{(Forward)} \\ Q = (I - C)(I + C)^{-1} = (I + C)^{-1}(I - C) & \text{(Inverse)} \end{cases} \tag{1}$$

where I indicates the identity matrix, and $C^{\mathrm{T}}C = I$, and $Q = -Q^{\mathrm{T}}$.

[*] Associate Professor of Aerospace Engineering, Department of Aerospace Engineering, 701 H.R. Bright Building, Room 741A, Texas A&M University, 3141 TAMU, College Station, Texas 77843-3141. AIAA and AAS Member.

Several applications of these equations can be found in attitude dynamic and control. In fact, if C represents the direction cosine matrix describing a given attitude (or rigid rotation), then the mapped Q matrix contains the three elements of the Gibbs vector, as the off-diagonal elements, a minimum number parameters to describe the attitude. Moreover, it is easy to derive the following relationships (See pp. 53-57 of Ref. [1]), which relate the angular velocity ω with the rate of change of C and Q

$$\tilde{\omega} = 2(I+Q)^{-1}\dot{Q}(I-Q)^{-1} = -\dot{C}C^{\mathrm{T}} = \begin{bmatrix} 0 & -\omega_3 & \omega_2 \\ \omega_3 & 0 & -\omega_1 \\ -\omega_2 & \omega_1 & 0 \end{bmatrix} \tag{2}$$

Another very important application of Cayley Transforms is for real Symmetric S matrices. In fact, any Symmetric matrix can be decomposed as

$$S = C\Lambda_S C^{\mathrm{T}} = (I-Q)(I+Q)^{-1}\Lambda_S(I-Q)^{-1}(I+Q) \tag{3}$$

that is, a minimal parametrization to describe Symmetric matrices[†]. Equations (1,2,3) can be specialized to any n-Dimensional Euclidean space. Finally, Eqs. (1) are extensively applied in all the fields where matrix transformation plays a key role.

An important aspect of the Cayley Transforms, easily to demonstrate, is the fact that C and Q share the same complex eigenvector matrix W. In fact, let $CW = W\Lambda_C$ be the spectral decomposition of C, then, since $W^\dagger W = I$, we can write

$$\begin{aligned} Q &= (I-C)(I+C)^{-1} = (WIW^\dagger - W\Lambda_C W^\dagger)(WIW^\dagger + W\Lambda_C W^\dagger)^{-1} = \\ &= W(I-\Lambda_C)W^\dagger W(I+\Lambda_C)^{-1}W^\dagger = W(I-\Lambda_C)(I+\Lambda_C)^{-1}W^\dagger = \\ &= W\Lambda_Q W^\dagger \end{aligned}$$

where $\Lambda_Q = (I-\Lambda_C)(I+\Lambda_C)^{-1}$, that in scalar form becomes

$$\lambda_Q = \frac{1-\lambda_C}{1+\lambda_C} \tag{4}$$

which shows that the eigenvalues of C and Q are related through a bilinear conformal transformation. Therefore, Cayley Transforms represent the extension to matrix field of the particular complex conformal mapping function $f_1(z) = \dfrac{1-z}{1+z}$. The mapping rule of this function, here called Cayley function $f_1(z)$, is shown in Fig. 1, where the presence of a mirror helps to understand the mapping geometry.

The relationship between Cayley mapping and the complex conformal function $f_1(z)$, arises the question whether it is possible to develop similar matrix mapping with different complex conformal functions. In particular, we will look to develop the

---

[†]Symmetric matrices are widely used in mechanics (as representing inertia and stress tensors, or as mass, stiffness, and damping matrices), as well as in dynamics and control (as representing covariance matrices in Kalman filter, gain matrices in optimal control, and as solution of the matrix Lyapunov and the Riccati differential equations).

missing similar matrix mapping to complete the mapping between the three most common real $n \times n$ matrices: the Orthogonal C, the Symmetric S, and the Skew-Symmetric Q matrices, respectively.

To this end, the problem to share the real eigenvector matrix P of S with the complex eigenvector matrix W of C and Q, arises. Spectral decomposition properties of C, Q, and S are summarized in Table 1.

Matrix	Definition	Eigenvalues	Eigenvectors
Orthogonal	$C^\mathrm{T}C = I$	Unit-Circle	Unitary
Skew-Symmetric	$Q = -Q^\mathrm{T}$	Imaginary Axis	Unitary
Symmetric	$S = S^\mathrm{T}$	Real Axis	Orthogonal

Table 1: Spectral decomposition properties for C, Q, and S matrices.

The eigenvector problem will be solved thanks to the Nucleus matrix M, while the eigenvalue problem (Which are the mapping functions to map the eigenvalues) will be solved by introducing the *Counter Clockwise* $f_2(z)$, and the *Clockwise* $f_3(z)$ complex mapping functions. However, prior to enter into these arguments, the *Ortho-Skew*, and the *Ortho-Sym* matrices, will be introduced.

The Ortho-Skew $\Im$, and the Ortho-Sym $\Re$ matrices

The intersection of the unit circle (eigenvalue locus of the Unitary matrices) and the imaginary axis (eigenvalue locus of the Skew-Hermitian matrices) yields to matrices which are, simultaneously, Unitary and Skew-Hermitian. These matrices can be, therefore, defined by the relationships

$$\{\, \Im^\dagger \Im = I, \; \Im = -\Im^\dagger \,\} \quad \Longrightarrow \quad \Im\Im = -I \tag{5}$$

The resulting matrix $\Im$ is, therefore, Unitary and Skew-Hermitian matrix set, here called improperly Ortho-Skew, which has been introduced in Ref. [3], presents very interesting properties. The general expression of these matrices, whose eigenvalues are only pure imaginary $\pm i$, is

$$\Im = i \sum_{k=1}^{p} w_k\, w_k^\dagger - i \sum_{k=p+1}^{n} w_k\, w_k^\dagger \qquad (w_i^\dagger\, w_j = \delta_{ij}) \tag{6}$$

where the complex directions w_k, which form the unitary eigenvector matrix W constitute an Orthogonal set, and where p indicates the number of positive eigenvalues "$+i$" ($0 \leq p \leq n$). Note that, since the w_k are Orthogonal, then if $p = n$ then $\Im = i\,I$, while if $p = 0$ then $\Im = -i\,I$.

The Ortho-Skew-Hermitian matrices are, in general, complex. However, in even dimensional spaces, it is possible to build real $\Im_e$ matrices as follows

$$\Im_e = \sum_{k=1}^{n/2} W_k \, \Im_2 \, W_k^\dagger = \sum_{k=1}^{n/2} [\, w_k \; \bar{w}_k \,] \begin{bmatrix} i & 0 \\ 0 & -i \end{bmatrix} [\, w_k \; \bar{w}_k \,]^\dagger \tag{7}$$

that can be simply called *Ortho-Skew*[‡].

Based on Eq. (5), subsequent powers of $\Im$ satisfy the following rule

$$\Im^k = \begin{cases} = +\Im & (k = 1, 5, 9, \cdots) \\ = -I & (k = 2, 6, 10, \cdots) \\ = -\Im & (k = 3, 7, 11, \cdots) \\ = +I & (k = 4, 8, 12, \cdots) \end{cases} \tag{8}$$

Analogously, the intersection of the unit-radius circle (eigenvalue locus of C) with the real axis (eigenvalue locus of S) yields to the definition of the *Ortho-Sym* $\Re$ matrices, which are, simultaneously, Orthogonal and Symmetric. The Ortho-Sym real matrices $\Re$, which have been introduced in Ref. [4], are defined by their own definitions

$$\Re^{\mathrm{T}} \Re = I \qquad \text{and} \qquad \Re = \Re^{\mathrm{T}} \tag{9}$$

Subsequent powers of these matrices obey to the simple rule

$$\Re^k = \begin{cases} = \Re & (k = \text{odd}) \\ = I & (k = \text{even}) \end{cases} \tag{10}$$

The general expression of $\Re$, whose eigenvalues are only "± 1", is

$$\Re = \sum_{k=1}^{p} p_k p_k^{\mathrm{T}} - \sum_{k=p+1}^{n} p_k p_k^{\mathrm{T}} \qquad (p_i^{\mathrm{T}} p_j = \delta_{ij}) \tag{11}$$

where the real directions p_k form an orthogonal set and where p indicates the number of positive eigenvalues "$+1$" ($0 \le p \le n$). Since the p_k are orthogonal, then if $p = n$ then $\Re = I$, while if $p = 0$ then $\Re = -I$.

These matrices, in particular the Ortho-Sym matrices (that can be built associated with a given C or S), will allow us to complete the one-to-one mapping between Orthogonal, Skew-Symmetric, and Symmetric matrices.

The Nucleus Matrix

The *nucleus* matrix is the 2×2 complex matrix

$$M = \frac{\sqrt{2}}{2} \begin{bmatrix} 1 & -i \\ 1 & i \end{bmatrix} \tag{12}$$

[‡]Also the complex $\Im$ matrices will be here improperly called Ortho-Skew.

This unitary matrix ($MM^\dagger = I_2$) allows to extract the information of the plane, provided by a complex-conjugate eigenvector pair $W_k = \frac{\sqrt{2}}{2}[\,\hat{p}_{1k} + i\hat{p}_{2k} \,\vdots\, \hat{p}_{1k} - i\hat{p}_{2k}\,] = [\,\hat{w}_k \,\vdots\, \hat{w}_k^\dagger\,]$, into a real plane defined by a real eigenvector pair $P_k = [\,\hat{p}_{1k} \,\vdots\, \hat{p}_{2k}\,]$. In fact

$$W_k\, M = \frac{\sqrt{2}}{2}[\,\hat{p}_{1k} + i\hat{p}_{2k} \,\vdots\, \hat{p}_{1k} - i\hat{p}_{2k}\,]\, \frac{\sqrt{2}}{2} \begin{bmatrix} 1 & -i \\ 1 & i \end{bmatrix} = [\,\hat{p}_{1k} \,\vdots\, \hat{p}_{2k}\,] = P_k \qquad (13)$$

This property allows us to map a complex unitary eigenvector matrix W into an associated real Orthogonal matrix P. In fact,

$$P = [\,P_1 \,\vdots\, P_2 \,\cdots\, \vdots\, p\,] = [\,W_1 \,\vdots\, W_2 \,\cdots\, \vdots\, p\,] \begin{bmatrix} M & 0 & \cdots & 0 \\ 0 & M & \cdots & 0 \\ \vdots & \vdots & \ddots & \vdots \\ 0 & 0 & \cdots & 1 \end{bmatrix} = W\, \mathcal{D}(M) \qquad (14)$$

where $\mathcal{D}(M)$ is a diagonal-block matrix containing, as a diagonal elements, M matrices associated with the $\lfloor n/2 \rfloor$ complex eigenvalue pairs, and a one associated to the real eigenvalue (if n is odd, only).

The nucleus matrix M allows us to build Ortho-Skew and Ortho-Sym matrices, $\Im(W)$ and $\Re(W)$, which are associated with a given unitary eigenvector matrix W. In fact, for a even dimensional space, we may write that

$$\begin{cases} \Im(W) = \sum_{k=1}^{n/2} W_k\, \Im_2\, W_k^\dagger = \sum_{k=1}^{n/2} P_k\, M^\dagger\, \Im_2\, M\, P_k^{\mathrm{T}} = \sum_{k=1}^{n/2} P_k\, \Im_2^{(\Re)}\, P_k^{\mathrm{T}} \\ \Re(W) = \sum_{k=1}^{n/2} P_k\, \Re_2 P_k^{\mathrm{T}} = \sum_{k=1}^{n/2} W_k\, M\, \Re_2\, M^\dagger\, W_k^\dagger = \sum_{k=1}^{n/2} W_k\, \Re_2^{(\Im)}\, W_k^\dagger \end{cases} \qquad (15)$$

where

$$\Im_2 = \begin{bmatrix} +i & 0 \\ 0 & -i \end{bmatrix} \qquad \text{and} \qquad \Im_2^{(\Re)} = M^\dagger\, \Im_2\, M = \begin{bmatrix} 0 & -1 \\ +1 & 0 \end{bmatrix} \qquad (16)$$

and where

$$\Re_2 = \begin{bmatrix} +1 & 0 \\ 0 & -1 \end{bmatrix} \qquad \text{and} \qquad \Re_2^{(\Im)} = M\, \Re_2\, M^\dagger = \begin{bmatrix} 0 & +1 \\ +1 & 0 \end{bmatrix} \qquad (17)$$

The nucleus matrix M has many important properties. First of all, it allows to build all the three *Pauli Spin Matrices* (See pp. 473-474 of Ref. [2])

$$\sigma_1 = \begin{bmatrix} 0 & 1 \\ 1 & 0 \end{bmatrix} = M\, M^{\mathrm{T}} \qquad \sigma_3 = \begin{bmatrix} 1 & 0 \\ 0 & -1 \end{bmatrix} = M^{\mathrm{T}} M \qquad (18)$$

and

$$\sigma_2 = M\, M\, M^{\mathrm{T}}\, M^{\mathrm{T}}\, M\, M^{\mathrm{T}} = \begin{bmatrix} 0 & -i \\ i & 0 \end{bmatrix} \qquad (19)$$

which are used in the complex 2×2 Cayley-Klein matrix to parameterize rigid rotation, and in quantum mechanics where they represent the angular-momentum matrices for spin ($\hbar$).

The second property (really unusual), is that M as well as M^T and $M^\dagger$, all present a power periodicity of 24

$$M^{24} = (M^T)^{24} = (M^\dagger)^{24} = I_2 \tag{20}$$

which allows us to write

$$M^{24\ell+k} = (M^T)^{24\ell+k} = (M^\dagger)^{24\ell+k} = M^k = (M^T)^k = (M^\dagger)^k \tag{21}$$

where ℓ and k represent any integer.

Conformal Complex Transformations

A conformal mapping, also called a conformal map, conformal transformation, angle-preserving transformation, or biholomorphic map, is a transformation $w = f(z)$ that preserves local angles. Let us consider the following interesting complex conformal mapping functions (See pp. 206-219 of Ref. [2]), together with their inverses

$$
\left\{
\begin{aligned}
w = f_1(z) = \frac{1-z}{1+z} &\iff z = f_1(w) = \frac{1-w}{1+w} \\[2mm]
w = f_2(z) = \frac{i-z}{i+z} &\iff z = f_3(w) = \frac{1-w}{1+w}\,i \\[2mm]
w = f_4(z) = \frac{i-z}{1+z} &\iff z = f_4(w) = \frac{i-w}{1+w} \\[2mm]
w = f_5(z) = \frac{1-z}{i+z} &\iff z = f_6(w) = \frac{1-iw}{1+w}
\end{aligned}
\right. \tag{22}
$$

Table 1 summarizes the mapping results for these functions when applied to the real axis ($z = \pm\Phi$), the imaginary axis ($z = \pm i\Phi$), and the unit-circle ($z = e^{\pm i\Phi}$).

	From Real axis	From Imaginary axis	From Unit-Circle
$f_1(z)$	Real	Unit-Circle	Imaginary
$f_2(z)$	Unit-Circle	Real	Imaginary
$f_3(z)$	Imaginary	Unit-Circle	Real
$f_4(z)$	$\mathcal{I} = \mathcal{R} + 1$	Circle at $[-1, +1]/2$	$\mathcal{I} = -\mathcal{R}$
$f_5(z)$	Circle at $[-1, -1]/2$	$\mathcal{I} = -\mathcal{R} - 1$	$\mathcal{I} = \mathcal{R}$
$f_6(z)$	$\mathcal{I} = \mathcal{R} - 1$	Circle at $[+1, -1]/2$	$\mathcal{I} = -\mathcal{R}$

Table 1: Mapping summary of $w = \mathcal{R} + i\mathcal{I} = f(z)$.

It is easy to demonstrate that the first four of these complex mapping functions obey to the following general rules

$$\begin{cases} z = f_1(f_1(z)) \\ z = f_4(f_4(z)) \end{cases} \quad \text{and} \quad \begin{cases} z = f_2(f_2(f_2(z))) \\ z = f_3(f_3(f_3(z))) \end{cases} \tag{23}$$

In the following sections, how to extend the first three complex conformal functions, $f_1(z)$, $f_2(z)$, and $f_3(z)$, to matrix mapping, is shown. The analysis for the remaining three functions, $f_4(z)$, $f_5(z)$, and $f_6(z)$, which are much more complicated, will be the subject of a future work.

The *Cayley* mapping function $f_1(z) = \dfrac{1-z}{1+z}$

As already stated, Cayley Transforms, given in Eq. (1), can be seen as the extension to matrix field of this mapping function. In fact, the eigenvalues of C and Q are related through the complex trigonometric identity

$$\mp i \tan\left(\frac{\Phi}{2}\right) = \frac{1 - e^{\pm i\Phi}}{1 + e^{\pm i\Phi}} \quad \Longleftrightarrow \quad e^{\pm i\Phi} = \frac{1 - \left[\mp i \tan\left(\frac{\Phi}{2}\right)\right]}{1 + \left[\mp i \tan\left(\frac{\Phi}{2}\right)\right]} \tag{24}$$

The geometry associated with this trigonometric identity or with the Cayley mapping function $f_1(z)$, is shown in Fig. 1, where the point $[-1, 0]$ works like a mapping reflecting point.

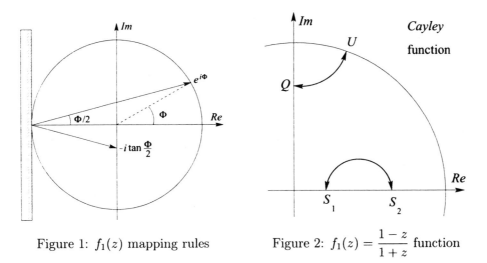

Figure 1: $f_1(z)$ mapping rules   Figure 2: $f_1(z) = \dfrac{1-z}{1+z}$ function

Figure 2 summarizes the mapping properties of the $f_1(z)$ function. In particular, $f_1(z)$ maps the real axis onto the real axis. This property can also be seen as the trigonometric identities

$$\cos \Phi = \frac{1 - \tan^2 \left(\dfrac{\Phi}{2}\right)}{1 + \tan^2 \left(\dfrac{\Phi}{2}\right)} \qquad \Longleftrightarrow \qquad \tan^2 \left(\frac{\Phi}{2}\right) = \frac{1 - \cos \Phi}{1 + \cos \Phi} \tag{25}$$

whose equivalent matrix mapping has the form

$$\begin{cases} S_1 = (I - S_2)(I + S_2)^{-1} = (I + S_2)^{-1}(I - S_2) \\ S_2 = (I - S_1)(I + S_1)^{-1} = (I + S_1)^{-1}(I - S_1) \end{cases} \tag{26}$$

where S_1 and S_2 are two Symmetric matrices. These two matrices have the same eigenvector matrix, and whose eigenvalues are $\lambda_k = \cos \Phi_k$, and $\tan^2 \left(\dfrac{\Phi}{2}\right)$, respectively.

Trigonometric/Matrix relationships

It is well known that any real square matrix can be decomposed as a sum of a Skew-Symmetric and a Symmetric matrices. Specified to an orthogonal matrix C, this decomposition is as follows

$$C = \left(\frac{C + C^\mathrm{T}}{2}\right) + \left(\frac{C - C^\mathrm{T}}{2}\right) = S + A \tag{27}$$

where S and A have the eigenvalue pairs $\cos \Phi_k$, and $\pm i \sin \Phi_k$, respectively. Therefore, it is possible to apply the first of Eq. (26) with $S_2 = S = \left(\dfrac{C + C^\mathrm{T}}{2}\right)$. The transformed matrix S_1, as it easy to demonstrate, can be set as $S_2 = QQ^\mathrm{T}$, which satisfies the two requested properties of having the same eigenvector matrix of C, and whose eigenvalue pairs are obtained as

$$\left[\mp i \tan \left(\frac{\Phi}{2}\right)\right]\left[\pm i \tan \left(\frac{\Phi}{2}\right)\right] = \tan^2 \left(\frac{\Phi}{2}\right) \tag{28}$$

Therefore, Eq. (26) becomes

$$\begin{cases} C + C^\mathrm{T} = 2(I - QQ^\mathrm{T})(I + QQ^\mathrm{T})^{-1} = 2(I + QQ^\mathrm{T})^{-1}(I - QQ^\mathrm{T}) \\ QQ^\mathrm{T} = (2I - C - C^\mathrm{T})(2I + C + C^\mathrm{T})^{-1} = (2I + C + C^\mathrm{T})^{-1}(2I - C - C^\mathrm{T}) \end{cases} \tag{29}$$

Analogously, the trigonometric property

$$\pm i \sin \Phi = -\frac{\mp 2i \tan \left(\dfrac{\Phi}{2}\right)}{1 + \tan^2 \left(\dfrac{\Phi}{2}\right)} \tag{30}$$

allows us to write the associated matrix identity

$$A = \frac{C - C^\mathrm{T}}{2} = -2\,Q\,(I + QQ^\mathrm{T})^{-1} = -2\,(I + QQ^\mathrm{T})^{-1}\,Q \tag{31}$$

(the sign "−" comes from the fact that, associated with Q we have the eigenvalues $\mp i \tan\left(\dfrac{\Phi}{2}\right)$, while associated with A, we have the eigenvalues $\pm i \sin\Phi$).

Analogously, the definition of the tangent trigonometric function, which can be written as

$$\pm i \tan\Phi = \frac{\pm i \sin\Phi}{\cos\Phi} = -2\,\frac{\mp i \tan\left(\dfrac{\Phi}{2}\right)}{1 - \tan^2\left(\dfrac{\Phi}{2}\right)} \tag{32}$$

allows us to introduce a Skew-Symmetric matrix T that can be seen as a *tangent matrix*

$$T = (C - C^\mathrm{T})(C + C^\mathrm{T})^{-1} = (C + C^\mathrm{T})^{-1}(C - C^\mathrm{T}) \tag{33}$$

that can also be expressed or decomposed as

$$T = -2\,Q\,(I - QQ^\mathrm{T})^{-1} = -2\,(I - QQ^\mathrm{T})^{-1}\,Q \tag{34}$$

Analogously, the expression of the tangent half-angle

$$\mp i \tan\left(\frac{\Phi}{2}\right) = -\frac{\pm i \sin\Phi}{1 + \cos\Phi} = \frac{1 - \cos\Phi}{\pm i \sin\Phi} \tag{35}$$

allows us to introduce the matrix identities

$$\begin{cases} Q = -(C - C^\mathrm{T})(2I + C + C^\mathrm{T})^{-1} = -(2I + C + C^\mathrm{T})^{-1}(C - C^\mathrm{T}) \\ Q = (2I - C - C^\mathrm{T})(C - C^\mathrm{T})^{-1} = (C - C^\mathrm{T})^{-1}(2I - C - C^\mathrm{T}) \end{cases} \tag{36}$$

Equations (29), (31), (33), (34), and (36) and the associated trigonometric relationships given in Eqs. (28), (30), (32), and (35), are, therefore, strictly connected. Note that, some of the above relationships represent a different way to write the Cayley Transforms. It is clear that matrix properties are substantially more complex than trigonometric properties. However, the above relationships allows us to see the Orthogonal, Symmetric, and Skew-Symmetric matrices as illuminated with an additional dimmer light. This small additional knowledge will certainly help to understand the nature, and to develop the "Matric Trigonometry", of which the known one (the scalar), only constitutes a small subset.

It is easy to demonstrate that the Ortho-Sym matrix $\Re$ represents the *key* matrix to map Symmetric onto Skew-Symmetric matrices, and viceversa. In fact, the relationship

$$S = \Re(W)\,A = -A\,\Re(W) \qquad \Longleftrightarrow \qquad A = \Re(W)\,S = -S\,\Re(W) \tag{37}$$

where $\Re(W)$ means a Ortho-Sym matrix $\Re$ which is built with the eigenvector matrix W of A, while the eigenvector $P = W\,\mathcal{D}(M)$ of S is obtained by Eq. (14).

The *Counter Clockwise* function $f_2(z) = \dfrac{i-z}{i+z} = \dfrac{1+iz}{1-iz}$

To extend the application of $f_2(z)$ to matrix field (see Fig. 3 for its mapping summary), it is possible to demonstrate that, for the Symmetric matrices S, the extension becomes

$$\begin{cases} C & = (I + \Re S)(I - \Re S)^{-1} = (I - \Re S)^{-1}(I + \Re S) \\ C^{\mathrm{T}} & = (I + \Re S)^{-1}(I - \Re S) = (I - \Re S)(I + \Re S)^{-1} \end{cases} \tag{38}$$

where $\Re = \Re(W)$ is the Ortho-Sym matrix associated with the eigenvector matrix of C. The previous equation can be written also in the form

$$\begin{cases} C & = (\Re - S)(\Re + S)^{-1} = (\Re - S)^{-1}(\Re + S) \\ C^{\mathrm{T}} & = (\Re + S)(\Re - S)^{-1} = (\Re + S)^{-1}(\Re - S) \end{cases} \tag{39}$$

these equations implies the trigonometric identities

$$\frac{i \mp \tan\left(\dfrac{\Phi}{2}\right)}{i \pm \tan\left(\dfrac{\Phi}{2}\right)} = \cos\Phi \pm i\,\sin\Phi \tag{40}$$

Just for example, let us to demonstrate Eq. (39). If, as it happens in Cayley Transforms (that is, the formal structure of the eigenvalue mapping is identical to the matrix mapping), then we need to look for an unknown matrix X such that

$$C = (X - S)(X + S)^{-1}$$

is Orthogonal. From this equation we can easily derive that the unknown matrix X has the expression

$$X = (I - C)^{-1}(I + C)\,S.$$

Setting $c \equiv \cos\Phi$, $s \equiv \sin\Phi$, and $t \equiv \tan\left(\dfrac{\Phi}{2}\right)$, we can write the expression for X in the 2-D case

$$X \equiv \begin{bmatrix} 1-c & -s \\ s & 1-c \end{bmatrix}^{-1} \begin{bmatrix} 1+c & s \\ -s & 1+c \end{bmatrix} \begin{bmatrix} -t & 0 \\ 0 & t \end{bmatrix} = \begin{bmatrix} 0 & 1 \\ 1 & 0 \end{bmatrix}$$

this matrix, which is independent from Φ, is Symmetric, and Orthogonal, with eigenvalues ± 1. This implies the conclusion that the unknown matrix X is the Ortho-Sym matrix $\Re$. Demonstration of Eq. (38) may be done on the same way.

The $f_2(z)$ mapping function, here named *Counter Clockwise* (because it moves from real axis, to imaginary axis, to unit circle, and back to real axis), cannot be applied to Orthogonal and to Skew-Symmetric matrices because $f_2(z) \neq -f_2(z^\dagger)$. In particular, this mapping function satisfies

$$f_2(e^{i\Phi})f_2(e^{-i\Phi}) = -1 \qquad \text{and} \qquad f_2(i\Phi)f_2(-i\Phi) = +1 \tag{41}$$

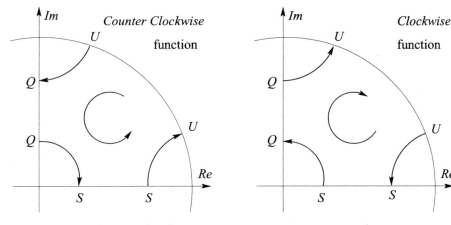

Figure 3: $f_2(z) = \dfrac{i-z}{i+z} = \dfrac{1+iz}{1-iz}$ function   Figure 4: $f_3(z) = \dfrac{1-z}{1+z} i$ function

The *Clockwise* function $f_3(z) = \dfrac{1-z}{1+z} i$

The mapping function $f_3(z)$ (see Fig. 4 for its mapping summary), can be used to find a way to map Orthogonal to Symmetric. The resulting mapping is

$$S = (I - C)(I + C)^{-1}\Re = (I + C)^{-1}(I - C)\Re \qquad (42)$$

whose demonstration is easily derived from Eq. (66) which, in turn, can be seen as the a way to map Skew-Symmetric to Symmetric and viceversa

$$S = Q\Re = -\Re Q \qquad \text{and} \qquad Q = S\Re = -\Re S \qquad (43)$$

The $f_3(z)$ mapping function, here named *Clockwise* (because it moves from real axis, to unit-circle, to imaginary axis, and back to real axis), cannot be applied to Symmetric matrices because $f_3(\Phi) \neq -f_3(-\Phi)$, and cannot be applied to Skew-Symmetric matrices because $f_3(i\,\Phi) \neq f_3(-i\,\Phi)^\dagger$. In particular, this mapping function satisfies

$$f_3(\Phi)f_3(-\Phi) = -1 \qquad \text{and} \qquad f_3(i\,\Phi)f_3(-i\,\Phi) = -1 \qquad (44)$$

Mapping Summary

All the above relationship are summarized below:

- From Orthogonal to Skew-Symmetric:

$$Q = (I - C)(I + C)^{-1} = (I + C)^{-1}(I - C) \qquad (45)$$

- From Skew-Symmetric to Orthogonal:

$$C = (I - Q)(I + Q)^{-1} = (I + Q)^{-1}(I - Q) \qquad (46)$$

- From Orthogonal to Symmetric:

$$S = (I - C)(I + C)^{-1}\Re = (I + C)^{-1}(I - C)\Re \qquad (47)$$

- From Symmetric to Orthogonal:

$$C = (\Re - S)(\Re + S)^{-1} = (\Re - S)^{-1}(\Re + S) \qquad (48)$$

- From Skew-Symmetric to Symmetric:

$$S = Q\Re = -\Re Q\Re \qquad (49)$$

- From Symmetric to Skew-Symmetric:

$$Q = S\Re = -\Re S \qquad (50)$$

The above set of equations constitute the complete matrix mapping relationships among Orthogonal, Skew-Symmetric, and Symmetric matrices.

From this set of equations it is possible to demonstrate the following

$$\begin{cases} \Re C\Re = C^{\mathrm{T}} \\ \Re Q\Re = -Q = Q^{\mathrm{T}} \\ \Re S\Re = -S \end{cases} \quad \text{and} \quad \begin{cases} \Im C\Im = -C \\ \Im Q\Im = -Q = Q^{\mathrm{T}} \\ \Im S\Im = S \end{cases} \qquad (51)$$

which yield to the relationships

$$\Re = -i\, W\, \mathcal{D}(\Im_2)\, W^{\dagger} \quad \text{and} \quad \Im = -i\, W\, \mathcal{D}(\Re_2)\, W^{\dagger} \qquad (52)$$

where $\mathcal{D}(\Im_2)$, and $\mathcal{D}(\Re_2)$ are the $n \times n$ diagonal matrices obtained with the 2×2 block elements $\Im_2$, and $\Re_2$, respectively.

Direct Product Mapping

The purpose of this section is to find the relationships allowing one-to-one mapping among Orthogonal, Skew-Symmetric, and Symmetric matrices, by a direct product. This implies, for instance, to find out which is the matrix G such that $S = GC$ maps a given orthogonal matrix C onto a symmetric matrix S, which has assigned eigenvalues. Here again, the problem to share the eigenvectors, is solved using the Nucleus matrix.

In fact, setting $c_k = \cos \Phi_k$ and $s_k = \sin \Phi_k$, the spectral decomposition of a given Orthogonal matrix C can be written as

$$C = \sum_{k=1}^{n_p} W_k \, \Lambda_{2k} \, W_k^\dagger + p \, p^{\mathrm{T}} \quad \text{where } \Lambda_{2k} = \begin{bmatrix} c_k + is_k & 0 \\ 0 & c_k - is_k \end{bmatrix} \tag{53}$$

which can also be written as

$$C = \sum_{k=1}^{n_p} [W_k M][M^\dagger \Lambda_{2k} M][M^\dagger W_k^\dagger] + p \, p^{\mathrm{T}} = \sum_{k=1}^{n_p} P_k C_{2k} P_k^{\mathrm{T}} + p \, p^{\mathrm{T}} \tag{54}$$

where

$$W_k \, M = P_k \quad \text{and} \quad M^\dagger \Lambda_{2k} \, M = \begin{bmatrix} \cos \Phi_k & \sin \Phi_k \\ -\sin \Phi_k & \cos \Phi_k \end{bmatrix} = C_{2k} \tag{55}$$

Therefore, the spectral decomposition of real Orthogonal matrices can also be written in the full real form

$$C = \sum_{k=1}^{n_p} P_k \, C_{2k} \, P_k^{\mathrm{T}} + p \, p^{\mathrm{T}} \quad \text{where} \quad C_{2k} = \begin{bmatrix} c_k & s_k \\ -s_k & c_k \end{bmatrix} \tag{56}$$

Similarly, Skew-Symmetric matrices can be expanded as

$$A = \sum_{k=1}^{n_p} P_k \, A_{2k} \, P_k^{\mathrm{T}} \quad \text{where} \quad A_{2k} = \begin{bmatrix} 0 & a_k \\ -a_k & 0 \end{bmatrix}. \tag{57}$$

while for the Symmetric matrices we have

$$S = \sum_{k=1}^{n_p} P_k \, S_{2k} \, P_k^{\mathrm{T}} \quad \text{where} \quad S_{2k} = \begin{bmatrix} r_k & 0 \\ 0 & -r_k \end{bmatrix} \tag{58}$$

Thanks to the eigenvector orthogonality

$$P_i^{\mathrm{T}} P_j = 0_2, \; i \neq j; \qquad P_i^{\mathrm{T}} P_i = I_2; \qquad P_i^{\mathrm{T}} p = 0_{2,1}; \qquad p^{\mathrm{T}} p = 1. \tag{59}$$

the expressions provided by Eqs. (56-58) allows us to obtain compact formulations for subsequent products of Orthogonal, Symmetric, and Skew-Symmetric matrices. For instance, using these expansions, the product between an Orthogonal and a Symmetric matrix becomes

$$C \, S = \left[\sum_{i=1}^{n_p} P_i \, C_{2i} \, P_i^{\mathrm{T}} + p \, p^{\mathrm{T}} \right] \left[\sum_{j=1}^{n_p} P_j \, S_{2j} \, P_j^{\mathrm{T}} \right] \tag{60}$$

where all the terms with $i \neq j$ vanish [because of Eq. (59)]. Thus,

$$C \, S = \sum_{k=1}^{n_p} P_k \, [\, C_{2k} \, S_{2k} \,] \, P_k^{\mathrm{T}} = \sum_{k=1}^{n_p} P_k \, F_{2k} \, P_k^{\mathrm{T}} \tag{61}$$

where $F_{2k} = C_{2k} S_{2k}$. This result provides us with the tool to develop general matrix mapping by means of a direct product. In fact, the G matrix that maps the matrix B onto the final matrix $F = G B$ can be built as follows

$$G = F B^{-1} = \sum_{k=1}^{n_p} P_k G_{2k} P_k^{\mathrm{T}} \qquad \text{where} \quad G_{2k} = F_{2k} B_{2k}^{-1} \tag{62}$$

This procedure holds provided that the matrices B_{2k} are not singular. The inverse mapping transformation (from F onto B) is simply

$$G^{-1} = B F^{-1} = \sum_{k=1}^{n_p} P_k G_{2k}^{-1} P_k^{\mathrm{T}} \qquad \text{where} \quad G_{2k}^{-1} = B_{2k} F_{2k}^{-1} \tag{63}$$

that holds if the matrices F_{2k} are not singular.

From the above, the following product properties

$$\begin{cases} A S = -S A \\ C S = S C^{\mathrm{T}} \\ C A = A C \end{cases} \implies \begin{cases} C S A = -C A S = A C S \\ A S C = -S A C = -S C A \end{cases} \tag{64}$$

are easily obtained.

Specifically to our end, omitting the k index, setting $s = \sin \Phi$ and $c = \cos \Phi$, and indicating with G_{BF} and G_{2k} the matrices used to map the B matrix onto the F matrix, we deal with the following nine cases:

From Symmetric onto Symmetric

$$G_{SS} = S_f S_i^{-1} = \begin{bmatrix} r_f & 0 \\ 0 & -r_f \end{bmatrix} \begin{bmatrix} r_i & 0 \\ 0 & -r_i \end{bmatrix}^{-1} = \frac{r_f}{r_i} I_2 \tag{65}$$

From Skew-Symmetric onto Symmetric

$$G_{AS} = S_f A_i^{-1} = \begin{bmatrix} r & 0 \\ 0 & -r \end{bmatrix} \begin{bmatrix} 0 & a \\ -a & 0 \end{bmatrix}^{-1} = -\frac{r}{a} \Re_2 \tag{66}$$

From Orthogonal onto Symmetric

$$G_{US} = S_f C_i^{-1} = \begin{bmatrix} r & 0 \\ 0 & -r \end{bmatrix} \begin{bmatrix} c & s \\ -s & c \end{bmatrix}^{-1} = r \begin{bmatrix} c & -s \\ -s & -c \end{bmatrix} \tag{67}$$

From Symmetric onto Skew-Symmetric

$$G_{SA} = A_f S_i^{-1} = \begin{bmatrix} 0 & a \\ -a & 0 \end{bmatrix} \begin{bmatrix} r & 0 \\ 0 & -r \end{bmatrix}^{-1} = -\frac{a}{r} \Re_2 \tag{68}$$

From Skew-Symmetric onto Skew-Symmetric

$$G_{AA} = A_f A_i^{-1} = \begin{bmatrix} 0 & a_i \\ -a_i & 0 \end{bmatrix} \begin{bmatrix} 0 & a_f \\ -a_f & 0 \end{bmatrix}^{-1} = \frac{a_i}{a_f} I_2 \tag{69}$$

From Orthogonal onto Skew-Symmetric

$$G_{UA} = A_f C_i^{-1} = \begin{bmatrix} 0 & a \\ -a & 0 \end{bmatrix} \begin{bmatrix} c & s \\ -s & c \end{bmatrix}^{-1} = a \begin{bmatrix} s & c \\ -c & s \end{bmatrix} \tag{70}$$

From Symmetric onto Orthogonal

$$G_{SU} = C_f S_i^{-1} = \begin{bmatrix} c & s \\ -s & c \end{bmatrix} \begin{bmatrix} r & 0 \\ 0 & -r \end{bmatrix}^{-1} = \frac{1}{r} \begin{bmatrix} c & -s \\ -s & -c \end{bmatrix} \tag{71}$$

From Skew-Symmetric onto Orthogonal

$$G_{AU} = C_f A_i^{-1} = \begin{bmatrix} c & s \\ -s & c \end{bmatrix} \begin{bmatrix} 0 & a \\ -a & 0 \end{bmatrix}^{-1} = \frac{1}{a} \begin{bmatrix} s & -c \\ c & s \end{bmatrix} \tag{72}$$

From Orthogonal onto Orthogonal

$$G_{UU} = C_f C_i^{-1} = \begin{bmatrix} c_1 & s_1 \\ -s_1 & c_1 \end{bmatrix} \begin{bmatrix} c_2 & s_2 \\ -s_2 & c_2 \end{bmatrix}^{-1} = \begin{bmatrix} c_3 & s_3 \\ -s_3 & c_3 \end{bmatrix} \tag{73}$$

where $c_3 = \cos(\Phi_1 - \Phi_2)$, and $s_3 = \sin(\Phi_1 - \Phi_2)$

Example

The above equations, from Eq. (65) through Eq. (73), can be used, for instance, to keep the information of rotation (contained in C), into Symmetric, Skew-Symmetric forms. An an example, let us to find the matrix mapping an Orthogonal matrix C onto Symmetric and Skew-Symmetric forms, whose eigenvalues are expressed as $\tan\left(\dfrac{\Phi_k}{4}\right)$[§]. S is obtained using Eq. (67), with $r_k = \tan\left(\dfrac{\Phi_k}{4}\right)$

$$G_{2k} = G_{US} = \tan\left(\frac{\Phi_k}{4}\right) \begin{bmatrix} \cos\Phi_k & -\sin\Phi_k \\ -\sin\Phi_k & -\cos\Phi_k \end{bmatrix} \tag{74}$$

thus, the matrix

$$T_{US} = \sum_{k=1}^{m} P_k \, G_{2k} \, P_k^{\mathrm{T}} \tag{75}$$

is such that the matrix

$$S = T_{US}C \tag{76}$$

is Symmetric and has eigenvalues $\pm\tan\left(\dfrac{\Phi_k}{4}\right)$.

Analogously, to obtain the Skew-Symmetric matrix, we should apply Eq. (70), with $a_k = \tan\left(\dfrac{\Phi_k}{4}\right)$

$$G_{2k} = G_{UA} = \tan\left(\frac{\Phi_k}{4}\right) \begin{bmatrix} \sin\Phi_k & \cos\Phi_k \\ -\cos\Phi_k & \sin\Phi_k \end{bmatrix} \tag{77}$$

thus, the matrix

$$T_{UA} = \sum_{k=1}^{m} P_k \, G_{2k} \, P_k^{\mathrm{T}} \tag{78}$$

is such that the matrix

$$A = T_{UA}C \tag{79}$$

is Skew-Symmetric and has eigenvalues $\pm i \tan\left(\dfrac{\Phi_k}{4}\right)$.

Conclusion

This paper shows that Cayley Transforms can be considered the extension to the matrix field of the complex conformal mapping function $f_1(z)$. This results has been extended to similar matrix transforms which map Orthogonal from/onto Symmetric, and Symmetric from/onto Skew-Symmetric matrices. This has been achieved

---

[§]Note that there is an explicit reference to the Modified Rodriguez Parameters (MRP), a minimum non-singular parameter set to describe Orientation

using a set of real matrices which are simultaneously Orthogonal, and Symmetric (the $Ortho - Sym$ matrices), and using two different complex conformal mapping functions (namely, the $f_2(z)$, and the $f_3(z)$, that are here called *clockwise*, and *counter clockwise* functions). Moreover, the close relationship between some trigonometric identities and some matrix mappings is shown. This paper ends with a description of a matrix mapping technique (always mapping Orthogonal, Symmetric, and Skew-Symmetric matrices) that is obtained by direct matrix product.

Acknowledgment

I would warmly like to thank Dr John L. Junkins and all the colleagues of the Aerospace Department of Texas A&M University for directly and indirectly encouraging me to keep going with this research.

References

[1] J.L. Junkins and Y. Kim. *Introduction to Dynamics and Control of Flexible Structures*. American Institute of Aeronautics and Astronautics, AIAA Education Series, Washington, D.C., 1993.

[2] G.A. Korn and T.M. Korn. *Mathematical Handbook for Scientists and Engineers*. McGraw-Hill Book Company, 1221 Avenue of the Americas, New York, N.Y. 10020, Second edition, 1968.

[3] D. Mortari. On the Rigid Rotation Concept in n-Dimensional Spaces. *The Journal of the Astronautical Sciences*, 49(3):401–420, July-September 2001.

[4] A.K. Sanyal. Geometrical Transformations in Higher Dimensional Euclidean Spaces. Master's thesis, Department of Aerospace Engineering, Texas A&M University, College Station, TX, May 2001.

SPACECRAFT ANGULAR RATE ESTIMATION ALGORITHMS FOR STAR TRACKER-BASED ATTITUDE DETERMINATION

Puneet Singla,[*] John L. Crassidis[†] and John L. Junkins[‡]

In this paper, two different algorithms are presented for the estimation of spacecraft body angular rates in the absence of gyro rate data for a star tracker mission. In first approach, body angular rates are estimated with the spacecraft attitude using a dynamical model of the spacecraft. The second approach makes use of a rapid update rate of star camera to estimate the spacecraft body angular rates independent of spacecraft attitude. Essentially the image flow of the stars is used to establish a Kalman filter for estimating the angular velocity. The relative merits of both the algorithms are then studied for the spacecraft body angular rates measurements. The second approach has an advantage of being free from any bias in attitude estimates.

Introduction

Spacecraft angular rate data plays an important role in attitude determination and attitude control. With the use of rate data, the attitude of spacecraft can be predicted between two different frames of star tracker data. Generally, three axis gyros are used on board to provide the body angular rate information. In the presence of densely measured rate data, the exact kinematic model can replace the dynamical model. But when rate data is not available, then estimation accuracy is obviously dependent on 1) accuracy of the star measurements, 2) their frequency in time, and 3) the accuracy of the dynamical model. An accidental gyro failure (e.g. failure of four of six rate gyros on the Earth Radiation Budget Satellite[1]) or intentionally omission of gyros (e.g. in Small Explorer (SAMPEX)[1]) due to their high cost can necessitate "gyroless" attitude estimation. The loss of gyro data can result in unacceptably high propagation errors.

The problems related to spacecraft attitude estimation in loss of gyro data has been discussed by Mook,[1] and he proposed an estimation algorithm, which can in principle take

---

[*] Graduate Student, Aerospace Engineering, Texas A&M University, College Station, Texas 77843-3141. E-mail: puneet@tamu.edu.

[†] Assistant Professor, Department of Aerospace and Mechanical Engineering, University at Buffalo, State University of New York, Amherst, New York 14260-4400. E-mail: john@eng.buffalo.edu.

[‡] George J. Eppright Chair Professor, Director of the Center for Mechanics and Control, Department of Aerospace Engineering, Texas A&M University, College Station, Texas 77843-3141. E-mail: junkins@tamu.edu. Fellow AAS.

care of both inaccurate rate data and inaccurate dynamic model. He suggested that the Minimum Model Error (MME) approach to be used to obtain the most accurate state estimate for poorly modelled dynamic systems by taking into account the error in model dynamics. Afterwards these MME estimates may be used for model identification. Crassidis and Markley[2] have obtained accurate estimates for the SAMPEX spacecraft by using only magnetometer sensor measurements. But the disadvantages with the MME based approach is that it falls into the category of batch estimation (i.e., all the data should be processed simultaneously), which significantly restricts its use in real time applications. The Kalman filter is an ideal choice for real time applications. However, the accuracy of the Kalman filter estimation depends in a complicated way upon the accuracy of dynamic model and tuning of the process noise covariance matrix. So between two sets of measurements, the estimates are still subjected to accumulation of model error. It should be noted that the Kalman filter does consider model error, but virtually all implementations consider the model error to be a "white noise random process". This model usually inadequate to represent colored (correlated) model errors; for the case of poorly known systematic (non-white) model errors, the MME and related algorithms are more attractive. While MME methods enjoy some theoretical advantages, both the Kalman filter and MME algorithms require tuning, however the required artistic tuning of the Kalman filter algorithms are in a more mature state of development. Fisher et. al.[3] discussed the use of the Kalman filter for attitude and angular rate determination using attitude sensor outputs alone.

Sufficient information about the body angular rates can be obtained from the attitude sensor measurements, if attitude sensor data frequency is fast enough to capture the spacecraft motion. Star cameras are very accurate and, in view of recent active pixel sensor cameras, star camera frame rates are increasingly high. This suggests the possibility of deriving angular velocity measurements from "star motion" on the focal plane. In this paper, two different sequential algorithms will be presented for spacecraft body angular rates estimation in the absence of gyro rate data for a star tracker mission.

1. In the first approach, body angular rates of spacecraft will be estimated with spacecraft attitude using the Kalman filter. This method uses a dynamical model in which, external torques acting on spacecraft are modelled by random walk process. So the performance of this algorithm will depend on the validity of assumed dynamical model for the given case.

2. The second approach makes use of the rapid update rate of the star camera to approximate body angular velocity vector independent of attitude estimation. A time derivative of star tracker body measurements is taken to establish "measurement equations" for estimating angular velocity. First order and second order finite difference approaches will be used to approximate the time derivative of body measurements. A sequential Least Squares algorithm is used to estimate the spacecraft angular rates. This algorithm works fine, if the sampling interval of star data is well within Nyquist's limit for the actual motion of the spacecraft.

The structure of this paper proceeds as follows. First a brief review of star tracker model is given. The subsequent section introduces the attitude dependent angular velocity

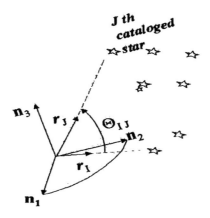

Figure 1 Catalog Star Pair

estimation followed by the attitude independent approach. Finally, computer simulation results are presented to test the algorithms.

Star Tracker Model

Star positions are a very accurate source as the reference system for the attitude determination problem as their position is fixed with respect to inertial system fixed to earth. Spacecraft attitude is determined by taking the photographs of the star by a star camera. Pixel formats of the order of 512×512 or larger are commonly used to provide good resolution pictures. The first stage in attitude determination is to identify the stars with reference to on-board catalog. Star catalog contain the spherical co-ordinate angles of the stars (α is the right ascension and δ is the declination, Figure 1) to a high accuracy.

Many algorithms have been developed for the star identification.[4] All star identification algorithms can be divided into three categories: 1) Direct match, 2) Angular separation match, and 3) Phase match. Star identification algorithm based upon the Angular separation approach are very popular .[5] After star identification is made, image plane coordinates of the stars are given by using a pinhole camera model for the star camera. Photograph image plane coordinates of jth star are given by following co-linearity equations:

$$x_j = f \frac{C_{11}r_{x_j} + C_{12}r_{y_j} + C_{13}r_{z_j}}{C_{13}r_{x_j} + C_{32}r_{y_j} + C_{33}r_{z_j}} + x_0 \tag{1}$$

$$y_j = f \frac{C_{21}r_{x_j} + C_{22}r_{y_j} + C_{23}r_{z_j}}{C_{13}r_{x_j} + C_{32}r_{y_j} + C_{33}r_{z_j}} + y_0 \tag{2}$$

where f is the effective focal length of the star camera and (x_0, y_0) are principal point offsets, determined by the ground or on-orbit calibration, C_{ij} are the direction cosines matrix elements, and the inertial vector $\mathbf{r}_j$ is given by

$$\mathbf{r}_j = \left\{ \begin{array}{c} r_{x_j} \\ r_{y_j} \\ r_{z_j} \end{array} \right\} = \left\{ \begin{array}{c} \cos \delta_j \cos \alpha_j \\ \cos \delta_j \sin \alpha_j \\ \sin \delta_j \end{array} \right\} \tag{3}$$

choosing the z-axis of the image coordinate system towards the boresight of the star camera as shown in Figure 2, the measurement unit vector $\mathbf{b}_j$ is given by following equation:

$$\mathbf{b}_j = \frac{1}{\sqrt{x_j^2 + y_j^2 + f^2}} \left\{ \begin{array}{c} -(x_j - x_0) \\ -(y_j - y_0) \\ f \end{array} \right\} \tag{4}$$

The relationship between measured star direction vector $\mathbf{b}_j$ in image space and their

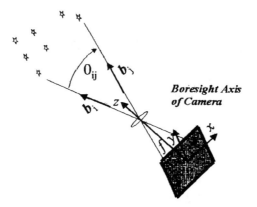

Figure 2 Image Plane Star Pair

projection $\mathbf{r}_j$ on the inertial frame is given by

$$\mathbf{b}_j = \mathbf{C} \mathbf{r}_j + \boldsymbol{\nu}_j \tag{5}$$

where $\mathbf{C}$ is the attitude direction cosine matrix denotes the mapping between image and inertial frame and $\boldsymbol{\nu}_j$ is a zero mean Gaussian white noise process with covariance $\mathbf{R}_j$.

Attitude Dependent Angular Velocity Estimation Algorithm

In this approach, the spacecraft body angular rate are estimated with spacecraft attitude using the Kalman filter. Actually, this algorithm has been derived from the attitude determination algorithm presented in ref.[6] The state vector of the Kalman filter consists of 3 component of angular velocity in place of the rate bias vector as in.[6]

$$\mathbf{x} = \left\{ \begin{array}{c} \delta q \\ \omega \end{array} \right\} \tag{6}$$

The angular acceleration of the spacecraft is modelled by first order random process.

$$\tau = \dot{\omega} = \eta_2 \tag{7}$$

where η_2 represents a Gaussian random variable with the following known statistical properties:

$$\mathbf{E}(\eta_2) = 0 \tag{8}$$
$$(\eta_2\eta_2^T) = \sigma_v^2\mathbf{I} \tag{9}$$

Therefore, the state equations for new state vector are given as:

$$\dot{x} = \mathbf{F}\mathbf{x} + \mathbf{G}\mathbf{w} \tag{10}$$

where $\mathbf{w}$ is a noise vector defined as:

$$\mathbf{w} = \left\{ \begin{array}{c} \eta_1 \\ \eta_2 \end{array} \right\} \tag{11}$$

and matrix $\mathbf{F}$ and $\mathbf{G}$ are given by following equations.[7]

$$\mathbf{F} = \left[\begin{array}{cc} -\tilde{\omega}(t) & -\frac{1}{2}\mathbf{I}_{3\times3} \\ \mathbf{O}_{3\times3} & \mathbf{O}_{3\times3} \end{array} \right] \tag{12}$$

$$\mathbf{G} = \left[\begin{array}{cc} \frac{1}{2}\mathbf{I}_{3\times3} & \mathbf{O}_{3\times3} \\ \mathbf{O}_{3\times3} & \mathbf{I}_{3\times3} \end{array} \right] \tag{13}$$

Adopting the procedure described in refs.,[6,7] the state propagation and update equations for Kalman filter can be written as:

Propagation Equations

$$\hat{q}_{k+1} = [\cos(\frac{\theta_k}{2})\mathbf{I}_{4\times4} + \sin(\frac{\theta_k}{2})\mathbf{\Omega}(\hat{n}_k)]\hat{q}_k \tag{14}$$

where

$$\theta_k = \omega_n * (t_{k+1} - t_k) \text{ and } \omega_n = \|\hat{\omega}_k\| = \sqrt{\hat{\omega}_{k_1}^2 + \hat{\omega}_{k_2}^2 + \hat{\omega}_{k_3}^2}$$

$$\mathbf{P}_{k+1} = \mathbf{\Phi}_k\mathbf{P}_k\mathbf{\Phi}_k + \mathbf{G}\mathbf{Q}_k\mathbf{G} \tag{15}$$

with

$$\mathbf{\Phi}_k = \left[\begin{array}{cc} \mathbf{\Phi}_{1_k} & \mathbf{\Phi}_{2_k} \\ \mathbf{O}_{3\times3} & \mathbf{I}_{3\times3} \end{array} \right] \tag{16}$$

$$\mathbf{\Phi}_{1_k} = \mathbf{I}_{3\times3} + \frac{\tilde{\omega}^T}{\omega_n}\sin\theta_k + (\frac{\tilde{\omega}^T}{\omega_n})^2(1 - \cos\theta_k) \tag{17}$$

$$\mathbf{\Phi}_{2_k} = \frac{1}{2}[\mathbf{I}_{3\times3}\Delta t + \frac{(\tilde{\omega}^T(1 - \cos\theta_k))}{\omega_n^2} + \frac{\tilde{\omega}\tilde{\omega}(\theta_k - \sin\theta_k)}{\omega_n^3}] \tag{18}$$

Update Equations

$$\hat{x}_k^+ = \hat{x}_k^- + \mathbf{K}_k(\bar{y}_k - \mathbf{H}_k\hat{x}_k^-) \tag{19}$$
$$\mathbf{P}_k^+ = (\mathbf{I} - \mathbf{K}_k\mathbf{H})\mathbf{P}_k^- \tag{20}$$
$$\mathbf{K}_k = \mathbf{P}_k^-\mathbf{H}_k^T(\mathbf{H}_k\mathbf{P}_k^-\mathbf{H}_k^T + \mathbf{R}_k)^{-1} \tag{21}$$

where

$$\mathbf{H}_k = \left[\begin{array}{cc} \mathbf{L} & \mathbf{O}_{3\times3} \end{array} \right] \tag{22}$$
$$\mathbf{L} = 2[\hat{b}\otimes] \tag{23}$$

1307

Simulations

Using the J-2000 star catalog with stars of visual magnitude brighter than $M_v \sim 6.4$, assuming $8^0 \times 8^0$ FOV star camera and 17μ radian (for 1024×1024 pixel array) of centroiding error, star data are simulated at a frame rate frequency of 10Hz. Three different test cases were considered. For the first case the spacecraft is assumed to be in LEO orbit with true angular velocity, $\omega = \left\{ \begin{array}{ccc} 0 & 0.0011 & 0 \end{array} \right\}^T$.

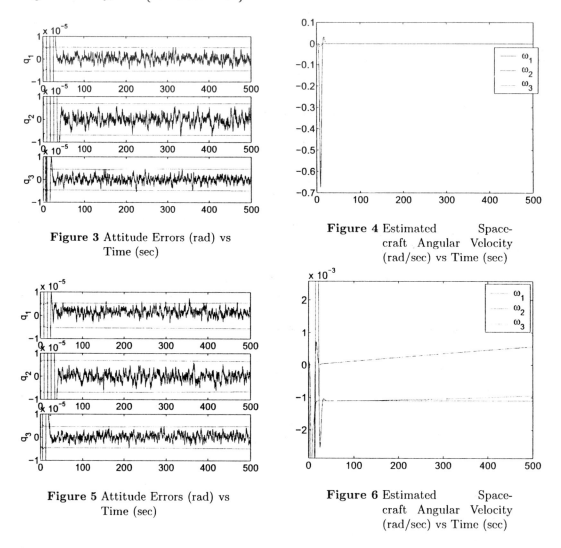

Figure 3 Attitude Errors (rad) vs Time (sec)

Figure 4 Estimated Spacecraft Angular Velocity (rad/sec) vs Time (sec)

Figure 5 Attitude Errors (rad) vs Time (sec)

Figure 6 Estimated Spacecraft Angular Velocity (rad/sec) vs Time (sec)

Figures 3 and 4 show the plots of estimated spacecraft attitude and spacecraft angular velocities, respectively. From Figure 4 it is clear that the algorithm is able to estimate the angular velocity and attitude precisely, but the attitude estimates are somewhat less accurate as compared to the estimates when angular rates are available.

For the second test case, it is assumed that spacecraft angular velocity has small slowly time varying oscillations about the x and z axes but has a constant velocity of 0.0011 rad/sec about the y-axis. Figures 5 and 6 show the plots of estimated angular velocities and attitudes of the spacecraft respectively. From the plots, we can infer that algorithm given in this section is able to estimate the attitude and angular velocities of the spacecraft successfully even in the absence of rate data. However, the estimates are sensitive to "dropout" when sparse star regions are encountered, where star-id confidence may be too low.

For the third test case, it is assumed that spacecraft angular velocity has fast time varying angular oscillations about the x and z axes but has a constant velocity of 0.0011 rad/second about the y-axis. But unfortunately, for third test case this algorithm fails, i.e., the estimated errors are significantly outside their 3σ bounds as the angular velocities are rapidly changing with time.

The main disadvantage associated with this approach is that angular rate estimates are subject to any error in spacecraft attitude estimates and vice-vera.

Attitude Independent Approach

In this section, a deterministic approach will be used to determine the spacecraft body angular rate vector from the star tracker body measurements. This approach makes use of the increasingly rapid update rates of modern star cameras and finite different analysis of "image flow" trajectories of the measured star line of sight vectors in the sensor coordinate system. Consider the following focal plane trajectory of line of sight vector for i^{th} star $\mathbf{b}_i(t)$:

$$\mathbf{b}_i(t) = \frac{1}{\sqrt{x_i(t)^2 + y_i(t)^2 + f^2}} \left\{ \begin{array}{c} -(x_i(t) - x_0) \\ -(y_i(t) - y_0) \\ f \end{array} \right\} \tag{24}$$

where $x_i(t)$ and $y_i(t)$ dictate the focal plane trajectory of ith star. Now, the true flow velocity measurement model is given by

$$\frac{d\mathbf{b}_i(t)}{dt} = \frac{d\mathbf{C}(t)}{dt} \mathbf{r}_i \tag{25}$$

Using the fact that $\frac{d\mathbf{C}(t)}{dt} = -[\boldsymbol{\omega}(t)\otimes]\mathbf{C}(t)$, we can rewrite equation (25) as:

$$\frac{d\mathbf{b}_i(t)}{dt} = -[\boldsymbol{\omega}(t)\otimes]\mathbf{b}_i(t) \tag{26}$$

The first order Taylor series expansion for line of sight vector for i^{th} star at time t_k is given by:

$$\mathbf{b}_i(k) = \mathbf{b}_i(k-1) + \frac{d\mathbf{b}_i(t)}{dt}|_{k-1}\Delta t + O(\Delta t^2) \tag{27}$$

Using the fact that $\bar{b}_i(k) = \mathbf{b}_i(k) + \boldsymbol{\nu}$ and substituting for $\frac{d\mathbf{b}_i(t)}{dt}$ from equation (26), we obtain the following finite difference approximation:

$$\mathbf{Y}_i(k) = \frac{1}{\Delta t}[\bar{b}_i(k) - \bar{b}_i(k-1)] = [\bar{b}_i(k-1)\otimes]\boldsymbol{\omega}(k-1) + \mathbf{w}_i(k) + O(\Delta t) \tag{28}$$

where the effective measurement error $\mathbf{w}_i(k)$ of $\mathbf{Y}_i(k)$ is:

$$\mathbf{w}_i(k) = \frac{1}{\Delta t}[\boldsymbol{\nu}_i(k) - \boldsymbol{\nu}_i(k-1)] + [\boldsymbol{\omega}(k-1)\otimes]\boldsymbol{\nu}_i(k-1) \tag{29}$$

From equation (29), it is clear that new noise vector $\mathbf{w}_i(k)$ is a function of angular rate vector $\boldsymbol{\omega}(k-1)$. Assuming a stationary noise process for $\boldsymbol{\nu}_i$, the following covariance relationship can be easily derived

$$\mathbf{E}(\mathbf{w}_i(k)\mathbf{w}_i^T(k)) = [\boldsymbol{\omega}(k-1)\otimes]\mathbf{R}_i[\boldsymbol{\omega}(k-1)\otimes]^T + \frac{1}{\Delta t}[\mathbf{R}_i[\boldsymbol{\omega}(k-1)\otimes] - [\boldsymbol{\omega}(k-1)\otimes]\mathbf{R}_i] + \frac{2}{\Delta t^2}\mathbf{R}_i \tag{30}$$

If the sampling interval is well within Nyquist's limit i.e. $\|\boldsymbol{\omega}\|\Delta t \leq \pi$, with a safety factor of 10 then $\|\boldsymbol{\omega}\|\Delta t \leq \frac{\pi}{10}$. Since, $\|[\boldsymbol{\omega}(k-1)\otimes]\| \approx \|\boldsymbol{\omega}\|$, then the following inequalities are true

$$\|[\boldsymbol{\omega}(k-1)\otimes]\mathbf{R}_i[\boldsymbol{\omega}(k-1)\otimes]^T\| \;<\; \frac{\pi^2}{100\Delta t^2}\|\mathbf{R}_i\| \;<<\; \frac{2}{\Delta t^2}\|\mathbf{R}_i\| \tag{31}$$

$$\left\|\frac{1}{\Delta t}[\mathbf{R}_i[\boldsymbol{\omega}(k-1)\otimes] - [\boldsymbol{\omega}(k-1)\otimes]\mathbf{R}_i]\right\| \;<\; \frac{2\pi}{10\Delta t}\|\mathbf{R}_i\| \;<\; \frac{2}{\Delta t^2}\|\mathbf{R}_i\| \tag{32}$$

It should be noted that second term of noise covariance matrix vanishes for the isotropic measurement errors (i.e. when $\mathbf{R}_i$ is scalar times identity matrix). Therefore the last term in equation (30) dominates the first two terms, which can effectively be ignored. Equation (28) can now be cast into a Kalman filter formulation, which leads to

$$\hat{\boldsymbol{\omega}}(k)^+ = \hat{\boldsymbol{\omega}}(k)^- + \mathbf{K}_k(\mathbf{Y}_i(k) - \mathbf{H}_k\hat{\boldsymbol{\omega}}(k)^-) \tag{33}$$

where

$$\mathbf{Y}_i(k) = \frac{1}{\Delta t}[\bar{\mathbf{b}}_i(k) - \bar{\mathbf{b}}_i(k-1)] \tag{34}$$

$$\mathbf{H}_k = [\bar{\mathbf{b}}_i(k-1)\otimes] \tag{35}$$

To take care of time varying angular rates we need to propagate the error covariance matrix according to the Ricatti equation. Therefore, again modelling the spacecraft angular rates by a first order statistical process, we get following propagation equations

$$\dot{\boldsymbol{\omega}} = \boldsymbol{\eta} \;\&\; \mathbf{E}(\boldsymbol{\eta}\boldsymbol{\eta}^T) = \sigma^2\mathbf{I} \tag{36}$$

$$\mathbf{P}_k^- = \mathbf{P}_{k-1}^+ + \mathbf{G}\mathbf{Q}\mathbf{G} \tag{37}$$

where

$$\mathbf{G} = \mathbf{I}_{3\times3} \tag{38}$$

$$\mathbf{Q} = \sigma^2\mathbf{I}_{3\times3}\Delta t \tag{39}$$

Therefore, only knowledge of the body vector measurements and sampling time interval is required to derive an angular velocity estimate.

We can improve the accuracy of spacecraft angular velocity estimates particularly for higher angular rates by considering a higher order finite-difference where the Taylor series truncation errors are of magnitude $O(\Delta t^2)$ as compared to $O(\Delta t)$ in equation (28)

$$\mathbf{Y}_i(k) = \frac{1}{2\Delta t}[4\bar{\mathbf{b}}_i(k-1) - 3\bar{\mathbf{b}}_i(k-2) - \bar{\mathbf{b}}_i(k)] = [\bar{\mathbf{b}}_i(k-2)\otimes]\boldsymbol{\omega}(k-2) + \mathbf{w}_i(k) + O(\Delta t^2) \tag{40}$$

where

$$w_i(k) = [\bar{b}_i(k-2)\otimes]\nu_i(k-2) + \frac{1}{2\Delta t}[4\nu_i(k-1) - 3\nu_i(k-2) - \nu_i(k)] \tag{41}$$

Again assuming a stationary noise process for $\nu_i(k)$, the following covariance relationship can be derived

$$E(w_i(k)w_i^T(k)) = [\omega(k-2)\otimes]R_i[\omega(k-2)\otimes]^T + \frac{3}{2\Delta t}[R_i[\omega(k-2)\otimes] - [\omega(k-2)\otimes]R_i] + \frac{13}{2\Delta t^2}R_i \tag{42}$$

If the sampling interval is well within Nyquist's limit i.e. $\|\omega\|\Delta t \le \pi$, with a safety factor of 10 then $\|\omega\|\Delta t \le \frac{\pi}{10}$. Since, $\|[\omega(k)\otimes]\| \approx \|\omega\|$, then the following inequalities are true

$$\|[\omega(k-2)\otimes]R_i[\omega(k-2)\otimes]^T\| \; < \; \frac{\pi^2}{100\Delta t^2}\|R_i\| << \frac{13}{2\Delta t^2}\|R_i\| \tag{43}$$

$$\|\frac{3}{2\Delta t}[R_i[\omega(k-2)\otimes] - [\omega(k-2)\otimes]R_i]\| \; < \; \frac{6\pi}{20\Delta t}\|R_i\| < \frac{13}{2\Delta t^2}\|R_i\| \tag{44}$$

again the second term vanishes for isotropic measurement errors and therefore the last term in equation (42) dominates the first two terms, which can be ignored. Equation (40) can also be cast into a Kalman filter formulation, which leads to

$$\hat{\omega}(k)^+ = \hat{\omega}(k)^- + K_k(Y_i(k) - H_k\hat{\omega}(k)^-) \tag{45}$$

where

$$Y_i(k) \;=\; \frac{1}{2\Delta t}[4\bar{b}_i(k+1) - 3\bar{b}_i(k) - \bar{b}_i(k+2)] \tag{46}$$

$$H_k \;=\; [\bar{b}_i(k)\otimes] \tag{47}$$

The covariance matrix P can be propagated according to equation (37).

It should be noted that the second order approach has a order of magnitude less Taylor series truncation error than the first order approach but at the price of an increased standard deviation in the effective measurement noise (by a factor of $\sqrt{13}/2$). But, the Kalman filter is inadequate to represent the truncation error whereas it can treat the effective measurement noise as "Gaussian white noise process". Therefore, we expect second order approach to give better results than the first order approach whenever truncation error dominate the increase in effective measurement error which is the case for high angular velocities.

Simulations

Using the J-2000 star catalog with stars of magnitude up to 6.0, assuming $8^0 \times 8^0$ field of view star camera and 17μ radian (for 1024×1024 pixel array) of centroiding error, star data are simulated at a frame rate frequency of 10Hz. Three different test cases defined above were considered.

Figures 7 and 9 show the plots of the true angular velocity and the estimated spacecraft angular velocities from the first order and second order approaches, respectively for the first test case. Figures 8 and 10 show the plots of the corresponding spacecraft angular velocity error with 3σ bounds for the first order and second order approaches, respectively.

From these plots, it is clear that the second order approach is able to estimate the angular velocity more accurately than the first order approach.

Similarly, Figures 11 and 13 show the plots of the true and estimated spacecraft angular velocity for the second test case, by first and second order approach respectively. Figures 12 and 14 show the corresponding angular velocity error and 3σ bounds plots for first and second order approach respectively. From these plots, it is again clear that the second order approach is able to estimate the angular velocity more accurately than the first order approach. This can be due to the fact that the extra information provided by the second order approach for the star trajectory compensate for an increased standard deviation in the estimate noise.

Figures 15 and 16 show the plots of the true and estimated spacecraft angular velocity and spacecraft angular velocity error with 3σ bound by first order approach for the third test case. From the plots, it is clear that for this test case the angular velocity errors are significantly outside their 3σ bound when the magnitude of fairly changing angular velocities is high. Where as, from figures 17 and 18, we can notice that second order attitude independent algorithm works fine even for fairly high magnitude rapidly changing angular velocities case. This can be due to the fact that the second order approach not only gives better approximation of the star trajectory for rapidly changing angular velocity case but also compensate for an increased standard deviation in the estimate noise.

From the plots, we can infer that algorithm based upon the second order approach gives more accurate results for higher angular rates but surprisingly by, at the price of an increased standard deviation in the estimate noise (by a factor of $\sqrt{13}/2$).

Concluding Remarks

The important issue in this paper is to address angular rate estimation for attitude determination in case of gyro failure which increases the domain of practical applicability of attitude estimation algorithms. As usual, the convergence of the Kalman Filter depends jointly upon: 1) the accuracy of the dynamical model and process noise representation, 2) the frequency and accuracy of the attitude measurements. For the attitude independent approach, approximations of the derivatives of the "image flow" trajectories of individual stars (imaged in high frame rate cameras) gives more accurate rate estimation results than the usual Kalman filter with typical modelling errors. The main advantage of the attitude independent algorithm is that now our spacecraft body rate estimates are free from any error in attitude estimates. The second order attitude independent algorithm gives better results than the first order approach for high magnitude angular rate case. However, the second-order approach requires two time steps ahead to estimate the angular velocity at the current time (i.e., knowledge of $\tilde{\mathbf{b}}_i$ at the $k + 2$ step is required to estimate ω at the k step). The first-order approach requires only one-time step ahead. For this reason the first-order approach should be used when possible.

REFERENCES

1. D. J. Mook, "Robust Attitude Determination without Rate Gyros," *Proceedings of the AAS/GSFC International Symposium on Space Flight Dynamics*, No. 93-299, NASA-Goddard Space Center, Greenbelt, MD, April 1993.

2. J. L. Crassidis and F. L. Markely, "A Minimum Model Error Approach For Attitude Estimation," *Journal of Guidance, Control and Dynamics*, Vol. 20, No. 6, Nov. -Dec. 1997, pp. 1241–1247.

3. H. L. Fisher, M. D. Shuster, and T. E. Strikwedra, "Attitude Determination For The Star Tracker Mission," *AAS/AIAA Astrodynamics Conference*, AAS, Stowe, VT, 7 Aug-10 Aug 1989.

4. J. M. Sidi, *Spacecraft Dynamics and Control*, Cambridge University Press, Cambridge, UK.

5. D. Mortari, J. L. Junkins, and M. Samaan, "Lost-In-Space Pyramid Algorithm for Robust Star Pattern Recognition," *Guidance and Control Conference*, No. 01-004, AAS, Breckenridge, Colorado, 31 Jan- 4Feb 2001.

6. P. Singla, T. D. Griffith, and J. L. Junkins, "Attitude Determination and On-Orbit Autonomous Calibration of Star Tracker For GIFTS Mission," *AAS/AIAA Space Flight Mechanics Meeting*, No. 02-101, AAS, San-Antonio, TX, 27 Jan- 30 Jan 2002.

7. P. Singla, "A New Attitude Determination Approach using Split Field of View Star Camera," *Masters Thesis report*, Aerospace Engineering, Texas A&M University, College Station, TX, USA.

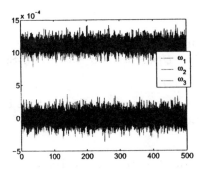

Figure 7 True and Estimated Spacecraft Angular Velocity (rad/sec) Obtained by 1st Order Approach for Test Case 1

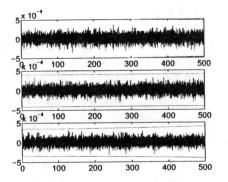

Figure 8 Estimated Spacecraft Angular Velocity Errors and 3σ Bound Obtained by 1st Order Approach for Test Case 1

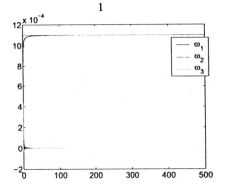

Figure 9 True and Estimated Spacecraft Angular Velocity (rad/sec) Obtained by 2nd Order Approach for Test Case 1

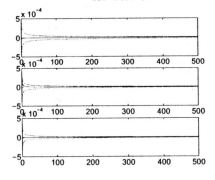

Figure 10 Estimated Spacecraft Angular Velocity Errors and 3σ Bound Obtained by 2nd Order Approach for Test Case 1

1314

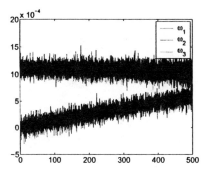

Figure 11 True and Estimated Spacecraft Angular Velocity (rad/sec) Obtained by 1st Order Approach for Test Case 2

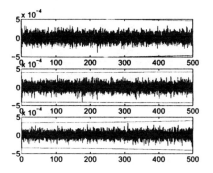

Figure 12 Estimated Spacecraft Angular Velocity Errors and 3σ Bound Obtained by 1st Order Approach for Test Case 2

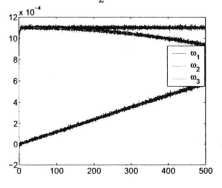

Figure 13 True and Estimated Spacecraft Angular Velocity (rad/sec) Obtained by 2nd Order Approach for Test Case 2

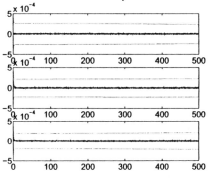

Figure 14 Estimated Spacecraft Angular Velocity Errors and 3σ Bound Obtained by 2nd Order Approach for Test Case 2

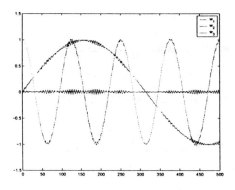

Figure 15 True and Estimated Spacecraft Angular Velocity (rad/sec) Obtained by 1st Order Approach for Test Case 3

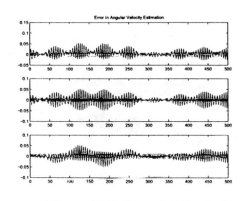

Figure 16 Estimated Spacecraft Angular Velocity Errors and 3σ Bound Obtained by 2nd Order Approach for Test Case 3

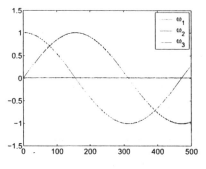

Figure 17 True and Estimated Spacecraft Angular Velocity (rad/sec) Obtained by 2nd Order Approach for Test Case 3

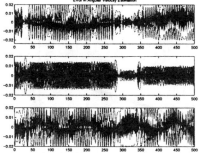

Figure 18 Estimated Spacecraft Angular Velocity Errors and 3σ Bound Obtained by 2nd Order Approach for Test Case 3

AUTONOMOUS ARTIFICIAL NEURAL NETWORK STAR TRACKER FOR SPACECRAFT ATTITUDE DETERMINATION

Aaron J. Trask[*] and Victoria L. Coverstone[†]

An artificial neural network based autonomous star tracker prototype for precise spacecraft attitude determination is developed. A new technique of star pattern encoding that removes the star magnitude dependency is presented. The convex hull technique is developed in which the stars in the field of view are treated as a set of points. The convex hull of these points is found and stored as line segments and interior angles moving clockwise from the shortest segment. This technique does not depend on star magnitudes and allows a varying number of stars to be identified and used in calculating the attitude quaternion. This technique combined with feed-forward neural network pattern identification creates a robust and fast technique for solving the "lost-in-space" problem. Night sky testing is used to validate a system consisting of a charged-coupled-device-based camera head unit and integrated control hardware and software. The artificial neural network star pattern match algorithm utilizes a sub catalog of the SKY2000 star catalog. The experimental results are real time comparisons of the star tracker observed motion with the rotational motion of the Earth. The time required to solve the "lost-in-space" problem for this star tracker prototype is on average 9.5 seconds.

INTRODUCTION

The work presented here consists of combining the concept of a star tracker with an Artificial Neural Network (ANN) to quickly determine the attitude of a spacecraft. The objectives of this research were to develop the hardware and software required to create an ANN based star tracker as well as to validate the star tracker using ground tests. Attitude determination is an integral part of spacecraft operations. Scientific imaging and other high precision pointing requirements have increased the need for precision attitude determination and the use of star trackers.

Star trackers use observations of stars compared with known stars to determine the spacecraft orientation in reference to the celestial sphere. The observation portion is performed by an optical sensor and control hardware. The actual star identification and attitude determination are left to the onboard computer or ground station. Only recently have autonomous star trackers been developed to output the pointing quaternion directly to the attitude control system without additional computation, thereby freeing up the onboard computer and communication system to perform other required tasks.

The star pattern identification and attitude determination algorithm developed uses an ANN that provided a fast and robust solution to the "lost-in-space" problem. The "lost-in-space" problem describes the problem of acquiring the initial attitude of a spacecraft. For spacecraft relying on star trackers for attitude knowledge, the probability of mission success is significantly increased by reliably solving this problem.

[*] Aerospace Engineer, Astrodynamics and Space Applications Office, Naval Research Laboratory, 4555 Overlook Avenue, S.W., Washington, DC 20375. Member AAS and AIAA.

[†] Associate Professor, Department of Aeronautical and Astronautical Engineering, University of Illinois at Urbana-Champaign, Urbana, Illinois 61801. Member AAS; Associate Fellow AIAA.

The field of Artificial Neural Networks (ANN) was pioneered by McCullock and Pitts in 1943 before the advent of computers. They showed that, in theory, the neuron models would compute any computable problem. In general, a problem is considered computable if an algorithm can be implemented that will give the correct output for any valid input. In this case, the algorithm would be an arrangement of trained neurons that would give the correct output for any valid input. The field of ANNs, as we know it today, emerged in the 1970s and research has recently increased. This has been motivated by studies of the human brain and its ability to compute an astounding amount of information in a short period. An ANN has the ability to perform complex recognition tasks in a small amount of time compared to conventional computer algorithms.[1] This trait is what makes ANNs ideal for star pattern identification.

Significant research has been performed in the field of star pattern identification for attitude determination. Multiple methods for solving the "lost-in-space" problem without *a priori* attitude knowledge have been developed. Bone[2] presents a comparison of various methods and modes of star identification and tracking algorithms as well as a discussion of on-board star catalog design.

The importance of robust algorithms that can handle erroneous events is important to star tracker reliability. Scholl[3] presented an algorithm capable of handling CCD images that might have stars not contained in the mission catalog as well as false stars (i.e. nearby debris and camera lens dust). This algorithm was based on magnitude-weighted star triangles.

Many different approaches have been taken in trying to solve the star identification problem. Udomkesmalee et al.[4] presented a stochastic star identification method. This method compares the statistics of the current field-of-view (FOV) with precomputed statistical patterns from a star catalog. The average probability of a correct match with this method was 0.976 using a 30° FOV with varying noise. Another approach was by Padgett and Kreutz-Delgado[5], Padgett et al.[6], and later Clouse and Padgett[7]. Their algorithm used a grid approach to star pattern encoding and matching. This was done by transforming the star field into a grid of "on" and "off" cells. This consisted of selecting a center star, a neighboring star, and aligning a grid. The cells containing a star were recorded and used to define patterns with neighboring stars, thereby forming a catalog of grid patterns. Chapel and Kiessig[8] developed a lightweight, low cost star camera that used the Stellar Compass™ software for pattern matching. It solved the "lost-in-space" problem and then switched to tracking mode for more efficient operation. Tracking mode selects stars in the FOV and compares them to the same stars in the prior FOV to determine change in star position and therefore change in spacecraft attitude. Liebe[9] presented an algorithm that used a star triple pattern. The parameters for the pattern were the angular distance to the first neighbor star, the angular distance to the second neighbor star, and the spherical angle between the two neighbor stars. Quine and Durrant-Whyte[10] then presented a binary tree search algorithm to perform star pattern identification. This algorithm used 15 binary questions to identify the star pattern and it performed with a 98.7% accuracy. van der Heide et al.[11] presented an algorithm combining the Liebe[9] and Quine-Durrant-Whyte[10] algorithms while maintaining a low memory requirement and a fast search strategy. This allowed a better than 99.99% reliability with one arcsec accuracy during software simulation testing. Pottech[12] presented an algorithm that used angular distance between stars, the sum of their magnitudes, and the absolute value of the difference of their magnitudes. This defines a three dimensional space, which the algorithm uses to determine close pairs of detected and catalog stars. Pottech[12] found this algorithm to perform at 92% reliability. Mortari[13], Mortari and Angelucci[14], and Mortari and Junkins[15] developed the Spherical-Polygon Search technique for wide- and multiple-FOV star trackers, which uses the fact that any vector representation of a star can always be expressed as a linear combination of two star vectors together with their cross product. Samann et al.[16] presented two algorithms for solving the recursive star identification problem. The first was the Spherical-Polygon Approach adapted from the previously developed algorithm by Mortari[13]. The second was the Star Neighborhood Approach, which used software pointers to a cone of neighboring stars in the catalog to reduce the search space.

Recently, researchers have started to apply ANNs to the "lost-in-space" problem with great success. Domeika et al.[17] applied a Hopfield network to a five star FOV simulation with only 50 stars in the catalog. Bardwell[18] used a Kohonen network for a 20° FOV and found 100% identification accuracy with no added noise to the simulation. With 5% random noise added, a 95% identification accuracy was observed. Domeika et al.[19] later applied Adaptive Resonance Theory 2 (ART2) to the problem with success. Lindsey et al.[20] applied a radial basis function (RBF) using star triangles as the pattern encoding method. Hong and Dickerson[21] used fuzzy neural logic networks (NLN) to find pattern matches with star triples generated from the Submillimeter Wave Astronomy Satellite (SWAS) run catalog. They achieved an accuracy of greater than 99% for a pattern match without the need of *a priori* attitude knowledge. Dickerson et al.[22] compared a RBF network with the fuzzy NLN on the same problem. They showed that the fuzzy NLN correctly identified more star patterns than the RBF network. They also found that the RBF needed an average of 10 hours to train whereas the fuzzy NLN only needed 0.5 hours on average.

The star tracker algorithm described within this paper consists of a multi-step method of star feature extraction, star pattern recognition, and pointing attitude determination through quaternion estimation. Each of these steps will be further discussed in the following sections.

STAR FEATURE EXTRACTION

The first step in extracting star features, after a star field image is acquired, is to calibrate the image by subtracting the dark frame. The dark frame is a measure of dark current, which are the thermal electrons that become trapped within the electron wells of the CCD. This increases the number of counts per pixel during both the exposure time and the read out time. A dark frame is acquired by taking an exposure with the shutter closed at the same CCD temperature as when the star field image was taken. To reduce the noise in the dark current calibration, multiple dark frames are averaged together and then subtracted from the star image. Figure 1 shows the star image after calibration.

Figure 1 Calibrated Star Image

The second step in feature extraction is to classify possible stars. This is done by performing a grid search for local maximums greater than the estimated peak intensity of the limiting instrument magnitude. A local maximum is classified as a star if it does not appear to be a radiation event (cosmic ray, terrestrial radiation, etc.) or dust particles. Radiation events create local pixel regions that appear as streaks or spots. The streaks are easy to identify and ignore while the point-like spots are slightly less obvious but can be distinguished from stars by their radial profile, which is point-like rather than a Gaussian distribution. Dust particles appear as out-of-focus rings and are easy to identify and ignore. Figure 2 shows what a classified star looks like after dark current calibration.

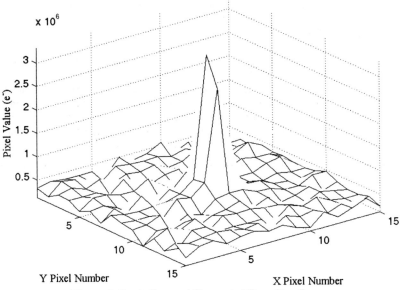

Figure 2 Dark Current Corrected Raw Star Image

After the stars are classified in the calibrated star image, the centroids and instrumental magnitude of each star are calculated. Instrumental magnitude is calculated using aperture photometry, which is shown in Eq. (1).[23]

$$M_{inst} = A - 2.5\log_{10}\left[\frac{\left(\sum_{i=1}^{N} S_i\right) - NB}{t}\right] \tag{1}$$

In Eq. (1), A is an arbitrary constant, N is the number of pixels in the aperture centered on the star, S_i is the signal from the ith pixel, B is the average sky background signal near the star, and t is the length of the exposure. If the instrumental magnitude is above the chosen threshold, the centroid of the star is calculated. This is done by using a center-of-mass centroid calculation method. The method mimics center-of-mass calculations in mechanics. Instead of density and volume of each object in the mechanics calculation, the pixel value and pixel area for each pixel containing the star is used. In order to decrease the error in this calculation, a small window of 3X3 pixels, centered on the peak pixel of the chosen star, is selected. The window size comes from the sensitivity of the centroid calculation to the edge of the star point spread function. The point-spread function (PSF) is a 2D Gaussian distribution and therefore has small values at the edges that can become overwhelmed by noise causing an offset of the centroid. A larger or smaller window increases the error in this combination of lens and CCD due to the noise added by a larger window or the loss of information caused by a smaller window. For clarity, a 15X15 pixel window is shown throughout the procedure example.

Another way of decreasing error is to sub-sample the selected star to create finer grid spacing using interpolation. This increases the number of pixels while decreasing their size, which allows a better estimate of the centroid. This is done using bilinear interpolation which is shown in Figure 3 and Eq. (2) for $x_1 \leq x \leq x_2$ and $y_1 \leq y \leq y_2$.[24] In Figure 3, t and u are the local interpolation variables and z is the value being interpolated. The result of sub-sampling by a factor of 10 of the star in Figure 2 is shown in Figure 4.

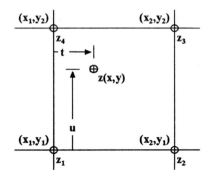

Figure 3 Bilinear Interpolation Notation

$$z(x,y) = (1-t)(1-u)z_1 + t(1-u)z_2 + tuz_3 + (1-t)uz_4 \tag{2}$$

where

$$t = \frac{(x-x_1)}{(x_2-x_1)} \tag{3}$$

and

$$u = \frac{(y-y_1)}{(y_2-y_1)} \tag{4}$$

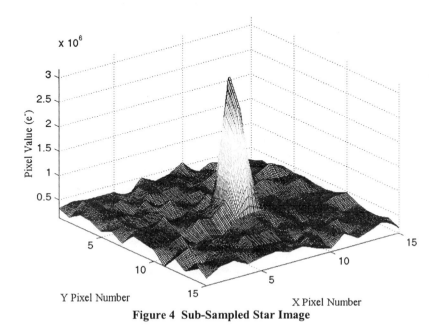

Figure 4 Sub-Sampled Star Image

To further decrease the error, the star can be smoothed to remove high frequency noise and discontinuities between pixels. This is done using a boxcar filter as shown in Eq. (5).

$$R_{ij} = \begin{cases} \dfrac{1}{w^2} \displaystyle\sum_{k=0}^{w-1}\sum_{l=0}^{w-1} A_{(i+k-w/2)(j+l-w/2)} & w/2 \le i \le N-w, w/2 \le j \le N-w \\ A_{ij} & \text{otherwise} \end{cases} \tag{5}$$

In Eq. (5), A is the square array of dimension N to be filtered, w is the width of the boxcar, and R is the resulting array with the same dimension as A. Figure 5 is the result of applying the boxcar filter to the subsampled star in Figure 4.

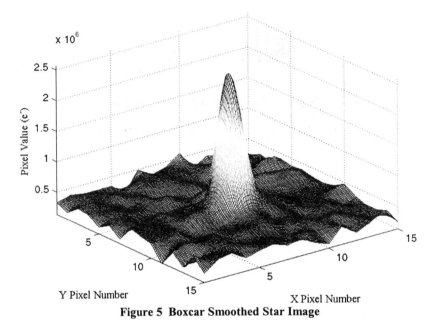

Figure 5 Boxcar Smoothed Star Image

To further reduce the noise in the centroid calculation, the background sky is subtracted. The background is determined by averaging the pixel values in an approximate annulus centered on the 3X3 window. The result of subtracting the background from Figure 5 is shown in Figure 6.

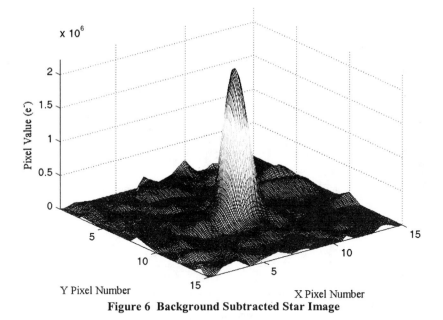

$\times 10^6$

Pixel Value (e⁻)

Y Pixel Number

X Pixel Number

Figure 6 Background Subtracted Star Image

After calibrating, sub-sampling, smoothing, and subtracting the background, the centroid is calculated using the center-of-mass method shown in Eqs. (6) and (7). In these equations, m is the sub-sampled pixel value, x,y are the sub-sampled pixel numbers, and x_{cm}, y_{cm} are the center of mass coordinates (centroid coordinates). Figure 7 shows the contours of the star with the initial centroid guess and the new calculated centroid.

$$x_{cm} = \frac{\sum_i \sum_j x_{ij} m_{ij}}{\sum_i \sum_j m_{ij}} \tag{6}$$

$$y_{cm} = \frac{\sum_i \sum_j y_{ij} m_{ij}}{\sum_i \sum_j m_{ij}} \tag{7}$$

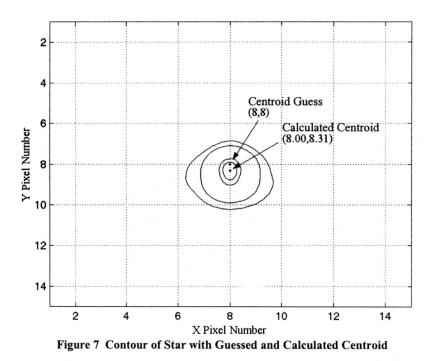

Figure 7 Contour of Star with Guessed and Calculated Centroid

Once the centroids have been calculated in pixel coordinates, they need to be converted to angular coordinates. This is ideally the inverse gnomonic projection, but due to various distortions, a mapping between CCD coordinates and angular coordinates on the celestial sphere is a better estimate (Eqs. (8) and (9)).[25,26]

$$\alpha_i = A * \arctan\left(\frac{x_i'}{f}\right) + B_1 \tag{8}$$

$$\beta_i = A * \arctan\left(\frac{y_i'}{f}\right) + B_2 \tag{9}$$

In Eqs. (8) and (9), α_i and β_i are the orthogonal separation angles from the star tracker bore axis (positive α about the y-axis and positive β about negative x-axis), x_i' and y_i' are the distances to the ith star in mm from the center of the CCD (bore axis), f is the focal length, A is a scale factor, and B_1 and B_2 are biases. A, B_1, and B_2 are found, prior to operation of the star tracker, from calibrations of measured relative star positions to catalog relative star positions.

Once the centroids are converted to angular coordinates, the encoding process for input to the ANN can be performed. The stars are considered a 2D point distribution in the FOV. The convex hull of these points is found and stored as line segments and angles. This method is independent of star magnitude, which alleviates the problem of large errors in magnitude calculations due to variations and degradation of the CCD.

The convex hull of a set of points is a convex polygon. In other words, if a set of points, P, has a convex hull, CH(P), where an edge of CH(P) is $\overline{pq}$, then all points in P must lie to the right of $\overline{pq}$ (Figure 8).[27]

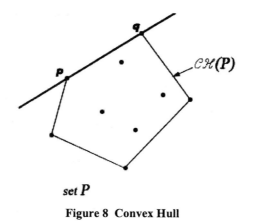

set P

Figure 8 Convex Hull

After the convex hull is found for the set of stars, any vertex with an angle, γ, that is within a half of a degree of 180° is deleted from the convex hull to reduce the error in the pattern encoding due to measured star position. The pattern is then stored as a vector of line segments and angles in order of the shortest segment first and then moving around the convex hull in a clockwise direction (Eq. (10)). Figure 9 shows the result of extracting the star features from the star image in Figure 1, computing the convex hull, and generating the pattern vector.

$$pattern = \{L_1 \quad \gamma_1 \quad L_2 \quad \gamma_2 \quad \cdots \quad L_n \quad \gamma_n\} \tag{10}$$

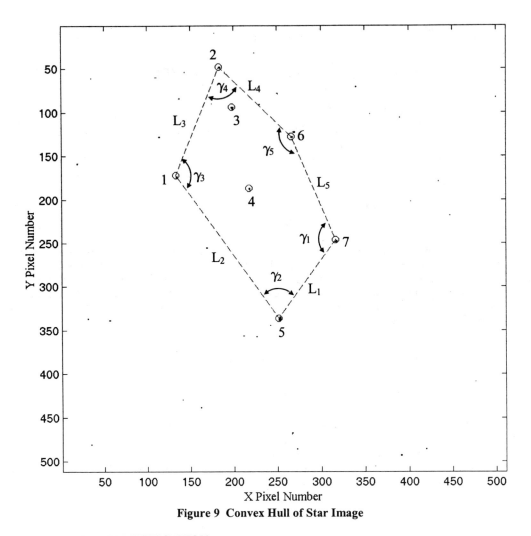

Figure 9 Convex Hull of Star Image

STAR PATTERN IDENTIFICATION

The ANN chosen for this research is the feed-forward neural network (FFNN). This ANN consists of multiple layers of neurons using a hyperbolic activation function. The neurons are fully connected between layers to form the FFNN.

The number of inputs is equal to the number of segment lengths and angles of the pattern. The inputs are real values and the outputs are the binary index of the recognized pattern. The number of hidden layers is optimized during training. The neural network is trained on the patterns generated in the run catalog. Once the pattern is fed in to the FFNN, the binary output is used to look up the pattern and corresponding catalog stars.

The ANN designed for the pattern recognition portion of the attitude determination algorithm, is a feed forward neural network (FFNN) also known as a multilayer perceptron (MLP). It consists of 28 inputs in the input layer, 30 neurons in the hidden layer, and 12 neurons in the output layer. The training set is all of the 3,828 star patterns in the run catalog generated from all stars with a limiting instrument magnitude of 5.0 in the SKY2000 Master Star Catalog. In order to improve the training set, mean removal was performed on each parameter after they were normalized. This creates a feature vector for the neural network that is less likely to cause saturation of the weights during training and allows each neuron to learn at approximately the same speed. Two methods of training the FFNN were attempted: the error back-propagation algorithm and the real-coded genetic algorithm.

The second training method attempted was the real-coded genetic algorithm (RCGA). The RCGA is based on concepts from biological evolutionary theory and is a stochastic optimization technique. This technique encodes multiple sets of parameter values as chromosomes (individuals), which are contained in the population. Various operators are applied to the population to create new generations. Each generation may contain previous generation chromosomes as well as new ones. These chromosomes are given a fitness, which is related to the value of the function being optimized.

The algorithm used here differs from other GA's in that it uses real valued alphabet to create the chromosomes instead of the typical discrete alphabet such as a binary or hexadecimal alphabet. The real coding was chosen because of the large number of parameters (weights) in the training of an ANN. If a binary alphabet was chosen, the chromosomes would become unmanageable and the GA would not converge. In the discrete coded genetic algorithm, the primary operator for taking steps in the problem space is the crossover operator. When parameters are coded in a discrete alphabet and concatenated to form a chromosome, crossover is performed anywhere along the chromosome. This allows pieces of a parameter's encoding to be swapped with another, effectively changing the parameter value, and introducing new information to the population. However, in the case of a real-coded chromosome, in which the chromosome is a vector of real values, crossover only swaps different parameter values instead of possibly introducing new information. Therefore, the mutation operator or a modified crossover operator becomes the primary operator in the RCGA. For more information of genetic algorithms, see Goldberg's[28] book titled, "Genetic Algorithms in Search, Optimization, and Machine Learning" or for genetic algorithm application to ANN training, see van Rooij et al.'s[29] book titled, "Neural Network Training using Genetic Algorithms."

The initial population was a random set of weights. A population size of 100 was used and each individual had a fitness corresponding to the mean square error of its output for all of the patterns in the run catalog. The mean square error of the individual is the mean of the square distance between the desired network output and the actual network output. The convergence criterion for the RCGA was based on the variance of the population. This training process was repeated with different values for the number of hidden elements in the FFNN. The FFNN's were trained and evaluated to minimize the number of hidden elements while maintaining 100% recognition. This resulted in the optimal number of hidden neurons for this problem of 30. The RCGA took 2,913 generations to train the FFNN to 100% accuracy in the recognition of the patterns in the run catalog. The parameters used in the RCGA are shown in Table 1. Due to the number of patterns, size of the network, and the number of individuals in each generation, the RCGA took approximately four hours to train the FFNN on an Intel Pentium III 450MHz computer. The results of the FFNN training using the RCGA are shown in Figure 10. The RCGA trained the FFNN to 100% recognition of the patterns in the run catalog. The output of the network is a set of real numbers. These are converted to a binary representation by any number less than zero becoming a zero and any number greater than zero becoming a one. Since the mean square error is a measure in the real domain, 100% accuracy in the binary domain doesn't necessarily represent a zero mean square error. This is the reason the results shown in Figure 10 do not converge to zero. The confidence of the result is calculated as the sum of the distance between real number result and the origin. When this number is high, the corresponding output has a high confidence. When the corresponding output is close to zero for each output neuron, the strength of the output neurons is weak and the confidence in the result is low.

Table 1

RCGA PARAMETERS

Parameter	Parameter Value
Population Size	100
Maximum Number of Generations	6000
Crossover Probability	0.6
Mutation Probability	0.005
Mutation Range (real valued step size)	±0.1

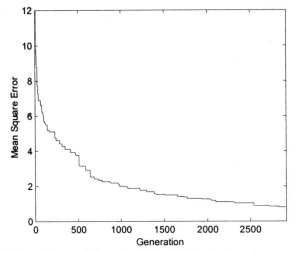

Figure 10 Mean Square Error of the Best Individual During RCGA Training

ATTITUDE DETERMINATION

The final step in the attitude determination algorithm is to calculate the pointing quaternion. After the FFNN returns the binary index number of the pattern, the coordinates of all the detected stars will be known on the celestial sphere. From these observations, the quaternions of each star are known in the star tracker frame and in the inertial frame. Therefore, the problem is to find a rotation matrix between frames that minimizes the cost function in Eq. (11). Wahba[30] first posed this problem and it is the basis for most attitude determination algorithms. Once this is found, the pointing quaternion aligned with the bore sight axis of the camera head unit can be calculated.

$$J = \sum_{j=1}^{n} \left\| \mathbf{v}_j^\star - \mathbf{M}\mathbf{v}_j \right\|^2 \tag{11}$$

In Eq. (11), $\mathbf{v}_j^\star$ is the unit vector of the jth star in the star tracker frame, $\mathbf{v}_j$ is the unit vector of the jth star in the inertial frame, and $\mathbf{M}$ is the least squares estimate of the rotation matrix between the inertial frame and the star tracker frame. In order to save computation time, QUEST[31], a proven algorithm for estimating $\mathbf{M}$, is used.

HARDWARE DEVELOPMENT

The camera head unit (CHU) consists of the AP7P CCD sensor unit with a parallel port interface from Apogee Instruments Inc., a Nikkor 50mm f1.2 lens from Nikon, and an infrared (IR) filter. The sensor is a SITe SI-502AB CCD, which has a 512X512 pixel array and a 24 μm pixel size. This was chosen for its high quantum efficiency and low read out noise. The quantum efficiency of a CCD describes its response to different wavelengths of light. The characteristics of the sensor are shown in Table 2. The AP7P is equipped with a Vincent 25mm blade shutter and a thermoelectric cooler. The read out time for the entire array is eight seconds, which is slower than desired for a flight model. The read out time limits the update rate of the star tracker attitude.

Table 2

SITE SI-502AB SENSOR CHARACTERISTICS

Parameter	Value
Format	512 × 512 pixels
Pixel Size	24 μm × 24 μm
Digital Resolution	16-bit at 35 kHz
Gain	5 e$^-$/ADU
Read Noise	7-11 e$^-$
Well	300k e$^-$
Dark Count	50 pA/cm^2 at 20°C
Dynamic Range	86 dB

The prototype CHU was tested during night sky observations. To minimize the effects of the atmosphere during night sky testing, a 092(89B) filter was required to only allow passage of the infrared (IR) spectral band.

The Nikkor 50mm f1.2 lens consists of seven elements in six groups. It has a 46° imaging angle and is 2.7in wide by 2.3in long. This lens was chosen for its high light gathering capability and low field distortion. This lens combined with the CCD gives the CHU a 14° field of view (FOV).

The resulting CHU has excellent properties for acquiring precision star frames. The only limitation is the read out time of the CCD. Future prototypes would benefit from selecting a CCD with a short read out time even though this increases the read out noise.

The prototype controller consists of a single board computer (SBC), the mediaEngine, by Bright Star Engineering. It is equipped with an Intel 200 MHz StrongARM processor and 256 MB of flash memory. Communication with the SBC is done through the ethernet port on the board. A PCMCIA socket parallel port adaptor was used to interface the single board computer with the CCD.

The operating system is embedded Linux, which allowed the use of the Random Factory's Linux drivers for the Apogee Instruments series of CCD cameras. The operating system is boot loaded and the attitude determination software and star catalog are stored in the flash memory. Commands are sent through a remote login to the SBC. Future prototypes of the controller would utilize RS-485 or RS-232 communication standards instead of the ethernet access to satisfy standard communication protocols onboard spacecraft. The completed CHU and CHU controller, which together make up the star tracker prototype, are shown in Figure 11.

Figure 11 Star Tracker Prototype

FIELD TRIALS

The field site at the Urbana Atmospheric Observatory (UAO) is located in Urbana, Illinois on High Cross Road at 40°10'1.2"N 88°10'1.2"W. The site consists of a climate controlled building with skyward looking windows. This allows zenith alignment and viewing for the entire field-of-view of the star tracker prototype.

During the attitude determination algorithm and control software debugging period, the star tracker prototype was run offline. This was achieved by acquiring images with the camera head unit and storing them in FITS format. These images were then used as inputs to the attitude determination algorithm to ease the debugging process.

Real-time night sky testing was performed at the UAO in order to test and validate the performance of the star tracker prototype. Occasional imaging errors occurred such as radiation events, as discussed previously, and processor load streaks. Abnormal processor load in the single board computer occasionally (once out of every 100 frames) caused a streak during read out of the CCD. These events were ignored by the software in most cases, since they do not appear the same as stars except when a streak crossed a star. In this case, more error was introduced in the attitude determination process due to the skewed estimate of the star centroid.

The results of one trial are shown in Figure 12. The right ascension, declination, and roll about the bore axis of the star tracker are shown. The attitude data from the star tracker had an output of once every 9.5 seconds on average. This corresponds to an attitude update frequency of 0.1 Hz. The slow update is due to the long read out time for the off-the-shelf CCD. The attitude determination software runs in 3-5 seconds depending upon the number of stars above the limiting magnitude in the field-of-view.

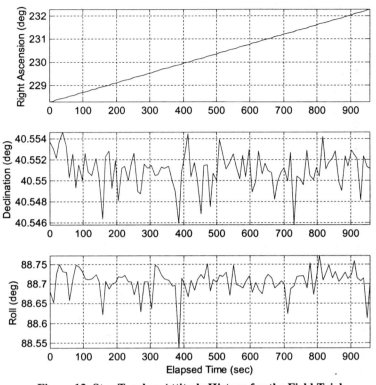

Figure 12 Star Tracker Attitude History for the Field Trial

In order to determine the error of the attitude measurement, a reference attitude history is needed. Since the star tracker was not mounted on an observatory telescope, the attitude history was estimated based on the Greenwich Hour Angle formula:[32]

$$GHA=280.46061837+360.98564736629*jd+0.000388*\left(\frac{jd}{36525}\right)^2 \qquad (12)$$

where jd is the Julian Date. The first measured attitude vector was used as the seed for the reference attitude calculation. This vector was transformed from the Earth-Centered-Inertial (ECI) frame to the Earth-Centered-Earth-Fixed (ECEF) frame. This was then used as the pointing vector to propagate forward due to the earth's rotation.

This estimate is only as good as the known time. The time stamps on the attitude data are only known to the second (an artifact of the time function used in the star tracker prototype). This creates an error in the right ascension estimate portion of the reference attitude of ±15 arcsec. Due to the fixed-mount test fixture, the roll and declination estimates are constant values and are not dependent on the time. The roll and declination estimates are taken to be the mean of each trial. The error of the measured attitude is shown in Figure 13. The mean of the absolute value of the error in right ascension is ±17 arcsec before considering the error introduced by the estimate. If the error introduced by the estimate is subtracted from the absolute value of the error and a new mean is calculated, then the mean of the absolute value of the error in right ascension becomes ±8 arcsec. This new error estimate represents the error of the measurement with respect to the maximum error in the reference attitude. In order to determine the star tracker performance with more certainty, night sky testing with a known pointing source is needed. The mean of the absolute value of the error in declination is ±5 arcsec and in roll is ±80 arcsec with respect to the constant mean values. Roll is the most error prone due to the sensitivity of calculating it from vector measurements.

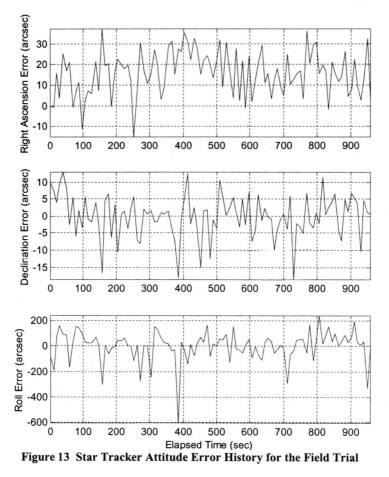

Figure 13 Star Tracker Attitude Error History for the Field Trial

An experiment was performed to see the effects of defocusing the stars on the CCD. This increases the number of pixels that a star covers and flattens the point spread function. The attitude estimate was slightly improved, but the increased chance of identifying non-celestial objects as stars outweighed the benefit. For the defocused trials, the mean of the absolute value of the error in right ascension is ±22 arcsec before considering the error introduced by the attitude estimate. If the error introduced by the estimate is considered, the mean of the absolute value of the error in right ascension becomes ±9 arcsec. The mean of the absolute value of the error in declination is ±4 arcsec and in roll is ±45 arcsec. The roll measurement showed the most improvement in error.

The results of the star tracker prototype during night sky testing are excellent. The "lost-in-space" problem is solved in less than twelve seconds for the worst case during night sky testing. In addition, the error in right ascension is approximately ±8 arcsec, the error in declination is approximately ±5 arcsec, and the error in roll is approximately ±80 arcsec. These errors are little worse than the current off-the-shelf autonomous star tracker by Ball Aerospace, the CT-633, which are 10 arcsec in right ascension and declination and 40 arcsec in roll.[33] However, the CT-633 solves the "lost-in-space" problem in less than 60 seconds 90% of the time and the star tracker prototype developed here does the same in 9.5 seconds on average.

CONCLUSIONS

The star tracker algorithm developed consists of a multi-step method for image calibration, star feature extraction, star pattern recognition, and pointing quaternion estimation. This research developed a unique method of encoding the star pattern with convex hulls. This technique combined with feed-forward neural network pattern identification creates a robust and fast technique for solving the "lost-in-space" problem. This allows all stars in the field of view (FOV), brighter than the limiting instrument magnitude, to be included in the attitude estimation. It also eliminates the dependency of the star magnitude during pattern recognition. In addition to the star tracker algorithm, prototype hardware was developed. This consisted of a camera head unit and controller designed for field trials of the star tracker algorithm.

Night sky testing was performed to test and evaluate the star tracker prototype. The performance of the star tracker is depicted in Table 3. The ability to solve the "lost-in-space" problem on average for this star tracker prototype is 9.5 seconds. This is a great improvement over the 60 seconds needed by the current off-the-shelf autonomous star tracker by Ball Aerospace, the CT-633.[33] Initial acquisition after launch as well as recovery from a loss of attitude knowledge during the mission would occur in much less time than the current off-the-shelf autonomous star tracker.

Table 3

STAR TRACKER CHARACTERISTICS AND PERFORMANCE

Parameter	Characteristic or Performance
Sensor	SITe 512X512 Scientific-Grade CCD
Format	511 × 511 pixels
Pixel Size	24 μm × 24 μm
Read Out	8 sec
f-Number	1.2
Focal Length	50 mm
Field of View	14° X 14°
Exposure	10 msec
Stellar Sensitivity	better than 5.0 instrument magnitude
Right Ascension Accuracy	±8 arcsec
Declination Accuracy	±5 arcsec
Roll Axis Accuracy	±80 arcsec
Attitude Update Frequency	0.1 Hz

FUTURE RESEARCH

The star tracker prototype has excellent performance, but suggestions for future research and development have been found during the design process. Obviously, the first suggestion is to trade exposure time for read out time. Increasing the clock rate during read out increases the amplifier noise but decreases the total read out time.[34] In order to overcome the increased noise level, the exposure time of the CCD must be increased. The read out time is the current limit on the attitude update rate.

In addition to the read out time, the issue of space qualifying the current mechanical shutter must be addressed. Mechanical shutters are not reliable in space. One possible solution to this is to use a CCD array with a masked region. The bare region is quickly clocked into the masked region and then read out slowly while the bare region gathers starlight. This creates a mechanical shutter effect without the need for moving parts.

The next suggestion would be to optimize the optics in order to provide a low f-number, large aperture, and an optimal FOV for the chosen sensor. In addition to these hardware changes, a custom controller should be developed with radiation tolerance in mind.

For the software aspect of the star tracker, a better model of the camera head unit is needed as well as better spectral information on the stars in the catalog. These are needed to improve the predicted instrument magnitude and develop an improved run catalog. In addition to the catalog improvements, the ANN could be redesigned to a group of ANN, each with its own portion of the run catalog patterns. This collective group could be trained faster as well as provide more insight into the confidence of each identified pattern.

Another suggestion would be to develop a faster centroiding and magnitude determination algorithm. Currently, other than the read out time, this takes the most time. One suggestion might be to develop an ANN that can identify stars in a FOV and return their instrument magnitude and centroid. An investigation into this would determine the feasibility and accuracy of such a concept.

The final suggestion would be to add a star-tracking mode. Once the "lost-in-space" problem is solved, the tracking mode would speed up the attitude update rate by looking for each star in a probability window. This would decrease the time spent identifying stars in the FOV as well as the time spent correlating those stars with the star catalog.

ACKNOLWEDGEMENT

This material is based on work supported by the University of Illinois at Urbana-Champaign.

REFERENCES

1. S. Haykin, *Neural Networks: A Comprehensive Foundation*, Prentice-Hall, Inc., Upper Saddle River, New Jersey, 1999.

2. J.W. Bone, "On-orbit star processing using multi-star star trackers," *Acquisition, Tracking, and Pointing VIII*, SPIE Proceedings 2221, 1994, pp. 6-14.

3. M.S. Scholl, "Star-field identification for autonomous attitude determination," *Journal of Guidance, Control, and Dynamics*, Vol. 18, No. 1, Jan.-Feb. 1995, p. 61-65.

4. S. Udomkesmalee, J.W. Alexander, and A.F. Tolivar, "Stochastic star identification," *Journal of Guidance, Control, and Dynamics*, Vol. 17, No. 6, Nov.-Dec. 1994, p. 1283-1286.

5. C. Padgett and K. Kreutz-Delgado, "A grid algorithm for autonomous star identification," *IEEE Transactions on Aerospace and Electronic Systems*, Vol. 33, No. 1, Jan. 1997, p. 202-213.

6. C. Padgett, K. Kreutz-Delgado, and S. Udomkesmalee, "Evaluation of star identification techniques," *Journal of Guidance, Control, and Dynamics*, Vol. 20, No. 2, Mar.-Apr. 1997, p. 259-267.

7. D.S. Clouse and C.W. Padgett, "Small field-of-view star identification using Bayesian decision theory," *IEEE Transactions on Aerospace and Electronic Systems*, Vol. 36, No. 3, July 2000, p. 773-783.

8. J.D. Chapel and R. Kiessig, "A Lightweight, Low-Cost Star Camera Designed for Interplanetary Missions," *Advances in the Astronautical Sciences*, Vol. 98, 1998, p. 345-355.

9. C.C. Liebe, "Star trackers for attitude determination," *IEEE Aerospace and Electronics Systems Magazine*, Vol. 10, No. 6, June 1995, pp. 10-16.

10. B.M. Quine and H.F. Durrant-Whyte, "A Fast Autonomous Star-acquisition Algorithm for Spacecraft," *Control Engineering Practice*, Vol. 4, No. 12, Dec. 1996, pp. 1735-1740.

11. E.J. van der Heide, M. Kruijff, S. Douma, S. Oude-Lansink, and C. de Boom, "Development and Validation of a Fast and Reliable Star Sensor Algorithm with Reduced Database," *49th IAF International Astronautical Congress*, Melbourne, Sept. 28 – Oct. 2, 1998.

12. S. Potteck, "A star recognition algorithm based on star groups and associated means," *Advances in the Astronautical Sciences*, Vol. 101, 1999, p. 87-94.

13. D. Mortari, "SP-Search: A New Algorithm for Star Pattern Recognition," *Advances in the Astronautical Sciences*, Vol. 102, pt. 2, 1999, p. 1165-1174.

14. D. Mortari and M. Angelucci, "Star Pattern Recognition and Mirror Assembly Misalignment for DIGISTAR II and III Multiple FOVs Star Sensors," *Advances in the Astronautical Sciences*. Vol. 102, pt. 2, 1999, p. 1175-1184.

15. D. Mortari and J.L. Junkins, "SP-Search Star Pattern Recognition for Multiple Fields of View Star Trackers," *Advances in the Astronautical Sciences*, Vol. 103, pt. 3, 2000, p. 2127-2143.

16. M.A. Samaan, D. Mortari, and J.L. Junkins, "Recursive Mode Star Identification Algorithms," *11th AAS/AIAA Space Flight Mechanics Meeting*, Santa Barbara, CA, February 11-14, 2001.

17. M. Domeika, E. Page, and G. Tagliarini, "Neural network approach to star field recognition," *Applications and Science of Artificial Neural Networks Proceedings*, SPIE Proceedings 2492, 1995, pp. 1007-1016.

18. G. Bardwell, "On-board artificial neural network multi-star identification system for 3-axis attitude determination," *Acta Astronautica*, Volume 35, 1995, Pages 753-761.

19. M.J. Domeika, C.W. Roberson, E.W. Page, and G.A. Tagliarini, "Adaptive resonance theory 2 neural network approach to star field recognition," *Applications and science of artificial neural networks II*, SPIE Proceedings 2760, 1996, p. 589-596.

20. C.S. Lindsey, T. Lindblad, and A. Eide, "A Method for Star Identification using Neural Networks," *Applications and science of artificial neural networks III*, SPIE Proceedings 3077, 1997, p. 471-478.

21. J. Hong and J.A. Dickerson, "Neural-Network-Based Autonomous Star Identification Algorithm," *Journal of Guidance, Control, and Dynamics*, Vol. 23, No. 4, July-August 2000, p. 728-736.

22. J.A. Dickerson, J. Hong, Z. Cox, and D. Bailey, "A comparison of radial basis function networks and fuzzy neural logic networks for autonomous star recognition," *Proceedings of the International Joint Conference on Neural Networks*, Vol. 5, 1999, pp. 3204–3207.

23. K.J. Mighell, "Algorithms for CCD stellar photometry," *Astronomical Data Analysis and Systems VIII, ASP Conference Series*, Vol. 172, 1999, pp. 317-328.

24. W.H. Press, S.A. Teukolsky, W.T. Vetterling, and B.P. Flannery, *Numerical Recipes in C: The Art of Scientific Computing*, Cambridge University Press, Cambridge, United Kingdom, 1999.

25. L.G. Taff, *Computational Spherical Astronomy*, Krieger Publishing Company, Malabar, Florida, 1991.

26. R. Laher, J. Catanzarite, T. Conrow, T. Correll, R. Chen, D. Everett, D. Shupe, C. Lonsdale, P. Hacking, N. Gautier, and K. Lebsock, "Attitude control system and star tracker performance of the Wide-Field Infrared Explorer spacecraft," Advances in the Astronautical Sciences, Vol. 105, pt. 2, 2000, pp. 723-752.

27. M. de Berg, M. van Kreveld, M. Overmars, and O. Schwarzkopf, *Computational Geometry: Algorithms and Applications*, Springer-Verlag, Berlin, Germany, 2000.

28. D.E. Goldberg, *Genetic Algorithms in Search, Optimization, and Machine Learning*, Addison-Wesley Publishing Company, Inc., Reading, Massachusetts, 1989.

29. A.J.F. van Rooij, L.C. Jain, and R.P. Johnson, *Neural Network Training using Genetic Algorithms*, World Scientific Publishing Co. Pte. Ltd., Singapore, 1996.

30. G. Wahba, "Problem 65-1: A Least Squares Estimate of Satellite Attitude," *SIAM Review*, Vol. 7, No. 3. (Jul., 1965), p. 409.

31. M.D. Shuster and S.D. Oh, "Three-axis attitude determination from vector observations," *Journal of Guidance and Control*, Vol. 4, No. 1, 1981, pp. 70-77.

32. *Astronomical Almanac 2001*, Nautical Almanac Office, United States Naval Observatory in cooperation with Her Majesty's Nautical Almanac Office, 2000.

33. Ball Aerospace, Web Site, http://www.ball.com.

34. J.R. Janesick, *Scientific Charge-Coupled Devices*, SPIE Press, Bellingham, Washington, 2001.

AN ANALYSIS OF
THE QUATERNION ATTITUDE DETERMINATION FILTER

Mark E. Pittelkau[*]

The full-quaternion attitude determination filter is analyzed to address questions of co-variance singularity and quaternion normalization. It is shown how nonsingularity of the covariance in the Extended Kalman Filter depends on the initial covariance, the process noise matrix, and implementation details of the filter. The covariance of a normalized quaternion estimate and the various means to achieve normalization are examined. The effect of a quaternion measurement update on the covariance and on the norm of the estimated quaternion is analyzed. It is also shown that the multiplicative and additive quaternion updates are equivalent. These are distinguished from an update called "rotational", which was proven elsewhere to be the constrained maximum-likelihood optimal update. It is demonstrated that the reduced-order body-referenced attitude determination filter is embedded in the full-quaternion filter.

Introduction

Attitude determination filters that include the full attitude quaternion in their state vector and that model the quaternion error covariance potentially have a singular covariance matrix when subject to the quaternion norm constraint or otherwise may exhibit problems related to the quaternion norm and its covariance. Singularity of the quaternion covariance matrix was mentioned in the landmark paper on attitude determination [1] and is often quoted as fact [2], although some maintain that the covariance is nonsingular [3]–[5]. Given that there is disagreement in the community, it would appear that the nature of the covariance matrix has not been fully analyzed in the literature and is not well understood. The following analysis should give greater insight into the nature of the covariance matrix and should clarify and resolve issues regarding singularity and quaternion normalization. This analysis is in the context of linearization in the Extended Kalman Filter (EKF): the true quaternion covariance is not necessarily singular (although it can be ill conditioned—see Appendix A); however, the EKF doesn't know anything about that. *The EKF is simply a first-order approximation* to the true probability density, and the computed EKF covariance may become singular even when the true covariance is nonsingular. The purpose of this paper is to show under what conditions the EKF covariance is singular, or nonsingular, or nonsingular but ill conditioned.

Another problem with the EKF in attitude determination is that the estimated quaternion may not maintain unity norm, depending on the implementation of the EKF. Various means of normalizing the estimated quaternion were addressed in [6, 7]. We reexamine the various means of normalizing the estimated quaternion and their effect on the covariance matrix in the EKF. It will be shown that normalization is not an unnatural or *ad hoc* step in the EKF, but that it is equivalent to a measurement update.

The process noise matrix is an important part of the full-quaternion EKF. A derivation of this matrix has not appeared in the literature. We treat the process noise matrix in detail here since it has direct influence on the conditioning of the EKF covariance matrix.

* The Johns Hopkins University, Applied Physics Laboratory, Space Department, 11100 Johns Hopkins Road, Laurel, Maryland 20723-6099. E-mail: Mark.Pittelkau@jhuapl.edu; mpittelkau@ieee.org.

Questions of singularity and quaternion normalization (for a rotational update) do not arise in the reduced-order body-referenced attitude determination filter [1, 2, 8, 9, 10]. It will be shown that the reduced-order filter is embedded in the full-quaternion attitude determination filter.

Definitions

First, some definitions are needed. Let be p, q, and r be quaternions corresponding to the direction cosine matrices T_p, T_q, and T_r. The quaternion operator $\otimes$ and operator matrix $[q\otimes]$ are defined such that the quaternion product $r = p \otimes q = [p\otimes]q$ corresponds to the direction cosine matrix product $T_r = T_p T_q$. The quaternion operator $\circledast$ and operator matrix $[q\circledast]$ are defined such that the product $r = q \circledast p = [q\circledast]p$ corresponds to the same direction cosine matrix product $T_r = T_p T_q$, i.e., the order of the quaternions is reversed in the quaternion product. It follows that $p \otimes q = q \circledast p$. The $[q\circledast]$ matrix can be partitioned such that $[q\circledast] = [\Xi \quad q]$, where Ξ is a 4×3 matrix. We will often use the identities $[q\otimes]^T = [q^*\otimes]$ and $[q\circledast]^T = [q^*\circledast]$, where q^* is the quaternion conjugate. We will write $\Xi(p)$ for the partition of $[p\circledast]$, but otherwise assume that Ξ without an argument depends on q in order to simplify the notation. Since $[q\circledast]$ is an orthonormal matrix for $|q| = 1$, we have that $\Xi\Xi^T + qq^T = I_4$ (the 4×4 identity matrix), $\Xi^T\Xi = I_3$ (the 3×3 identity matrix), and $\Xi^T q = 0$. The Ξ matrix is defined in [1] below and the elements of the $\otimes$ operator matrix are shown in [8]. We will also need to use the fact that for $|p| = 1$,

$$[p\circledast]^T[p\otimes] = [p\otimes][p\circledast]^T = \begin{bmatrix} T_p & 0 \\ 0^T & 1 \end{bmatrix} . \tag{1}$$

Dynamics Model

Let ω^b be the body angular rate in body coordinates governing the evolution of the true attitude quaternion q_t, and let $\omega_g = \omega^b - \eta$ be the measured angular rate, where η is a 3×1 vector of white measurement noise with covariance $\sigma_g^2 I_3$. For simplicity we will ignore angle white noise in the gyro measurement. The dynamics model for the attitude determination filter is given by the quaternion differential equation

$$\begin{aligned} \frac{dq_t}{dt} &= \tfrac{1}{2}\bar{\omega} \otimes q_t \\ &= \tfrac{1}{2}\bar{\omega}_g \otimes q_t + \tfrac{1}{2}\Xi(q_t)\eta , \qquad \bar{\omega}_g = \begin{bmatrix} \omega_g \\ 0 \end{bmatrix} \end{aligned} \tag{2}$$

subject to the quaternion norm constraint

$$q_t^T q_t = 1 . \tag{3}$$

The norm of the initial quaternion $q_t(t_0)$ at some initial time t_0 is naturally maintained by the solution to equation (2) because $d(q_t^T q_t)/dt = 2q_t^T(dq_t/dt) = 0$, so the constraint implies that $|q_t(t_0)| = 1$.

We will derive the quaternion and covariance update and propagation equations of the attitude determination filter in a manner similar to [1], but with only the quaternion as the state in order to simplify this analysis. Gyro bias, calibration, and misalignment states are not relevant to the discussion.

Measurement Model

A vector measurement v^s in sensor coordinates and a corresponding reference vector v^i in inertial coordinates are related through the attitude matrix $A(q_t)$ and the body-to-sensor transformation (mounting) matrix T_b^s by

$$v^s = T_b^s A(q_t)v^i , \qquad |q_t| = 1 . \tag{4}$$

A vector attitude sensor produces a measurement $y = h(v^s) + \mathbf{v}$, where $\mathbf{v}$ is white measurement noise with covariance R. The measurement sensitivity matrix is given by [1]

$$H = \frac{\partial h}{\partial q_t^T} = \frac{\partial h}{\partial (v^s)^T} \frac{\partial v^s}{\partial q_t^T} , \tag{5}$$

where $(v^s)^T$ is the transpose of the column matrix v^s. Define

$$u^T = \frac{\partial h}{\partial (v^s)^T} \mathbf{T}_b^s . \tag{6}$$

Taking into account that the elements of the unit quaternion q_t are not independent, we get

$$\frac{\partial v^s}{\partial q_t^T} = [v^b \times] \Xi^T , \tag{7}$$

where $\Xi = \Xi(q_t)$. Substitution of equations (6) and (7) into (5), yields the measurement sensitivity matrix

$$\begin{aligned} H &= 2u^T [v^b \times] \Xi^T \\ &= 2H_3 \Xi^T , \end{aligned} \tag{8}$$

where $v^b = \mathbf{A}(q_t)v^i$.

For a kinematics model with no quaternion norm constraint, the measurement model would have to be

$$v^s = \mathbf{T}_b^s A(q_t/|q_t|)v^i = |q_t|^{-2} \mathbf{T}_b^s A(q_t)v^i \tag{9}$$

for which the unconstrained measurement sensitivity matrix is identical to equation (8).

The measurement model (4) is invalid if q_t is unconstrained. The unconstrained partial derivative is

$$H = 2u^T \left([v^b \times] \Xi^T + v^b q^T \right) , \tag{10}$$

where $\Xi = \Xi(q_t)$ and $q = q_t$. The sensitivity matrix in [3, 5] can be written succinctly as

$$H = 2\Xi^T [\bar{v}^i \circledast] , \qquad \bar{v}^i = \begin{bmatrix} v^i \\ 0 \end{bmatrix} . \tag{11}$$

With a bit of quaternion algebra, this can be shown to be equal to the unconstrained measurement sensitivity matrix (10) with $h(v^s) = v^b$, for which $u = \mathbf{I}$.

Quaternion measurements are considered in Appendix B.

Covariance Update

The quaternion estimation error is $q_{ae} = q_t - q$. As shown in Appendix A, the *true* error covariance $\mathcal{E}\left\{ q_{ae} q_{ae}^T \right\}$ is generally nonsingular but possibly ill conditioned, in particular when the estimate q is constrained to unit norm. However, *the quaternion EKF computes only a first order approximation* P_4 *to the true covariance*. As remarked in the Introduction, the EKF covariance might be expected to be singular [1] but this may not always the case in practice [3, 4]. Our investigation into the nature of the EKF covariance begins with an orthogonal transformation of P_4,

$$[q \circledast]^T P_4 [q \circledast] = \begin{bmatrix} \frac{1}{4} P_3 & \frac{1}{2} w \\ \frac{1}{2} w^T & \sigma_n^2 \end{bmatrix} , \tag{12}$$

where the transformed matrix is partitioned into the 3×3 matrix $\frac{1}{4} P_3$, the 3×1 vector $\frac{1}{2} w$, and the scalar σ_n^2. It will become obvious as we proceed that P_3 is the body-referenced attitude error covariance matrix, that w is the cross-correlation between attitude error and quaternion norm error, and that σ_n^2 is the quaternion norm error variance. (See Appendix A for an independent derivation.) It is not required that $|q| = 1$ in this transformation as long as we are consistent in its definition. By pre- and post-multiplying equation (12) by the orthogonal transformations $[q\circledast] = [\Xi \quad q]$ and $[q\circledast]^T$, we see that P_4 has the general structure

$$P_4 = \begin{bmatrix} \Xi & q \end{bmatrix} \begin{bmatrix} \frac{1}{4}P_3 & \frac{1}{2}w \\ \frac{1}{2}w^T & \sigma_n^2 \end{bmatrix} \begin{bmatrix} \Xi^T \\ q^T \end{bmatrix}$$

$$= \tfrac{1}{4}\Xi P_3 \Xi^T + \tfrac{1}{2}\Xi w q^T + \tfrac{1}{2}q w^T \Xi^T + \sigma_n^2 q q^T . \tag{13}$$

Observe that the accuracy of the attitude error in body coordinates is given by the diagonal of P_3, *not* by the diagonal of P_4.

In the landmark Lefferts-Markley-Shuster paper [1], it was assumed a priori that the covariance is singular, so only the term $\frac{1}{4}\Xi P_3 \Xi^T$ was considered in their derivation of the attitude determination filter. The analysis here is general.

Using equation (8), the measurement residual covariance matrix is given by

$$\begin{aligned} V &= H P_4 H^T + R \\ &= 4 H_3 \Xi^T P_4 \Xi H_3^T + R \\ &= 4 H_3 \Xi^T \left[\tfrac{1}{4}\Xi P_3 \Xi^T + \sigma_n^2 q q^T \right] \Xi H_3^T + R \\ &= H_3 P_3 H_3^T + R \end{aligned} \tag{14}$$

and the Kalman gain $K = P_4 H^T (H P_4 H^T + R)^{-1}$ simplifies to

$$\begin{aligned} K &= P_4 H^T V^{-1} \\ &= 2 \begin{bmatrix} \Xi & q \end{bmatrix} \begin{bmatrix} \frac{1}{4}P_3 & \frac{1}{2}w \\ \frac{1}{2}w^T & \sigma_n^2 \end{bmatrix} \begin{bmatrix} \Xi^T \\ q^T \end{bmatrix} \Xi H_3^T V^{-1} \\ &= \begin{bmatrix} \Xi & q \end{bmatrix} \begin{bmatrix} \frac{1}{2}P_3 \\ w^T \end{bmatrix} H_3^T V^{-1} \end{aligned} \tag{15a}$$

$$\begin{aligned} &= \left(\tfrac{1}{2}\Xi P_3 + q w^T \right) H_3^T V^{-1} \\ &= \tfrac{1}{2}\Xi K_3 + q K_n , \end{aligned} \tag{15b}$$

where

$$K_3 = P_3 H_3^T V^{-1} \tag{16}$$

is the gain matrix of the reduced-order body-referenced attitude determination filter [1] and where

$$K_n = w^T H_3^T V^{-1} \tag{17}$$

is the quaternion-norm gain.

The covariance update can be computed by the quadratic form $P_4^+ = (I - KH)P_4(I - KH)^T + KRK^T$, which for the Kalman gain $K = P_4 H^T (H P_4 H^T + R)^{-1}$ only, is identical to $P_4^+ = (I - KH)P_4$ in the absence of numerical errors. Using equation (13) and equation (15a), the covariance update is computed as follows:

$$P_4^+ = (I_4 - KH)P_4$$

$$= P_4 - KVK^T$$

$$= \begin{bmatrix} \Xi & q \end{bmatrix} \begin{bmatrix} \frac{1}{4}P_3 & \frac{1}{2}w \\ \frac{1}{2}w^T & \sigma_n^2 \end{bmatrix} \begin{bmatrix} \Xi^T \\ q^T \end{bmatrix} - \begin{bmatrix} \Xi & q \end{bmatrix} \begin{bmatrix} \frac{1}{2}P_3 \\ w^T \end{bmatrix} H_3^T V^{-1} H_3 \begin{bmatrix} \frac{1}{2}P_3 & w \end{bmatrix} \begin{bmatrix} \Xi^T \\ q^T \end{bmatrix}$$

$$= \begin{bmatrix} \Xi & q \end{bmatrix} \begin{bmatrix} \frac{1}{4}(I_3 - K_3 H_3)P_3 & \frac{1}{2}(I_3 - K_3 H_3)w \\ \frac{1}{2}w^T(I_3 - K_3 H_3)^T & \sigma_n^2 - w^T H_3^T V^{-1} H_3 w \end{bmatrix} \begin{bmatrix} \Xi^T \\ q^T \end{bmatrix}$$

$$= \begin{bmatrix} \Xi & q \end{bmatrix} \begin{bmatrix} \frac{1}{4}P_3^+ & \frac{1}{2}w^+ \\ \frac{1}{2}(w^T)^+ & (\sigma_n^2)^+ \end{bmatrix} \begin{bmatrix} \Xi^T \\ q^T \end{bmatrix}, \tag{18}$$

where $P_3^+ = (I_3 - K_3 H_3)P_3$ is the updated covariance of the reduced-order filter [1]. Observe that the updated covariance, equation (18), is identical in structure to the a priori covariance, equation (13).

Suppose the covariance matrix is initialized to be positive definite and that the process noise in the norm-error subspace is zero. Since the updated covariance matrix must be at least positive semi-definite (in the absence of numerical error) the term $\sigma_n^2 - w^T H_3^T V^{-1} H_3 w$ will converge to a non-negative value. Thus w must converge to zero. This is also evident from the fact that $I - K_3 H_3$ is contractive. In the presence of process noise in the norm-error variance subspace, the norm-error variance will eventually diverge. In practice we have no a priori knowledge of w, so it is generally set to zero in the initial covariance matrix and will remain zero in the absence of cross-correlated process noise. If σ_n^2 is initialized to zero, then w necessarily has to be initialized to zero also so that the initial P_4 is nonnegative-definite. For an initial quaternion estimate of unit norm (zero norm error), it would be natural to set $\sigma_n^2 = 0$ and $w = 0$.

If the quaternion covariance matrix P_4 is initialized to a value such as $\frac{1}{4}\sigma_a^2 I_4$ with σ_a^2 equal to the variance of the initial attitude error in each axis, then the covariance P_3 in the attitude subspace will decrease as measurement updates occur and the variance in the norm-error subspace will remain at $\sigma_n^2 = \sigma_a^2$ or will grow in the presence of norm-error process noise. The quaternion covariance matrix can become ill-conditioned in applications where the attitude estimation error covariance becomes very small (because $\sigma_n^2 = \sigma_a^2$ is relatively large compared to $\text{tr}(\frac{1}{4}P_3)$), thus potentially causing numerical problems, even with the quadratic form of the covariance update. In practice there can be some "leakage" of the attitude covariance into the norm-error variance and vice versa due to numerical errors, thus reducing the performance of the filter or causing divergence of σ_n^2. If P_4 is initialized to respect the quaternion norm constraint (to first order, so that $\sigma_n^2 = 0$), then it is singular and, in the absence of numerical errors and in the absence of process noise in the norm-error subspace, will remain singular as measurement updates occur. In practice the filter can become numerically unstable, even if the quadratic form of the covariance update is used.

In [5, Figure 6] we see that one eigenvalue of the covariance is 1000 times larger than the other three. Because the largest eigenvalue decreased, it is evident that $w \neq 0$ in the initial covariance P^- at time $k = 1$, although it can't be determined from the figure whether this eigenvalue converged to a non-zero steady-state.

The analysis in equations (14) through (18) can be repeated for the sensitivity matrix equation (10) (or equivalently (11)). It can then be seen that the norm-error variance converges to zero, so the covariance becomes *singular*. Singularity is avoided by adding fictitious process noise. Following the procedure in the next section, it can be seen that the norm gain K_n goes to zero as σ_n^2 and w to to zero, thus leading to divergence in the norm.

State Update

We now examine the state update. The estimated quaternion perturbation correction is

$$\boldsymbol{\delta} = K\boldsymbol{\nu}$$
$$= \tfrac{1}{2}\Xi K_3\boldsymbol{\nu} + qK_n\boldsymbol{\nu}$$
$$= \tfrac{1}{2}\Xi\boldsymbol{\phi} + q\lambda, \tag{19}$$

where $\boldsymbol{\nu} = \boldsymbol{y} - \boldsymbol{h}(\boldsymbol{v}^s)$ is the measurement residual vector and where $\boldsymbol{\phi} = K_3\boldsymbol{\nu}$ and $\lambda = K_n\boldsymbol{\nu}$. Clearly the $q\lambda$ term affects the norm of the updated quaternion. The quaternion update is

$$\boldsymbol{q}^+ = \boldsymbol{q} + \boldsymbol{\delta} \tag{20a}$$

$$= \begin{bmatrix} \Xi & \boldsymbol{q} \end{bmatrix} \begin{bmatrix} \tfrac{1}{2}\boldsymbol{\phi} \\ 1 + \lambda \end{bmatrix}$$

$$= \boldsymbol{q} \circledast \begin{bmatrix} \tfrac{1}{2}\boldsymbol{\phi} \\ 1 + \lambda \end{bmatrix}$$

$$= \begin{bmatrix} \tfrac{1}{2}\boldsymbol{\phi} \\ 1 + \lambda \end{bmatrix} \otimes \boldsymbol{q} . \tag{20b}$$

This shows that the additive update (20a) and multiplicative update (20b) are identical. This is not a new result; see [1]. Since $\boldsymbol{q}^+$ is nonnormalized, a forced normalization is generally inserted after the update. A corresponding adjustment of the EKF covariance should be performed as discussed in the Normalization section below.

When $\boldsymbol{w} = \boldsymbol{0}$, we get $\lambda = 0$ and the filter gives the same update as in [1] in which the norm of the quaternion estimate diverges from unity. If $\boldsymbol{w}$ is valid, it contains norm-error information when $\boldsymbol{w} \neq \boldsymbol{0}$. Then $\lambda \neq 0$ corrects the norm of the quaternion estimate. The norm of the updated quaternion estimate diverges if $\boldsymbol{w}$ is not a valid cross-covariance (for example, as a result of numerical error in the covariance or process noise). Furthermore, after $\boldsymbol{w} \to \boldsymbol{0}$ we see from equation (19) with $\lambda = 0$ that $|\boldsymbol{q}^+|^2 = |\boldsymbol{q} + \boldsymbol{\delta}|^2 = |\boldsymbol{q}|^2(1 + \tfrac{1}{4}|\boldsymbol{\phi}|^2)$. Thus the norm of the quaternion estimate eventually will grow without bound monotonically if it is not normalized after each update, which will adversely affect the attitude represented by the estimated quaternion. A comparison of Figures 5 and 8 of [5] illustrates this; the attitude error for the nonnormalized filter[5, Figure 5] is significantly larger than for the normalized filter [5, Figure 8]. The norm error in [5, Figure 5] was in a transient phase since $\boldsymbol{w} \to \boldsymbol{0}$ there and had not yet begun a monotonic divergence. Nevertheless the damage was already done. ·

Note that $\boldsymbol{\phi} = K_3\boldsymbol{\nu}$ is identical to the attitude correction produced by the reduced-order body-fixed attitude determination filter [1]. It is shown in [11] that the constrained maximum likelihood estimate (MLE) is obtained by treating $\boldsymbol{\phi}$ as a rotation. The MLE optimal update is

$$\boldsymbol{q}^+ = \boldsymbol{r} \otimes \boldsymbol{q} \tag{21}$$

where

$$\boldsymbol{r} = \begin{bmatrix} \tfrac{1}{2}\boldsymbol{\phi} \dfrac{\sin(\varphi/2)}{\varphi/2} \\ \cos(\varphi/2) \end{bmatrix} , \qquad \varphi = |\boldsymbol{\phi}| . \tag{22}$$

This update, although multiplicative, should be called *rotational* to distinguish it from the nonnormalized multiplicative update in equation (20b) where the left multiplicand is not a pure rotation. Quaternion normalization is not needed with the rotational update except occasionally to correct accumulated machine roundoff error. Ostensibly the EKF covariance for the rotational update is singular because the norm error is zero, provided the initial estimate is of unit norm and the covariance is initialized with $\sigma_n^2 = 0$.

Normalization

The quaternion estimate must be normalized to prevent divergence and to maintain its significance as a representation of attitude. One means of normalizatin is "brute force" normalization by $q := q/|q|$. It is said in [6, 7] that brute force normalization is an "outside interference" in the EKF. Various means of normalizing the quaternion estimate and their effect on the covariance are analyzed in this section.

"MPM" Method [6, 7] An update of the quaternion estimate using the norm constraint $c(q) = 1 - q^T q$ as a pseudo-measurement residual with a small pseudo-measurement error variance will result in convergence of the covariance to singularity if there is no process noise added to the norm-error variance. A zero pseudo-measurement error variance results in the covariance update instantly rendering P_4^+ singular. The posterior covariance is then

$$P_4^+ = (I_4 - P_4 q q^T / q^T P_4 q) P_4$$
$$= \tfrac{1}{4} \Xi (P_3 - w w^T / \sigma_n^2) \Xi^T$$
$$= \tfrac{1}{4} \Xi P_3^+ \Xi^T . \qquad (23)$$

where $P_3^+ = P_3 - w w^T$. In the usual case that $w = 0$ we get $P_4^+ = \tfrac{1}{4} \Xi P_3 \Xi^T$. Although the constrained EKF covariance is singular, the true covariance from equations (A–10) and (A–11) is

$$P_{4\text{true}}^+ = [q_t \circledast] \begin{bmatrix} \tfrac{1}{4} P_3^+ & 0 \\ 0^T & \tfrac{1}{32} \big(\text{tr}(P_3^+) \big)^2 \end{bmatrix} [q_t \circledast]^T , \qquad (24)$$

which is nonsingular but possibly ill conditioned with condition number approximately equal to $8/\text{tr}(P_3^+)$. We assume that P_3^+ itself is not ill conditioned, which is a separate problem due to incomplete observability of attitude.

When the pseudo-measurement error variance is zero and $w = 0$, the pseudo-measurement update is equivalent to a quaternion normalization step in the EKF with a corresponding adjustment of the covariance. It is easy to show that $q^+ = q + K(1 - q^T q) = q + \tfrac{1}{2} q (1 - q^T q) \simeq q/\sqrt{q^T q}$. The approximation is based on a Taylor series expansion for $x^{-1/2}$. The approximation error is on the order of $\tfrac{3}{8}(1 - |q|^2)^2 \simeq \tfrac{3}{8}(1 - |q|)^4$, that is, of order four in the norm error. When $w \neq 0$, we also get an attitude correction that results in $|q| \neq 1$ (which means that "brute force" normalization by $q/|q|$ is not optimal when $w \neq 0$). Conceivably the pseudo-measurement update can be applied a second time since $w = 0$ after the first update. For efficiency one would naturally prefer to normalize the quaternion directly via $q/|q|$ in place of the second update. The update of the covariance is a numerical hazard because the computed result may have small negative eigenvalues. Process noise may be added to the norm-error subspace of the covariance to alleviate numerical problems, but this tactic is ill-advised and can be avoided.

"LOM" Method [6, 7] In the LOM method the measurement equation (4) is replaced with equation (9), for which the measurement sensitivity matrix is the same, equation (8). The behavior of the covariance and the Kalman gain is the same as in the previous sections, but the residual $\nu = y - |q|^{-2} T_b^s A(q) v^i$ is unaffected by the norm of the estimated quaternion. Therefore the norm diverges, as explained previously, but in this case the norm cannot converge when $w \neq 0$ because it is unobservable in this measurement model.

Remark 1. It is shown in [3, Eqs. (39) and (51)] that normalization of the updated quaternion $q^+ = q + \delta$ can be approximated *to first order* by $q^+/|q^+| = (q + \delta)/|q + \delta| \simeq q + \delta - q q^T \delta$, where $-q q^T \delta$ is a "norm correction" term added to $q + \delta$. However, $q + \delta - q q^T \delta = q + \Xi \Xi^T \delta = q + \Xi \delta'$ does not correct the norm to unity. The limitation is that it is first order, like the EKF itself. Any norm correction in the EKF has to be at least to second order. In [3] the author takes a circuitous route to conclude that $-q q^T \delta$ is zero, and then concludes that normalization does not affect the quaternion estimation error or the covariance. It is easily seen from the gain equation that $-q q^T \delta$ is zero only when $w = 0$ for equation (8) and when $\sigma_n^2 = 0$ and $w = 0$ for equation (10) or (11).

Remark 2. In [3, 5] it is concluded that because the estimation error is unchanged by normalization (which is incorrect as discussed in Remark 1 above), the quaternion normalization constraint has no effect on the quaternion error covariance and therefore that the constrained EKF covariance is identical to the unconstrained EKF covariance, which is nonsingular by construction. (In [3], the unconstrained covariance is nonsingular because it is initialized as $P_4 = \frac{1}{4}\sigma_a^2 I$ and the process noise was adjusted to prevent divergence. In [5], the initial covariance is $P_4 = \frac{1}{4}\sigma_a^2 I$ and the measurement covariance $R = \sigma^2 I_4$ is also nonsingular. The "partial reset" filter in [5, Table 1] is algebraically the same as the unconstrained additive EKF.) In [12] it is claimed that the covariance of the constrained quaternion estimate is the unconstrained covariance *increased* by a positive semidefinite amount, and similarly in [13] for the constrained estimate of the direction cosine matrix. Thus, [3] and [5] are not in agreement with [12] and [13] and these do not agree with equation (23). Obviously if the constraints were linear, the covariance of the constrained estimate would be singular. For the non-linearly constrained quaternion, the true covariance (24) may be close to singular and the first-order EKF approximation in equation (23) is singular.

Remark 3. The quaternion kinematics equation could be augmented with a normalizing feedback so that $dq/dt = \frac{1}{2}\bar{\omega} \otimes q + \frac{1}{2}\kappa(1 - q^T q)q$, where κ usually ranges from 1 to 10 depending loosely on the number of mantissa bits. (This is useful for numerical integration of dq/dt.) This modification results in P_4 converging to singularity in the norm-error variance subspace in the absence of numerical error or process noise adding to the norm error variance.

Covariance Prediction

Let us now look at the effect of process noise on P_4. Does the norm-error variance in P_4 grow as a result of process noise? If so, this could be numerically disastrous since this variance does not decrease with measurement updates.

The covariance prediction to measurement time t_k is $P_4^-(t_k) = \Phi(t_k, t_{k-1})P_4^+(t_{k-1})\Phi^T(t_k, t_{k-1}) + Q_k$, where the state transition matrix $\Phi(t_k, t)$ is given by

$$\Phi(t_k, t) = [p_t^k \otimes] \tag{25}$$

and where p_t^k is the solution to the quaternion differential equation

$$\dot{p}(t) = \frac{1}{2}\bar{\omega}(t) \otimes p(t), \qquad \bar{\omega}(t) = \begin{bmatrix} \omega^b(t) \\ 0 \end{bmatrix} \tag{26}$$

integrated from time t to time t_k with initial condition $p(t) = [0, \ 0, \ 0, \ 1]^T$ and body angular rate vector $\omega^b(t)$. The quaternion p_t^k represents the attitude at time t_k relative to the attitude at time t. Thus p_{k-1}^k is the change in attitude from t_{k-1} to t_k and $\Phi(t_k, t_{k-1}) = [p_{k-1}^k \otimes]$ is the corresponding state transition matrix. This change in attitude can be parameterized by a rotation vector θ. It usually suffices to assume that ω^b is constant over the interval $[t_{k-1}, t_k]$ so that $\theta = \omega^b(t_{k-1})T$, where $T = t_k - t_{k-1}$. Then

$$p_{k-1}^k = \begin{bmatrix} \frac{1}{2}\theta \dfrac{\sin(\theta/2)}{\theta/2} \\ \cos(\theta/2) \end{bmatrix}, \qquad \theta = |\theta| \tag{27a}$$

and

$$[p_{k-1}^k \otimes] = \exp\left(\tfrac{1}{2}[\bar{\theta} \otimes]\right)$$
$$= \cos(\theta/2)I_4 + \frac{\sin(\theta/2)}{\theta}[\bar{\theta} \otimes], \qquad \bar{\theta} = \begin{bmatrix} \theta \\ 0 \end{bmatrix}. \tag{27b}$$

The predicted attitude is $q_k^- = p_{k-1}^k \otimes q_{k-1}^+$. The process noise η and its covariance $\sigma_g^2 I_3$ were defined previously in the Dynamics Model section. From (2), the 4×4 process noise matrix Q_k at time t_k is given by

$$
\begin{aligned}
Q_k &= \int_{t_{k-1}}^{t_k} \int_{t_{k-1}}^{t_k} \tfrac{1}{4}\Phi(t_k,t)\,\Xi(\boldsymbol{q}(t))\,\mathcal{E}\left\{\boldsymbol{\eta}(t)\boldsymbol{\eta}^T(\tau)\right\}\Xi^T(\boldsymbol{q}(\tau))\,\Phi^T(t_k,\tau)\,\mathrm{d}t\,\mathrm{d}\tau \\
&= \int_{t_{k-1}}^{t_k} \tfrac{1}{4}\Phi(t_k,t)\,\Xi(\boldsymbol{q}(t))\,\sigma_{\mathrm{g}}^2 I_3\,\Xi^T(\boldsymbol{q}(t))\,\Phi^T(t_k,t)\,\mathrm{d}t \\
&= \tfrac{1}{4}\sigma_{\mathrm{g}}^2 \int_{t_{k-1}}^{t_k} \Phi(t_k,t)\,\Xi(\boldsymbol{q}(t))\,\Xi^T(\boldsymbol{q}(t))\,\Phi^T(t_k,t)\,\mathrm{d}t \\
&= \tfrac{1}{4}\sigma_{\mathrm{g}}^2 \int_{t_{k-1}}^{t_k} [\boldsymbol{p}_t^k\otimes]\left[I_4 - \boldsymbol{q}(t)\boldsymbol{q}^T(t)\right][\boldsymbol{p}_t^k\otimes]^T\,\mathrm{d}t \\
&= \tfrac{1}{4}\sigma_{\mathrm{g}}^2 \int_{t_{k-1}}^{t_k} \left[I_4 - \boldsymbol{q}_k\boldsymbol{q}_k^T\right]\mathrm{d}t \\
&= \tfrac{1}{4}T\sigma_{\mathrm{g}}^2\left[I_4 - \boldsymbol{q}_k\boldsymbol{q}_k^T\right] \tag{28a}\\
&= \tfrac{1}{4}T\sigma_{\mathrm{g}}^2\,\Xi(\boldsymbol{q}_k)\,\Xi^T(\boldsymbol{q}_k)\,. \tag{28b}
\end{aligned}
$$

This is the exact expression for the discretized process noise matrix, except that Ξ necessarily has to be computed from the estimated quaternion and Φ is computed from noisy gyro measurements. Observe that there is no process noise in the norm error subspace defined by $\boldsymbol{q}_k$. We have from equation (28a),

$$
\begin{aligned}
Q_k &= \tfrac{1}{4}T\sigma_{\mathrm{g}}^2\left[I_4 - (\boldsymbol{p}_{k-1}^k\otimes\boldsymbol{q}_{k-1})(\boldsymbol{p}_{k-1}^k\otimes\boldsymbol{q}_{k-1})^T\right] \\
&= \tfrac{1}{4}T\sigma_{\mathrm{g}}^2\,[\boldsymbol{p}_{k-1}^k\otimes]\left[I_4 - \boldsymbol{q}_{k-1}\boldsymbol{q}_{k-1}^T\right][\boldsymbol{p}_{k-1}^k\otimes]^T \\
&= \tfrac{1}{4}T\sigma_{\mathrm{g}}^2\,[\boldsymbol{p}_{k-1}^k\otimes]\,\Xi(\boldsymbol{q}_{k-1})\,\Xi^T(\boldsymbol{q}_{k-1})\,[\boldsymbol{p}_{k-1}^k\otimes]^T\,. \tag{29}
\end{aligned}
$$

Note that the Kalman gain used to compute the updated covariance P_4^+ at time $k-1$ is computed from $\Xi(\boldsymbol{q}_{k-1}^-)$ so that P_4^+ is of the form of equation (13) with $[\Xi\ \ \boldsymbol{q}] = [\boldsymbol{q}_{k-1}^-\circledast]$. Let us use the a priori estimate $\boldsymbol{q}_{k-1} = \boldsymbol{q}_{k-1}^-$ to evaluate the process noise matrix in equation (29). Substituting equation (29) into the covariance prediction equation yields

$$
\begin{aligned}
P_4^-(t_k) &= \Phi(t_k,t_{k-1})P_4^+(t_{k-1})\Phi^T(t_k,t_{k-1}) + Q_k \\
&= [\boldsymbol{p}_{k-1}^k\otimes]\left[P_4^+(t_{k-1}) + \tfrac{1}{4}T\sigma_{\mathrm{g}}^2\,\Xi(\boldsymbol{q}_{k-1})\,\Xi^T(\boldsymbol{q}_{k-1})\right][\boldsymbol{p}_{k-1}^k\otimes]^T \\
&= [\boldsymbol{p}_{k-1}^k\otimes]\Big[\tfrac{1}{4}\Xi(\boldsymbol{q}_{k-1})P_3^+(t_{k-1})\Xi^T(\boldsymbol{q}_{k-1}) \\
&\qquad + \tfrac{1}{2}\Xi(\boldsymbol{q}_{k-1})\boldsymbol{w}\boldsymbol{q}_{k-1}^T + \tfrac{1}{2}\boldsymbol{q}_{k-1}\boldsymbol{w}^T\Xi^T(\boldsymbol{q}_{k-1}) \\
&\qquad + \sigma_{\mathrm{n}}^2\boldsymbol{q}_{k-1}\boldsymbol{q}_{k-1}^T + \tfrac{1}{4}T\sigma_{\mathrm{g}}^2\,\Xi(\boldsymbol{q}_{k-1})\,\Xi^T(\boldsymbol{q}_{k-1})\Big][\boldsymbol{p}_{k-1}^k\otimes]^T \\
&= [\boldsymbol{p}_{k-1}^k\otimes]\Xi(\boldsymbol{q}_{k-1})\left[\tfrac{1}{4}P_3^+(t_{k-1}) + \tfrac{1}{4}T\sigma_{\mathrm{g}}^2 I_3\right]\Xi^T(\boldsymbol{q}_{k-1})[\boldsymbol{p}_{k-1}^k\otimes]^T \\
&\qquad + \tfrac{1}{2}\Xi(\boldsymbol{q}_k)\boldsymbol{w}\boldsymbol{q}_k^T + \tfrac{1}{2}\boldsymbol{q}_k\boldsymbol{w}^T\Xi^T(\boldsymbol{q}_k) + \sigma_{\mathrm{n}}^2\boldsymbol{q}_k\boldsymbol{q}_k^T \\
&= \Xi(\boldsymbol{q}_k)\left[\tfrac{1}{4}\mathbf{T}_{k-1}^k P_3^+(t_{k-1})\mathbf{T}_k^{k-1} + \tfrac{1}{4}T\sigma_{\mathrm{g}}^2 I_3\right]\Xi^T(\boldsymbol{q}_k) \\
&\qquad + \tfrac{1}{2}\Xi(\boldsymbol{q}_k)\boldsymbol{w}\boldsymbol{q}_k^T + \tfrac{1}{2}\boldsymbol{q}_k\boldsymbol{w}^T\Xi^T(\boldsymbol{q}_k) + \sigma_{\mathrm{n}}^2\boldsymbol{q}_k\boldsymbol{q}_k^T\,, \tag{30}
\end{aligned}
$$

where $\mathbf{T}_{k-1}^k$ is the transformation matrix corresponding to $\boldsymbol{p}_{k-1}^k$ and where $\boldsymbol{q}_k = \boldsymbol{p}_{k-1}^k\otimes\boldsymbol{q}_{k-1}^-$. The derivation of the last line from the previous in equation (30) involves a bit of quaternion matrix algebra utilizing equation (1). The result is intuitive: the quantity in brackets is the covariance prediction $P_3^-(t_k)$ for the reduced-order filter [1] and the entire expression is of the form of equation (13). Equation (30) indicates that the process noise matrix Q_k is singular in the norm-error variance subspace of the predicted covariance $P_4^-(t_k)$, so the variance σ_n^2 in the norm-error subspace does not grow; the process noise contributes only to the attitude covariance P_3.

Remark 4. The structure of the process noise matrix Q_k in equation (29) is dictated by the quaternion kinematics. To assume any other form, e.g., $Q_k = \sigma_g^2 I_4$, is incorrect and will result in suboptimal performance of the filter, even though it will keep the a priori covariance positive definite. The a posteriori covariance may still become singular under conditions noted in the previous section and in Appendix B.

Remark 5. The process noise matrix will render the predicted covariance matrix nonsingular if the process noise matrix is not computed with $q_{k-1} = q_{k-1}^-$ in equation (29) or if it is computed by some means of approximation, e.g., numerical integration in [4]. This is similar to the "leakage" problem mentioned earlier. The norm error variance will grow if it is not reduced by the covariance update, thereby eventually making the covariance matrix ill-conditioned. Although covariance leakage may prevent singularity, an algorithm that depends on computational error to make it work is not a robust algorithm.

The predicted covariance equation (30) was derived by using the a priori estimate q_{k-1}^- in the process noise matrix equation (29). In a typical implementation of the filter, the process noise matrix in equation (28b) is evaluated at the a priori (predicted) estimate q_k^-, so we must have $q_{k-1} = q_{k-1}^+$ in equation (29) so that $[p_{k-1}^k \otimes] \Xi(q_{k-1}^+) = \Xi(q_k^-)$. Since P_4^+ is computed with q_{k-1}^- (because q_{k-1}^+ is unavailable when the gain is computed), we have to adjust P_4^+ before prediction to time k to ensure that $\Phi(t_k, t_{k-1}) P_4^+ \Phi^T(t_k, t_{k-1})$ and Q_k are in the same coordinate system. This adjustment is simply propagating P_4^+ at time $k - 1$ with $[r_{k-1} \otimes] P_4^+ [r_{k-1} \otimes]^T$, where r_{k-1} is the quaternion correction defined by equation (22) at time $k - 1$. The resulting predicted covariance is then of the form of equation (30) with $q_k = q_k^-$. In the reduced-order filter, this corresponds to updating P_3^+ at time $k - 1$ with $\mathbf{T}(r_{k-1}) P_3^+ \mathbf{T}^T(r_{k-1})$, where $\mathbf{T}(r_{k-1})$ is the transformation matrix corresponding to r_{k-1}. To the author's knowledge this covariance propagation is not mentioned in the literature and is never done in practice, though it should be done in order to maintain the correct frame of reference for the covariance matrix and to prevent "leakage" in the covariance of the full quaternion filter.

Remark 6. In [3] it was reported that the filter diverged quickly when the a priori quaternion estimate was used in the measurement sensitivity matrix in the covariance update equation, and that divergence did not appear when the a posteriori quaternion estimate was used. A plausible explanation is "leakage" of the attitude covariance into the norm-error covariance that prevented singularity. Ficitious process noise was also added to prevent a subsequent slow divergence.

Remark 7. Observe from the gain, state update, covariance update, and covariance prediction equations (15b), (18), (20b) (or (21)), and (30) that the reduced-order body-referenced attitude determination filter is embedded in the full-quaternion attitude determination filter. As pointed out earlier, the rotational update defined by Eqs. (21) and (22) is the maximum-likelihood optimal and does not require quaternion normalization. An infrequent normalization may be used to compensate for machine roundoff error.

Conclusion

The covariance of the full-quaternion Extended Kalman Filter is analyzed in this paper and it is shown under what conditions the EKF covariance is singular, nonsingular, or otherwise ill conditioned. It particular, it was shown that singularity is often skirted in practice because the process noise matrix and the predicted covariance matrix may be computed in slightly different coordinate frames. Singularity may also be avoided by approximations to the process noise matrix that contribute to the norm-error variance or by intentionally adding fictitious process noise to prevent divergence. In the case of a quaternion measurement, singularity is avoided when the measurement covariance matrix is positive definite (which is artificial, not real), though the covariance asymptotically goes to zero in the absence of process noise. Singularity can also be avoided, ostensibly, by not updating the covariance in correspondence with normalization of the quaternion estimate. It was shown also that a nonsingular initial covariance can result in the updated covariance being ill conditioned.

Normalization of the quaternion estimate is also addressed in this paper. The norm of the estimated quaternion is shown to diverge from unity if the estimated quaternion is not normalized, and this can also affect the accuracy of the attitude represented by the quaternion estimate. Although it is said in the literature that the exact quaternion normalization $q/|q|$ is an "outside interference" in the EKF, it is shown in this paper that it is equivalent to the pseudo-measurement update up to order four in the norm error. It is also shown that the EKF covariance becomes singular when updated to account for the normalization of the estimate. Like any other measurement update, the pseudo-measurement update, and therefore quaternion normalization, is not an unnatural part of the EKF.

The attitude determination filter that includes the full quaternion as a state and that includes the full quaternion error covariance matrix as part of the filter covariance can potentially exhibit suboptimal performance or outright failure as a result of numerical problems due to either singularity or ill-conditioning. The reduced-order body-referenced attitude determination filter is immune from the covariance singularity and ill-conditioning issue addressed herein, requires much less computation, and is simpler to implement than the full-quaternion filter. The rotational quaternion update as defined herein is the maximum-likelihood optimal update and does not require normalization. The additive update and the equivalent multiplicative update should be avoided because they are not optimal and require normalization.

Although the structure of the covariance of the full-quaternion Extended Kalman Filter has been revealed, it should not be exploited to find *ad hoc* means of ensuring numerical stability. The reduced-order body-referenced attitude determination filter, which was shown to be embedded in the full-quaternion filter equations, should be the first and only choice for attitude determination filter design for the aforementioned reasons. However, for those who choose to use the full-quaternion filter, *caveat emptor*.

Acknowledgement

The author thanks F. Landis Markley of the Goddard Space Flight Center for reviewing and commenting on the manuscript, and especially for pointing out early in the development that the cross-covariance w between attitude and norm error had to be considered for complete generality and for bringing the pseudo-measurement update to my attention. The author is further indebted to Dr. Markley for revealing the existence of a second term in the unconstrained measurement sensitivity equation. Thanks also to Quang Lam of Swales Aerospace for helpful comments on the paper. Special thanks go to I. Y. Bar-Itzhack and J. Thienel for the invitation to pursue the covariance singularity and quaternion normalization issues.

References

1. LEFFERTS, E. J., MARKLEY, F. L., and SHUSTER, M. D., "Kalman Filtering for Spacecraft Attitude Estimation", *AIAA Journal of Guidance, Control, and Dynamics*, Vol. 5, No. 5, 1982, pp. 417–429.

2. PITTELKAU, M. E., "Kalman Filtering for Spacecraft System Alignment Calibration," *AIAA Journal of Guidance, Control, and Dynamics*, Vol. 24, No. 6, 2001, pp. 1187–1195.

3. BAR-ITZHACK, I. Y., and OSHMAN, Y., "Attitude Determination from Vector Measurements: Quaternion Estimation", *IEEE Transactions on Aerospace and Electronic Systems*, Vol. AES-21, No. 1, 1985, pp. 128–136.

4. DEUTSCHMANN, J. and BAR-ITZHACK, I. Y., "Extended Kalman Filter for Attitude Esti-

mation of the Earth Radiation Budget Satellite," NASA CP-3050, *Flight Mechanics and Estimation Theory Symposium*, Greenbelt MD, May 10–11, 1989, pp. 333–345.

5. BAR-ITZHACK, I. Y. and THIENEL, J., "On the Singularity in the Estimation of the Quaternion-of-Rotation", Paper Number AIAA-2002-4831, AIAA Astrodynamics Specialists Conference, Monteray CA, 7 August 2002, pp. ???–???.

6. BAR-ITZHACK, I. Y., DEUTSCHMANN, J., and MARKLEY, F. L., "Quaternion Normalization in Additive EKF for Spacecraft Attitude Determination," AIAA Paper No. 91-2706, *AIAA Guidance, Navigation, and Control Conference*, New Orleans, Aug 1991, pp. 908–916.

7. DEUTSCHMANN, J., MARKLEY, F. L., and BAR-ITZHACK, I. Y., "Quaternion Normalization in Spacecraft Attitude Determination," *Flight Mechanics and Estimation Theory Symposium*, NASA Conference Publication CP-3186, NASA Goddard Space Flight Center, May 1992, pp. 441–454.

8. PITTELKAU, M. E., "Spacecraft Attitude Determination Using The Bortz Equation", *AAS/AIAA Astrodynamics Specialists Conference*, Paper No. AAS 01-310, 30 July 2001.

9. PITTELKAU, M. E., "Everything is Relative in Spacecraft System Alignment Calibration", *AIAA Journal of Spacecraft and Rockets*, Vol. 39, No. 3, May–June 2002, pp. 460—466.

10. MARKLEY, F. L., "Attitude Error Representations for Kalman Filtering", *AIAA Journal of Guidance, Control, and Dynamics*, (to be published in 2003).

11. SHUSTER, M. D., "The Quaternion in the Kalman Filter", *Advances in the Astronautical Sciences*, Paper No. AAS-93-553, Vol. 85, 1993, pp. 25–37.

12. CHOUKROUN, D., BAR-ITZHACK, I. Y., and OSHMAN, Y., "A Novel Quaternion Kalman Filter", Paper Number AIAA-2002-4460, AIAA Astrodynamics Specialists Conference, Monteray CA, 7 August 2002, pp. ???–???.

13. BAR-ITZHACK, I. Y. and REINER, J., "Recursive Attitude Determination from Vector Observations: Direction Cosine Matrix Identification", *AIAA Journal of Guidance*, Vol. 7, No. 1, 1984, pp. 51–56.

A   Quaternion Covariance

An expression for a second-order approximation to the covariance of the quaternion error is derived in this appendix. The additive quaternion error q_{ae} is defined by $q_{\mathrm{ae}} = q_{\mathrm{t}} - q$, where q_{t} is the true attitude quaternion with $|q_{\mathrm{t}}| = 1$ and q is the estimated attitude quaternion. We will allow a random error in $|q|$. Let $\mathbf{1}$ represent the identity quaternion. The additive error can be written in terms of a rotational error $q_{\mathrm{re}} = q \otimes q_{\mathrm{t}}^*$ by

$$
\begin{aligned}
q_{\mathrm{ae}} &= q_{\mathrm{t}} - q \\
&= (\mathbf{1} - q \otimes q_{\mathrm{t}}^*) \otimes q_{\mathrm{t}} \\
&= (\mathbf{1} - q_{\mathrm{re}}) \otimes q_{\mathrm{t}} \\
&= q_{\mathrm{t}} \circledast (\mathbf{1} - q_{\mathrm{re}}) \\
&= [q_{\mathrm{t}} \circledast](\mathbf{1} - q_{\mathrm{re}}) .
\end{aligned}
\tag{A-1}
$$

The expected value of the additive error is

$$
\begin{aligned}
\bar{q}_{\mathrm{ae}} &= [q_{\mathrm{t}} \circledast](\mathbf{1} - \mathcal{E}\{q_{\mathrm{re}}\}) \\
&= [q_{\mathrm{t}} \circledast](\mathbf{1} - \bar{q}_{\mathrm{re}}) .
\end{aligned}
\tag{A-2}
$$

Thus we have

$$
q_{\mathrm{ae}} - \bar{q}_{\mathrm{ae}} = -[q_{\mathrm{t}} \circledast](q_{\mathrm{re}} - \bar{q}_{\mathrm{re}}) .
\tag{A-3}
$$

The additive error covariance is thus written in terms of the rotational error covariance as

$$
\mathcal{E}\left\{(q_{\mathrm{ae}} - \bar{q}_{\mathrm{ae}})(q_{\mathrm{ae}} - \bar{q}_{\mathrm{ae}})^T\right\} = [q_{\mathrm{t}} \circledast]\,\mathcal{E}\left\{(q_{\mathrm{re}} - \bar{q}_{\mathrm{re}})(q_{\mathrm{re}} - \bar{q}_{\mathrm{re}})^T\right\}[q_{\mathrm{t}} \circledast]^T .
\tag{A-4}
$$

The rotational error can be represented in general as

$$
\boldsymbol{q}_{\mathrm{re}} = (1 + \mu + \eta) \begin{bmatrix} \frac{1}{2}\boldsymbol{\phi} \dfrac{\sin(\varphi/2)}{\varphi/2} \\[2mm] \cos(\varphi/2) \end{bmatrix}, \tag{A-5}
$$

where $\boldsymbol{\phi}$ is the rotation error vector, $\varphi = |\boldsymbol{\phi}|$, μ is a mean offset of the norm from unity, and η is a zero mean random variable that represents the norm error in $\boldsymbol{q}$. (We have $1 + \mu + \eta = |\boldsymbol{q}_{\mathrm{re}}| = |\boldsymbol{q} \otimes \boldsymbol{q}_{\mathrm{t}}^*| = |\boldsymbol{q}|$ since the norm is invariant under an orthogonal transformation.) A series expansion yields

$$
\boldsymbol{q}_{\mathrm{re}} = (1 + \mu + \eta) \begin{bmatrix} \frac{1}{2}\boldsymbol{\phi}\left(1 - \dfrac{\varphi^2}{4\cdot 3!} + \dfrac{\varphi^4}{16\cdot 5!} - \cdots\right) \\[3mm] 1 - \dfrac{\varphi^2}{4\cdot 2!} + \dfrac{\varphi^4}{16\cdot 4!} - \cdots \end{bmatrix}. \tag{A-6}
$$

The rotation (attitude) error is assumed to be zero mean with covariance P_3, that is, $\mathcal{E}\{\boldsymbol{\phi}\} = \boldsymbol{0}$ and $\mathcal{E}\{\boldsymbol{\phi}\boldsymbol{\phi}^T\} = P_3$. We then have $\sigma^2 = \mathcal{E}\{\varphi^2\} = \mathcal{E}\{\boldsymbol{\phi}^T\boldsymbol{\phi}\} = \mathrm{tr}(P_3)$. Note that since $\boldsymbol{\phi}$ is unbiased, $\mathcal{E}\{\boldsymbol{\phi}\varphi^{2k}\} = \boldsymbol{0}$ for $k = 1, 2, \ldots$. We will initially consider $\mu = 0$ and later consider nonzero μ and biased $\boldsymbol{\phi}$. The norm error has zero mean and variance σ_{n}^2, that is, $\mathcal{E}\{\eta\} = 0$ and $\mathcal{E}\{\eta^2\} = \sigma_{\mathrm{n}}^2$. We also assume that the norm error and the rotational error are correlated so that $\mathcal{E}\{\eta\boldsymbol{\phi}\} = \boldsymbol{w}$. For the sake of calculation we assume that the random variables are Gaussian so that $\mathcal{E}\{\varphi^4\} = 3\sigma^4$ and $\mathcal{E}\{\eta\boldsymbol{\phi}\boldsymbol{\phi}^T\eta\} = \mathcal{E}\{\eta^2\}\mathcal{E}\{\boldsymbol{\phi}\boldsymbol{\phi}^T\} + 2\mathcal{E}\{\eta\boldsymbol{\phi}\}\mathcal{E}\{\boldsymbol{\phi}^T\eta\} = \sigma_{\mathrm{n}}^2 P_3 + 2\boldsymbol{w}\boldsymbol{w}^T$.

From these definitions we easily obtain

$$
\bar{\boldsymbol{q}}_{\mathrm{re}} = \mathcal{E}\{\boldsymbol{q}_{\mathrm{re}}\} \simeq \begin{bmatrix} \frac{1}{2}\boldsymbol{w} \\[2mm] 1 - \frac{1}{8}\sigma^2 + \frac{3}{384}\sigma^4 \end{bmatrix} \tag{A-7}
$$

and

$$
\tilde{\boldsymbol{q}}_{\mathrm{re}} = \boldsymbol{q}_{\mathrm{re}} - \bar{\boldsymbol{q}}_{\mathrm{re}} \simeq \begin{bmatrix} \frac{1}{2}\boldsymbol{\phi} + \frac{1}{2}(\eta\boldsymbol{\phi} - \boldsymbol{w}) \\[2mm] \frac{1}{8}(\sigma^2 - \varphi^2) + \frac{1}{384}(\varphi^4 - 3\sigma^4) + \eta(1 - \frac{1}{8}\varphi^2 + \frac{3}{384}\varphi^4) \end{bmatrix}
$$

$$
\simeq \begin{bmatrix} \frac{1}{2}\boldsymbol{\phi} + \frac{1}{2}(\eta\boldsymbol{\phi} - \boldsymbol{w}) \\[2mm] \frac{1}{8}(\sigma^2 - \varphi^2) + \eta \end{bmatrix}. \tag{A-8}
$$

Thus the covariance of the rotational error quaternion is easily calculated to be, to within an error on the order of σ^6,

$$
P_{\mathrm{re}} = \mathcal{E}\{\tilde{\boldsymbol{q}}_{\mathrm{re}}\tilde{\boldsymbol{q}}_{\mathrm{re}}^T\} = \begin{bmatrix} \frac{1}{4}(1 + \sigma_{\mathrm{n}}^2)P_3 + \frac{1}{4}\boldsymbol{w}\boldsymbol{w}^T & \frac{1}{2}\boldsymbol{w} \\[2mm] \frac{1}{2}\boldsymbol{w}^T & \frac{1}{32}\sigma^4 + \sigma_{\mathrm{n}}^2 \end{bmatrix}. \tag{A-9}
$$

We expect that $\sigma_{\mathrm{n}}^2 \ll 1$ when $\boldsymbol{q}$ is a "good" estimate of $\boldsymbol{q}_{\mathrm{t}}$, so that $\boldsymbol{w}\boldsymbol{w}^T$ is small compared to the attitude error covariance P_3. We can then further approximate P_{re} by

$$
P_{\mathrm{re}} = \begin{bmatrix} \frac{1}{4}P_3 & \frac{1}{2}\boldsymbol{w} \\[2mm] \frac{1}{2}\boldsymbol{w}^T & \frac{1}{32}\sigma^4 + \sigma_{\mathrm{n}}^2 \end{bmatrix}. \tag{A-10}
$$

From equation (A–4), the additive quaternion error covariance, in terms of the rotational quaternion error covariance, is given by

$$
P_{\mathrm{ae}} = [\boldsymbol{q}_{\mathrm{t}}\circledast]P_{\mathrm{re}}[\boldsymbol{q}_{\mathrm{t}}\circledast]^T. \tag{A-11}
$$

It is easy to show that the covariance P_{re} and therefore P_{ae} can be singular for certain values of $\boldsymbol{w}$. For example, let $P_3 = \sigma_\phi^2 I_3$ and $\boldsymbol{w} = \rho\sigma_\phi\sigma_{\mathrm{n}}[1,0,0]^T$, where ρ is the correlation coefficient. Now consider a quaternion of unit norm. For $\sigma_n^2 = 0$ we have $\boldsymbol{w} = \boldsymbol{0}$ and the reciprocal condition number is σ_ϕ^2, which for low accuracy estimation is not too small compared to a typical double precision machine roundoff of $2.22(10^{-16})$. For single precision arithmetic, the covariance is ill conditioned even for low accuracy estimation, $\sigma_\phi = 0.001$ for example. For arcsecond accuracy, $\sigma_\phi = 5.4(10^{-6})$ and the covariance is close to being ill-conditioned with respect to double precision. Note that high-order terms that were neglected in the covariance are typically very small and lost in computer roundoff error, so for all practical purposes can be neglected in this analysis.

From equation (A–2) and equation (A–7) we get the mean value of the additive quaternion error in terms of the mean value of the rotational quaternion error,

$$\bar{\boldsymbol{q}}_{\mathrm{ae}} = [\boldsymbol{q}_\mathrm{t} \circledast] \begin{bmatrix} -\frac{1}{2}\boldsymbol{w} \\ \frac{1}{8}\sigma^2 \end{bmatrix} . \tag{A–12}$$

Considering now the norm offset μ in equation (A–5), equation (A–12) becomes

$$\bar{\boldsymbol{q}}_{\mathrm{ae}} = [\boldsymbol{q}_\mathrm{t} \circledast] \begin{bmatrix} -\frac{1}{2}\boldsymbol{w} \\ 1 - (1 + \mu)(1 - \frac{1}{8}\sigma^2) \end{bmatrix} . \tag{A–13}$$

The mean value of the additive quaternion error is minimized when $\mu = \frac{1}{8}\sigma^2/(1 - \frac{1}{8}\sigma^2)$, which yields

$$\bar{\boldsymbol{q}}_{\mathrm{ae}} = [\boldsymbol{q}_\mathrm{t} \circledast] \begin{bmatrix} -\frac{1}{2}\boldsymbol{w} \\ 0 \end{bmatrix} . \tag{A–14}$$

Recall that the preceeding derivations are under the assumption that the attitude error is unbiased. Thus equation (A–14) shows that the quaternion estimate can be biased even of the rotational attitude error (which is of primary interest) is unbiased.

Let's consider a bias in the attitude error. This can be modeled in either of two ways: let $\mathcal{E}\{\boldsymbol{\phi}\} = \boldsymbol{b}$ or multiply equation (A–5) by a bias quaternion

$$\boldsymbol{q}_\mathrm{b} = \begin{bmatrix} \frac{1}{2}\boldsymbol{b} \dfrac{\sin(b/2)}{b/2} \\ \cos(b/2) \end{bmatrix} , \qquad b = |\boldsymbol{b}| \tag{A–15}$$

such that $\boldsymbol{q}_{\mathrm{bre}} = \boldsymbol{q}_\mathrm{b} \otimes \boldsymbol{q}_{\mathrm{re}}$ is the biased rotational error quaternion. We choose the latter route. To a little better than first order we have

$$\bar{\boldsymbol{q}}_{\mathrm{bre}} = \begin{bmatrix} \frac{1}{2}(\boldsymbol{w} + \boldsymbol{b} + \frac{1}{2}\boldsymbol{w} \times \boldsymbol{b}) \\ 1 - \frac{1}{8}\sigma^2 - \frac{1}{4}\boldsymbol{b}^T\boldsymbol{w} \end{bmatrix} . \tag{A–16}$$

and

$$\bar{\boldsymbol{q}}_{\mathrm{bae}} = [\boldsymbol{q}_\mathrm{t} \circledast] \begin{bmatrix} -\frac{1}{2}(\boldsymbol{w} + \boldsymbol{b} + \frac{1}{2}\boldsymbol{w} \times \boldsymbol{b}) \\ \frac{1}{8}\sigma^2 + \frac{1}{4}\boldsymbol{b}^T\boldsymbol{w} \end{bmatrix} . \tag{A–17}$$

The additive quaternion error is minimized when $\boldsymbol{b} = -\boldsymbol{w}$, which causes the rotational error to be biased. A mean offset μ can be introduced as before. The quaternion can then be made unbiased (zero mean) when $\boldsymbol{b} = -\boldsymbol{w}$ and $\mu = (\frac{1}{8}\sigma^2 + \frac{1}{4}\boldsymbol{b}^T\boldsymbol{w})/(1 - \frac{1}{8}\sigma^2 - \frac{1}{4}\boldsymbol{b}^T\boldsymbol{w})$, which also causes the rotational error to be biased.

B Quaternion Measurement Update

We have already examined the state and covariance update when a three-axis attitude measurement is available. Now suppose we have a quaternion "measurement". Of course there is no device that measures a quaternion because it is not a physical quantity. The quaternion measurement is simply a representation of attitude as measured by other means, for example a set of focal plane measurements in a star tracker. We will nevertheless treat the quaternion measurement as a physical quantity. A first-order linear approximation to the quaternion measurement is

$$q_{\mathrm{m}} = Hq + \mathbf{v}\,, \tag{B-1}$$

where $H = I_4$ is the linear observation matrix, q is the attitude quaternion, and $\mathbf{v}$ is measurement noise with

$$\mathcal{E}\left\{\mathbf{v}\mathbf{v}^T\right\} = R_4 = [\Xi \quad q]\begin{bmatrix}\frac{1}{4}R_3 & 0 \\ 0^T & \sigma_1^2\end{bmatrix}\begin{bmatrix}\Xi^T \\ q^T\end{bmatrix}$$

$$= \tfrac{1}{4}\Xi R_3 \Xi^T + \sigma_1^2 q q^T\,. \tag{B-2}$$

In this expression, R_3 is the covariance of error in the measured attitude and σ_1^2 is the covariance of error in the norm of q_{m}. Since any proper attitude quaternion is of unit norm, it is natural that $\sigma_1^2 = 0$. However, we will allow $\sigma_1^2 \neq 0$ for the purpose of discussion. We have assumed in equation (B-2) that there is no correlation between the attitude measurement error and the norm error in the measurement q_{m}. Note that even though we have allowed σ_1^2 to be non-zero in the filter, we assume that the norm of the quaternion measurement is unity (the true norm error variance of the measurement is zero), otherwise the quaternion estimate will be biased.

Remark A-1. Observe that if $R_4 = \sigma^2 I_4$, then we have $R_3 = \sigma^2 I_3$ and $\sigma_1^2 = \sigma^2$. We shall not impose this restriction on R_4, instead we will allow the values of R_3 and σ_1^2 to be defined individually.

We will now examine the update of the state and covariance. If the covariance matrix is initialized with $w = 0$, then w will remain zero in the absence of numerical error and in the absence of process noise in the norm-error subspace. We will assume that $w = 0$. The analysis for $w \neq 0$ yields the same conclusion but is significantly more involved to show.

From equation (13) with $w = 0$ and equation (B-2), the Kalman gain is

$$K_4 = P_4 H^T (H P_4 H^T + R_4)^{-1}$$

$$= [\Xi \quad q]\begin{bmatrix}\frac{1}{4}P_3 & 0 \\ 0^T & \sigma_n^2\end{bmatrix}\begin{bmatrix}(\frac{1}{4}P_3 + \frac{1}{4}R_3)^{-1} & 0 \\ 0^T & 1/(\sigma_n^2 + \sigma_1^2)\end{bmatrix}\begin{bmatrix}\Xi^T \\ q^T\end{bmatrix}$$

$$= \Xi(P_3(P_3 + R_3)^{-1})\Xi^T + \frac{\sigma_n^2}{\sigma_n^2 + \sigma_1^2}qq^T$$

$$= \Xi K_3 \Xi^T + \frac{\sigma_n^2}{\sigma_n^2 + \sigma_1^2}qq^T\,, \tag{B-3}$$

where $K_3 = P_3(P_3 + R_3)^{-1}$ is the gain for the rotational part of the update.

Examine first the covariance update. With a little matrix manipulation it is easily shown that

$$P_4^+ = (I_4 - K_4 H_4)P_4$$

$$= [\Xi \quad q]\begin{bmatrix}\frac{1}{4}(I_3 - K_3)P_3 & 0 \\ 0^T & \dfrac{\sigma_1^2}{\sigma_n^2 + \sigma_1^2}\sigma_n^2\end{bmatrix}\begin{bmatrix}\Xi^T \\ q^T\end{bmatrix}\,, \tag{B-4}$$

where $K_3 = P_3 V^{-1}$. This result is the same as (18) with $H_3 = I_3$ and $w = 0$ except that now the norm error variance is reduced in equation (B-4). A more general result with $w \neq 0$ requires significantly more matrix algebra, but it does not affect our conclusions below and therefore is omitted.

Now we examine the state update. Let us assume that $|q| = \alpha \simeq 1$, that is, the quaternion state estimate is not of unit length. The state update is

$$q^+ = q + K_4(q_{\mathrm{m}} - q)$$

$$= q + \Xi K_3 \Xi^T (q_{\mathrm{m}} - q) + \frac{\sigma_{\mathrm{n}}^2}{\sigma_{\mathrm{n}}^2 + \sigma_{\mathrm{l}}^2} qq^T(q_{\mathrm{m}} - q)$$

$$= q + \Xi K_3 [I_3 \quad 0] \left(q^* \circledast q_{\mathrm{m}} - \begin{bmatrix} 0 \\ \alpha \end{bmatrix} \right) + \frac{\sigma_{\mathrm{n}}^2}{\sigma_{\mathrm{n}}^2 + \sigma_{\mathrm{l}}^2} q(q^T q_{\mathrm{m}} - \alpha^2)$$

$$= q + \tfrac{1}{2}\Xi\theta + q\lambda \tag{B-5}$$

where the vector $\theta = 2K_3[I_3 \quad 0](q^* \circledast q_{\mathrm{m}}) = 2K_3[I_3 \quad 0](q_{\mathrm{m}} \otimes q^*)$ is the attitude correction and where the scalar $\lambda = (\sigma_{\mathrm{n}}^2/(\sigma_{\mathrm{n}}^2 + \sigma_{\mathrm{l}}^2))(q^T q_{\mathrm{m}} - \alpha^2)$ is the norm-error correction. The analysis of the update for θ is identical to the previous analysis. We need only to examine the update for λ. Since $|q_{\mathrm{m}}| = 1$, $q^T q_{\mathrm{m}} = \alpha + \eta$, where η is a residual error. Letting $\sigma_{\mathrm{n}} \to \infty$ (infinite norm-error estimate variance) or $\sigma \to 0$ (zero norm-error measurement variance) we get

$$q^+ = \tfrac{1}{2}\Xi\theta + q(1 + \alpha - \alpha^2) + q\eta \tag{B-6}$$

By writing $\alpha = 1 + \varepsilon$ it is easy to show that $|q(1 + \alpha - \alpha^2)| = 1 + 2\varepsilon^2$. Therefore if the rotational part of the update were zero and either $\sigma_{\mathrm{n}}^2 \to \infty$ or $\sigma_{\mathrm{l}}^2 \to 0$, then q^+ would be of unit length to within $2\varepsilon^2$ (which is due to linearization) plus a noise term. However, the norm error will be larger than this "noise floor" due to the presence of the attitude correction term. The update sequence can be arranged so that the attitude and covariance update are performed first, followed by the update of the norm and the covariance (the residual has to be recomputed). This sequential update is possible because the attitude measurement error and the norm measurement error are independent in equation (B-2). For finite σ_{n}^2 and $\sigma_{\mathrm{l}}^2 > 0$, the norm of q^+ would never naturally reach the noise floor. In particular, for an initial $\sigma_{\mathrm{n}}^2 \to \infty$, we get $\sigma_{\mathrm{n}}^2 = \sigma_{\mathrm{l}}^2$ after the first update, as can be seen in equation (B-4), so the norm correction on subsequent updates would be small unless sufficient process noise is added to σ_{n}^2. The need for this is called "filter tuning" in [3], but the reason why it was needed was not stated. The process noise results in the quaternion norm approaching but never reaching unity. Forced normalization of the quaternion is therefore necessary.

Setting σ_{l}^2 too small will make the covariance ill conditioned and setting $\sigma_{\mathrm{l}}^2 = 0$ will make the covariance singular. Creating a large process noise to feed σ_{n}^2 (to keep σ_{n}^2 large) when $\sigma_{\mathrm{l}}^2 = 0$ will not necessarily alleviate numerical problems in the covariance update calculation because the updated covariance is supposed to be singular. The inevitable conclusion is that a filter designed to process the quaternion "measurement" offers no advantage in producing a normalized quaternion estimate over a rotational update of the reduced-order body-referenced filter.

ATTITUDE INTERPOLATION

Sergei Tanygin[*]

Interpolation may be required for attitude data when repropagation is either impossible or impractical. Nonlinearity of kinematics and potential aliasing between revolutions are among the unique challenges presented by the attitude motion. Selections of the reference frame and attitude parameterization become very important. It is possible to apply standard polynomial interpolation techniques to the attitude data and, also, include the angular velocity data in order to achieve a better accuracy and to reduce potential for aliasing between revolutions. It is also possible to employ interpolated polynomial trajectories for fixed duration near optimal maneuver design and to achieve additional optimization for spinners.

INTRODUCTION

A need for interpolation may generally arise given a set of grid points, where each point contains corresponding values of independent and dependent variables. Interpolation is one of the ways to produce values between the points in the absence of other information relating dependent and independent variables. Interpolation also does that while ensuring that values at the grid points are matched. Interpolation methods vary according to what type of functions they use, how many grid points they use and how many derivatives they can take advantage of at each point. One of the simplest and most common types of functions used for interpolation is a polynomial. Table 1 includes a classification of polynomial interpolation methods according to the number of grid points and derivatives.[1]

TABLE 1 POLYNOMIAL INTERPOLATION METHODS

	Two points	>Two points
No derivatives	Linear Lagrange	Lagrange
1st derivatives	1st Order Taylor or 2-point Osculating	Osculating
1st and higher derivatives	Taylor	Hermite

---

[*] Lead Engineer, Analytical Graphics, Inc., 40 General Warren Blvd., Malvern, Pennsylvania 19355. E-mail: stanygin@stk.com. Member AAS and AIAA.

In practice, ephemeris interpolation has been used extensively when position and optionally velocity data are available at a discrete set of times. The data may originate from simulations or from telemetry and estimation. Re-propagation of ephemeris can be an alternative to interpolation, but it requires knowledge of force models. Even if force models are known, the amount of extra data and computations may become unnecessary when desired accuracy can be satisfied with interpolation. Conceptually, attitude interpolation parallels ephemeris interpolation: it enables relatively fast and accurate computation of attitude and angular velocity at any time spanned by a discrete set of points. The differences come from non-linear nature of attitude composition operations and attitude kinematics. Results of interpolation will depend on both attitude parameterization and reference frame selected for data points. The most straightforward approach calls for independent interpolation of each element of parameterization. At the same time, this places significant restrictions on types of acceptable parameterizations: attitude parameterization becomes unsuitable for interpolation if it imposes constraints on its elements or if it exhibits discontinuities or singularities.

TABLE 2 ATTITUDE PARAMETERIZATIONS

	Constraints	Singularities	Discontinuities
Direction cosine matrix	X		
Unit quaternion	X		
Cayley-Klein parameters	X		
3 subsequent angle-axis rotations		X	
Eigen-axis and function of eigen-angle		X	X
• **Rotation vector**		X	

According to Table 2, these restrictions eliminate from the consideration every possible parameterization. However, a closer examination of the rotation vector parameterization reveals that its singularity at the origin can be easily resolved. The general definition and kinematics are described below:[2]

$$\mathbf{q} = \begin{bmatrix} \hat{\boldsymbol{\varphi}} \sin \phi / 2 \\ \cos \phi / 2 \end{bmatrix},$$ (1)

$$\boldsymbol{\omega} = [1 - \sin \phi / \phi](\hat{\boldsymbol{\varphi}}^T \dot{\boldsymbol{\varphi}})\hat{\boldsymbol{\varphi}} \\ + [\sin \phi / \phi]\dot{\boldsymbol{\varphi}} + [(\cos \phi - 1)/\phi](\hat{\boldsymbol{\varphi}} \times \dot{\boldsymbol{\varphi}})'$$ (2)

$$\dot{\boldsymbol{\omega}} = \ddot{\boldsymbol{\varphi}} - \dot{\phi}[\sin \phi / \phi + 2(1 - \cos \phi)/\phi^2](\hat{\boldsymbol{\varphi}} \times \ddot{\boldsymbol{\varphi}}) \\ + (\dot{\phi}/\phi^2)[(1 - \cos \phi) - 3(1 - \sin \phi / \phi)] \\ [\phi \hat{\boldsymbol{\varphi}} \times (\hat{\boldsymbol{\varphi}} \times \ddot{\boldsymbol{\varphi}}) + \dot{\boldsymbol{\varphi}} \times (\hat{\boldsymbol{\varphi}} \times \dot{\boldsymbol{\varphi}})]$$ (3)

where $\mathbf{q}$ is the four-parameter vector representing the attitude in terms of the unit quaternion, $\boldsymbol{\omega}$ and $\dot{\boldsymbol{\omega}}$ are the body angular velocity and acceleration in the body fixed frame, $\boldsymbol{\varphi} = \hat{\boldsymbol{\varphi}} \phi$ is the rotation vector with the direction $\hat{\boldsymbol{\varphi}}$ along the eigen-axis of rotation relative to the reference frame and with the magnitude ϕ equal to the eigen-angle of rotation. However, near the origin, the definition is simplified to

$$\mathbf{q} \rightarrow \begin{bmatrix} \boldsymbol{\varphi}/2 \\ 0 \end{bmatrix}$$ (4)

and the kinematical relationships reduce to direct correspondence of the rotation vector velocity to the body angular velocity and the rotation vector acceleration to the body angular acceleration:

$$\boldsymbol{\omega} \rightarrow \dot{\boldsymbol{\varphi}},$$ (5)

$$\dot{\boldsymbol{\omega}} \rightarrow \ddot{\boldsymbol{\varphi}}.$$ (6)

These results are important not only for interpolation, but also for design of near optimal attitude maneuvers later in this paper. Finally, note that norm of the rotation vector corresponds to the eigen angle, which, in turn, represents a "distance" measure in attitude space.

ROTATION VECTOR INTERPOLATION

A method for attitude interpolation using rotation vector parameterization is presented in this section. The method combines useful properties of this parameterization described in the introduction with standard polynomial interpolation techniques. The method contains the following steps:

1. Find N attitude grid points centered around time of interest

2. Redefine these N points with respect to attitude of grid point nearest in time

3. Convert resulting points to rotation vector parameterization

4. Perform N point Lagrange interpolation on each rotation vector element independently

5. Convert resulting rotation vector to desired attitude parameterization

The procedure outlined above does not require knowledge of the angular velocity: Step 4 is performed for the attitude alone. The angular velocity can be interpolated separately by also using Lagrange interpolation provided that angular velocity grid points are available. However, de-coupling the attitude from the angular velocity ignores kinematical relationship between the two and may result in a significant loss of accuracy. Note that interpolation of rotations presents a special challenge: without the angular velocity data it may not be possible for the interpolation to distinguish between attitudes separated by complete revolutions, the effect often referred to as aliasing of revolutions. This effect may ultimately result in a sign error causing the apparent interpolated motion to proceed in the direction opposite to the actual motion. These problems provide strong arguments for incorporating the angular velocity data into the attitude interpolation. This can be done by replacing the two separate Lagrange interpolations, one for the attitude and the other for the angular velocity, with the single osculating interpolation that operates on the rotation vector and its first derivative. The derivative needs to be related to the angular velocity via rotation vector kinematics (Eq.(2)). Hence, the interpolation method incorporating angular velocity contains the following steps:

1. Find N attitude and angular velocity points centered around time of interest

2. Redefine these N points with respect to attitude of point nearest in time

3. Convert resulting attitude and angular velocity points to rotation vector parameterization and its velocity

4. Perform N point osculating interpolation on each pair of rotation vector element and its derivative independently; each element of rotation vector velocity is interpolated using derivative of interpolating polynomial

5. Convert resulting rotation vector and its velocity to desired attitude parameterization and angular velocity

Note that N-point Lagrange interpolation employs polynomials of degree N-1, whereas N-point osculating interpolation employs polynomials of degree 2N-1. It is instructive to consider 2-point interpolations in more details in order to gain a better insight into the construction of interpolating polynomials as well as in order to establish the mathematical foundation for the next section.

2-point osculating interpolation employs cubic polynomial that passes between two grid points leaving and arriving at specified slopes in specified time (Fig. 1). Without loss of generality, the time of the first grid point can be set to 0, so that the second grid point simply occurs at the time elapsed between the two points, T, and so that the time along the polynomial progresses between 0 and T.

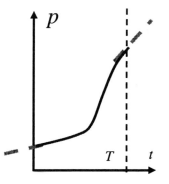

Figure 1 Cubic polynomial $p(t)$ in 2-point osculating interpolation

2-point osculating cubic polynomial is constructed as a linear combination of four other cubic polynomials. These polynomials, often referred to as basis, have coefficients that depend only on the time elapsed between the grid points and do not depend on either values or slopes at the two grid points. Each of the four basis polynomials is then multiplied by one of four constants: two values and two slopes at the grid points. These four scaled polynomials added together compose the 2-point osculating cubic polynomial. Hence, the interpolation of the rotation vector $\varphi(t)$ and its derivatives $\dot{\varphi}(t)$, $\ddot{\varphi}(t)$ results in cubic, quadratic and linear polynomials, $\overline{\varphi}(t)$, $\dot{\overline{\varphi}}(t)$ and $\ddot{\overline{\varphi}}(t)$, respectively. All of them can be formulated as a linear combination of the four other polynomials in the vector form:

$$\overline{\varphi}(t) = \varphi_0 \overline{p}_0(t) + \dot{\varphi}_0 \overline{r}_0(t) + \varphi_T \overline{p}_T(t) + \dot{\varphi}_T \overline{r}_T(t) , \tag{7}$$

$$\dot{\overline{\varphi}}(t) = \varphi_0 \dot{\overline{p}}_0(t) + \dot{\varphi}_0 \dot{\overline{r}}_0(t) + \varphi_T \dot{\overline{p}}_T(t) + \dot{\varphi}_T \dot{\overline{r}}_T(t) , \tag{8}$$

$$\ddot{\overline{\varphi}}(t) = \varphi_0 \ddot{\overline{p}}_0(t) + \dot{\varphi}_0 \ddot{\overline{r}}_0(t) + \varphi_T \ddot{\overline{p}}_T(t) + \dot{\varphi}_T \ddot{\overline{r}}_T(t) , \tag{9}$$

where $\varphi(0) = \varphi_0$, $\dot{\varphi}(0) = \dot{\varphi}_0$, $\varphi(T) = \varphi_T$, $\dot{\varphi}(T) = \dot{\varphi}_T$ and

$$\overline{p}_0(t) = \left[1 + 2\frac{t}{T}\right]\frac{(t-T)^2}{T^2} , \tag{10}$$

$$\overline{p}_T(t) = 1 - \overline{p}_0(t) = \left[3 - 2\frac{t}{T}\right]\frac{t^2}{T^2} , \tag{11}$$

$$\overline{r}_0(t) = \frac{t(t-T)^2}{T^2} , \tag{12}$$

$$\overline{r}_T(t) = \frac{(t-T)t^2}{T^2} , \tag{13}$$

$$\dot{\overline{p}}_0(t) = -\dot{\overline{p}}_T(t) = 6\frac{t(t-T)}{T^3} , \tag{14}$$

$$\dot{\overline{r}}_0(t) = \frac{(t-T)(3t-T)}{T^2} , \tag{15}$$

$$\dot{\overline{r}}_0(t) = \frac{t(3t-2T)}{T^2} , \tag{16}$$

$$\ddot{\overline{p}}_0(t) = -\ddot{\overline{p}}_T(t) = 6\frac{2t-T}{T^3} , \tag{17}$$

$$\ddot{\overline{r}}_0(t) = 2\frac{3t-2T}{T^2} , \tag{18}$$

$$\ddot{\overline{r}}_T(t) = 2\frac{3t-T}{T^2} . \tag{19}$$

As stated above, the essential properties of 2-point osculating interpolation are $\overline{\varphi}(0) = \varphi(0) = \varphi_0$, $\dot{\overline{\varphi}}(0) = \dot{\varphi}(0) = \dot{\varphi}_0$, $\overline{\varphi}(T) = \varphi(T) = \varphi_T$, $\dot{\overline{\varphi}}(T) = \dot{\varphi}(T) = \dot{\varphi}_T$. In turn, these properties can be deduced from observing the following properties of the basis polynomials:

$$\overline{p}_0(0) = \overline{p}_T(T) = \dot{\overline{r}}_0(0) = \dot{\overline{r}}_T(T) = 1 , \tag{20}$$

$$\begin{aligned}\overline{p}_0(T) = \overline{p}_T(0) &= \overline{r}_0(0) = \overline{r}_T(0) = \overline{r}_0(T) = \overline{r}_T(T) \\ &= \dot{\overline{p}}_0(0) = \dot{\overline{p}}_T(0) = \dot{\overline{p}}_0(T) = \dot{\overline{p}}_T(T) = 0\end{aligned} \tag{21}$$

The construction of 2-point osculating polynomials can be somewhat simplified if the grid points are re-defined with respect to the attitude of one of them. For example, re-defining both grid points to be relative to the attitude of the first point, makes that point the origin and makes its rotation vector parameterization a zero vector, $\boldsymbol{\varphi}_0 = \mathbf{0}$. Thus, 2-point osculating polynomials become a linear combination of only three basis polynomials. Another benefit of placing the origin at one of the grid points is the simplification of kinematical relationships of the rotation vector velocity and acceleration with angular velocity and acceleration: near the origin, the the rotation vector velocity and acceleration approximately become the angular velocity and acceleration, respectively (Eqs.(5,6)).

This and previous sections presented the attitude interpolation methods that utilize the attitude and, optionally, the angular velocity data. The angular acceleration along the interpolating polynomials was also formulated here, but its significance is the subject of the next section.

NEAR OPTIMAL FIXED DURATION MANEUVER

Given an attitude maneuver with specific duration and specific initial and target attitudes (and angular velocities), minimizing the overall torque spent during the maneuver is certainly one of most desirable objectives. The challenge lies in relating the shape of the attitude trajectory to the torque along that trajectory: both kinematics and dynamics must be accounted for and they are generally non-linear. Hence, one reason to be interested in the angular acceleration along the interpolating polynomial is because it relates to the rotation vector acceleration on one side and to the body fixed torque on the other side. The 2-point osculating polynomial using rotation vector parameterization quickly emerges as a good candidate for the desired trajectory because of its two properties:

1. Trajectory passes between initial and final attitudes leaving and arriving with specified angular velocities in specified time

2. Trajectory is least curved in rotation vector parameterization

The first property was fully discussed in the previous section. The second property comes from using calculus of variations to minimize the following objective functions:

$$J(\phi_i) = \int_0^T \ddot{\phi}_i^2 \, dt, i = 1,2,3,$$

(22)

where $\phi_i \in \mathfrak{R}^1$ is the ith element of the rotation vector $\boldsymbol{\varphi}$. The Euler-Lagrange equation[4,5] yields the following condition for ϕ_i:

$$\ddddot{\phi}_i \equiv 0, i = 1,2,3,$$

(23)

which is clearly satisfied if ϕ_i is any cubic polynomial.[6] Hence, the 2-point osculating polynomials $\overline{\phi}_i$ developed in the previous section as part of $\overline{\boldsymbol{\varphi}} = [\overline{\phi}_1 \quad \overline{\phi}_2 \quad \overline{\phi}_3]^{\mathrm{T}}$ minimize the objective functions (Eq.(22)):

$$\min_{\phi_i} J(\phi_i) = J(\overline{\phi}_i), i = 1,2,3 , \tag{24}$$

This is the first step towards minimization of the ultimate objective function:

$$J(\mathbf{M}) = \int_0^T \mathbf{M}^{\mathrm{T}}\mathbf{M} dt , \tag{25}$$

where $\mathbf{M} \in \mathfrak{R}^3$ is the applied torque.

The results below follow directly from the properties of rotation vector kinematics presented in the previous sections (Eqs.(5,6)):

$$\lim_{\boldsymbol{\varphi} \to 0} \ddot{\boldsymbol{\varphi}} = \dot{\boldsymbol{\omega}} , \tag{26}$$

$$\lim_{\dot{\boldsymbol{\varphi}} \to 0} \ddot{\boldsymbol{\varphi}} = \dot{\boldsymbol{\omega}} , \tag{27}$$

$$\lim_{\substack{\angle(\boldsymbol{\varphi},\dot{\boldsymbol{\varphi}}) \to 0 \\ \angle(\boldsymbol{\varphi},\ddot{\boldsymbol{\varphi}}) \to 0}} \ddot{\boldsymbol{\varphi}} = \dot{\boldsymbol{\omega}} . \tag{28}$$

In other words, the rotation vector acceleration approaches the angular acceleration under either one of the following three conditions:

1. Trajectory is small

2. Trajectory is slow

3. Trajectory is close to maintained pure spin

This also means that the objective functions based on the angular acceleration components is approximately minimized under the same conditions:

$$\min_{\dot{\omega}_i} \int_0^T \dot{\omega}_i^2 dt \approx \int_0^T \ddot{\overline{\omega}}_i^2 dt, i = 1,2,3 , \tag{29}$$

where $\dot{\omega}_i, \ddot{\overline{\omega}}_i \in \mathfrak{R}^1$ and where $\ddot{\overline{\omega}}_i$ is computed as part of the angular acceleration vector $\dot{\overline{\boldsymbol{\omega}}} = [\dot{\overline{\omega}}_1 \quad \dot{\overline{\omega}}_2 \quad \dot{\overline{\omega}}_3]^{\mathrm{T}}$ along the 2-point osculating rotation vector trajectory.

The next step towards minimization of the applied torque considers the time rate of change of the rigid body angular momentum in the body fixed frame, $\mathbf{I}\dot{\boldsymbol{\omega}} \in \mathfrak{R}^3$, where

$0 < \mathbf{I}^T = \mathbf{I} \in \Re^{3\times3}$ is the body fixed inertia matrix. The following relationship between the objective functions dealing with every component of the angular acceleration, $\dot{\omega}_i$, and with the objective function dealing with $\mathbf{I}\dot{\omega}$ can be established:

$$\min_{\dot{\omega}_i} \int_0^T \dot{\omega}_i^2 dt, i = 1,2,3 \Rightarrow \min_{\dot{\omega}_i, i=1,2,3} \int_0^T (\mathbf{I}\dot{\omega})^T \mathbf{I}\dot{\omega} dt, \tag{30}$$

where the relationship is, of course, preserved in any body fixed frame, but is particularly evident in the principal frame.

The final step must include attitude dynamics in order to relate the time rate of change of the rigid body angular momentum in the body frame to the applied torque:

$$\mathbf{I}\dot{\omega} + \omega \times \mathbf{H} = \mathbf{M}, \tag{31}$$

where $\mathbf{H} \in \Re^3$ is the body total angular momentum in the body fixed frame, which includes the angular momentum of the rigid body itself and the internal angular momentum due to parts moving with respect to the rigid body. The effect of cross coupling, $\omega \times \mathbf{H}$, can be negligible under either one of these two conditions:

1. Trajectory is slow

2. Trajectory is close to pure spin about one of principal axes and internal angular momentum, if any, is close to same axis

This means that under these conditions, the ultimate objective function (Eq.(25)) is approximately minimized whenever $J(\mathbf{I}\dot{\omega}) = \int_0^T (\mathbf{I}\dot{\omega})^T \mathbf{I}\dot{\omega} dt$ is minimized.

In summary, under the sets of kinematical and dynamical conditions, the maneuver trajectory defined by the 2-point osculating polynomial in the rotation vector parameterization approximately minimizes the overall magnitude of the torque applied during the maneuver. While these conditions are satisfied for a large group of maneuvers, e.g. spin-up/-down, slow large angle, etc., the most challenging agile large angle maneuvers may become far from optimal when using the proposed trajectory. The rest of this section presents modifications to the trajectory that may improve its optimality even for agile large angle maneuvers.

The conditions imposed by the attitude dynamics are especially difficult to relax, because of the significant and highly non-linear contribution of the cross coupling during agile maneuvers. However, because of this difficulty, the cross coupling is often compensated for agile spacecraft via the closed-loop feedback linearization in order to ensure a better predictability of attitude trajectories. Accepting this penalty on the applied torque leaves only the kinematical conditions to be relaxed. The attitude trajectories that minimize $J(\mathbf{I}\dot{\omega})$ will be referred to as kinematically optimal in the rest of this paper.

Consider the maneuver trajectory defined in the rotation vector parameterization relative to the attitude of the initial point using the 2-point osculating interpolation. As stated in this and previous sections, this selection of the reference frame makes the trajectory depart from the origin, which means that, at this point, the rotation vector velocity and acceleration are equal to the angular velocity and acceleration, respectively. If this equivalence were maintained throughout the trajectory, the trajectory would be kinematically optimal, but, generally, this is not the case. Generally, the linear progression of the rotation vector acceleration throughout the trajectory does not correspond to the linear progression of the angular acceleration. The latter can become more and more divergent and curved as the trajectory moves further away from the origin, because the effect of kinematical non-linearity can become more and more pronounced. One of the most intuitive approaches to countering this effect is to somehow introduce periodic corrections to the interpolation polynomials. For example, while the same target attitude and angular velocity are sought, the entire interpolation problem can be re-cast using the current trajectory point as the initial point and using the remaining maneuver time to update interpolation polynomials (Fig. 2). This procedure can be repeated as often as necessary depending on the significance of the non-linear kinematical effect. The method of periodic corrections becomes effectively a closed-loop guidance method with periodic updates. For example, the rotation vector, its velocity and acceleration during the period of $0 \leq t \leq t' \leq T$ are governed by the following set of equations:

$$\bar{\varphi}(t) = \omega_0 \bar{r}_0(t) + \varphi_T \bar{p}_T(t) + \dot{\varphi}_T \bar{r}_T(t),$$ (32)

$$\dot{\bar{\varphi}}(t) = \omega_0 \dot{\bar{r}}_0(t) + \varphi_T \dot{\bar{p}}_T(t) + \dot{\varphi}_T \dot{\bar{r}}_T(t),$$ (33)

$$\ddot{\bar{\varphi}}(t) = \omega_0 \ddot{\bar{r}}_0(t) + \varphi_T \ddot{\bar{p}}_T(t) + \dot{\varphi}_T \ddot{\bar{r}}_T(t),$$ (34)

where all terms multiplied by $\varphi_0 \equiv 0$ are removed and where $\dot{\varphi}_0$ is replaced with ω_0, because the two are equal at the origin. At time t' the problem can be re-cast: the target rotation vector and its velocity are redefined relative to $\bar{\varphi}(t')$ resulting in φ_T' and $\dot{\varphi}_T'$, respectively; the basis polynomials are rebuilt using the remaining maneuver time of $T - t'$ instead of T (all updated polynomials are indicated by "prime" in the next set of equations); the original angular velocity ω_0 is replaced with its current counterpart, $\omega_0' = \omega'(t')$. Hence, the following set of equations governs the rotation vector, its velocity and acceleration after this correction and until the end of the maneuver or until the next correction:

$$\bar{\varphi}'(t) = \omega_0' \bar{r}_0'(t) + \varphi_T' \bar{p}_T'(t) + \dot{\varphi}_T' \bar{r}_T'(t),$$ (35)

$$\dot{\bar{\varphi}}'(t) = \omega_0' \dot{\bar{r}}_0'(t) + \varphi_T' \dot{\bar{p}}_T'(t) + \dot{\varphi}_T' \dot{\bar{r}}_T'(t),$$ (36)

$$\ddot{\bar{\varphi}}'(t) = \omega_0' \ddot{\bar{r}}_0'(t) + \varphi_T' \ddot{\bar{p}}_T'(t) + \dot{\varphi}_T' \ddot{\bar{r}}_T'(t).$$ (37)

Note that the time t in the equations above is also reset: it starts at 0 and, in the absence of other corrections, continues until the end of the maneuver at time $T - t'$.

Recall that, in general, more frequent corrections result in smaller contributions from the non-linear kinematics and, thus, result in less curved angular velocity and acceleration trajectories. Therefore, it is natural to seek a transition from periodic corrections to continuous correction throughout the trajectory. The transition becomes clear if the angular acceleration at the initial point of any of the periodic corrections is examined:

$$\dot{\overline{\omega}}'(0) = \ddot{\overline{\varphi}}'(0) = -\omega'_0 \frac{4}{T-t'} + \varphi'_T \frac{6}{(T-t')^2} - \dot{\varphi}'_T \frac{2}{T-t'}. \tag{38}$$

This equation can be rewritten as a continuous function of the original time t assuming that the trajectory includes an infinite number of corrections, each applied over an infinitely small period of time. In other words, any point on this trajectory can be considered the initial point of some correction. The resulting equation after straightforward re-grouping takes form of the closed-loop guidance law. The law specifies the angular acceleration at any point on the trajectory as a function of the current time, total maneuver duration, the current attitude and angular velocity as well as the target attitude and angular velocity:

$$\dot{\overline{\omega}}(t) = \ddot{\overline{\varphi}}(t) = 2 \frac{3\varphi_T(t) - [2\omega(t) + \dot{\varphi}_T(t)](T-t)}{(T-t)^2}, \tag{39}$$

where $\varphi_T(t)$ and $\dot{\varphi}_T(t)$ are the target rotation vector and its velocity relative to the current attitude along the trajectory. Note the apparent singularity as $t \to T$, however, the numerator in this expression becomes proportional to $(T-t)^2$ just as $t \to T$, so that

$$\lim_{t \to T} \dot{\overline{\omega}}(t) = \textbf{const} . \tag{40}$$

In practice, as $t \to T$, the guidance law may be turned off and replaced with a simple attitude control law tracking the target attitude and angular velocity. The guidance law uses continuous local linearization and effectively "slides" interpolating polynomials along the trajectory towards the target (Fig. 2). The resulting trajectory reaches the target at the specified time in a near kinematically optimal manner.

The importance of the appropriate selection of the reference frame for interpolation and for maneuver design is clearly evident. The next section demonstrates the case when not only the initial attitude, but also the target attitude of the maneuver is utilized as the reference frame during the maneuver design.

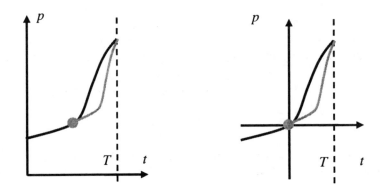

Figure 2 "Sliding" 2-point osculating interpolation of *p(t)*

CLOCK-ANGLE OPTIMIZATION FOR SPINNERS

Consider the attitude trajectory for which the initial point, the target point or both specify only the angular velocity vector and its orientation in some reference frame. For example, if the target trajectory point only specifies the angular velocity, the target attitude itself is not fully defined and possesses an additional degree of freedom: the clock-angle α_T about the angular velocity vector (Fig. 3).

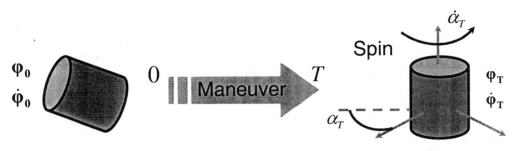

Figure 3 Target clock-angle for spinners

Additional optimization may be performed with respect to this parameter, α_T, assuming that the attitude trajectory takes the form of the 2-point osculating polynomial described in the previous sections. The manner in which the clock-angle enters the cost function strongly depends on the selection of the reference frame for the interpolated rotation vector. In general, the composition operation in the rotation vector parameterization is non-linear and includes trigonometric functions. However, the deliberate selection of the inertial reference frame that aligns one of its axes, e.g. the third axis, with the target angular velocity can linearize the composition operation and limit the clock-angle dependency to only one component of the rotation vector, e.g. $\bar{\phi}_3$ (Fig. 4).

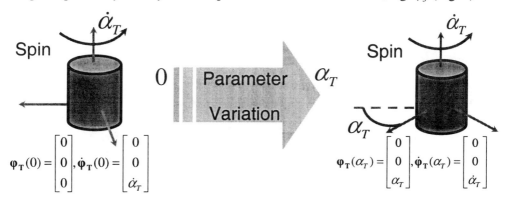

Figure 4 Clock-angle variation

Then, the resulting cost function for the parametric optimization is straightforward:

$$J_p(\alpha_T) = \int_0^T \ddot{\bar{\phi}}_3^2(\alpha_T)dt \tag{41}$$

and so is the solution to the parameter optimization problem[4,5] $\min_{\alpha_T} J_p(\alpha_T) = J_p(\bar{\alpha}_T)$:

$$\bar{\alpha}_T = \phi_{03} + \frac{\dot{\phi}_{03} + \dot{\alpha}_T}{2}T, \tag{42}$$

where ϕ_{03} and $\dot{\phi}_{03}$ are third components of the initial rotation vector and its velocity relative to the inertial reference frame that aligns its third axis with the target angular velocity; $\dot{\alpha}_T$ is the magnitude of the target angular velocity in that same frame. Note that in that frame both the angular velocity and the rotation vector velocity are aligned with the third axis. As expected, the optimal clock-angle value $\bar{\alpha}_T$ is the one that minimizes curvature of the rotation vector velocity, which ultimately leads to the reduction by one of the polynomial degrees for the third component of the rotation vector, its velocity and acceleration (Fig. 5):

$$\bar{\phi}_3(t) = \phi_{03} + \dot{\phi}_{03}t + (\dot{\alpha}_T - \dot{\phi}_{03})\frac{t^2}{2T}, \tag{43}$$

$$\dot{\bar{\phi}}_3(t) = \dot{\phi}_{03} + (\dot{\alpha}_T - \dot{\phi}_{03})\frac{t}{T}, \tag{44}$$

$$\ddot{\overline{\phi}}_3(t) = \frac{\dot{\alpha}_T - \dot{\phi}_{03}}{T} .$$ (45)

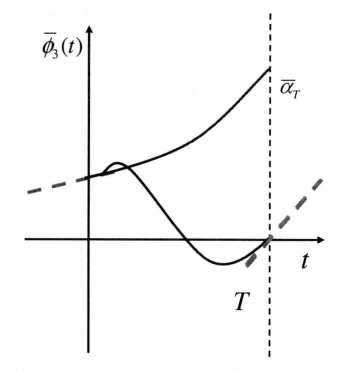

Figure 5 Reduction of curvature with optimal clock-angle

EXAMPLES

· This section illustrates the performance of the closed-loop guidance law. The following conditions are defined for the maneuver:

- Initial quaternion and angular velocity:

$$\mathbf{q}(0s) = \begin{bmatrix} 0 & 0 & 0 & 1 \end{bmatrix}^T$$

$$\omega(0s) = \begin{bmatrix} 0 & 0 & 0.055 \end{bmatrix}^T \text{ deg/s}$$

- Target quaternion and angular velocity:

$$\mathbf{q}(180s) = \begin{bmatrix} 0.3829 & 0.6621 & -0.4139 & -0.4936 \end{bmatrix}^T$$

$$\omega(180s) = \begin{bmatrix} -0.314 & -0.6947 & 0.0226 \end{bmatrix}^T \text{ deg/s}$$

The kinematical performance measure related to the original cost function is defined as follows:

$$Jr(t) = \sqrt{\int_0^t \dot{\omega}^T \dot{\omega} \, d\tau} \, . \tag{46}$$

The error eigen-angle is computed relative to the frame aligned with the target attitude at the final time. Similarly, the error for the rotation vector velocity $\dot{\phi}(t)$ is computed relative to the frame aligned and rotating with the target frame at the final time. The error for the angular velocity $\omega(t)$ is computed as a direct difference between its current vector and the target vector. As expected, all of the errors reach zero at the final time and the difference between the errors in the rotation vector velocity and the angular velocity tend to become equivalent as the attitude trajectory approaches the target.

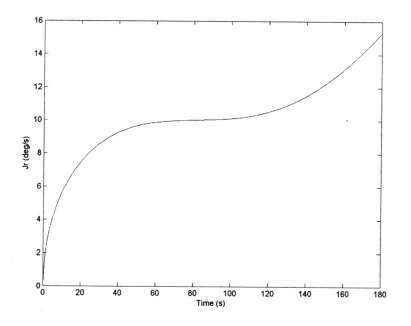

Figure 6 Performance measure

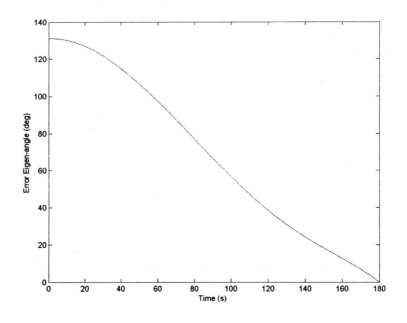

Figure 7 Error eigen-angle relative to target attitude

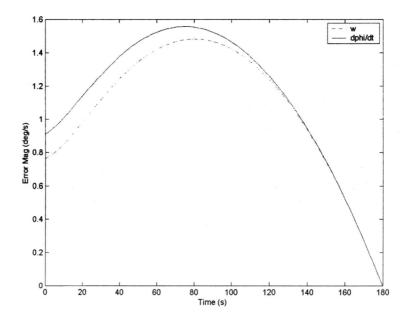

Figure 8 Error magnitudes in terms of angular velocity and rotation vector velocity relative to target attitude

CONCLUSIONS

The paper demonstrated that the attitude interpolation can be implemented using standard polynomial interpolation techniques and the rotation vector parameterization. The angular velocity data can be incorporated in the attitude interpolation improving the accuracy and removing possible aliasing of multiple revolutions.

The paper also demonstrated how the optimality of acceleration along 2-point osculating polynomials can be utilized in the design of near-optimal fixed duration maneuvers. The near-kinematically optimal closed-loop guidance law was designed based on successive and continuously updated interpolations.

The paper specifically addressed maneuver design for spinners, for which additional optimization is possible if the target clock-angle of the spin can be adjusted.

REFERENCES

1. P.J. Davis, *Interpolation and Approximation*, Dover Publications Inc., New York, 1975, pp.24-55.

2. M.D. Shuster, "A Survey of Attitude Representations," *Journal of the Astronautical Sciences*, Vol.41, No.4, 1993, pp.439-517.

3. K. Shoemake, "Animation Rotation with Quaternion Curves," *Computer Graphics,* Vol.19(3), 1985, pp. 245-254.

4. D. Zwillinger, *Handbook of Differential Equations,* 2nd Ed., Academic Press Inc., New York, 1992, pp. 5, 7, 88-93.

5. D. A. Pierre, *Optimization Theory with Applications,* Dover Publications Inc., New York, 1986, pp. 45-46, 62-75.

6. A.H. Barr, B. Currin, S. Gabriel, and J.F. Hughes, "Smooth Interpolation of Orientations with Angular Velocity Constraints using Quaternions," *SIGGRAPH' 92* Proceedings, 1992, pp. 313-320.

NAVIGATION AND ORBIT DETERMINATION - OPERATIONS I

SESSION 14

Chair: Kim Luu
U.S. Air Force Research Laboratory

RECONSTRUCTION OF THE VOYAGER SATURN ENCOUNTER ORBITS IN THE ICRF SYSTEM

Robert A. Jacobson[*]

The Voyager 1 and Voyager 2 spacecraft visited the Saturnian system in November 1980 and August 1981, respectively. Campbell et al.[1] discussed the determination of the spacecraft orbits, and Campbell and Anderson[2] used data from the encounters to improve knowledge of the Saturnian gravity field. In anticipation of the Cassini tour, we have re-examined the results from the Voyager mission. We obtain Voyager trajectories in the International Celestial Reference Frame, and we revise the gravity field taking advantage of improvements made in modelling and data processing since the previous work. We also incorporate a full dynamical model for the Saturnian satellites into the analysis for the first time.

INTRODUCTION

The Saturnian system was visited by the Voyager 1 and Voyager 2 spacecraft in November 1980 and August 1981, respectively. Campbell et al.[1] provided a detailed discussion of the determination of the orbits of both spacecraft. Using tracking data acquired during the Voyager encounters, Campbell and Anderson[2] subsequently improved the knowledge of the Saturnian gravity field. The Voyager imaging data also aided in the determination of the orbits of the Saturnian satellites[3,4,5]. As a part of the preparation for the Cassini tour of the Saturnian system, beginning in July 2004, we re-examined the results from the Voyager mission. The objectives of our analysis were:

- to repeat the gravity field investigation taking advantage of improvements made in modelling and data processing since the original analysis.
- to utilize all of the Voyager imaging data in the development of the Saturnian satellite ephemerides; the archival Voyager reconstructed trajectories do not span the entire data interval.
- to obtain Voyager trajectories in the International Celestial Reference Frame (ICRF) which will facilitate combined Voyager and Cassini data analyses; the original Voyager analysis was in the B1950 system and Cassini uses the ICRF.

[*] Member of the Technical Staff, Jet Propulsion Laboratory, California Institute of Technology, 4800 Oak Grove Drive, Pasadena, California 91109-8099. Associate Fellow AIAA.

Campbell *et al.* performed separate reconstructions for each spacecraft. In the Voyager 1 case they used a 42 day data arc from 11 October 1980 to 22 November 1980. The data included noncoherent one-way Doppler, coherent two-way and three-way Doppler, range, and imaging (pictures of the satellites against a stellar background acquired with vidicon system). For Voyager 2 they used a 55 day data arc from 1 August 1981 to 24 September 1981. The data types were identical to those of Voyager 1 with the exception that no one-way or three-way Doppler was used. The Saturnian satellite ephemerides during the Voyager 1 encounter, where a high accuracy Titan orbit was needed, were produced with numerical integration; during the Voyager 2 encounter they were based on less accurate analytical theories.

In their gravity work Campbell and Anderson used a combination of the Voyager data and coherent Doppler obtained from the earlier Pioneer 11 Saturn encounter[6]. Besides the gravity parameters, they obtained reconstructed trajectories for all three spacecraft. In this case the reconstructions were 'simultaneous' in the sense that estimates of the common parameters such as those of the gravity field were determined from the combined data set. The Voyager 2 data also included 6 hours of noncoherent one-way Doppler just after the Tethys flyby. A single set satellite ephemerides, produced with the analytical theories, covered the three encounters[3]. The data spans, however, were limited to the periods from 10 days before to 5 days after each Saturn closest approach. Consequently, the trajectories applied only the close encounter time frame.

Our analysis relies on data arcs of 105 days (7 August 1980 to 20 November 1980) for Voyager 1 and 106 days (8 June 1981 to 22 September 1981) for Voyager 2. The arcs begin at the time of the earliest useable imaging data; they terminate slightly earlier than those of Campbell *et al.* because we elected to stop at the end of the availability of the calibrations for the effects of interplanetary plasma. To enhance our gravity parameter solutions, we include the Voyager 2 noncoherent one-way Doppler and the Pioneer 11 Doppler used by Campbell and Anderson. We cover the encounters with a single set satellite ephemerides obtained from a dynamically complete numerical integration of the satellite orbits. We have also extended the ephemerides to include the Lagrangian satellites Helene, Telesto, and Calypso. Observations of these satellites provide valuable information on the masses of Tethys and Dione[4].

ANALYSIS

Data

We relied on three basic types of data in the analysis: spacecraft radiometric tracking, spacecraft optical navigation observations, and satellite astrometry. The Pioneer Doppler and Voyager Doppler and range tracking data are essentially identical to that used in the earlier analyses. The only differences, as noted in the previous section, were the in lengths of the Voyager data arcs. A complete description of the spacecraft data appears in the references. We calibrated the tracking data for the effects of the Earth's troposphere and ionosphere and the interplanetary plasma (the Pioneer Doppler was calibrated only for the troposphere). The Voyager imaging data were originally referenced to star catalogs in B1950 system; we modified them, replacing the reference star locations with ICRF positions from the Tycho 2 star catalog.

The satellite astrometry is derived from telescopic observations made at several astronomical observatories over the time period 1966 to 2003. It includes both satellite to satellite relative positions, satellite to planet relative positions, and absolute satellite positions. References for the pre-1990 observations of the major satellites may be found in Strugnell and Taylor[7]. The post-1990 observation set contains relative positions measured photographically[8-11], relative positions measured with CCDs[12-19], and ICRF positions measured with CCDs[20-25]. Photographic relative positions of the Lagrangian satellites may be found in Ref. 26-36, and CCD relative positions appear in Ref. 17,18,37,38. Augmenting the data from Earthbased observatories are satellite to satellite relative positions of the major and Lagrangian satellites obtained with the Hubble Space Telescope[39,40].

Dynamical Model

Our dynamical model contains both gravitational and non-gravitational forces; the former affect the motion of the spacecraft, planet, and satellites whereas the latter affect only the spacecraft.

Sources of the gravitational forces are the Sun, the solar system planets, and the Saturnian satellites. JPL planetary ephemeris DE405[41] provides the positions and masses of the Sun and planets. The satellite positions are from ephemerides based on high precision numerical integration and are significantly more accurate than those from the analytical theories[3] used previously. The increased accuracy not only improves the dynamical modelling of the spacecraft trajectories but also enhances the contribution of the optical navigation data to the determination of the Voyager trajectories. Moreover, the integrated ephemerides, unlike the analytical theories, are sensitive to the values of the gravity parameters; the sensitivity aids in the determination of those parameters.

The gravity field of the planet is represented by the standard spherical harmonic expansion of its gravitational potential. We use the same degree and order harmonic coefficients as did Campbell and Anderson. The orientation of the pole of Saturn (needed for the gravity field model) cannot be accurately determined from spacecraft and satellite data. Campbell and Anderson adopted the pole direction found by Simpson et al.[42] using Voyager occultation data. We replaced that pole with the revised one from French et al.[43] and have included the pole precession rate of Nicholson et al.[44].

Solar radiation pressure effects on the spacecraft are modelled with the formulation of Georgevic[45]. The values of the parameters in the models were determined during the Earth-Jupiter cruise period of each spacecraft. We retained those values.

The Voyager spacecraft are three-axis stabilized. Attitude is changed and maintained by groups of thrusters which are unbalanced, i.e., they do not fire in pairs separated from the center of mass on opposite moment arms. Consequently, there is a net translational velocity imparted to the spacecraft each time a thruster is fired. In addition, due to a design flaw, the exhaust plumes from the pitch thrusters strike the spacecraft adding to the translational velocity when they are fired. We included impulses along the spacecraft axes at the times of a number of the larger attitude changes to account for the translations. The remaining attitude control pulses were modelled as the sum of constant and stochastic accelerations along the spacecraft axes. This model also absorbed the effects of non-isotropic thermal radiation from the RTGs (Pu^{238} radioactive thermal generators which provide electrical power) and solar pressure mis-modelling. The non-gravitational accelerations were a major source of Voyager navigation error; Ref. 46 contains an excellent discussion of them. During the time frame of our analysis each Voyager spacecraft made two trajectory correction maneuvers (TCM). We modelled them with the finite burn (rocket equation) model.

The Pioneer spacecraft is spin stabilized(5 rpm). To maintain the high-gain antenna pointing toward the Earth the spin axis had to be re-oriented (precessed) periodically. Two precession maneuvers occurred within our data arc, and we modelled them as velocity impulses along the spacecraft axes. Pioneer is also subject to non-gravitational accelerations due to gas leaks in the attitude control system, thermal radiation from the RTGs, and mis-modelling of the solar pressure acceleration. We accounted for these effects with a constant acceleration along the spin axis; the spacecraft's rotation causes the effects of accelerations normal to the spin axis average out.

Method of Solution

We determined the orbits of the spacecraft, the planet, and the satellites by adjusting parameters in the dynamical model to obtain a weighted least-squares fit to the observational data. The fundamental adjustable parameters were:

- epoch position and velocity of each spacecraft and satellite
- elements of the Saturn orbit
- GM's of the Saturnian system and the satellites
- gravitational harmonics of Saturn
- thrust magnitude and direction for large spacecraft maneuvers
- impulsive velocity changes for small spacecraft maneuvers
- non-gravitational accelerations

In order to obtain an adequate fit to the observations we also had to adjust the following parameters in the observation model:

- one-way Doppler biases and drift rates
- station dependent range biases
- spacecraft camera pointing angles

Unlike all previous analyses we did not need to account for errors in the locations of the Earth tracking stations because they are well known in the ICRF system. For the same reason we ignored possible errors in the ephemeris of the Earth.

We processed the observations with a batch-sequential, square-root information filter, treating the stochastic non-gravitational accelerations as colored noise with a 1 day correlation time and the range biases and camera pointing angles as white noise. The accelerations were batched at 1 day intervals, the range biases were batched by tracking pass, and the pointing angles were batched by picture. As did Campbell and Anderson, we included a priori information on Saturn's zonal harmonics from the ringlet constraint devised by Nicholson and Porco[47]. A priori information on the Saturn orbit, based on the data used in the development of DE405, was provided by E. M. Standish[48].

With multiple data types data weights balance the information provided by each type as well as represent the accuracy of the type. Assigning the weights is as much an art as a science. Our selections were guided by knowledge of the potential accuracy of the type coupled with an examination of the data residuals.

We set separate Doppler weights for each DSN pass to correspond to an accuracy of 2.5 times the root-mean-square (rms) of the residuals for that pass. However, no two-way Doppler was weighted tighter than 1.0 mm sec^{-1}, and no one-way or three-way Doppler was weighted tighter than 2.0 mm sec^{-1}. The accuracies represented by the range weights were 400 m for Voyager 1 and 200 m for Voyager 2 (the interplanetary plasma had a higher effect on Voyager 1). The range was deweighted from its actual accuracy of a few tens of meters because fitting it is sensitive to the non-gravitational accelerations. The stochastic range biases account for range calibration errors and further deweight the range data. The biases had 10 m a priori uncertainties.

The accuracy assumed for most of the imaging data was 0.5 pixels for the stars and major satellites and 1.0 pixels for the Lagrangian satellites (their locations were not measured as carefully as were the locations of the major satellites). Near the encounter the weights for the major satellites were decreased to represent a larger uncertainty of 1.0 pixel. The change accounted for the increased difficulty in finding the image centers as their size grew with decreasing spacecraft range. Moreover, the close encounter images were overexposed further complicating the centerfinding. The post-encounter Voyager 2 images, taken with the wide angle camera, were deweighted (7.0 pixel uncertainty) because of centerfinding problems introduced by their high phase angles.

The satellite astrometric data were grouped according to data type, observatory, and the observing period in which they were acquired. The accuracy of each group was taken to be equal to the rms of residuals of the group.

RESULTS

Our estimated gravity field parameter values and those obtained by Campbell and Anderson appear in Table 1. We are in close agreement on the GMs of the system, Rhea, and Titan, and the J_2 of Saturn and, except for the Dione and Iapetus GMs, agree within the uncertainties on the other parameters. Our analysis included a massive Hyperion based on an assumed density of 1.1 gm/cm^3 but omitted the effects of the masses of the Lagrangian satellites (their GMs are estimated to be less than 0.002 km^{3}sec^{-2}). We ignored the Saturnian C_{22} and S_{22}; Campbell and Anderson's determination of those two parameters was marginal at best suggesting that they have little effect on the spacecraft trajectories.

Our GMs of the inner four satellites are determined primarily from the Earthbased observations and the satellite dynamics. The Lagrangian satellite data are the primary source of the information on the Tethys and Dione GMs; the earlier values were found by Kozai[49] based on observations of Mimas and Enceladus and the Mimas-Tethys and Enceladus-Dione orbit resonances. Those resonances coupled with observations of Tethys and Dione (and the Lagrangian satellites) also lead to the GMs of Mimas and Enceladus. We match Kozai's GMs surprisingly well considering our differing observation sets and orbital motion model (he used an approximate analytical theory).

Our Iapetus GM is about 25% larger than Campbell and Anderson's result and is somewhat less certain. However, it is close to the values they found from the Pioneer data only (136 km^{3}sec^{-2}) and the Voyager 1 data only (131 km^{3}sec^{-2}). Their final result was apparently dominated by the contribution of the Voyager 2 data; we did not find that domination.

We estimated corrections to the Saturn ephemeris of the order of 800 km in the in-orbit direction and 250 km in the radial and out-of-plane directions. The in-orbit correction is slightly larger than the 700 km 1-σ uncertainty associated with DE405; it should be noted that no Voyager Saturn data were used in the development of DE405.

Our corrected satellite ephemerides differ from those described in Ref. 5 by less than 150 km. A thorough discussion of the satellite orbit determination will appear in a future paper.

Table 2 provides the distances and times of the Pioneer and Voyager spacecraft close approaches to Jupiter and its satellites. The values are in good agreement with those given by Campbell and Anderson. Differences are mainly due to differences in the satellite ephemerides.

Table 1

GRAVITY PARAMETERS

Parameter[†]	Campbell and Anderson	Reconstruction
GM_{system}	37940630. ±200.	37940672. ±100.
GM_{Mimas}	2.50± 0.06[‡]	2.56± 0.04
$GM_{\text{Enceladus}}$	4.9 ± 2.4[‡]	5.77± 1.19
GM_{Tethys}	45. ± 10.	41.21± 0.04
GM_{Dione}	70.2 ± 2.2[‡]	73.13± 0.02
GM_{Rhea}	154. ± 4.	154.59± 3.87
GM_{Titan}	8978.2 ± 1.	8978.03± 0.92
GM_{Hyperion}		0.72± 0.35[‡]
GM_{Iapetus}	106. ± 10.	131.72± 15.00
J_2	16298. ±10.	16294.6± 6.0
J_4	−915. ±40.	−919.8±26.1
J_6 ⎫ × 10^6	103. ±50.	99.7±27.6
J_8 ⎬	−10.[‡]	−10.[‡]
C_{22} ⎭	0.7± 1.	
S_{22}	−0.2± 1.	
α_{p}	40.580±0.016[‡]	40.5955±0.0036[‡]
δ_{p}	83.540±0.002[‡]	83.5381±0.0002[‡]
$\dot{\alpha}_{\text{p}}$		−0.04229[‡]
$\dot{\delta}_{\text{p}}$		−0.00444[‡]

[†]units: GM(km^{3}sec^{-2}), $\alpha_{\text{p}}, \delta_{\text{p}}$(deg), $\dot{\alpha}_{\text{p}}, \dot{\delta}_{\text{p}}$(deg century^{-1})
[‡]not estimated

Table 2

CLOSE APPROACH DISTANCES AND TIMES

Object	Pioneer 11		Voyager 1		Voyager 2	
	R (km)	T_{CA}	R (km)	T_{CA}	R (km)	T_{CA}
Saturn	80930	E^a	184141	E^b	161126	E^c
Mimas	104210	E- 0^h1	88407	E+ 2^h0	309764	E- 0^h8
Enceladus	222311	E+ 2^h0	201934	E+ 2^h1	87021	E+ 0^h4
Tethys	329398	E+ 1^h9	415532	E- 1^h5	93018	E+ 2^h8
Dione	291292	E- 0^h5	161499	E+ 3^h9	502289	E- 2^h3
Rhea	345723	E+ 6^h0	73985	E+ 6^h6	645320	E+ 3^h1
Titan	362910	E+25^h5	6498	E-18^h1	666096	E-17^h8
Hyperion	665977	E-82^h5	870823	E+17^h0	472737	E-26^h0
Iapetus	1033186	E-76^h5	2476556	E+43^h5	908487	E-74^h0

[a] 1 Sep. 1979 16:30:34 GMT,   [b]12 Nov. 1980 23:45:43 GMT
[c]26 Aug. 1981 03:24:05 GMT

Tables 3 and 4 constrast the navigation results found by the previous analysis and our reconstruction. The tables contain the standard B-plane coordinates, the error ellipse semi-major (SMAA) and semi-minor (SMIA) axes and orientation angle (θ) measured from the T axis, and the time of closest approach (TCA) and the error in that time. The agreement between the results is quite good considering differing reference frames, models, and estimation procedures. The most striking differences are our larger error ellipse semi-major axis and smaller arrival time uncertainty for Voyager 2. We speculate that the differences are caused by Campbell *et al.*'s use of the non-dynamic satellite ephemerides versus our use of the integrated ephemerides for Voyager 2.

The estimated velocity changes imparted by the spacecraft maneuvers appear in Table 5. Except for the TCMs all changes are referred to the spacecraft coordinate axes. For Pioneer 11 the Z axis is along the spin axis which was maintained within 3 degrees of the Earth direction; the other two axes were normal to Z and to each other. For the Voyagers the Z axis is the axis of symmetry of the spacecraft bus and the X and Y axes are the pitch and yaw axes, respectively. The centerline of the high gain antenna is aligned with the Z axis and was normally pointed toward the Earth. For the TCMs, modelled as finite burns, the table contains the ICRF coordinates of the velocity change accumulated during the maneuver.

Table 3

VOYAGER 1 NAVIGATION PERFORMANCE - TITAN B-PLANE

Source	B·R (km)	B·T (km)	SMAA (km)	SMIA (km)	θ (deg)	TCA (H:M:S)	σ_{TCA} (sec)
Campbell *et al.*	1875	6250	2.	1.	67.7	5:41:14	0.03
This work	1879	6252	1.3	0.2	110.1	5:41:14	0.04

Table 4

VOYAGER 2 NAVIGATION PERFORMANCE - JUPITER B-PLANE

Source	B·R (km)	B·T (km)	SMAA (km)	SMIA (km)	θ (deg)	TCA (H:M:S)	σ_{TCA} (sec)
Campbell *et al.*	-17797	364021	1.5	1.4	48.2	3:24:57	0.1
This work	-17802	364024	13.4	1.1	87.6	3:24:57	0.01

The values of the constant non-gravitational accelerations are given in Table 6. The Pioneer acceleration is less than the level found by Null. The Voyager constant accelerations are consistent with those found during Voyager mission operations. Figures 1 and 2 show the Voyager stochastic accelerations along the Z axis; the accelerations along the other axes are an order of magnitude smaller. The stochastic accelerations are near the expected levels of 5×10^{-12} km sec^{-2}; the root-mean-square values are less than 3×10^{-12} km sec^{-2}.

Figures 3—10 show the spacecraft data residuals and give an idea of the data spans, noise levels, and quality of the data fit. The residuals confirm that our fit to the data is as good if not slightly better than the fits done for the earlier navigation and gravity analyses.

Some of the variation in the noise in the Doppler data is the result of differing sample times. All of the close encounter data had 1 min sample times. At the start of the arc the Voyager 1 data was compressed to a 1 hour sample time; the Voyager 2 data used 20 min samples. Near the encounter times the compression varied between 5 min and 20 min for both spacecraft. Solar conjunction is responsible for the Voyager 1 data gap in late September. Solar plasma effects related to the conjunction caused the high data noise in early September and October. In general both the Voyager 1 range and Doppler are noisier than that of Voyager 2; probably reflecting higher activity in the interplanetary plasma during the Voyager 1 encounter period.

In the optical residual figures the symbols (somewhat difficult to see), indicate which satellite was being imaged: Mimas(M), Enceladus(E), Tethys(t), Dione(D), Rhea(R), Titan(T), Hyperion(H), Iapetus(I), Helene(h), Telesto(+), Calypso(C). The residuals are well within the observational uncertainties. The degradation of fit to the close encounter images and post-encounter Voyager 2 images, attributed to centerfinding errors, can clearly be seen.

CONCLUDING REMARKS

In this paper we have reported a new reconstruction of the Voyager Saturn encounter trajectories. The reconstruction was done as part of an investigation of the Saturnian system gravity field and the orbits of the Saturnian satellites. The new trajectories are needed in order to properly process the Voyager tracking and optical navigation data for that investigation. Because of improvements in our models and data processing procedures and our use of the modern ICRF reference frame, we believe the new trajectories to be the most accurate descriptions produced thus far for the spacecraft motion through the Saturnian system.

Table 5

MANEUVERS (mm sec^{-1})

Time(TDB)		$\Delta \dot{X}$	$\Delta \dot{Y}$	$\Delta \dot{Z}$	Event
Pioneer 11					
28-Aug-1979	13:37:09	-0.031	0.030	2.622	precession
28-Aug-1979	17:47:22	-0.035	0.031	1.347	precession
Voyager 1					
19-Aug-1980	11:45:00	-1.826	-1.669	-3.009	Ant. & Sun sensor cal.
23-Aug 1980	09:00:00	-1.742	-1.397	-3.705	Cruise science maneuver
10-Oct-1980	19:09:51	-1360.091	1183.871	-274.575	TCM8
07-Nov-1980	03:39:58	-550.861	1229.997	-640.510	TCM9
13-Nov-1980	05:30:00	85.892	-130.782	-3.861	Science turns
13-Nov-1980	07:30:00	-56.890	110.370	-2.380	Science turns
13-Nov-1980	21:30:00	0.536	-3.260	-0.862	Science turns
Voyager 2					
19-Jul-1981	11:16:25	-469.648	882.510	648.463	TCM8
01-Aug-1981	00:00:00	-32.485	129.635	-5.098	Vertical system scan
13-Aug-1981	08:00:00	15.890	-4.254	-1.215	Roll to Procyon
15-Aug-1981	09:30:00	-14.243	27.886	-0.660	Roll to Canopus
18-Aug-1981	21:26:16	-230.724	1205.798	-534.272	TCM9
24-Aug-1981	07:50:00	-45.208	80.182	-0.821	Roll to Miaplacidus
25-Aug-1981	12:40:00	-31.082	-10.951	3.606	Science turns
26-Aug-1981	02:39:11	-2.299	-0.946	4.741	Science turns
26-Aug-1981	02:47:31	-2.147	-0.879	4.714	Science turns
26-Aug-1981	03:08:47	-16.523	-6.714	4.671	Science turns
26-Aug-1981	03:26:06	-14.960	-6.070	4.685	Science turns
26-Aug-1981	05:25:39	-1.346	-0.569	7.688	Science turns
26-Aug-1981	05:41:53	-1.288	-0.543	-1.239	Science turns
26-Aug-1981	07:38:00	-0.215	-0.034	1.315	Science turns
26-Aug-1981	08:09:10	34.058	23.408	1.946	Roll to Vega
04-Sep-1981	01:46:00	-0.508	-2.048	-0.781	Roll to Canopus
05-Sep-1981	03:10:00	0.272	0.961	-1.330	Roll to Miaplacidus
09-Sep-1981	22:30:00	0.613	0.470	-3.519	Scale factor test

Table 6

NON-GRAVITATIONAL ACCELERATIONS
$$(\text{km sec}^{-2} \times 10^{-12})$$

Axis	Pioneer 11	Voyager 1	Voyager 2
X		-3.589	-1.242
Y		6.361	1.592
Z	-3.825	-5.353	-4.039

ACKNOWLEDGEMENT

We would like to thank Bill Owen for converting the Voyager optical data to the ICRF system. Thanks to George Null, Jim Campbell, Steve Synnott, and Ed Riedel for their informative discussions. The research described in this paper was performed at the Jet Propulsion Laboratory, California Institute of Technology, under contract with the National Aeronautics and Space Administration.

REFERENCES

1. Campbell, J. K., Jacobson, R. A., Riedel, J. E., Synnott, S. P., and Taylor, A. H. (1982) 'Voyager I and Voyager II Saturn Encounter Orbit Determination', AIAA Paper No. 82-0419, presented at the AIAA 20th Aerospace Sciences Meeting, Orlando, Flordia.
2. Campbell, J. K. and Anderson, J. D. (1989) 'Gravity Field of the Saturnian System from Pioneer and Voyager Tracking Data', *The Astronomical Journal*, **97**, pp. 1485-1495.
3. Jacobson, R. A., Campbell, J. K., and Synnott, S. P. (1982) 'Satellite Ephemerides for Voyager Saturn Encounter', AIAA Paper No. 82-1472, presented at the AIAA/AAS Astrodynamics Converence, San Diego, California.
4. Jacobson, R. A. (1995) 'The Orbits of the Minor Saturnian Satellites' *Bulletin of the American Astronomical Society*, **27**, No.3, pp. 1202-1203.
5. Jacobson, R. A. (1996) 'The Orbits of the Saturnian Satellites from Earthbased and Voyager Observations, *Bulletin of the American Astronomical Society*, **28**, No.3, pp. 1185.
6. Null, G. W., Lau, E. L., Biller, E. D., and Anderson, J. D. (1981) 'Saturn Gravity Results Obtained from Pioneer 11 Tracking Data and Earth-based Saturn Satellite Data', *The Astronomical Journal*, **86**, pp. 456-468.
7. Strugnell, P. R., and Taylor, D. B. (1990) 'A catalogue of ground-based observations of the eight major satellites of Saturn, 1874-1989', *Astronomy and Astrophysics Supplement Series*, **83**, pp. 289-300.
8. Standish, E. M. (1996) 'Astrographic Positions of the Satellites of Jupiter, Saturn, and Uranus', JPL Internal Document IOM 312.1-96-020, Jet Propulsion Laboratory, Pasadena, Ca.
9. Veiga, C. H. and Vieira Martins, R. (1999) 'Photographic positions for the first eight satellites of Saturn', *Astronomy and Astrophysics Supplement Series*, **139**, pp. 305-310.

10. Whipple, A. L. (1993) 'McDonald observations in 1992 and 1993', personal communication.
11. Whipple, A. L. (1995) 'McDonald observations in 1994', personal communication.
12. Harper, D., Murray, C. D., Beurle, K., Williams, I. P., Jones, D. H. P., Taylor, D. B., and Greaves, S. C. (1997) 'CCD astrometry of Saturn's satellites 1990-1994', *Astronomy and Astrophysics Supplement Series*, **121**, pp. 65-69.
13. Harper, D., Beurle, K., Williams, I. P., Murray, C. D., Taylor, D. B., Fitzsimmons, A., and Cartwright I. M. (1999) 'CCD astrometry of Saturn's satellites in 1995 and 1997', *Astronomy and Astrophysics Supplement Series*, **136**, pp. 257-259.
14. Peng, Q. Y., Vienne, A., and Shen, K. X. (2002) 'Positional measuring procedure and CCD observations for Saturnian satellites', *Astronomy and Astrophysics*, **383**, pp. 296-301.
15. Nicholson, P. D. (1992) 'Palomar observations in 1990', personal communication.
16. Nicholson, P. D. (1994) 'Palomar observations in 1993', personal communication.
17. Rohde, J. R. and Pascu, D. (1993) 'Astrometric Observations of Helene (SXII), Telesto(SXIII), and Calypso (SXIV)', *Bulletin of the American Astronomical Society*, **25**, pp. 1235.
18. Rohde, J. R. and Pascu, D. (1994) 'CCD astrometry of Helene (SXII), Telesto(SXIII), and Calypso (SXIV): 1993 observations', *Bulletin of the American Astronomical Society*, **26**, pp. 1024.
19. Vienne, A., Thuillot, W., Veiga, C. H., Arlot, J.-E., and Vieira Martins, R. (2001) 'Saturnian satellite observations made in Brazil during the 1995 opposition with an astrometric analysis', *Astronomy and Astrophysics*, **380**, pp. 727-733.
20. Owen, W. M. (2002) 'Table Mountain observations', personal communication.
21. Qiao, R. C., Shen, K. X., Liu, J. R., and Harper, D. (1999) '1994-1996 CCD astrometric observations of Saturn's satellites and comparison with theories', *Astronomy and Astrophysics Supplement Series*, **137**, pp. 1-5.
22. Rapaport, M., Teixeira, R., Le Campion, J. F., Ducourant, C., Camargo, J. I. B., and Benevides-Soares, P. (2002) 'Astrometry of Pluto and Saturn with the CCD meridian instruments of Bordeaux and Valinhos', *Astronomy and Astrophysics*, **383**, pp. 1054-1061.
23. Stone, R. C. and Harris, F. H. (2000) 'CCD positions determined in the international celestial reference frame for the outer planets and many of their satellites in 1995-1999', *Astronomical Journal*, **119**, pp. 1985-1998.
24. Stone, R. C. (2000) 'CCD positions for the outer planets and many of their satellites. IV. FASTT observations taken in 1999-2000', *Astronomical Journal*, **120**, pp. 2124-2130.
25. Stone, R. C. (2001) 'Positions for the outer planets and many of their satellites. V. FASTT observations taken in 2000-2001', *Astronomical Journal*, **122**, pp. 2723-2733.
26. Lamy, P. and Mauron, N. (1980) IAU Circ. No. 3491
27. Laques, P. and Lecacheux, J. (1980) IAU Circ. No. 3483
28. Leliévre (1980) IAU Circ. No. 3545
29. Retisema, H. J. (1980) IAU Circ. No. 3466
30. Veillet, C. (1980) IAU Circ. No. 3470
31. Larson, S. M. and Fountain, J. W. (1981) IAU Circ. No. 3602
32. Pascu, D. and Seidelmann, P. K. (1981) IAU Circ. No. 3619
33. Reitsema, H. J. (1981) 'The libration of the Saturnian satellite Dione B', *Icarus*, **48**, 23.
34. Seidelmann, P. K., Harrington, R. S., Pascu, D., Baum, W. A., Currie, D. G., Westphal, J. A., and Danielson, G. E. (1981) 'Saturn Satellite Observations and Orbits from the 1980 Ring Plane Crossing', *Icarus*, **47**, pp. 282-287.

35. Oberti, P., Veillet, C. and Catullo, V. (1989) 'Lagrangian satellites of Tethys and Dione. I. Reduction of observations', *Astronomy and Astrophysics Supplement Series*, **80**, pp. 289-297.

36. Veiga, C. H. and Vieira Martins, R. (2000) 'Astrometric photographic observations of Helene', *Astronomy and Astrophysics Supplement Series*, **143**, pp. 405-407.

37. Martinka, S. and Pascu, D. (1998) data held at the U. S. Naval Observatory, personal communication

38. Veiga, C. H., Vieira Martins, R., Vienne, A., Thuillot, W., and Arlot, J.-E. (2003) 'CCD astrometric observations of Saturnian satellites', *Astronomy and Astrophysics*, to appear.

39. French, R. G. (2001) 'HST observations', personal communication.

40. McGhee, C. A., Nicholson, P. D., French, R. G., and Hall, K. J. (2001) 'HST observations of Saturnian satellites during the 1995 ring plane crossings', *Icarus*, **152**, pp. 282-315.

41. Standish, E. M. (1998) 'JPL Planetary and Lunar Ephemerides,DE405/LE405', JPL Internal Document IOM 312.F-98-048, Jet Propulsion Laboratory, Pasadena, Ca.(available electronically at http://ssd.jpl.nasa.gov/iau-comm4)

42. Simpson, R. A., Tyler, G. L., and Holberg, J. B. (1983) 'Saturn's Pole: Geometric Correction based on Voyager UVS and Radio Occultations', *The Astronomical Journal*, **88**, pp. 1531-1536.

43. French, R. G., Nicholson, P. D., Cooke, M. L., Elliot, J. L., Matthews, K., Perkovic, O., Tollestrup, E., Harvey, P., Chanover, N. J., Clark, M. A., Dunham, E. W., Forrest, W., Harrington, J., Pipher, J., Brahic, A., Grenier, I., Roques, F., and Arndt, M. (1993) 'Geometry of the Saturn System from the 3 July 1989 Occultation of 28 Sgr and Voyager Observations', *Icarus*, **103**, pp. 163-214.

44. Nicholson, P. D., French, R. G., and Bosh, A. S. (1999) 'Ring plane crossings and Saturn's pole precession', *Bulletin of the American Astronomical Society*, **31**, pp. 1140.

45. Georgevic, R. M. (1971) 'Mathematical Model of the Solar Radiation Force and Torques Acting on the Components of a Spacecraft', JPL TM 33-494, Jet Propulsion Laboratory, Pasadena, Ca.

46. Campbell, J. K., Synnott, S. P., Riedel, J. E., Mandell, S., Morabito, L. A., and Rinker, G. C. (1980) 'Voyager 1 and Voyager 2 Jupiter Encounter Orbit Determination', AIAA Paper No. 80-0241 presented at the AIAA 18th Aerospace Sciences Meeting, Pasadena, California.

47. Nicholson, P. D., and Porco, C. C. (1988) 'A New Constraint on Saturn's Zonal Gravity Harmonics from Voyager Observations of an Eccentric Ringlet', *Journal of Geophysical Research*, **93**, pp. pp. 10209-10224.

48. Standish, E. M. (2002) 'A covariance for Saturn', personal communication.

49. Kozai, Y. (1957) 'On the Astronomical Constants of the Saturnian Satellites Systen', *Annals of the Tokyo Astronomical Observatory Second Series*, **5**, pp. 73-106.

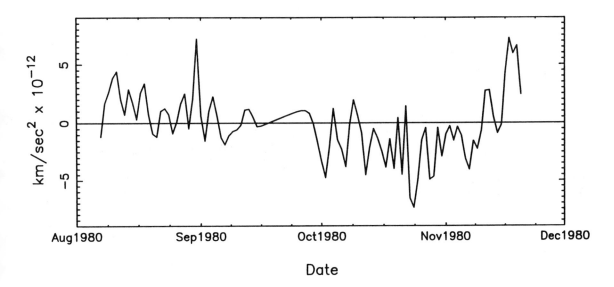

Figure 1: Voyager 1 Z Stochastic Acceleration

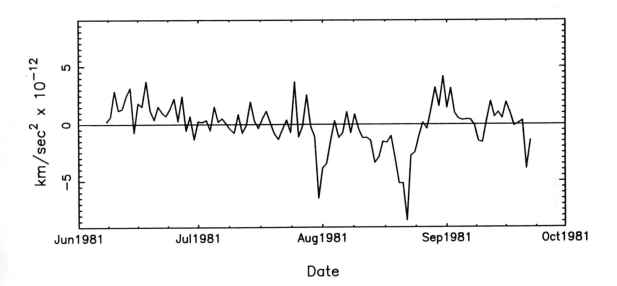

Figure 2: Voyager 2 Z Stochastic Acceleration

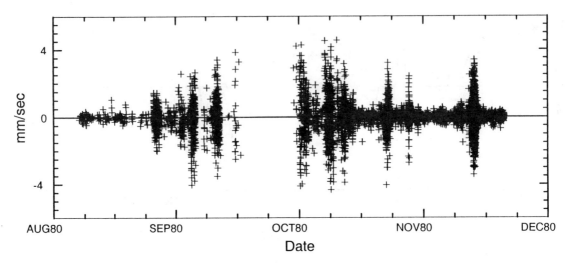

Figure 3: Voyager 1 Doppler Residuals

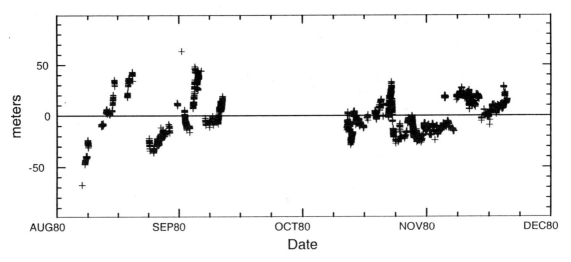

Figure 4: Voyager 1 Range Residuals

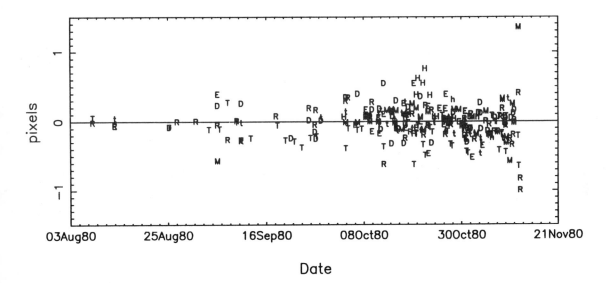

Figure 5: Voyager 1 Pixel Residuals

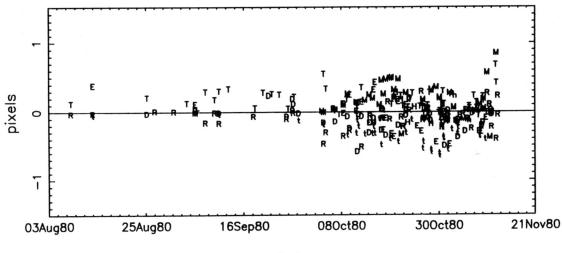

Figure 6: Voyager 1 Line Residuals

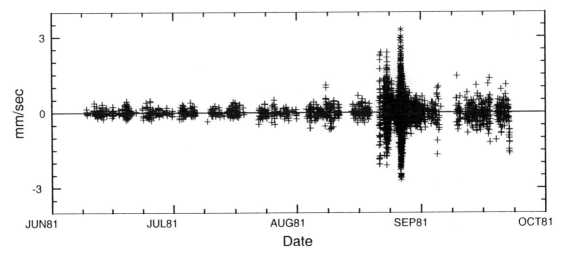

Figure 7: Voyager 2 Doppler Residuals

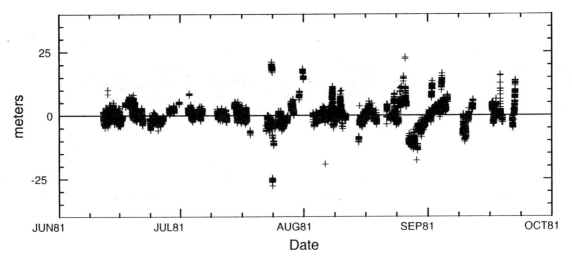

Figure 8: Voyager 2 Range Residuals

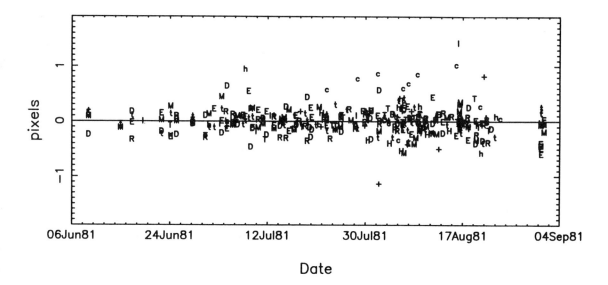

Figure 9: Voyager 2 Pixel Residuals

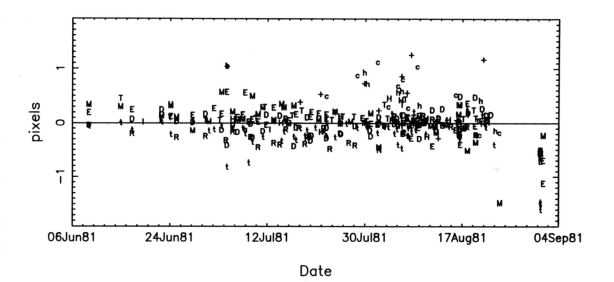

Figure 10: Voyager 2 Line Residuals

SATELLITE EPHEMERIDES UPDATE SCHEDULE FOR THE CASSINI MISSION

Ian Roundhill[*] and Duane Roth[*]

The Cassini mission will arrive at Saturn in July 2004 to explore the Saturnian system. The navigation team will update the ephemerides of 9 satellites to improve spacecraft navigation and provide information on satellite location for pointing of the spacecraft. This paper outlines the approach used to predict pointing uncertainty when pointing information is generated with a spacecraft prediction and older satellite ephemerides. This modeling is then used to choose times when the satellite ephemerides need to be updated to meet requirements.

* Jet Propulsion Laboratory, California Institute of Technology, 4800 Oak Grove Drive, Pasadena, California 91109-8099.

Introduction

The purpose of this paper is to describe the satellite ephemerides update schedule for the Cassini mission and the analysis method used. When Cassini enters the Saturnian system in July 2004, the navigation team will be responsible for providing the ephemerides for nine of the natural satellites. Throughout the mission, new sets of satellite and spacecraft ephemerides will be computed by the navigation team and provided to users within the project. One major user is the science planning team, which uses the ephemerides to compute pointing information.

The motivation for this study is the effect of updating the satellite ephemerides. An update requires work by the navigation team to re-compute the spacecraft reference trajectory. This new reference trajectory is then used to update the spacecraft sequences. Updating the spacecraft sequences is not a trivial task and should only be done when necessary. This analysis will help define when the spacecraft sequences need to be updated with a new reference trajectory to satisfy pointing requirements. Updates to small segments of the spacecraft sequence are possible but were projected to consume five days when this study was initiated. The project would like to update large amounts of the sequences at once and minimize local updates to reduce work.

This paper is composed of three parts. First, the pointing uncertainty algorithm is described. This algorithm includes a previously utilized method that combines spacecraft and satellite uncertainties from the same data set and a new algorithm for combining information from different data sets. Second, the filter setup used for uncertainty predictions is outlined. This includes the filter timing, data schedule and weights, parameters used, and filter outputs.

The final section of the memo describes the results of applying the filter and pointing uncertainty algorithm throughout the mission. An output of this process is a proposed list of times when the satellite ephemerides need to be updated. The pointing uncertainty validity throughout the tour is shown in a series of plots that indicate when requirements are not being met.

Computation of Pointing Uncertainty

The orbit determination program SIGMA was used to predict position uncertainties for the spacecraft and the Saturnian satellites. This information can then be used to compute the pointing uncertainty due to navigation errors. Two separate algorithms were used and are described below.

When the satellite ephemerides and the spacecraft ephemeris are the result of the same SIGMA filter run, the correlated information is used to predict the pointing. This case would be used when a late spacecraft update is to be computed and both the spacecraft and satellite ephemerides are updated at the same time. The filter outputs in the RTN frame are used for this computation. These outputs include the relative position uncertainty of the spacecraft with respect to the satellite and this computation was done using the cross correlation of the two bodies in addition to their individual position uncertainties. This relative uncertainty is expressed in a matrix:

$$\Gamma^{sc,sat} = \begin{bmatrix} \Gamma_{rr} & \Gamma_{rt} & \Gamma_{rn} \\ \Gamma_{tr} & \Gamma_{tt} & \Gamma_{tn} \\ \Gamma_{nr} & \Gamma_{nt} & \Gamma_{nn} \end{bmatrix} \tag{1}$$

The square root of Γ_{rr} is the one-sigma uncertainty in the radial direction:

$$\sigma_{radial} = \sqrt{\Gamma_{rr}} . \tag{2}$$

The 2×2 lower right-hand submatrix of Γ contains information about the uncertainty ellipse perpendicular to the radial direction. The length of the semi-major axis of this ellipse is the maximum uncertainty in the perpendicular direction and is computed as the square root of the maximum eigenvalue of the ellipse.

$$\sigma_{perpendicular} = \sqrt{max\left(eigenvalue\left(\begin{bmatrix} \Gamma_{tt} & \Gamma_{tn} \\ \Gamma_{nt} & \Gamma_{nn} \end{bmatrix}\right)\right)} . \tag{3}$$

The pointing uncertainty is then computed as the ratio of the perpendicular uncertainty to an adjusted range value. This adjusted range is the reference range minus the satellite radius and one-sigma radial uncertainty:

$$pointing\ error_{sc,sat} = \frac{\sigma_{perpendicular}}{range_{sc,sat} - radius_{sat} - \sigma_{radial}} . \tag{4}$$

Figure 1 shows a pictorial representation of this ratio and shows how the pointing error is the perpendicular error divided by an adjusted range value.

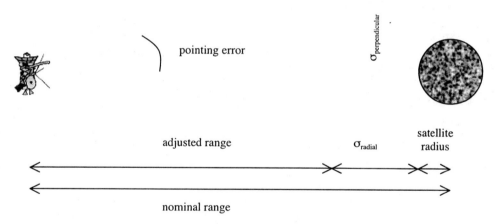

Figure 1: Diagram of Pointing Uncertainty Calculation

When the satellite and spacecraft ephemerides are produced with different data cutoffs, the cross correlations are not defined. This is the more common instance that will occur during the tour. Perhaps only a handful of satellite updates will be made but many spacecraft updates will be computed. For this algorithm, the barycentric position uncertainties of the spacecraft and satellite are combined. The SIGMA run for a chosen maneuver solution to update the satellite ephemerides produces the position uncertainty of a satellite in Saturn barycentric EME 2000 coordinates at a given map or predicted time:

$$\Gamma^{sat} = \begin{bmatrix} \Gamma^{sat}_{xx} & \Gamma^{sat}_{xy} & \Gamma^{sat}_{xz} \\ \Gamma^{sat}_{yx} & \Gamma^{sat}_{yy} & \Gamma^{sat}_{yz} \\ \Gamma^{sat}_{zx} & \Gamma^{sat}_{zy} & \Gamma^{sat}_{zz} \end{bmatrix} \tag{5}$$

A separate SIGMA run for the maneuver being considered produces the position uncertainty of the spacecraft in the same barycentric EME 2000 coordinates at the same map time:

$$\Gamma^{sc} = \begin{bmatrix} \Gamma^{sc}_{xx} & \Gamma^{sc}_{xy} & \Gamma^{sc}_{xz} \\ \Gamma^{sc}_{yx} & \Gamma^{sc}_{yy} & \Gamma^{sc}_{yz} \\ \Gamma^{sc}_{zx} & \Gamma^{sc}_{zy} & \Gamma^{sc}_{zz} \end{bmatrix} \tag{6}$$

Note that no cross correlations between the spacecraft and satellite are present. These two position uncertainties can then be combined and rotated to a radial, transverse, and normal (RTN) coordinate frame:

$$\Gamma^{sc,sat} = R\left(\Gamma^{sc} + \Gamma^{sat}\right)R^T \tag{7}$$

where the rotation matrix R is defined as:

$$R = \begin{bmatrix} \hat{r} \\ \hat{t} \\ \hat{n} \end{bmatrix} = \begin{bmatrix} \dfrac{\left(\vec{r}_{sc} - \vec{r}_{sat}\right)}{\left|\vec{r}_{sc} - \vec{r}_{sat}\right|} \\ \hat{n} \times \hat{r} \\ \dfrac{\left(\vec{r}_{sc} - \vec{r}_{sat}\right) \times \left(\vec{v}_{sc} - \vec{v}_{sat}\right)}{\left|\left(\vec{r}_{sc} - \vec{r}_{sat}\right) \times \left(\vec{v}_{sc} - \vec{v}_{sat}\right)\right|} \end{bmatrix} \tag{8}$$

This matrix $\Gamma^{sc,sat}$ can then be used in Eqs (1)-(4). These equations compute a radial and perpendicular error. A conservative range value is computed by subtracting the uncertainty in the range and the satellite radius from the nominal range. The pointing error is the ratio of the perpendicular error and the adjusted range value.

Pointing Requirements

The navigation pointing requirements used throughout this analysis are 1.02 mrad for ranges between 20,000 and 30,000 km and 0.79 mrad for ranges above 30,000 km[1]. These values will typically be shown as a stair step line as the requirements change over time. These Cassini project requirements are used for all 9 satellites during all phases of the mission.

Filter Setup

Tour Filter Times

A separate filter run was conducted for every maneuver. During the tour of Saturn these filter runs will be required for maneuver design and will be available for pointing updates. The data spanned from the beginning of the filter to a data cutoff point before the maneuver. The filter then continued past the data cutoff to compute the predicted uncertainty of the spacecraft and satellites. The end filter time varies depending on the maneuver. For maneuvers for which a satellite ephemeris update will be computed, the end filter time was beyond the beginning time of the next satellite ephemeris update. For maneuvers without a satellite ephemeris update, the end filter time was the time of the second subsequent maneuver. The time step chosen for predictions was typically 1 day and 15 minutes near close approaches to satellites.

Table 1: Filter Timing

Timing Item	Tour Analysis
Begin Filter Span	Apoapse time before previous Saturn periapse
Data cutoff	2 days before the maneuver
End Filter Span	If maneuver OD is not being used for satellite ephemeris update time of second subsequent maneuver plus 2 days, otherwise time of next satellite ephemeris update implementation.
Start of uncertainty predictions	4 days after data cutoff
Prediction Time Step	15 minutes near flybys and 1 day otherwise

Data Schedule and Weights

Radiometric doppler and ranging data was simulated for the entire tour and weighted at 0.2 mm/s and 100 m respectively. The tracking schedule included daily passes alternating between Goldstone and Madrid. All data within 12 hours of a targeted flyby was removed because of science observations. During solar conjunction, data was deweighted and removed based on the Sun-Earth-Probe (SEP) angle. This information is summarized in Table 2.

Optical data was included using a Picture Sequence File (PSF) prepared by the navigation team. The data includes pictures of nine Saturnian satellites (Mimas, Enceladus, Tethys, Dione, Rhea, Titan, Hyperion, Iapetus, and Phoebe) and background stars. The optical data was weighted at 0.5 pixels for the satellites except for Titan, which was weighted at 1 pixel. The background stars were weighted at 0.25 pixels. One pixel is approximately 6 micro radians. This information is also presented in Table 2.

Table 2: Data Weights

	Doppler	Ranging	Satellites other than Titan	Titan	Background Stars
Normal weights	0.2 mm/s	100 m			
SEP angle: 7.5° – 15.0°	1 mm/s	100 m			
SEP angle: 3.0° – 7.5°	5 mm/s	Deleted	0.5 pixels	1.0 pixels	0.25 pixels
SEP angle: 0.0° – 3.0°	Deleted	Deleted			
Within 12 hours of targeted flyby	Deleted	Deleted			

Estimated, Stochastic, and Considered Parameters

The filter includes various estimated, stochastic, and considered parameters. The spacecraft dynamical model is adjusted by estimating the spacecraft state, antenna reflectivity, maneuvers (OTMs), and reaction control system (RCS) usage. The state includes the position and velocity of the spacecraft at the start of the filter. The RCS usage is designed to model the use of RCS instead of reaction wheels during targeted flybys to attain higher turn rates. This event is modeled as an impulsive maneuver at closest approach. The spacecraft dynamical model also includes uncorrelated stochastic accelerations computed every 8 hours that are intended to model forces caused by the radioisotope thermoelectric generators (RTG).

The dynamical model for the Saturnian system is adjusted by estimating various parameters for Saturn and the nine satellites. The Saturn parameters include the Saturn state, GM, J2, J4, and pole orientation. The satellite parameters are the states and GMs.

The observation model includes stochastic and considered parameters. For every optical picture three uncorrelated stochastic parameters, (RA, DEC and TWIST), are computed to adjust the direction and rotation of the picture to fit the satellite and background stars. The radiometric observation model includes considered parameters for the Earth-Moon barycenter, Earth polar motion and timing, troposphere, ionosphere, and station locations.

The apriori uncertainties for these parameters come from various sources. The spacecraft state, maneuvers, targeted flyby RCS usage, and antenna reflectivity apriori uncertainties are unchanged for each filter. The maneuver apriori matrices are the result of simulations using the nominal OD solution and are intended to model the expected variations from the nominal maneuver design[2,3]. The station location correlated matrix is the result of VLBI measurements and conventional and GPS surveys[4]. The Saturn and satellite ephemeris, GM, J2, J4, and pole orientation terms apriori uncertainties are the post-fit uncertainties of the data used for a previous fit. The first fit uses the apriori values from SAT077 and a custom Phoebe solution[5].

Table 3: Estimated, Stochastic, and Considered Parameter Uncertainties

Estimated Parameters	Tour Apriori uncertainty
Spacecraft state	150 m and 100 m/s
Maneuvers (3-axis magnitudes)	3×3 Covariance Matrices
Targeted Flyby RCS usage (3-axis magnitudes)	160 mm/s diagonal
Antenna reflectivity (specular and diffuse reflectivity coefficients)	0.0213 and 0.0158
Saturn state	DE405 planetary solution scaled by 5^2
Saturn GM, J2, J4, pole orientation	Uncertainty at apoapse as described below.
Satellite states & GMs	
Stochastic Parameters	
Spacecraft acceleration	4.5×10^{-12} km/s^2 in 3 axes
Camera pointing	1° in Right Ascension, Declination, and Twist angles.
Considered Parameters	
Earth-Moon barycenter	DE405 planetary solution formal covariance scaled by 5^2
Earth Polar Motion and Timing	10 cm per axis
Troposphere	1 cm for dry & 1 cm for wet
Ionosphere	5 cm for daytime & 1 cm for nighttime
Station Locations	Covariance determined from VLBI solutions and conventional and GPS surveys accurate to about 10 cm.

Satellite Uncertainty Processing

For tour analysis, the filter span for each maneuver begins at a previous apoapse and not at the beginning of the tour. This means that the filter does not include any optical or radiometric data before the beginning of the filter span. For the Saturnian dynamical model, this information needs to be available in the filter. The information is provided by the apriori covariance of the associated parameters. Obviously this covariance is not constant for each apoapse and improves throughout the tour as more data is applied. To provide these covariance matrices, a separate set of filter runs were conducted. For these auxiliary runs the filter setup was the same but with a different filter span and outputs. These filters began at apoapse and ended at the next apoapse and did not provide predicted uncertainty information. The output instead was the covariance for the Saturnian parameters at the end apoapse. This chain of filters began at Saturn Orbit Insertion (SOI) and proceeded with each output Saturnian covariance being used as the input Saturnian covariance for the next orbit revolution. The Saturnian covariance at each apoapse was saved and used as the apriori covariance for the primary filter runs described previously[6]. The initial Saturnian covariance used at SOI is the product of a previous Saturn Approach analysis[7].

Additional Filter Inputs

The trajectory and partial derivatives used for the spacecraft is intended to match the T2002-01 reference trajectory. To account for numerical errors in the integrator, the reference trajectory during the tour was integrated in pieces from each apoapse for 2 orbit revolutions. This short length is acceptable because the spacecraft trajectory is only mapping forward through two maneuvers, which is typically less than one orbit. For the cases when mapping through two maneuvers is beyond 2 orbits (e.g. OTM-24), the mapping ends at the end of the second orbit. The trajectory and partial derivatives used for the Earth and Saturn is the DE405 solution and the trajectory and partial derivatives used for the Saturnian satellites is SAT077.

Filter Output

The filter was configured to provide two sets of uncertainties. The first set was the relative uncertainty of the spacecraft with respect to eight satellites in the RTN frame. The eight satellites used were Mimas, Enceladus, Tethys, Dione, Rhea, Titan, Hyperion, and Iapetus; Phoebe was removed due to a limitation in the number of mappings that SIGMA can compute. The radial direction is away from the satellite towards the spacecraft. The transverse direction is perpendicular to the radial direction and along the relative velocity vector. The normal direction is perpendicular to the orbit plane along the angular momentum direction and is also perpendicular to the radial and transverse directions.

The second set was the position uncertainty of the spacecraft and satellites with respect to the Saturn Barycenter oriented along the Earth Mean Equator of 2000. These two sets of position uncertainties were then used to compute the pointing uncertainty outside of SIGMA.

Tour Results

The previous sections of this memo have described the filter setup that is used and the pointing uncertainty algorithm that transforms the filter outputs into pointing errors. For every maneuver the filter was run with the appropriate data start and stop times. The filter produces both correlated RTN uncertainties and uncorrelated barycentric position uncertainties and these outputs are saved. A MATLAB script was prepared to read these filter outputs and apply the pointing uncertainty algorithm.

A set of satellite ephemerides update times was chosen and modified throughout the study and these times dictated which pointing algorithm was used for each maneuver filter. The satellite ephemeris update times used are shown in Table 4. For each update the table indicates the maneuver name, data cutoff time, and when this update was implemented. For the first four update times the impact of the update was considered to be instantaneous. This means that satellite and spacecraft ephemerides are to be updated within the four days and uploaded to the spacecraft and the correlated pointing uncertainty algorithm was used. For the final two satellite ephemerides updates a 15 week dwell period was used to allow time for replanning of the reference trajectory and observation planning. For all maneuvers other than the first four shown in Table 4, the uncorrelated pointing uncertainty was used. For these maneuvers the spacecraft position uncertainty for that maneuver was combined with the latest previous implementation date of a satellite ephemerides update. For example, a maneuver on January 12, 2006 would produce a spacecraft prediction uncertainty that would be combined with a satellite prediction uncertainty from the E2-6 filter run.

Table 4: Proposed Satellite Ephemerides Update Times for Tour

Maneuver Name	Data cutoff date	Implementation Date
SOI+2 days	July 1, 2004	July 5, 2004
TA-3	October 21, 2004	October 25, 2004
TB-3	December 8, 2004	December 12, 2004
TC+2	January 14, 2005	January 19, 2005
E2-6	July 6, 2005	November 3, 2005
Post T24 Apoapse	February 5, 2007	May 26, 2007

The result of this algorithm is the pointing uncertainty for each satellite that is a function of time. An example plot of these results is shown in Figure 2. For this plot the satellite ephemerides are updated at the SOI+2 and TA-3 maneuvers but the spacecraft ephemeris is updated at each of the five listed maneuvers (SOI+2, SOI+16, PRM, PRMCU, and TA-3). The pointing requirement is also depicted in the figure as the solid horizontal line.

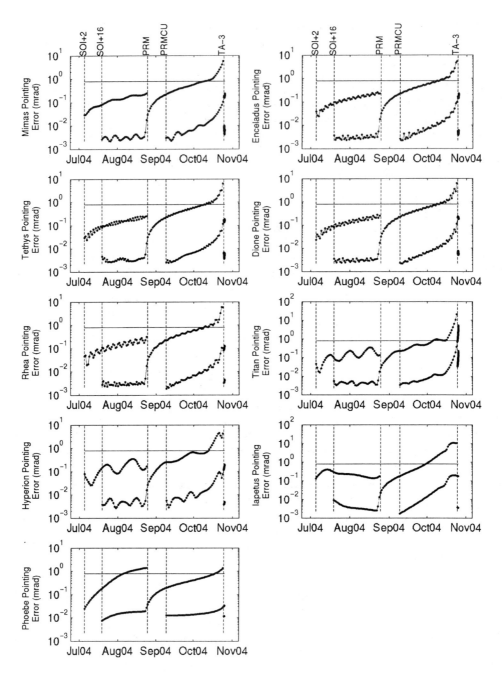

All x–ticks for first of indicated month

Figure 2: Predicted Pointing Uncertainty During First Orbit of Saturn

All of the pointing uncertainties and requirements can be combined to produce a chart for when the requirements are not met. These charts are shown in Figures 3-8 and have the following characteristics:

- o The x-axis indicates time during the tour
- o The y-axis indicates a maneuver number.
- o The shaded bar shows the time period for which the pointing requirement is met. A bar may be discontinuous during times of close approaches to satellites.
- o The shaded bar for each maneuver has been collapsed to bottom of the chart to show a combined interval of when the requirements are being met.
- o Some of the maneuver names are provided at the beginning of the bar. The maneuver names are typically defined as relative to a satellite flyby. A letter defines the satellite flyby (Enceladus, Dione, Rhea, Hyperion, or Iapetus), a number or letter defines which flyby, and then the number of days is given. The nomenclature has undergone various alterations, but a '-3' maneuver is three days before a flyby.
- o The tick marks attached to the shaded bar that extend vertically denote endpoints of the shaded bar. Areas with multiple tick marks are accurate but difficult to interpret on this time scale.
- o The diamond symbol indicates a maneuver OD that will be used for updating of satellite ephemeris.
- o The square symbol indicates when the satellite ephemeris is implemented and the time is indicated.

The results indicate that the pointing requirements are being met for the vast majority of the tour. The updates for the first two Titan flybys (TA and TB) are required because of the pointing requirements during the Titan flyby. Some highlights of these figures include:

- o The pointing may fail after a flyby and before the information from the next maneuver is available. This can be seen after many of the Titan flybys.
- o The probe mission pointing is not met by using the OD delivery from the ODM maneuver. This fact is already known and probe mission pointing will be addressed as a separate issue by the navigation team.
- o During extended periods of no maneuvers, the pointing fails. This can be seen in the time after Titan 5, Titan 20, and Titan 36. This problem can be easily remedied by providing additional spacecraft solutions during these times.
- o The first four satellite ephemeris updates are considered instantaneous. This was done to come close to meeting the pointing requirements for flybys early in the tour; Titan A and Titan B will require late satellite ephemeris updates. The first update at SOI was chosen for convenience of analysis and will most likely be performed prior to SOI.

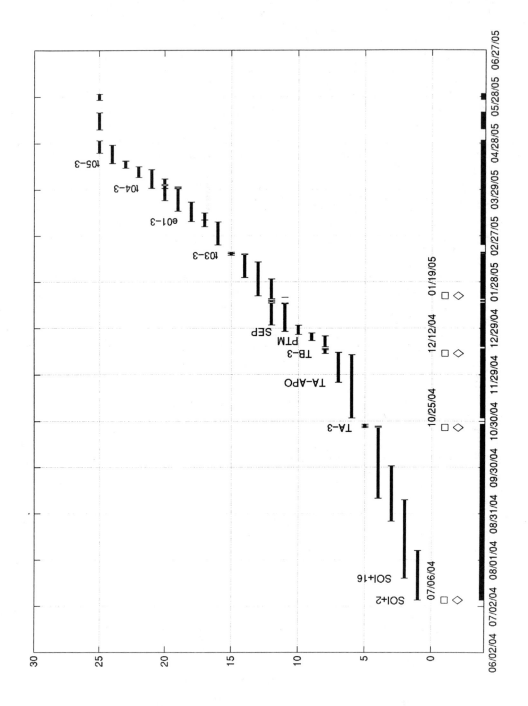

Figure 3: Time Periods of Meeting Navigation Pointing Requirements (6/2/04 – 6/27/05)

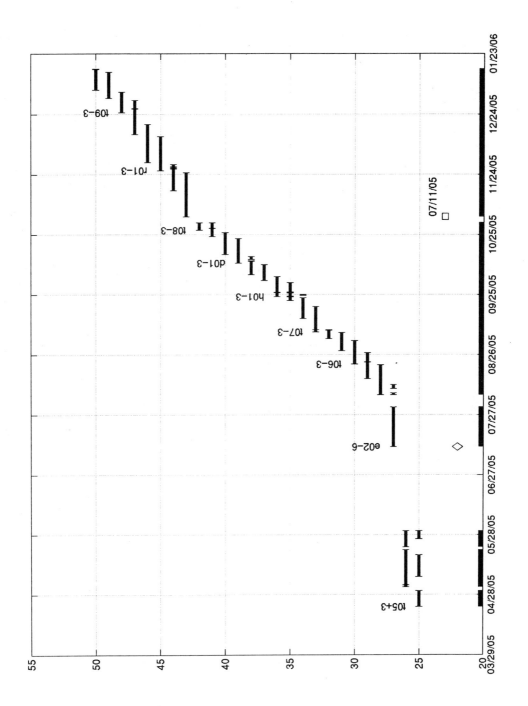

Figure 4: Time Periods of Meeting Navigation Pointing Requirements (3/29/05-1/23/06)

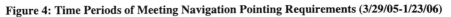

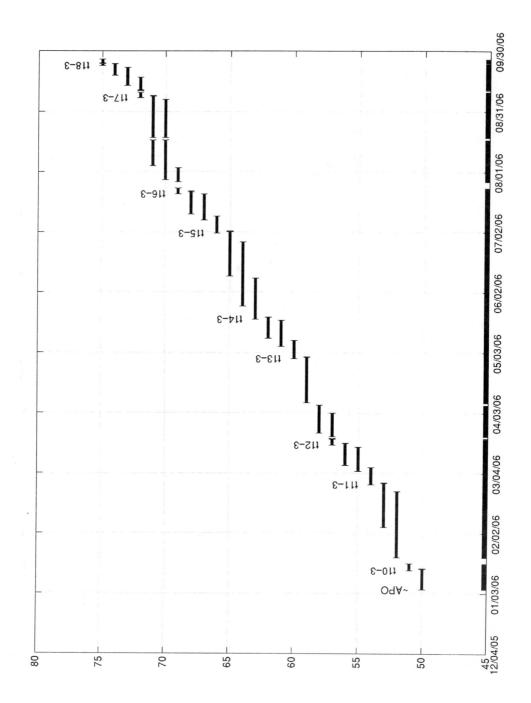

Figure 5: Time Periods of Meeting Navigation Pointing Requirements (12/4/05-9/30/06)

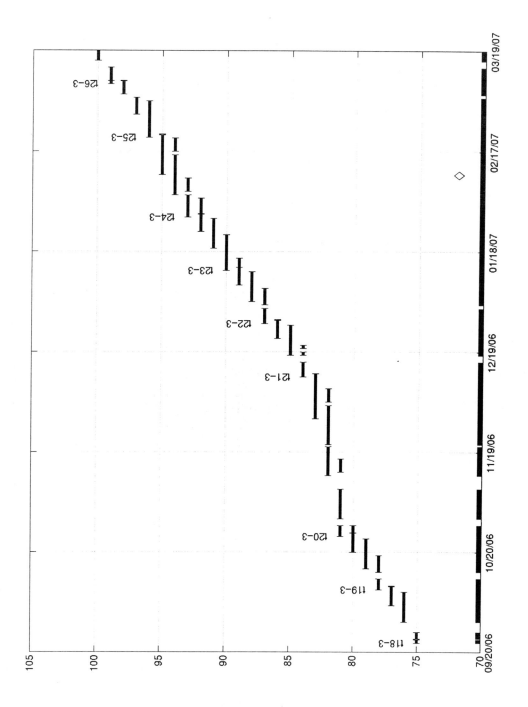

Figure 6: Time Periods of Meeting Navigation Pointing Requirements (9/20/06-3/19/07)

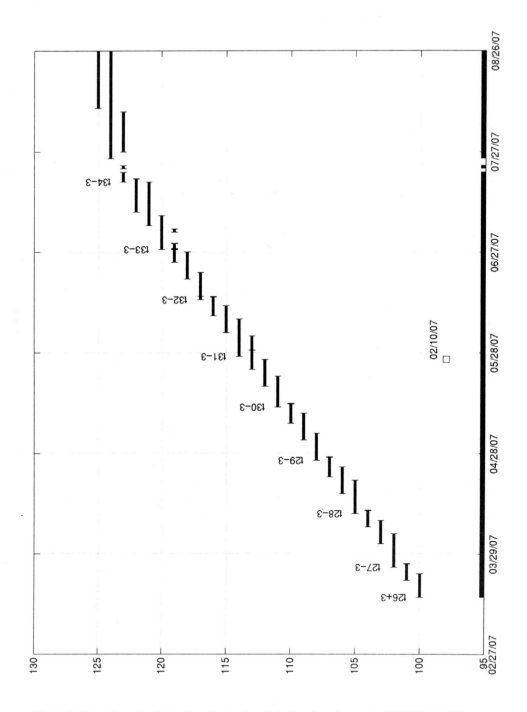

Figure 7: Time Periods of Meeting Navigation Pointing Requirements (2/27/07-8/26/07)

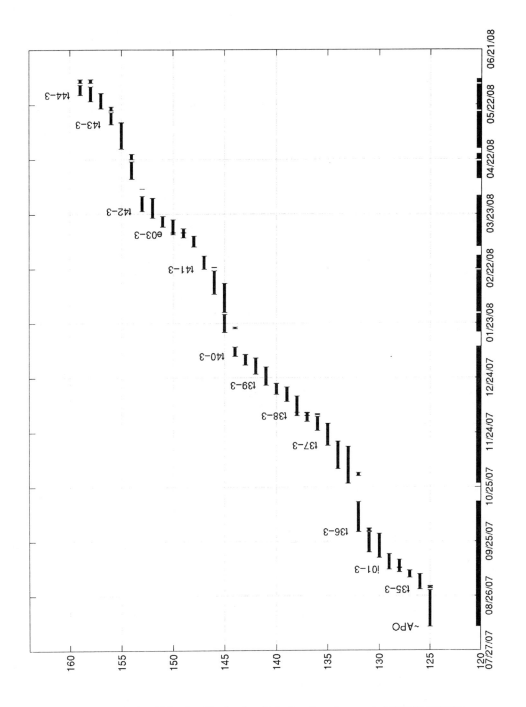

Figure 8: Time Periods of Meeting Navigation Pointing Requirements (7/27/07-6/21/08)

Conclusions

The pointing requirements due to navigation error can be met for most of the tour. The current set of maneuvers already defines a set of orbit determination solutions that can be used to meet requirements for a large portion of the tour. In addition, areas of non-compliance during times of a low frequency of maneuvers can easily be covered with additional solutions. Unfortunately, some time periods after Titan flybys may have uncorrectable violations of requirements. The spacecraft trajectory is perturbed significantly during the Titan flyby and only after adequate radiometric tracking can the pointing update process begin. This paper also outlines a preliminary set of satellite ephemeris update times. These times will be used for further iteration based on the expectations and abilities of the science and spacecraft teams.

Acknowledgements

The research described in this paper was carried out at the Jet Propulsion Laboratory, California Institute of Technology, under a contract with the National Aeronautics and Space Administration.

References

1. "SCO FRD 427-1 Target Pointing," Cassini Internal Document.
2. Goodson, Troy, "Maneuver Simulation Results for Saturn Approach through Huygens Probe Delivery," JPL IOM 312.G-02-010, August 29, 2002.
3. Hahn, Yungsun, Private Communication.
4. Folkner, William M., JPL internal web site http://epic/nav/eop/stations.html.
5. Hildebrand, C., "Revised SAT077 Covariance", email dated 30 March 2000.
6. Ionasescu, Rodica, "Predicted Uncertainties for the Cassini Tour T2002-01: Baseline Case," JPL IOM312.B/17-02, October 11, 2002.
7. Bordi, John, "Saturn Approach Orbit Determination," JPL IOM 312.B/14-02, September 3, 2002.

CASSINI NAVIGATION DURING SOLAR CONJUNCTIONS VIA REMOVAL OF SOLAR PLASMA NOISE

P. Tortora,[*] L. Iess,[†] J. J. Bordi,[‡] J. E. Ekelund[‡] and D. Roth[‡]

The Cassini spacecraft and its ground segment are currently testing a novel radio frequency multilink technology to perform radio science experiments. During solar conjunctions, this allows the complete removal of the solar plasma noise from the Doppler observables, with benefits also for deep space navigation. This is obtained combining the carrier frequencies of three independent down-links: two of them, a X- and a Ka-band (Ka1), are coherent with a X-band up-link, while an additional Ka-band down-link (Ka2), is coherent with a Ka-band up-link. During the June-July 2002 Cassini solar conjunction, this procedure was fully tested for the first time. We show that, using the adopted multifrequency plasma calibration scheme, the standard deviation of the Doppler frequency residuals is reduced up to a factor of 200 over the uncalibrated X-band data. This large improvement in the data quality, revealed by values of the frequency stability previously achieved only during solar oppositions, makes the navigation accuracy of deep space probes nearly independent of the solar elongation angle.

INTRODUCTION

Current deep space navigation systems rely on X-band radio-links, used for both range and range-rate measurements. The frequency stability of these links (measured by the Allan deviation) is generally on the order of 10^{-13} at 1000s integration time[1], yielding range-rate accuracies on the order of 1.5×10^{-3} cm/s. Near solar oppositions, when the sun-earth-probe (SEP) angle is close to 180°, higher stabilities (on the order of 2×10^{-14}) can be achieved, since the solar wind velocity is nearly parallel to the line-of-sight[2]. The lowest stabilities are obtained, for missions in the ecliptic plane, during solar conjunctions (SEP angles close to zero) lasting up to two weeks. The dramatic decay in the attainable navigation accuracy is caused by the signal phase scintillation due to the solar plasma. Radio metric data collected when the line of sight falls within 40 solar radii from the sun are generally not used for the orbit determination process, due to the high measurement errors introduced by the solar corona, leading to long time spans during which navigation cannot rely on actual data.

[*]   Università di Bologna, II Facoltà di Ingegneria, Via Fontanelle 40, I-47100 Forlì (FC), Italy.

[†]   Università di Roma "la Sapienza", Dipartimento di Ingegneria Aerospaziale ed Astronautica, Via Eudossiana 18, I-00184, Rome, Italy.

[‡]   Jet Propulsion Laboratory, California Institute of Technology, Navigation and Mission Design Section, 4800 Oak Grove Drive, Pasadena, California 91109-8099.

More than 30 days of tracking data, across the 2000 and 2001 solar conjunctions, were removed for the orbit reconstruction of the Cassini spacecraft, currently in cruise flight to Saturn[3-5]. This strategy is widely accepted and proven during the cruise flight, but it is not recommended during critical mission phases, when frequent ground-commanded maneuvers are executed. An example of this is the Cassini Saturn Orbit Insertion (SOI) maneuver, scheduled on July 1st 2004, only seven days before a solar conjunction[6]. In order to reduce the orbit determination uncertainties, the whole maneuver would greatly benefit from using all radio metric data collected leading up to and following SOI.

The Cassini spacecraft and ground segment are currently used to test a novel multilink radio frequency (RF) system to perform radio science experiments[7] (RSE). The on-board configuration provides, in addition to the standard X-band up-link/X-band down-link (X/X), an exciter (KEX) which generates a Ka-band down-link signal (at 32.5 GHz), coherent with the received X-band up-link, resulting in a X/Ka link. Moreover, a Ka/Ka link is obtained using a coherent frequency translator (KaT), which transmits to the ground a Ka-band signal (at 32.5 GHz) coherent with a Ka-band up-link (at 34 GHz). The primary goals of RSE are the measurement of the solar gravitational deflection during the Solar Conjunction Experiments (SCE)[8-9], and the search for low-frequency gravitational waves (due, for example, to the coalescence of massive black hole binaries) during solar oppositions[10].

The multifrequency plasma calibration scheme[11], originally proposed to calibrate the Doppler observables used for the estimation of the post-Newtonian parameter γ during the SCE, has also shown to be very effective in improving the accuracy of the orbit determination process. This has been demonstrated during the test of the Cassini ground and on-board systems, performed on May-June 2001[12], and during the June-July 2002 solar conjunction[13]. With this method, the sky frequencies, reconstructed using data from a wideband open loop receiver (OL) in the three bands (X/X, X/Ka, Ka/Ka) are coherently combined to remove the effects of the solar plasma, the major noise source in the Doppler observable. However, the observables used by the orbit determination program (ODP)[14] developed at the Jet Propulsion Laboratory, are obtained from the block V receivers (BVR) which digitally lock and track the carrier in a closed loop (CL). Thus, using the OL plasma calibrated sky frequencies for the orbit determination process requires the additional computation of Doppler observables compatible with the data format required by the ODP.

The analysis of the 2001 Cassini solar conjunction data[12], has shown that the use of the multifrequency plasma calibration scheme can lower the Allan deviation to values of 2×10^{-14}, at integration times of 1000 s, when the impact parameter is about 25 solar radii. This corresponds to a range rate accuracy of about 3×10^{-4} cm/s. At an impact parameter of about 6 solar radii, the Allan deviation is on the order of 4×10^{-14} (1000 s integration time), still well below the corresponding original uncalibrated X-band value, and a factor of 25 better than the Ka-band value.

For the 2002 Cassini solar conjunction a much wider data set has been analyzed[13]. The results show that, while the stability of the uncalibrated links degrades as the line-of-sight get closer to the Sun, the Allan deviation of the plasma calibrated frequency residuals has nearly constant values of $1\text{-}2\times10^{-14}$ at integration times of 1000 s. Similar values were previously achieved only near solar oppositions.

Moreover, at the Deep Space Network (DSN) complex located at Goldstone (CA) – DSS 25, an advanced media calibration (AMC) system has been developed and implemented to perform RSE. It consists of water vapor radiometers, digital pressure sensors and microwave temperature profilers[15-16] providing a precise calibration of the frequency shifts due to the dry and wet components of the Earth troposphere. The analysis of the calibration data collected during the 2002 Cassini solar conjunction has shown that the frequency stability is improved by about a factor of 3, when the AMC, rather than the standard seasonal tropospheric models, are applied to the navigation data.

The remainder of this paper is organized as follows: first a brief description of the June-July 2002 Cassini Solar Conjunction is given, pointing out the geometry of the Cassini trajectory, as seen from the Earth, and summarizing the data acquired. The algorithm to reconstruct the sky frequencies from wideband OL data is then illustrated, followed by a description of the solar effects on signal properties. The plasma calibration scheme, used to generate the "plasma-free" observables, is described and, using a simple model of the orbital dynamics, the stability of the residuals is characterized in terms of Allan deviations. In the following Section, the techniques for the reduction of the non-dispersive tropospheric effects, using the advanced media calibration system, are analyzed. Finally, we compare the frequency residuals and orbital solutions obtained using uncalibrated and calibrated data in the ODP. Concluding remarks are given in the last Section.

GEOMETRY OF THE 2002 CASSINI SOLAR CONJUNCTION

During June-July 2002 Cassini first solar conjunction experiment (SCE1), the radio science instrumentation on-board the spacecraft and at the DSN ground stations was operated with 24 hour coverage. At the DSS 25 antenna, the only one with Ka-band uplink capability, the OL receivers acquired and sampled the down-link carrier in three bands (X/X, X/Ka and Ka/Ka), while at the DSS 45 (Canberra, Australia) and DSS 65 (Madrid, Spain) the OL receivers acquired only the X/X signal. Table 1 shows a summary of both the expected and actually acquired data for the days of year (DOY) 157/2002 to 186/2002. The amount of data acquired for the X/Ka link at DSS 25 is identical to that of the X/X link.

Table 1: Summary of OL data acquired at DSN stations during 2002 Cassini solar conjunction

Band	DSN station	Cumulative Expected Pass Duration (hh.mm)	Cumulative Actual Pass Duration (hh.mm)	% Acquired
X/X	DSS 25	359.00	340.08	95 %
Ka/Ka	DSS 25	262.40	188.15	72 %
X/X	DSS 45	129.20	125.08	97 %
X/X	DSS 65	254.36	241.21	95 %

The relatively low percentage of Ka/Ka data acquired at DSS-25 is due to a malfunctioning of the Ka-band up-link transmitter, caused by a heat exchanger problem. This resulted in either the complete absence of data for some passes or some tracks being shorter than expected.

Since the plasma calibration is made possible by the simultaneous acquisition of the down-link signals in the three bands, Table 2 summarizes the amount of multifrequency data expected and actually acquired at DSS 25; in practice it represents the overlap between the X/X and Ka/Ka data sets (rows 1 and 2 in Table 1). In Table 2, actual tracks shorter than 5 hrs have not been considered and DOY 172 has been discarded due to the impossibility to reliably estimate the X/X and X/Ka sky frequency.

Table 2: Summary of multifrequency OL data acquired at DSS 25 during 2002 Cassini solar conjunction

Cumulative Expected Pass Duration (hh.mm)	Cumulative Actual Pass Duration (hh.mm)	% Acquired
262.40	166.35	63 %

The geometry of the 2002 Cassini solar conjunction and the relative distance of the line-of-sight vector from the center of the Sun (impact parameter) are shown respectively in Figures 1 and 2. The minimum SEP angle was reached on DOY 172/2002 (June 21st) at about 1:14 pm UTC, when the impact parameter was 1.6 solar radii. The multifrequency link data closest to conjunction were acquired on DOY 173/2002, at an impact parameter of about 3.5 solar radii.

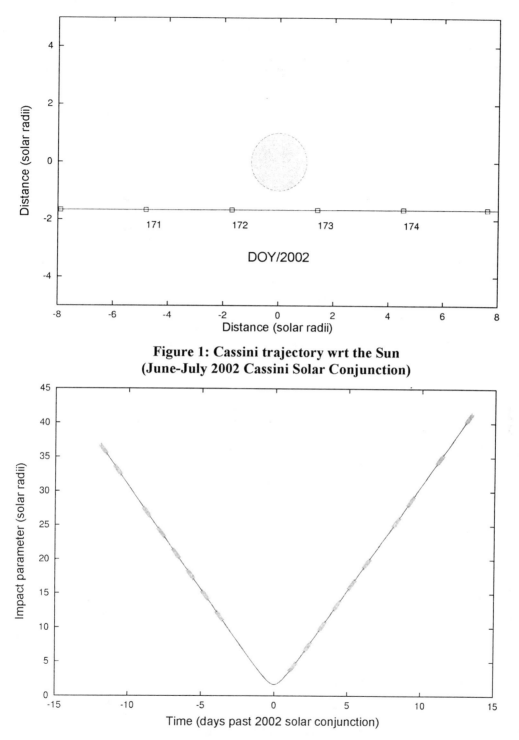

Figure 1: Cassini trajectory wrt the Sun
(June-July 2002 Cassini Solar Conjunction)

Figure 2: Cassini impact parameter vs. time
(The dots represent times when DSS 25 multifrequency tracking is available)

RECONSTRUCTION AND CHARACTERIZATION OF THE OPEN LOOP SKY FREQUENCIES

The radio science receiver (RSR) files, acquired at the DSN complexes, consist of records of the I and Q (in phase and in quadrature) components of the down-converted carrier, plus a header containing ancillary information needed for a full reconstruction of the signal.

During conjunctions, the radio waves received at the ground station are corrupted by the solar and interplanetary plasma. This noise is inherently non-stationary and its effects are strongly correlated among the three bands. As the refractive index of the plasma differs from unity by a quantity proportional to the inverse square of the carrier frequency, Ka-band signals are much less affected by the corona than X-band signals. However, even a Ka-band carrier shows strong scintillation when the beam is very close to the sun (less than about 5 solar radii). In the RSR, the carrier is down-converted to nearly zero frequency using an accurate model of the signal dynamics, and then sampled at the pre-selected rate (1 KHz in our case). To allow for a full reconstruction, a complex representation is used, where the incoming carrier is beaten against two 90° phase-shifted reference signals.

The procedure for the reconstruction of the sky frequency from open loop data acquired at solar conjunction has been described in Ref. 12. It has been shown that the typical algorithms used to reconstruct the signal frequency from digital samples, involving a digital Phase-Locked Loop (PLL) implemented on a computer, are not suitable at impact parameters below 6-8 solar radii. The strong phase scintillation causes frequent loss of lock, especially for the X/X and X/Ka carriers, leading to a significant loss of data. This is why a different frequency reconstruction algorithm, consisting of a frequency estimator which processes sequentially (and independently) fixed-length data intervals, has been implemented for the Cassini conjunction experiment.

The reconstructed carrier exhibits amplitude fluctuations due to the solar corona[13]. At Ka-band, additional, minor amplitude variations along each pass were induced by the pointing system of the ground antenna. Amplitude scintillation is particularly strong on DOY 168 and 173, especially at X-band.

The X/X, X/Ka and Ka/Ka observables have also been characterized by analyzing their frequency residuals[13]. To this end, the sky frequencies have been fitted using a simple orbital model[12], based on the assumption that the signal dynamics, over time scales of a few hours (a tracking pass), is affected only by the Earth rotation and linear drifts of the spacecraft angular coordinates (right ascension and declination) and radial velocity. Approaching and leaving conjunction, the stability of the three links is significantly degraded. In such conditions the solar plasma noise affecting the radio metric data causes a dramatic decay in the attainable accuracy of the orbit solution. As a matter of fact, radio data collected when the line of sight falls within 40 solar radii from the Sun are usually discarded; for the Cassini orbit reconstruction, about 30 days of tracking data were removed across the 2000 and 2001 solar conjunctions[5].

COMPUTATION OF PLASMA-FREE DOPPLER OBSERVABLES
USING THE MULTIFREQUENCY LINK

The output of the sky frequency reconstruction algorithm, summarized in the previous section, is a set of three independent observables at each time instant:

$$\left(f_{sky}\right)^{obs}_{X/X}, \quad \left(f_{sky}\right)^{obs}_{X/Ka}, \quad \left(f_{sky}\right)^{obs}_{Ka/Ka} \tag{1}$$

where the subscript specifies the up-link and down-link band and the superscript identifies that they are observed quantities.

Assuming that each observed sky frequency contains three independent contributions, due respectively to the spacecraft orbital motion and to the crossing of the solar corona in the up-link and in the down-link, we can define y as the relative frequency shift and write the following set of equations[11]:

$$
\begin{aligned}
y^{obs}_{X/X} &= y_{nd} + y_{pl\_up} + \frac{1}{\alpha^2_{X/X}} y_{pl\_dn} \\[2mm]
y^{obs}_{X/Ka} &= y_{nd} + y_{pl\_up} + \frac{1}{\alpha^2_{X/Ka}} y_{pl\_dn} \\[2mm]
y^{obs}_{Ka/Ka} &= y_{nd} + \frac{1}{\beta^2} y_{pl\_up} + \frac{1}{\beta^2}\frac{1}{\alpha^2_{Ka/Ka}} y_{pl\_dn}
\end{aligned}
\tag{2}
$$

The observable frequency shifts have been computed from the observed sky frequencies as:

$$
\begin{aligned}
y^{obs}_{X/X} &= \frac{\left(f_{sky}\right)^{obs}_{X/X}}{\alpha_{X/X}\left(f_X\right)_\uparrow} - 1 \\[2mm]
y^{obs}_{X/Ka} &= \frac{\left(f_{sky}\right)^{obs}_{X/Ka}}{\alpha_{X/Ka}\left(f_X\right)_\uparrow} - 1 \\[2mm]
y^{obs}_{Ka/Ka} &= \frac{\left(f_{sky}\right)^{obs}_{Ka/Ka}}{\alpha_{Ka/Ka}\left(f_{Ka}\right)_\uparrow} - 1
\end{aligned}
\tag{3}
$$

where $\alpha_{XX} = 880/749$, $\alpha_{X/Ka}=3344/749$ and $\alpha_{Ka/Ka} = 14/15$ are the turnaround ratios for the three links and $\beta = \left(f_{Ka}\right)_\uparrow / \left(f_X\right)_\uparrow$ is the ratio between the X and Ka-band up-link frequencies. In Eq. (2), y_{nd} is the orbital (*non dispersive*) contribution, while y_{pl_up} and y_{pl_dn} are respectively the *plasma up-link* and *plasma down-link* contributions to the relative frequency shift (referred to a $\left(f_X\right)_\uparrow$ carrier).

The set of equations (2) is easily solved for y_{nd}, y_{pl_up}, and y_{pl_dn}:

$$y_{pl\_dn} = \left(\frac{1}{\alpha_{X/X}^2} - \frac{1}{\alpha_{X/Ka}^2} \right)^{-1} \left(y_{X/X}^{obs} - y_{X/Ka}^{obs} \right) \tag{4}$$

$$y_{pl\_up} = \frac{y_{X/Ka}^{obs} - y_{Ka/Ka}^{obs} - y_{pl\_dn} \left(\frac{1}{\alpha_{X/Ka}^2} - \frac{1}{\beta^2} \frac{1}{\alpha_{Ka/Ka}^2} \right)}{1 - \frac{1}{\beta^2}} \tag{5}$$

$$y_{nd} = y_{Ka/Ka}^{obs} - \left(y_{pl\_up} + \frac{1}{\alpha_{Ka/Ka}^2} y_{pl\_dn} \right) \frac{1}{\beta^2} \tag{6}$$

Once the non-dispersive relative frequency shift y_{nd} is solved for, one can compute the *non dispersive* sky frequency for each band. So, for example, the non dispersive sky frequency for the X/X band can be written as:

$$\left(f_{sky} \right)_{X/X}^{nd} = \left(f_{sky} \right)_{X/X}^{obs} - \alpha_{X/X} \left(\Delta f_X \right)^{pl\_up} - \frac{\left(\Delta f_X \right)^{pl\_dn}}{\alpha_{X/X}} \tag{7}$$

where $\left(\Delta f_X \right)^{pl\_up} = y_{pl\_up}(f_X)_\uparrow$ and $\left(\Delta f_X \right)^{pl\_dn} = y_{pl\_dn}(f_X)_\uparrow$.
Substituting in Eq. (7) and dividing by $\alpha_{X/X}(f_X)_\uparrow$ we get:

$$\frac{\left(f_{sky} \right)_{X/X}^{obs}}{\alpha_{X/X}(f_X)_\uparrow} = \frac{\left(f_{sky} \right)_{X/X}^{nd}}{\alpha_{X/X}(f_X)_\uparrow} + y_{pl\_up} + \frac{y_{pl\_dn}}{\alpha_{X/X}^2}$$

which offers, by comparison with the first row of the set of equations (2), and using Eq. (3), the result:

$$\left(f_{sky} \right)_{X/X}^{nd} = (1 + y_{nd}) \alpha_{X/X}(f_X)_\uparrow \tag{8}$$

Eq. (8) shows that the so-called "plasma-free" (non-dispersive) sky frequency (X/X band) is obtained as a linear combination of the three X/X, X/Ka and Ka/Ka observables since, as shown in Eq. (6), they all contribute to the non-dispersive frequency shift y_{nd}. The stability of the plasma-free link can be compared to the corresponding uncalibrated links by computing the Allan deviation of the frequency residuals obtained by fitting the sky frequency with the simple orbital model described in the previous section.

Figure 3 shows a cumulative plot of the Allan deviations (at 1000s int. time) for the available multifrequency link passes. For each pass the stability of the raw X/X, X/Ka and Ka/Ka links are directly compared to the corresponding plasma-free one. To avoid excessive contamination from tropospheric noise and systematic errors, only data acquired above 20° of elevation have been considered.

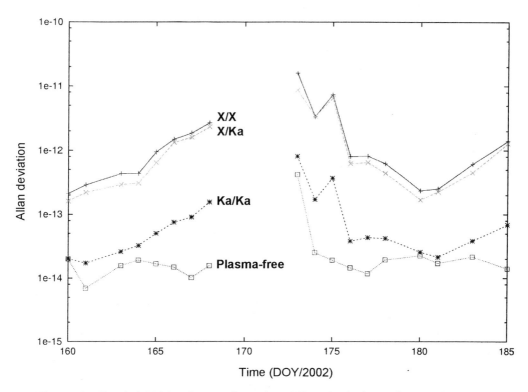

Figure 3: Cassini 2002 solar conjunction Allan deviations @1000s int. time

While the Allan deviation of the X/X, X/Ka and Ka/Ka links degrades with smaller SEP angles, the plasma-free signal exhibits a nearly constant stability at levels of $1\text{-}2\times10^{-14}$, at integration times of 1000 s, a value previously achieved only near solar oppositions. The largest improvement in the signal stability is obtained for the DOY 175/2002, where the Allan deviation of the uncalibrated X/X signal is reduced by about 3 orders of magnitude. However, on DOY 173/2002 the plasma-free signal stability is about 50 times worse than its average value. This can be explained by considering that three main effects limit the applicability of the plasma calibration scheme (Eq. 2) based upon multi-frequency links:

- Diffraction and physical optics effects

- Magnetic corrections to the refractive index

- Spatial separation between the ray paths at different frequencies

In Ref. 12 it was thoroughly explained that at small impact parameters the diffraction effects due to physical optics are responsible for the occasional signal fading at small SEP angles. Moreover, when the impact parameter is smaller than 5 solar radii (as it was on DOY 173/2002 – see Figure 2) both magnetic corrections to the refractive index and density gradients in the solar corona (which cause a change in the impact parameter) become non-negligible.

In conclusion, Figure 3 shows that for the 30 days of the SCE1, the quality of the plasma calibrated Doppler observables is nearly independent of the SEP angle, with the exception of the pass on DOY 173/2002 where the impact parameter of the signal beam was smaller than 5 solar radii.

THE ADVANCED MEDIA CALIBRATION SYSTEM

To perform the RSE, the DSS 25 complex has been equipped with an outstanding media calibration system[15-16] capable of providing a full calibration of the dry and wet path delays due to the Earth troposphere. It consists of two independent systems, where water vapor radiometers, digital pressure sensors and microwave temperature profilers have been installed and symmetrically located a short distance from the main 34m antenna at the Goldstone complex. This system was tested for the first time during the first gravitational waves experiment (GWE1), on November 2001-January 2002[10].

In Ref. 9, the data reduction of the AMC acquired during the SCE1 has been described in detail. By comparison with the corresponding data acquired during the GWE1, it has been shown that for many passes, the dry component of the zenith path delay was much noisier than expected. It turned out that the larger zenith path delay fluctuations levels were, with high probability, due to the surface wind speed which, during many passes of the SCE1, exceeded the 15-20 mph. The increased zenith path delay "noise" levels then, should be due to the atmosphere and not some wind-induced "jiggle" in the pressure sensor mechanism. This can be seen in the substantial agreement of the zenith path delays, as read from the systems 1 and 2. An easy way to filter out the high frequency dry zenith path delay fluctuations is to average, through a moving window (a 100 s interval has been used for the SCE1), the raw data acquired by the instruments, in order to remove the local, small scale effects.

The parallel analysis of the GWE1 and SCE1 advanced media calibrations data has revealed that the use of the water vapor radiometers and digital pressure sensors is more effective during solar conjunctions. This is mainly due to two concurrent reasons: first, SCE are carried out during daytime while GWE take place during the night; second, due to Cassini's present orbital position, for the northern hemisphere ground stations, the SCE observations are made in summer while the GWE ones are in winter. As a result, the tropospheric noise levels, which can be reduced making use of the AMC, are significantly higher during SCE than GWE. As shown in the next section, the frequency stability is improved by about a factor of 3, when the AMC, rather than the standard seasonal tropospheric models, are applied to the SCE1 navigation data.

CASSINI ORBIT DETERMINATION USING PLASMA AND TROPOSPHERE CALIBRATED NAVIGATION OBSERVABLES

In the previous sections we pointed out the procedures needed to compute a plasma free sky frequency using three independent signals simultaneously acquired in the bands X/X, X/Ka and Ka/Ka. These calibrated sky frequencies can now be used to form a plasma-free Doppler observable compatible with the data format usually handled by the ODP. For two-way Doppler observables[§] the ODP uses the received sky frequency in the following manner to compute the observable[17]:

$$obs = \alpha_{Up/Dn} \cdot f_{REF} - f_{sky} \tag{9}$$

where f_{REF} is the Doppler reference frequency and $\alpha_{Up/Dn}$ is the turnaround ratio, which depends only upon the down-link band, since it is always referred to an S-band up-link carrier. For an X-band down-link signal, $\alpha_{S/X} = 880/221$. Thus, at each timetag, the plasma calibrated sky frequency can be directly substituted to the corresponding uncalibrated one to form a plasma calibrated observable:

$$\left(obs\right)_{X/X}^{nd} = \alpha_{S/X} \cdot f_{REF} - \left(f_{sky}\right)_{X/X}^{nd} \tag{10}$$

The result of this procedure is a new navigation file where the Doppler observables, to be processed by the ODP, are free from solar plasma noise. This new observable has been computed using OL data (RSR files and a digital frequency estimator), which have been processed for the first time for orbit determination.

For a full comparison between the plasma-free and raw observables, we processed (in the ODP) the original, plasma free and plasma free with the AMC navigation files, where the observables are compressed at 300s.

To compare the results of using the three different tracking data sets in terms of spacecraft navigation, a short arc, spanning just the solar conjunction period (between June 9th and July 5th, 2002), was processed. The initial conditions on June 9th, 2002 were obtained from a long arc solution that has an epoch of February 28th, 2001. The only parameters estimated in the short arc solutions were the spacecraft state and the radioisotope thermoelectric generator (RTG) radiation accelerations. For all three cases, only station DSS 25 two-way Doppler data were used. The weights assigned to the tracking data were assigned on a pass-by-pass basis, where the a priori uncertainty assigned to each pass is equal to the 1-σ value of the frequency residuals.

[§] One-way, two-way and three-way Doppler observables are formed respectively when:
- the spacecraft (S/C) generates a signal which is received at the ground station
- the same ground station generates the up-link signal to the S/C and receives the down-link signal from it
- the up-link signal to the S/C is generated in a ground station but the down-link signal from the S/C is received in a different ground station

Figure 4 shows the two-way Doppler frequency residuals obtained by running the ODP and using the original (uncalibrated), X-band tracking data file.

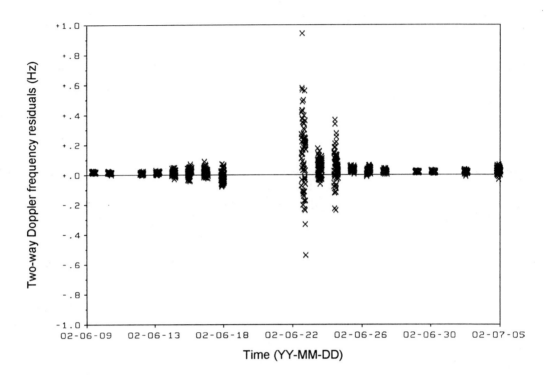

Figure 4: Cassini 2002 solar conjunction X/X Doppler frequency residuals obtained using plasma uncalibrated data and the standard troposphere corrections

Note the increasing variation of the frequency residuals when the spacecraft gets close to conjunction (which occurred on June 21st). On June 24th, 2002, when the average level of the residuals was already getting back to lower values, there is an evident signature in the data, already noticed in terms of increased Allan deviations in Figure 3. This was probably caused by a strong solar event in a direction perpendicular to the Earth-to-Cassini line of sight. The overall 1-σ value of these residuals is 7.16×10^{-2} Hz.

Figure 5 shows exactly the same time interval of Figure 4, but now the plasma calibrated two-way Doppler observables have been processed. The y-axis scale reveals that the noise reduction is huge with nearly constant levels for the 17 passes shown. The June 22nd, 2002, pass has not been included in the plot since it is the only one where the multifrequency plasma calibration scheme did not offer reliable results – see Figure 3. The 1-σ value of the plasma calibrated frequency residuals is 3.5×10^{-4} Hz, more than 200 times better than the corresponding uncalibrated ones.

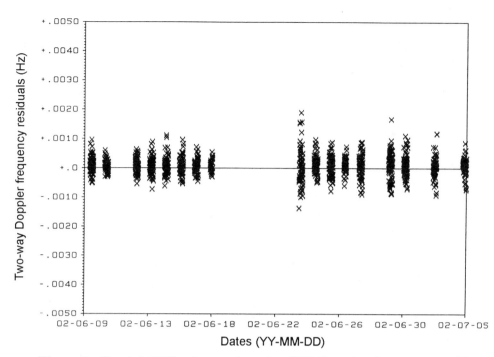

Figure 5: Cassini 2002 solar conjunction X/X Doppler frequency residuals obtained using plasma calibrated data and the standard troposphere corrections

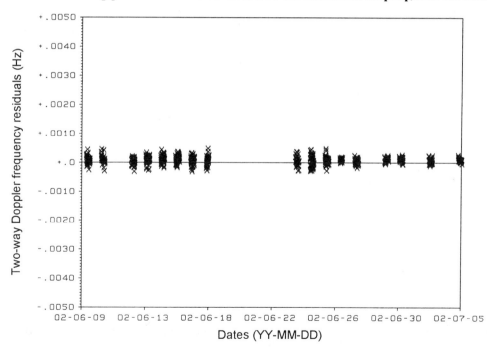

Figure 6: Cassini 2002 solar conjunction X/X Doppler frequency residuals obtained using plasma calibrated data and the advanced media calibrations

Figure 6 shows the additional improvement in the stability of the frequency residuals, which can be obtained once the AMC are applied to the plasma calibrated navigation observables. For some passes, especially after conjunction, the noise reduction is close to a factor of 5, while for some others the effectiveness of the advanced troposphere calibration is less evident. The 1-σ value of the plasma calibrated frequency residuals is 1.2×10^{-4} Hz, respectively a factor of 3 and 600 better than the corresponding plasma calibrated and uncalibrated ones.

In order to help visualize the differences in the solutions and the uncertainties in the estimates, the results are mapped forward to the Saturn B-plane in Figure 7. For reference, the long arc solution and error ellipse is also shown in this figure. The long arc solution is considered the best estimate as of July 2002 of the Saturn B-plane location, so it provides a measure of the accuracy of the short arc solutions. The uncalibrated solution is clearly the poorest, with a large offset in the B.T direction relative to the long arc solution. The solution made with the plasma calibrated data shows a significant improvement in the level of agreement with the long arc solution. Furthermore, the addition of the advanced media calibrations results in even better agreement with the long arc solution.

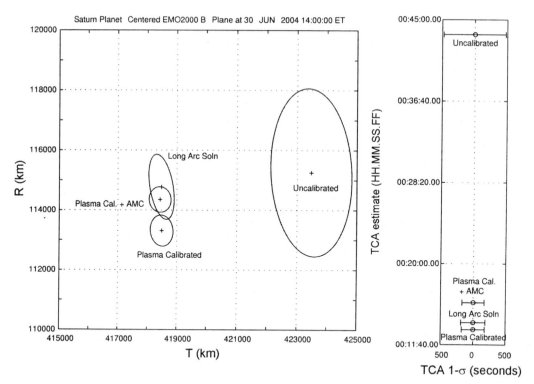

Figure 7: Saturn centered B-plane plot of the Cassini orbital solutions obtained using different sets of X-band data collected during the 2002 solar conjunction. On the right the different estimates of the time of closest approach

These results are impressive when considering that the short arc solutions rely on roughly 30 days of Doppler data from only one ground station during solar conjunction. It should be pointed out that the error ellipses and associated mean values in this plot are for comparison purpose only and do not represent current best knowledge of Cassini's ephemeris. In all cases, no attempt was made to model the errors associated with future maneuvers or thrusting events, which would normally dominate the size of the error ellipse. Furthermore, the actual mean values will differ because future spacecraft maneuvers and orientation changes (that will impact the direction of the radiation accelerations) are not modeled.

Another way to assess the quality of the solutions is to compare the value of the RTG acceleration estimate in the spacecraft z-direction. A very accurate estimate of this value was made during the GWE1. The RTG acceleration estimate is very precise since during the GWE the spacecraft orientation was not changed and was maintained using the reaction wheels rather than the reaction control thrusters. Additionally, continuous 24 hour tracking of the spacecraft was available during the GWE. The z-direction of the RTG acceleration was well determined since the spacecraft was oriented such that the z-axis was pointed towards Earth. Table 3 shows the values of the estimates of this parameter for the three different short arc cases, as well as the estimate from the long arc solution. Once again, the results show a dramatic improvement in comparison to the long arc solution when the solar plasma calibrations are included in the Doppler data. However, little improvement is seen in this parameter when the advanced media calibrations are included in the tracking data.

Table 3: Summary of the RTG acceleration estimates obtained using different sets of X-band data and ODP setup

Case	RTG acceleration Km/s^2	1-σ uncertainty Km/s^2
Long Arc Solution	-2.98×10^{-12}	0.05×10^{-12}
Uncalibrated Short Arc	-7.44×10^{-12}	1.12×10^{-12}
Plasma Calibrated Short Arc	-2.89×10^{-12}	0.13×10^{-12}
Plasma Calibrated + AMC Short Arc	-3.07×10^{-12}	0.08×10^{-12}

CONCLUSIONS

The Cassini spacecraft and its ground segment are equipped with an advanced radio frequency system (RF) to perform radio science experiments. In addition to the standard X-band down-link, coherent with a X-band up-link, the spacecraft on board configuration allows two additional down-links in the Ka-band, one coherent with the X-band up-link and the other coherent with a Ka-band up-link. The simultaneous acquisition of the three down-link carriers is possible only at the Goldstone (CA) complex – DSS 25 – which is the only DSN station with a Ka-band up-link and down-link capability. This configuration, originally devised to perform an accurate test of the general relativity during solar conjunctions, allows the complete removal of the solar plasma noise from the Doppler observables. Moreover, the DSS 25 has been equipped with an advanced media calibration system which allows the full calibration of the dry and wet path delay components of the Earth troposphere.

During the 2002 Cassini solar conjunction experiment, the RF system was operated continuously for 30 days, from June 6th through July 5th. The multifrequency plasma calibration scheme has been applied to all those DSS 25 passes where the three independent down-links were available (18 passes in total). Then, the plasma-free Doppler observables, derived from radio science open loop receivers, have been fitted using the orbit determination program (ODP) to test the capabilities of the new system for precision spacecraft navigation. In addition, the advanced troposphere calibrations have been applied to the plasma calibrated data resulting in a data set of the highest quality. The analysis of the ODP frequency residuals reveals that the application of the plasma and troposphere calibrations to the Doppler observables yields a global improvement of a factor of 600 over the corresponding uncalibrated data. Thus, with this new technique, the data acquired near solar conjunctions can be successfully calibrated in order to gain frequency stabilities and orbital solution accuracies usually recorded when the spacecraft is at solar oppositions.

During the 2003 Cassini solar conjunction experiment (SCE2), a real time test of the plasma calibration technique proposed in this paper will be performed. The 30 days of continuous multifrequency tracking from DSS 25 will have the main goal of a second precise test of General Relativity, but will also allow to gain further confidence in the capabilities of this novel method for spacecraft navigation. This experience is valuable in view of the potential applications for the Saturn orbit insertion maneuver (SOI) (which occurs only 7 days before a solar conjunction, on July 1st, 2004) and during the four solar conjunctions which will occur during the Cassini Saturn tour (2004-2008).

ACKNOWLEDGMENTS

The work of P.T. and L.I has been funded by the Italian Space Agency (ASI). The work of J.J.B, J.E.E. and D.R. was performed at the Jet Propulsion Laboratory, California Institute of Technology, under a contract with the National Aeronautics and Space Administration.

REFERENCES

1. Thornton, C.L., Border, J.S., "Radiometric Tracking Techniques for Deep-Space Navigation", *Deep-Space Communications and Navigation Series*, JPL Publication 00-11, Oct. 2000.

2. Armstrong, J.W., Woo, R., Estabrook, F.B., "Interplanetary Phase Scintillation and the Search for Very Low Frequency Gravitational Radiation", *The Astrophysical Journal*, Vol.230, 1979, pp. 570-574.

3. Roth, D.C., Guman, M.D., Ionasescu, R., Taylor, A.H., "Cassini Orbit Determination From Launch to the First Venus Flyby", *Paper AIAA 98-4563, AIAA/AAS Astrodynamics Specialist Conference*, August 10-12 1998, Boston, MA, USA.

4. Guman, M.D., Roth, D.C., Ionasescu, R., Goodson, T.D., Taylor, A.H., Jones, J.B., "Cassini Orbit Determination From First Venus Flyby to Earth Flyby", *Advances in the Astronautical Sciences, Spaceflight Mechanics 2000*, Vol.105, pp.1053-1072.

5. Roth, D.C., Guman, M.D., Ionasescu, R., "Cassini Orbit Reconstruction From Earth To Jupiter", *Advances in the Astronautical Sciences, Spaceflight Mechanics 2002*, Vol.112, pp.693-704.

6. Cassini Mission Plan, Rev. N, JPL D-5564, May 2002.

7. Kliore, A.J., et al. "Cassini Radio Science", *Accepted for publication on Space Science Reviews*.

8. Iess, L., Giampieri, G., Anderson, J.D., Bertotti, B., "Doppler measurement of the solar gravitational deflection", *Classical and Quantum Gravity*, Vol.16, 1999, pp. 1487-1502.

9. Iess, L. et al., "The Cassini Solar Conjunction Experiment: a New Test for General Relativity", *Proceedings of the IEEE Aerospace Conference*, March 8-15, 2003, Big Sky, Montana, USA.

10. Abbate, S.F., et al. "The Cassini Gravitational Waves Experiment", *Proceedings of the SPIE Conference 4856 on "Astronomy Outside the EM Spectrum"*, August 2002, Waikoloa HI, USA.

11. Bertotti, B., Comoretto, G., Iess, L., "Doppler Tracking of Spacecraft with Multifrequency Links", *Astronomy & Astrophysics*, Vol.269, 1993, pp. 608-616.

12. Tortora, P., Iess, L., Ekelund, J.E., "Accurate Navigation of Deep Space Probes using Multifrequency Links: the Cassini Breakthrough during Solar Conjunction Experiments", *Proceedings of The World Space Congress*, October 10-19, 2002, Houston, USA.

13. Tortora, P., Iess, L., Herrera, R.G., "The Cassini Multifrequency Link Performance during 2002 Solar Conjunction", *Proceedings of the IEEE Aerospace Conference*, March 8-15, 2003, Big Sky, Montana, USA

14. Ekelund, J.E. et al., "Orbit Determination Program (ODP) Reference Manual", JPL D-18671 (JPL internal document), Volume 1-4, Pasadena, CA, Feb. 2000.

15. Resch, G. M. et al. "The Media Calibration System for Cassini Radio Science: Part III", *IPN Progress Report*, 42-148, 1-12, 2002.

16. Keihm, S. J., "Water Vapor Radiometer Measurements of the Tropospheric Delay Fluctuations at Goldstone over a Full Year", *TDA Progress Report*, 42-122, 1-11, 1995.

17. Moyer, T.D., "Formulation for Observed and Computed Values of Deep Space Network Data Types for Navigation", *Deep-Space Communications and Navigation Series*, JPL Publication 00-7, Oct. 2000

INTERPLANETARY NAVIGATION DURING ESA'S
BEPICOLOMBO MISSION TO MERCURY

Rüdiger Jehn,[*] Juan L. Cano,[†] Carlos Corral[‡] and Miguel Belló-Mora[**]

Presently, the use of Solar Electric Propulsion (SEP) as the principal propulsion and navigation system for interplanetary missions is reaching its maturity level after decades of consideration only on the design boards. NASA's *Deep Space 1* successfully demonstrated the use of its ion propulsion system, while ESA is about to launch its *SMART-1* mission to the Moon based on the use of a Hall thruster. Among other low-thrust missions, ESA is also planning to send a probe to Mercury using a SEP system supported by a series of flybys at Venus and Mercury. The navigation based on a low-thrust propulsion system prior to the flybys at Venus and Mercury is the subject of the present paper. This includes the orbit determination accuracy assessment of the interplanetary trajectory during coast and thrust arcs, as well as the performance of the propulsion system and the ranging devices. The cases studied in the paper demonstrate the observability of the low-thrust vector components, especially its out-of-plane component. A brief description is given of the software tool SEPNAV that was developed for the analysis of low-thrust navigation.

INTRODUCTION

The European Space Agency (ESA) is currently planning to send a probe to Mercury at the end of this decade or at the beginning of the next one. The proposed mission is called *BepiColombo* in recognition of the work performed by the Italian scientist Giuseppe Colombo in the sixties, who explained the 3:2 resonant motion of Mercury. A mission based on low-thrust propulsion together with a series of planetary swingbys at Venus and Mercury is the current baseline to reach Mercury. In the fastest option (transfer time of a bit more than two years) four swingbys and a number of long low-thrust arcs will be needed to arrive at Mercury and to obtain low delta-V insertion conditions.

ESA awarded a study contract to DEIMOS Space S.L. for the creation of a software tool, called SEPNAV, to adequately address the trajectory determination and guidance problems associated to the low-thrust trajectories as well as to perform a series of application cases for the *BepiColombo* mission.

This paper presents a description of the features of SEPNAV, and some main trajectory determination results.

---

[*] ESA Staff Member at the European Space Operations Centre (ESOC), Robert Bosch Str. 5, D-64293 Darmstadt, Germany. E-mail: Ruediger.Jehn@esa.int.

[†] Staff Member of DEIMOS Space S.L., Ronda de Poniente 19, Edificio Fiteni VI, Portal 2, 2°, 28760 Tres Cantos (Madrid), Spain. E-mail: Juan-Luis.Cano@deimos-space.com.

[‡] Staff of GMV S.A. located at ESA/ESOC, Robert Bosch Str. 5, D-64293 Darmstadt, Germany. E-mail: Carlos.Corral.van.Damme@esa.int.

[**] General Director of DEIMOS Space S.L., Ronda de Poniente 19, Edificio Fiteni VI, Portal 2, 2°, 28760 Tres Cantos (Madrid), Spain. E-mail: Miguel.Bello@deimos-space.com.

THE SEPNAV NAVIGATION TOOL

SEPNAV (*Solar Electric Propulsion NAVigation*) is a software tool developed to simulate low-thrust navigation problems combined with planetary flybys.

The tool consists of a set of modules, which allow to solve trajectory estimation problems and guidance problems from two different perspectives. On one side, the tool can be run in a so called *one-shot mode*, which allows to obtain theoretical knowledge covariance and dispersion covariance analysis along a nominal trajectory. On the other hand the tool can be executed in a *full simulation mode*, which allows carrying out *Monte Carlo* simulations and comparing the results obtained with the ones in the *one-shot mode*.

SEPNAV is not meant to be used operationally but to be used as a navigation tool for preliminary assessment purposes.

The main modules of SEPNAV are the following:

1. Trajectory Propagation Module

2. Measurements Generation Module

3. Trajectory Determination Module

4. Trajectory Guidance Module

They are described in the following subsections.

Trajectory Propagation Module

This module allows propagating the spacecraft state vector including the effects of different forces like third body perturbations from the main solar system bodies, solar radiation pressure, low-thrust propulsion, chemical propulsion and other non-gravitational accelerations.

The propagation is done by direct integration of the spacecraft state vector by means of an adaptive step Runge-Kutta-Fehlberg algorithm of 7th-8th order.

This module is used as a service package by the rest of the main modules, providing the spacecraft state vector at any epoch from an initial time and state and a given control law (in terms of chemical propulsion and/or low-thrust propulsion).

The module can be run in a deterministic mode, to simulate nominal and estimated world dynamics, or in a stochastic mode, to simulate a real-world propagation.

Measurements Generation Module

Several types of measurements are simulated in the tool. They will allow to carry out the trajectory determination process.

The following types of measurements are considered:

1. Radiometric measurements from Earth based ground stations (range, range-rate, DOR and ΔDOR)

2. Optical measurements of target bodies (basically target azimuth and declination in a spacecraft reference frame)

3. Accelerometer measurements of the spacecraft motion (mainly to measure the solar electric propulsion performance and the solar radiation pressure interaction)

The system can actually model perfect measurements and measurements with stochastic components in order to simulate the real-world observables. The selected measurements will feed the estimation filter to allow the computation of the spacecraft state.

Trajectory Determination Module

The implemented solution for trajectory determination is based on the use of a *Square Root Information Filter* (SRIF), following the work of Bierman[1]. Such denomination refers to the type of formulation utilised in the filtering process, which is posed in terms of the Square Root Information Matrix (SRIM). The SRIM is the square root of the inverse of the covariance matrix. It was shown by Bierman and others that the estimation process based on SRIF is at the same time more accurate and more stable than other equivalent approaches (e.g. Kalman).

The formulation of the proposed approach with SRIF allows to include in the estimation process not only the modelling of the dynamic variables as defined by their equations, but also the effect of exponentially correlated random variables (ECRVs) and consider biases.

The implementation of the information filter allows the processing of a number of measurements in so called *mapping time intervals*, a concept which is equivalent to the selection of a time interval to process a batch of measurements, if one speaks in terms of minimum least squares information processing. The use of SRIF allows obtaining an estimated deviation in the state vector at the beginning of the mapping time interval. This is done by combining the *a priori* information with the information provided by the associated dynamics and the measurements in a mapping time interval. Then, the augmented state and the SRIM are propagated to the next mapping time.

Both covariance analysis and full trajectory estimation processes can be performed with the tool. In the first case, the update and propagation of the knowledge and dispersion covariances is carried out without actually determining the state. Whereas in the second case, Monte Carlo simulations can be executed on simulated real world cases with random parameters.

On one hand, results are available in terms of knowledge and dispersion parameter covariance evolution. On the other hand, results are provided in terms of statistical analysis of the Monte Carlo results.

Trajectory Guidance Module

Interplanetary guidance has been traditionally based on the use of chemical thrusters at given points in the trajectory to steer the spacecraft towards the nominal trajectory. Models for the thrusters and the errors introduced in their performance are the input to the guidance module. In the case of low-thrust trajectories it is advantageous to use the low thrust itself for guidance of the spacecraft. This can be done by modulation of:

1. The modulus of the thrust vector

2. The thrust vector angles

3. The switching times of the thrust

In the case of SEPNAV a combination of changes in the thrust modulus and the thrust vector angles can be selected for guidance purposes.

From the possible guidance options for low-thrust trajectories, the *Trajectory Re-optimisation* choice was considered out of the scope of the application. Thus, a *Feedback Guidance* choice was selected to find the corrections that would need to be introduced in the control variables to bring the spacecraft back to its nominal trajectory. This was done assuming a linear approximation of the dynamic and the measurement equations close to the nominal scenario and in some cases iterating over the solution to account for non-linearities. Two options were implemented. One based on a *Minimum Least Squares* (MLS) approach and one on a *Linear-Quadratic Control* (LQC) approach. The formulation for both methods was presented in Cano[4] *et al.*

In the MLS approach, a solution similar to the one of NASA's *Deep Space 1* by Bhaskaran[2] *et al.*, was implemented. This method allows reaching the target point by introducing changes in the thrust modulus and the thrust angles in a number of discretised time intervals. Those changes are computed by solving a minimum least squares problem where the deviations in the final state are the objective function to minimise and the changes in thrust the variables to adjust.

In the case of the LQC approach a quadratic cost function is defined based on the deviations in the final state and discretised deviations of the guidance variables in time. Such method is complemented by a linear representation of the dynamic equations, which together define an optimisation problem. This problem is then translated into a Ricatti equation by using the Pontryagin principle, as demonstrated by Bryson and Ho[3]. This equation needs to be solved to obtain the change in the thrust variables which minimise the cost function.

ESA'S BEPICOLOMBO MISSION

The *BepiColombo* mission to Mercury is one of ESA's *Cornerstone Missions*. The present trajectory design foresees the use of a SEP system combined with a number of planetary swingbys at Venus and Mercury.

The trajectory was originally proposed by Langevin[8] and then optimised both with direct optimisation methods, Vasile[10] *et al.*, and with indirect methods, Katzkowski[6] *et al.*

In the mission opportunity of 2010 the launch with a Soyuz rocket takes place on 22 July 2010 (Figure 1 shows an ecliptic projection of this mission). The departure excess velocity is 1.9 km/s and the launch mass 1,390 kg. The maximum thrust of 340 mN is only reached after the two first thrust arcs when the spacecraft is well inside the Venus orbit where sufficient solar energy is available for full thrust. Arrival will be 2.3 years later on 27 Oct 2012. After jettison of the SEP system a mass of 620 kg can be inserted into a 400 x 12,000 km Mercury orbit.

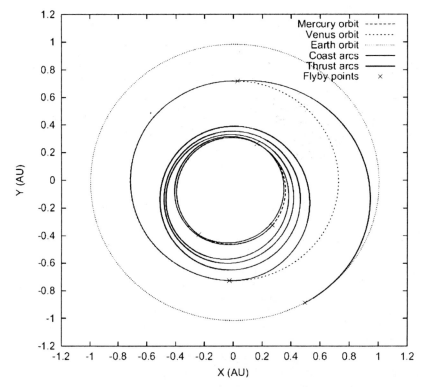

Figure 1: Ecliptic projection of the BepiColombo transfer from Earth to Mercury. Thrust arcs are marked with thick lines, coast arcs with thin lines and flybys with asterisks

For navigation studies the trajectory can be split in the following segments:

1. Earth-Venus segment: with a 97-day thrust arc followed by a 30-day coast arc,

2. Venus-Venus segment: 111-day coast arc of 180° where the spacecraft travels all the time below (wrt the ecliptic) Venus,

3. Venus-Mercury segment: 4.5 heliocentric orbits with alternating thrust and coast arcs until the first Mercury flyby (for 435 days),

4. Mercury-Mercury segment: 40-day coast arc of 180°,

5. Final approach: final segment from second Mercury flyby to Mercury orbit insertion (including a 19-day thrust arc).

SEPNAV was designed for the independent analysis of each of these segments, although the analysis of several segments together is not excluded.

TRAJECTORY DETERMINATION RESULTS

SEPNAV was applied to the covariance analysis of the *BepiColombo* trajectory. Several cases were defined and executed, first during coast arcs to allow comparison with the results obtained with other tools, and finally during arcs with thrust to observe the tool performances. The following assumptions were taken:

<table>
<tr><td colspan="2" align="center">**Table 1**
INITIAL 1σ KNOWLEDGE COVARIANCE
(AND AUTOCORRELATION TIME, IF ECRV)</td></tr>
<tr><td>Position</td><td>500 km</td></tr>
<tr><td>Velocity</td><td>5 m/s</td></tr>
<tr><td>Solar radiation pressure</td><td>10 % (10 days)</td></tr>
<tr><td>Non-gravitational accelerations</td><td>10^{-11} km/s^2 (1 day)</td></tr>
</table>

Table 2
MEASUREMENT ASSUMPTIONS

Ground stations		Madrid and Perth
Stations position bias (X,Y,Z) .		1 m, 1 m, 2 m
Range	Sampling	1 per hour
	Bias	2 m
	Noise (white)	10 m
Doppler	Sampling	6 per hour
	Bias	0 mm/s
	Noise (white)	0.3 mm/s

Orbit determination during coast arcs

These test cases over coast arcs, were meant to be compared with results of a navigation analysis during coast phases by Corral and Jehn[5], which were obtained with INTNAV, ESA's tool for interplanetary navigation assessments (this tool does not allow to consider low-thrust propulsion).

Plots are presented for the evolution of 1-σ position and velocity knowledge covariance during 30 days before the first Venus flyby and for the following Venus-Venus coast arc. The initial date MJD2000 = 3954 corresponds to 29/10/2010. Mapping time intervals of 0.5 days were used in the computations. It can be reminded (see Table 2) that radiometric measurements were collected from two ground stations in all the passes available.

Plots in Figure 2 show how the knowledge is improved from the initially pessimistic 500 km in position and 5 m/s in velocity in each component. It then remains approximately constant at around 25 km uncertainty in total position while the velocity uncertainty keeps decreasing to a value of 10 mm/s before the Venus flyby. In both plots the flyby at Venus can be seen around day 3984. The expected focusing effect for the position uncertainty in contrast to the velocity uncertainty magnifying effect can be observed. If the obtained orbit knowledge is kept for the next Venus-Venus coast arc, the plots in Figure 3 are obtained. The position uncertainty rises to a level around 40 km before decreasing again just before the next Venus flyby. The velocity knowledge covariance remains more or less constant after the first flyby and increases slightly before the second flyby.The results agree quite notably with those computed with INTNAV[5].

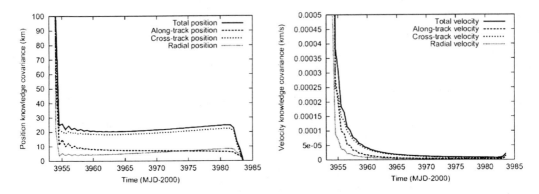

Figure 2: Position and velocity knowledge covariance before the first Venus flyby

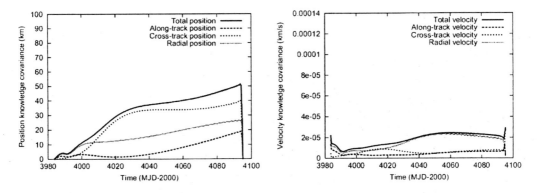

Figure 3: Position and velocity knowledge covariance before the second Venus flyby

Orbit determination during low-thrust arcs

Simulations were made taking the same assumptions as for the previous cases (see Table 1 and Table 2) and testing the effect of the thrust parameters (thrust magnitude and thrust angles) assuming different dynamic properties for them.

Figure 4 shows the definition used for the thrust angles: the *angle of attack*, α, is the angle between the thrust vector and its projection on the local tangential plane to the trajectory, positive outside the orbit, and the *side-slip angle* is the angle between this projection and the velocity vector (in the figure T is along the velocity, Z perpendicular to the orbital plane, and N so that the TZN system is right-handed).

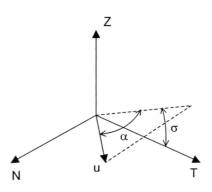

Figure 4: Thrust angles definition

1430

Depending on the case, the thrust parameters were modelled as

- Estimated constant deviations
- Consider biases
- ECR variables

The simulations were run over a 95-day leg of the trajectory prior to the first Mercury flyby, composed of a 47-day thrust arc followed by a 48-day coast arc, see Figure 5, and assuming the values proposed in Table 1 and Table 2. The thrust vector direction forms a 65° angle with the line-of-sight from the Earth at the beginning (see Figure 6), which increases to 150° at the moment when it is switched off (thrust stops at around day 4480). It is also important in the analysis of the results to show the evolution of the radial, the along-track and cross-track directions with respect to the line of sight (LOS). The cross-track component (mainly out-of–ecliptic) is almost perpendicular to the LOS, thus making the behaviour of the other two components to be complementary. Whenever the along-track component or the radial component is parallel to the LOS, the other is normal and vice versa. The radial component is almost perpendicular to the line of sight when the thrust is switched off.

Figure 8 presents the evolution of the knowledge in position and velocity when the thrust parameters were assumed to have constant deviations in their performance which were estimated. The thrust modulus will be usually determined by direct measurements of the current and voltage supplied by the power system to the engine and their relation to the thrust variables. From those available parameters, and possibly others (as for example working temperatures, number of hours of operation, etc.) the mass flow rate, the specific impulse, the engine efficiency and the thrust force will be obtained from the calibrated operational tables. However, this could be verified by radiometric measurements as done in this case.

An initial 1-σ uncertainty of 5 % in thrust modulus (which for this arc is 340 mN) and 3° in the thrust angles were assumed for those variables. Figure 9 presents the uncertainty associated to the estimated thrust parameters. The plot shows that the thrust modulus and the thrust angle related to the thrust in-plane component (the angle of attack) can be estimated to a value two orders of magnitude better that the initial values. In other words, constant deviations in thrust modulus are observable to a level of 0.1 % and constant angular changes in thrust are observable to an accuracy of 0.05°. However, the thrust modulus is affected by the uncertainty in the knowledge of the spacecraft mass, which for this case was assumed to be 1 kg. With a better knowledge of the spacecraft mass, the thrust modulus can also be better determined.

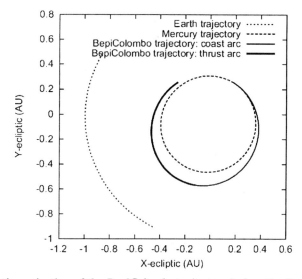

Figure 5: Ecliptic projection of the *BepiColombo* trajectory before the first Mercury flyby

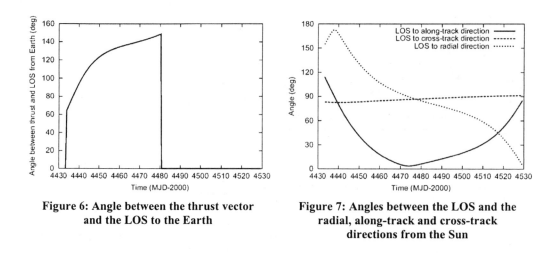

Figure 6: Angle between the thrust vector
and the LOS to the Earth

Figure 7: Angles between the LOS and the
radial, along-track and cross-track
directions from the Sun

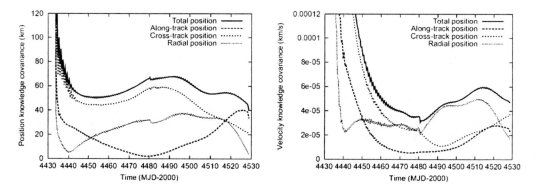

Figure 8: Position and velocity knowledge covariance for systematic estimated deviations in the
thrust parameters

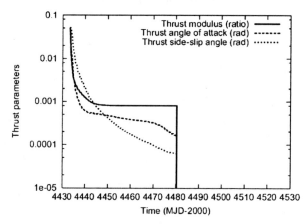

Figure 9: Estimated thrust parameters

Regarding the out-of-plane component of the thrust (the side-slip angle), systematic changes in its value will be even more observable, as its knowledge improves in this case even one order of magnitude more than e.g. the thrust modulus. This actually means that the out of plane SEP forces (changing inclination and RAAN) are better observed than the in-plane components whenever they are estimated.

It has to be noted that the knowledge is not only based on the processing of the measurements available (which would give a very poor value for the cross-track component for a single ground station pass) but the information from the past dynamics is also used and combined by the information filter with the information provided by the observables.

It can also be noted that the three components of the thrust vector are quickly estimated to a good accuracy as seen in Figure 9, which explains why the position and velocity knowledge in Figure 8 are not very much affected by the thrust uncertainty.

The second simulation case was set up to show the effect of unmodelled constant deviations in the performance of the thrust system. This was done by assuming the thrust parameters as consider biases and by analysing the resulting evolution of the knowledge covariance. For this case, consider biases in thrust modulus of 0.15 %, in thrust angle of attack of 0.1° and side-slip angle of 0.01° were taken. Results are presented in Figure 10.

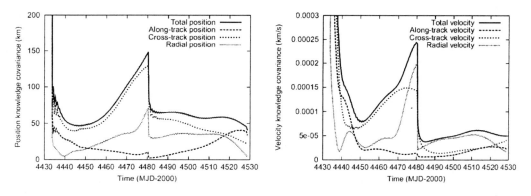

Figure 10: Position and velocity knowledge covariance assuming consider biases in the thrust parameters

To explain the behaviour of the curves in Figure 10, it is needed to correlate the evolution of each position and velocity components with the angles shown in Figure 7. In the case of the velocity vector, it can be observed that the least observable component is the cross-track component, as one might expect. After the initial improvement in the velocity knowledge down to the value compatible with the system uncertainties, the knowledge is not improved further, but degraded little by little due to the lack of thrust biases acknowledgement in the dynamics. As the moment of thrust switch off approaches, the radial component becomes normal to the LOS from the Earth and thus poorly observable. Therefore the uncertainty in this component increases considerably even exceeding the uncertainty in the cross-track component.

The evolution of the uncertainty in the position vector is somewhat similar but the effects are smaller. The uncertainty in the knowledge of the cross-track component drives the error in position, slightly modulated by the other two components. In summary, unexpected deterministic deviations in the thrust parameters will have a large impact on the state accuracy if they are not determined. Whenever the thrust stops, the knowledge quickly improves to values compatible with ballistic segments.

The last simulation case involved the assumption that the deviations of the thrust parameters from their nominal behaviour were modelled as exponentially correlated random variables. The assumed 1-σ steady state covariance for the three thrust parameters was 5 % for the modulus and 3° for the angles. Autocorrelation times of 1 day were assumed for all three thrust parameters.

Figure 11 shows the evolution of the knowledge position and velocity covariance along the simulation time. The oscillation in the results in the thrusting arc are introduced by the different number of measurements available in the mapping time intervals (the mapping time step was taken as 0.5 days) and by the dispersion introduced by the noisy thrust components.

Again, it can be seen that the cross track component drives the behaviour of the determination process. Covariances are larger than in the case of the estimated deviations as expected, and the covariances are seen to clearly improve after the thrust switch off, as also expected.

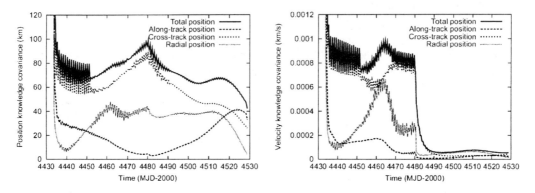

Figure 11: Position and velocity knowledge covariance with noisy thrust parameters

The effect of solar plasma

The effect that the solar plasma introduces in the estimation process based on radiotracking, was studied for a trajectory segment starting on day 4138 (40 days after the second Venus flyby) and ending on day 4188, during which the spacecraft passed behind the Sun (solar conjunction). The moment of smallest elongation from the Sun occurs around day 4167. The spacecraft is thrusting throughout the whole segment. The thrust parameters were assumed as consider biases with the following values: 0.15 % in modulus, 0.1° in angle-of-attack and 0.01° in side-slip angle. An X-band uplink combined with a Ka-band downlink was selected, assuming a 600 s doppler integration time.

Figure 12 shows the evolution of the 1-σ covariance in range and doppler measurement errors as given by Eqs. (1) and (2), obtained from Timm[9]. For range, the error depends on a constant error σ_{ran0} (10 m, see Table 2), the Sun-Earth-spacecraft angle α, and the Earth-Sun-spacecraft angle β:

$$\sigma_{ran} = \sqrt{\sigma_{ran_0}^2 + \sigma_{ran_p}^2}$$

$$\sigma_{ran_p} = \frac{2.13}{\sin \alpha} \beta \quad (m)$$

(1)

For doppler, the error depends on a constant error σ_{dop0} (0.3 mm/s, see Table 2), the same angle α, the speed of light c, the downlink frequency f_c, a constant k_1 depending on the transmission frequencies and the doppler integration time T:

$$\sigma_{dop} = \sqrt{\sigma_{dop_0}^2 + \sigma_{dop_p}^2}$$

$$\sigma_{dop_p} = \frac{0.73\,c\,\sqrt{k_1}\,(\sin\alpha)^{-1.225}}{f_c\,T^{0.175}} \quad (m/s) \tag{2}$$

$$k_1 = \begin{cases} 5.5\cdot 10^{-6} & X-up\,/\,X-down \\ 5.2\cdot 10^{-5} & X-up\,/\,Ka-down \\ 2.3\cdot 10^{-7} & Ka-up\,/\,Ka-down \end{cases} \tag{3}$$

$$f_c = \begin{cases} 8.4\cdot 10^9\ Hz & X-downlink \\ 32.0\cdot 10^9\ Hz & Ka-downlink \end{cases}$$

The error in range rises from an initial value of around 10 m up to 5.5 km at the point of closest conjunction. In the doppler case, the error grows from 0.3 mm/s up to 60 mm/s. In any case, measurements are not considered when the solar elongation of the spacecraft is below 1.3°.Figure 13 presents the evolution of position and velocity covariance accounting for the solar plasma effect. As expected, the uncertainty levels are higher than the cases where the errors in the measurements were assumed constant. The knowledge evolution reaches a peak just a couple of days after the solar conjunction. The knowledge in position decreases by 65 %, whereas the knowledge in velocity decreases by 35 %.

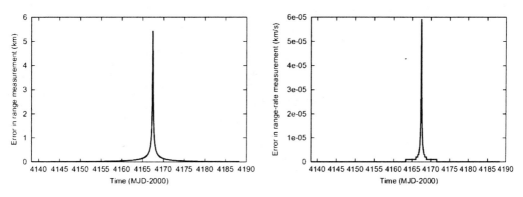

Figure 12: Range and doppler noise as modelled by equations (1) to (3)

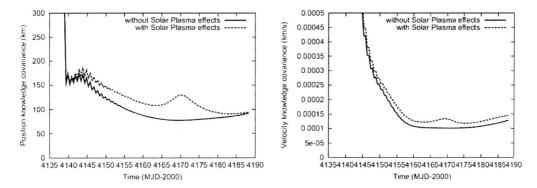

Figure 13: Effect of solar plasma on the position and velocity knowledge covariance (solar conjunction occurs on day 4167).

Covariance analysis before a planetary flyby

Finally, a case is shown where the thrust system is operated close to a flyby to demonstrate the degradation in the estimation performance when a noisy thrust system is in operation. A Mercury to Mercury segment of the trajectory similar to the one considered before is now analysed. The time between the two flybys at Mercury is 40 days, where the first 33 days are a coast arc and the last 7 days a thrust arc.

It was assumed that during thrusting a noisy system behaviour is present with the following 1-σ steady state covariance of the Gauss-Markov noises: 5 % in thrust modulus, 3° in each of the thrust angles. The autocorrelation times were taken to be 1 day for the three variables and the mapping time interval also as 1 day. The assumptions in Table 1 and Table 2 where also taken hereafter for the rest of the system parameters. The estimation process was carried out once per day.

In addition, it was assumed that the guidance system is able to correct the dispersion after each trajectory determination process to the level of the knowledge (by computing the required deviations in the rest of the thrust law that would drive the spacecraft to its nominal trajectory).

Figure 14 shows the results obtained in this case in terms of 1-σ knowledge and dispersion standard deviation of the flyby altitude. If one compares the knowledge when the noisy thrust system is on with the case of perfect thrust system behaviour (corresponding to the case with no thrust), it is easily observed how the knowledge degrades from a value close to 40 km during the thrust arc to a value close to 100 km. The final improvement in the knowledge when approaching Mercury is due to the focussing effect in position produced by the flyby itself (opposite effect in velocity). Regarding the dispersion, after each guidance process is introduced, the trajectory covariance increases up to the moment where new knowledge is available and the retargeting process is carried out. The 1-σ final dispersion at the moment of the flyby would be around 75 km in this case.

This case clearly shows the effects that a noisy thrust system induces in knowledge and dispersion close to a planetary flyby. With the SEPNAV tool it is possible now to analyse the noise tolerances that can be accepted in the solar electric propulsion to guarantee an accurate flyby. If it turns out that the requirements on the propulsion system are too stringent, it is recommended to introduce coast arcs prior to the flyby.

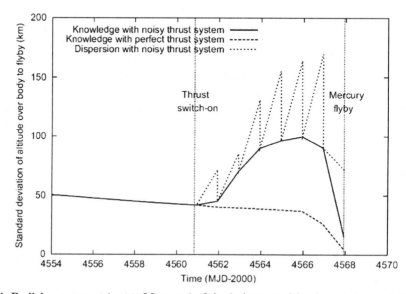

Figure 14: Radial component (wrt to Mercury) of the 1-sigma position knowledge and dispersion. At the time of pericentre passage it gives the flyby altitude knowledge and dispersion.

CONCLUSIONS

The results in this paper show that any estimated systematic changes in the thrust variables seem to be highly observable. In particular, the out-of-plane thrust component can be better determined than the in-plane component and the thrust modulus.

On the other hand, if those deviations are unnoticed (thus, not included in the estimation process), the position and velocity uncertainty will increase considerably. This applies in particular to the out-of-plane component of the trajectory.

If the thruster performance is noisy, the position and velocity estimation will be notably degraded. Therefore, it may become necessary to introduce coast arcs before critical events like flybys to allow for a better orbit determination. With the SEPNAV tool it can be analysed what levels of noise in the solar electric propulsion system can be tolerated.

The solar plasma effects on the trajectory estimation were shown to degrade the accuracy in the estimates whenever solar conjunctions occur. However, the effects are only temporary and may become only important if a flyby takes place at the same time.

ACKNOWLEDGEMENT

The development of the SEPNAV software tool and the related study for the *BepiColombo* mission was carried out under ESA contract, reference 15760/01/D/HK(SC).

REFERENCES

1. Bierman G.J., *Factorisation Methods for Discrete Sequential Estimation*, Mathematics in Science and Engineering, Vol. 128, Academic Press, 1977.

2. Bhaskaran S. *et al.*, *Orbit Determination Performance Evaluation of the Deep Space 1 Autonomous Navigation System*, AAS 98-193, Vol. 99, Spaceflight Mechanics, 1998.

3. Bryson A.E., Ho Y., *Applied Optimal Control*, Hemisphere Publishing Corp., 1975.

4. Cano J.L., Belló-Mora M., Jehn R., *Low-Thrust Navigation for BepiColombo Mission*, IAF-02-A.1.03, 53rd International Astronautical Congress, The World Space Congress - 2002, Houston, 10-19 October 2002.

5. Corral C., Jehn R., *BepiColombo Mercury Cornerstone Mission Analysis: Orbit Determination and Guidance*, MAS WP No. 439, ESA/ESOC, Darmstadt, October 2002.

6. Katzkowski M., Jehn R., *Optimum Trajectories to Mercury Combining Low-Thrust with Gravity Assists*, IAF-01-Q.2.05, 52nd International Astronautical Congress, Toulouse, 1-5 October 2001.

7. Katzkowski M., Corral C., Jehn R., *et al.*, *BepiColombo Mercury Cornerstone Mission Analysis: Input to Definition Study*, MAS WP No. 432, ESA/ESOC, Darmstadt, April 2002.

8. Langevin Y., *Chemical and Solar Electric Propulsion Options for a Mercury Cornerstone Mission*, IAF Paper No. IAF-99-Q.2.07, IAF Amsterdam, 1999.

9. Timm R., *BepiColombo: Solar Plasma Influence on Ranging Accuracy*, internal ESA document, July 2002.

10. Vasile M., Bernelli F., Jehn R. and Janin G., *Optimal Interplanetary Trajectories Using a Combination of Low-Thrust and Gravity Assist Manoeuvres*, IAF-00-A.5.07, October 2000.

GENESIS TRAJECTORY AND MANEUVER DESIGN STRATEGIES DURING EARLY FLIGHT

Roby S. Wilson[*] and Kenneth E. Williams[†]

As the fifth Discovery mission, the Genesis spacecraft was launched on August 8, 2001 with a science objective to collect solar wind samples for a period of approximately two and a half years while in orbit in the vicinity of the Sun-Earth L_1 libration point. These samples will eventually be delivered back to the Earth for analysis, posing a formidable challenge in terms of both mission design and navigation. This paper discusses trajectory and maneuver design strategies employed during the early phases of flight to accommodate spacecraft and instrument design constraints, while achieving the science objectives of the mission. Topics to be discussed include: mission overview, spacecraft design and constraints, maneuver analyses and trajectory re-optimization studies, and operational flight experience to date.

INTRODUCTION

Mission Overview

Genesis is the fifth mission selected by NASA under its low cost Discovery program. The primary goal of the Genesis mission is to collect solar wind particles over a 2.5-year period and return them safely back to Earth. As such, Genesis will be the first mission to return extra-terrestrial samples since the Apollo missions nearly 30 years ago. After years of development, Genesis was successfully launched from Kennedy Space Center on August 8, 2001. The trajectory of the spacecraft takes it out to the vicinity of the Sun-Earth L_1 libration point (along the line between the Sun and the Earth). After insertion into a large amplitude Lissajous (or halo orbit) about the L_1 point, the collection arrays were deployed to begin sampling the solar wind. Genesis will collect samples for five revolutions, or approximately 2.5 years, about the libration point. After completing its time near L_1, Genesis follows a free return trajectory home via a looping return about the Sun-Earth L_2 point to set up a daylight entry and mid-air capture over the western United States in September 2004. After a safe return to Earth, the samples will be

---

[*] Jet Propulsion Laboratory, California Institute of Technology, MS 301-140L, 4800 Oak Grove Drive, Pasadena, California 91109-8099. E-mail: Roby.Wilson@jpl.nasa.gov.

[†] Jet Propulsion Laboratory, California Institute of Technology, MS 264-370, 4800 Oak Grove Drive, Pasadena, California 91109-8099. E-mail: Kenneth.Williams@jpl.nasa.gov.

curated at Johnson Space Center and made available for study to scientists throughout the world. This type of sample return has never been attempted before and presents a challenge to both mission design and navigation that will be discussed in this paper.

The Genesis mission is lead by Principal Investigator Dr. Donald Burnett of the California Institute of Technology. The Genesis team consists of members from the Jet Propulsion Laboratory, Lockheed Martin Astronautics, Los Alamos National Laboratory, and the Johnson Space Center. Project management resides at JPL along with mission planning, navigation, and sequencing. The spacecraft and operations teams are headed by Lockheed Martin, and the science team primarily resides at Los Alamos. The handling and curation of the returned samples will be lead by the Johnson Space Center.

Science Overview

The science objective for the Genesis mission is to precisely determine the elemental and isotopic composition of the solar wind. The solar wind is thought to be compositionally identical to the Sun's photosphere. Furthermore, it is believed that the photosphere is representative of the solar nebula from which the solar system was formed. By studying the solar wind then, scientists can, in essence, study the very material that formed the sun and all the planets, moons, asteroids, and comets.

The solar wind is not constant in speed or composition. It is affected by activity on the surface of the Sun, such as coronal mass ejections. The Genesis spacecraft has electron and ion monitors to help classify the nature of the solar wind at any given instant. Based up this classification, the spacecraft deploys various collector arrays that have been specifically designed for that particular solar wind regime. Genesis also has an electrostatic concentrator instrument that is tuned to optimize collection of O, N, and C ions, while rejecting most of the H ions from the solar wind. A portion of the samples collected will be made available to scientists immediately, while the rest is held in reserve for future, perhaps as yet undeveloped, analyses.

Because the science objective is to collect pristine solar wind samples, it is critical that the collection be performed away from any interaction with the Earth's electromagnetic environment. This requirement drove the selection of a Lissajous or halo orbit about the Sun-Earth L_1 point. Trajectories in this region remain in front of the bowshock interaction of the solar wind with the Earth's magnetic field and suffer no eclipsing issues.

Spacecraft Overview

The Genesis spacecraft is depicted in Figure 1. The spacecraft consists of basically two parts. The first part is the bus that comprises the solar panels, propulsion system, attitude control system, avionics system, communications systems, as well as the electron and ion monitors. The spacecraft is spin-stabilized with a nominal spin rate of about 1.6 RPM. Due to the need to minimize sample contamination on the top of the spacecraft, all of the thrusters are on the aft side (as shown in the right portion of Figure 1). This configuration produces unbalanced thrusting that must be accounted for in terms of both attitude control and propulsive maneuver design. The implications to trajectory and maneuver design are discussed in detail later in the paper.

The second part of the spacecraft is the Sample Return Capsule or SRC. The SRC is the portion of the spacecraft that is designed to re-enter the atmosphere and separates from the spacecraft bus prior to re-entry. The SRC consists of a hinged capsule (as shown in Figure 2) that contains the sample collection canister, including the collection arrays and the electro-static concentrator. The SRC backshell is covered with an ablative material to protect the capsule during re-entry, while the foreshell contains a parafoil to help slow down the capsule during the atmospheric portion of its descent. After the SRC has reached its terminal descent velocity, it will be captured in mid-air by a helicopter and then taken safely to the processing facility.

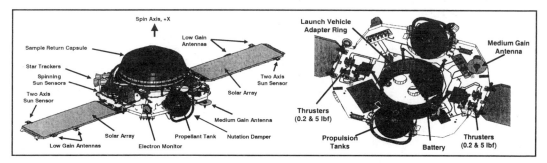

Figure 1. Fore and Aft Views of the Genesis Spacecraft

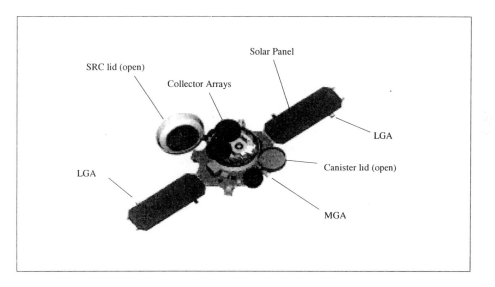

Figure 2. Genesis Spacecraft in Science Collection Configuration

Trajectory Overview

The Genesis trajectory was designed using Dynamical Systems Theory coupled with a two-level differential corrections process. This combination of techniques allowed all of the various top-level mission constraints to be satisfied, such as: no interference from the Earth's electro-magnetic field, at least 23 months of science collection, and daylight entry over Utah for mid-air capture. These top-level constraints have driven the design and re-design of the trajectory. Earlier papers provide details on procedures employed to determine suitable trajectory solutions (see references).

The Genesis trajectory for the July/August 2001 launch opportunity is shown in Figure 3. This view of the trajectory is in a frame that rotates with the Earth about the Sun. Thus, the Sun is always to the left along the –X axis.

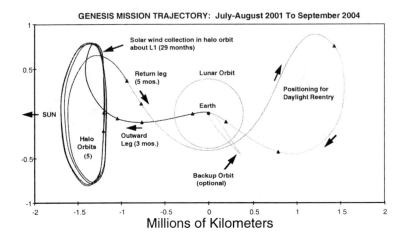

Figure 3. Genesis Trajectory in Sun-Earth Rotating Frame

The trajectory launches in the sunward direction towards the Sun-Earth L_1 point and injects onto the Lissajous or halo orbit near the first downward crossing of the Sun-Earth line. This maneuver is called the Lissajous Injection Maneuver or LOI. Genesis successfully performed its LOI maneuver on November 16, 2001 and is currently about half way through its planned stay near L_1. After completing five revolutions in the Lissajous orbit, the natural dynamics of the motion cause the trajectory to depart L_1, pass close to the lunar orbit, and make a loop around L_2. (Note that the Moon actually only plays a small role in this return trajectory; the closest approach is over 300,000 km away from the Moon. The primary dynamics at work is the Sun-Earth-Spacecraft three body problem.) The L_2 loop positions the spacecraft for a daylight entry over Utah from the northwest. It is of note that the Earth's terminator roughly coincides with the X=0 line in the plot. Trajectories returning from L_2 cross the terminator into daylight as they re-enter, while trajectories returning from L_1 cross into darkness. This fact necessitated the return from L_2 to allow the helicopter pilots to be able to adequately see the SRC for retrieval.

EARLY TRAJECTORY CORRECTIONS

Design

The launch period for Genesis opened on July 30, 2001 and extended through August 14. In all cases, each launch date assumed a short-coast arc launch trajectory leading to the same LOI point on November 16, 2001. In order to accommodate COLA (launch collision avoidance) and minimize the effect on the LOI delta-V, the length of each launch opportunity was limited to no more than 2 minutes on each launch date. For convenience, a common launch vehicle trajectory could be assumed for 2-3 day periods, known as launch blocks, centered on July 30, August 2, August 5, August 8, August 11 and August 14, respectively. Injection into the transfer orbit was achieved with a Delta II 7326 launch vehicle with a Star 37 third stage.

Up to five trajectory correction maneuver (TCM) opportunities were scheduled between launch and LOI to correct transfer orbit injection errors and set up the proper conditions for performing the LOI maneuver. These maneuvers were designed in accordance with an early TCM strategy, summarized in Table 1, which attempted to account for limitations in spacecraft capabilities and to minimize operational complexity in the critical period immediately following launch.

Maneuver	Primary Location	Backup Location(s)	Conditions/Explanations
All Transfer	See below	See below	Maneuver strategy expanded to accommodate ACS delta-design and ground s/w development. All transfer maneuvers targeted to LOI location near L1 (minor targeting variations possible).
TCM-0 (Contingency)	L+24h	Same as TCM-1 primary/backup(s)	Emergency TCM used in place of TCM-1 only if maximum maneuver size allowed by spacecraft power and thermal constraints exceeded at L+48h (> 110 m/s anti-sunward or > 130 m/s sunward design same as TCM-1; highly unlikely contingency.
TCM-1	L+48h	L+3d, L+4d, L+5d, L+6d... (see below)	Performed only if required maneuver at L+48h > 5 m/s but <110 m/s if anti-sunward or <130 m/s if sunward (option to not perform if as high as 30 m/s, based on operational assessment of cost of TCM-3 without earlier maneuvers); fixed aimpoints in plane containing Sun and injection (TIP) attitude near Ecliptic Plane. Two-maneuver optimization with TCM-3 assumed. Maneuver on spinning sun sensors (SSS) only, assuming injection attitude near Earth-based velocity vector (Boeing's PMA and DTO data provide the actual injection attitude); must observe ACS-defined keepout zones per flight rule 0001-A-ACS. Use one of two fixed aimpoints in either sunward or anti-sunward direction; only three inertial aimpoints sufficient to cover nominal blocks. Daily backup opportunities (if delayed before execution); minimum 48 hour delay if execution aborted to reestablish OD and verify s/c health and safety; switch to TCM-2 on L+7d if TCM-1 was required but was unable to be executed.
TCM-2 (Backup Only)	N/A	L+7d and L+18d	Highly unlikely maneuver (e.g., if TCM-1 required but not executed by L+7d or extremely large injection error could not be corrected sufficiently by TCM-0/1). In the very unlikely event that TCM-2 is needed, consider the following: • Two-maneuver optimization with TCM-3. • Employ best possible design (i.e., greatest flexibility in burn direction that can be afforded) based on SSS-only attitude control with appropriate observation of KOZs. • Backup opportunity at L+18d avoid periods of low target adjustment capability. • Only need to use OD-supported determination of initial attitude when clock angle knowledge has been lost or severely degraded by previous attitude maneuvers.
TCM-3	L+35d	L+49d	Performed only if > 1.25 m/s to avoid use of smaller thruster sun maneuver modes (relief to maneuver decomposition software development schedule pre-launch. Spin Track must be operational by this point (otherwise, drastic mission re-design anticipated). Prime location at L+35d and backup opportunity at L+49d avoid OD quiet period; if backup used, may require replanning of small forces calibration and science start activities.
TCM-4 (Contingency)	L+65d	L+72d	Contingency maneuver (no optimization), highly unlikely given transfer phase minimum maneuver size constraint. Retain opportunity as backup to TCM-3. Prime location at L+65d to afford better target adjustment capability. Backup opportunity at L+72d; prior to start of science (opening of canister, deployment of arrays, etc.) to avoid possible additional cycle on mechanisms.

Table 1. Summary of Early TCM Strategy

During the first few days after launch, the attitude control subsystem (ACS) was effectively limited only to spinning sun sensors to support spacecraft attitude determination. Star trackers, required to support the spin track mode with full three-axis attitude determination, could not be relied upon to support the ACS until a calibration was performed a few days after launch. For a majority of interplanetary missions, this would not be a major issue, since the first TCM is typically performed weeks after launch, allowing plenty of time for checkout and calibration of various spacecraft systems. However, as shown in Figure 4, the assumed injection covariance was large enough for Genesis to necessitate a potentially large post-launch TCM. The data shown in the figure are statistical delta-v magnitudes, including mean and 95% probable maximum delta-v, based on monte-carlo simulation results. These magnitudes grow rapidly as time past injection increases. Such potentially exorbitant delta-v costs precluded delaying the first TCM more than a few days after launch.

On the other hand, the TCM could not be planned until tracking data from two, or preferably three, ground stations had been collected and processed to determine the actual spacecraft orbit resulting from launch injection errors. Also, it was considered prudent to allow sufficient time to deal with a variety of potential operational difficulties that could readily arise during this period. Consequently, the first TCM (designated TCM-1) was scheduled to be performed nominally at 48 hours after the post-injection target interface point (TIP), which occurred about 40 minutes after launch. As a contingency, if launch errors were extremely large, owing to about a 1% probability of early shutdown of the second stage of the launch vehicle, then obtaining highly accurate orbit determination information was less critical than performing some sort of trajectory correction as quickly as possible. In this case, the first TCM would be executed at 24 hours past TIP. This contingency TCM (designated TCM-0) would replace TCM-1 in such an event.

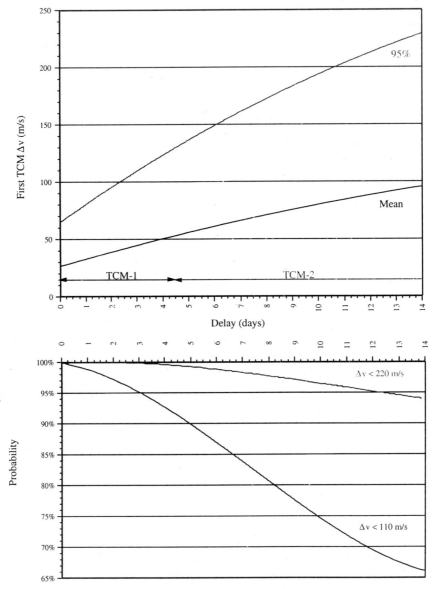

Figure 4. Potential ΔV Costs for First TCM

Due to the limited ACS capability, the implementation of TCM-0 or 1 had the potential to be quite complex with considerable ground-spacecraft interaction. Fortunately, this first TCM was needed mainly to correct the launch energy or C_3, so the required maneuver orientation would be ideally in the direction of, or opposite to, the spacecraft velocity. Also, because Genesis was injected from the launch vehicle in a spinning state, an injection attitude in the direction of the velocity prior to injection could be assumed. So if burn directions were selected to lie in a plane which included the injection attitude and the Sun, the spacecraft could be oriented in a pre-planned, fixed pointing direction merely by dead-reckoning with sun sensors. Two burn directions particular to each launch block were chosen, one sunward to compensate for an injection underburn, the other anti-sunward in the event of an injection overburn.

The sun sensors, primarily the spinning sun sensors (SSS) operating at post-injection spin rates of 9.5 RPM and higher, impose constraints on spacecraft attitude relative to sunward and anti-sunward directions. These constraints, known as keep-out zones (KOZ) are designed to prevent a potentially mission-fatal situation where the ACS cannot determine spin rate due to false Sun crossing indications in the presence of nutation and allowing for the possibility of single thruster failures. Consequently, the fixed aimpoints selected for various launch blocks were always on the edge or beyond the KOZ's, albeit as close to the ideal velocity or anti-velocity direction as allowed operationally. For operational simplicity, the number of inertial orientations for the entire launch period could be reduced to just three directions. These were specified in separate Maneuver Profile Files (MPF's), one sunward and two anti-sunward (the second providing an update for the first starting with the August 8 launch block). Each MPF was processed into a specific maneuver sequence before launch. After launch, the Navigation Team needed only to determine which of three cases (sunward or anti-sunward TCM-1 or sunward TCM-0) would apply and provide a simple burn magnitude update in accordance with the estimated injection C_3 error. An illustration of the guideline used in this process for the August 8 launch block is indicated in Figure 5.

Following the aforementioned launch energy correction, a TCM at 35 days after launch (designated TCM-3) was scheduled to correct pointing errors arising from injection and earlier TCM's. Further TCM's were included for contingency purposes only, at 7 days after launch (TCM-2) and 65 days after launch (TCM-4). These TCM's would only be needed in the event of severe spacecraft anomalies associated with abort or delay of TCM-0/1 and/or TCM-3.

Execution

Because of hardware concerns on July 30 and bad weather on subsequent launch opportunities, the actual launch did not occur until August 8. In reality, the injection provided by the Delta Star 37 third stage was so accurate that only a small anti-sunward burn of about 5 m/s was needed to correct a slight injection overburn. This was achieved with an overall magnitude of about 8 m/s, including turns and spin changes, with an error of about +4% (overburn).

During the period leading up to the scheduled TCM-3 on September 12, there was an indication of excessive temperature of the batteries needed for SRC recovery at the end the mission. This was most probably due to contamination of adjacent surfaces internal to the SRC itself, which had the potential to greatly increase the temperature of these batteries over the course of the mission and compromise the recovery of the SRC. This anomaly necessitated that scheduled activities be delayed or postponed to allow for remedial actions to be carried out. Fortunately, it was determined that the direction of the LOI maneuver could be improved, from the standpoint of Earth visibility, by canceling TCM-3 and other transfer TCM's altogether. This led to some re-optimization of the post-LOI trajectory as well. These considerations are explained in more detail in the following section on LOI re-design.

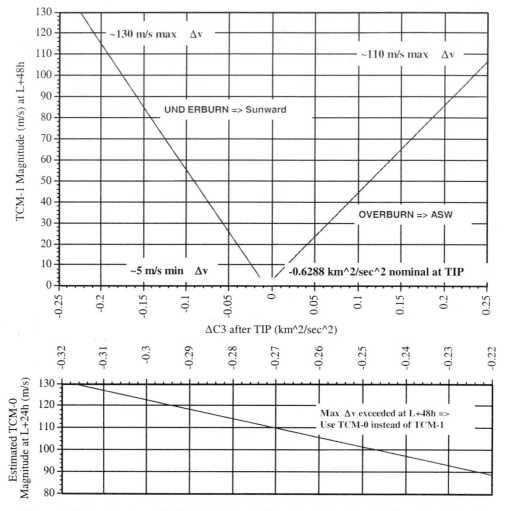

Figure 5. Determination of TCM-0/1 Delta-V Magnitude based on C_3 from Post-Injection Orbit Determination

LOI MANEUVER

Re-Design

After reconstructing TCM-1, as discussed above, the TCM-3 maneuver size and direction relative to the Sun were estimated based on a monte-carlo simulation with 5000 samples, as shown in Figure 6. The data shown assume execution of a delayed TCM-3 on September 18 or about 41 days after launch. However, the contamination issue discussed earlier complicated potential execution of TCM-3. To deal with this problem, it was decided that the SRC should be partially closed to avoid further exposure to the Sun that might result in further annealing of contaminant particles to SRC interior surfaces. This partial closure would also permit contaminants to be baked off and vented away using all available spacecraft heaters for a short enough period to minimize risk to the SRC batteries. Differences in mass properties among SRC backshell configurations produced different maneuver decompositions required to achieve the total desired delta-v in the presence of unbalanced thrusters. With the SRC backshell open, this decomposition would yield a single turn-burn-turn sequence with the burn direction outside the KOZ's prescribed for the sun sensors. However, now that the SRC backshell would be closed, the directions indicated in Figure 6 would produce a violation of the anti-sunward KOZ if performed as a single burn. To avoid such a constraint violation would require execution of TCM-3 as a double or "dogleg" maneuver in two parts on consecutive days. Such an implementation would entail considerable operational complexity and could interfere with efforts to resolve the contamination problem.

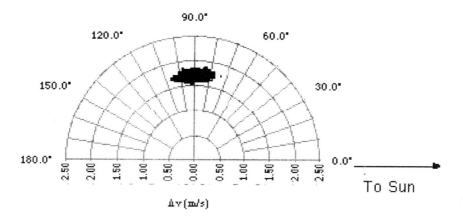

Figure 6. TCM-3 Magnitude and Direction per TCM-1 Reconstruction

Happily, an alternative arose which provided additional operational benefits for LOI execution. From pre-launch analysis, the LOI had a large prospect of being executed at ~90 deg from the Earth-Sun line. In this attitude, the burn would be completely out of view of the Earth since the orientation of the spacecraft would be such that both LGA coverage patterns would be blocked effectively by the solar panels. However, if TCM-3 were canceled, the required LOI maneuver would be affected as shown in Figure 7. Although the magnitude of the maneuver is not affected significantly, the view angle from Earth now appeared to be most likely in the range of 65-80 deg such that the LOI burn would have a better chance of being visible in real time. Due to all of these factors, it was decided to cancel TCM-3.

In light of operational concerns, an additional sensitivity analysis was performed to determine the impact delaying the LOI maneuver without TCM-3. As indicated in Figure 8, the direction of LOI remains favorable, but the magnitude would grow considerably from around 25 to 43 m/s over the course of four weeks. More alarmingly, as indicated in Figure 9, the effect on the first LOI cleanup maneuver, designated SKM-1A, is quite significant. (SKM stands for Station Keeping Maneuver.) SKM-1A, scheduled nominally on December 12 or 26 days after LOI, could increase from less than 10 m/s to around 45 m/s, making it potentially larger than LOI itself.

To avoid these problems, re-optimization of the Genesis trajectory was needed. As shown in Figures 10 and 11, recomputing the trajectory limits the potential growth of delta-v costs as a consequence of LOI execution delay more than the baseline case. Such re-optimization also provided a better Earth view angle for LOI; in fact, the view angle actually improves with delays. The implications of this re-optimization for later SKM's will be discussed in a later section.

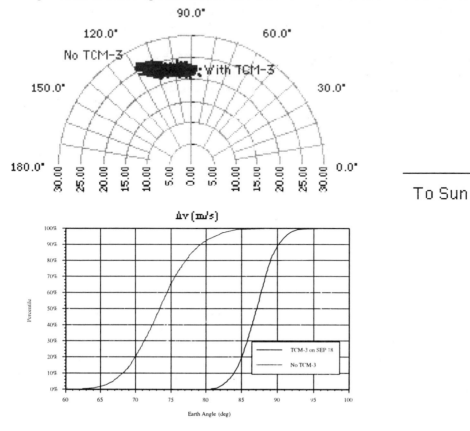

Figure 7. Impact of Excluding TCM-3 on LOI Magnitude and Direction

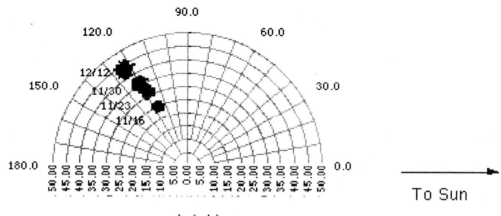

Figure 8. Sensitivity of LOI Magnitude and Direction to Delay

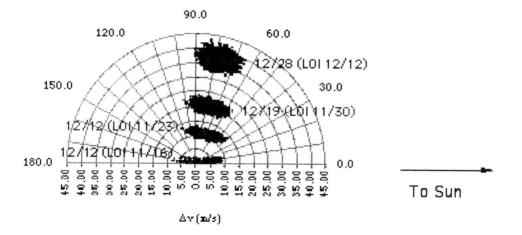

Figure 9. Sensitivity of SKM-1A Magnitude and Direction to Delays

Execution and Cleanup of LOI

LOI was executed as planned on November 16, 2001 with an overall magnitude of about 25 m/s and a −1.2% error (underburn) within 0.8 deg of the planned direction. After completion of this large maneuver, and with concerns about exacerbating the contamination problem alleviated, science collection could commence with opening of the SRC backshell. The resultant SKM-1A after LOI reconstruction was a mere 1.1 m/s at about 18 deg off Sun. This meant that SKM-1A would be small enough and in the right direction to allow for a single burn sequence. The primary benefit of this type of maneuver is that the SRC could be kept open with minimal interruption to solar wind collection and no risk of damaging the SRC backshell and concentrator with an additional close-open cycle. An assessment of SKM-1B, the first regular SKM, suggested that further re-biasing of SKM-1A would be beneficial for SKM-1B. Redirecting SKM-1A to about 25 deg off Sun would pull the direction of SKM-1B farther away from the Sun to maximize the prospects of a simple maneuver sequence. This minimizes the chances of a double maneuver driven by a sunward KOZ of 12.5 deg. It also avoids as much as possible going farther than 28 deg off Sun, which would degrade the accuracy of the maneuver by requiring intermediate turns to the burn attitude supported entirely via dead-reckoning on sun sensors. Uncertainties in SKM-1B magnitude and direction are shown in Figure 12, based on a post-LOI reconstruction with re-biasing of SKM-1A assumed. The boundaries shown reflect the effect on total delta-v of combining turns and spin adjustments with the burn itself.

SKM-1A was executed as scheduled on December 12 with a magnitude of only 1.115 m/s and a +2.30% error within about 0.5 deg of the planned direction. SKM-1B on January 16, 2002 ended up as a near-Sun single maneuver with a magnitude of 1.328 m/s with a 1.44% magnitude and 0.17 deg direction error.

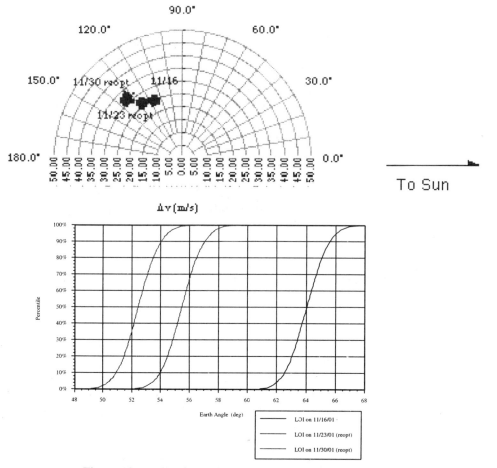

Figure 10. Implications of Trajectory Re-Optimization for Nominal and Delayed LOI

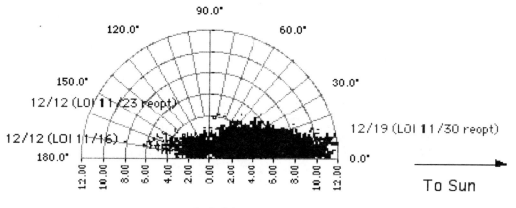

Figure 11. Implications of Trajectory Re-Optimization on SKM-1A for Nominal and
Delayed LOI-SKM-1A Cases

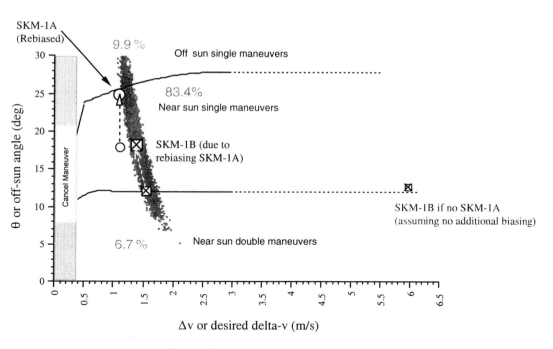

Figure 12. Re-Biasing Strategy for SKM-1A and 1B

STATION-KEEPING MANEUVERS

Initial Design

Due to modeling deficiencies inherent in any design and the sensitivity of trajectories in the three-body problem to small perturbations, it is necessary to periodically perform station-keeping maneuvers (SKM) to maintain the desired Lissajous trajectory. Although trajectories in the three-body problem are sensitive to perturbing forces, the time constant to instability is relatively long (on the order of weeks to months). Thus, statistically, the size of any given SKM is generally less than 0.5 m/s and is fairly constant over a few weeks period. Typical libration point missions perform SKM's once or twice per revolution (see references). This is acceptable since these missions typically do not need to be in a precise orbit; any orbit satisfying the mission constraints will suffice.

For Genesis, however, the Lissajous portion of the trajectory is critical to set up the "free-return" trajectory that brings the spacecraft and its samples back to Earth. In fact, the LOI maneuver at launch plus three months that placed the spacecraft into the Lissajous orbit is actually the maneuver that started Genesis on its path home. Pre-launch analyses performed at JPL and Purdue University suggested that, for a Genesis-type trajectory, SKM's should be performed every two months, or three times per revolution. Thus, the mission has planned 15 SKM's denoted by a number corresponding to the revolution and a letter A, B, or C. Hence, the order of the SKM's is 1A, 1B, 1C, 2A, 2B, 2C, etc. (Note that, as discussed, SKM-1A is actually the cleanup maneuver for LOI.) This station-keeping plan keeps the spacecraft on its intended path without allowing the magnitude of the maneuvers to grow beyond the typical 1 m/s or so.

Statistically, any given SKM may be pointed in any direction with varying magnitudes. The Genesis spacecraft is capable of performing arbitrary maneuvers (with some limitations). However as has been pointed out earlier, the execution of certain maneuver types can become very operationally complex. For example, small (less than 0.5 m/s) anti-sunward pointing maneuvers are difficult to execute. To help mitigate the operational complexities associated with these maneuvers, it was decided pre-launch to bias each station-keeping maneuver in a regular manner. The initial bias chosen for each SKM was 1.5 m/s in a near-sunward pointing direction. When the statistical variations were overlaid with these biases, the resulting maneuvers were generally sunward pointing with magnitudes between 0.5 and 2.5 m/s. The upper magnitude coincides with a pre-launch guideline for maximum maneuver size on the smaller (0.2 lbf) thrusters. This is significant, since switching to larger (5 lbf) thrusters poses a contamination risk and would require an additional close-open cycle for the SRC backshell and concentrator cover.

These biases were incorporated into the design of the Lissajous trajectory using the same techniques used to design the original un-biased solution. In the original design, the LOI maneuver was the only post-launch deterministic impulsive event. This design gave rise to the notion of the "free return" trajectory that has been mentioned earlier. By designing the biases at each SKM (and subsequently at each return TCM) into the trajectory, the solution is no longer a "free" return since each maneuver *must* be performed in order to return to Earth. This is the trade-off that was made to ensure that each SKM and TCM is pointed in a favorable direction with an acceptable magnitude.

SKM Redesign

After it was decided to cancel the rest of the transfer TCM's, an effort was undertaken to redesign the trajectory to achieve a more favorable LOI maneuver, as was discussed in the previous section. The goal of the redesign was to improve LOI without affecting the return portion of the trajectory. As part of this redesign, the pre-launch SKM biases were re-examined in light of the developing operational experience with the spacecraft.

The pre-launch Navigation Plan allowed for execution errors as large as 6% at the 3-sigma level. When such errors were modeled in monte-carlo simulation runs, large dispersions with respect to sun angle had been evident. Consequently, the 1.5 m/s biases were set close to the Sun at a 5 deg off-sun cone angle and a 0 deg clock angle, the prevailing attitude in the direction of expected maximum solar wind flux, as required for science collection. By allowing the biases to float in both magnitude and direction, a more optimal LOI was determined that placed the spacecraft into a slightly different Lissajous orbit than the pre-launch design. The resulting SKM biases are shown in Figure 13. Note that the cone angles range from 4 to 7 deg off sun with varying clock angles, and the magnitudes range from 1.509 to 1.524 m/s. The added flexibility in the biases aided in redesigning the trajectory to return on essentially the same path as the pre-launch solution, with no added cost after LOI. In retrospect, designing biases into the Lissajous orbit allows redesign flexibility that can only be achieved through additional maneuvers in an unbiased solution.

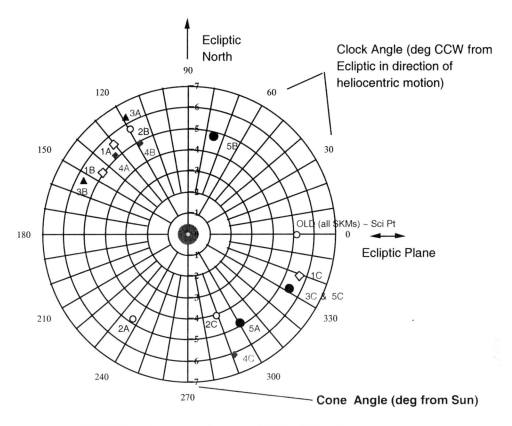

NOTE: Magnitudes are in range 1.509-1.524 m/s.

Figure 13. Bias Shift as Designed for First Post-Launch Trajectory Re-Optimization

SKM Re-Redesign

As further experience with the spacecraft and maneuver design process was gained during 2002 flight operations, it became apparent that a further change to the SKM biases would be beneficial to simplify maneuver operations. By shifting the off-sun cone angles to between 12.5 and 28 deg, all future SKM's after 1C could statistically be expected to be the most benign maneuver type, namely the single leg near-Sun variety. In this redesign, SKM-2A was allowed to vary in magnitude and direction (within reason) and in fact ended up being about 0.75 m/s at 22 deg off sun. The remaining 11 SKM's were then reoptimized to satisfy the new pointing requirements, while still targeting the same return trajectory back to Utah. The reoptimized SKM's are shown in Figure 14, where SKM's 2A through 3B are the actual maneuvers performed to date and 3C through 5C are the estimated maneuvers as determined by the redesign. Here again, the flexibility of the biases in the Lissajous for redesign of the trajectory is proven to be useful.

1453

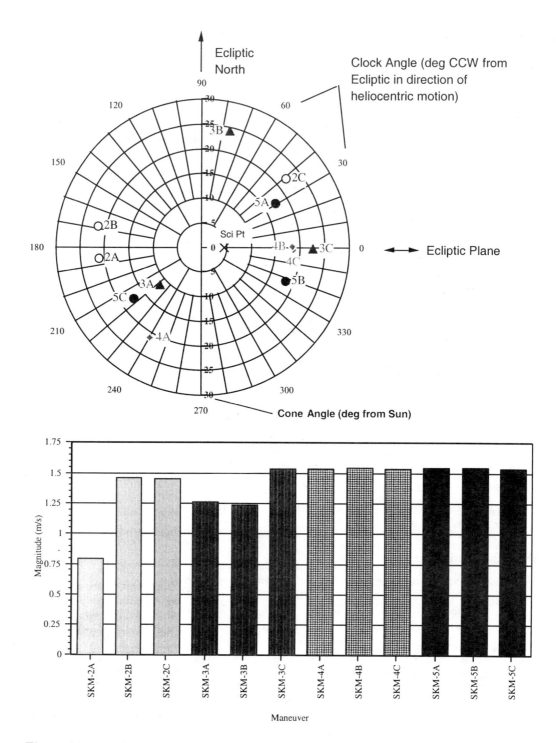

Figure 14. Actual Biases to Date for SKM-2A through SKM-3A with Revised Estimates
for SKM-3B through SKM-5C

1454

CONCLUSIONS

The Genesis spacecraft has proven to be a very reliable platform and the operational experience to date has been nominal. The flexibility of the Lissajous portion of the trajectory to redesign efforts has been shown to be quite beneficial to simplifying the operations, while not compromising the return of the samples. Future papers will discuss the operations through the remainder of the Lissajous and during the return to Earth and subsequent entry leading up to the mid-air recovery in Utah.

ACKNOWLEDGEMENTS

This research was carried out at the Jet Propulsion Laboratory, California Institute of Technology, under a contract with the National Aeronautics and Space Administration. Reference herein to any specific commercial product, process, or service by trade name, trademark, manufacturer, or otherwise, does not constitute or imply its endorsement by the United States Government or the Jet Propulsion Laboratory, California Institute of Technology.

REFERENCES

[1] Farquhar, R., "Trajectories and Orbital Maneuvers for the First Libration Point Satellite", *Journal of Guidance & Control*, Vol. 3, No. 6, Nov-Dec 1980.

[2] E. Hirst, P. Hong, Project Document GN 61000-200, Genesis Planning, Control, Analysis, and Recovery Document, Volume 2, *Mission Plan*, July 13, 2001 *(JPL internal document)*.

[3] Ken Williams, Phil Hong, Don Han, "Maneuver Design and Calibration for the Genesis Spacecraft," AAS 99-399, AIAA/AAS Astrodynamics Conference, August 16-19, 1999.

[4] K. C. Howell, B. T. Barden, R. S. Wilson, M. Lo, "Trajectory Design Using a Dynamical Systems Approach with Application to Genesis", AAS 97-709, AAS/AIAA Astrodynamics Specialists Conference, August 4 – 7, 1997.

[5] Martin Lo, et al, "Genesis Mission Design," AIAA 98-4468, AIAA/AAS Astrodynamics Conference, August 10-12, 1998.

[6] Julia Bell, Martin Lo, Roby Wilson, "Genesis Trajectory Design", AAS 99-398, AAS/AIAA Astrodynamics Specialists Conference, August 16 – 19, 1999.

[7] K. C. Howell, B. T. Barden, and M. W. Lo, "Application of Dynamical Systems Theory to Trajectory Design for a Libration Point Mission," The Journal of Astronautical Sciences, Vol. 45, No. 2, April-Jun3 1997, pp. 161-178.

[8] B. T. Barden, R. S. Wilson, K. C. Howell, B. G. Marchand, "Summer Launch Options for the Genesis Mission", AAS 01-306, AAS/AIAA Astrodynamics Specialists Conference, July 30 – August 2, 2001.

[9] D. Han, K. Williams, Project Document GN 61000-200-19, Genesis Planning, Control, Analysis, and Recovery Document, Volume 19, *Navigation Plan*, July 23, 2001 *(JPL internal document)*.

[10] G. Brown, R. Haggard, R. Corwin, "Parafoil Mid-Air Retrieval for Space Sample Return Missions," AIAA-01-2018, AIAA Aerodynamic Decelerator Systems Technology Conference, May 21-24, 2001.

[11] P. Hong, G. Carlisle, N. Smith, "Look, Ma, No HANS," IEEE-02-7803-7231, 2002 IEEE Aerospace Conference, March 9-16, 2002.

[12] Kenneth E. Williams, "Overcoming Genesis Mission Design Challenges," IAA-L-0603P, Fourth IAA International Conference on Low-Cost Planetary Missions, Laurel, MD, May 2-5, 2000.

[13] Don Han, Genesis Maneuver Reconstruction Reports, dated 30 August 2001, 28 November 2001, 19 December 2001, 30 January 2002, 1 April 2002, 3 June 2002, 6 August 2002, 3 October 2002 *(JPL internal documents)*.

[14] K. C. Howell, B. T. Barden, "Investigation of Biasing Options for the Genesis Mission," Purdue Interoffice Memo AAE-9840-007, June 1998.

[15] Kenneth Williams, Martin Lo, Roby Wilson, Kathleen Howell, Brian Barden, "Genesis Halo Orbit Station Keeping Design," Ref. 53, CNES 15th International Symposium on Spaceflight Dynamics, Biarritz, France, June 26-30, 2000.

[16] K.C. Howell and B.T. Barden, "Station-keeping Study for Genesis," Purdue IOM AAE-9840-007, June 9, 1998, West Lafayette, Indiana.

[17] K.C. Howell and B.T. Barden, "Summary of the Endstate Variations for a Finite Set of Different Controllers and Maneuver Strategies," Purdue IOM AAE-9840-014, December 1998, West Lafayette, IN.

[18] K.C. Howell and B.T. Barden, "Investigation of the Effects of Maneuver Uncertainties on a Finite Set of Different Controllers and Maneuver Strategies," Purdue IOM AAE-9840-009, July 1998, West Lafayette, IN.

[19] R.W. Farquhar, D.P. Muhonen, C.R. Newman, and H.S. Heuberger, "Trajectories and Orbital Maneuvers for the First Libration-Point Satellite," *Journal of Guidance and Control*, Vol. 3, No. 6, 1980, pp. 549-554.

[20] K.C. Howell and H.J. Pernicka, "Station-keeping Method for Libration Pont Trajectories," *Journal of Guidance, Control, and Dynamics*, Vol. 16, No. 1, January-February 1993, pp. 151-159.

[21] K.C. Howell and S.C. Gordon, "Orbit Determination Error Analysis and a Station-keeping Strategy for Sun-Earth L1 Libration Point Orbits," *Journal of the Astronautical Sciences*, Vol. 42, No. 2, April-June 1994, pp. 207-228.

[22] N. Smith, K. Williams, R. Wiens and C. Rasbach, "Genesis - The Middle Years," IEEE-03-7803-7651, to be presented at 2003 IEEE Aerospace Conference, March 7-15, 2003.

ORBIT DETERMINATION SUPPORT FOR
THE MICROWAVE ANISOTROPY PROBE (MAP)

Son H. Truong,* Osvaldo O. Cuevas*
and Steven Slojkowski†

NASA's Microwave Anisotropy Probe (MAP) was launched from the Cape Canaveral Air Force Station Complex 17 aboard a Delta II 7425-10 expendable launch vehicle on June 30, 2001. The spacecraft received a nominal direct insertion by the Delta expendable launch vehicle into a 185-km circular orbit with a 28.7° inclination. MAP was then maneuvered into a sequence of phasing loops designed to set up a lunar swingby (gravity-assisted acceleration) of the spacecraft onto a transfer trajectory to a lissajous orbit about the Earth-Sun L2 Lagrange point, about 1.5 million km from Earth. Because of its complex orbital characteristics, the mission provided a unique challenge for orbit determination (OD) support in many orbital regimes. This paper summarizes the premission trajectory covariance error analysis, as well as actual OD results. The use and impact of the various tracking stations, systems, and measurements will be also discussed. Important lessons learned from the MAP OD support team will be presented. There will be a discussion of the challenges presented to OD support including the effects of delta-Vs at apogee as well as perigee, and the impact of the spacecraft attitude mode on the OD accuracy and covariance analysis.

INTRODUCTION

The Microwave Anisotropy Probe (MAP) is the second Medium Class Explorer (MIDEX) mission of the National Aeronautics and Space Administration (NASA). The main goal of the MAP observatory is to measure the temperature fluctuations, known as anisotropy, of the cosmic microwave background (CMB) radiation over the entire sky and to produce a map of the CMB anisotropies with an angular resolution of approximately 3 degrees. This map of the anisotropy distribution will help determine how structures formed in the early universe, will determine the ionization history of the universe, and will refine estimates of key cosmological parameters. In particular, these data will be used to shed light on several key questions associated with the Big

---

*   Flight Dynamics Engineer, NASA Goddard Space Flight Center, Greenbelt, Maryland 20771.

†   Member of the Technical Staff, Computer Sciences Corporation, 10110 Aerospace Drive, Lanham-Seabrook, Maryland 20706.

Bang theory and to expand on the information gathered from the Cosmic Background Explorer (COBE) mission, flown in the early 1990s. The L2 lissajous orbit was selected by the MAP program to minimize environmental disturbances, maximize observing efficiency, and to provide instrument thermal stability. A lissajous trajectory is considered as a three-dimensional quasi-periodic orbit. [2] The science mission minimum lifetime is two years of observations at L2 with a desired lifetime of 5 years. [1]

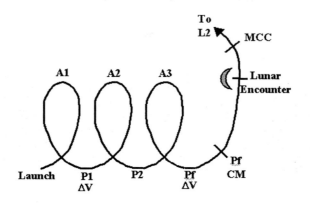

Figure 1 MAP (3.5-loop) Trajectory Schematic

MAP used a lunar gravity assist strategy since it reduced the fuel required to achieve the desired lissajous orbit. The MAP transfer orbit consisted of 3.5 phasing loops. Figure 1 shows the MAP trajectory schematic all the way through L2. [3] The first loop had a period of 7.5 days, the second and third loops were 9 days long, and the last half loop was 4.5 days long. A correction maneuver at the third perigee (P_f or Pfinal) was planned for approximately 18 hours after the last perigee maneuver to accurately achieve the targeted lissajous orbit. The lunar encounter took place approximately 30 days after launch. After the lunar encounter, the spacecraft cruised for 60 days before it arrived in the vicinity of the L2 libration point. Two mid-course correction (MCC) maneuvers were performed a week after the periselene (i.e., lunar encounter or swingby) to refine MAP's post-launch trajectory. Now that MAP is at its operational L2 lissajous orbit, the MAP satellite is commanded to perform occasional station-keeping (SK) maneuvers in order to maintain its orbit around L2. At L2, MAP will maintain a lissajous orbit with a MAP-Earth vector between 0.5° and 10.5° off the Sun-Earth vector in order to satisfy its communications requirements while avoiding eclipses. [6,7]

Telemetry, tracking and command is provided by the NASA Deep Space Network (DSN). The Tracking and Data Relay Satellite System (TDRSS) was also used during launch and early orbit operations.

The NASA GSFC Flight Dynamics Analysis Branch (FDAB) performed premission covariance analysis in order to determine MAP orbit determination requirements for maneuver planning and calibration for each phase of the mission. Once the satellite was launched, the Navigation Team from Honeywell and the Computer Sciences Corporation (CSC) performed definitive orbit determination in support of the mission. This paper presents the results of the premission orbit analysis, the technique and results of the post-launch OD process, and evaluates the OD accuracy. Important lessons learned from the MAP Navigation team support are also presented.

PREMISSION ORBIT ERROR ANALYSIS

The purpose of the orbit error analysis was to provide definitive and predicted ephemeris accuracy estimates to help the MAP Project plan for orbit control. Another purpose of this analysis was to determine if the orbit accuracy and tracking requirements as specified in the MAP Detailed Mission Requirements (DMR)[9] document could be met, and to propose other tracking scenarios if necessary. The FDAB's Orbit Determination Error Analysis System (ODEAS) was used to perform the covariance analysis and it was based on a nominal trajectory provided by the trajectory design group. The nominal trajectory did not include a maneuver at perigee 2 (P2). Several tracking scenarios were investigated for each phase of the mission.[5] Estimated ephemeris accuracies were derived for different post-maneuver and pre-maneuver tracking scenarios. Results from the ODEAS runs show that required orbit accuracy can be satisfied if tracking support includes both range and range rate measurements from at least two of three DSN 26-meter stations (Goldstone, California; Canberra, Australia; and Madrid, Spain) under a proposed tracking schedule as follows:

- For maneuver recovery: 6 to 18 hours of continuous tracking support after each planned maneuver (M).
- For maneuver planning:
  From the transfer trajectory injection (TTI) to the first perigee (P1): three 1-hour passes/day (alternating northern (N) and southern (S) hemisphere DSN stations) then continuous tracking from M-16hrs to M-12hr.
  From P1 to Pfinal: three 1-hour passes/day (alternating N&S hemisphere DSN stations) then continuous tracking from M-16hr to M-12hr.
  From Pfinal to periselene (Ps): three 1-hour passes/day (alternating N&S hemisphere DSN stations) then continuous tracking from M-16hr to M-12hr.
  From Ps to L2 insertion: one 37-min pass/day (alternating N&S hemisphere DSN stations)
  L2 Nominal: one 37-min pass/day (alternating N&S hemisphere DSN stations)

Table 1 shows post-maneuver definitive and predicted ephemeris accuracies under different tracking scenarios. For each scenario in Table 1, the maneuver is supported by continuous tracking followed by no tracking support until the next planned maneuver (e.g., from TTI to P1). Three post-maneuver tracking data arcs were evaluated: 6 hours, 12 hours and 18 hours (for Pfinal only). After 12 hours of continuous tracking, RSS position error (3σ) was on the order of 500 m and RSS velocity error (3σ) was on the order of 2 cm/s. With Delta-V magnitudes of 3 km/s at TTI, 22 m/s at P1, and 7 m/s at Pfinal, the estimated velocity error is not a significant fraction of the burn magnitude. For this paper, the phrase "definitive ephemeris" is used for a post-processed trajectory generated by an orbit determination process (i.e., OD with tracking data) and the word "predicted ephemeris" is used for trajectories generated by orbit propagation (i.e., without tracking data). It should be noted that errors on the predicted ephemerides (i.e., no tracking support after 6 or 12 hours) are quite high, which is typical.

Table 1

POST-MANEUVER EPHEMERIS ACCURACY ESTIMATES

Epoch	Tracking Support	Definitive Ephemeris Accuracy (3σ)	Predicted Ephemeris Accuracy (3σ)
TTI	Maneuver+ 6 hours	Pos: 1.039km Vel: 2.72 cm/s	Pos: 63.419 km (at P1) Vel: 9.06 m/s
	Maneuver+ 12 hours	Pos: 559 m Vel: 1.14 cm/s	Pos: 35.062 km (at P1) Vel: 5.01 m/s
P1	Maneuver+ 6 hours	Pos: 300 m Vel: 1.05 cm/s	Pos: 167.19 km (at Pf) Vel: 19.47 m/s
	Maneuver+ 12 hours	Pos: 241 m Vel: 0.35 cm/s	Pos: 146 km (at Pf) Vel: 17 m/s
Pfinal	Maneuver + 6 hours	Pos: 416 m Vel: 3.5 cm/s	Pos: 4.85 km (at Ps) Vel: 33.81 cm/s
	Maneuver + 12 hours	Pos: 886 m Vel: 2.57 cm/s	Pos: 4.340 km (at Ps) Vel: 29.44 cm/s
	Maneuver + 18 hours	Pos: 564 m Vel: 1.14 cm/s	Pos: 2.729 km (at Ps) Vel: 21.88 cm/s

The orbit determination process used for this error analysis is a batch least-squares method and the orbit error depends on many factors including spacecraft position on the orbit. It is expected that more tracking coverage would improve the overall post-processed and predicted orbit accuracy but not necessarily at any specific time (e.g., the estimated definitive orbit error at the end of the 12-hr arc from Pf is 886 m while that of the 6-hr arc is only 416 m).

Table 2

DEFINITIVE EPHEMERIS ACCURACY ESTIMATES
FOR MANEUVER PLANNING

Epoch	Definitive Ephemeris Accuracy (3σ) at Maneuver – 24 hours	Definitive Ephemeris Accuracy (3σ) at Maneuver – 12 hours	Definitive Ephemeris Accuracy (3σ) at Maneuver
TTI (To plan P1 maneuver)	Pos: 233 m Vel: 0.18 cm/s	Pos: 149 m Vel: 0.27 cm/s	Pos: 33 m Vel: 0.89 cm/s
P1 (to plan Pfinal maneuver)	Pos: 554 m Vel: 0.24 cm/s	Pos: 243 m Vel: 0.38 cm/s	Pos: 53 m Vel: 0.93 cm/s
Pfinal (to plan periselene maneuver)	Pos: 194 m Vel: 0.15 cm/s	Pos: 161 m Vel: 0.14 cm/s	Pos: 90 m Vel: 1.06 cm/s
L2 – 3 weeks (to plan SK)	Pos: 2.376 km Vel: 0.16 cm/s	Pos: 2.364 km Vel: 0.16 cm/s	Pos: 2.354 km Vel: 0.16 cm/s

Epoch	Definitive Ephemeris Accuracy (3σ) at Epoch + 1 week	Definitive Ephemeris Accuracy (3σ) at Epoch + 2 weeks	Definitive Ephemeris Accuracy (3σ) at Epoch + 3 weeks
Periselene (to plan MCC)	Pos: 1.514 km Vel: 0.11 cm/s	Pos: 1.872 km Vel: 0.07 cm/s	Pos: 1.990 km Vel: 0.06 cm/s
L2 Insertion (to plan MCC)	Pos: 2.634 km Vel: 0.08 cm/s	Pos: 2.644 km Vel: 0.08 cm/s	Pos: 2.593 km Vel: 0.1 cm/s

Table 2 shows the results of the error analysis for maneuver planning for a range of pre-maneuver tracking data cut-off times. For all phases of the mission, the estimated RSS definitive position

1460

error is ≤ 2.644 km and the estimated RSS definitive velocity error is ≤ 1.06 cm/s at the start time of the planned maneuver. It should be noted that where there is uncertainty about the exact time of the planned maneuver at the time of the analysis, only estimated accuracy at the end of the tracking arc (e.g., 1 week) was given.

Table 3 describes the proposed orbit accuracy and tracking requirements for the MAP mission. It should be noted that the estimated accuracies are three-sigma (3σ) values. The error covariance analysis showed that these orbit requirements could be met if the existing GSFC operational orbit related systems (e.g., the Goddard Trajectory Determination System (GTDS)) are used to support the mission under the specified tracking support from DSN. Based on the covariance analysis, this tracking support was different from the original DMR document.

Table 3

ORBIT ACCURACY AND TRACKING REQUIREMENTS

Mission Phase	Service	Data Type	Pass Frequency	Definitive Ephem Requirements (3σ)	Data Sample Rate	Predicted Ephem Requirements (3σ)
LEO (L+0 to L + 1 day)	26-m or 34-m	Range and Doppler	Continuous 2 stations for first 12 hours, one station second 12 hours	Best Obtainable	Doppler:1/10s Range:1/10s	Position: 25 km
Transfer Trajectory Phase-nominal (2-F+3 days) Nominal Support	26-m or 34-m	Doppler, range, and angles from 26-m	2 – 4 one hour passes/day	Position: 5 km Velocity: 5 cm/s	Doppler:1/10s Range: 1/10s	Position: 50 km Velocity: 10 cm/s (1-week arc)
Transfer Trajectory Phase-maneuvers & lunar gravity Assist (phasing loops)	26-m or 34-m	Doppler, range	Near Continuous M – 16h to M– 12h (4 hour span) and M start to M+8 hours	Position: 5 km Velocity: 5 cm/s	Doppler:1/10s Range: 1/60s	Position: 50 km Velocity: 10 cm/s (1-week arc)
Cruise (Gravity Assist to L2 Insertion) (~70 days)	70-m or 34-m	Doppler, range	Nominal: One 37-min pass/day*	Position: 5 km Velocity: 5 cm/s	Doppler:1/10s Range: 1/60s	Position: 50 km Velocity: 10 cm/s (1-week arc)
Cruise-maneuvers	70-m or 34-m	Doppler, range	Near Continuous M-16h to M-12h (4 h span) and M start through M+8 hr	Position: 5 km Velocity: 5 cm/s	Doppler:1/10s Range: 1/60s	Position: 50 km Velocity: 10 cm/s (1-week arc)
L2-nominal (2 years)	70-m or 34-m	Doppler, range	One 37 min pass/day*	Position: 5 km Velocity: 5 cm/s	Doppler:1/10s Range: 1/60s	Position: 100 km Velocity: 40 cm/s (1-week arc)
L2-maneuvers Delta V/Delta H	70-m or 34-m	Doppler, range	Near Continuous from M-4h to M+8h	Position: 5 km Velocity: 5 cm/s	Doppler:1/10s Range: 1/60s	Position: 100 km Velocity: 40 cm/s (1-week arc)

*alternating N & S hemisphere DSN stations

POSTLAUNCH ORBIT DETERMINATION SUPPORT

MAP was launched on June 30, 2001 at 19:46 UTC from Kennedy Space Center aboard a Delta II rocket. Over the course of the following four weeks, MAP executed 3.5 phasing loops about the Earth prior to periselene on July 30, 2001.

The maneuver team conducted thruster firings at every apogee and perigee of the phasing loop period. A thruster calibration was performed at apogee 1 (A1), and engineering burns were executed at apogee 2 (A2) and apogee 3 (A3). Perigee maneuvers (P1, P2, P3) and a final correction maneuver (P_{fc}) at 18 hours after P3 were executed to give MAP necessary energy to reach periselene, correct errors, and fine tune the trajectory.

As a result of these thruster events conducted during the phasing loops, the maximum tracking data arc available for orbit determination (OD) was one half of an orbit. Other perturbations affecting OD were also present during the phasing loops. Additional thruster testing was performed between spacecraft separation and A1. Furthermore, the solar radiation force on MAP is strongly dependent on the spacecraft attitude mode, which changed frequently during the phasing loop period. A timeline of MAP orbit events is shown in Table 4.

Table 4

MAP ORBIT EVENTS

Event	Start Time	Duration (s)	Delta–V (m/s)
S/C Sep	2001/06/30 21:12		
Apogee 1 Delta-V	2001/07/04 13:22	106.0	1.92
Perigee 1 Delta-V	2001/07/08 04:43	1274.4	20.19
Apogee 2 Delta-V	2001/07/12 16:11	40.6	0.25
Perigee 2 Delta-V	2001/07/17 03:36	1777.2	2.51
Apogee 3 Delta-V	2001/07/21 18:54	43.4	0.29
Perigee 3 Delta-V	2001/07/26 10:29	546.2	7.41
P3 Correction Maneuver	2001/07/27 04:30	23.9	0.31
Lunar Gravity Assist	2001/07/30 16:37		
Mid-course Correction 1	2001/08/06 16:37	18.0	0.10
Mid-course Correction 2	2001/09/14 16:37	6.6	0.04
StationKeeping # 1	2002/01/16 16:51	72.9	0.43
StationKeeping # 2	2002/05/08 16:03	53.8	0.35
StationKeeping # 3	2002/07/30 16:39	71.8	0.47
StationKeeping # 4	2002/11/05 19:21	94.3	0.56

MAP orbit determination was performed using GTDS, running on a Windows-NT platform. GTDS employs a batch least-squares estimator with the option to solve for additional force parameters such as the solar radiation pressure coefficient (C_r).

During the phasing loop period, MAP tracking predominantly consisted of range and Doppler measurements from DSN 34-meter antennas, with DSN 26-meter antennas providing some additional support. Near perigee, TDRS 2-way Doppler tracking was received. For all the solutions presented in this report, Doppler observations were sampled down to roughly equal the number of range observations to balance the effects of both data types in the solutions. Angle data, when available, was only used in short-arc solutions, due to the noise of angle observations and their increased uncertainty at larger radial distances.

Intensive orbit determination analysis was performed to evaluate ephemeris accuracy, orbit perturbations (e.g., due to changes in spacecraft attitude), major systematic errors, and performance of short post-maneuver tracking data arcs and of L2 operations.

For most of the OD solutions presented in this paper, the 1-sigma standard deviation of post-fit range residuals varied between 2 m and 10 m. For those solutions over tracking data arcs without attitude or other disturbances, the 1-sigma standard deviation was usually below 5 m. Large residual divergences, like that illustrated in Figure 2, do not significantly drive up the residual standard deviation because most of the observations in the "residual tail" are edited out of the solution. The 1-sigma standard deviation of range-rate residuals was typically between 0.1 cm/sec and 2 cm/sec, depending on the attitude mode of the spacecraft.

Ephemeris Consistency

The ideal method of determining the accuracy of OD solutions is by comparison to precision definitive ephemeris; however, these methods are not available for most missions. Thus it is common to assess OD accuracy through the examination of overlapping definitive ephemeris spans. When using this method, one takes the ephemeris consistency to be a measure of definitive ephemeris accuracy, under the assumption that no systematic errors bias both solutions.

For an operational mission, it is not difficult over time to collect a statistical set of overlapping well-determined solutions. In the case of MAP, the presence of delta-Vs executed at all apogees and perigees made the task of generating overlapping definitive ephemeris difficult during the phasing-loop period since it reduced the maximum tracking data arc to a half of an orbit. The approach used here to generate a set of overlapping half-orbit definitive ephemerides was to modify the tracking data spans and other input parameters in the orbit determination process. This procedure is additionally complicated by the perturbations induced by attitude dependent changes in the solar radiation force.

A set of orbit solutions was generated using all available data between each apogee and perigee. A second set was also generated by deleting 24 hours of arc from each half-orbit span. In order to mitigate the potential for solution degradation caused by the loss of advantageous geometry, the data was deleted from the middle of the half-orbit spans. The ephemerides generated from these solutions were then compared over their common definitive span or overlapping timespan. The results are shown in Table 5. In addition to position and velocity, C_r was determined for all of these solutions.

Table 5

DEFINITIVE OVERLAP COMPARES

Definitive Overlap Comparison Span	Maximum Definitive Position Difference (m)	Maximum Definitive Velocity Difference (cm/s)
S/C Sep to A1	530	34.8
A1 to P1	32	0.3
P1 to A2	64	0.4
A2 to P2	389	1.5
P2 to A3	871	11.2
A3 to P3	46	0.1

While Table 5 shows a significant dispersion in ephemeris position and velocity differences, the largest differences are associated with spans where thruster firings, deployments, or attitude changes are present in the arc. In the case of the separation-to-A1 arc, initial deployments and thruster testing which occurred during the first 3 days of the mission make solutions very sensitive to the tracking data arc. In addition, many attitude changes took place during the flight to A1 that impacted the force of solar radiation experienced by the spacecraft. Solutions over the separation-to-A1 arc generally exhibit higher range residual standard deviations than over other, less perturbed arcs. Cleaner solutions were obtained by cutting the tracking data arc right after the last thruster testing activity performed, but this resulted in the loss of useful perigee tracking.

Orbit Perturbations Due to Changes in Spacecraft Attitude Mode

Attitude mode changes are likely responsible for larger differences in the A2 to P2 and P2 to A3 tracking arcs. Those with the lowest differences, A1 to P1, P1 to A2, and A3 to P3, are largely unperturbed by attitude mode changes.

Throughout the phasing loops, for the purposes of OD considerations, MAP's attitude could be characterized generically in two modes. The first is a group of states, identified here as "flat spin" modes, in which the spacecraft rotates about its Z-axis while the Z-axis is on or close to (generally within 2 degrees of) the spacecraft to Sun line. The other was the operational science "observing mode" configuration in which the spacecraft rotates about its Z-axis, while the Z-axis precesses along a 22.5-degree cone about the Sun line. In each of these two modes the spacecraft presents a different area to the Sun. As a result, MAP experienced a different force of solar radiation during each mode. When solving for C_r as part of the OD process, a value of 1.45 was typically found for observing mode arcs, while a value of 1.78 was characteristic of flat spin mode. When tracking data arcs contained data from both modes, the OD solutions determined a value somewhere between these two.

The effect of the difference in solar radiation force in each mode is clearly visible in Figure 2. This plot displays final vector residuals computed in an A2 to P2 solution using all available data. This solution features a prominent divergence when the spacecraft transitioned from a flat spin mode to observing mode. From the range residuals, it can be seen that the solution solved to the data taken when the spacecraft was in observing mode, due to the greater proportion of data in this mode. This effect is characteristic of the least-squares estimation process. In fact, most of the range data prior to Day 194 as seen in Figure 2 was excluded from the solution by the differential corrector sigma edit criteria. The 80-meter residuals at the beginning of the tail are at the 30-sigma level. This resulted in an effective reduction in tracking data due to the loss of nearly all usable data in flat spin mode. In both the A2 to P2 and P2 to A3 arcs, the difference in solar radiation force in both modes is too big to allow the least-squares estimation to average well over the attitude change, resulting in states which solve to one or the other set of data.

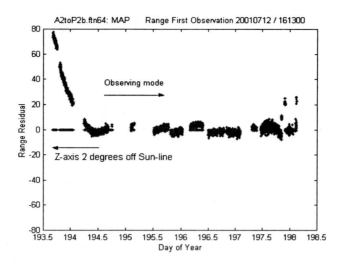

Figure 2   Post-fit Apogee 2 to Perigee 2 Range Residuals
(Range in meters)

Systematic Errors

An effort was made to examine two potential sources of systematic errors that may bias the overlap comparisons; that due to station location uncertainty and that due to uncertainties in a-priori range biases. In operational GTDS runs that apply given station locations and a-priori range biases instead of solving for these parameters, no consideration of their uncertainties is incorporated in the covariance of the estimated state.

Table 6

**DEFINITIVE EPHEMERIS DIFFERENCES
DUE TO VARIATIONS IN DSN STATION LOCATIONS**

Solution Arc and Comparison Span	Maximum Definitive Position Difference (m)	Maximum Definitive Velocity Difference (cm/s)
S/C Sep to A1	185	1.4
A1 to P1	140	0.5
P1 to A2	130	1.6
A2 to P2	205	0.7
P2 to A3	105	0.7
A3 to P3	89	0.1
P3 correction to MCC	35	0.03

To examine OD uncertainty due to station geo-location, a set of runs were made using a set of DSN station locations obtained from Telecommunications and Data Acquisition (TDA) Progress

1465

Report by Folkner,[8] which differs slightly from the operational set. The operational geodetics employed in Flight Dynamics Facility (FDF) for these stations are currently being evaluated against those in the TDA report. MAP orbit solutions employing the TDA report locations have lower weighted root-mean-squares and observation residual standard deviations than solutions using the operational locations. Again, two sets of runs were made, the first using FDF operational station locations and the second using the station locations reported in the TDA report. The same tracking data was employed in each series of runs, and each solution solved for position, velocity, and C_r. The results are presented in Table 6.

Finally, a set of solutions was gathered to assess the effects of station range bias uncertainties. Historical analysis employing WIND and SOHO observations indicates that the sequential ranging assembly (SRA) range data received from DSN 34-meter stations and processed in FDF is observed to have a 100 ± 30 meter bias. In operations, this bias is typically solved-for, when possible. However, due to the predominance of 34-meter tracking in the MAP solutions, it was not possible to accurately determine the biases during MAP phasing-loop support, therefore a-priori value of 100 m was applied. Since MAP has been on station at L2, a series of parametric solutions have been performed to establish a new set of a-priori biases for MAP support, 88 meters for DS24, 98 meters for DS34, and 90 meters for DS54.

A series of runs were also made applying a 100-meter bias on all 34-meter DSN ranging, and another set using the new biases as a-priori. The same tracking data were used in each series of runs, and each solution solved for position, velocity, and C_r. The results are presented in Table 7.

As Tables 5-7 show, in all cases the definitive ephemeris uncertainty is less than 900 m in position, and less than 35 cm/s in velocity. In fact, most cases show position differences of less than 400 m and velocity differences of less than 5 cm/s. The largest position and velocity differences are associated with spans where thruster tests, deployments, or attitude mode changes were present. Comparison of Table 5 and Table 7 shows that in the absence of orbit disturbances, the largest consistent component of definitive ephemeris uncertainty appears to be due to uncertainties in range biases.

Table 7

DEFINITIVE EPHEMERIS DIFFERENCES
INDUCED BY RANGE BIAS DIFFERENCES

Solution Arc and Comparison Span	Maximum Definitive Position Difference (m)	Maximum Definitive Velocity Difference (cm/s)
S/C Sep to A1	366	4.1
A1 to P1	344	0.9
P1 to A2	550	1.6
A2 to P2	490	0.9
P2 to A3	402	1.1
A3 to P3	247	0.7
P3 correction to MCC	703	0.2

1466

Performance of Short Post-maneuver Tracking Data Arcs

Table 8 displays definitive and predictive differences between solutions that employ just 6 hours of post-maneuver tracking versus baseline definitive ephemeris. All of the baseline solutions used for the short-arc comparisons employed the JPL TDA Progress Report 42-128 station locations for the DSN sites, the parametrically determined 34-meter range biases, and solved for a coefficient of solar radiation. Baseline and short-arc runs apply a Doppler bias to compensate for the spacecraft's rotation about its Z-axis. The 6-hour and 12-hour short-arc solutions did not solve for solar radiation, but applied the value computed in the baseline solution. The 6- and 12-hour solutions included angle data, when available, but angle data was excluded from the baseline solutions. The tracking data arcs used in the 6-hour solutions were cut at six hours after the maneuver end time, regardless of whether six hours of tracking data had been received by that time.

It is apparent from Table 8 that post-perigee solutions performed much better than post-apogee solutions. This better performance is due to advantageous orbit geometry provided by the TDRS tracking at perigee. Excluding TDRS data from the post-P1 solution results in a single-station solution (DS46 only) with definitive error of 95 km in position and 39 m/s in velocity, and a predictive error of 1549 km in position and 8 m/s in velocity. The errors are primarily in the cross-track direction. Without TDRS data, the post-P2 definitive solution accuracy degraded to 13 km in position and 19 m/s in velocity and 77 km in position and 47 cm/s in velocity when predictive.

The separation +6-hour solution is worse than other near-perigee short arc solutions due to thruster testing, deployments, and attitude disturbances, which occurred later in the arc. In the case of the post-A1 six-hour solution, only approximately 4 hours of usable tracking were available during the first 6 hours. To obtain convergence with that amount of tracking, it was necessary to de-weight the Doppler observations from their nominal values. A 7-hour solution with nominal data weights gave significantly better results. No angle data was available in this arc. The post-A3 solution is degraded as a result of being a single-station solution.

Table 8

6-HOUR POST-MANEUVER TRACKING DATA ARC PERFORMANCE

Predictive Span	6-hour Maximum Definitive Position Difference	6-hour Maximum Definitive Velocity Difference	6-hour Solution Maximum Predictive Position Difference	6-hour Solution Maximum Predictive Velocity Difference
S/C Sep to A1	1.7 km	8.9 cm/s	17.8 km	8.9 cm/s
A1 to P1	90.7 km	1.3 m/s	927 km	397 m/s
P1 to A2	332 m	2.5 cm/s	1.9 km	1.3 cm/s
A2 to P2	820 m	6.6 cm/s	49.0 km	24.5 m/s
P2 to A3	624 m	33.4 cm/s	1.9 km	0.9 cm/s
A3 to P3	5.4 km	13.4 cm/s	140 km	53.2 m/s
P3 correction to periselene	17.6 km	28.4 cm/s	62.6 km	3.4 m/s

For orbits lacking good observability of the orbit plane, post-maneuver short-arc OD accuracy has been shown to benefit from constraining the plane of the post-maneuver state to be very close to that of the a-priori estimate. This technique is implemented operationally by performing the solution in Keplerian coordinates and applying tight a-priori covariances for the inclination and right ascension of ascending node. Typically, covariances of 1.0×10^{-12} degrees^2 are applied. Table 9 shows the results of reprocessing the post-apogee 2 and post-apogee 3 solutions under these constraints. For each of these solutions, the a-priori vector employed was taken from a longer, definitive post-maneuver solution.

The technique of tightening a-priori covariances on the right ascension and inclination likely represented a better estimate of the plane than would have been available for the 6-hour solution operationally, especially in the case of maneuvers with out-of-plane components. Both cases show considerable improvement under the constrained plane scheme.

Performance of the 12-hour post-maneuver tracking data arc was also studied. Nearly all 12-hour solutions showed improvement over the 6-hour solutions. Definitive overlap compares were 254 m position and 2 cm/s velocity for the post-P1 arc; 639 m position and 0.2 cm/s velocity for the post-P3 arc. The worst 12-hour definitive overlap compares were 2.7 km position for the post-A1 12-hour arc and 33.3 cm/s velocity for the post-P2 12-hour arc. The worst predictive compares were 48.1 km and 24 m/s, both for the post-A2 12-hour solution. In the case of the post-A2 12-hour solution, definitive overlap compares of 2.4 km and 8.1 cm/s were obtained, both worse than the 6-hour solution. This result may be an indication that the definitive ephemeris uncertainty in the first 12 hours of post-apogee tracking is truly on the order of 2 km.

Table 9

6-HOUR TRACKING DATA ARC WITH CONSTRAINED PLANE

Predictive Span	6-hour Maximum Definitive Position Difference (km)	6-hour Maximum Definitive Velocity Difference (cm/s)	6-hour Solution Maximum Predictive Position Difference (km)	6-hour Solution Maximum Predictive Velocity Difference (m/s)
A2 to P2	1.034	1.2	8.6	4.3
A3 to P3	2.251	8.9	101	38.6

L2 Operations

During routine operations at L2, MAP typically receives 45 minutes of range and Doppler tracking per day, almost exclusively from the DSN 70-meter antennas. L2 station-keeping requirements are for a minimum of 4 hours pre-burn and 4-hours post-burn near-continuous tracking.[9] MAP will execute approximately 4 station-keeping maneuvers per year.
MAP reached the proximity of L2 in early October 2001. The first station-keeping (SK1) maneuver occurred on January 16, 2002 and the second on May 8, 2002. A series of solutions were generated using 6-week tracking data arcs between October 3, 2001 until SK1 and then between SK1 and SK2, with 3 weeks of data overlap from span to span.

Table 10 shows the definitive and predictive ephemeris differences between these solutions. The predictive compares are taken from the end of the definitive arc until the next maneuver, compared against the final OD prior to the maneuver. All solutions employed the JPL TDA Progress Report 42-128 geodetics for DSN station locations, solved for C_r, and applied optimal range observation biases.

The high predictive and definitive compares prior to the 5 December solution are due to an extreme solar storm that occurred on November 6, 2001. The effect of this storm is very evident, when compared to the ephemeris differences seen in the unperturbed cases. In all other cases, both predictive and definitive ephemeris differences are less than 5 km in position and 1 cm/s in velocity. In this orbit regime, definitive ephemeris differences induced by differences between JPL TDA Progress Report geodetics and the operational FDF geodetics are about 3 km position and 0.1 cm/s velocity. The definitive ephemeris difference between application of the optimal range biases and nominal (100 meters) range biases is about 2 km and 0.1 cm/s.

Table 10

6-WEEK SOLUTION ARCS WITH 3-WEEK DEFINITIVE OVERLAP (L2 OPERATIONS)

Solution Tracking Data Span	3-Week Maximum Definitive Overlap Position Difference With Prior Ephemeris (km)	3-Week Maximum Definitive Overlap Velocity Difference With Prior Ephemeris (cm/s)	Maximum Predicted Position Difference With Final Pre-manuever Ephemeris (km)	Maximum Predicted Velocity Difference With Final Pre-manuever Ephemeris (cm/s)
Arrived at L2, approximately 1 Oct 2001				
10/03/01 To 10/14/01	-	-	145	9
10/24/01 To 12/05/01	30	3	45	2
10/14/01 To 12/26/01	26	2	2.4	0.1
12/05/01 To 01/16/02	1.7	0.1	-	-
SK1 16 Jan 2002				
01/17/02 To 02/28/02	-	-	3.5	0.2
02/07/02 To 03/21/02	1.5	0.1	1.6	0.1
02/28/02 To 04/11/02	1.2	0.1	3.4	0.1
03/21/02 To 05/08/02	2.7	0.1	-	-
SK2 8 May 2002				

To date MAP has successfully executed a total of 4 L2 station-keeping maneuvers. No significant problems have been observed concerning definitive orbit determination at L2.

PREMISSION AND POST-PROCESSED OD COMPARISON

Comparing the premission covariance analysis results with those from the post-processed OD is a complicated and delicate task. The comparison of predictive ephemeris accuracies is not necessary since the two processes use the same algorithms and force models for orbit propagation. The problem is therefore limited to the comparison of definitive ephemeris accuracies.

As discussed in the previous section, in the absence of an independent precision ephemeris, the common method of accessing the post-processed OD accuracy is through definitive ephemeris overlap compares, under the assumption that no perturbations and systematic errors bias the solutions. The covariance analysis provides the estimated ephemeris accuracy under certain tracking scenarios to plan for maneuver control, whereas post-processed maximum overlaps provide information about the orbit consistency. Results from the covariance analysis and those from post-processed OD are, therefore, not meant to be equivalent. However, by leaving aside the data that are dominated by perturbations and systematic errors, a general correlation between results from the two processes, as displayed in Table 11, can be seen. This correlation suggests that orbit errors estimated by the premission covariance analysis are relatively close to the actual OD errors, enough to assist mission analysts with maneuver planning and recovery. It also shows that even with changes in tracking scenarios due to actual tracking support availability, the requirements for definitive and predicted ephemeris accuracies to be performed by the post-launch OD support (Table 3) can be met.

Table 11

PREMISSION ANALYSIS AND POST-PROCESSED OD COMPARISON

Event	Estimated Definitive Ephemeris Accuracy (3σ) from Premission Analysis (From Tables 1 & 2)	Post-processed Maximum Definitive Overlap Compares (From Tables 5 & 8 and 12-h performance)
At P1	Pos: 33 m Vel: 0.89 cm/s	Pos: 32 m Vel: 0.3 cm/s
At P3 (Pfinal)	Pos: 53 m Vel: 0.93 cm/s	Pos: 46 m Vel: 0.1 cm/s
6 hours from P1	Pos: 300 m Vel: 1.05 cm/s	Pos: 332 m Vel: 2.5 cm/s
12 hours from P1	Pos: 241 m Vel: 0.354 cm/s	Pos: 254 m Vel: 2 cm/s
12 hours from P3	Pos: 886 m Vel: 2.57 cm/s	Pos: 639 m Vel: 0.2 cm/s
3-week arc after L2 insertion	Pos: 2.593 km Vel: 0.1 cm/s	Pos: 2.7 km Vel: 0.1 cm/s

ORBIT DETERMINATION SUPPORT LESSONS LEARNED

A number of lessons for OD may be taken from this analysis. The advantageous orbit geometry provided by TDRS tracking at perigee passage gave 6-hour and 12-hour post-perigee solutions greater predictive accuracy than post-apogee solutions. Application of tight a-priori covariances on right ascension and inclination can improve the predictive accuracy of short-arc solutions by reducing the state-space of the estimation to those solutions in the vicinity of the expected orbital plane. This technique requires that the a-priori state represent a good estimate of the post-maneuver plane. In the case of in-plane maneuvers, the last pre-maneuver state may be used, but for plane-changing maneuvers, the accuracy of the constrained prediction will be dependent upon the accuracy of the predicted post-maneuver state. This technique may be especially helpful when TDRS tracking is not available.

Attitude mode changes impact the definitive ephemeris uncertainty by effectively reducing the usable tracking data span. This is primarily a consequence of least-squares estimation and could be ameliorated by the implementation of an attitude-dependent area model or potentially by sequential filter estimation. This effect did not have a negative impact on predicted ephemeris accuracy as these errors did not have time to grow appreciably before the next delta-V, but did notably degrade definitive accuracy. The magnitude of this effect depends on the spacecraft structure. In the case of MAP, the spacecraft solar arrays and Sun shade system act similarly to a solar sail in the flat-spin modes.

Studies on MAP OD accuracy are impacted by the apogee-perigee maneuver scheme during the phasing loops, which makes obtaining consecutive overlapping full-orbit solutions almost impossible. The magnitude of the ephemeris differences can be driven by the choice of comparison end time, especially at apogee and perigee, where radial and along-track differences change rapidly. Future analysis of deep-space mission ephemeris accuracy would benefit from the generation of independent precision orbit ephemeris whose inherent accuracy is well known.

Finally, results from the premission covariance analysis are used as a baseline for mission and maneuver planning only. They could be significantly different from position overlap obtained from postlaunch OD data, especially when involving major perturbations, biases, and differences in tracking scenarios.

CONCLUSION

Intensive premission covariance analysis and actual postlaunch orbit determination were performed in support of MAP. The premission analysis provided valuable information on frequency of tracking, definitive arc length, and measurement types in supporting various mission phases. Several challenges were presented to the postlaunch orbit determination accuracy analysis including perturbations due to changes in spacecraft attitude, systematic errors due to uncertainties in station location and a-priori range bias, and the lack of independent precision definitive ephemeris. Even though results from the covariance analysis and the actual OD are not equivalent, they show a general correlation. They suggest that errors estimated from the covariance analysis can be close to actual errors, enough to benefit the mission to plan for maneuver control. Many important lessons learned from the MAP orbit determination support would help improve future mission support. The technique of tightening a-priori covariances on right ascension and inclination can be especially helpful when TDRS tracking is not available.

ACKNOWLEDGMENTS

The authors would like to acknowledge the outstanding contributions of the MAP Navigation team of Computer Sciences Corporation (CSC) and Honeywell for their outstanding support during the MAP Launch and early orbit activities.

REFERENCES

1. K. Richon and M. Matthews, "An Overview Of The MAP Trajectory Design", AAS 97-728, August 1997.
2. O. Cuevas, et. al, "MAP Trajectory Design Peer Review", presentation to MAP Project, December 1999.
3. O. Cuevas, et al, "MAP Trajectory Status", presentation to MAP Project, March 2000.
4. D. Fink and S. Slojkowski, "Draft MAP Navigation Post-launch Report," memorandum to S. Coyle, March 2002.
5. S. Truong, "MAP Orbit Covariance Analysis Final Report," memorandum to the Flight Dynamics Analysis Branch, July 2000.
6. L. Newman and D. Rohrbaugh, "Trajectory Design for the Microwave Anisotropy Probe (MAP)", 16th International Symposium on Space Flight Dynamics, Pasadena, CA, December 2001.
7. L. Newman and D. Rohrbaugh, "MAP Trajectory Post-launch Report," November 2001.
8. Joseph H. Yuen, Ed., "The Telecommunications and Data Acquisition Progress Report 42-128," The Jet Propulsion Laboratory, October-December 1996.
9. S. Coyle, et al, "Detailed Mission Requirements (DMR) for the Microwave Anisotropy Probe (MAP)", Goddard Space Flight Center, October 1999.

AAS 03-204

NAVIGATING CONTOUR USING
THE NONCOHERENT TRANSCEIVER TECHNIQUE

Eric Carranza, Anthony H. Taylor,
Dongsuk Han, Cliff E. Helfrich, Ramachand Bhat and Jamin S. Greenbaum[*]

The successful navigation of the Comet Nucleus Tour spacecraft was performed at the Jet Propulsion Laboratory and was conducted with the use of the new noncoherent transceiver technique developed by the Applied Physics Laboratory. Descriptions of the mission and trajectory are provided, followed by a summary of the challenges to navigation. After launch, about six weeks of noncoherent tracking data were acquired while the spacecraft was in the initial phasing orbits about Earth. Unfortunately, radio contact with CONTOUR could not be re-established after the solid rocket motor (SRM) burn that sent the spacecraft onto its interplanetary trajectory. Following the SRM, ground based optical measurements indicated the spacecraft had broken into pieces and was presumed lost. Discussions include the conditioning performed on the 2-way noncoherent Doppler data, the orbit determination process, and the post SRM trajectory reconstruction.

INTRODUCTION

The COmet Nucleus TOUR (CONTOUR) spacecraft was launched on July 3, 2002 from Cape Canaveral, Florida. CONTOUR was the sixth mission flown in the National Aeronautics and Space Administration's Discovery Program, a program which aims at achieving highly focused planetary science investigations at a low cost. The objective of the $159 million CONTOUR mission was to conduct scientific flyby studies of comets, Encke and Schwassmann-Wachmann 3, with the option of changing targets in-flight or visiting an additional target after the Schwassmann-Wachmann 3 encounter. The flyby science goals included the following: image nucleus parts at a resolution of 4 m/pixel to reveal details of morphology and comet processes; determine nucleus size, shape, rotation state, albedo/color heterogeneity and activity; map composition of nucleus surface and coma; obtain detailed compositional measurements of gas and dust in the near-nucleus environment; and assess the level of outgassing[1].

[*] The authors are Members of the Technical Staff, Jet Propulsion Laboratory, California Institute of Technology, 4800 Oak Grove Drive, Pasadena, California 91109-8099.

This paper will describe the mission's planned trajectory, the navigation challenges, as well as the navigation tasks that were involved in achieving mission objectives. The challenges include transceiver issues, orbit determination of the spacecraft during the six-week Earth orbit phase, maneuver design, and post solid rocket motor burn reconstruction.

TRAJECTORY AND SPACECRAFT DESCRIPTIONS

CONTOUR was designed, built, operated and managed by the Johns Hopkins University's Applied Physics Laboratory (APL), while the California Institute of Technology's Jet Propulsion Laboratory (JPL) navigated it. The Deep Space Network (DSN), managed by JPL, provided spacecraft telecommunication support. APL successfully launched its spacecraft aboard a three-stage Boeing Delta II rocket on July 3, 2002 at 02:41 A.M. (EDT). It was placed into a highly eccentric Earth parking orbit with an apogee radius of 115,000 km, a perigee radius of 6,620 km (240 km in altitude), an inclination of 30.5 degrees, and an orbital period of 1.73 days (see Fig. 1 & 2). CONTOUR remained in this parking orbit from launch until August 15, 2002, when it fired its STAR-30 solid rocket motor (SRM) to leave Earth orbit and enter a heliocentric orbit. CONTOUR was to remain in this heliocentric orbit through an Earth flyby in the summer of 2003, followed by a flyby of its first target (comet Encke) in November of the same year. However, due to an anomaly during the SRM burn, radio contact with the spacecraft could not be re-established and it was presumed lost.

CONTOUR was designed in the shape of an octagon cylinder, approximately 3 m in diameter and 2 m in height, and had a 1 m mast located at the center of its aft end (see Fig. 3). Its total weight was approximately 970 kg; this includes a dry weight of 397 kg, 70 kg of Hydrazine fuel, and 503 kg for the STAR-30 SRM[1]. (The SRM was built into the spacecraft; it was not a stage that could be jettisoned.) The spacecraft's sides and aft end were covered with solar array panels. A seven-layered dust shield covered the craft's front end; its purpose was to protect CONTOUR and the science instruments from dust particles during the comet flybys. CONTOUR had four science instruments on board: the CONTOUR Remote Imager/Spectrograph (CRISP), the CONTOUR Forward Imager (CFI), the CONTOUR Dust Analyzer (CIDA), and the Neutral Gas Ion Mass Spectrometer (NGIMS). CONTOUR also carried four antennas, three on the aft end and one on the forward end. The Aft Low Gain and Pancake antennas were located on the mast, which was on the spacecraft spin axis, while the High Gain antenna was located 50 cm from the spin axis on the body of the craft's aft end; the Forward Low Gain antenna was located 40 cm from the spin axis on the front end. Each antenna used the noncoherent transceiver system for communications; the system used one of APL's long line of ultra-stable oscillators[2]. Figure 4 shows CONTOUR's oscillator specifications.

NAVIGATION CHALLENGES

The greatest technical challenge to navigation was to develop robust procedures to deal with the uncertainties introduced by the noncoherent navigation technique. Since this was the first operational use of the technique, it had to be validated before launch, and new software and procedures had to be incorporated into the orbit determination process for post-launch operations. Navigation usage of both Doppler and ranging data were affected by this technique since the usually trusted DSN data types had to be examined more carefully during the orbit determination process to see if they were adversely affected.

The relatively small CONTOUR navigation team had a full-time equivalent staffing of 5 people from six months before launch until 3 months after launch. Two of these people were dedicated to the noncoherent transceiver processing, leaving 3 to deal with the traditional navigation activities of preparation for launch and operations after launch. Some pre-launch analysis which is normally performed was delayed to concentrate on developing the new processing technique and contingencies; e.g., there was no covariance analysis done for the Earth orbit phase. The team was able to partially test some parts of the non-coherent navigation technique on a previously launched spacecraft (TIMED), although differences between this low Earth orbiter and the CONTOUR mission profile still left many aspects untested at launch. By the time of launch, navigation team activities peaked as last minute interface tests on the small forces and attitude predicts files were performed while the team also prepared for post-launch contingency scenarios. The team uncertainty and concern before launch quickly faded once the CONTOUR spacecraft was safely in orbit and it became obvious the noncoherent processing technique was working properly.

NONCOHERENT TRANSCEIVER

CONTOUR was the first interplanetary spacecraft to use a transceiver instead of a transponder. This resulted in a number of departures in the task of obtaining Doppler and range data for use in orbit determination. To use transceiver-based tracking data, new procedures and software were established and exercised prior to launch[4,5].

The primary data type was 2-way X-band noncoherent Doppler received by the Deep Space Network (DSN) almost continuously during the Earth phasing orbits. Because 1-way Doppler is usually much too inaccurate for navigation because of spacecraft oscillator frequency drift, this data required correction in order to be usable in the standard JPL orbit determination process. The correction process was developed by APL, and used both non-coherent Doppler and spacecraft telemetry data to obtain nearly coherent 2-way Doppler data; therefore, the successful navigation of the CONTOUR spacecraft relied not only on the radiometric data, but on the telemetry data, as well. (This later dependence was a first in deep space navigation.) The corrected Doppler data were treated as equivalent to coherent, 2-way Doppler data, the standard data type used in navigation. The details of this correction are described in detail in a paper written by Dr. J. R. Jensen[6].

The use of a transceiver onboard the spacecraft meant that the downlink frequency, driven by the onboard oscillator, was independent of the uplink frequency. This independence introduced errors in the 2-way Doppler data, which were caused by bias and drift of the spacecraft frequency reference. In order to support radiometric accuracy requirements, CONTOUR carried additional hardware that, in essence, measured the difference between the received uplink frequency and the transmitted downlink frequency and placed these measurements into the telemetry. On the ground, the correction software then used this information to convert the Doppler observables to those that would have been obtained had a transponder been used[2]. Figure 5 shows both the corrected and uncorrected 2-way Doppler observables for a pass that occurred toward the end of July 18, 2002 UTC (Day Of Year 199); the X represents the uncorrected observable and the O represents the corrected observable. Although it is difficult to see in this pass, the corrected data has a near-constant offset of 3 Hz from the uncorrected data after DOY 199.8602. Prior to this, it can be seen that the corrected data has a sharp positive slope. This is caused by the uplink frequency sweep the station performs when attempting to acquire the spacecraft. The uncorrected observable is not affected during the sweep because the downlink frequency observed is independent of the uplink signal. This typical figure validates what is expected from the noncoherent transceiver system. Approximately 96% of all 2-way noncoherent Doppler data was corrected during flight.

The ranging system used by the DSN was originally designed to use coherent Doppler information to "rate aid" the correlation of the received ranging modulation code with the local code. Without rate aiding, the received and local waveforms drift with respect to each other and the correlation fails. In order to achieve correlation using CONTOUR's noncoherent transceiver system, it was necessary to continually ramp the uplink frequency from the DSN so that the signal arriving at the spacecraft was pre-compensated for the Doppler shift on the uplink leg. Thus, the spacecraft oscillator's downlink frequency approximated the frequency that would result from a transponder-equipped spacecraft receiving the same ramped signal and turning it around to the downlink. This technique required precise control of the uplink frequency. It introduced a vulnerability to frequency errors in the uplink signal or the spacecraft oscillator, which resulted in failed correlations whenever the received and local waveforms drifted more than 1/4 wavelength relative to each other.

In order to minimize the potential for range correlation errors due to frequency prediction errors, a nonstandard set of ranging tones were used during the first 8 days of the mission. (See Table 1.) This choice of ranging components had the unexpected effect of causing errors in the sequential ranging assembly (SRA) and the loss of 50% of the range measurements. (See Fig 6, this plot shows the ranging anomalies associated with miscorrelations. The "good" (useable) residuals are near 0, while the anomalous residuals due to miscorrelations are spread over several million meters in distinct layers.) This problem was eliminated later when a standard set of components were used. The SRA correlation errors were due to the choice of ranging tones employed, rather than directly due to the use of a transceiver to support range measurements. The source of these correlation errors is not yet understood. Following the change of range components, 82% of the range measurements were good, except during one interval following the spin-up from 20 to 60 rpm where a change in the spacecraft oscillator that was not included in the uplink predictions caused range correlation errors.

ORBIT DETERMINATION MODELING

Orbit Determination (OD) for CONTOUR was performed using corrected 2-way X-band noncoherent Doppler and 2-way noncoherent ranging data collected by the DSN's 34 m network. Doppler data were collected and corrected at 1-second intervals, then compressed to 60 second intervals (except on launch day, when 1-second data were used). Ranging data were collected at sample rates of 29 and 37 seconds, depending on the specific ranging system parameter values used. When using the compressed data during flight, the 2-way Doppler data were weighted at a sigma from 2 mm/sec to 10 mm/sec and the range data were weighted at a sigma from 5 m to 100 m. Invalid data were removed from both data sets, and both sets were calibrated for media effects (troposphere and ionosphere). Additionally, the Doppler data were calibrated for spin-polarization biases, and the ranging data were calibrated for the timing delay at the spacecraft. The orientation of the spacecraft's - Z-axis (the spin pole) relative to each DSN station was modeled but the spin signature was not removed from the Doppler tracking data. (CONTOUR spin rate varied from 20 to 60 rpm)

The dynamic models used in determining the CONTOUR orbits were the Newtonian point-mass model, the relativity model, the Earth and Lunar oblateness and solid tide models, the impulsive maneuver model for unintentional delta-V events, the finite maneuver model for intentional delta-V events, the atmospheric drag model, and the solar radiation pressure model. The Newtonian point-mass model computed the gravitational acceleration of the spacecraft due to the nine planets, the Sun and the Moon by treating the bodies as point-masses. The relativity model computed the relativistic perturbative acceleration caused by the Sun. The oblateness and solid tide models computed the acceleration of the spacecraft due to Earth and Moon's oblateness and the tidal acceleration effect of the Earth and Lunar solid tides caused by the Sun, respectively. The impulsive maneuver model was used to model unintentional delta-Vs (e.g., incidentally caused by turns), and the finite maneuver model was used to model intentional delta-Vs, i.e., the Orbit Change Maneuvers (OCMs) and the Solid Rocket Motor (SRM) maneuver. The atmospheric drag model computed the deceleration of CONTOUR caused by Earth's atmosphere. The solar radiation pressure models computed the acceleration of the spacecraft due to solar radiation.

The Earth gravity model used in the OD was the JGM-3 gravity model[7], truncated to the 50th degree and order. The Earth atmospheric model used for drag was the Jacchia-Roberts atmosphere density model[8]. This model was used in place of the DTM model[9,10], which was used for initial studies, because the former model was built into certain software APL used. Figure 7 shows a comparison between the Jacchia-Roberts model and the DTM model during the first perigee pass of CONTOUR; it can be seen that both models show reasonable agreement to each other. The figure also shows the large drag pressure exerted on the spacecraft over the brief period CONTOUR passed through perigee. The CONTOUR spacecraft shape model for drag was basic and was composed of three components: (1) a 1-sided flat plate for the aft side (the side that had the high gain antenna), (2) an open-ended cylinder to represent the sides of the spacecraft, and (3) a 1-sided plate for the forward side (the side with the SRM nozzle).

Other models that were used in the OD process include the solid Earth tide correction model, the continental plate motion and ocean loading models, the precession and nutation models, and the relativistic light time correction model. The precession and nutation models used the JPL Earth Orientation Parameters[11]. The Earth-fixed coordinate system used in the OD was consistent with the International Earth Rotation Service (IERS) terrestrial reference frame labeled ITRF93[12]. The locations of the DSN stations were consistent with the ITRF93 reference frame[13]. The ephemerides of the Sun, the Moon, and the planets were defined by the JPL DE405 planetary ephemeris[14,15]. The inertial coordinate system used for orbit integration was Earth centered, Earth mean equator and vernal equinox system at J2000. The *a priori* sigma values for each component of the spacecraft's position were 100 km and 0.1 m/sec for the spacecraft's velocity components. See Table 2 for a listing of *a priori* sigma values for other parameters estimated in the filter, as well as a listing of the consider parameters.

The software used to process the tracking data was JPL's Orbit Determination Program (ODP)[16]. The ODP solves for spacecraft position, velocity, and other requested parameters using a square root information (SRIF) weighted least squares filter[17,18]. The Doppler data corrections and the orbit fits were performed on a 700 MHz Intel Pentium II workstation running RedHat Linux 6.1.

ORBIT DETERMINATION RESULTS

The concern for launch was that the new noncoherent transceiver technique would either fail to work, or that there would be significant but subtle undetected biases in the corrected data, or that the correction process would encounter unplanned delays during the first critical pass over Goldstone, and consequently there would be no useful coherent Doppler data available for early orbit determination. Early OD deliveries were to begin at Launch + 2 hours (one hour into the initial Goldstone pass) in order to support acquisition at the next DSN track at Canberra beginning at L + 7 hours. Failure to deliver an accurate orbit could have meant a loss of the Canberra pass, and perhaps serious difficulties acquiring the spacecraft after that if there were launch anomalies. Another OD delivery was needed no later than L + 11 hours to support a decision to do a periapsis raise maneuver at the first apoapsis (L + 22 hours) to prevent losing the spacecraft in the event of a serious launch error.

Since coherent Doppler is the workhorse data type of JPL navigation during launch phase, it was necessary to put more emphasis on other data types in case of its failure. The other data types would either replace the corrected Doppler data if necessary, or serve as independent validation of it. Ranging would ordinarily be the next data type of choice, but the same feature that made it necessary to correct the Doppler data (i.e., non-coherence) also put the ranging data at risk because of the trajectory-induced uncertainty in the frequency that the DSN needed to transmit in order to insure that the noncoherent ranging technique (without the conventional rate aiding) would work. Figure 8 shows the a priori errors in the knowledge of the spacecraft downlink frequency. In order to achieve valid ranging, the DSN would have to transmit at a frequency which, after being Doppler shifted on the uplink and turned around at the spacecraft by the same ratio that a transponder would have used, would be within a given error of the actual downlink frequency. In the case of the DSN ranging setup used (and optimized) for launch, the maximum allowable frequency error would be about 1300 Hz. Comparing this to the limits in Figure 8 indicates that there was a reasonable (but not excellent) chance that ranging would be usable beginning about two hours after launch, but that the conditions would then worsen later in the pass unless the actual downlink frequency was determined and the DSN changed its uplink frequency accordingly.

Only one DSN data type would be totally unaffected by noncoherence during launch, and that was the tracking angle data of the acquisition antennas, one of them at Goldstone and one at Canberra. These data are used only at launch, and are not as powerful as coherent Doppler and ranging, so there was an accuracy issue as well as a reliability issue associated with their use.

Fortunately, on launch day the Doppler correction technique and the prototype processing system worked with timely deliveries of calibrated data to the navigation team. Ranging data also appeared mostly successful, as did the angle data. Experiments were tried in near real-time with all three data types, individually and in combinations, and the solutions were consistent with each other to within the errors expected of each type, so that the worry of a bias being introduced into the Doppler by the correction process was alleviated.

The next major concern for launch was over the signature modulated onto the Doppler data by the spin of the spacecraft in combination with the offset of the forward low gain antenna from the spin axis. The forward low-gain antenna was in use from the initial Goldstone acquisition into the beginning of the first Canberra pass (from about L + 1 to L + 8.6 hours). After this initial period of forward low-gain antenna use, most of the rest of the orbital phase (except for short periods near periapsis) was flown on either the pancake antenna or the aft low-gain antenna, both of which had no nominal offset from the spin axis, and thus induced no nominal spin modulation on the Doppler.

The initial spin of the spacecraft was 49.5 RPM, a much higher rate than has been seen by JPL navigators on any other spacecraft. At about L + 2.5 hours, the spacecraft was spun down to 22.8 RPM, still faster than ever seen by JPL navigators. At L + 13 hours (by which time the spacecraft had switched to the pancake antenna), the rate was cleaned up to 20 RPM, and left close to that rate for most of the rest of the orbit phase until it was spun up again to 60 RPM in preparation for the SRM burn.

The modulation on the Doppler during the early part of the launch is seen in Figure 9. The modulation as a function of antenna offset (ρ), spin axis angle from the tracking station (α), Doppler compression time (Δt), and spin rate (ω in rad/s) is

$$D = \frac{2\rho \sin \alpha}{\Delta t} \sin\left(\omega \frac{\Delta t}{2}\right) \cos \omega (t - t_0)$$

where D is the 1-way Doppler modulation (in terms of range rate, not Hz). This function explains the larger envelopes at about 0.38 day and 0.42 day when the compression time was changed briefly from 1 second to 0.1 second, and why the envelope widened at 0.4 day when the spin rate actually decreased from 49.5 to 22.8 RPM (i.e., a sampling effect).

The modulation in Figure 9 is quite large, exceeding half a meter per second after 0.4 day. The best way of treating this data is not obvious. One method would be to "de-modulate" it by determining the exact parameter values in the above equation and calibrating the signature out. The danger is that if the rate is not known precisely, a large low-frequency signature could be left in the signal, the worst possible situation if the filter should allow it to masquerade as a change in orbital or other parameters. Another way would be to not model the modulation at all, but simply deweight the data to the approximate envelope of the signal, and count on the high-frequency averaging of the filter to provide an accurate mean. This was the method the navigation team used, not because of any stringent analysis, but because it was the easiest, fastest, cheapest way to approach the problem. The worry was that with longer Doppler compression times, such as 1 minute, and short spans of data (e.g., a short DSN pass), odd sampling effects and incomplete averaging might leave subtle undesirable signatures in the data. Fortunately, except for the launch, most of the orbit phase was spent on the aft and pancake low gain antennas so that there was no spin modulation to worry about. Unfortunately, there was never enough time or resources to do a thorough analysis on the best way to treat the spin-modulated data, so that this question may continue to haunt navigators of future spinning spacecraft for which antenna placement is not benign.

The last topic of this section concerns the modeling and estimation of spacecraft drag. The spacecraft periapsis altitude was very low, ranging between a low of 194 and a high of 249 km, and neither the Jacchia-Roberts nor the DTM atmosphere density models are particularly accurate in this range, particularly at the lower end, even with weekly updates of the solar flux inputs. The drag model we used in the navigation software used the same spacecraft components (a flat plate at each end of a cylinder) as the solar pressure model. Because of time and resource limitations, only a drag force was modeled, based on the combined cross-section of the three components, rather than a more complicated model which was available to model lift and side-slip forces. An overall drag coefficient of 2.8 was used. A constant scale factor of the area was estimated to remove bias in the model.

Despite the simplicity of the model, the estimate of the scale factor and the size of the data residuals were quite reasonable through the first two periapses. However, when the data arc was extended from launch through the third periapsis, the residuals no longer were well behaved, and the area scale factor estimate began to take on implausible values. It became obvious that estimating the scale factor as a constant throughout all periapses was the wrong thing to do. The estimation procedure was changed so that the scale factor was estimated stochastically at each periapsis, with zero correlation between periapses. This had the desired effect of improving the residuals and returning the scale factors to believable values, as well as decoupling the orbit-to-orbit error.

MANEUVER DESIGN

There were 23 spacecraft maneuvers performed during the Earth orbit phase of the mission, which lasted 43 days. They were designed by APL to achieve proper conditions at the Earth departure maneuver. Because of the large DeltaV (1.9 km/s) and the resonant nature of the CONTOUR trajectory, the SRM had to be executed at the exact conditions determined by JPL for optimizing the interplanetary trajectory. Failure to achieve these conditions would require large clean-up maneuvers following Earth departure.

POST SRM RECONSTRUCTION

The Solid Rocket Motor burn on August 15, 2002 was intended to terminate CONTOUR's Earth orbit phase and inject it into a heliocentric orbit, and was performed at a point where the spacecraft was below the horizon for all DSN stations. The spacecraft was last seen, in apparently good order, by DSS 65 at Madrid at the nominal end of its pass about an hour before the burn. About 40 minutes after the burn, CONTOUR would have risen for DSS 25 at Goldstone and DSSs 34 and 46 at Canberra. Radio contact could not be re-established.

On the evening of August 16, the Spacewatch 1.8 m telescope at Kitt Peak, Arizona observed two objects, denoted A and B, in the general vicinity of the nominal post-maneuver trajectory. Later, in some of the Spacewatch images of the same date, a third object, denoted C, was discovered. Thereafter additional images of all three objects were obtained by Spacewatch, the LINEAR facility in New Mexico, the University of Hawaii's 2.24 m telescope at Mauna Kea, JPL's Table Mountain 1 m telescope in California, and at the Farpoint Observatory's 0.3 m telescope in Kansas. Table 3 is a summary of these optical tracks.

These observations dating to August 21, 2002, plus radiometric data from 13 August until the SRM burn, were used by CONTOUR Navigation to estimate separate SRM burns and accelerations due to solar radiation pressure for each of the three objects. These results (along with their 1-sigma uncertainties) are presented in Tables 4 and 5, respectively. It was determined that one of the components, B, is apparently (with high probability) a very low mass object of between 3.3 and 3.4 kg. Component A could contain 190 kg of the nominal spacecraft or be as small as 108 kg for a 1-sigma deviation. There was too little data for C (only on August 16) to determine anything meaningful about its mass. All three objects will return to the vicinity of Earth almost exactly a year after the SRM maneuver. Earth-centered B-plane results for these encounters are given in Table 6, along with their respective 1-sigma uncertainties. Figure 10 shows the pointing uncertainties of component A for ground observers during the cruise back to Earth. Interestingly, the pointing uncertainties show a dip close to the flyby time, implying that it might be possible to find the object with a fairly narrow-field (and sensitive) telescope.

CONCLUSIONS

The Jet Propulsion Laboratory undertook the task of navigating the CONTOUR spacecraft, which used the successful transceiver system for telecommunications. The use of this system resulted in a number of departures from standard JPL and DSN procedures for obtaining Doppler and ranging data for use in orbit determination. Calibration of the noncoherent transceiver Doppler data using the down link telemetery was necessary before it could be used as tracking data in the navigation solutions. A new process was jointly developed by the DSN, the developers of the JHU/APL transceiver, and JPL navigation to accomplish the calibration in near real time with minimal delays. The new process was successful in calibrating the noncoherent transceiver Doppler data so that its use in the orbit determination process, except for the extra steps involving the calibration itself, was essentially equivalent to the use of standard DSN Doppler tracking.

One of the lessons learned from the experience of CONTOUR is that the prototype calibration system, which was operated by members of the CONTOUR navigation team, should instead be operated by those more familiar with the details of the transceiver electronics, its operation and its telemetry to avoid unnecessary delays in processing the calibrations. This would also provide immediate feedback to the project for validating (and debugging, if necessary) the proper operation of the transceiver. Improvements in file handling and automation of the prototype process could also be made to reduce the risk of data delivery delays or outages that could impact critical navigation deliveries on future projects. Any loss of telemetry during a noncoherent Doppler pass would, of course, render that data unusable for navigation. Improvements should be made to the prototype process to minimize the possibility of loss of telemetry, especially during critical periods, to further reduce navigation risks.

Examining the post-processing orbit determination results, it does not appear necessary to model out the spin signature on the Doppler data for a spacecraft like CONTOUR, at least during the Earth orbital phase. Adequate orbit determination was achieved by de-weighting the data to the approximate envelope of the modulation with little degradation of accuracy. It does appear necessary to stochastically model the drag at every periapsis of a CONTOUR-like orbit.

CONTOUR's launch was one of the most challenging launches supported by JPL navigation because of the uncertainty surrounding the operation of the new noncoherent processing technique. The navigation team had to prepare for contingencies (that thankfully never occurred) by providing backup tracking techniques in case the prototype system had failed. The successful completion of the Earth orbit phase navigation proved that even a relatively small team of dedicated individuals from the DSN, the transceiver developers at JHU/APL, and the JPL navigation team could accomplish their goal of making the noncoherent tracking technique work for space navigation.

ACKNOWLEDGMENTS

CONTOUR images, stats, and cost information courtesy of the official CONTOUR Project web page, CONTOUR animation, and a presentation given at JPL by personnel from APL [19,20,21]. We appreciate the advice Tomas Martin-Mur gave us to complete the task at hand, as well as the tremendous support given to us by the personnel at the Deep Space Network and The Johns Hopkins University, Applied Physics Laboratory. The work described in this paper was carried out at the Jet Propulsion Laboratory, California Institute of Technology, under contract with the National Aeronautics and Space Administration.

REFERENCES

1.   Farquhar, R. W., and D. W. Dunham, Launch and Trajectory Alternatives for the CONTOUR Comet Mission, paper presented at the International Astronautical Congress, Houston, TX, Oct 10-19, 2002.

2.   Jensen, J. R., K. B. Fielhauer, M. J. Reinhart, D. K. Srinivasan, In-Flight CONTOUR Radiometric Performance, IEEEAC paper # 1084, updated November 25, 2002.

3.   Carranza, E., A. H. Taylor, and B. G. Williams, CONTOUR Navigation, Interplanetary Network Directorate Deep Space Mission System (DSMS), Presentation at the Delta Readiness Review for CONTOUR, June 4, 2002.

4.   Jensen, J. R., and R. S. Bokulic, Experimental Verification of Noncoherent Doppler Tracking at the Deep Space Network, IEEE Transactions on Aerospace and Electronic Systems, Vol. 36, pp. 1401-1406, 2000.

5.   Fielhauer, K. B., CONTOUR Non-Coherent Navigation End-to-End and Post Environmental Performance Results, JHU/APL Memo SER-02-020, April 17, 2002.

6.   Jensen, J. R., and R. S. Bokulic, Highly Accurate Noncoherent Technique for Spacecraft Doppler Tracking, IEEE Transactions on Aerospace and Electronic Systems, Vol. 35, No. 3, pp. 963-973, July 1999.

7.   Tapley, B.D., M. M. Watkins, J. C. Ries, *et al.*, The Joint Gravity Model 3, Journal of Geophysical Research, Vol. 101, No. B12, Dec 10, 1996.

8.   Roberts, C., An Analytic Model for Upper Atmosphere Densities based on Jacchia's 1970 Models, Celestial Mechanics Vol. 4, 1971, pp 368-377.

9.   Shum, C. K., J.C. Reies, B. D.Tapley, P. Escudier, and E. Delaye, Atmospheric Drag model for Precision Orbit Determination, CSR-86-02, Center for Space Research, The University of Texas at Austin, Austin, Texas, January 1986.

10.   Barlier, F., C. Berger, J. L. Falin, G. Kockarts, and G. Thuillier, A Thermospheric Model based on Satellite Drag Data, Aeronomica Acta, A- No. 185, 1977.

11.   Folkner, W. M., J. A. Steppe, and S. H. Oliveau, Earth Orientation Parameter file description and usage, *JPL Internal Interoffice Memo. 335.1-11-93*, Jet Propulsion Laboratory, California Institute of Technology, Pasadena, CA, 1993.

12.   Boucher C., Z. Altamimi, and L. Duhem, Results and analysis of the ITRF93, *IERS Tech. Note 18*, Obs. de Paris, 1994.

13.   Folkner, W. M., DSN station locations and uncertainties, *Progress Rep. 42-128*, pp. 1-34, Jet Propulsion Laboratory, California Institute of Technology, Pasadena, CA, 1996.

14.   Standish, E. M., and Newhall, X X, New accuracy levels for solar system ephemerides, in *IAU Symp. 172: Dynamics, Ephemerides and Astrometry of the Solar System*, edited by S. Ferraz-Mello et al., pp. 29-36, IAU, Paris, 1996.

15.   Newhall, X X, J. G. Williams, and E. M. Standish, Planetary and Lunar Ephemerides, Lunar laser ranging, and Lunar physical librations, in *IAU Symp. 172: Dynamics, Ephemerides and Astrometry of the Solar System*, edited by S. Ferraz-Mello et al., pp. 37-44, IAU, Paris, 1996.

16.   Moyer, T. D., Mathematical formulation of the Double-Precision Orbit Determination Program, *Technical Report 32-1527*, Jet Propulsion Laboratory, California Institute of Technology, Pasadena, CA, 1971.

17.   Lawson, C. L., and R. J. Hanson, *Solving Least Squares Problem*, Society for Industrial and Applied Mathematics, Philadelphia, PA, 1995.

18.   Bierman, G. J., *Factorization Methods for Discrete Sequential Estimation*, Academic Press, New York, 1977.

19.   Web site for the Comet Nucleus Tour Project, http://www.contour2002.org.

20.   Maas, Dan, CONTOUR The Movie Animation.

21.   Briefing by APL to the DSN on Noncoherent Navigation Technologies and Flight Results, November 13, 2002.

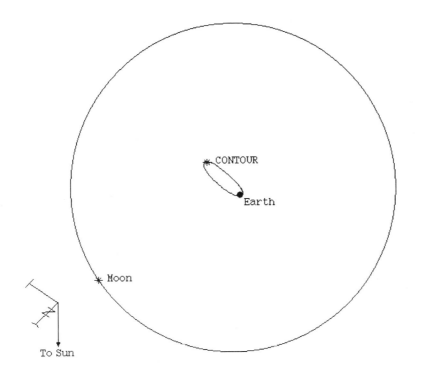

Figure 1: CONTOUR Orbit as viewed from north ecliptic.

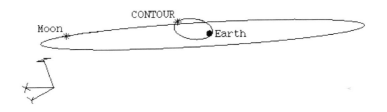

Figure 2: CONTOUR Orbit as viewed from the Sun.

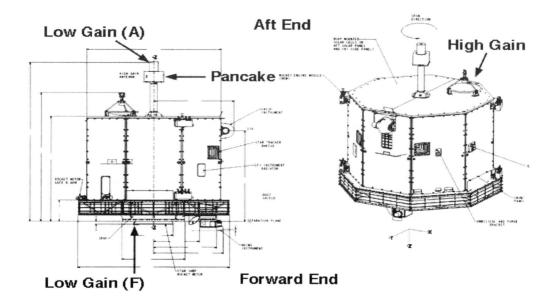

Figure 3: CONTOUR Spacecraft.

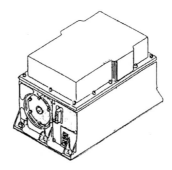

Output	Two 10 dBm @ 30.6 MHz
Drift Rate	5×10^{-10} per 24 hours
Temperature Stab.	1×10^{-11} per °C (-5 to +25 °C)
Phase Noise	-90 dBc/Hz @ 1 Hz offset
	-130 dBc/Hz @ 1 KHz offset
Allan Deviation	1×10^{-12} in 1, 10, or 100 sec. Periods

Figure 4: CONTOUR Oscillator.

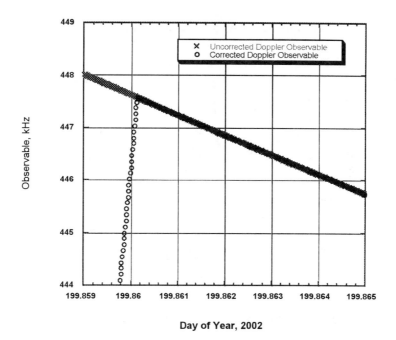

Figure 5: 2-way Doppler Observables for July 18, 2002.

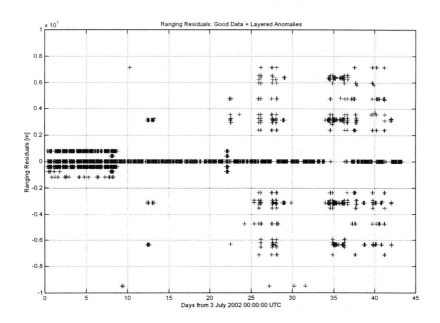

Figure 6: Miscorrelated Range Residuals.

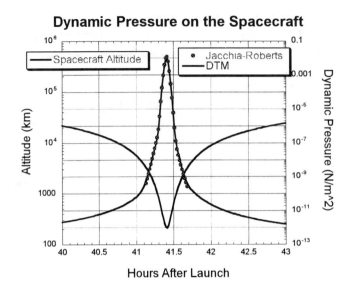

Figure 7: Dynamic Pressure on the Spacecraft.

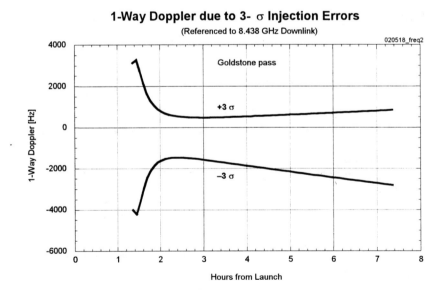

Figure 8: Errors in Best Ranging Frequency After Launch.

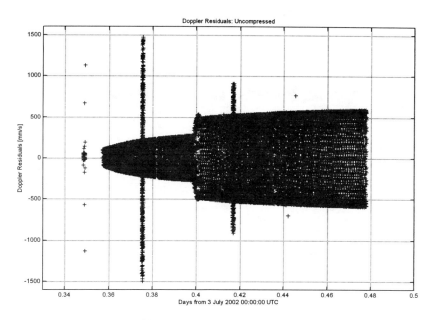

Figure 9: Doppler Residuals.

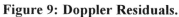

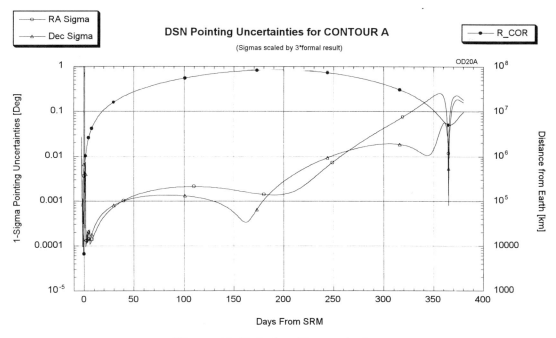

Figure 10: Pointing Uncertainties.

Table 1: Ranging Parameter Settings

	Launch Settings	After July 11 17:38
Clock	5	4
Chop	8	4
High	17	20
T1 (sec)	2	2
T2 (sec)	1	1
Cycle Time (sec)	29	37
Ambiguity (km)	1187	9494

Table 2: Filter Parameters

Parameters	A Priori Uncertainty (1-sigma)	Comments
Estimated as bias:		
Epoch state (6)	100 km, 0.1 m/s per axis	
Solar pressure acceleration (3)	50% of a priori value for each surface	Scale factors for cylinder and two flat plates (A priori values = 1)
Drag coefficients, constant part (3)	50% of a priori value for each surface	Scale factors for cylinder and two flat plates. (A priori values =1)
Impulsive events	20 mm/s per axis	All thrusting events other than planned maneuvers. (A priori values = 0)
Planned maneuvers (3)	5° RA, 5° Dec, 5% ΔV	
Ranging bias, constant part	5 m	Constant bias, all ranging points.
Estimated as stochastic:		
Drag coefficients, stochastic part (3)	0.5 each surface	Scale factors for cylinder and two flat plates. (A priori values = 0). t = 0, updated each apoapsis.
Ionosphere (day)	0.05	
Ionosphere (night)	0.01	
Ranging bias, stochastic part	2 m	Independent bias each ranging pass.
Considered:		
DSN station locations (9)	15 cm per axis	Sufficiently conservative to cover media errors as well.
Earth ephemeris		
Station locations		
Troposphere (dry)		
Troposphere (wet)		
Ionosphere (day)		
Ionosphere (night)		
Polar motion		
UT1		

Table 3: Summary of Optical Data

	16 Aug	17 Aug	18 Aug	19 Aug	20 Aug	21 Aug	Total
A	2S,2L	7M,4T	4M	2M	1M	4M	26
B	2S,3L	3M,3T	4M	2M	0	0	17
C	2M	0	0	0	0	0	2

S = Space Watch, L = Linear, M = Mauna Kea, T = Table Mountain

Table 4: Estimated SRM

	CONTOUR-A	CONTOUR-B	CONTOUR-C	NOMINAL
RA (deg.)	259.046 ±0.006	258.780 ±0.006	259. ±1.0	258.731
Dec (deg.)	29.350 ±0.001	29.583 ±0.001	28.8 ±0.2	29.295
ΔV (m/sec)	1849.8 ±0.1	1850.3 ±0.1	1896. ±20	1920.1
ΔV (% SRM)	96	96	98	100

Table 5: Solar Pressure Results

	Relative acceleration	1-sigma uncertainty	Mass limits for 1-sigma errors and 4.3 m^2 area [kg]
NOMINAL	1	-----	-----
A	3.6	1.0	190/108
B	149	2.0	3.4/3.3
C	2.9	100	495/5

Table 6: Earth Flyby B-Plane Results

	B•R [km]	B•T [km]	Date/Time [UTC]
A	435,000 ± 1,400	4,947,000 ± 16,000	16 Aug 13:47:39 ± 330 sec
B	580,000 ± 2,700	6,985,000 ± 32,800	17 Aug 01:36:55 ± 670 sec
C	93,700 ± 169,000	721,000 ± 1,564,000	15 Aug 13:24:41 ± 9.0 hour

ESTIMATING GENERAL RELATIVITY PARAMETERS
FROM RADIOMETRIC TRACKING
OF HELIOCENTRIC TRAJECTORIES

Ryan S. Park,[*] **Daniel J. Scheeres,**[†] **Giacomo Giampieri,**[‡]
James M. Longuski[**] **and Ephraim Fischback**[††]

The theory of GR can be tested by precisely measuring small changes in the trajectory of a spacecraft as it travels near the Sun. An important question is with what accuracy can the relativistic coefficients γ and β be estimated from such a trajectory. We present a detailed covariance analysis of this question, analyzing uncertainties in the spacecraft state and in the GR parameters. The measurement data types simulated in our analysis are range, Very Long Baseline Interferometry (VLBI), and Doppler measurements. Also included are the effects of different phase angles between the Earth and the spacecraft trajectory as well as the realistic error sources. The worst case analysis shows that estimates of these parameters should be obtainable to the order of 0.001 or better, assuming modest improvement in measurement capabilities.

1 Introduction

It has been shown [1] that the Theory of GR can be tested by precisely measuring small changes in the position of a spacecraft as it escapes from the sun on a hyperbolic trajectory. But how well can the relativistic coefficients, also known as the Parameterized Post Newtonian (PPN) parameters, γ and β be estimated from such trajectory (Figure 1)? By forming a state transformation from the initial state and GR parameters to the data measurements, and computing proper partial derivatives, we can estimate the order of accuracy to which we should be able to determine these coefficients, using covariance analysis. A preliminary study based on an analytic approach [2] showed the possibility of measuring the PPN parameters to the order of accuracy 0.001 or better with a foreseeable improvement in spacecraft tracking system (K-band system). This paper presents more detailed numerical analysis including various error sources, such as process noise and station location errors. Also, elliptic orbits are studied to analyze the sensitivity of the uncertainty distribution on the increased data arc and repeated radiometric measurements. The GR effect is at a maximum in the proximity of the Sun and more useful data can be obtained by placing the spacecraft on an elliptic orbit. Moreover, closed orbits allow multiple periapsis passages, which allow increased number of radiometric measurements.

[*] Department of Aerospace Engineering, The University of Michigan, FXB Building, 1320 Beal Avenue, Ann Arbor, Michigan 48109-2140. E-mail: sanghp@umich.edu.

[†] Department of Aerospace Engineering, The University of Michigan, 3048 FXB Building, 1320 Beal Avenue, Ann Arbor, Michigan 48109-2140. E-mail: scheeres@umich.edu.

[‡] Space and Atmospheric Physics, The Blackett Laboratory Imperial College, London, SW7 2BW. E-mail: g.giampieri@ic.ac.uk.

[**] School of Aeronautics and Astronautics, Purdue University, 1282 Grissom Hall, West Lafayette, Indiana 47907-1282. E-mail: longuski@ecn.purdue.edu.

[††] Physics Department, Purdue University, 525 Northwestern Avenue, West Lafayette, Indiana 47907-2036. E-mail: ephraim@physics.purdue.edu.

2 The Transient Effect of γ and β

2.1 General Relativistic Perturbation

The perturbing relativistic acceleration, to first Post-Newtonian (PN) order [3], can be written as:

$$\delta \vec{a} = \frac{m}{r^3} \left[2(\gamma + \beta) \frac{m\vec{r}}{r} - \gamma v^2 \vec{r} + 2(\gamma + 1)(\vec{r} \cdot \vec{v})\vec{v} \right] \tag{1}$$

where $\vec{r}$ and $\vec{v}$ are the spacecraft state vectors and m is the normalized gravitational constant [i.e., $m = $ (gravitational constant, μ)/(speed of light, c)]. It can be seen that the relativistic perturbation is present only in the orbital plane. The acceleration components decomposed into the radial R, transverse S, and out-of-plane W directions are:

$$R = \frac{m^2(1 + e\cos f)^2}{a^3(e^2 - 1)^3} \left[\left(1 - e^2\right)\gamma + 2\beta(1 + e\cos f) + 2(\gamma + 1)e^2 \sin^2 f \right], \tag{2}$$

$$S = \frac{m^2(1 + e\cos f)^2}{a^3(e^2 - 1)^3} \left[2(\gamma + 1)(1 + e\cos f)e\sin f \right], \tag{3}$$

$$W = 0. \tag{4}$$

2.2 Hyperbolic Lagrange Equations

The Hyperbolic Lagrange Planetary Equations [6], with proper changes (i.e., $e^2 > 1, a \to -a$) can be represented as:

$$\frac{da}{dt} = -\frac{2a^{3/2}}{\sqrt{m(e^2 - 1)}} \left[Re\sin f + S(1 + e\cos f) \right], \tag{5}$$

$$\frac{de}{dt} = \sqrt{\frac{a(e^2 - 1)}{m}} \left[R\sin f + \frac{S}{e}\left(\frac{p}{r} + \frac{r}{a}\right) \right], \tag{6}$$

$$\frac{di}{dt} = \frac{r}{h}W\cos(\omega + f), \tag{7}$$

$$\frac{d\Omega}{dt} = \frac{rW\sin(\omega + f)}{h\sin i}, \tag{8}$$

$$\frac{d\omega}{dt} = \frac{1}{e}\sqrt{\frac{a(e^2 - 1)}{m}} \left[-R\cos f + S\frac{2 + e\cos f}{1 + e\cos f}\sin f \right] - \frac{d\Omega}{dt}\cos i, \tag{9}$$

$$\frac{dM}{dt} = n - \frac{1}{na}\left[\frac{2r}{a} - \frac{e^2 - 1}{e}\cos f \right]R + \frac{e^2 - 1}{nae}\left[\frac{r}{p} \right]S. \tag{10}$$

where a is semi-major axis, e is eccentricity, i is inclination, ω is argument of perigee, Ω is longitude of ascending node, M is mean anomaly, f is true anomaly, n is mean motion, and F is hyperbolic eccentric anomaly.

After substitution of the perturbing relativistic acceleration components into the Hyperbolic Lagrange Equations [6], the change in the orbital elements from periapsis passage to some value of true anomaly can be approximated by keeping the elements on the right-hand-side constant and varying the true anomaly.

$$\Delta a = \frac{2em}{(e^2-1)^2}\left[(2+2\beta+3\gamma+2e^2+\gamma e^2)\,\Delta\cos f - (2+\beta+2\gamma)\,e\Delta\sin^2 f\right] \tag{11}$$

$$\Delta e = -\frac{m}{a(e^2-1)}\left[(\gamma+2\beta+4e^2+3\gamma e^2)\,\Delta\cos f - (2+\beta+2\gamma)e\Delta\sin^2 f\right] \tag{12}$$

$$\Delta\omega = \frac{m}{a(e^2-1)}\left[(2-\beta+2\gamma)\,\Delta f - \left(\frac{2\beta+(1-e^2)\gamma}{e}\right)\Delta\sin f - (2+\beta+2\gamma)\,\Delta(\sin f\cos f)\right] \tag{13}$$

$$\Delta M = \frac{3m}{a(e^2-1)^2}\left[\gamma+\beta+(2+\gamma)e^2+(2+3\gamma+2\beta+(2+\gamma)e^2)e+(2+2\gamma+\beta)e^2\right]n\Delta t$$
$$-\frac{m}{a\sqrt{e^2-1}}\left[(2+2\gamma+\beta)\Delta(\sin f\cos f)+\frac{\gamma+2\beta+(4+3\gamma)e^2}{e}\Delta\sin f\right]-(2+\gamma)\frac{m}{a}\Delta F \tag{14}$$

Figures 2 and 3 show the change in the orbital elements due to the relativistic effect. The initial epoch is at the periapsis with $R_P = 4R_\odot$ and $V_\infty = 39$ km/s, where R_P and $R_\odot$ are the perihelion distance and radius of the Sun, respectively. It is important to note that the largest change in orbital elements occur very early in the trajectory and essentially disappear after a few days.

Of most interest are the partial derivatives of orbital elements with respect to the relativistic constants, γ and β, based on the equations found above. If we let E be the set of orbital elements, a change in E due to relativistic effects can be represented as $E = E_o + \Delta E$, where E_o is the set of initial orbital elements. Taking partials with respect to the GR parameters yields:

$$\frac{\partial E}{\partial(\gamma,\beta)} = \frac{\partial(E_o+\Delta E)}{\partial(\gamma,\beta)} = \frac{\partial\Delta E}{\partial(\gamma,\beta)} \tag{15}$$

The partial derivatives of orbital elements with respect to γ are

$$\frac{\partial a}{\partial\gamma} = -\frac{2em}{(e^2-1)^2}\left[(3+e^2)(1-\cos f)+2e\sin^2 f\right], \tag{16}$$

$$\frac{\partial e}{\partial\gamma} = \frac{m}{a(e^2-1)}\left[(1+3e^2)(1-\cos f)+2e\sin^2 f\right], \tag{17}$$

$$\frac{\partial\omega}{\partial\gamma} = \frac{m}{a(e^2-1)}\left[2f-2\sin f\cos f+\frac{(e^2-1)}{e}\sin f\right], \tag{18}$$

$$\frac{\partial M}{\partial\gamma} = \frac{3m}{a(e^2-1)^2}\left[1+3e+3e^2+e^3\right]M-\frac{m}{a\sqrt{e^2-1}}\left[2\sin f\cos f+\frac{(1+3e^2)}{e}\sin f\right]-\frac{m}{a}F. \tag{19}$$

The partial derivatives of orbital elements with respect to β are

$$\frac{\partial a}{\partial\beta} = -\frac{2em}{(e^2-1)^2}\left[2(1-\cos f)+e\sin^2 f\right], \tag{20}$$

$$\frac{\partial e}{\partial\beta} = \frac{m}{a(e^2-1)}\left[2(1-\cos f)+e\sin^2 f\right], \tag{21}$$

$$\frac{\partial\omega}{\partial\beta} = \frac{m}{a(e^2-1)}\left[-f-\frac{2\sin f}{e}-\sin f\cos f\right], \tag{22}$$

$$\frac{\partial M}{\partial\beta} = \frac{3m}{a(e^2-1)^2}\left[1+2e+e^2\right]M-\frac{m}{a\sqrt{e^2-1}}\left[\sin f\cos f+\frac{2}{e}\sin f\right]. \tag{23}$$

Figure 4 shows the ratios of the partial derivatives of the orbital elements with respect to the GR parameters as the spacecraft travels on the hyperbolic trajectory. The initial conditions are the same as used above. It is interesting to note that the partials of ω show different behavior between γ and β. This plot clearly shows that there is a possibility of solving for γ and β separately by tracking the spacecraft.

3 The Covariance Analysis and the Least Square Approximation

3.1 Measurement Data Types

For our analysis, three different measurement data types are considered. The first data type is two-way radar range measurement (Z_ρ). This measures the distance between the spacecraft and the tracking station based on the travel time of the uplink and downlink signals. If we let T be signal travel time, the range value $|\rho| \sim 0.5cT$. The second data type we consider is Very Long Baseline Interferometry (VLBI) measurement $(Z_{m,n})$. VLBI measures the longitudinal and latitudinal angles of the spacecraft trajectory in the plane of sky of the tracking station. Combined with range measurements, the 3-dimensional position of the spacecraft can be obtained. The final data type we consider are Doppler measurements, $Z_{\dot\rho}$, which measure the frequency shift (Doppler Effect) [7] in the transmitted signals which gives range-rate and, because of the Hamilton-Melbourne effect, provides angular information on the trajectory. Doppler measurements can most simply be represented as range-rate, $\dot\rho \approx (\rho_{i+1} - \rho_i)/(t_{i+1} - t_i)$.

3.2 State to be Estimated

At epoch, the spacecraft is located at perihelion $(R_P = 4R_s)$ of its heliocentric hyperbolic trajectory with $a_o = 0.58$ AU, $e_o = 1.03$, and $i_o = \omega_o = \Omega_o = M_o = 0$. The trajectories of spacecraft and Earth were assumed to be coplanar and the spacecraft escapes the Sun with $V_\infty = 39$ km/s, which corresponds to periapsis velocity $V_p = 311$ km/s. This nominal trajectory was solved using the following modified Kepler's Equation,

$$\sqrt{\frac{\mu}{|a|^3}}(t - \tau) = e\sinh(F) - F \tag{24}$$

where τ is the periapsis passage time and F is the hyperbolic eccentric anomaly. It is shown [1] that the general relativistic effect is at a maximum when the spacecraft is on a parabolic trajectory. The hypothetical trajectory from Ref. [1] will most likely fly into perihelion as an elliptic orbit and then boost to a hyperbolic escape trajectory. Hence, the above initial state is considered to be the actual condition at epoch. All of these measurement data types are analyzed using a range of different initial phase angles between the Earth and the spacecraft trajectory (i.e., the initial Earth-Sun-spacecraft angle, ϕ, shown in Figure 1.). At epoch for the initial covariance matrix, the conservative initial uncertainties were, $\sigma_x = \sigma_y = \sigma_z = 1$ km, $\sigma_u = \sigma_v = \sigma_w = 1$ m/s, and $\sigma_\gamma = \sigma_\beta = 1$, where σ represents how accurately we know the state components of the spacecraft initially.

3.3 Computation of the Information and Covariance Matrix

Define the function J as follows:

$$J = \frac{1}{2}\sum_i^N \frac{1}{\sigma_i^2}\left[\overline{Z}_i - Z_i\right]^2, \tag{25}$$

where N is the number of data measurements, σ_i is the data noise uncertainty, $\overline{Z}_i$ is the actual data measurement, and Z_i is the predicted data measurement. Also, let Y_o be a column vector composed of initial state and GR parameters (i.e., $Y_o = [\vec{r}_o\ \vec{v}_o\ \gamma_o\ \beta_o]^T$). We want to minimize the function J in order to fit the GR parameters to the data. The specific types of data measurement considered were range, VLBI, and Doppler as mentioned earlier. Now minimizing J and linearizing the nominal values of γ and β, we obtain the following expressions.

$$\sum_i^N \frac{1}{\sigma_i^2}\left(\overline{Z}_i - Z_{i,o}(Y_o)\right)\left(\frac{\partial Z_i(Y_o)}{\partial Y_o}\right)_o = \left[\sum_i^N \frac{1}{\sigma_i^2}\left(\frac{\partial Z_i(Y_o)}{\partial Y_o}\right)_o\left(\frac{\partial Z_i(Y_o)}{\partial Y_o}\right)_o^T\right]\partial Y_o. \tag{26}$$

1496

where we define the information matrix as

$$\Lambda = \left[\sum_i^N \frac{1}{\sigma_i^2} \left(\frac{\partial Z_i(Y_o)}{\partial Y_o} \right)_o \left(\frac{\partial Z_i(Y_o)}{\partial Y_o} \right)_o^T \right]. \tag{27}$$

The Gaussian Probability Density function is then defined as

$$F(\vec{r}_o, \vec{v}_o, \gamma, \beta) = \frac{\sqrt{\det(\Lambda)}}{16\pi^4} e^{-\frac{1}{2}(Y_o^T \Lambda Y_o)}. \tag{28}$$

The covariance matrix of initial state and GR parameters, P, is then:

$$P = \Lambda^{-1}. \tag{29}$$

4   Model Description

4.1   Partial Derivatives of Data Measurements with respect to State Vector, $\left(\frac{\partial Z}{\partial X} \right)$

Now, we want to analyze the effect of a given initial condition on the GR parameters, and in order to compute the information matrix, Λ, we must first compute the partial derivatives of the measurement partials, $\left(\frac{\partial Z}{\partial X} \right)$, where $X = \begin{bmatrix} \vec{r} & \vec{v} \end{bmatrix}^T$.

The first data type we consider are range measurements,

$$Z_\rho = |\vec{r} - \vec{r}_E - \vec{r}_{TS}| = |\vec{\rho}|, \tag{30}$$

which measure the distance between the spacecraft and the tracking station (TS). Here, $\vec{r}_{TS}$ is the vector representing the location of the Earth tracking station (Goldstone, CA) with origin at Earth center and its analytic representation is

$$\vec{r}_{TS}(t) = \begin{bmatrix} 1 & 0 & 0 \\ 0 & \cos\psi & -\sin\psi \\ 0 & \sin\psi & \cos\psi \end{bmatrix} \cdot \begin{bmatrix} R_E \cos(\alpha + \omega_E t) \sin\delta \\ R_E \sin(\alpha + \omega_E t) \sin\delta \\ R_E \cos\delta \end{bmatrix} = \Psi \cdot \vec{r}_{TS}(0), \tag{31}$$

where ψ is the Earth obliquity (23.45^o), R_E is Earth mean radius (6378 km), α is the right ascension (243.17^o), and δ is the declination (54.67^o). Now, we take the partial derivative of Z_ρ with respect to X to find,

$$\frac{\partial Z_\rho}{\partial X} = \begin{bmatrix} \dfrac{\partial Z_\rho}{\partial \vec{r}} \\[2mm] \dfrac{\partial Z_\rho}{\partial \vec{v}} \end{bmatrix}^T = \begin{bmatrix} \hat{\rho}(t) \\ 0 \end{bmatrix}_{6 \times 1}^T \tag{32}$$

where $\hat{\rho}$ is the unit position vector of the spacecraft from the Earth tracking station. We consider the precision of the range measurement, σ_i, to range over 10^{-2}, 10^{-3}, and 10^{-4} km.

An alternative data measurement type considered is VLBI, which yields accurate angular measurements of the spacecraft relative to a radio source. We represent this measurement as a set of angles,

$$Z_{(m,n)} = \begin{bmatrix} Z_m & Z_n \end{bmatrix}^T, \tag{33}$$

where Z_m and Z_n are the longitudinal and the latitudinal angular measurements, respectively. Taking partials with respect to X yields,

$$\frac{\partial Z_{(m,n)}}{\partial X} = \begin{bmatrix} \dfrac{\hat{m}_o^T}{\rho} & 0 \\[2ex] \dfrac{\hat{n}_o^T}{\rho} & 0 \end{bmatrix}_{2\times 6}, \tag{34}$$

where we define

$$\hat{l}_o = \hat{\rho}, \tag{35}$$

$$\hat{m}_o = \hat{l}_o \times \hat{n}_o, \tag{36}$$

$$\hat{n}_o = \frac{\hat{z} - (\hat{z} \cdot \hat{l}_o)\hat{l}_o}{|\hat{z} - (\hat{z} \cdot \hat{l}_o)\hat{l}_o|}, \tag{37}$$

with $\hat{z} = \begin{bmatrix} 0 & 0 & 1 \end{bmatrix}^T$. ρ is the range from Earth to the spacecraft as we defined earlier. The precision used for the angular measurements were 5, 1, and 0.1 nrad. These angular measurements give us the angular position of the spacecraft.

The final data measurement type we consider is Doppler,

$$Z_D = \frac{d}{dt}|\vec{r} - \vec{r}_E - \vec{r}_{TS}| = \hat{\rho} \cdot \dot{\vec{\rho}}, \tag{38}$$

which is widely used for interplanetary missions. The Doppler measurements measure the shift in frequency due to the Doppler effect, and contains both the range and angular information. The partial derivative of Z_D results in,

$$\frac{\partial Z_D}{\partial X} = \begin{bmatrix} \dfrac{\partial \hat{\rho}}{\partial \vec{r}} \dot{\vec{\rho}} \\[2ex] \hat{\rho} \end{bmatrix}_{6\times 1}^T, \tag{39}$$

where

$$\frac{\partial \hat{\rho}}{\partial \vec{r}} = \frac{1}{\rho}\left[I_3 - \hat{\rho}\,\hat{\rho}^T \right]. \tag{40}$$

The precision, σ_i, used for the Doppler measurement were 10^{-6}, 10^{-7}, and 10^{-8} km/s.

4.2 Numerical Model Description

Define the state vector as

$$Y = \begin{bmatrix} \vec{r} & \vec{v} & \gamma & \beta \end{bmatrix}^T. \tag{41}$$

The total time derivative of Y is then expressed as,

$$\dot{Y} = F(Y, \gamma, \beta) = \begin{bmatrix} \dot{\vec{r}} & \dot{\vec{v}} & \dot{\gamma} & \dot{\beta} \end{bmatrix}^T = \begin{bmatrix} \vec{v} & \vec{a} & 0 & 0 \end{bmatrix}^T, \tag{42}$$

where

$$\vec{a} = -\frac{\mu}{r^3}\vec{r} + \delta\vec{a}. \tag{43}$$

$\delta\vec{a}$ is the relativistic perturbing acceleration given in Section 2 and its non-dimensionalized form is

$$\delta\vec{a} = \frac{\mu}{c^2 r^3}\left[2(\gamma + \beta)\frac{\mu\vec{r}}{r} - \gamma v^2\vec{r} + 2(\gamma + 1)(\vec{r}\cdot\vec{v})\vec{v}\right]. \tag{44}$$

Now, the measurement partial can be expressed as,

$$\frac{\partial Z}{\partial Y_o} = \frac{\partial Z}{\partial Y}\frac{\partial Y}{\partial Y_o}, \tag{45}$$

where the state transformation matrix, $\Phi = \frac{\partial Y}{\partial Y_o}$,

$$\Phi = \begin{bmatrix} \Phi_{\vec{r}\vec{r}_o} & \Phi_{\vec{r}\vec{v}_o} & \Phi_{\vec{r}(\gamma_o,\beta_o)} \\ \Phi_{\vec{v}\vec{r}_o} & \Phi_{\vec{v}\vec{v}_o} & \Phi_{\vec{v}(\gamma_o,\beta_o)} \\ \Phi_{(\gamma,\beta)\vec{r}_o} & \Phi_{(\gamma,\beta)\vec{v}_o} & \Phi_{(\gamma,\beta)(\gamma_o,\beta_o)} \end{bmatrix} = \begin{bmatrix} \left(\frac{\partial X}{\partial X_o}\right)_{6\times 6} & \left(\frac{\partial X}{\partial(\gamma,\beta)}\right)_{6\times 2} \\ 0_{2\times 6} & I_{2\times 2} \end{bmatrix}_{8\times 8}, \tag{46}$$

with $\Phi_o = I_{8\times 8}$. Now taking the time derivative of the state transformation matrix, Φ, yields,

$$\dot{\Phi} = \begin{bmatrix} \left(\frac{\partial X}{\partial X_o}\right)'_{6\times 6} & \left(\frac{\partial X}{\partial(\gamma,\beta)}\right)'_{6\times 2} \\ 0_{2\times 6} & 0_{2\times 2} \end{bmatrix}_{8\times 8}. \tag{47}$$

The superscript $'$ denotes the total time derivative of a partial derivative matrix. The time derivative of the state partial with respect to initial state yields,

$$\left(\frac{\partial X}{\partial X_o}\right)' = \left(\frac{\partial \dot{X}}{\partial X}\right)_{X(t)} \cdot \left(\frac{\partial X}{\partial X_o}\right) \tag{48}$$

where

$$\left(\frac{\partial \dot{X}}{\partial X}\right)_{X(t)} = \begin{bmatrix} \frac{\partial \vec{v}}{\partial \vec{r}} & \frac{\partial \vec{v}}{\partial \vec{v}} \\ \frac{\partial \vec{a}}{\partial \vec{r}} & \frac{\partial \vec{a}}{\partial \vec{v}} \end{bmatrix} = \begin{bmatrix} 0_{3\times 3} & I_{3\times 3} \\ \frac{\partial}{\partial \vec{r}}\left(-\frac{\mu}{r^3}\vec{r} + \delta\vec{a}\right) & \frac{\partial}{\partial \vec{v}}\left(-\frac{\mu}{r^3}\vec{r} + \delta\vec{a}\right) \end{bmatrix}_{6\times 6}, \tag{49}$$

and the partial with respect to state vectors,

$$\frac{\partial}{\partial \vec{r}}\left(-\frac{\mu}{r^3}\vec{r} + \delta\vec{a}\right) = -\frac{\mu}{r^3}\left[I_{3\times 3} - \frac{3}{r^2}\vec{r}\vec{r}^T\right] + \frac{2\mu^2(\gamma + \beta)}{c^2 r^4}\left[I_{3\times 3} - \frac{4}{r^2}\vec{r}\vec{r}^T\right] - \frac{\mu\gamma}{c^2}\frac{(\vec{v}\cdot\vec{v})}{r^3}\left[I_{3\times 3} - \frac{3}{r^2}\vec{r}\vec{r}^T\right]$$

$$+ \frac{2\mu(\gamma + 1)}{c^2 r^3}\left[\vec{v}\vec{v}^T - \frac{3}{r^2}(\vec{r}\cdot\vec{v})\vec{r}\vec{v}^T\right], \tag{50}$$

$$\frac{\partial}{\partial \vec{v}}\left(-\frac{\mu}{r^3}\vec{r} + \delta\vec{a}\right) = -\frac{2\mu\gamma}{c^2 r^3}\vec{v}\vec{r}^T + \frac{2\mu(\gamma + 1)}{c^2 r^3}\left[\vec{r}\vec{v}^T + I_{3\times 3}(\vec{r}\cdot\vec{v})\right]. \tag{51}$$

The time derivative of state partial with respect to the GR paramters can be expressed as,

$$\left(\frac{\partial X}{\partial(\gamma,\beta)}\right)' = \left(\frac{\partial \dot X}{\partial X}\right)_{X(t)} \cdot \left(\frac{\partial X}{\partial(\gamma,\beta)}\right) + \frac{\partial \dot X}{\partial(\gamma,\beta)}. \tag{52}$$

These partials are defined as:

$$\frac{\partial \dot X}{\partial(\gamma,\beta)} = \left[\begin{array}{cc} \dfrac{\partial \vec v}{\partial \gamma} & \dfrac{\partial \vec v}{\partial \beta} \\[2mm] \dfrac{\partial \vec a}{\partial \gamma} & \dfrac{\partial \vec a}{\partial \beta} \end{array}\right] = \left[\begin{array}{cc} 0 & 0 \\[2mm] \dfrac{\partial(\delta \vec a)}{\partial \gamma} & \dfrac{\partial(\delta \vec a)}{\partial \beta} \end{array}\right], \tag{53}$$

where

$$\frac{\partial(\delta \vec a)}{\partial \gamma} = \frac{\mu}{c^2 r^3}\left[\frac{2\mu \vec r}{r} - v^2 \vec r + 2(\vec r \cdot \vec v)\vec v\right] \tag{54}$$

$$\frac{\partial(\delta \vec a)}{\partial \beta} = \left[\frac{2\mu^2}{c^2 r^4}\vec r\right]. \tag{55}$$

5   Filter Model and Error Sources

5.1   Square Root Information Filter

The inversion of the Information Matrix usually entails loss of numerical precision. The Square Root Information Filter (SRIF) is used to obtain a more numerically precise Covariance Matrix, P. Define

$$\Lambda = R^T R \tag{56}$$

where R is the SRIF Matrix [4]. Let T_H be an orthogonal householder transformation matrix such that

$$T_H\left[\begin{array}{c} R \\ \dfrac{\partial Z}{\partial Y_o} \end{array}\right] = \left[\begin{array}{c} R' \\ 0 \end{array}\right]. \tag{57}$$

The updated information matrix is then,

$$\Lambda' = (R')^T(R') \tag{58}$$

and the updated covariance matrix, P', becomes

$$P' = (\Lambda')^{-1} = (R')^{-1}(R')^{-T}. \tag{59}$$

5.2 Stochastic Acceleration

The effect of time correlated random accelerations can be included in the SRIF matrix based on Ref. [4],

$$\dot{R} = R_o \dot{\Phi}^{-1} = -R_o \Phi^{-1} A = -RA + F(R) \tag{60}$$

where $F(R)$ is the stochastic acceleration given as follows:

$$F(R) = -\frac{1}{2} R B P_\omega B^T R^T R \tag{61}$$

where

$$B = \begin{bmatrix} 0_{3\times3} & I_{3\times3} & 0_{2\times3} \end{bmatrix}^T, \tag{62}$$

$$P_\omega = 2T\sigma_a^2 I_{3\times3}. \tag{63}$$

Here, $\sigma_a = 10^{-12}$ km/s^2 is the steady state acceleration noise and $T = 0.5$ days is the correlation time. Between measurements the SRIF matrix is propagated by solving this differential equation.

5.3 Station Location Error

Define the consider station location vector, $\vec{r}_{SL}$ as,

$$\vec{r}_{SL} = \begin{bmatrix} R_E \sin\delta \\ R_E \cos\delta \\ \alpha \end{bmatrix}. \tag{64}$$

Then taking partials of range measurement with respect to the station location vector results in

$$\frac{\partial Z_\rho}{\partial \vec{r}_{SL}} = -\left(\Psi \frac{\partial \vec{r}_{TS_o}}{\partial \vec{r}_{SL}} \right)^T \hat{\rho} \tag{65}$$

where

$$\frac{\partial \vec{r}_{TS_o}}{\partial \vec{r}_{SL}} = \begin{bmatrix} \cos(\alpha + \omega_E t) & 0 & -R_E \sin(\alpha + \omega_E t)\sin\delta \\ \sin(\alpha + \omega_E t) & 0 & R_E \cos(\alpha + \omega_E t)\sin\delta \\ 0 & 1 & 0 \end{bmatrix}. \tag{66}$$

Taking partials of VLBI measurement with repect to the station location vector yields,

$$\begin{bmatrix} \dfrac{\partial Z_m}{\partial \vec{r}_{SL}} \\[3mm] \dfrac{\partial Z_n}{\partial \vec{r}_{SL}} \end{bmatrix} = \begin{bmatrix} -\left(\Psi \dfrac{\partial \vec{r}_{TS_o}}{\partial \vec{r}_{SL}} \right)^T \dfrac{\hat{m}_o}{\rho} \\[3mm] -\left(\Psi \dfrac{\partial \vec{r}_{TS_o}}{\partial \vec{r}_{SL}} \right)^T \dfrac{\hat{n}_o}{\rho} \end{bmatrix}. \tag{67}$$

Finally, the partial derivatives of Doppler measurements result in,

$$\frac{\partial Z_D}{\partial \vec{r}_{SL}} = -\frac{1}{\rho} \left(\Psi \frac{\partial \vec{r}_{TS_o}}{\partial \vec{r}_{SL}} \right)^T \left(I_3 - \hat{\rho}\hat{\rho}^T \right)^T \dot{\vec{\rho}} - \left(\Psi \frac{\partial \vec{v}_{TS_o}}{\partial \vec{r}_{SL}} \right)^T \hat{\rho}. \tag{68}$$

5.4 Solar Occultation Effects

When the spacecraft passes in front of or behind the Sun (Figure 5), we cannot obtain radiometric measurements. Since the trajectory originates close to the Sun, this can be an important effect in the early stage of the experiment.

Define,

$$\chi = \cos^{-1} \frac{\vec{\rho} \cdot (-\vec{r}_E)}{\rho r}, \tag{69}$$

where $\vec{\rho} = \vec{r} - \vec{r}_E$.

Then, χ is simply the spacecraft-Earth-Sun angle. Based on the geometry of the Earth and Sun, assuming the Earth is in circular orbit about the Sun, the angle between $\vec{r}_E$ and the tangent vector from center of the Earth to outer radius of the Sun, ξ, can be computed and its value is approximately $0.267°$. We assume no measurements are taken if $\chi \leq \xi + 0.5°$ for Doppler and VLBI measurements, and $\chi \leq \xi + 5°$ for Range measurements.

6 Results

6.1 Baseline Results

For our analysis, the trajectory condition given in [1,2] was first verified as the baseline case. The spacecraft was initially located at the periapsis of the heliocentric hyperbolic trajectory with $R_p = 4R_\odot$ and $V_\infty = 39$ km/s. All of the data measurements considered were analyzed with different initial phase angles (ϕ) in order to find the optimum position of the Earth (i.e., phase angle ϕ that gives the most accurate GR estimates).

Figure 6 shows the uncertainties when the Range, VLBI, and Doppler measurements are combined. Current technology can provide noise factors of $\sigma_{i,\rho} = 10^{-3}$ km, $\sigma_{i,\dot{\rho}} = 10^{-7}$ km/s , and $\sigma_{i,(m,n)} = 1$ nrad. These noise factors are directly related to how much information we can obtain from the spacecraft trajectory. In other words, one can simply consider the noise factor as a scaling quantity since the uncertainties are linearly proportional to it. We assume that, in the future, each of these will be decreased by an order of magnitude by adoption of K-band measurement systems. Figure 7 shows the correlations between σ_γ and σ_β (i.e., $\sigma_{\gamma\beta}/\sqrt{\sigma_{\gamma\gamma}\sigma_{\beta\beta}}$). It can be noticed that the GR parameters become more correlated as the spacecraft moves away from periapsis. This is an indication that we must obtain the maximum number of measurements when in proximity of the Sun to separately estimate γ and β.

NOTE: Range $\sigma_i = \sigma_{i,\rho}$, VLBI $\sigma_i = \sigma_{i,(m,n)}$, and Doppler $\sigma_i = \sigma_{i,\dot{\rho}}$

Table 1: Timespan of 30 days with $\Delta t = 15$ minutes.

N = 2881	σ_γ	σ_β	Worst σ_γ	Worst σ_β
Medium Accuracy	$1.35 \cdot 10^{-4}$	$8.75 \cdot 10^{-4}$	$7.26 \cdot 10^{-3}$	$4.70 \cdot 10^{-2}$
High Accuracy	$1.35 \cdot 10^{-5}$	$8.75 \cdot 10^{-5}$	$7.26 \cdot 10^{-4}$	$4.70 \cdot 10^{-3}$

All of the values of σ_γ and σ_β shown in the above table are the final ones taken at the end of the timespan. "Medium Accuracy" consists of combined measurements with currently provided noise factors, whereas "High Accuracy" consists of combined measurements with values one order of magnitude lower than the "Medium Accuracy" case. The uncertainties σ_γ and σ_β tell us how accurately we can estimate the GR parameters. Two obvious ways to increase the accuracy of these parameters are either taking more measurements or by improving the value of the data noise. It is important to note that the GR parameter γ can be estimated more accurately than β. The worst case estimates are obtained by multiplying the formal uncertainties by $\sqrt{N}$, where N is the number of measurements. This shows that some improvements to current measurement technology must be made to estimate γ and β to the proposed order 10^{-3} or better.

Figure 8 shows the uncertainties and correlations when the range, Doppler, and VLBI measurements are combined with different initial phase angles; considering the solar occultation effect. The estimates are taken at the end of a 10 day timespan, and each measurement was updated every 15 minutes. We notice that the uncertainties are relatively sensitive to the initial phase angle, ϕ, which tells that the measurements taken in early stage of the trajectory significantly enhance the overall estimates of GR parameters.

As mentioned earlier, longer data arcs and repeated measurements of a closed orbit may improve results. Figure 9 shows the sensitivity of uncertainties in GR parameters as a function of escape velocity, $V_\infty^2 = -\mu/a$, where negative V_∞ represents orbits with negative orbital energies (i.e., elliptic orbits). This shows that the overall change in the uncertainty distribution is not significant, which means that there is a possibility of conducting this new test of general relativity based on elliptic orbits. An important item to notice is that the estimation of β becomes more accurate as specific orbital energy decreases, whereas the estimate of γ is better off with hyperbolic orbits. Considering the repetition in tracking measurements and increased data arc in the vicinity of Sun, elliptical orbits provide more accurate measurements of γ and β, assuming the spacecraft is capable of enduring multiple close passes of the Sun.

6.2 Effects of Consider Parameters

The effect of stochastic acceleration on the state variables (i.e., position and velocity components) are usually negligible for a short period of time; however, its effect on the GR variables was rather significant. Figures 10 and 11 show how the process noise affects the overall tracking performance. This error alone degraded the covariance of γ and β by an order of magnitude. This with the occultation effect will cause a significant problem in estimating GR parameters as well as separately measuring them. If the steady-state uncertainty, σ_a, decreases by an order of magnitude ($\sigma_a = 0.1$ nm/s^2), it will be possible to obtain the desired GR accuracy of 0.001. Figures 10 and 11 also show how exactly we should know the station location to achieve this accuracy. As it can be seen, station location uncertainty less than 0.2 meter will ensure the required accuracy. Fortunately, the current technology provides location uncertainties less than 0.2 meters, and we can disregard the effect of station location error. Figure 12 presents time response of GR estimates when station location error of 0.8, 0.2, and 0.1 meters were considered. This error does not affect the overall result of the uncertainties, but the oscillation of the GR estimates indicates that there is a possibility of estimating the station location errors from this test.

6.3 Comparison of Baseline and Solar Probe Trajectories

At the moment, the most feasible NASA mission to carry out this new test of General Relativity is the Solar Probe mission. The semi-major axis of this mission is approximately 2.6 AU with the periapsis distance of $4R_s$ as before. The Solar Probe will perform a Jupiter gravity assist to obtain required speed and will have the periapsis velocity $V_p \approx 308$ km/s, which is slightly less than the baseline case. Figure 13 indicates that the resulting covariance in GR parameters, as well as their correlation, will not be degraded by flying in an elliptic orbit. The hyperbolic trajectory provides a slightly better result; however, Solar Probe's multiple perihelion passages will increase the level of accuracy. This indicates that there is a strong possibility in carrying out this new GR experiment as a part of the Solar Probe mission objectives.

7 Conclusions/Future Work

The purpose of this paper was to show the feasibility of carrying out the new, unique test of general relativity proposed in Ref. [1] and [2]. The spacecraft was considered as an idealized particle and we applied covariance analysis to study the uncertainty distributions based on hyperbolic and elliptic orbits. Also, various error sources were included to study their impact on the overall performance of uncertainty dynamics. Although this is a preliminary-level analysis, it captures most of the fundamentals for performing the actual experiment. Several crucial characteristics of uncertainties in PPN parameters were obtained from this analysis. The first important fact is that placing a spacecraft on a heliocentric elliptic orbit performs essentially equivalent to a hyperbolic orbit, plus the increased data arc and multiple orbits will enhance the estimation accuracy of these coefficients. This led us to consider carrying out this test as a part of the Solar Probe mission, which is currently under development by NASA. To do so, some improvements in spacecraft tracking technology will be required. The stochastic effect significantly degrades the accuracy of the measurements, and hence, it would be necessary to decrease the steady-state uncertainty, σ_a, to less than 10^{-13} km/s^2. The station location error does decrease the accuracy of PPN parameters, however, the current technology provides station location error less than 0.2 meters.

Under ideal conditions, we have demonstrated that GR can be tested to unprecedented accuracy by tracking spacecraft trajectories near the Sun. To show the feasibility of an actual experiment we need to perform a more detailed analysis to take into account various nongravitational forces that are known to be important due to solar radiation, solar dust, and solar wind. These disturbances will be addressed in a future work.

8 References

[1] James M. Longuski, Ephraim Fischbach, and Daniel J. Scheeres, "Deflection of Spacecraft Trajectories as a New Test of General Relativity," *Physical Review Letters*, Vol. 86, No. 14, 2001, pp. 2942-2945.

[2] James M. Longuski, Ephraim Fischbach, Daniel J. Scheeres, Giacomo Giampieri, and Ryan S. Park , "Measurement of the Deflection of Spacecraft Trajectories as a New Test of General Relativity," *Physical Review Letters* (To be submitted).

[3] C.M. Will, "Theory and Experiment in Gravitational Physics," Cambridge, 1993.

[4] D.J. Scheeres, D. Han, Y. Hou "Influence of Unstable Manifolds on Orbit Uncertainty," *Journal of Guidance, Control, and Dynamics*, Vol. 24, No. 3, 2001, pp. 573-585.

[5] K. D. Mease, J. D. Anderson, L. J. Wood, and L. K. White "Tests of General Relativity Using Starprobe Radio Metric Tracking Data," *Journal of Guidance, Control, and Dynamics*, Vol. 7, No. 1, 1983, pp. 36-44.

[6] J. Prussing and B. Conway, "Orbital Mechanics," Oxford University Press, Inc. 1993, New York.

[7] T.W. Hamilton and W.G. Melbourne, "Information Content of a Single Pass of Doppler Data from a Distant Spacecraft," *JPL Space Programs Summary*, Vol. 3, No. 37-39, 1966, pp. 18-23.

[8] R. A. Brouke, "On the Matrizant of the Two-Body Problem," *Astron. and Astrophys.*, Vol. 6, 1970, pp. 173-182.

9   Figures

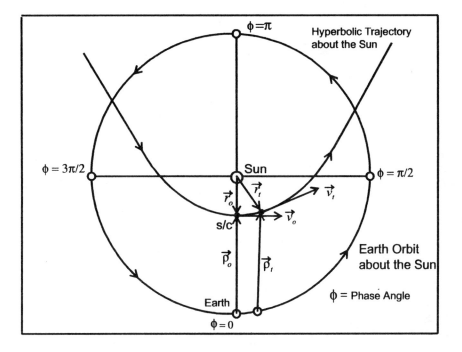

Figure 1: Hyperbolic Flyby of a Spacecraft near the Sun.

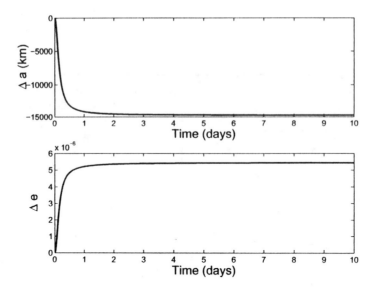

Figure 2: Change in the semi-major axis and eccentricity due to the relativistic effect.

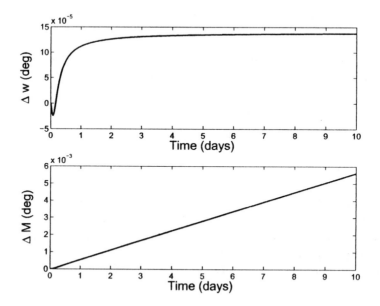

Figure 3: Change in the argument of perigee and mean anomaly due to the relativistic effect.

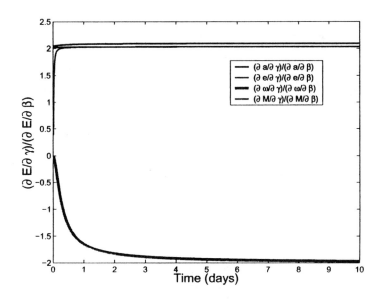

Figure 4: Ratios of orbital element partial derivatives with respect to GR parameters. Time variation of argument of perigee shows the potential to separately estimate γ and β.

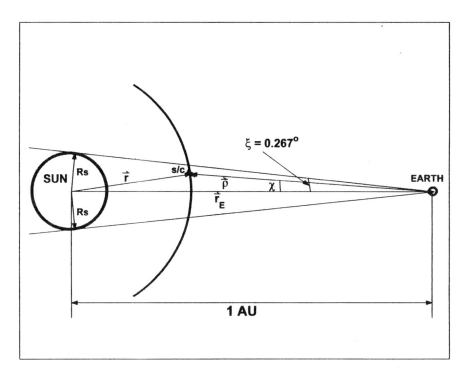

Figure 5: Occultation Effect due to Sun.

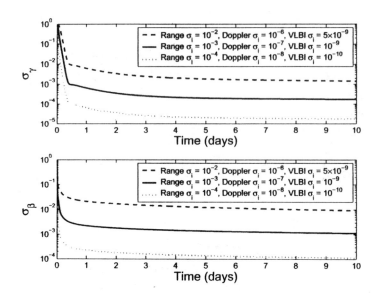

Figure 6: Uncertainties of γ and β measurements as a function of time for different tracking accuracies, σ_i. (No error sources included)

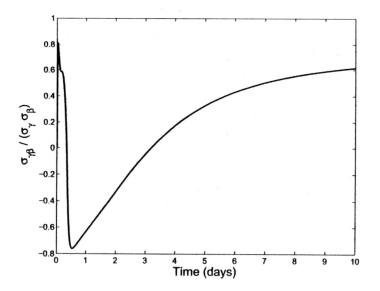

Figure 7: Correlation of γ and β for range, VLBI, and Doppler measurements, with accuracies $\sigma_\rho = 10^{-3}$ km, $\sigma_{m,n} = 10^{-9}$ rad, and $\sigma_{\dot{\rho}} = 10^{-7}$ km/s. (No error sources included)

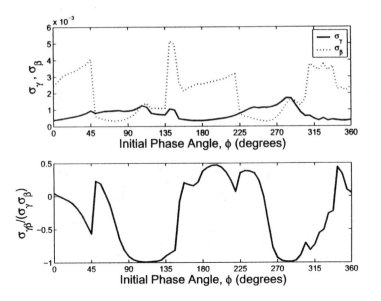

Figure 8: Accuracy and correlation of γ and β estimates as a function of initial phase angle with the Earth. Covariance are taken at the end of 10-day timespan with $\Delta t = 15$ minutes. Range, VLBI, and Doppler measurement accuracies are $\sigma_\rho = 10^{-3}$ km, $\sigma_{m,n} = 10^{-9}$ rad, and $\sigma_{\dot{\rho}} = 10^{-7}$ km/s.

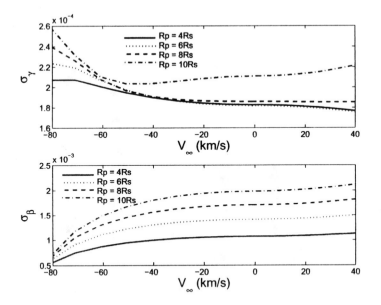

Figure 9: Accuracy of γ and β estimates as a function of escape velocity, V_∞. Covariance are taken at the end of 10-day timespan with $\Delta t = 15$ minutes. Measurement accuracies are $\sigma_\rho = 10^{-3}$ km, $\sigma_{m,n} = 10^{-9}$ rad, and $\sigma_{\dot{\rho}} = 10^{-7}$ km/s. (No error sources included)

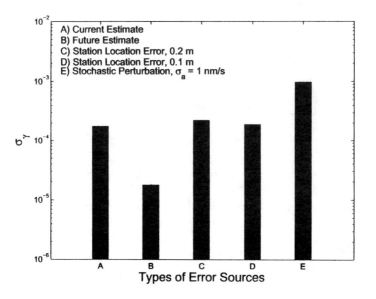

Figure 10: Effect of various errors sources on the accuracy of γ. Estimates are taken at the end of 10-day timespan with $\Delta t = 15$ minutes. Measurement accuracies are $\sigma_\rho = 10^{-3}$ km, $\sigma_{m,n} = 10^{-9}$ rad, and $\sigma_{\dot\rho} = 10^{-7}$ km/s.

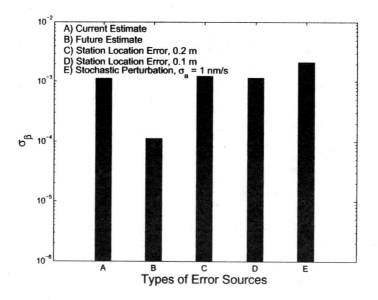

Figure 11: Effect of various errors sources on the accuracy of β. Estimates are taken at the end of 10-day timespan with $\Delta t = 15$ minutes. Measurement accuracies are $\sigma_\rho = 10^{-3}$ km, $\sigma_{m,n} = 10^{-9}$ rad, and $\sigma_{\dot\rho} = 10^{-7}$ km/s.

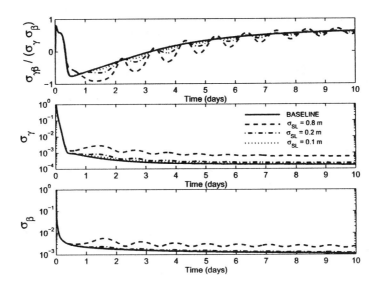

Figure 12: Effect of station location errors on the accuracy of γ and β. Measurement accuracies were $\sigma_\rho = 10^{-3}$ (km), $\sigma_{m,n} = 10^{-9}$ (rad), and $\sigma_{\dot{\rho}} = 10^{-7}$ (km/s) with measurement update for every 15 minutes.

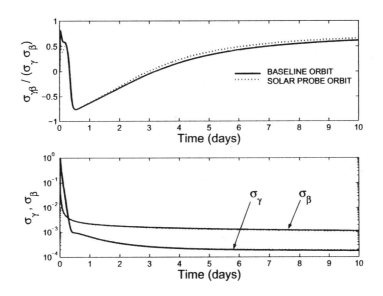

Figure 13: Comparison of GR parameter estimates of baseline and Solar Probe trajectories. Measurement accuracies are $\sigma_\rho = 10^{-3}$ km, $\sigma_{m,n} = 10^{-9}$ rad, and $\sigma_{\dot{\rho}} = 10^{-7}$ km/s with measurement update for every 15 minutes. (No error sources are included)